Name Reactions in Heterocyclic Chemistry II

Wiley Series on Comprehensive Name Reactions

Jie Jack Li, *Series Editor*

Name Reactions in Heterocyclic Chemistry
Edited by Jie Jack Li

Name Reactions of Functional Group Transformations
Edited by Jie Jack Li

Name Reactions for Homologation, Part 1 and Part 2
Edited by Jie Jack Li

Name Reactions for Carbocyclic Ring Formations
Edited by Jie Jack Li

Name Reactions in Heterocyclic Chemistry II
Edited by Jie Jack Li

Name Reactions in Heterocyclic Chemistry II

Edited by
Jie Jack Li
Bristol-Myers Squibb Company

Foreword by
E. J. Corey
Harvard University

WILEY

A JOHN WILEY & SONS, INC., PUBLICATION

Published by John Wiley & Sons, Inc., Hoboken, New Jersey
Published simultaneously in Canada

Library of Congress Cataloging-in-Publication Data:

Name reactions in heterocyclic chemistry / edited by Jie Jack Li.
p. cm.
Includes bibliographical references and index.
ISBN 978-0-470-08508-0
1. Heterocyclic chemistry. 2. Chemical reactions. I. Li, Jie Jack.
QD400.N34 2005
547'

oBook ISBN: 978-1-118-09282-8
ePDF ISBN: 978-1-118-09285-9
ePub ISBN: 978-1-118-09284-2

10 9 8 7 6 5 4 3 2 1

Contents

Appendixes

Foreword

Part of the charm of synthetic organic chemistry derives from the vastness of the intellectual landscape along several dimensions. First, there is the almost infinite variety and number of possible target structures that lurk in the darkness waiting to be made. Then, there is the vast body of organic reactions that serve to transform one substance into another, now so large in number as to be beyond credibility to a nonchemist. There is the staggering range of reagents, reaction conditions, catalysts, elements, and techniques that must be mobilized to tame these reactions for synthetic purposes. Finally, it seems that new information is being added to that landscape at a rate that exceeds the ability of a normal person to keep up with it. In such a troubled setting any author, or group of authors, must be regarded as heroic if through their efforts, the task of the synthetic chemist is eased.

This last volume on heterocyclic chemistry fills the holes left behind in Volume 1 and brings to the attention of practicing synthetic chemists and students of chemistry a wide array of tools for the synthesis of new and useful molecules. It is a valuable addition to the literature by any measure and surely will prove its merit in years to come. The new knowledge that arises with its help will be impressive and of great benefit to humankind.

E. J. Corey
February 1, 2011

Preface

This book is the sixth and the last volume of the *Comprehensive Name Reactions* series, an ambitious project conceived by Professor E. J. Corey of Harvard University in the summer of 2002. Volume 1, *Name Reactions in Heterocyclic Chemistry*, was published in 2005. Volume 2, *Name Reactions for Functional Group Transformations came out in* 2007. Volumes 3 and 4 *Name Reactions for Homologations Part I and Part II* were published in 2009. And Volume 5, *Name Reactions on Ring Formations* came out in 2010. Continuing the traditions of the first five volumes, each name reaction in Volume 6 is also reviewed in seven sections:

1. Description
2. Historical Perspective
3. Mechanism
4. Variations and Improvements
5. Synthetic Utility
6. Experimental
7. References.

I also introduce a symbol [R] to highlight review articles, book chapters, and books dedicated to the respective name reactions.

I have incurred many debts of gratitude to Professor E. J. Corey. He once told me "The desire to learn is the greatest gift from God," which has been a true inspiration. Furthermore, it has been my great privilege and a pleasure to work with a collection of stellar contributing authors from both academia and industry. Some of them are world-renowned scholars in the field; some of them have worked intimately with the name reactions that they have reviewed; some of them even discovered the name reactions that they authored in this series. As a consequence, this book truly represents the state-of-the-art for name reactions in heterocyclic chemistry.

I welcome your critique.

Jack Li
February 1, 2011

Jie Jack Li and E. J. Corey, circa 2000.

Contributing Authors:

Dr. Nadia M. Ahmad
Eli Lilly and Company
Erl Wood Manor
Windlesham
Surrey, GU20 6PH
United Kingdom

Dr. Narendra B. Ambhaikar
Dr. Reddy's Laboratories
CPS
Bollaram Road, Miyapur
Hyderabad - 500 049
A. P., India

Professor Alexandru T. Balaban
Texas A&M University at Galveston
200 Seawolf Parkway
Galveston, TX 77553-1675, United States

Dr. Ke Chen
Process Research and Development
Bristol-Myers Squibb Company
New Brunswick, NJ 08901, United States

Dr. David A. Conlon
Process Research and Development
Bristol-Myers Squibb Company
New Brunswick, NJ 08901, United States

Dr. Daniel P. Christen
3345 Winter Brook Pl.
Jamestown, NC 27282, United States

Michael T. Corbett
Department of Chemistry
Caudill and Kenan Laboratories, CB 3290
The University of North Carolina at Chapel Hill
Chapel Hill, NC 27599-3290, United States

Dr. Timothy T. Curran
Chemical Development
Vertex Pharmaceuticals
130 Waverly Street
Cambridge, MA 02139, United States

Dr. Kyle J. Eastman
Discovery Chemistry
Bristol-Myers Squibb Company
5 Research Parkway
Wallingford, CT 06492, United States

Dr. Matthew J. Fuchter
Department of Chemistry
Imperial College London
London SW7 2AZ
United Kingdom

Professor Micheal Fultz
Department of Chemistry
West Virginia State University
Institute, WV 25112, United States

Professor Brian Goess
Department of Chemistry
Furman University
3300 Poinsett Highway
Greenville, SC 29613, United States

Dr. Xiaojun Han
Discovery Chemistry
Bristol-Myers Squibb Company
5 Research Parkway
Wallingford, CT 06492, United States

Dr. Ying Han
Discovery Chemistry
Bristol-Myers Squibb Company
5 Research Parkway
Wallingford, CT 06492, United States

Dr. Richard A. Hartz
Discovery Chemistry
Bristol-Myers Squibb Company
5 Research Parkway
Wallingford, CT 06492, United States

Dr. Martin E. Hayes
Medicinal Chemistry
Abbott Bioresearch Center
381 Plantation Street
Worcester, MA, 01605, United States

Dr. Matthew D. Hill
Discovery Chemistry
Bristol-Myers Squibb Company
5 Research Parkway
Wallingford, CT 06492, United States

Dr. James Kempson
Discovery Chemistry
Bristol-Myers Squibb Company
P.O. Box 5400
Princeton, NJ 08543-5400, United States

Dr. Brian A. Lanman
Amgen, Inc.
Mailstop 29-1-B
One Amgen Center Drive
Thousand Oaks, CA 91320, United States

Dr. Jie Jack Li
Discovery Chemistry
Bristol-Myers Squibb Company
5 Research Parkway
Wallingford, CT 06492, United States

Noha S. Maklad
Department of Chemistry
Pfizer Global Research and Developme
Eastern Point Road
Groton, CT 06340, United States

Professor Richard J. Mullins
Department of Chemistry
Xavier University
3800 Victory Parkway
Cincinnati, OH 45207-4221, United Sta

Dr. Jennifer Xiaoxin Qiao
Discovery Chemistry
Bristol-Myers Squibb Company
P.O. Box 5400
Princeton, NJ 08543-5400, United State

Dr. Jeremy Richter
Discovery Chemistry
Bristol-Myers Squibb Company
P.O. Box 5400
Princeton, NJ 08543-5400, United State

Professor Nicole L. Snyder
Hamilton College
198 College Hill Road
Clinton, NY 13323, United States

Dr. Gerald J. Tanoury
Chemical Development
Vertex Pharmaceuticals Incorporated
130 Waverly Street
Cambridge, MA 02139-4242, United States

Dr. Stephen W. Wright
Department of Chemistry
Pfizer Global Research and Developmen
Eastern Point Road
Groton, CT 06340, United States

Dr. Thomas A. Wynn
Medicinal Chemistry
Alkermes Inc.
852 Winter St
Waltham MA, 02451-1420, United States

Dr. Bingwei Vera Yang
Discovery Chemistry
Bristol-Myers Squibb Company
P.O. Box 5400
Princeton, NJ 08543-5400, United States

Dr. Ian S. Young
Process Research and Development
Bristol-Myers Squibb Company
New Brunswick, NJ 08901, United States

Dr. Ji Zhang
Process Research and Development
Bristol-Myers Squibb Company
New Brunswick, NJ 08901, United States

Professor Alexandros Zografos
Department of Chemistry
Aristotle University of Thessaloniki
Thessaloniki 54124, Greece

1.1 Blum Aziridine Synthesis

Jeremy M. Richter

1.1.1 Description

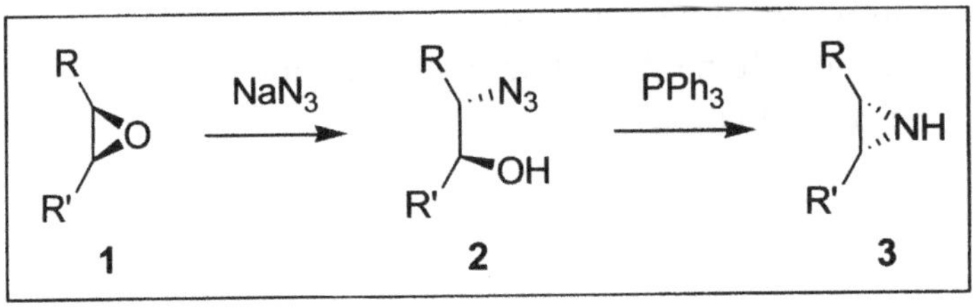

The Blum aziridine isomerization describes the net coversion of epoxides **1** into N–H aziridines **3** *via* an intermediate azido–alcohol **2**. The reaction proceeds by opening of the epoxide with an azide source followed by the Staudinger reduction and cyclization of the intermediate azido–alcohol to give the aziridine.[1] The reaction proceeds with net inversion of stereochemistry around the epoxide.

1.1.2 Historical Perspective

While investigating chemical carcinogens, Jonathan Blum and co-workers[1] hypothesized that arene imines were chemical intermediates in carcinogenesis and therefore sought to prepare several phenanthrene imines to test this hypothesis. However, they found that unsubstituted arene imines could not be prepared by the current methods and required alternate means by which to access these structures. Starting from the stable arene oxides (**4**), Blum and co-workers discovered that these epoxides could be opened by the action of sodium azide, forming the intermediate azido–alcohol **5**. They then found that further heating with triphenylphosphine provided the desired N–H aziridine **6** in good yield.

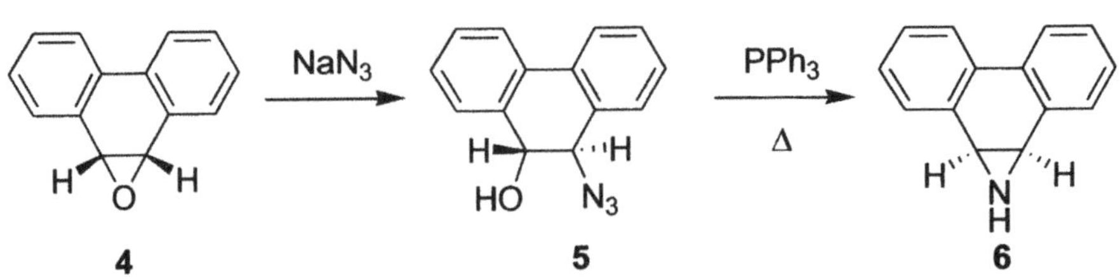

1.1.3 Mechanism

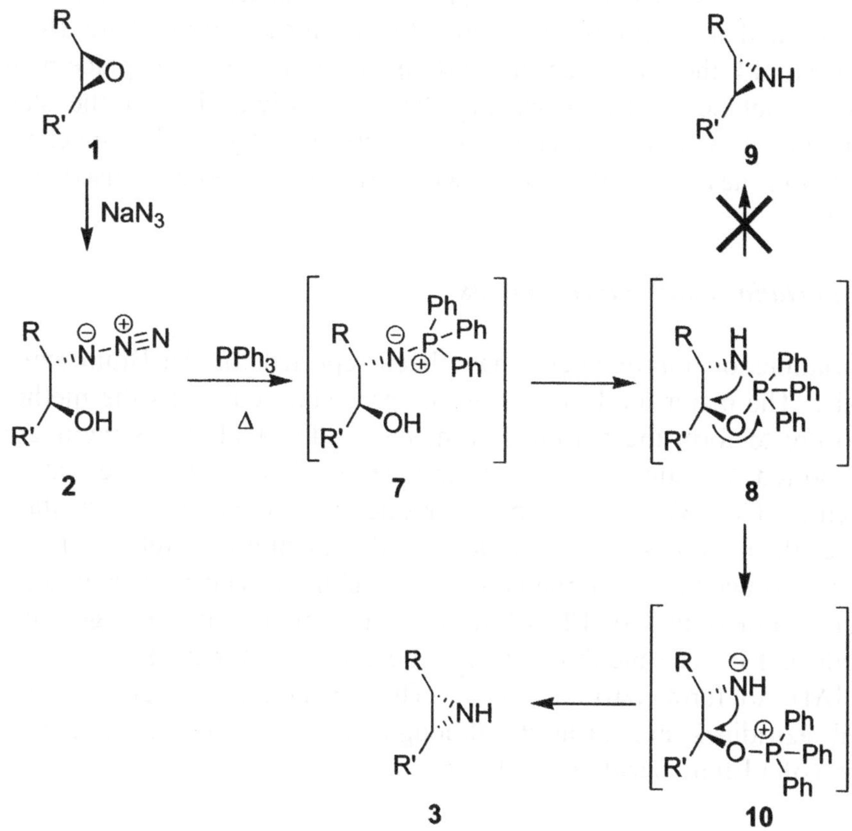

In Blum's original publication on the reaction that bears his name, he postulated a mechanistic interpretation of this process.[1] While investigating sterically flexible systems, he astutely noticed that the *cis/trans* nature of the double bond is conserved during the course of the reaction. That is, the reaction proceeds with complete inversion of stereochemistry but no *cis/trans* isomerisation is observed. In spite of this observation, two mechanistic interpretations could be put forth to explain the overall transformation. Little controversy exists surrounding the initial reaction intermediates and the preparation of the isolable azido–alcohol **2**. Good evidence also exists for the azide reduction proceeding through a standard Staudinger sequence, as the amino-alcohol can be easily isolated if no heat is applied to the system.[2] From this point, the mechanism can be explained by two competing pathways, diverging from the cyclic intermediate **8**. It would certainly seem reasonable to propose that the cyclization occurs in a concerted fashion (*cf.* Wittig Reaction), leading to direct expulsion of triphenylphosphine oxide from intermediate **8** (as shown by the arrows below). However, this reaction would proceed with retention of stereochemistry at the oxygen-containing carbon, leading to *cis/trans* isomerization. Experimental observations are in contrast to this mechanistic explanation (*vide supra*). Alternatively, the cyclic intermediate **8** could decompose to the linear intermediate **10**. The nitrogen would then be free to participate in an S_N2 displacement of triphenylphosphine oxide, preserving the *cis* configuration of the starting material, albeit with net inversion of stereochemistry. This mechanistic interpretation has been generally accepted and is widely used in the literature.[3]

1.1.4 Variations and Improvements

Few variations or improvements have been reported for the Blum aziridine synthesis. The major modification to the procedure concerns the method of ring closure to form the aziridine from the azido–alcohol. Several groups have reported variations in which the azido–alcohol **2** is activated for displacement (X = Ms, Ts, *etc.*) before reduction of the azide. This alternate sequence then allows for a milder and potentially higher efficiency cyclization to occur.[4,5] Alternatively, the Staudinger can be performed first to generate amino-alcohol **12**, followed by acylation of the nitrogen to give intermediate **13**. The alcohol can then be activated for displacement (*e.g.*, PPh_3/DIAD) to form aziridine **14**.[6,7] This method is especially useful if acylated aziridines are desired, although it does not satisfy the strict definition of a Blum aziridine synthesis.

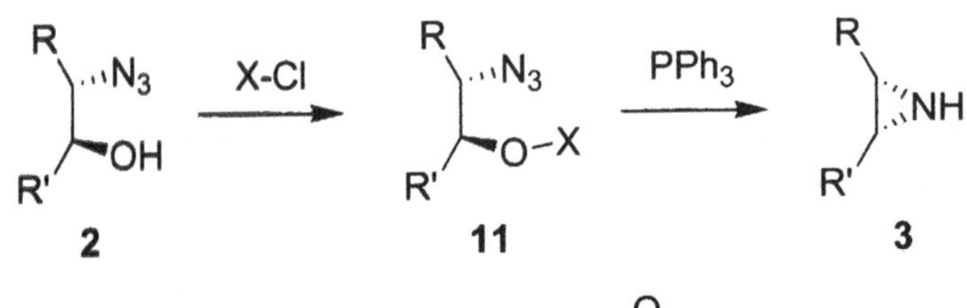

2 **11** **3**

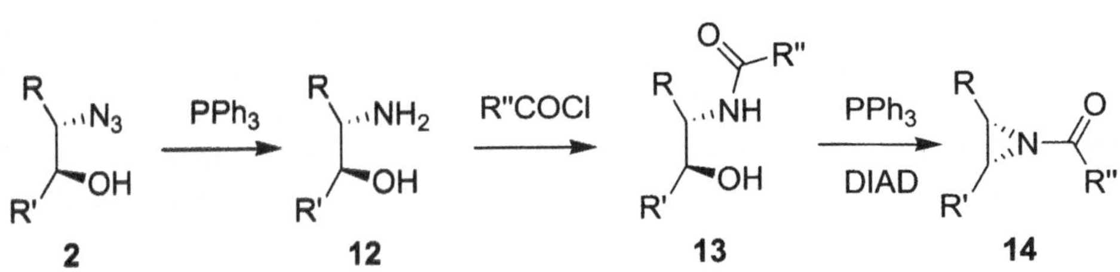

2 **12** **13** **14**

1.1.5 Synthetic Utility

Total Synthesis

The Blum aziridine synthesis has found widespread utility in the synthetic community. The field of total synthesis has especially benefited from the power of this transformation. In one example of how this reaction can be applied in complex natural product synthesis, Hanessian and co-workers applied a Blum reaction in the total synthesis of chlorodysinosin A (**17**).[8] In this sequence, epoxide **15** was opened with sodium azide and the primary alcohol protected as the silyl ether. The azide was then treated with triphenylphosphine and heat, resulting in concomitant reduction and aziridine formation to give intermediate **16**. Aziridine **16** was eventually processed to chlorodysinosin A (**17**).

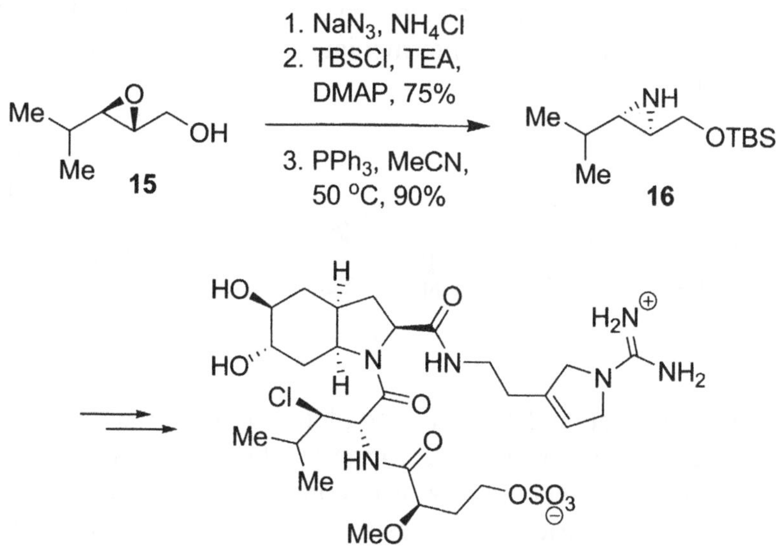

chlorodysinosin A (**17**)

Somfai and Åhman have applied the Blum aziridine synthesis to the total synthesis of indolizidine 209D (**20**).[9,10] Epoxide **18** was opened with sodium azide and the primary alcohol protected as the silyl ether. The azide was then treated with triphenylphosphine and heat, resulting in concomitant reduction and cyclization to give intermediate **19**. Aziridine **19** was eventually processed to indolizidine 209D (**20**).

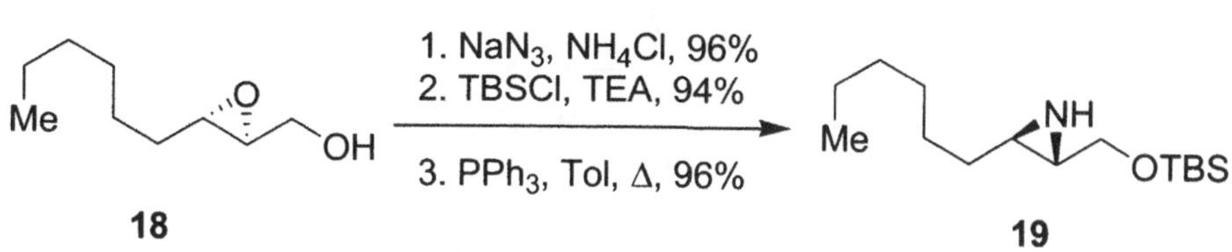

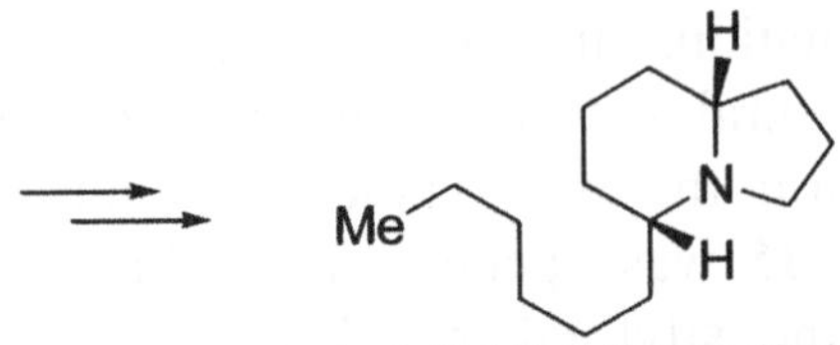

indolizidine 209D (**20**)

Bäckvall and co-workers used the Blum reaction in the total synthesis of ferruginine (**23**).[11] Epoxide **21** was opened with sodium azide and the resultant azido–alcohol was reduced and cyclized with triphenylphosphine in good yield to give aziridine **22**. Intermediate **22** was eventually processed to ferruginine (**23**).

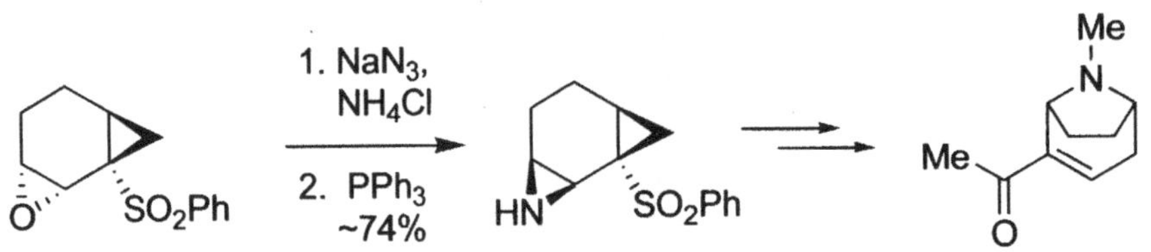

21 **22** ferruginine (**23**)

Finally, Tanner and Somfai completed a formal total synthesis of thienamycin (**26**) using the Blum aziridine synthesis as a key step.[12] As in the previous examples, epoxide **24** was converted to aziridine **25** in good yield using a Blum aziridine synthesis. Intermediate **25** was eventually processed to thienamycin (**26**).

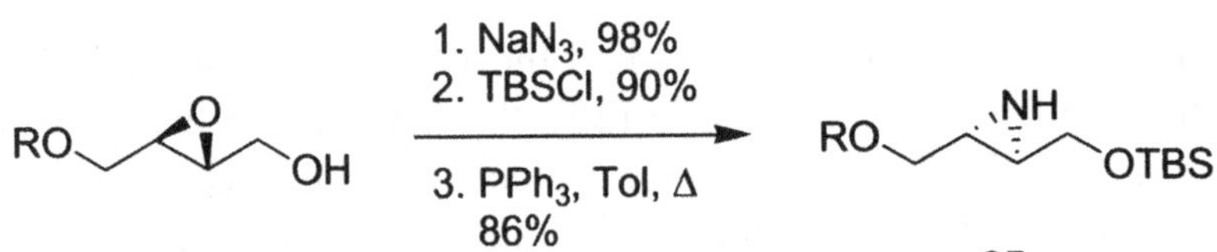

24 **25**

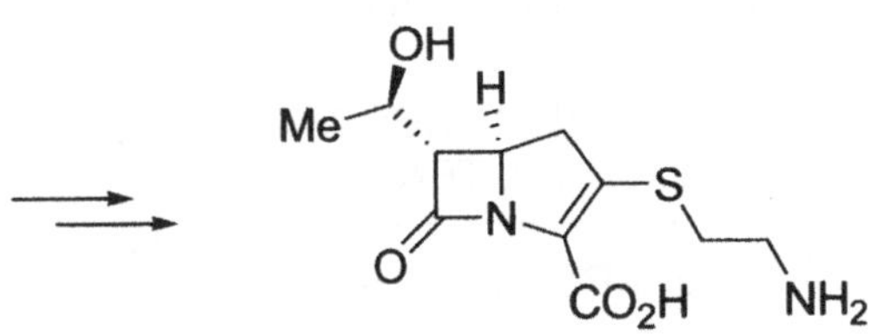

thienamycin (**26**)

Medicinal and Process Chemistry

Robinson and co-workers reported the preparation of aromatase inhibitors that used the Blum aziridine synthesis as a key step.[13] Epoxy–steroid **27** was opened with sodium azide to form the azido-alcohol, which was then reductively cyclized with triphenylphosphine and heat to provide the desired aziridine **28**. Analog **28** exhibited modest inhibitory activity toward human placental aromatase.

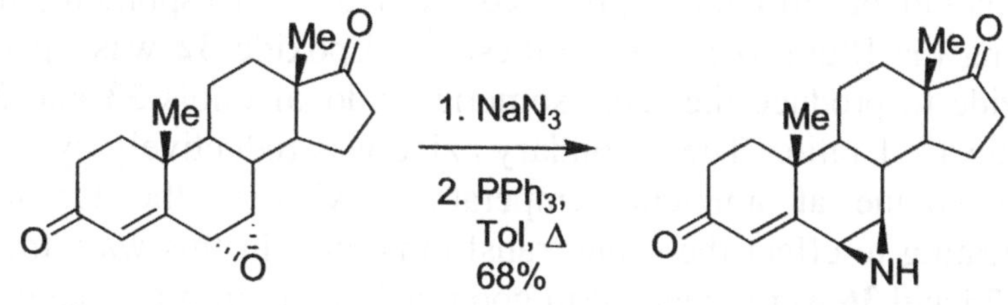

27 **28**

Researchers at Bristol-Myers Squibb reported the preparation of an epothilone analog using the Blum aziridine synthesis as a key step.[14] Epothilone C (**29**) was epoxidized to provide intermediate **30**. Opening of the epoxide with sodium azide followed by reductive cyclization forged the desired N–H aziridine analogue **31** in good yield, especially in light of the complex setting for this transformation.

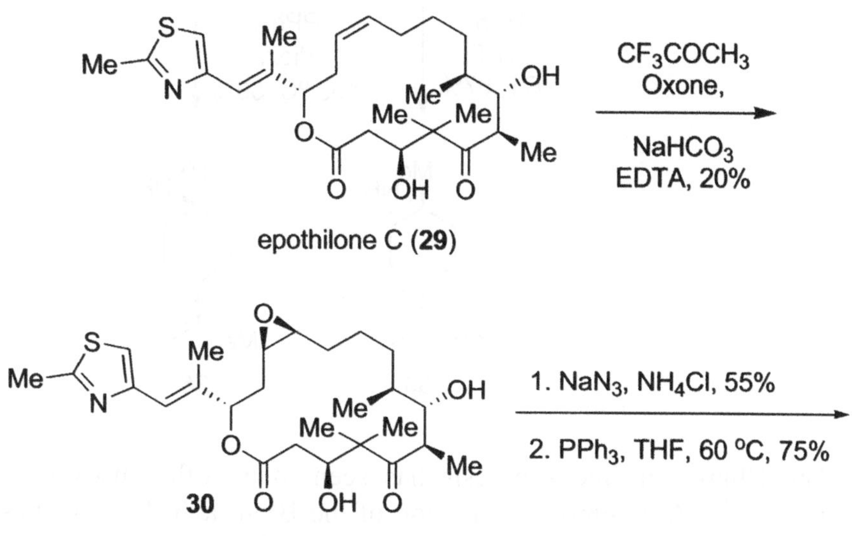

30

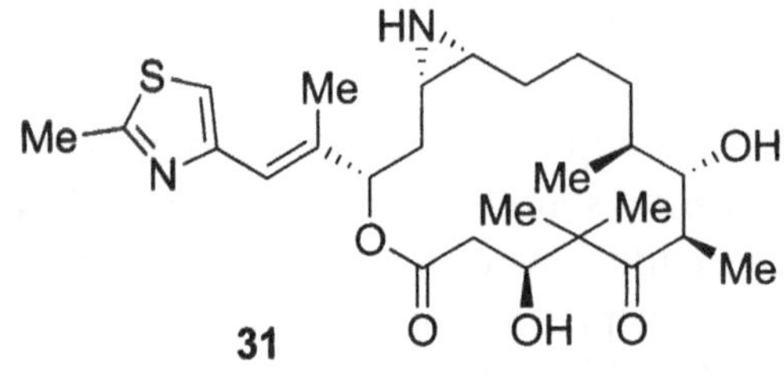

31

Researchers at Lexicon Pharmaceuticals found that limonene aziridines could be efficiently prepared from the corresponding limonene oxides using the Blum aziridine synthesis.[15] Epoxide **32** was opened with sodium azide to produce the regioisomeric azido–alcohols **33** and **34** in an approximate 1 : 1 ratio. The secondary azide was reductively cyclized with triphenylphosphine at ambient temperature, whereas the tertiary azide required heating to effect the same transformation. In this way, the desired aziridines **35** and **36** were prepared in good yield on multigram scale.

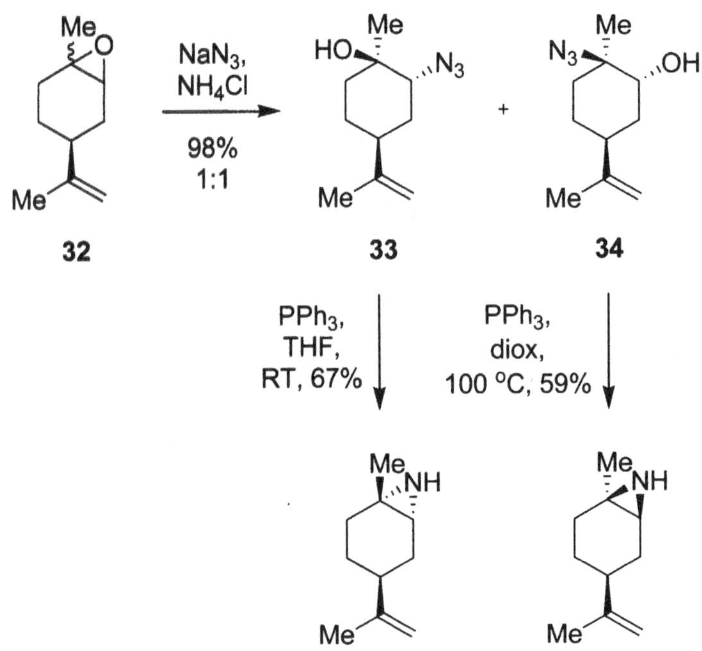

The Blum aziridine synthesis has seen many other uses since its development.[16,17] A common application of the Blum aziridine synthesis is in the preparation of authentic standards while developing novel reactions or synthetic routes.[18–20] The reaction has also found broad utility in the preparation of starting materials or intermediates in novel reaction development.[21–31]

1.1.6 Experimental

Blum aziridine synthesis to prepare a chlorodysinosin A intermediate 16[8]

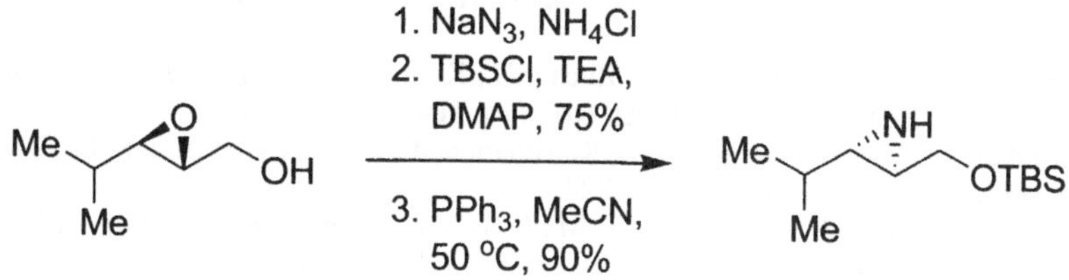

15 **16**

To a solution of **15** in $CH_3OCH_2CH_2OH$ (0.7 M) was added NaN_3 (10 equiv), NH_4Cl (2 equiv), and water (4 equiv). The mixture was heated to reflux for 24 h, cooled to room temperature, and the solvent removed under vacuum. The residue was then taken up in water, extracted with EtOAc, dried over Na_2SO_4, and concentrated. After drying under reduced pressure for 18 h, the resulting amber oil was taken up in CH_2Cl_2, cooled to 0 °C, and treated sequentially with Et_3N (1.5 equiv), DMAP (cat.), and TBSCl (1.1 equiv). The solution was stirred at 0 °C for 2 h, diluted with CH_2Cl_2, washed with 1 M HCl, dried over Na_2SO_4, and concentrated under vacuum. The crude residue was purified by flash chromatography over silica gel (8% EtOAc/hexanes) to give the azido–alcohol as a 1:1 mixture of regioisomers (75%).

The mixture of azido–alcohols was dissolved in MeCN (0.5 M) and treated with PPh_3 (1.1 equiv). The solution was stirred at RT for 2 h, then heated to 50 °C for 18 h. After removal of MeCN under vacuum, the mixture was taken up in Et_2O, filtered through a pad of Celite, and the filtrate concentrated under vacuum. The crude residue was purified by flash chromatography over silica gel (25% EtOAc/hexanes) to give **16** as a colorless liquid (90% yield).

1.1.7 References

1 Ittah, Y.; Sasson, Y.; Shahak, I.; Tsaroom, S.; Blum, J. *J. Org. Chem.* **1978,** *43*, 4271–4273.
2 Blum, J.; Yona, I.; Tsaroom, S.; Sasson, Y. *J. Org. Chem.* **1979,** *44*, 4178–4182.
3 Fürmeier, S.; Metzger, J. O. *Eur. J. Org. Chem.* **2003,** 649–659.
4 Tanner, D.; Birgersson, C.; Gogoll, A. *Tetrahedron* **1994,** *50*, 9797–9824.
5 Bromfield, K. M.; Gradén, H.; Hagberg, D. P.; Olsson, T.; Kann, N. *Chem. Commun.* **2007,** 3183–3185.
6 Page, M. F. Z.; Jalisatgi, S. S.; Maderna, A.; Hawthorne, M. F. *Synthesis* **2008,** 555–563.
7 Hirose, T.; Sunazuka, T.; Tsuchiya, S.; Tanaka, T.; Kojima, Y.; Mori, R.; Iwatsuki, M.; Ōmura, S. *Chem. Eur. J.* **2008,** *14*, 8220–8238.

8 Hanessian, S.; Del Valle, J. R.; Xue, Y.; Blomberg, N. *J. Am. Chem. Soc.* **2006,** *128*, 10491–10495.

9 Åhman, J.; Somfai, P. *Tetrahedron Lett.* **1995,** *36*, 303–306.

10 Åhman, J.; Somfai, P. *Tetrahedron* **1995,** *51*, 9747–9756.

11 Bäckvall, J.-E.; Jonsson, S. Y.; Löfström, C. M. G. *J. Org. Chem.* **2000,** *65*, 8454–8457.

12 Tanner, D.; Somfai, P. *Tetrahedron Lett.* **1987,** *28*, 1211–1214.

13 Njar, V. C. O.; Hartmann, R. W.; Robinson, C. H. *J. Chem. Soc., Perkin Trans. 1* **1995,** 985–991.

14 Regueiro-Ren, A.; Borzilleri, R. M.; Zheng, X.; Kim, S.-H.; Johnson, J. A.; Fairchild, C. R.; Lee, F. Y. F.; Long, B. H.; Vite, G. D. *Org. Lett.* **2001,** *3*, 2693–2696.

15 Voronkov, M. V.; Gontcharov, A. V.; Kanamarlapudi, R. C.; Richardson, P. F.; Wang, Z.-M. *Org. Process Res. Dev.* **2005,** *9*, 221–224.

16 Zhao, Y.-J.; Tan, L.-J. S.; Li, B.; Li, S.-M.; Loh, T.-P. *Chem. Commun.* **2009,** 3738–3740.

17 Metzger, J. O.; Fürmeier, S. *Eur. J. Org. Chem.* **1999,** 661–664.

18 Hirose, T.; Sunazuka, T.; Tsuchiya, S.; Tanaka, T.; Kojima, Y.; Mori, R.; Iwatsuki, M.; Omura, S. *Chem. Eur. J.* **2008,** *14*, 8220–8238.

19 Serafin, S. V.; Zhang, K.; Aurelio, L.; Hughes, A. B.; Morton, T. H. *Org. Lett.* **2004,** *6*, 1561–1564.

20 Jin, X. L.; Sugihara, H.; Daikai, K.; Tateishi, H.; Jin, Y. Z.; Furuno, H.; Inanaga, J. *Tetrahedron* **2002,** *58*, 8321–8329.

21 Wipf, P.; Venkatraman, S.; Miller, C. P. *Tetrahedron Lett.* **1995,** *36*, 3639–3642.

22 Mordini, A.; Sbaragli, L.; Valacchi, M.; Russo, F.; Reginato, G. *Chem. Commun.* **2002,** 778–779.

23 Crotti, P.; Di Bussolo, V.; Favero, L.; Macchia, F.; Renzi, G.; Roselli, G. *Tetrahedron* **2002,** *58*, 7119–7133.

24 Pulipaka, A. B.; Bergmeier, S. C. *Synthesis* **2008,** 1420–1430.

25 Marples, B. A.; Toon, R. C. *Tetrahedron Lett.* **1999,** *40*, 4873–4876.

26 Oh, K.; Parson, P. J.; Cheshire, D. *Synlett* **2004,** 2771–2775.

27 Yadav, J. S.; Bandyopadhyay, A.; Reddy, B. V. S. *Synlett* **2001,** 1608–1610.

28 Watson, I. D. G.; Yudin, A. K. *J. Org. Chem.* **2003,** *68*, 5160–5167.

29 Yu, L.; Kokai, A.; Yudin, A. K. *J. Org. Chem.* **2007,** *72*, 1737–1741.

30 Tsang, D. S.; Yang, S.; Alphonse, F.-A.; Yudin, A. K. *Chem. Eur. J.* **2008,** *14*, 886–894.

31 O'Brien, P.; Pilgram, C. D. *Org. Biomol. Chem.* **2003,** *1*, 523–534.

1.2 Gabriel–Heine Aziridine Isomerization

Jeremy M. Richter

1.2.1 Description

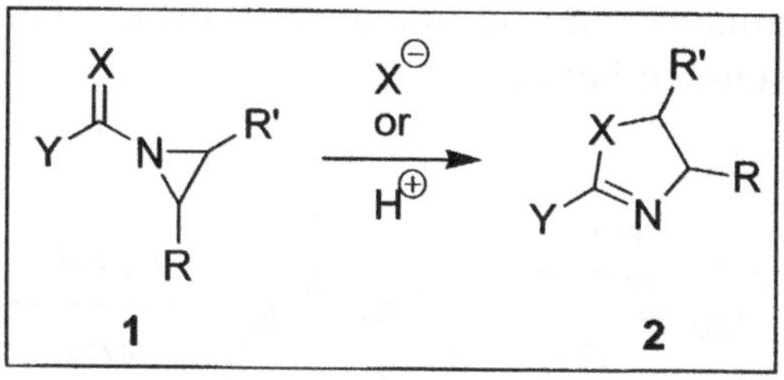

The Gabriel–Heine aziridine isomerization describes the rearrangement of an acylaziridine **1** into an oxazoline **2**.[1–5]

1.2.2 Historical Perspective

Harold Heine and co-workers were working with ethylenimines, and derivatives thereof, when they became intrigued by the reaction originally reported by Gabriel and coworkers in which thioacyl aziridine **3** was isomerized to thiazoline **4** upon attempted distillation.[1] This report went virtually unnoticed until Heine and co-workers decided to investigate the reaction further.[2] They found that upon exposure of **5** in refluxing heptane to small amounts of aluminum halides, oxazoline **6** was isolated in nearly quantitative yield. They also discovered that the reaction does not occur under purely thermal conditions, but catalysis is required.

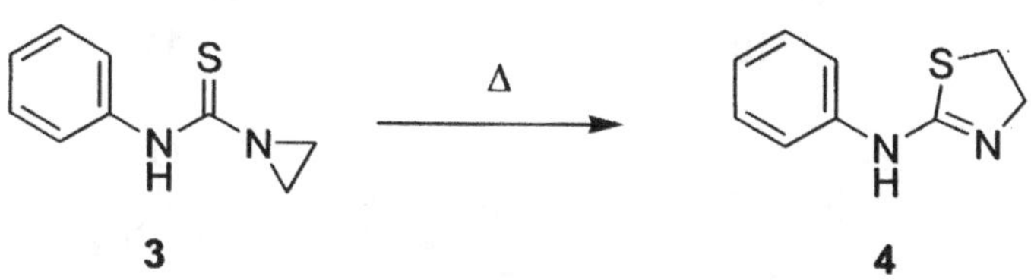

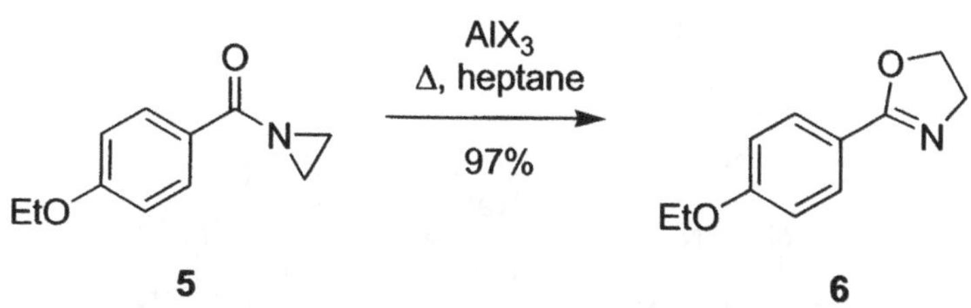

Further work from the Heine group revealed that the reaction could be accomplished under milder conditions and that the choice of catalyst determined the regiochemical outcome of the reaction.[3] For example, in the presence of iodide, **7** rearranges to **8**, wherein the least substituted carbon atom migrates. Alternatively, upon exposure to acid, **7** rearranges to **9**, wherein the most substituted carbon atom migrates. These results suggest that alternate mechanisms may be operable in these two transformations, a point that will be discussed below.

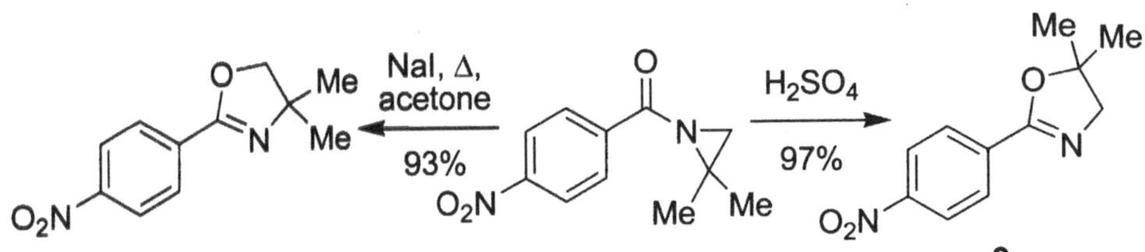

Heine also demonstrated that the rearrangement can occur on differentially acylated aziridines to give rise to different heterocycles such as imidazolines (**11**), imidazolones (**13**), and complex heterocycles such as **15**.[4,5] Futhermore, this reaction has been the subject of reviews by Heine and others.[6,7]

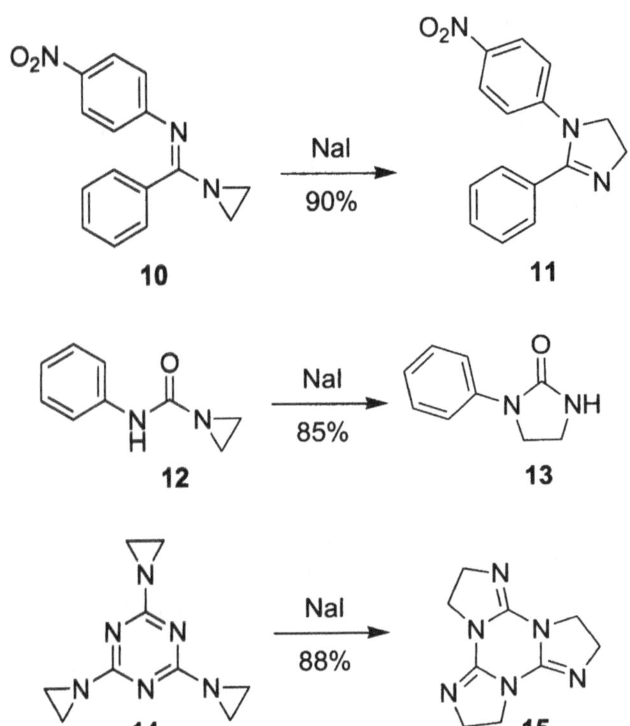

1.2.3 Mechanism

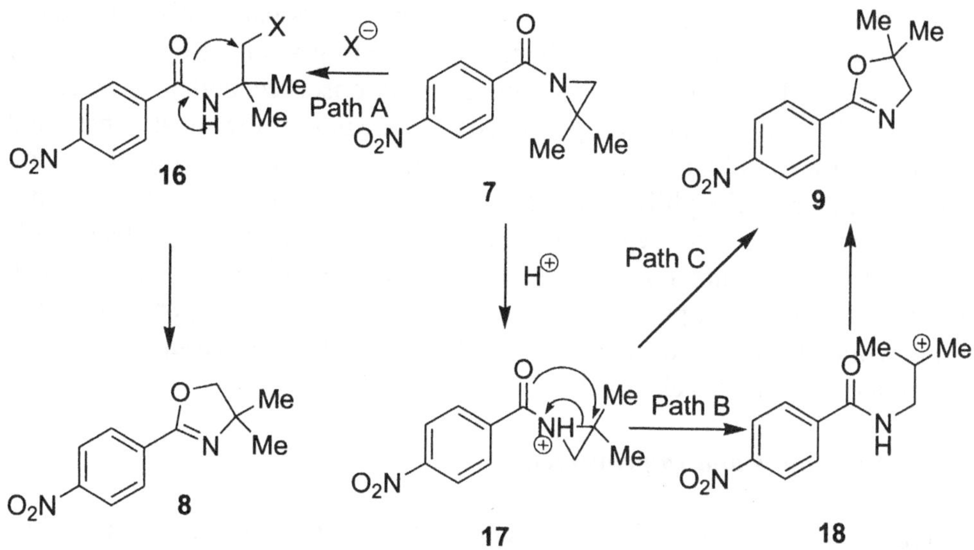

In Heine's seminal publications, he astutely noticed that different products are observed under acidic and nucleophilic catalysis.[3] Heine put forth a mechanistic explanation for the formation of both of these products.[6] He proposed that the rearrangement of **7** under nucleophilic catalysis proceeds through aziridine ring opening with the halide followed by displacement with oxygen to form **8** (two sequential S_N2 reactions, Path A). Alternatively, he proposed that the acid-catalyzed rearrangement of **7** to **9** proceeds through protonation of the aziridine nitrogen, formation of a tertiary carbocation, and subsequent attack by the oxygen on this carbocation (Path B). This mechanistic explanation of the nucleophilic catalysis (*i.e.,* **7** to **8**) is still generally accepted and has received further support as more examples have been presented in the literature.[8] His explanation for the acid-catalyzed rearrangement has come under scrutiny, though, and has been modified through further studies. Shortly after Heine put forth his mechanistic interpretation, Nishiguchi and co-workers proposed that the acid catalyzed rearrangement could potentially proceed through an S_Ni reaction (*i.e.,* Path C).[9] Convincing clarity in terms of both nucleophilic and acidic catalysis was not gained until the computational and experimental studies of Hori and co-workers, who concluded that two sequential S_N2 reactions (Path A) were likely operable for nucleophilic catalysis and an S_Ni pathway (Path C) accounted for the observed product under acidic catalysis.[10]

Further evidence exists to support these mechanistic interpretations, in that retention of configuration is observed if the aziridine is chiral in nature.[11] The mechanism is thus limited to either a double inversion or front-side attack to account for this retention of stereochemistry.

Any mechanistic discussion of reactions of this nature also inherently requires comments on the regioselectivity of the migration. Regioselectivity in the Gabriel–Heine aziridine isomerization is observed under a variety of conditions.[12,13] As mentioned previously (*vide supra*), the most substituted carbon migrates under acidic catalysis and the least substituted carbon migrates under nucleophilic catalysis. This trend holds for acyl shifts, but thioacyl shifts are less regioselective, with more scrambling observed.[14] Finally, Eastwood and co-workers showed that aziridines substituted with electron-donating groups form 2,4-substituted oxazolines upon rearrangement, while those substituted with electron-withdrawing groups form 2,5-substituted oxazolines selectively.[15]

1.2.4 Variations and Improvements

Several variations and improvements of the Gabriel–Heine aziridine isomerization have been reported, primarily surrounding the catalyst selection and/or reaction conditions. An electrochemical rearrangement has been reported, which gave the desired oxazolines in moderate yield.[16] In addition to the catalysts initially reported, several other catalysts have been used in this reaction, including TfOH,[17] Yb(biphenol)OTf, Ti(O*i*-Pr)$_4$, Zr(Cp)$_2$(SbF$_6$)$_2$,[18] Cu(OTf)$_2$, Zn(OTf)$_2$, BF_3·OEt_2, $MgBr_2$·OEt_2,[19] Sn(OTf)$_2$,[20,21] Mn(salen),[22] P_2S_5,[23] and TBAI.[24] The reaction has also been performed in the microwave in good yields and rapid reaction times.[25,26] Finally, one major variation of the Gabriel–Heine aziridine isomerization is the use of similar reaction conditions to convert acyl–aziridines into oxazolidinones using BF_3·OEt_2.[4,27,28]

1.2.5 Synthetic Utility

Total Synthesis

The Gabriel–Heine aziridine isomerization has been used only twice in the context of total synthesis, despite finding widespread utility in alternate contexts. The Vogel group has prepared 3-amino-3-deoxy-L-talose (**21**) using the Gabriel–Heine reaction as a key step.[29–31] The synthesis commences with readily available aziridine **19**, which undergoes a Gabriel–Heine rearrangement under triflic acid catalysis at 80 °C in hexafluoroisopropanol to generate oxazoline **20**. This intermediate was further transformed into 3-amino-3-deoxy-L-talose (**21**).

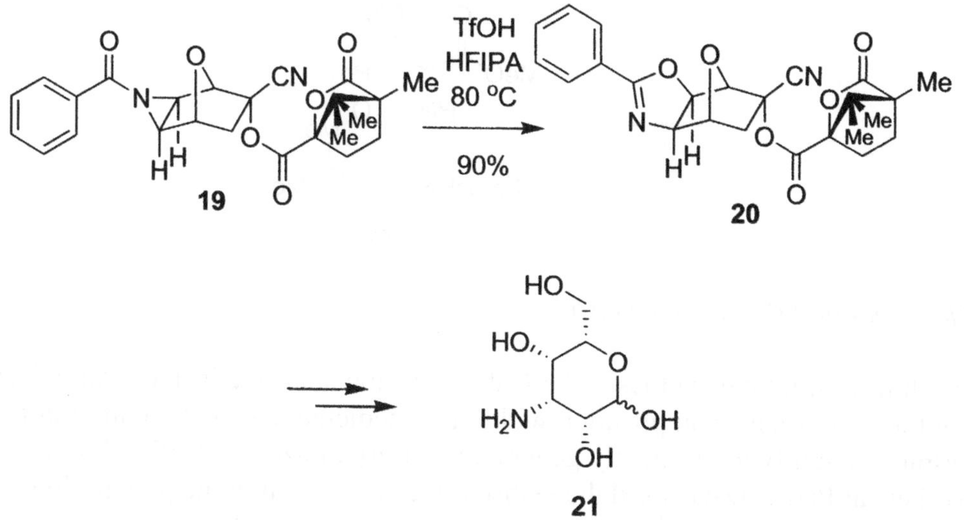

Cardillo and co-workers used the Gabriel–Heine aziridine isomerization during the preparation of a dipeptide fragment of lysobactin.[32–36] Both enantiomers of the fragment were prepared using this reaction. The group first prepared the incorrect diastereomer of the target fragment by ring expansion of aziridine **22** to oxazoline **23** using $BF_3 \cdot OEt_2$ in nearly quantitative yield. It is surprising that the diastereomer (**24**) corresponding to the natural product was rearranged solely by the action of triethyl amine and dimethylaminopyridine to give **25** in 75% yield. This compound was further processed to the desired intermediate for lysobactin (**26**).

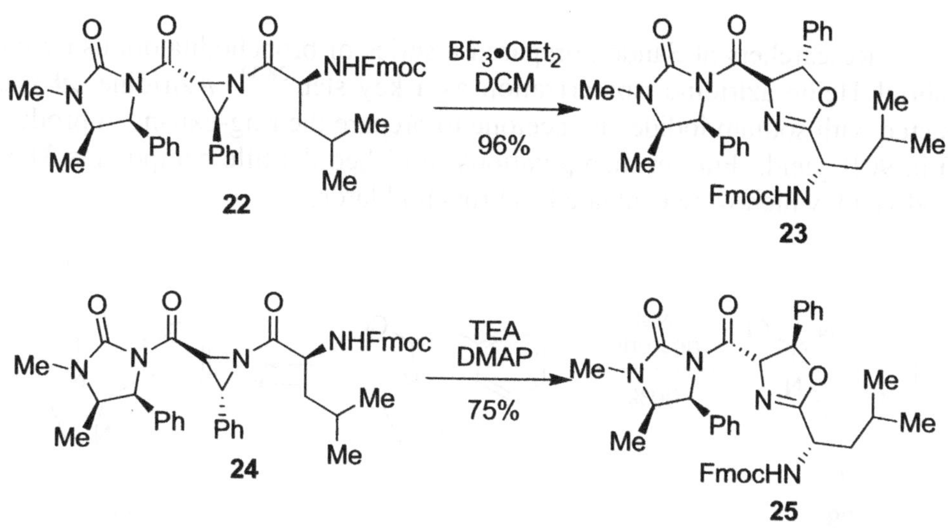

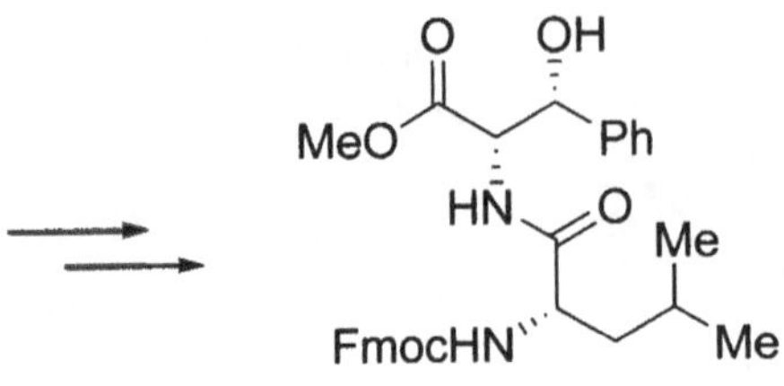

26

Medicinal and Process Chemistry

Much like with total synthesis, the Gabriel–Heine aziridine isomerization has not found widespread application within the medicinal chemistry community, despite its ability to efficiently generate a variety of azoles. DeWald and co-workers at Parke–Davis used the Gabriel–Heine reaction in the preparation of potential antipsychotics.[37] Compound **27** was rearranged upon exposure to sodium iodide in acetone to provide compounds of type **28** in moderate to good yield, which were evaluated as nondopamine-binding antipsychotics.

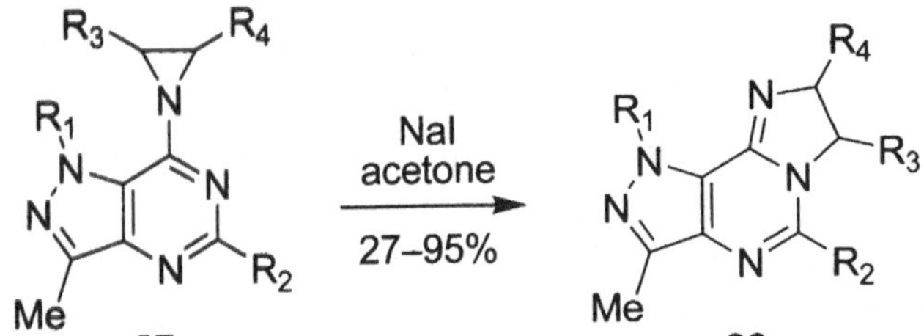

27 28

Researchers at Sandoz prepared a series of bronchodilators using the Gabriel–Heine aziridine isomerization as a key step.[38,39] Aziridine **29** was reacted with sodium iodide and acetone to prepare the ring-expanded product **30** in 90% yield. Further manipulations furnished the title compounds **31** in good yield, which were evaluated as bronchodilators.

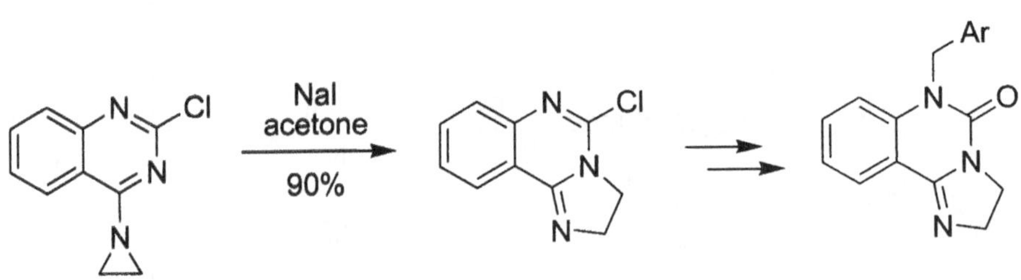

29 30 31

Iqbal and co-workers used the Gabriel–Heine aziridine isomerization to confirm the stereochemistry of an intermediate for preparation of an HIV protease inhibitors.[40] Aziridine **32** was rearranged with sodium iodide in acetonitrile to give the desired oxazoline **33**, which enabled verification of the stereochemistry in the previous step.

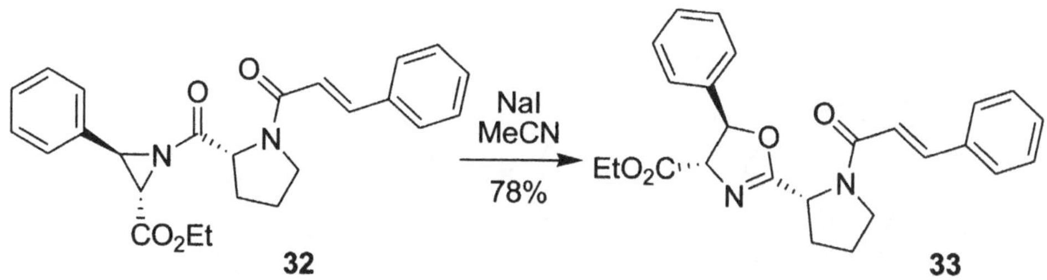

Miscellaneous Examples

Kohn and Jung reported the conversion of alkenes into vicinal diamines using the Gabriel–Heine aziridine isomerization as a key step during this sequence.[41] Alkene **34** was converted into the aziridine **35** using standard chemistry. Rearrangment under the action of sodium iodide provided the desired imidazolone, which was hydrolyzed with barium hydroxide to give the vicinal diamine **36**.

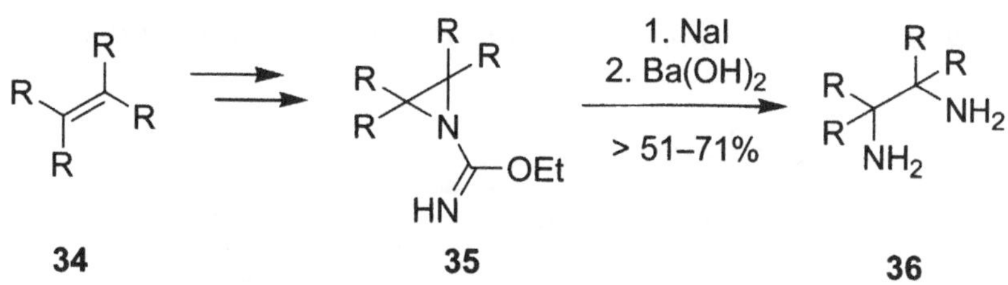

Bonini and co-workers reported the use of the Gabriel–Heine aziridine isomerization to prepare ferrocenyl-oxazolines.[42] Bis-aziridine **37** was rearranged with sodium iodide in acetonitrile to provide the desired bisoxazoline **38** in excellent yield.

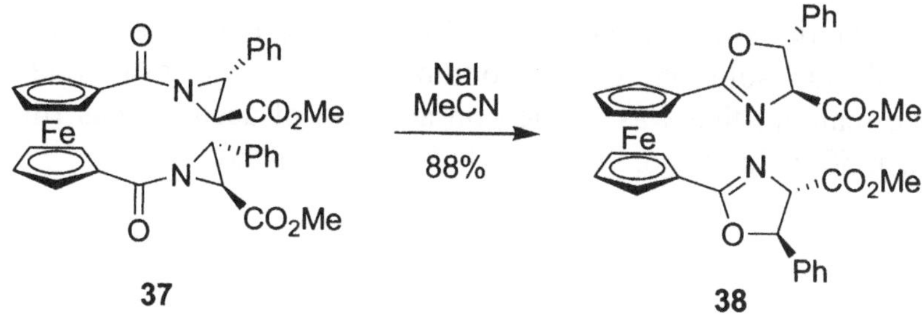

The Banks and Mattay groups independently reported the use of the Gabriel–Heine aziridine isomerization in the context of fullerenes.[43,44] Acylaziridinofullerine **39** was rearranged simply by heating to provide the oxazininofullerine **40** in quantitative yield.

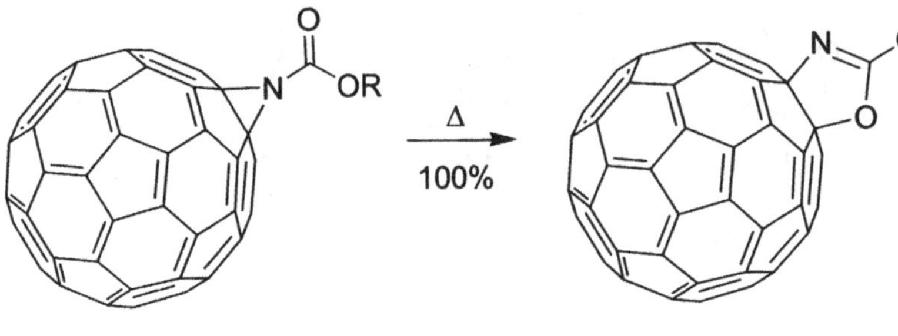

39 **40**

The Gabriel–Heine aziridine isomerization has seen many other uses since its development, whether in the preparation of starting materials or for studies of fundamental reactivities.[45–54]

1.2.6 Experimental

Gabriel–Heine Aziridine Isomerization to Prepare Bronchodilators[39]

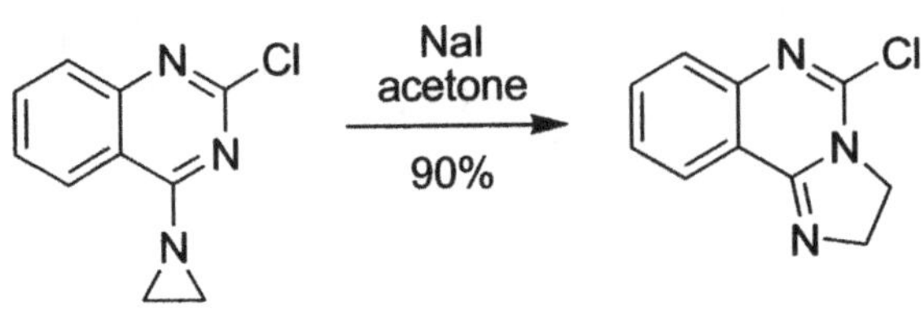

29 **30**

Compound **29** (32 g, 0.16 mol) was dissolved in anhydrous acetone (500 mL) and stirred with 3.2 g (0.021 mol) of NaI for 90 min. The solvent was evaporated *in vacuo* and the residue treated with methylene chloride. Insoluble material was filtered off, and the filtrate extracted with saturated sodium chloride solution, dried (sodium sulfate), and evaporated. The crystalline material that was formed on addition of acetone was filtered off and dried at 50 °C in vacuo, yielding 28.7 g of **30** (90%).

Gabriel–Heine Aziridine Isomerization to Prepare the Lysobactin Fragment[32]

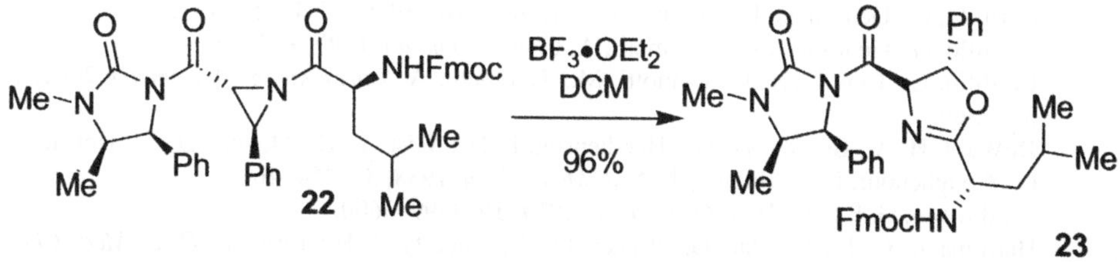

A solution of **22** (0.24 g, 0.36 mmol) in CH_2Cl_2 (5 mL) was treated with $BF_3{\cdot}Et_2O$ (0.045 mL, 0.36 mmol) at room temperature for 6 h. The reaction was quenched with sat. $NaHCO_3$ (3 mL), extracted three times with CH_2Cl_2, and dried over Na_2SO_4. The solvent was evaporated at reduced pressure to afford **23** (0.23 g, 96%), which required no further purification.

1.2.7 References

1. Gabriel, S.; Stelzner, R. *Ber.* **1895,** *28*, 2929.
2. Heine, H. W.; Proctor, Z. *J. Org. Chem.* **1958,** *23*, 1554–1556.
3. Heine, H. W.; Fetter, M. E.; Nicholson, E. A. *J. Am. Chem. Soc.* **1959,** *81*, 2202–2204.
4. Heine, H. W.; Bender, H. S. *J. Org. Chem.* **1960,** *25*, 461–463.
5. Heine, H. W.; Kenyon, W. G.; Johnson, E. M. *J. Am. Chem. Soc.* **1961,** *83*, 2570–2574.
6. Heine, H. W. *Angew. Chem., Int. Ed.* **1962,** *1*, 528–532.
7. McCoull, W.; Davis, F. A. *Synthesis* **2000,** 1347–1365.
8. Jung, S.-H.; Kohn, H. *J. Am. Chem. Soc.* **1985,** *107*, 2931–2943.
9. Nishiguchi, T.; Tochio, H.; Nabeya, A.; Iwakura, Y. *J. Am. Chem. Soc.* **1969,** *91*, 5835–5841.
10. Hori, K.; Nishiguchi, T.; Nabeya, A. *J. Org. Chem.* **1997,** *62*, 3081–3088.
11. Cardillo, G.; Gentilucci, L.; Mohr, G. P. *Eur. J. Org. Chem.* **2001,** 3545–3551.
12. El Shafei, Z. M.; Guthrie, R. D. *J. Chem. Soc. C* **1970,** 843–846.
13. Deutsch, A. S.; Fanta, P. E. *J. Org. Chem.* **1956,** *21*, 892–895.
14. Nishiguchi, T.; Tochio, H.; Nabeya, A.; Iwakura, Y. *J. Am. Chem. Soc.* **1969,** *19*, 5841–5846.
15. Eastwood, F. W.; Perlmutter, P.; Yang, Q. *Tetrahedron Lett.* **1994,** *35*, 2039–2042.
16. Archier-Jay, D.; Besbes, N.; Laurent, A.; Laurent, E.; Stamm, H.; Tardivel, R. *Tetrahedron Lett.* **1989,** *30*, 2271–2272.
17. Eastwood, F. W.; Perlmutter, P.; Yang, Q. *J. Chem. Soc., Perkin Trans. 1.* **1997,** 35–42.
18. Ferraris, D.; Drury, W. J. III; Cox, C.; Lectka, T. *J. Org. Chem.* **1998,** *63*, 4568–4569.
19. Cardillo, G.; Gentilucci, L.; Gianotti, M.; Tolomelli, A. *Synlett.* **2000,** 1309–1311.
20. Lucarini, S.; Tomasini, C. *J. Org. Chem.* **2001,** *66*, 727–732.
21. Luppi, G.; Tomasini, C. *Synlett* **2003,** 797–800.
22. Nishimura, M.; Minakota, S.; Takahashi, T.; Oderaotoshi, Y.; Komatsu, M. *J. Org. Chem.* **2002,** *67*, 2101–2110.
23. Vingiello, F. A.; Rorer, M. P.; Ogliaruso, M. A. *Chem. Commun.* **1971,** 329.
24. Tomalia, D. A. *J. Heterocyclic Chem.* **1967,** *4*, 419–421.
25. Cardillo, G.; Gentilucci, L.; Gianotti, M.; Tolomelli, A. *Tetrahedron* **2001,** *57*, 2807–2812.
26. Samimi, H. A.; Mamaghani, M.; Tabatabaeian, K. *J. Heterocyclic Chem.* **2008,** *45*, 1765–1770.
27. Tomasini, C.; Vecchione, A. *Org. Lett.* **1999,** *1*, 2153–2156.
28. Olofsson, B.; Somfai, P. *J. Org. Chem.* **2002,** *67*, 8574–8583.
29. Allemann, S.; Vogel, P. *Synlett* **1993,** 801–803.
30. Hünenberger, P.; Allemann, S.; Vogel, P. *Carbohydrate Res.* **1994,** *257*, 175–187.
31. Allemann, S.; Vogel, P. *Synthesis* **1991,** 923–928.

32. Cardillo, G.; Gentilucci, L.; Gianotti, M.; Tolomelli, A. *Eur. J. Org. Chem.* **2000,** 2489–2494.
33. Cardillo, G.; Gentilucci, L.; Tolomelli, A.; Tomasini, C. *Tetrahedron Lett.* **1997,** *38*, 6953–6956.
34. Cardillo, G.; Gentilucci, L.; Tolomelli, A. *Tetrahedron* **1999,** *55*, 15151–15158.
35. Cardillo, G.; Gentilucci, L.; Tolomelli, A. *Chem. Commun.* **1999,** 167–168.
36. Cardillo, G.; Gentilucci, L.; Gianotti, M.; Tolomelli, A. *Tetrahedron: Asymmetry* **2001,** *12*, 563–569.
37. DeWald, H. A.; Beeson, N. W.; Hershenson, F. M.; Wise, L. D.; Downs, D. A.; Heffner, T. G.; Coughenour, L. L.; Pugley, T. A. *J. Med. Chem.* **1988,** *31*, 454–461.
38. Hardtmann, G. E.; Ott, H. *J. Org. Chem.* **1974,** *39*, 3599–3600.
39. Hardtmann, G. E.; Koletar, G.; Pfister, O. R.; Gogerty, J. H.; Iorio, L. C. *J. Med. Chem.* **1975,** *18*, 447–453.
40. Saha, B.; Nandy, J. P.; Shukla, S.; Siddiqui, I.; Iqbal, J. *J. Org. Chem.* **2002,** *67*, 7858–7860.
41. Kohn, H.; Jung, S.-H. *J. Am. Chem. Soc.* **1983,** *105*, 4106–4108.
42. Bonini, B. F.; Fochi, M.; Comes-Franchini, M.; Ricci, A.; Thijs, L.; Zwanenburg, B. *Tetrahedron: Asymmetry* **2003,** *14*, 3321–3327.
43. Banks, M. R.; Cadogan, J. I. G.; Gosney, I.; Hodgson, P. K. G.; Langridge-Smith, P. R. R.; Millar, J. R. A.; Taylor, A. T. *Tetrahedron Lett.* **1994,** *35*, 9067–9070.
44. Averdung, J.; Mattay, J.; Jacobi, D.; Abraham, W. *Tetrahedron* **1995,** *51*, 2543–2552.
45. Hassner, A.; Burke, S. S.; Cheng-Fan, I, J. *J. Am. Chem. Soc.* **1975,** *97*, 4692–4700.
46. Han, Y.; Xie, Y.-X.; Zhao, L.-B.; Fan, M.-J.; Liang, Y.-M. *Synthesis* **2008,** 87–93.
47. Gero, S. D.; Hildesheim, J.; Walczak, E.; Guthrie, R. D.; Smith, C. W. *J. Chem. Soc.* **1970,** 1402–1404.
48. Lu, Z.; Zhang, Y.; Wulff, W. D. *J. Am. Chem. Soc.* **2007,** *129*, 7185–7194.
49. Murai, N.; Komatsu, M.; Tagii, T.; Nishihara, H.; Ohchiro, Y.; Agawa, T. *J. Org. Chem.* **1977,** *42*, 847–850.
50. Claudi, F.; Franchetti, P.; Grifantini, M.; Martelli, S. *J. Org. Chem.* **1974,** *39*, 3508–3511.
51. Campbell, M. M.; Campbell, A. C.; Peace, A.; Pick, J.; Woods, G. F. *J. Chem. Soc., Chem. Commun.* **1985,** 1164–1165.
52. Keen, B. T.; Krass, D. K.; Paudler, W. W. *J. Heterocyclic Chem.* **1976,** *13*, 807–811.
53. Tomalia, D. A.; Ojha, N. D.; Thill, B. P. *J. Org. Chem.* **1969,** *34*, 1400–1405.
54. Prabhakaran, E. N.; Nandy, J. P.; Shukla, S.; Tewari, A.; Das, S. K.; Iqbal, J. *Tetrahedron Lett.* **2002,** *43*, 6461–6466.

1.3 Shi Epoxidation

Bingwei Vera Yang

1.3.1 Description

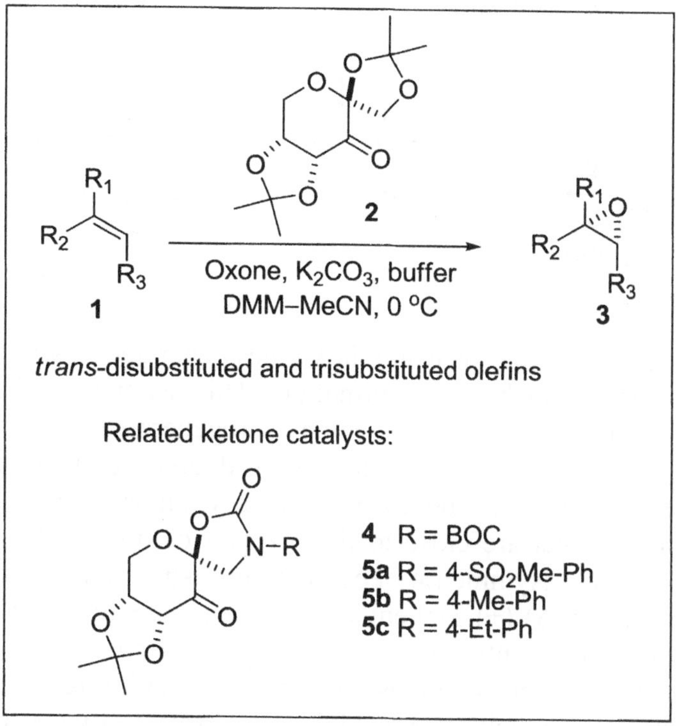

The Shi epoxidation refers to the asymmetric epoxidation of alkenes **1** using Oxone (potassium peroxymonosulfate, $2KHSO_5 \cdot KHSO_4 \cdot K_2SO_4$) as the primary oxidant and a fructose-derived chiral ketone catalyst **2**.[1,2] This procedure generates epoxides **3** with high enantiomeric excesses from *trans*-disubstituted and trisubstituted olefins. *cis*-Disubstituted olefins and styrenes are asymmetrically expoxidized under similar conditions using glucose-derived catalysts **4**[3–6] or **5**.[7,8]

1.3.2 Historical Perspective

Professor Yian Shi at the Colorado State University first reported the use of a fructose-derived chiral ketone **2** for the asymmetric epoxidation in 1996.[1] This ketone is conveniently synthesized from an inexpensive chiral starting material D-fructose *via* ketalization and oxidation.[9–11] The enantiomer of ketone **2**, *ent*-**2**, can be prepared by the same methods from L-fructose, which is derived from L-sorbose.[12,13]

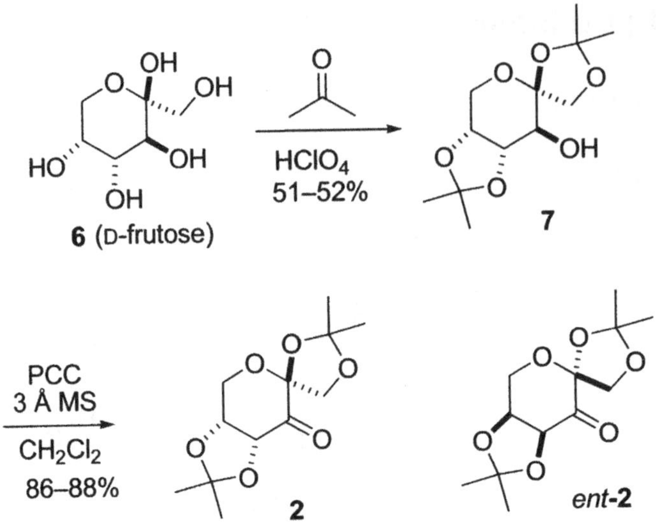

6 (D-frutose) 7 2 *ent*-2

The epoxidation of olefins using dioxiranes generated *in situ* from Oxone and ketones is an established transformation. This reaction can be performed with catalytic amount of ketone, which is regenerated after the reactive intermediate dioxirane delivers oxygen to the double bond. Furthermore, a chiral ketone catalyst could be used for an asymmetric epoxidation if the chiral control elements are close to the reacting carbonyl. In 1984, Curci reported the first asymmetric epoxidation using chiral ketone (+)-isopinocamphone or (*S*)-(+)-3-phenylbutan-2-one (maximum *ee* = 12.5%).[14] However, it was only until the discovery of Shi's fructose-derived chiral ketone **2** that the organocatalytic asymmetric epoxidation received extensive attention, particularly notable for its high enantioselectivity, broad generality and green chemistry advantages. The wide scope of olefin substrates, especially the unactivated alkenes, makes Shi epoxidation one of the most powerful methods for converting olefins to chiral epoxides.

1.3.3 Mechanism

The epoxidation with *in situ* generated dioxiranes often requires careful control of the reaction pH. Since Oxone rapidly autodecomposes at high pH, early epoxidations were usually carried out at pH 7–8. In contrast, higher pH was found to be beneficial to epoxidation with ketone **2**.[9,15] For example, conversion of *trans*-β-methylstyrene **11** to its epoxide **12** increased from ~ 5% at pH 7–8 to > 80% at pH > 10 while a high enantioselectivity (90–92% *ee*) was retained. Analysis of the reaction cycle implied that a Baeyer–Villiger oxidation from intermediate **8** could be one of the possible decomposition pathways for ketone **2**. A higher pH would facilitate the formation of anion **9** and subsequent formation of dioxirane **10**, thus suppressing the competing Baeyer–Villiger oxidation. The catalytic procedure at pH 10 requires substantially less Oxone, < 30% of the amount at pH 7–8, suggesting that ketone **2** can react with Oxone fast enough to avoid the autodecomposition of Oxone. The epoxidation is typically carried out around pH 10.5 by adding either K_2CO_3 or KOH as the reaction proceeds. Performing the reaction at higher pH greatly reduces the required amount of ketone catalyst, 30 mol% in most cases, leading to a catalytic process of epoxidation.

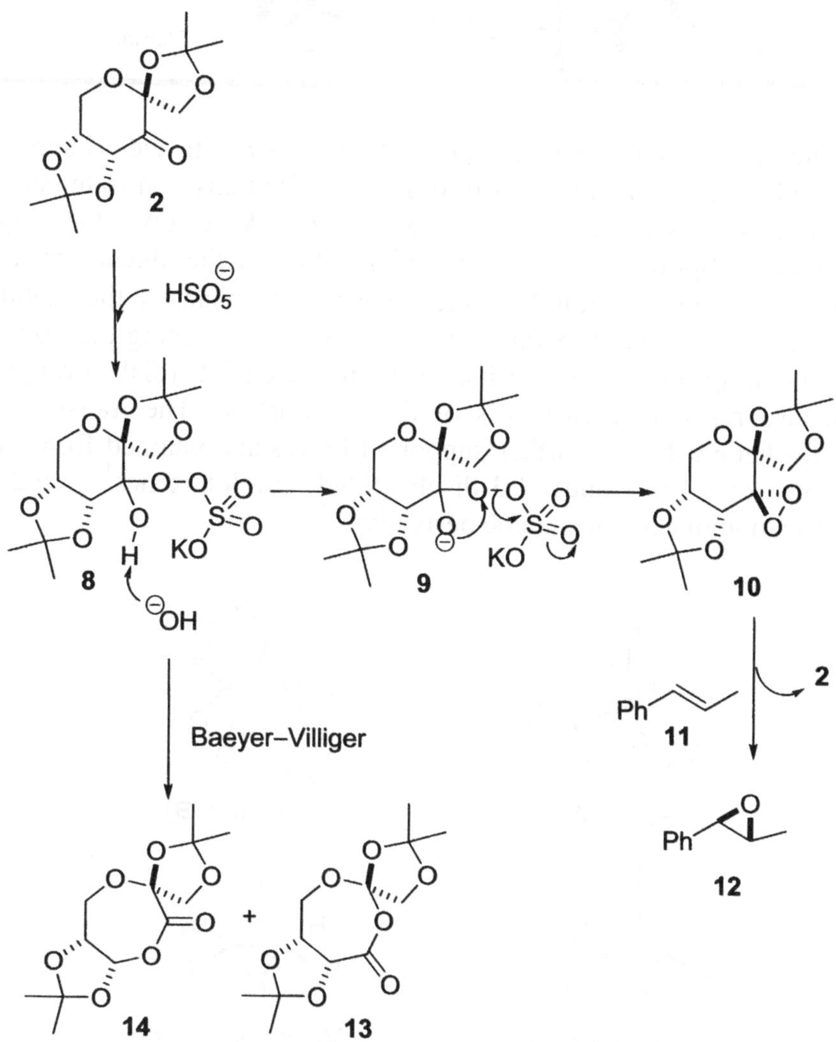

The stereochemical outcome of the epoxidation can be rationalized by a spiro transition state model. Two extreme epoxidation modes, spiro and planar, are known for epoxidation with dimethyldioxirane, and the spiro transition state is the optimal transition state for oxygen atom transfer from dimethyldioxirane to alkene, presumably due to the stabilizing interactions between the oxygen nonbonding orbital and the alkene π* orbital in the spiro transition state.[16]

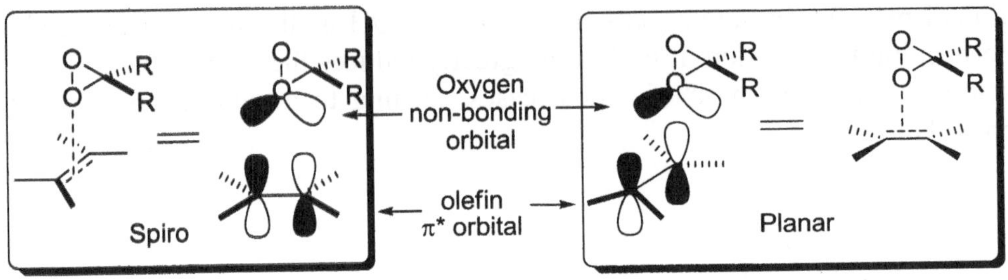

Studies have shown that the epoxidation of *trans*-di and tri-substituted olefins with ketone **2** mainly goes through the spiro transition state (spiro **A**). Planar transition state **B** competes with spiro **A** to give the opposite enantiomer.[1,9] Spiro **A** is favored by conjugation of the alkene that lowers the energy of the π* orbital of the alkene and enhances the stabilizing interaction between the dioxirane and the olefin. Decreasing the size of R_1 (further favoring spiro **A**) and/or increasing the size of R_3 (disfavoring planar **B**) can also result in higher *ee*'s of the epoxidation. The transition state modes for ketone **2** were further supported by results obtained from kinetic resolution of 1,6- and 1,3-disubstituted cyclohexenes[17] and de-symmetrization of cyclohexadiene derivatives.[18]

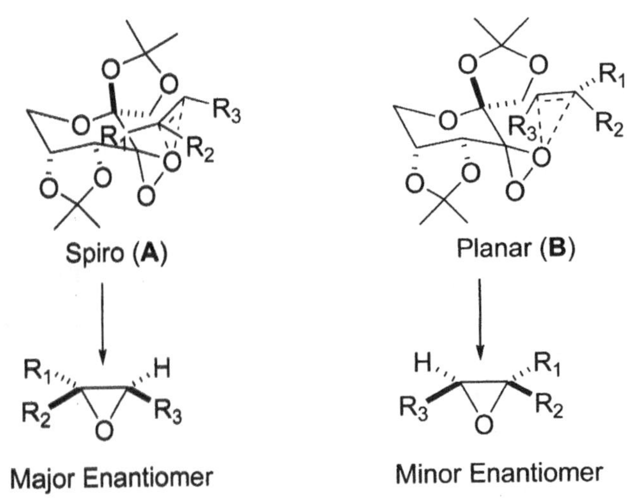

1.3.4 Variations and Improvements

Variation of Chiral Ketones

The asymmetric epoxidation with chiral ketone **2** has achieved high enantionselectivity with a wide range of unfunctionalized *trans*-disubstituted and trisubstituted olefins (selected examples are listed below).[9] 2,2-Disubstituted vinyl silanes are epoxidized in high *ee*'s and enantiomerically enriched 1,1-disubstituted epoxides can be obtained *via* the desilylation of these epoxides (*e.g.*, **18**).[19] Allylic alcohols and conjugated dienes and enynes are effective substrates (e.g. **19** and **20**).[20–23] The epoxidations of enol ethers and enol esters were also studied.[24] The resulting epoxides (*e.g.*, **21**) from enol esters can undergo stereoselective rearrangement to give optically active α-acyloxy ketones, (*S*)-**22** or (*R*)-**23**, under different acidic conditions.[25–27]

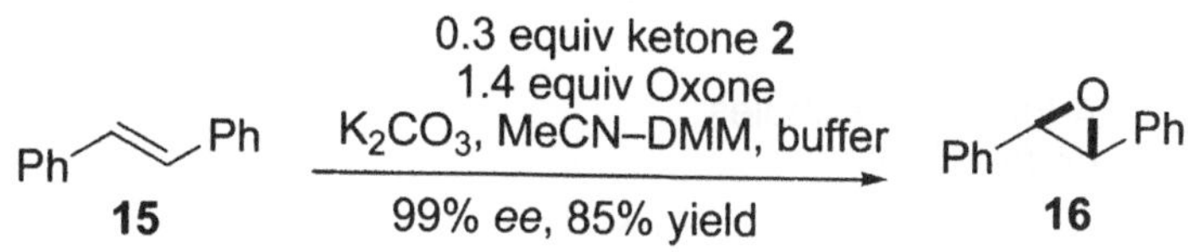

Additional examples, *ee* (yield%):

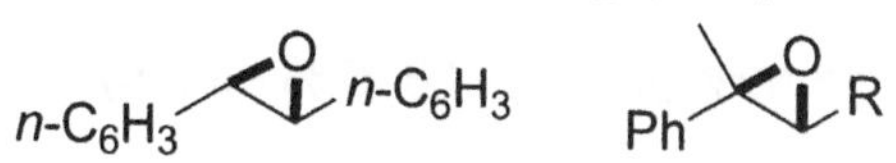

17 95% *ee* (89) **18a** R = TMS, 94% *ee* (74)
18b R = H

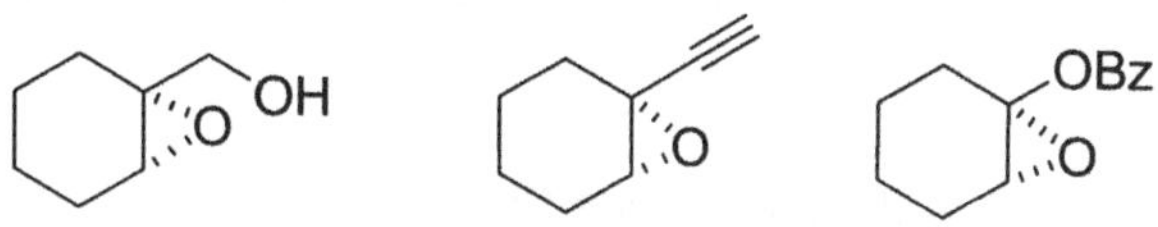

19 94% *ee* (93) **20** 93% *ee* (78) **21** 93% *ee* (82)

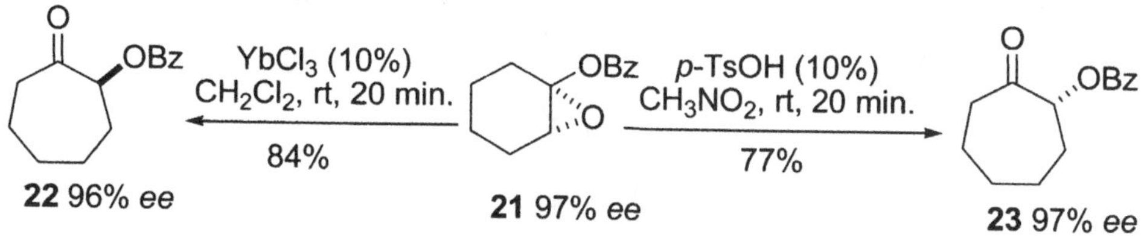

To effectively suppress the decomposition of ketone catalyst *via* Bayer–Villiger oxidation (see the mechanism scheme), Shi replaced the fused ketal moiety in ketone **2** by a more electron-withdrawing oxazolidinone (**24**) and acetates (**25**).[28,29] Only 5 mol% (1 mol% in some cases) of ketone **24** was needed to get comparable reactivity and enantioselectivity for 20–30 mol% of ketone **2**.[28]

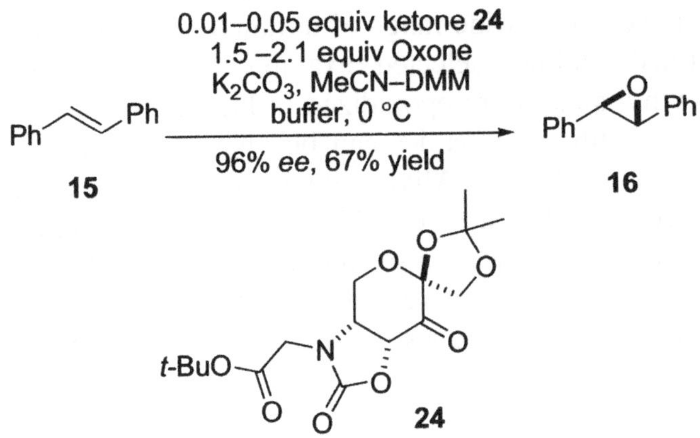

Additional examples
ee (yield%):

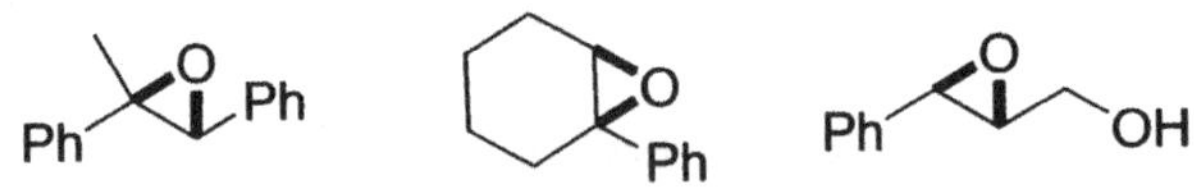

26 93% *ee* (80) **27** 92% *ee* (74) **28** 87% *ee* (89)

Ketone **25** has shown to provide high *ee*'s and good yields for epoxidation of a number of electron-deficient α,β-unsaturated esters,[29] whereas ketone **2** epoxidizes α,β-unsaturated esters sluggishly due to the low reactivity of its dioxirane as an electrophilic reagent toward electron-deficient olefins.

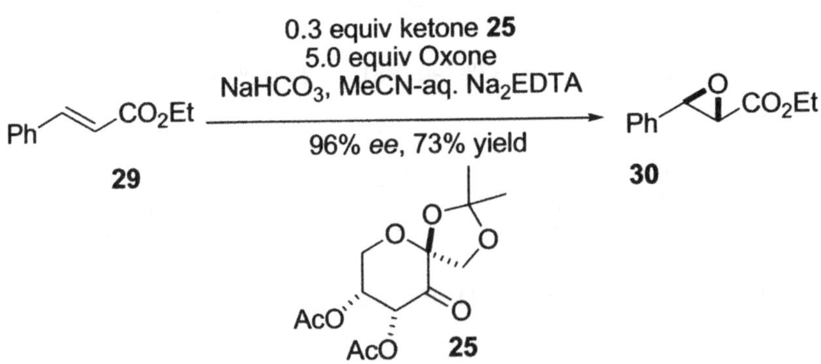

Additional examples
ee (yield%):

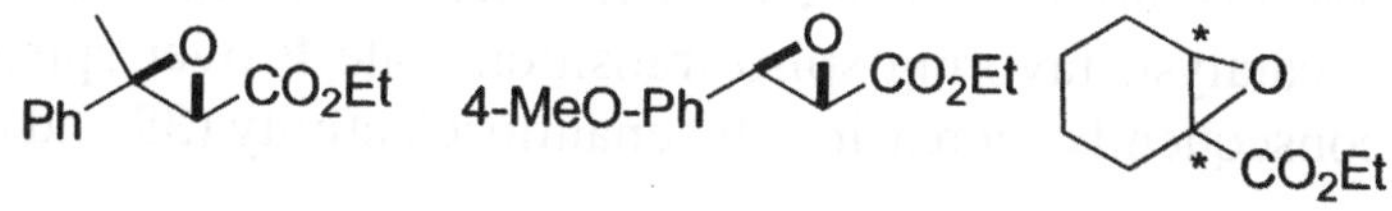

31 96% *ee* (93) **32** 90% *ee* (57) **33** 93% *ee* (77)

The substrate scope was expanded to *cis*-olefins and certain terminal olefins when a new series of glucose-derived ketones **4** and **5** were developed. In 2000, Shi reported an *N*-Boc oxazolidinone-bearing ketone **4** to be a highly enantioselective catalyst for the epoxidation of various *cis*-olefins conjugated with an aromatic or alkyne group.[3] The stereopreference of the olefin appears to be directed by an attraction from the oxazolidinone moiety of ketone **4**. In the transition state, Rπ substituent on the substrate prefers to be proximal to the oxazolidinone of ketone **4** (spiro **C** favored over spiro **D**).[3–6]

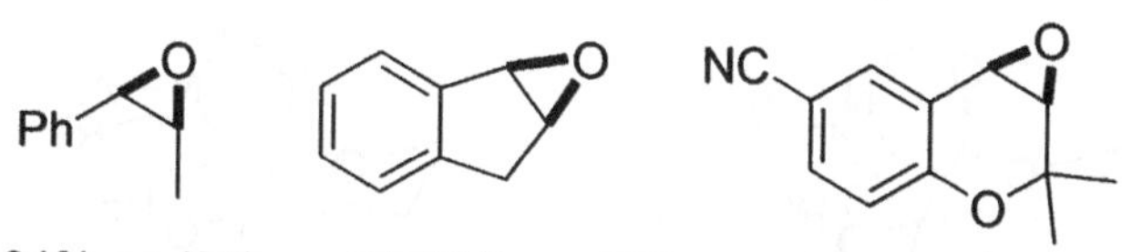

34 91% *ee* (87) **35** 83% *ee* (88) **36** 91% *ee* (61)

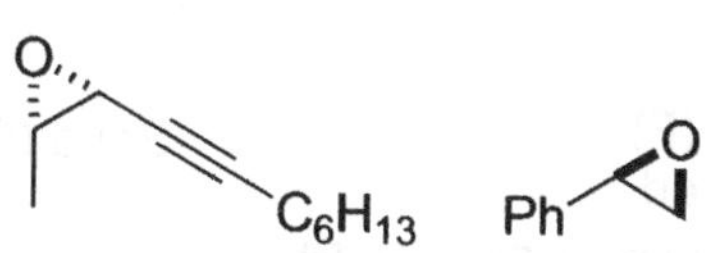

37 87% *ee* (77) **38** 81% *ee* (92)

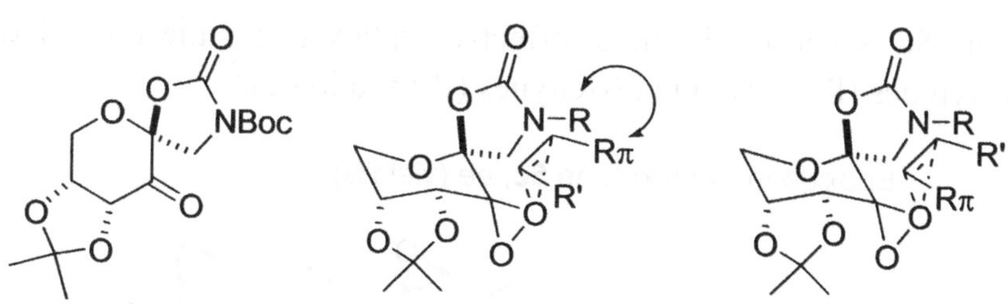

4 Spiro (**C**) Favored Spiro (**D**)

The *N*-aryl oxazolidinone-bearing ketone catalysts **5** are readily prepared in large quantities from glucose and inexpensive anilines in four steps.[30] Phenyl groups substituted with methylsulfonyl (**5a**) or alkyls (**5b**, **5c**) consistently provide high enantioselectivity for a variety of *cis*-olefins and certain terminal olefins.[7,8] *cis*-β-Methylstyrenes can be epoxidized with ketones **5a** and **5b** in high conversions and *ee*'s.[31] Substituents on the phenyl group of the olefins further enhance its (Rπ) interaction with the *N*-aryl group of the ketone catalyst, favoring spiro transition state **F** over spiro transition state **G** and consequently increasing the enantioselectivity (**39** and **40** *vs.* **34**).

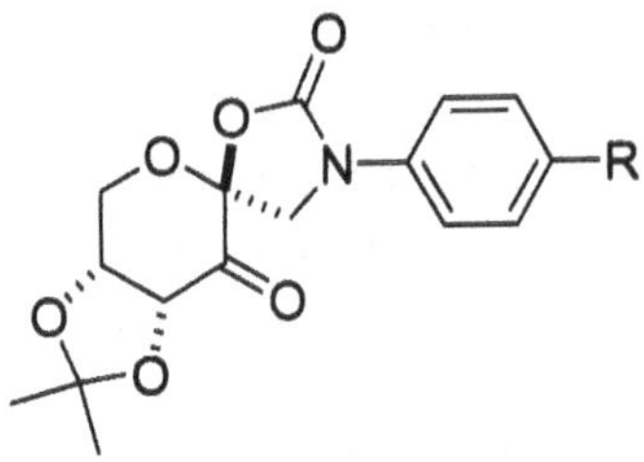

5a R = SO_2Me
5b R = Me
5c R = Et

Epoxidation with ketones **5a** or **5b**, *ee* (yield %):

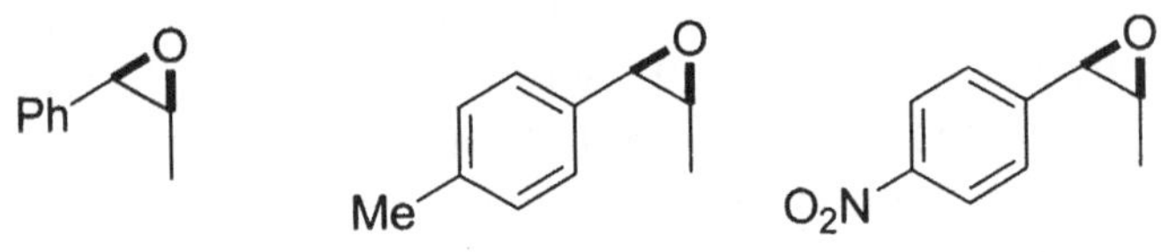

34	**39**	**40**
90% *ee* (100)/**5a**	92% *ee* (96)/**5a**	97% *ee* (91)/**5a**
84% *ee* (99)/**5b**	88% *ee* (100)/**5b**	98% *ee* (86)/**5b**

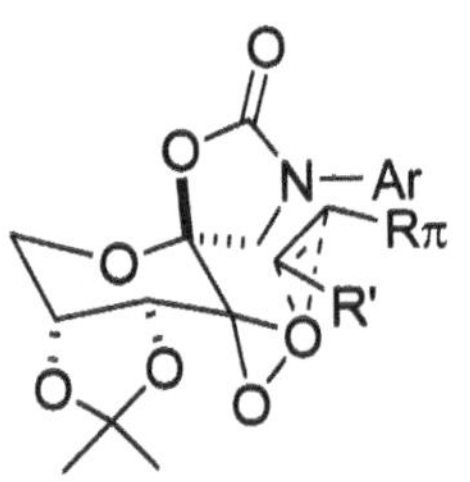

Spiro (**F**)
Favored

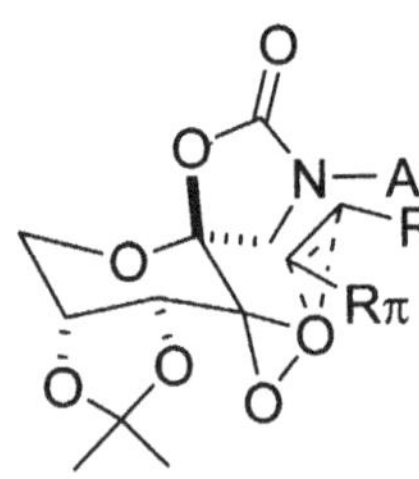

Spiro (**G**)

Ar = 4-alkyl-Ph, 4-SO_2Me-Ph;
Rπ = alkene, alkyne, Ph, substituted Ph; R' = H, alkyl.

Ketone **5c** is one of the most effective catalysts for the epoxidation of various styrenes (R′ = H)[8] and *cis*-enynes (Rπ = alkyne).[32]

Epoxidation with ketone **5c**, *ee* (yield%):

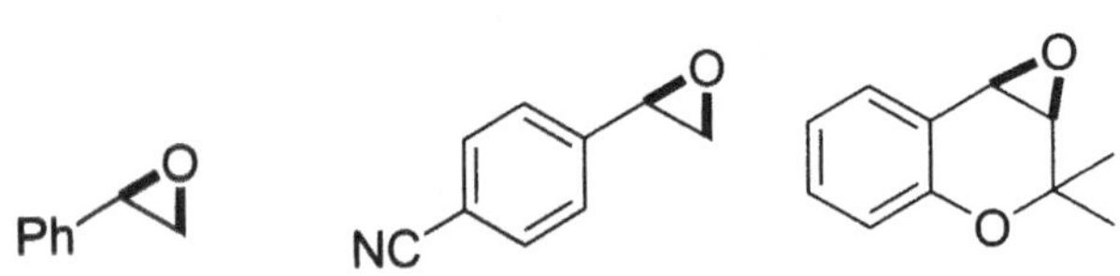

38 86% *ee* (72) **41** 90% *ee* (86) **42** 84% *ee* (100)

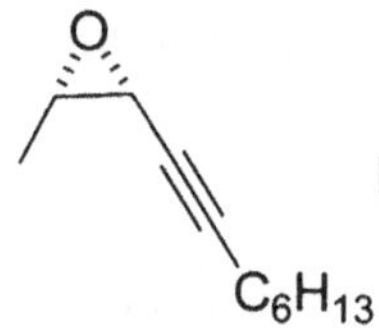

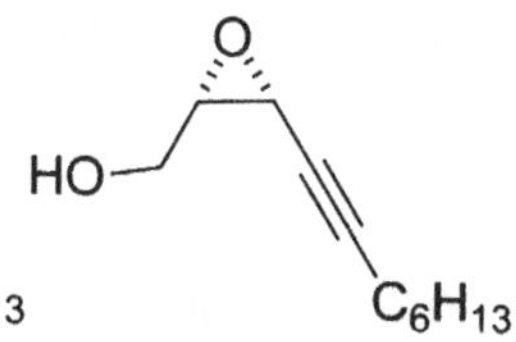

37 90% *ee* (84) **43** 97% *ee* (68)

Ketone **5b** demonstrates a broader substrate scope than that displayed by any other *N*-aryl-oxazolidinone-containing catalyst. Conjugated *cis*-dienes **44** can be epoxidized with **5b** in high *ee*'s with no *cis–trans* isomerization.[33]

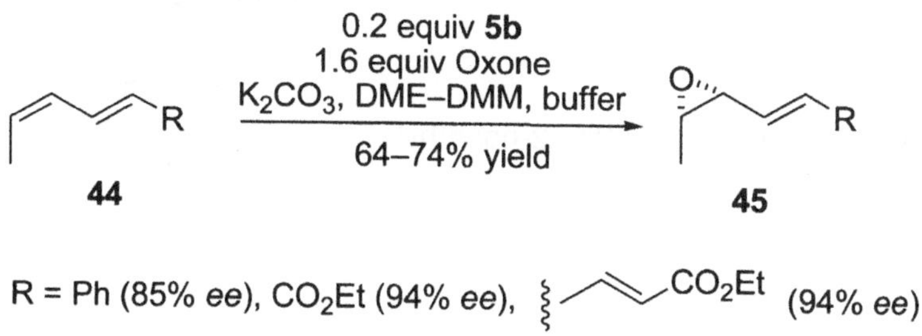

Trisubstituted and tetrasubstituted benzylidenecyclobutanes **46** undergo highly enantioselective epoxidation with ketone **5b** followed by epoxide rearrangement upon treatment with Et_2AlCl or LiI to afford optically active 2-aryl cyclopentanones (**48**, **49** or **50**).[34,35]

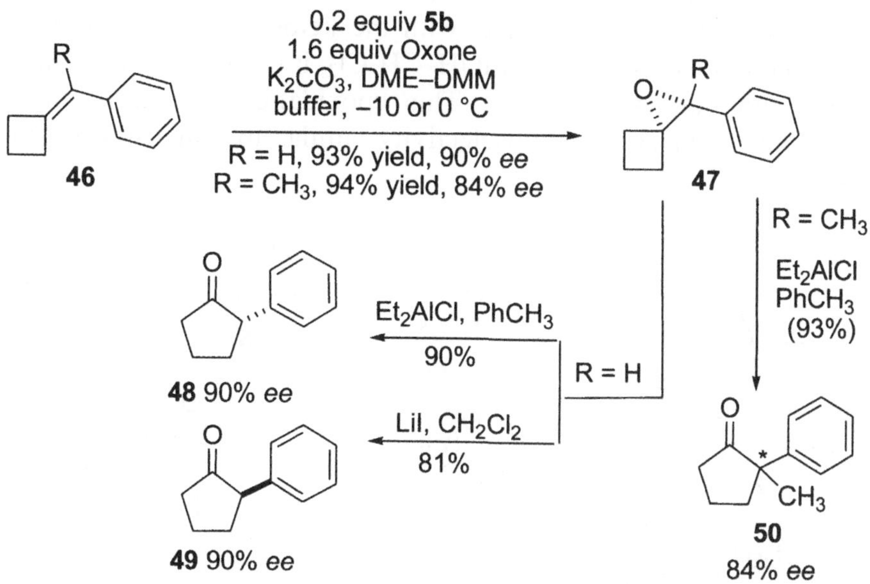

When benzyliden-cyclopropanes **51** are subjected to epoxidation with **5b**, optically active γ-aryl-γ-butyrolactones **53** can be obtained in moderate yield and good enantioselectivity *via* an *in situ* epoxide rearrangement and a Baeyer–Villiger oxidation.[36]

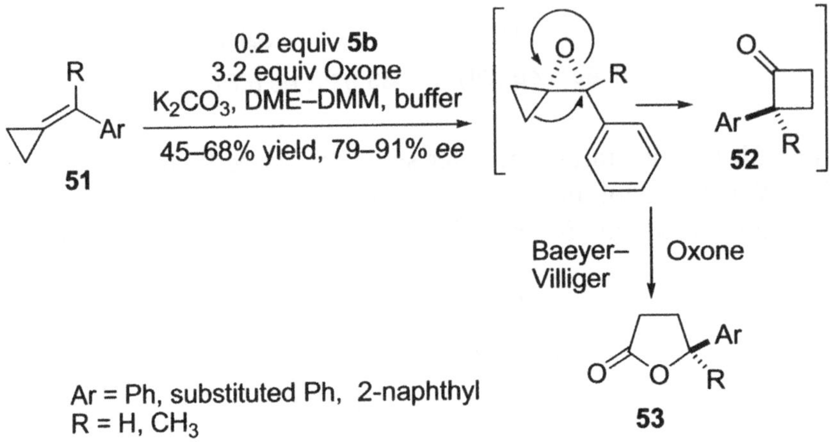

Variations of Oxidant: Hydrogen Peroxide as Primary Oxidant

While Oxone has been commonly used to generate dioxiranes from ketones, Shi's studies have shown that epoxidation with ketone **2** or **5c** can be carried out with a nitrile and H_2O_2 as the primary oxidant, giving high enantioselectivity for a variety of olefins.[37–39] Peroxyimidic acid **55** is likely to be the active oxidant that reacts with the ketone to form dioxirane **10**. Mixed solvents, such as CH_3CN–EtOH–CH_2Cl_2, improve the conversions for substrates with poor solubilities. This epoxidation system is mild and provides conversion and enantioselectivity similar to that using Oxone as oxidant.

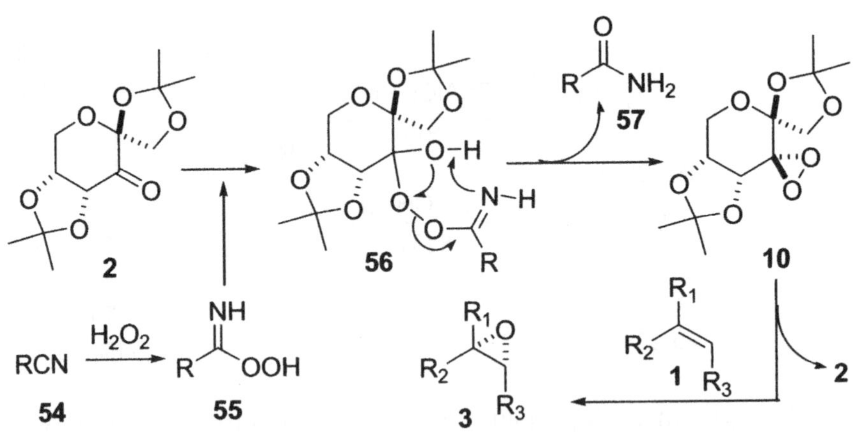

Epoxidation with ketones **2** or **5c**, *ee* (yield %):

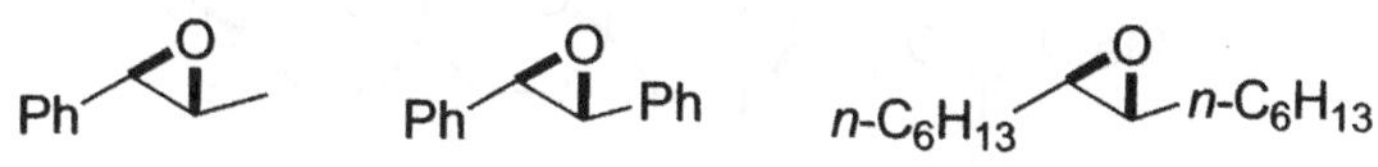

11 92% *ee* (93)/**2** **15** 98% *ee* (90) /**2** **17** 92% *ee* (97) /**2**

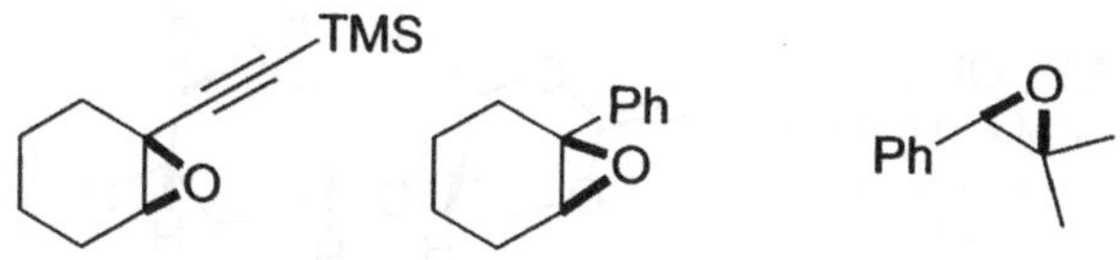

58 95% *ee* (93)/**2** **59** 96% *ee* (90)/**2** **60** 92% *ee* (82)/**5c**

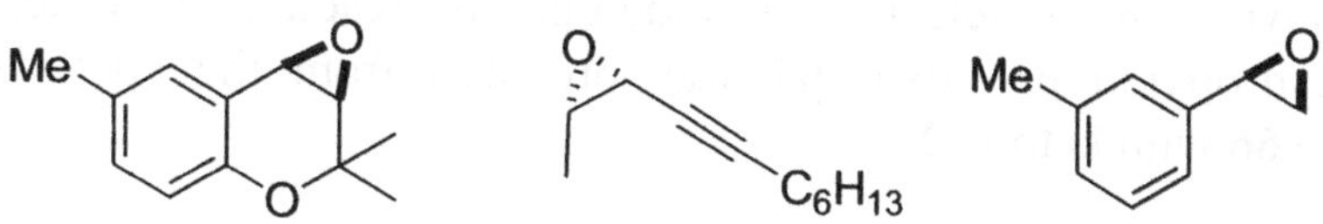

61 91% *ee* (89)/**5c** **17** 90% *ee* (65)/**5c** **62** 83% *ee* (93)/**5c**

1.3.5 Synthetic Utility

The availability of ketone **2** and its effectiveness toward a wide variety of *trans*-disubstituted and trisubstituted olefins make Shi epoxidation a widely applicable method in many syntheses published over the past decade. Selected examples are highlighted in this section.

A group of polycyclic polyether natural products are of special interest owing to their fascinating structure and biological activities. One of the proposed biosynthetic origins of these molecules features an epoxide-opening cascade pathway. Shi asymmetric epoxidation of un-activated alkenes has been frequently employed in the preparation of polyepoxide intermediates. McDonald and co-workers studied a series of tandem *endo*-selective and stereospecific oxacyclization of polyepoxides mediated by Lewis acid. Polyepoxides, such as **64**, can be obtained from the epoxidation of triene **63** with ketone **2**.[40] Furthermore, a cascade cyclization, initiated by a Lewis acid-promoted epoxide opening of **64**, furnished the desired polyether **65**.

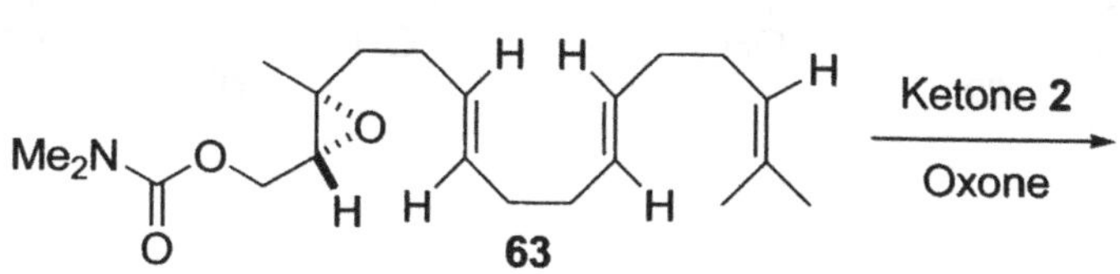

63

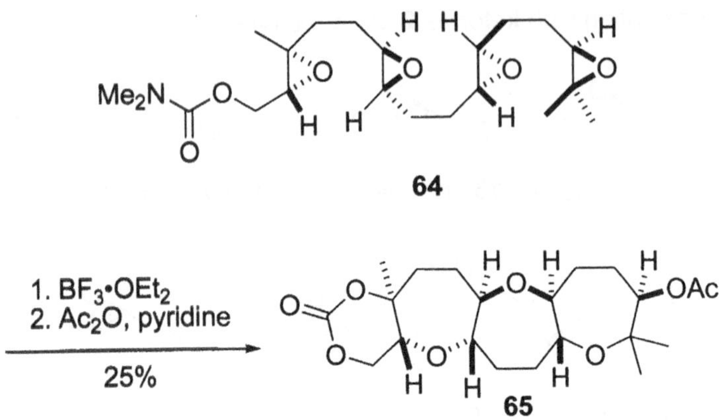

In recent studies, Jamison and co-workers reported the formation of tetrahydropyran **68** *via* selective epoxide-opening reactions in water.[41] The polytetrahydropyran precursor **67** was prepared from the epoxidation of polyalkene **66** with ketone **2**.

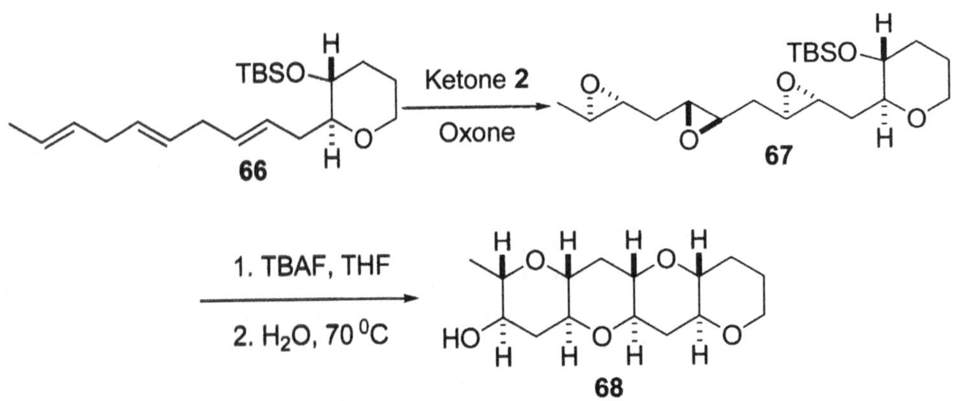

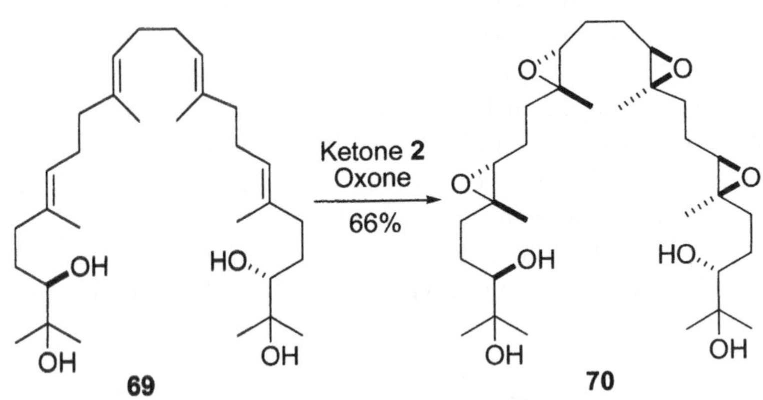

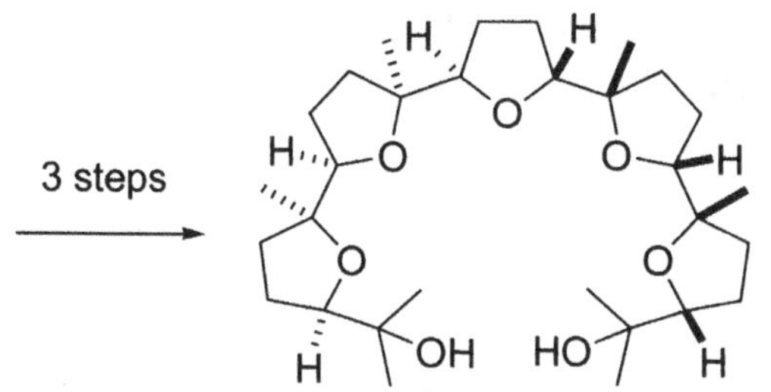

71, glabrescol

In 2000, in an effort to verify the structure of a polycyclic oxasqualenoid, glabrescol **71**, Corey and co-workers applied the Shi epoxidation in the conversion of tetraene **69** to tetra-epoxide **70**, which was subsequently transformed to glabrescol **71** in three steps.[42,43]

The high specificity of the Shi epoxidation permits the regioselective epoxidation in some polyene compounds. McDonald and co-workers employed ketone **2** in the total synthesis of nakorone and abudinol.[44] Triene–yne **72** was selectively epoxidized on the two more electron-rich double bonds, leaving the olefin next to the electron-withdrawing sulfone group intact. Bis-epoxide **73** was transformed into both *ent*-nakorone **75** and *ent*-abudinol B **76**.

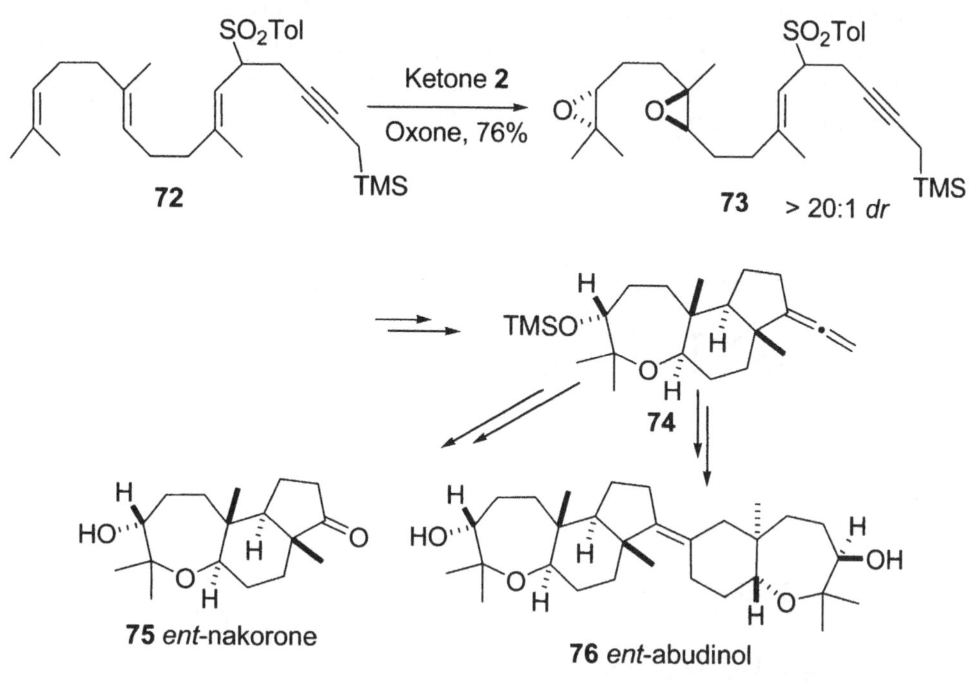

In 2006, Ready and co-workers reported that compound **77**, which contains three double bonds, was regio- and stereo-selectively epoxidized at the desired C7–C8 double bond. The resulting epoxide was converted into (+)-nigellamine A_2 **78**.[45]

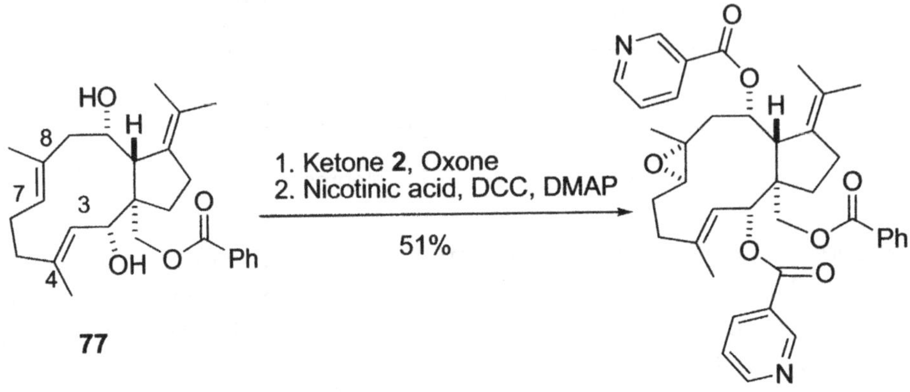

77

(+)-nigellamine A_2 **78**

Selective epoxidation of polyene compound has also achieved with *ent*-**2**, the enantiomer of ketone **2**. In Morimoto's total synthesis of polyether (+)-aurilol, Shi epoxidation was utilized twice, with ketone **2** and *ent*-**2**, respectively.[46] Epoxidation of **79** with ketone **2** gave epoxide **80** with high diastereoselectivity. Epoxide **80** underwent acid catalyzed 5-*exo-tet* cyclization to produce tetrahydrofuran **81** with the desired stereochemistry. Subsequently, diene **82** was selectively epoxidized with *ent*-**2** only at the trisubstituted olefin to give epoxide **83**. Epoxides **80** and **83** played important roles in setting stereocenters in the final product.

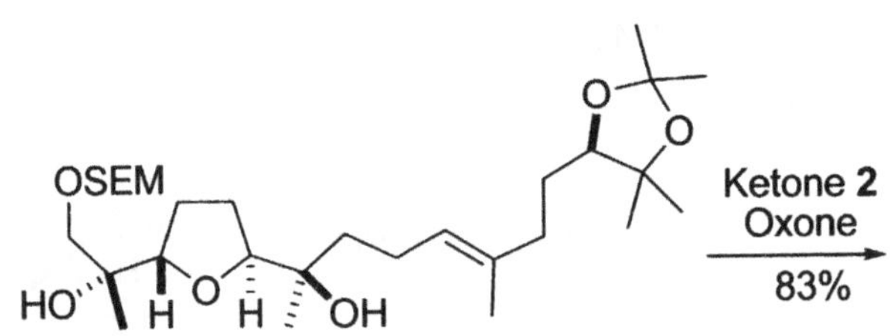

79

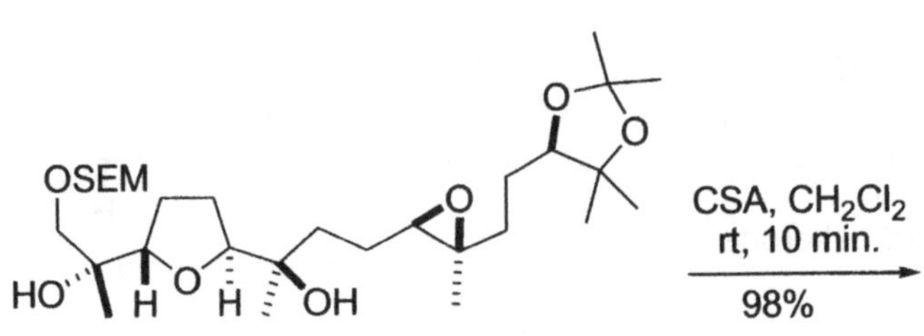

80 (> 15 : 1)

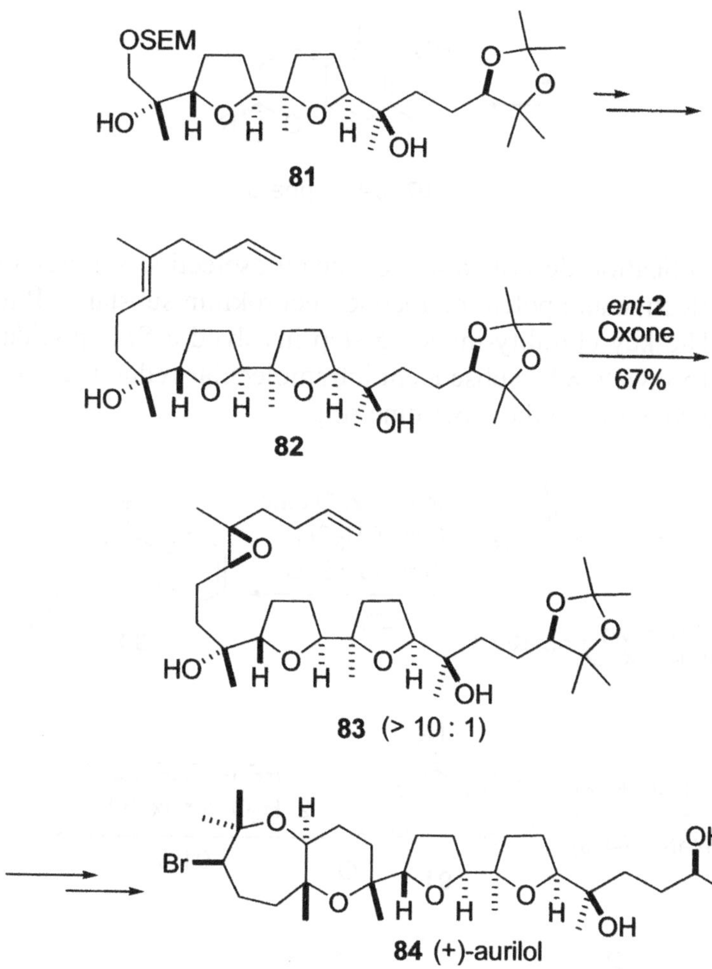

A diastereoselective synthesis of α-tocopherol **87** features a Shi epoxidation with *ent*-**2** and a carefully controlled intramolecular epoxide opening cyclization for the formation of the chromanol ring. Good conversion and high enantioselectivity have been achieved in the epoxidation step.[47]

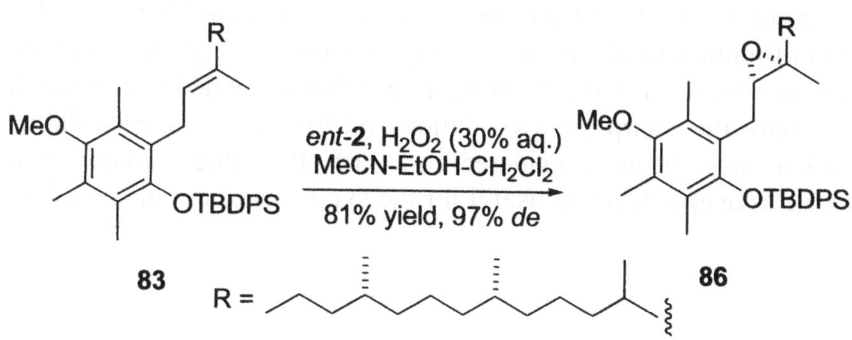

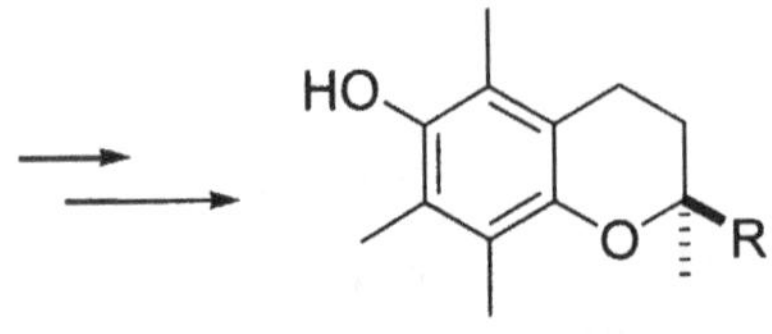

87 α-tocopherol

A recent publication described a short enantioselective synthesis of (+)-L-733,060, a selective and potent nonpeptide neurokinin substance P receptor antagonist.[48] The key chirality-inducing step involved a Shi epoxidation of homoallylic carboxylate **89**. Subsequent intramolecular reductive cyclization of azidolactone constructed the piperidine ring.

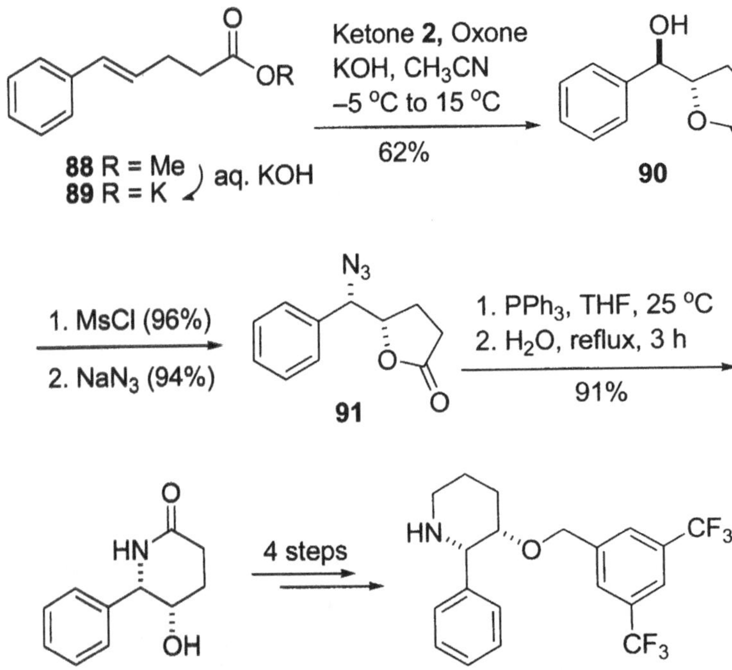

In summary, the broad application of the Shi epoxidation in the total synthesis of natural products and in drug discovery is a good indication that the reaction will receive more attention and find extended use in the future. Together with the Sharpless epoxidation and the Jacobsen epoxidation, Shi epoxidation has been considered one of the three major catalytic enantioselective epoxidations useful for the synthesis of chiral epoxides.

1.3.6 Experimental

Standard Conditions

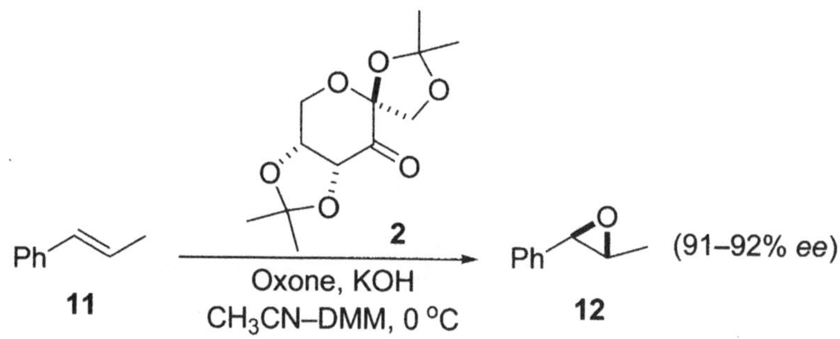

(R,R)-*trans*-β-Methylstyrene oxide (12).[38]

A 2-L, three-necked, round-bottomed flask equipped with a 5-cm, egg-shaped, Teflon-coated stir bar and two addition funnels is cooled in an ice bath. The flask is charged with *trans*-β-methylstyrene **11** (5.91 g, 50.0 mmol), 500 mL of a 2 : 1 mixture of dimethoxymethane (DMM) and acetonitrile (CH_3CN), 300 mL of potassium carbonate–acetic acid buffer solution, tetrabutylammonium hydrogen sulfate (0.375 g, 1.1 mmol), and the chiral ketone **2**[11] (4.52 g, 17.5 mmol, 35 mol%). One addition funnel is charged with a solution of Oxone (46.1 g, 75.0 mmol) in 170 mL of aqueous 4×10^{-4} M disodium ethylenediaminetetraacetate (Na_2EDTA) solution, and the other addition funnel is charged with 170 mL of 1.47 M aqueous potassium hydroxide (KOH) solution. The two solutions in the addition funnels are added dropwise at the same rate over 2.5 h to the cooled reaction mixture, which is stirred vigorously at 0 °C. The resulting suspension is stirred at 0 °C for an additional hour, and then 250 mL of pentane is added. The aqueous phase is separated and extracted with two 250-mL portions of pentane, and the combined organic phases are dried over Na_2SO_4, filtered, and concentrated by rotary evaporation at 0 °C. The resulting oil is loaded onto 50 g of Whatman 60 Å (230–400 mesh) silica gel packed in a 5-cm-diameter column. The silica gel is first washed with 200 mL of hexane to remove trace amounts of unreacted olefin, then the product is eluted with 200 mL of 10 : 1 hexane:ether to afford 6.02–6.31 g (90–94%) of *trans*-β-methylstyrene oxide **12**.

*Asymmetric Epoxidation Using Ketone **2** and H_2O_2 as Primary Oxidant*

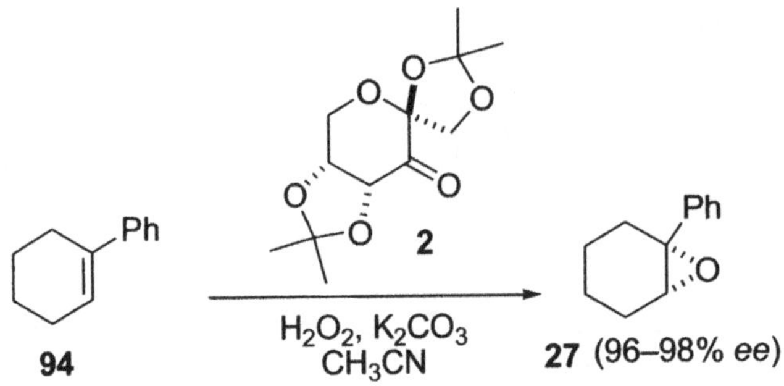

(*R*,*R*)-1-Phenylcyclohexene oxide (27).[38]
A 250-mL, round-bottomed flask equipped with a 4.5-cm, egg-shaped Teflon-coated magnetic stirbar is charged with 1-phenylcyclohexene **94** (7.91 g, 50.0 mmol) and the chiral ketone **2**[11] (1.29 g, 5.00 mmol, 10 mol%). The flask is cooled in an ice bath, and 75 mL of CH_3CN and 75 mL of a solution of 2.0 M potassium carbonate and 4×10^{-4} M EDTA are added. The reaction mixture is cooled to 0 °C, and 20 mL (200 mmol) of 30% hydrogen peroxide (H_2O_2) is added. The resulting mixture is vigorously stirred at 0 °C for 6 h, then diluted with 50 mL of hexane. The aqueous phase is separated and extracted with three 200-mL portions of hexane, and the combined organic phases are washed with two 50-mL portions of 1 M aqueous sodium thiosulfate solution and 100 mL of brine, dried over Na_2SO_4, filtered, and concentrated by rotary evaporation at 0 °C. The resulting oil is applied to 180 g of Whatman 60 Å (230–400 mesh) silica gel packed in a 5-cm-diameter column; then the product is eluted with 600 mL of hexane and finally 1 L of 20 : 1 hexane/Et_2O to afford 6.88–8.01 g (79–92%) of (*R*,*R*)-1-phenylcyclohexene oxide (**27**) as a colorless oil.

1.3.7 References

1. Tu, Y.; Wang, Z.-X.; Shi, Y. *J. Am. Chem. Soc.* **1996,** *118*, 9806–9807.
2. [R] For leading reviews on the Shi epoxidation, see: (a) Frohn, M.; Shi, Y. *Synthesis* **2000,** *14*, 1979–2000. (b) Shi, Y. *Acc. Chem. Res.* **2004,** *37,* 488–496. (c) Wong, O. A.; Shi, Y. *Chem. Rev.* **2008,** 108, 3958–3987. (d) Wong, O. A.; Shi, Y. *Top. Curr. Chem.* **2010,** *291*, 201–232.
3. Tian, H.; She, X.; Shu, L.; Yu, H.; Shi, Y. *J. Am. Chem. Soc.* **2000,** *121,* 11551–11568.
4. Tian, H.; She, X.; Xu, J.; Shi, Y. *Org. Lett.* **2001,** *3*, 1929–1931.
5. Tian, H.; She, X.; Yu, H.; Shu, L.; Shi, Y. *J. Org. Chem.* **2002,** *67*, 2435–2446.
6. Shu, L, Shen, Y. M., Burke, C, Goeddel, D.; Shi, Y. *J. Org. Chem.* **2003,** *68*, 4963–4965.
7. Shu, L.; Wang, P.; Gan.; Shi, Y. *Org. Lett.* **2003,** *5*, 293–296.
8. Goeddel, D.; Shu, L.; Yuan, Y.; Wong, O. A.; Wang, B.; Shi, Y. *J. Org. Chem.* **2006,** *71*, 1715–1717.
9. Wang, Z.-X.; Tu, Y.; Frohn, M.; Zhang, J.-R.; Shi, Y. *J. Am. Chem. Soc.* **1997,** *119*, 11224–11235.
10. Mio, S.; Kumagawa, Y.; Sugai, S. *Tetrahedron* **1991,** *47*, 2133–2144.

11. Tu, Y.; Frohn, M.; Wang, Z.-X.; Shi, Y. *Org. Synth.* **2003,** *80,* 1–9; *Coll. Vol. 11,* **2009,** 177–182.
12. Chen, C.-C.; Whistler, R. L. *Carbohydr. Res.* **1988,** *175*, 265–271.
13. Zhao, M.-X.; Shi, Y. *J. Org. Chem.* **2006,** *71*, 5377–5379.
14. Curci, R.; Fiorentino, M.; Serio, M. R. *J. Chem. Soc., Chem. Commun.* **1984,** 155–156.
15. Wang, Z.-X.; Tu, Y.; Frohn, M.; Shi, Y. *J. Org. Chem.* **1997,** *62*, 2328–2329.
16. [R] For general leading references on dioxiranes see (a) Murray, R. W. *Chem. Rev.* **1989,** *89*, 1187. (b) Adam, W.; Curci, R.; Edwards, J. O. *Acc. Chem. Res.* **1989,** *22*, 205. (c) Curci, R.; Dinoi, A.; Rubino, M. F. *Pure Appl. Chem.* **1995,** *67*, 811. (d) Clennan, E. L. *Trends Org. Chem.* **1995,** *5*, 231. (e) Adam, W.; Smerz, A. K. *Bull. Soc. Chim. Belg.* **1996,** *105*, 581.
17. Frohn, M.; Zhou, X.; Zhang, J. R.; Tang, Y.; Shi, Y. *J. Am. Chem. Soc.* **1999,** *121*, 7718–7719.
18. Lorenz, J. C.; Frohn, M.; Zhou, X.; Zhang, J. R.; Tang, Y.; Burke, C.; Shi, Y. *J. Org. Chem.* **2005,** *70*, 2904–2911.
19. Warren, J. D.; Shi, Y. *J. Org. Chem.* **1999,** *64*, 7675–7677.
20. Wang, Z. X.; Shi, Y. *J. Org. Chem.* **1998,** *63*, 3099–3104.
21. Frohn, M.; Dalkiewicz, M.; Tu, Y.; Wang, Z. X.; Shi, Y. *J. Org. Chem.* **1998,** *63*, 2948–2953.
22. Cao, G. A.; Wang, Z. X.; Tu, Y.; Shi, Y. *Tetrahedron Lett.* **1998,** *39*, 4425–4428.
23. Wang, Z. X.; Cao, G.; Shi, Y. *J. Org. Chem.* **1999,** *64*, 7646–7650.
24. Zhu, Y.; Tu, Y.; Yu, H.; Shi, Y. *Tetrahedron Lett.* **1998,** *39*, 7819–7822.
25. Zhu, Y.; Manske, KJ.; Shi, Y. *J. Am. Chem. Soc.* **1999,** *121*, 4080–4081.
26. Feng, X.; Shu, L.; Shi, Y. *J. Am. Chem. Soc.* **1999,** *121,* 11002–11003.
27. Zhu, Y.; Shu, L.; Tu, Y.; Shi, Y. *J. Org. Chem.* **2001,** *66,* 1818–1826.
28. Tian, H.; She, X.; Shi, Y. *Org. Lett.* **2001,** *3*, 715–717.
29. Wu, X.-Y.; She, X.; Shi, Y. *J. Am. Chem. Soc.* **2002,** *124,* 8792–8783.
30. For large-scale synthesis of ketones **5**, see: Zhao, M.-X.; Goeddel, D.; Li, K.; Shi, Y. *Tetrahedron* **2006,** *62*, 8064–8068.
31. Shu, L.; Shi, Y. *Tetrahedron Lett.* **2004,** *45*, 8115–8117.
32. For *cis*-enyne, see Burke, C. P.; Shi, Y. *J. Org. Chem.* **2007,** *72*, 4093–4097.
33. Burke, C. P.; Shi, Y. *Angew. Chem., Int. Ed.* **2006,** *45*, 4475–4478.
34. Shen, Y.-M.; Wang, B.; Shi, Y. *Angew. Chem., Int. Ed.* **2006,** *45*, 1429–1432.
35. Shen, Y.-M.; Wang, B.; Shi, Y. *Tetrahedron Lett.* **2006,** *47*, 5455–5458.
36. Wang, B.; Shen, Y.-M.; Shi, Y. *J. Org. Chem.* **2006,** *71*, 9519–9521.
37. Shu, L.; Shi, Y. *Tetrahedron Lett.* **1999,** *40*, 8721–8724.
38. Wang, Z.-X.; Shu, L.; Frohn, M.; Tu, Y.; Shi, Y. *Org. Synth.* **2003,** *80,* 9–17; *Coll. Vol. 11,* **2003**, 183–188.
39. Burke, C. P.; Shu, L.; Shi, Y. *J. Org. Chem.* **2007,** *72*, 6320–6323.
40. Valentine, J. C.; McDonald, F. E.; Neiwert, W. A.; Hardcastle, K. I. *J. Am. Chem. Soc.* **2005,** *127*, 4586–4593.
41. Vilotijevic, I.; Jamison, T. F. *Science* **2007,** *317*, 1189–1192.
42. Corey, E. J.; Xiong, Z. *J. Am. Chem. Soc.* **2000,** *122*, 4831–4834.
43. Corey, E. J.; Xiong, Z. *J. Am. Chem. Soc.* **2000,** *122*, 9328–9329.
44. Tong, R.; Valentine, J. C.; McDonald, F. E.; Cao, R.; Fang, X.; Hardcastle, K. I. *J. Am. Chem. Soc.* **2007,** *129*, 1050–1051.
45. Bian, J.; Van Wingerden, M.; Ready, J. M. *J. Am. Chem. Soc.* **2006,** *128*, 7428–7429.
46. Morimoto, Y.; Nishikawa, Y.; Takashi, M. *J. Am. Chem. Soc.* **2005,** *127*, 5806.
47. Julien, C.; Axel, B.; Antoinette, C.; Wolf, D. W. *Org. Lett.* **2008,** *10*, 512–516.
48. Emmanuvel, L.; Sudalai, A. *Tetrahedron Lett.* **2008,** *49*, 5736–5738.

2.1 Clauson–Kaas Pyrrole Synthesis

Thomas Andrew Wynn

2.1.1 Description

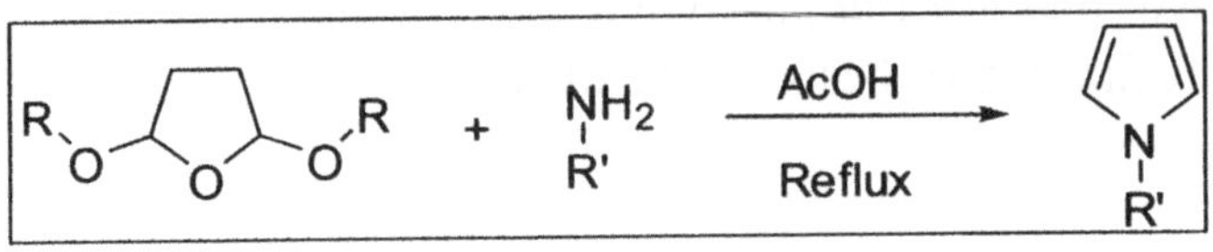

The Clauson–Kaas pyrrole synthesis involves the acid-catalyzed cyclization of a primary amine and a dialkoxytetrahydrofuran to form an *N*-substituted pyrrole.[1]

2.1.2 Historical Perspective

The key biological relevance of pyrroles has long been recognized because they are in several biological building blocks such as heme, the amino acids proline and hydroxyl proline (in a reduced form), and a host of natural products. Studies of pyrrole stretch back to the very beginning of synthetic organic chemistry, with the first description appearing in 1834 by Runge.[2] Pyrrole was the first characterized in 1858[3] and the structural determination reported in 1870 by Bayer.[4] The importance of pyrrole derivatives has driven the development of a host of different methods for generating the ring system.[5] At the time of Clauson–Kaas's original report the major method for generating pyrroles was the Paal–Knorr reaction. The ease of generation and stability of 2,5-dialkoxytetrahydrofurans[6] has greatly increased the utility of the Paal–Knorr reaction.

2.1.3 Mechanism

Although no formal investigations into the reaction mechanism has been undertaken, the most likely proposal involves acid-catalyzed acetal hydrolysis that releases succindialdehyde. The succindialdehyde then enters into the Paal–Knorr reaction pathway.[7] Once the dialdehyde is revealed the acid catalyst aids the attack of the nucleophilic nitrogen one of the carbonyls forming a hemiaminal. The nitrogen is then free to attack the second carbonyl after which two dehydrations generate the pyrrole ring system. A variety of nitrogen nucleophiles have shown to undergo the Clauson–Kaas reaction and the nucleophilicity of the nitrogen does have an effect on the rate of the reaction. The less nucleophilic nitrogens typically take longer to go to completion, and many studies have been reported that look at developing efficient catalytic systems for less reactive systems.

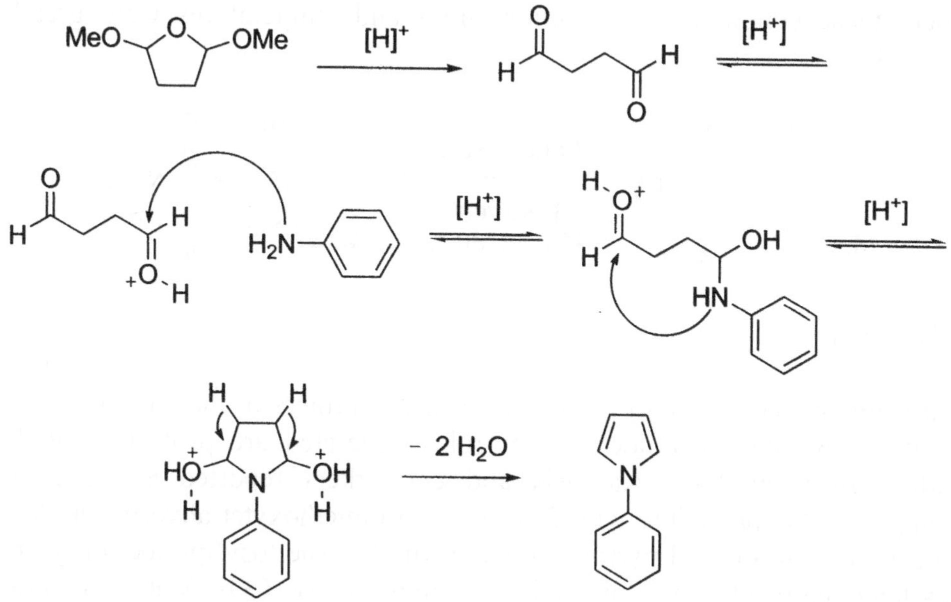

2.1.4 *Variations and Improvements*

A variety of nucleophilic nitrogens can undergo pyrrole formation with dialkoxytetrahydrofurans. Anilines, amides, sulphonamides, and alkyl amines all react cleanly to form substituted pyrroles. With judicious choice of amides or sulfonamide a variety of N–H protected pyrroles can be generated in short order.

Amides

Menger and Donohue made use of commercially available 2,5-dimethoxytetrahydrofuran (**1**) and benzamide to generate Cbz-protected pyrrole in their studies on the base-catalyzed hydrolysis rate of *N*-acylpyrroles.[8] A slight excess of **1** and long reaction times at reflux were needed to generated the product in modest (47%) yield.

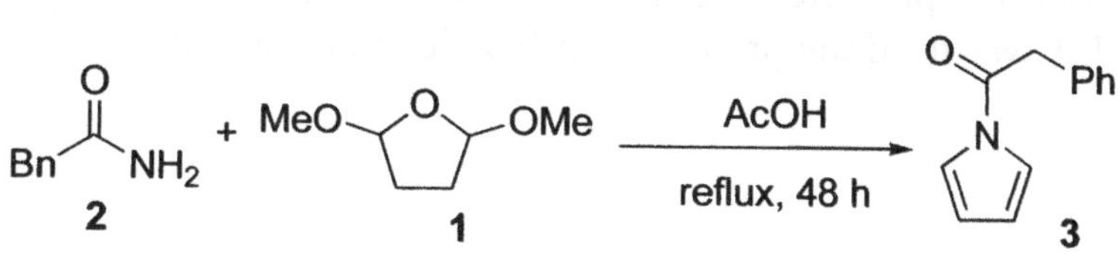

Ekkati and Bates reacted more functionalized amides using 2,5-dimethoxytetrahydrofuran as the solvent and thionyl chloride as the acid source to generate a variety of *N*-acyl pyrroles similar to **5** with good yields.[9] Under these conditions the reaction times and temperatures were greatly reduced.

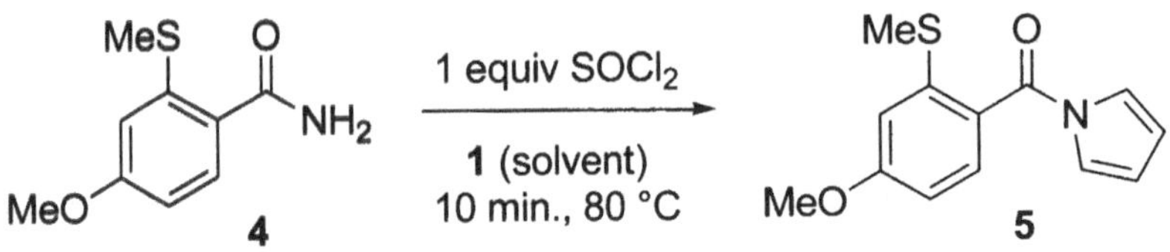

4 **5**

Sulfonamides

Sulfonamides have also seen great success as partners in the Clauson–Kaas reaction. Similar to amides the pyrroles generated are protected by the starting sulfonamides. Karousis and co-workers reported a successful example of this procedure with 3-formyl-2,5-dimethoxytetrahydrofuran **7** to generate tosyl-protected pyrrole carbaldehyde **8**. The tosyl-protecting group was later removed under mildly basic conditions (K_2CO_3, MeOH at room temperature).[10]

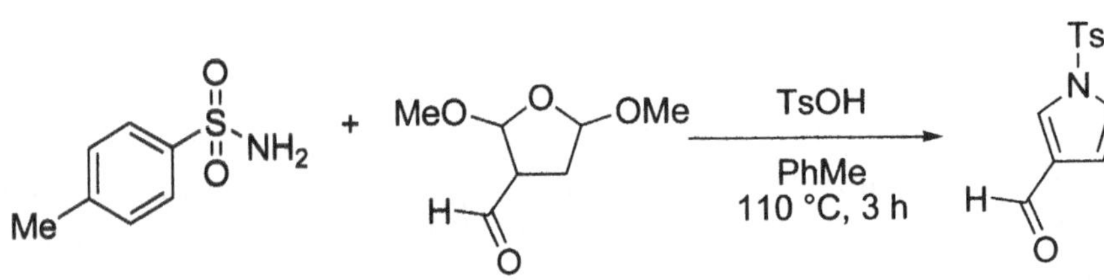

6 **7** **8**

Abid and co-workers reported that using reaction conditions that vary only in the amount of trifluoroacetic acid protected pyrroles, indoles and carbazoles are generated.[11] The reaction conditions involved reacting an aryl sulfonamide with 5 equiv of 2,5-dimethoxytetrahydrofuran in the dichloromethane and different amounts of TFA. When a catalytic amount of TFA was added the expected pyrrole (**10**) resulted. When a full equivalent of TFA was used the protected indole was produced (**11**) and finally when 3.5 equiv of TFA was used the protected carbazole (**12**) resulted.

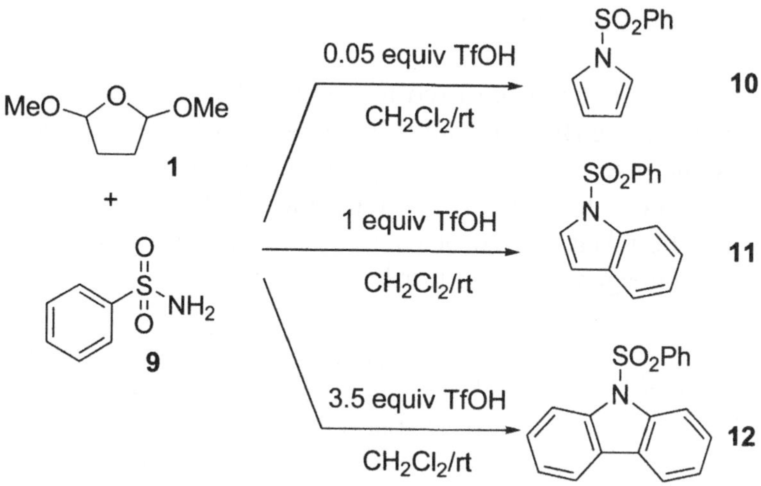

The authors propose that the reaction initially proceeds through the Paal–Knorr reaction of the succindialdehyde with the sulfonamide nitrogen to generate the pyrrole. In the presence of more than a catalytic amount of TFA the excess succindialdehyde can then undergo acid-catalyzed Friedel–Crafts annulation onto the pyrrole followed by elimination of water to generate the indole. The water released in the Paal–Knorr and annulation reactions are postulated to attenuate the acidity of the TFA[12] thus the reaction halts at the indole with one equivalent of TFA. The authors found that lower pH achieved with 3.5 equiv of TFA was needed the produce the carbazole in high yields.

Alkyl Amines

Kashima and co-workers reported the reaction of aliphatic amine hydrochloride salts (**13**) with 2,5-dimethoxytetrahydrofuran in a benzene/water mixture also generated *N*-alkyl substituted carbazoles (**14**) in modest (34–69%) yields depending on the amine chosen.[13] In this case the hydrochloride salt was used in excess to the 2,5-dimethoxytetrohydrofuran.

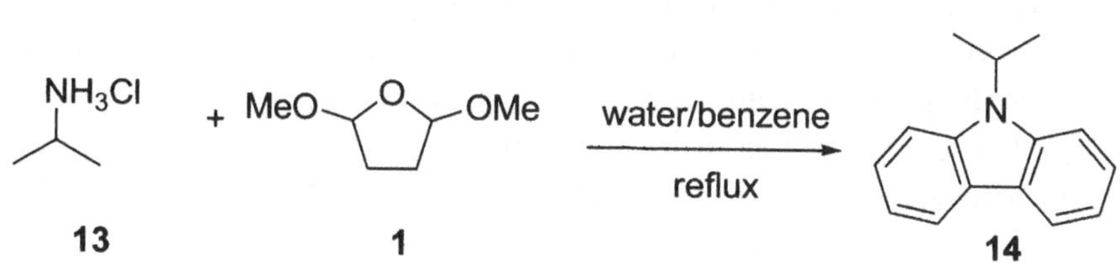

Alternate Reaction Conditions

A variety of catalysts has been used to promote the Clauson–Kaas reaction. The most common are Brønsted acids but a variety of other Lewis acids have also seen some use. An interesting report by Fang used phosphorus pentoxide in toluene to shorten reaction times.[14] Others have commented that care must be taken to rigorously remove water from this reaction with amides to avoid the generation of nitriles.[9]

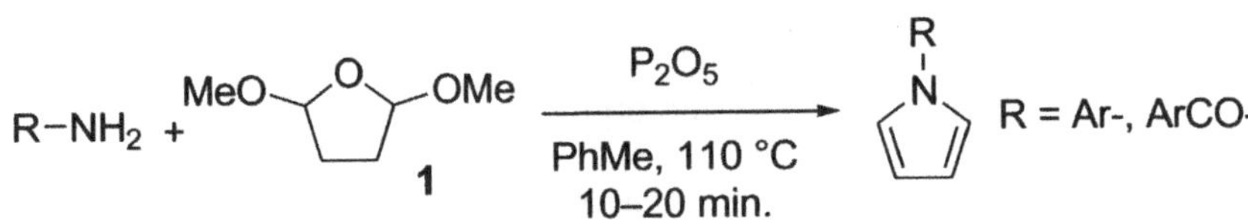

The quite versatile Lewis acid, scandium triflate has a been reported to act as a mild catalyst for Clauson–Kaas reactions. Zuo and co-workers investigated a battery of Lewis acids and found the greatest success with $Sc(OTf)_3$.[15] In their report they used these conditions to generate a variety of aryl and heteroaryl pyrroles. In this case, deactivated nitrogen nucleophiles (**15**) worked well under these reaction conditions to yield heteroaryl substituted pyrroles similar to (**16**).

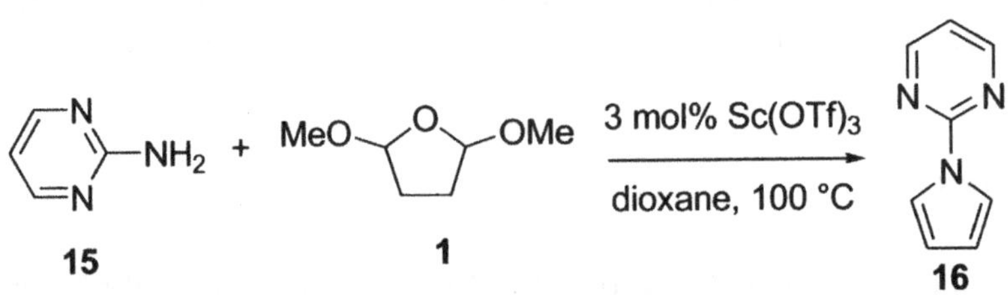

As with most thermal reactions, microwave heating has been applied to the Clauson–Kaas reaction. Polshettiwar and co-workers have added in interesting twist by running the reaction in water and using a magnetic nanoparticle supported glutathione organocatalyst (Nano-FGT).[16] The combination of these various technologies yielded a very green process where the solvent was water and the catalyst was readily removed and recycled. These reaction conditions were shown to be quite general with most types nitrogen nucleophiles, aliphatic and aromatic amines, amides, and sulfonamides all reacting well.

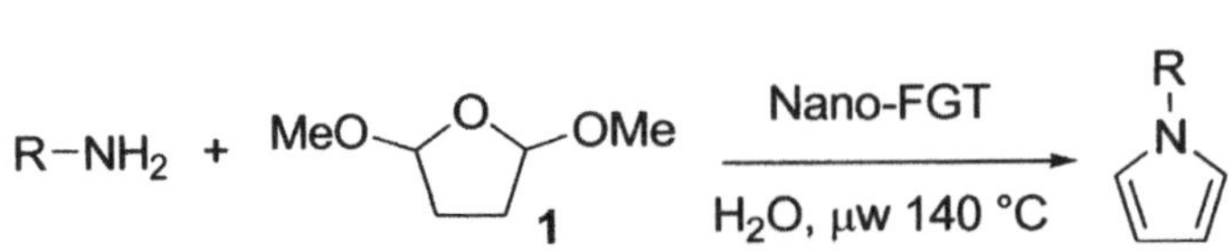

Abid and co-workers reported another green reaction system in which amides or sulfonamides along with 2,5-dimethoxytetrahydrofuran were adsorbed to K-10 montmorillonite clay and then irradiated for short times at 100 °C.[17] The resulting pyrroles were then isolated by simply diluting with ether and filtrating from the insoluble clay.

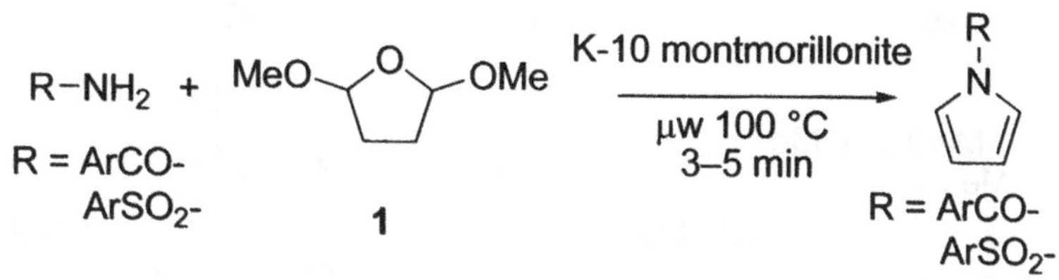

2.1.5 *Synthetic Utility*

Although the readily available 2,5-dimthoxytetrahydrofuran is by far the preferred precursor for most examples of the Clauson–Kaas reaction several examples of more highly functionalized tetrahydrofurans have been reported. Merz and co-workers generated 3,4-dialkoxy pyrroles (**19**) from the corresponding 3,4-dialkoxy-2,5-dimethoxytertohydrofurans (**18**) under mildly acidic conditions.[18] The tetra-substituted tetrahydrofurans were access through permanganate oxidation of 2,5-dimethoxydihydrofuran **17** followed by alkylation of the corresponding diol.

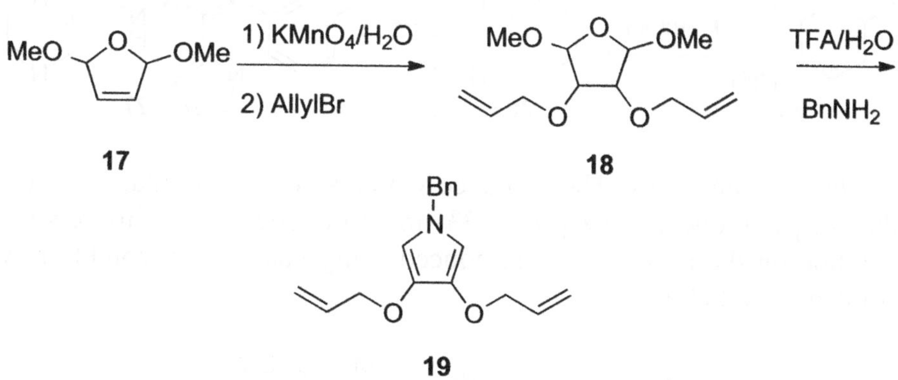

Eiden and Grusdt synthesized an even more complicated tetrahydrofuran to generate trisubstituted pyrroles (**24**) as bronchodilators.[19] In this case, the authors used the Michael addition of 2-thiomethylacetate onto the unsaturated ester **21**. The resulting highly functionalized tetrahydrofuran then underwent the Clausen–Kaas reaction to yield the desired protected pyrrole in modest (42%) yield.

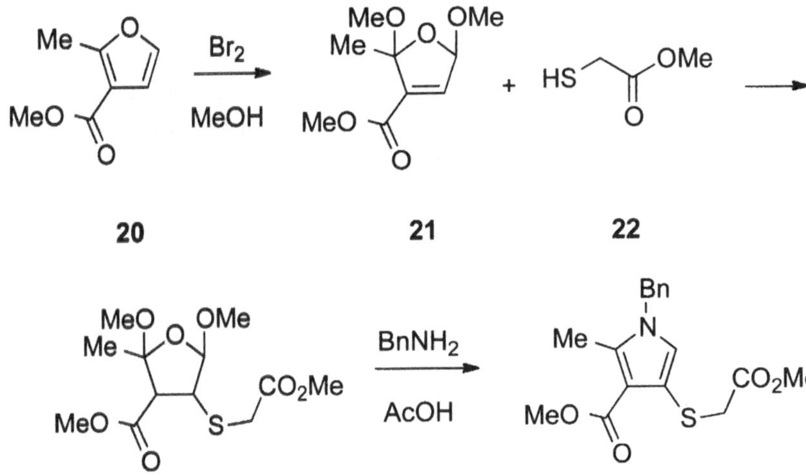

20 21 22

23 24

The Clauson–Kaas reaction has also seen some utility in the synthesis of complex heterocycles where the pyrrole is generally installed early and then further functionalized. Plasencia and co-workers have a fine example of this strategy in their reported synthesis of quinoxalinhydrazides (**27**), which showed some anticancer activity.[20]

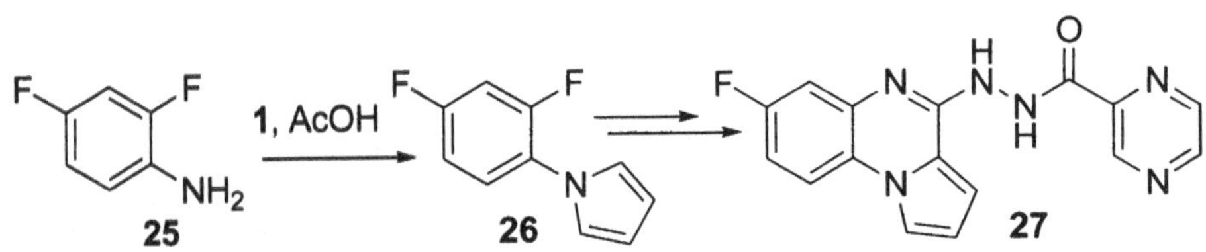

25 26 27

In the late 1970s Hara and co-workers used a similar strategy to synthesize pyrrolobenzodiazepines (**33**) in short order.[21] In this report 3,4 substitution on the pyrrole was introduced using Paal–Knorr condition with substituted 1,4-diketones.

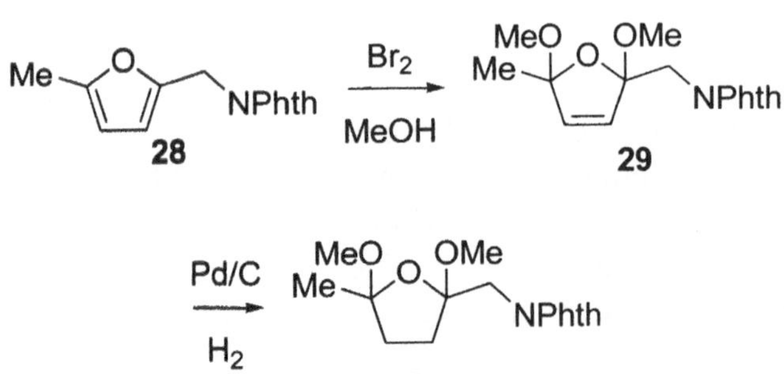

28 29 30

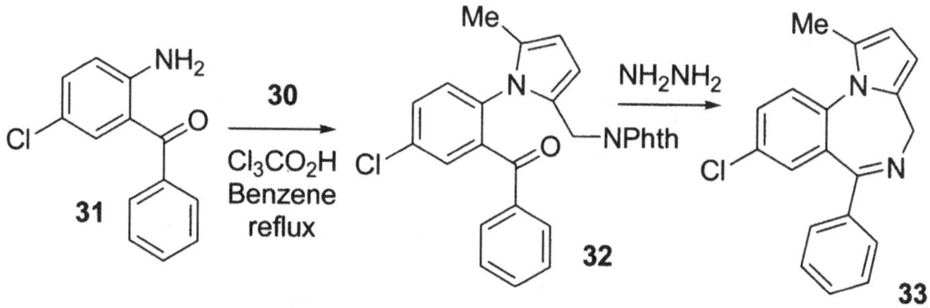

Park and co-workers reported a slightly more involved synthesis of pyrrolobenzazepines (**38**), starting from properly substituted anilines (**34**) and 2,5-dimethoxytetrohydrofuran.[22] Following this method a series of pyrrolobenzazepines substituted with halogens, nitro- and methoxy-groups were generated.

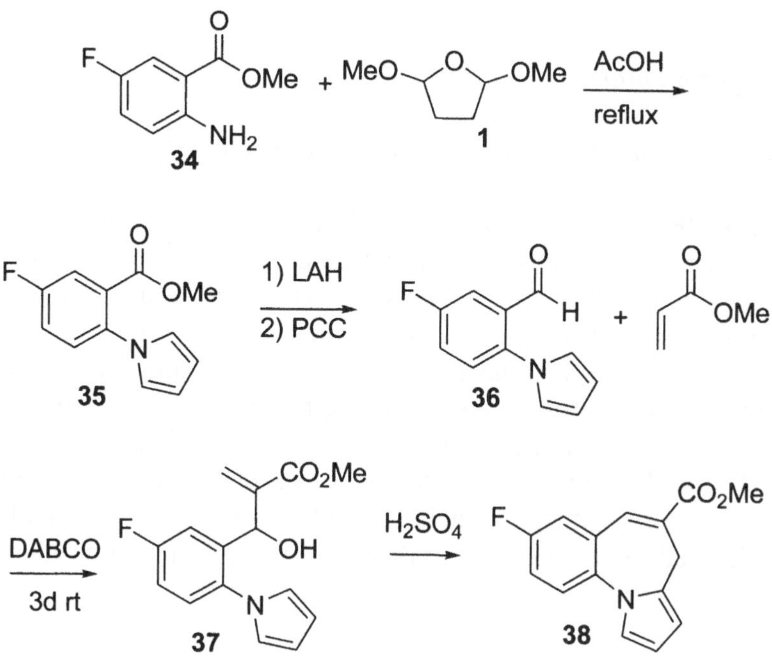

An impressive example of the functional group compatibility of the Clauson–Kaas reaction was reported by Frontata-Uribe and co-workers in the synthesis of optically active pyrrole D-glucosamine derivatives **41**.[23] One drawback to this synthetic route was the large amounts of elimination by-product **42** observed after the removal of the acetate protecting groups. The low yield of the desired product was thought to be due to the general instability of the unprotected pyrrole D-glucosamine.

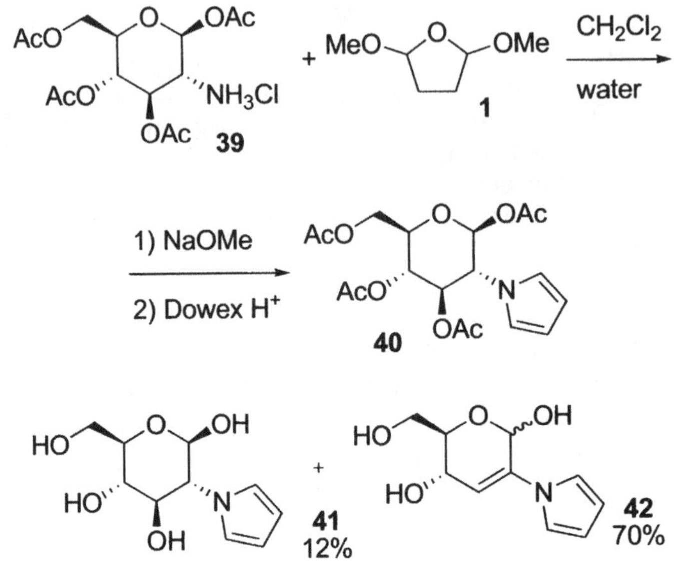

2.1.6 *Experimental*

Synthesis of 1-(2-methoxycarbonylphenyl)pyrrole 43[24]

A solution of 90 g (0.59 mol) of methyl anthranilate in 265 mL of glacial acetic acid is placed in a 1-L round-bottomed flask equipped with a reflux condenser and a magnetic stirrer. The stirrer is started, and 78 g (0.59 mol) of 2,5-dimethoxytetrahydrofuran is added over 10–15 min. The solution is heated under reflux for 1 h, during which time the solution turns deep red to black in color. The heating is discontinued, the condenser is replaced with a Vigreux column, and the acetic acid is removed by distillation at aspirator pressure. The dark residue is distilled under reduced pressure through a 25-cm column packed with glass helices, and 84–96 g (70–80%) of slightly yellow 1-(2-methoxycarbonylphenyl)pyrrole is collected, bp 90–95 °C (2 mm).

Synthesis of *N*-Benzoylpyrrole 3[8]

2,5-Dimethoxytetrahydrofuran **3** (8.5 g, 0.064 mol) and benzamide (6.0 g, 0.050 mol) in 50 mL of glacial acetic acid were boiled under reflux for 48 h. The mixture was cooled, poured onto ice, neutralized with sodium bicarbonate, and extracted with ether. Removal of the solvent from the ether extract left a thick brown oil that was purified first by steam distillation and then by vacuum distillation, bp 125° C (1.8 mm). The colorless product (4.0 g, 47%) possessed a carbonyl stretching band at 1698 cm^{-1} and a characteristic nmr sextet (triplets at 6.0 and 7.1). GLC analysis proved that the product was greater than 95 pure.

General Procedure for Clay-Adsorbed Microwave Reactions[11]

Amines or sulfonamides (1.0 mmol) and 2,5-dimethoxytetrahydfuran (0.2 g, 15 mmol) were mixed in 3 mL of ether in a round-bottomed flask, after which 500 mg of montmorillonite K-10 was added. After 5 min stirring, the solvent was evaporated *in vacuo* to produce the dry mixture of reactants adsorbed on the catalyst surface. The dry mixture was transferred to a reaction tube (10 cm long and 1 cm in diameter) and irradiated in a focused microwave reactor (CEM Discover Benchmate) at standard temperature (100 °C). The reaction temperature was determined and maintained by a built-in infrared temperature detector-controller. After satisfactory conversion, ether was added to the cold mixture, and the product was separated from catalyst by gravity filtration. The products were isolated as crystals or oils and purified by flash chromatography.

2.1.7 *References*

1 (a) Clauson–Kaas, N., Tyle, Z. *Acta Chem. Scand.* **1952,** *6,* 667. (b) Clauson–Kaas, N., Elming, N. *Acta Chem. Scand.* **1952,** *6,* 867–874.
2 Runge, R. *Ann. Physik.* **1834,** *31,* 67.
3 Anderson, T. *Ann.* **1858,** *105,* 349.
4 Bayer, A.; Emmerling, H. *Ber.* **1870,** *3,* 517.
5 For reviews on pyrrole synthesis see [R] (a) Fisher, H.; Orth, H. in *Die Chermie des Pyrrols,* Akademische Verlag, Leipzig, 1934. [R] (b) Corwin, A. H. in *Heterocyclic Chemistry,* vol. 1, Wiley, New York, 1950. [R] (c) Jones, R. A. *Pyrroles* In *Chemistry Heterocyclic Compounds* v 48 pt 1, Wiley, New York, **1990.**
6 2,5-Dimethoytetrahydrofuran (**1**) is available from a variety of commercial sources.
7 For detailed studies on the mechanism of the Paal–Knorr reaction see (a) Katritzky, A. R.; Oetercamp, D. L.; Yousaf, T. I. *Tetrahedron* **1987,** *43,* 5171, (b) Amarnath, V.; Anthony, D. C; Amarnath, K.; Valentine, W. M.; Wetterau, L. A.; Graham, D. G. *J. Org. Chem.* **1991,** *56,* 6924–6931.
8 Menger, F. M.; Donohue, J. A. *J. Am. Chem. Soc.* **1973,** *95,* 432–437.
9 Ekkati, A. R.; Bates, D. K. *Synthesis* **2003,** 1959–1961.
10 Karousis, N.; Liebscher, J.; Varvounis, G. *Synthesis* **2006,** 1494–1498.
11 Abid, M.; Teixeira, L.; Toeroek, B. *Tetrahedron Lett.* **2007,** *48,* 4047–4050.
12 (a) Saito, S.; Sato, Y.; Ohwada, T.; Shudo, K. *J. Am. Chem. Soc.* **1994,** *116,* 2312. (b) Olah, G. A.; Batamack, P.; Deffieux, D.; Torok, B.; Wang, Q.; Molnar, A.; Prakash, G. K. S. *Appl. Catal. A* **1996,** *146,* 107.
13 Kashima, C.; Hibi, S.; Maruyama, T.; Omote, Y. *Tetrahedron Lett.* **1986,** *27,* 2131–2134.
14 Fang, Y.; Leysen, D.; Ottenheijm, H. C. J. *Synth. Commun.* **1995,** *25,* 1857–1861.
15 Zuo, B.; Chen, J.; Liu, M.; Ding, J.; Wu, H.; Su, W. *J. Chem. Res.* **2009,** 14–16.
16 (a) Polshettiwar, V.; Baruwati, B.; Varma, R. S. *Chem. Commun.* **2009,** 1837–1839. (b) Polshettiwar, V.; Varma, R. S. *Tetrahedron* **2010,** *66,* 1091–1097.
17 Abid, M.; Landge, S. M.; Torok, B. *Org. Prep. Proced. Int.* **2006,** *38,* 495–500.
18 Merz, A.; Meyer, T. *Synthesis* **1999,** 94–99.
19 Eiden, F.; Grusdt, U. *Arch. Pharm.* **1989,** *322,* 807–810.

20 Plasencia, C.; Grande, F.; Oshima, T.; Cao, X.; Yamada, R.; Sanchez, T.; Aiello, F.; Garofalo, A.; Neamati, N. *Cancer Biol. Ther.* **2009,** *8*, 458–465.

21 Hara, T.; Kayama, Y.; Mori, T.; Itoh, K.; Fujimori, H.; Sunami, T.; Hashimoto, Y.; Ishimoto, S. *J. Med. Chem.* **1978,** *21*, 263–268.

22 Park, S. P.; Song, Y. S.; Lee, K.-J. *Tetrahedron* **2009,** *65*, 4703–4708.

23 Frontana-Uribe, B. A.; Escarcega-Bobadilla, M. V.; Juarez-Lagunas, J.; Toscano, R. A.; Garcia, d. l. M. G. A.; Salmon, M. *Synthesis* **2009,** 980–984.

24 Josey, A. D. *Org. Synth.* **1967,** *47*, 81.

2.2 Houben–Hoesch Acylation of Pyrroles

Richard J. Mullins and Kenneth E. Schwieter

2.2.1 Description

The Houben–Hoesch acylation of pyrroles describes the electrophilic substitution of an activated nitrile onto an electron-rich pyrrole ring. Typically requiring either a Lewis acid or protic acid, the resulting imine is immediately hydrolyzed to yield the corresponding ketone.

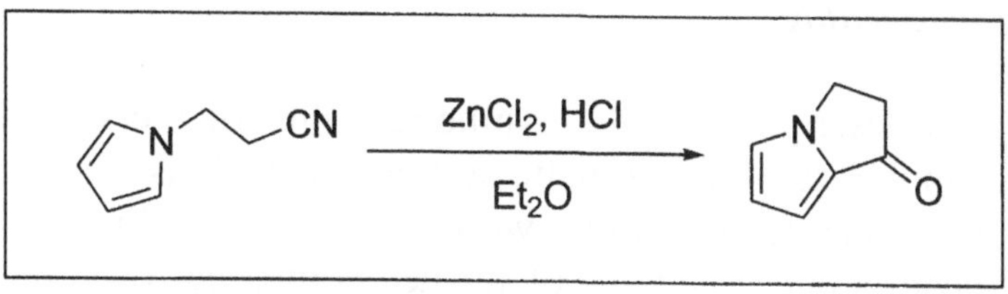

2.2.2 Historical Perspective

In a reaction similar to the Friedel–Crafts acylation of an aromatic ring, Ludwig Gatterman, in 1898, reported the Lewis acid-promoted reaction of hydrocyanic acid and benzene to produce aromatic aldehydes in a reaction that now bears his name.[1] The related electrophilic substitution of an activated nitrile onto an electron-rich aromatic ring was first reported in 1915 by the German chemist Kurt Hoesch,[2] who later served as the biographer of the legendary chemist Emil Fischer.[3] The reaction was extended and generalized by Josef Houben while at the Biologische Reichsanstalt in Berlin.[4] It was later extended to other π-excessive heterocycles, such as pyrroles, which are the focus of this chapter. For a more thorough discussion of the history of the Houben–Hoesch reaction as it applies to traditional aromatic systems, the reader is directed to an excellent review[5] as well as to a chapter in a previous volume in this series.[6]

2.2.3 Mechanism

The Houben–Hoesch reaction proceeds via a straightforward electrophilic aromatic substitution mechanism. Following protonation or Lewis acid activation of the alkyl nitrile, nucleophilic attack by the electron-rich pyrrole selectively at C(2) produces the resonance stabilized intermediate **1**. Elimination of H^+ reestablishes the aromaticity of the pyrrole, resulting in imine **2**, which is rapidly hydrolyzed to produce the ketone **3**.[7]

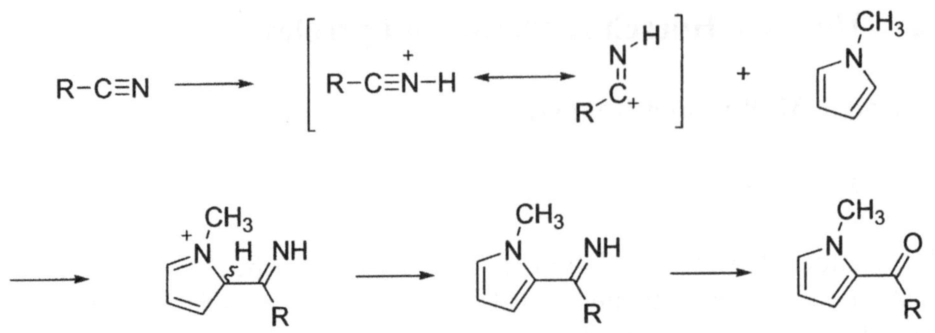

1 2 3

2.2.4 *Synthetic Utility*

For a complete description of the synthetic utility of the Houben–Hoesch reaction as it applies to other aromatic systems, as well as some mechanistic discussions, the reader is directed to two reviews on the subject.[5,6] Due to the diminished electrophilic reactivity of nitriles compared to other carboxylic acid derivatives as well as the broad number of Friedel-Crafts substrates, most Houben–Hoesch pyrrole acylation reactions are conducted intramolecularly. However, an impressive intermolecular example was delineated by Chang and co-workers in efforts directed toward the synthesis of novel 2-[5-aroylpyrrolo]alkanoic acids, for evaluation of their potential analgesic and anti-inflammatory activities.[8] Treatment of substituted pyrrole **4** and 3-cyanopyridine (**5**) with acid in dry chloroform resulted in the preparation of **6** in good yield.

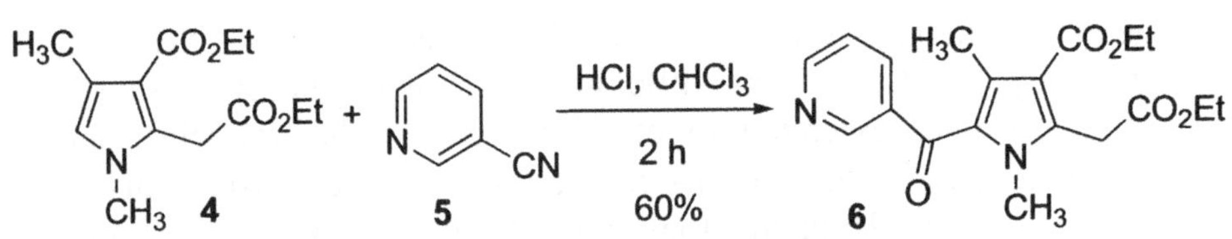

The first intramolecular Houben–Hoesch acylation of pyrrole was reported by Clemo and Ramage,[9] and later used by Adams and co-workers[10] in their efforts to prepare 1-hydroxypyrrolizidine and related compounds. Treatment of cyanopyrrole **7** under standard conditions provided **8**, albeit with inconsistent yields presumably due to a nonproductive polymerization reaction that accompanies product formation. A short time later, Gabel provided a modified procedure, using $BF_3 \cdot OEt_2$ as the solvent.[11] Formation of a somewhat stable BF_3·pyrrole complex is suggested to prevent polymerization, while still allowing for cyclization.

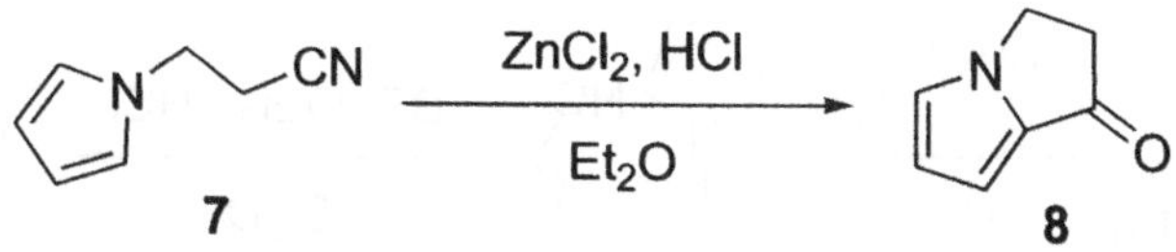

The intramolecular Houben–Hoesch acylation of pyrroles has been used for the synthesis of a number of compounds with interesting properties.[12,13] An intramolecular acylation was reported by the Meinwald group to synthesize the hairpencil secretion of the butterfly *Lycorea ceres ceres*.[14] *N*-Cyanoethyl-3-methyl pyrrole (**9**) undergoes intramolecular acylation to selectively yield the 2,3-product (**10**) as opposed to the 3,5-product. Although two regioisomers are expected from attack at C(2) and C(5), the methyl group at C(3) selectively activates C(2) for nucleophilic attack.

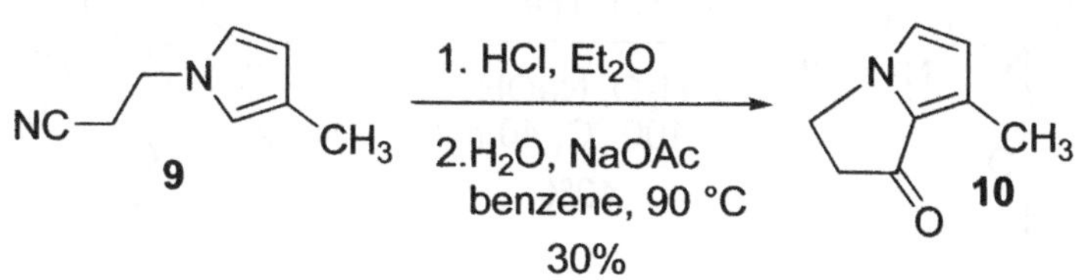

Not just limited to five-membered rings, the intramolecular Houben–Hoesch can be used to form rings of various sizes. The Patterson group synthesized cyclopentano, cyclohexano, and cycloheptano[*a*]pyrroles under Houben–Hoesch conditions using the Lewis acid $BF_3{\cdot}OEt_2$ followed by a Wolf–Kishner reduction.[15] The yield of the acylation step seemed to depend on the stability of the ring size of the product, providing the cyclohexano compound in highest yield.

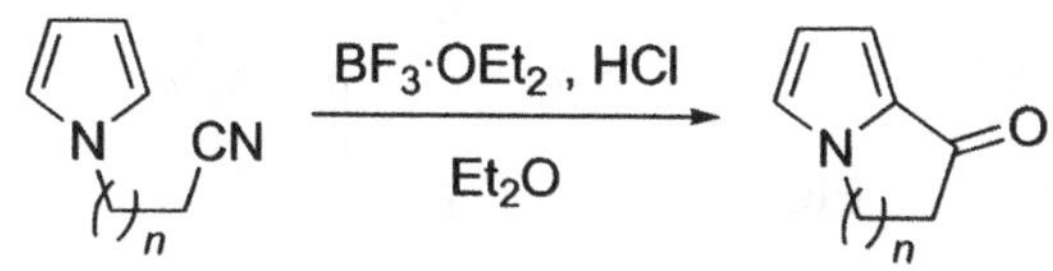

(*n* = 1, 33%; *n* = 2, 59%; *n* = 3, 31%)

An interesting variant of the above reaction, in which unsaturation was present in the newly formed ring, was used by Flitsch and co-workers.[16] Following a low-yielding Wittig reaction to produce the *E*-isomer **12** as the separable minor component, cyclization was effected under standard Houben–Hoesch conditions to give **13**.

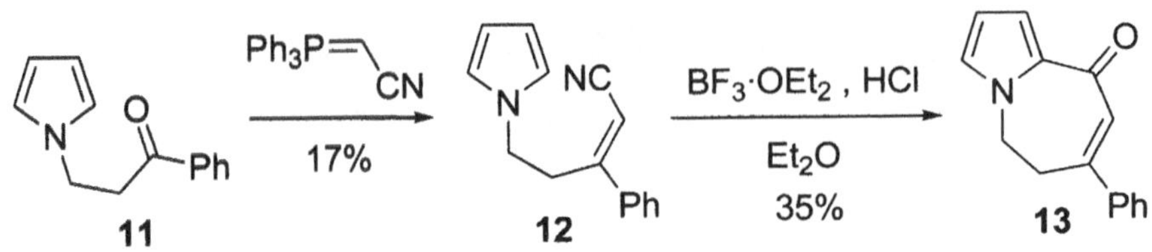

The intramolecular acylation has proven to be quite general, regardless of substituents attached on the pyrrole ring. As demonstrated in the synthesis of **15**, the presence of a protected amine at C(2) is tolerated.[17] Similarly, a large number of 5-aryl-1,2-dihydro-1-pyrrolizinones (**17**), compounds with anti-inflammatory and analgesic properties, have been synthesized using this method.[18]

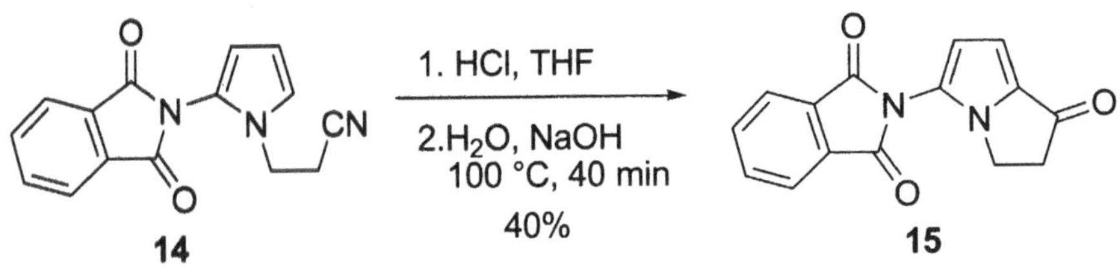

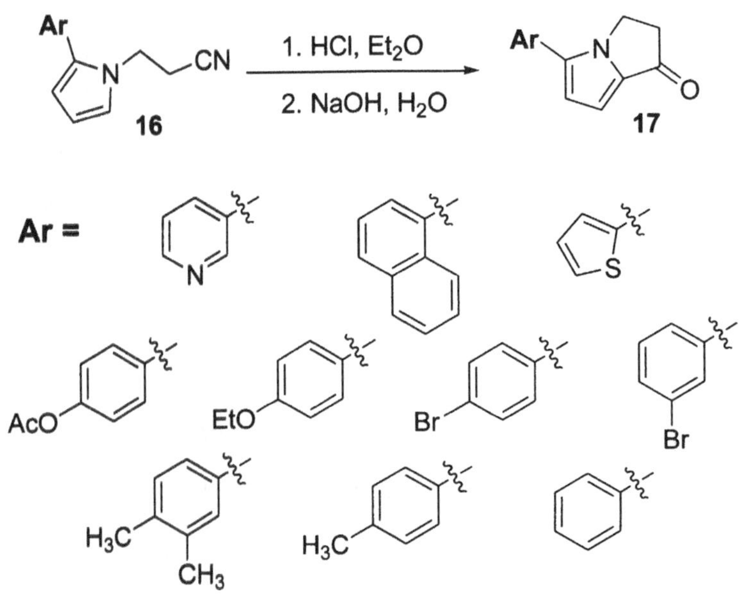

The Houben–Hoesch has also been used in reactions with indoles.[19] Ottoni and co-workers recently developed a mild and highly efficient acylation method to obtain 3-acylindoles with high regioselectivity.[20] Precomplexation between indole (**18**) and $SnCl_4$ in CH_2Cl_2 is followed by treatment with acetonitrile in nitromethane to give the 3-acylindole **19** in very high yield. The high level of regioselectivity that results is attributed to the increased nucleophilicity of the 3-postion once coordinated to the Lewis acid. This method makes the use of protecting groups unnecessary while avoiding the dimerization and trimerization sometimes seen with unsubstituted indole under Lewis acidic conditions. These results find precedence in the Gabel work described above, as it applied to the pyrrole ring.[11]

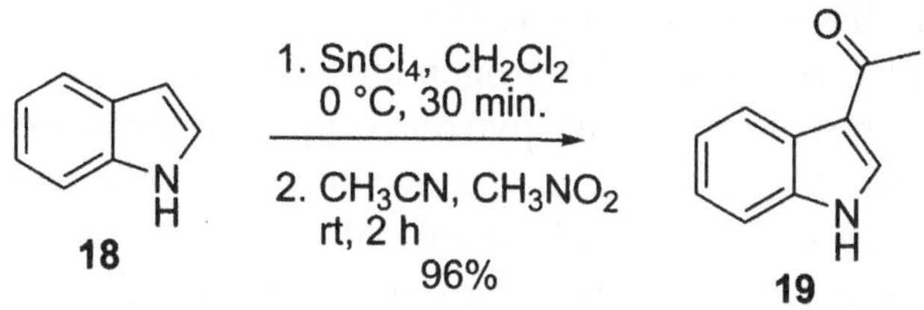

If the 3-position of indole is substituted, the Houben–Hoesch acylation can proceed at other positions on the ring. As a representative example, dimethoxyindole **20** is reacted with benzylcyanide under acidic conditions to produce the indole **21**, which has been acylated at the 7-position.[21]

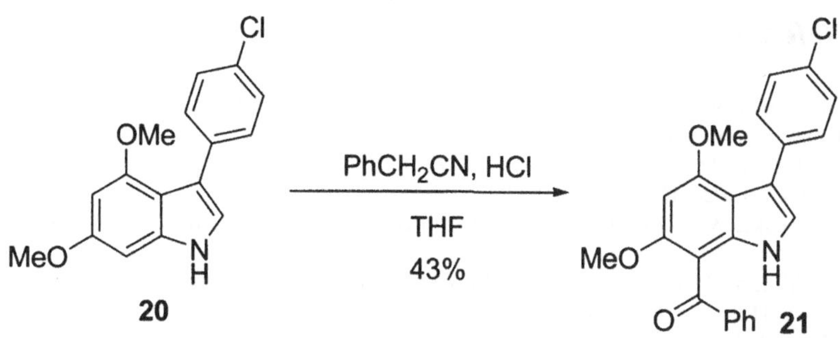

In a similar manner, Houben–Hoesch acylation was effected at the 4-position when indole **22** was treated with nitrile **23** in the presence of $AlCl_3$.[22]

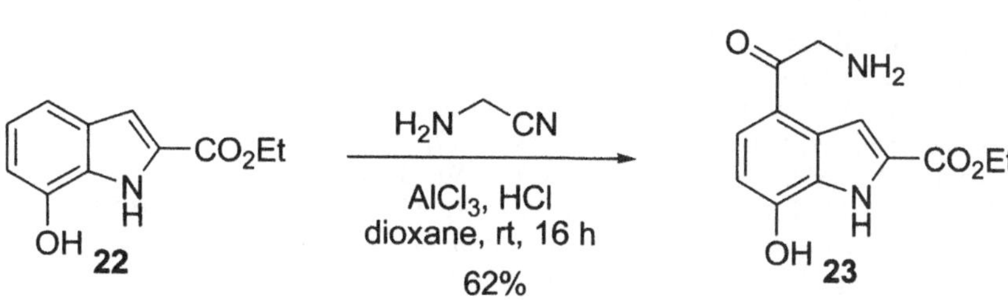

2.2.5 *Experimental*

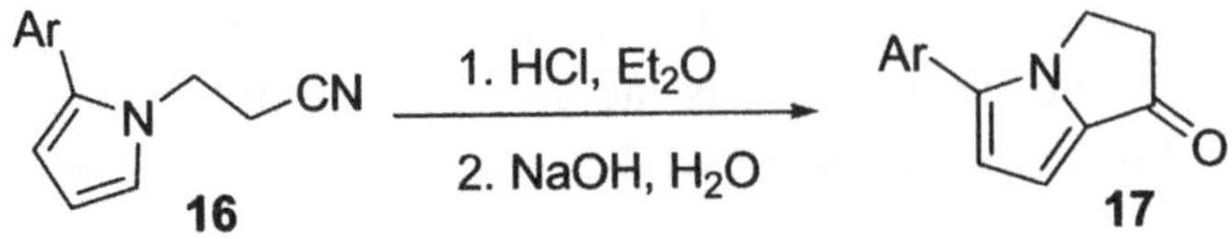

5-Aryl-2,3-dihydro-1-pyrrolizinone (17)[18]

General procedure: A stream of hydrogen chloride was gently passed through a solution of 4.1 mmol of 3-(2-arylpyrrol-yl) propanenitrile (**16**) in 30 mL of ether at 0–5 °C for 3 h. The solvent was decanted, the precipitate was washed with dry ether (30 mL × 2), and 40 mL of water was added. Then 10% aqueous sodium hydroxide was added until pH 4–4.5. The mixture was stirred at 30–40 °C for 1 h, and then heated at 85–90 °C for 2 h. After cooling to room temperature, the mixture was extracted with methylene chloride. The combined extracts were washed with water and dried over $MgSO_4$. After removing the solvent, the crude product was obtained as a solid. The crude product was purified by recrystallization from ethanol, or column chromatography on silica gel.

2.2.6 *References*

1. Gatterman, W.; Berchelmann, W. *Ber.* **1898,** *31*, 1765–1769.
2. Hoesch, K. *Ber.* **1915,** *48*, 1122–1133.
3. [R] Fischer, H. O. L. *Annu. Rev. Biochem.* **1960,** *29*, 1–14.
4. Houben, J. *Ber.* **1926,** *59*, 2878–2891.
5. [R] Ruske, W. In *Friedel–Crafts and Related Reactions*, Interscience: New York, **1964,** 383–497.
6. [R] Mullins, R. J.; O'Reilly, M. C. In *Name Reactions for Carbocyclic Ring Formations*; Wiley: Hoboken, NJ **2010,** 675–687.
7. [R] Li, J. J. *Name Reactions: A Collection of Detailed Reaction Mechanisms*, Springer-Verlag, Telos, NY, 2002, p 200.
8. Chang, M. N.; Biftu, T.; Boulton, D. A.; Finke, P. E.; Hammond, M. L.; Pessolano, A. A.; Zambias, R. A.; Bailey, P.; Goldenberg, M.; Rackham, A. *Eur. J. Med. Chem., Chim. Ther.* **1986,** *21*, 363–369.
9. Clemo, G. R.; Ramage, G. R. *J. Chem. Soc.* **1931,** 49.
10. Adams, R.; Miyano, S.; Fleš, D. *J. Am. Chem. Soc.* **1960,** *82*, 1466–1468.
11. Gabel, N. W. *J. Heterocyclic Chem.* **1967,** *4*, 627
12. Braunholtz, J. T.; Mallion, K. B.; Mann, F. G. *J. Chem. Soc.* **1962**, 4346–4353.
13. Schnekenburger, J.; Breit, E. *Arch. Pharm.* **1977,** 152–160.
14. Meinwald, J.; Meinwald, Y. C. *J. Am. Chem. Soc.* **1966,** *88*, 1305–1310.
15. Patterson, J. M.; Brasch, J.; Drenchko, P. *J. Org. Chem.* **1962,** *27*, 1652–1659.
16. Flitsch, W.; Kappenberg, F.; Schmitt, H. *Chem. Ber.* **1978,** *111*, 2407–2422.
17. Sun, G.; Yang, Z.; Zhang, S. F. *J. Indian Chem. Soc.* **2003,** *80*, 851–852.
18. Yu, H.; Wang, F.; Zhang, S. F. *Chinese Chem. Lett.* **2003,** *14*, 565–568.
19. Albrecht, R.; Heindl, J.; Loge, O. *Eur. J. Med. Chem.* **1985,** *20*, 57–60.
20. Ottoni, O.; Neder, A., Dias, A. K. B.; Cruz, R. P. A.; Aquino, L. B. *Org. Lett.* **2001,** *3*, 1005–1007.
21. Black, D. S.; Kumar, N.; Wahyuningsih, T. D. *ARKIVOC* **2008,** *6*, 42–51.

22. Norcini, G.; Allievi, L.; Bertolini, G.; Casagrande, C.; Miragoli, G.; Santangelo, F.; Semeraro, C. *Eur. J. Med. Chem.* **1993,** *28*, 505–511.

2.3 Overman Pyrrolidine Synthesis

Matthew D. Hill

2.3.1 Description

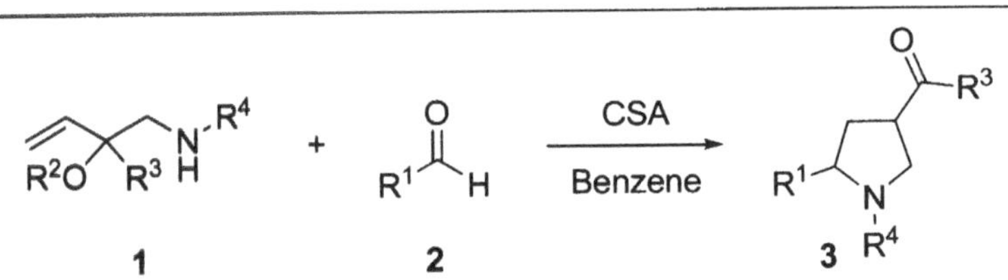

The Overman pyrrolidine synthesis is a tandem reaction, or cascade, used to generate acylpyrrolidine derivatives. This process begins with condensation of an allylic alcohol/ether-containing secondary homoallylic amine with an aldehyde, followed by an aza-Cope rearrangement and subsequent Mannich reaction. Commonly, this reaction is run in refluxing benzene with an acidic additive, such as *d*-10-camphorsulfonic acid (CSA).

2.3.2 Historical Perspective

Professor Larry E. Overman first reported the titled reaction as a young member of the University of California, Irvine faculty. Since his initial report in 1979,[1] the Overman group has used this cascade approach to a variety of natural products, including (+)-strychnine,[2] (–)-pancracine,[3] and (±)-gelsemine.[4]

2.3.3 Mechanism

The Overman pyrrolidine synthesis is a tandem process that harnesses the natural reactivity of an iminium species formed after acid-promoted and reversible condensation of an allylic alcohol/ether-containing secondary homoallylic amine **1** and an aldehyde **2**.[5] Once formed, the productive pathway requires the iminium species to undergo a reversible [3,3]-sigmatropic rearrangement, or aza-Cope rearrangement, affording a second iminium species **4**. A Mannich reaction closes the pyrrolidine ring to give the desired product **3**, and serves as the cascade terminator.[6]

Relative configuration of the stereogenic centers created during the cascade can be predicted using a transition state analysis of the Mannich step. In a study conducted within the Overman lab, (*E*)-alkenes were found to furnish 4,5-*cis*-pyrrolizidines, while the corresponding (*Z*)-alkenes gave 4,5-

trans-pyrrolizidines **5**.[7] Not only was the selectivity lower with (*Z*)-alkenes but also to a minor extent the Mannich addition occurred by way of a boat transition state, giving pyrrolizidine **5**.

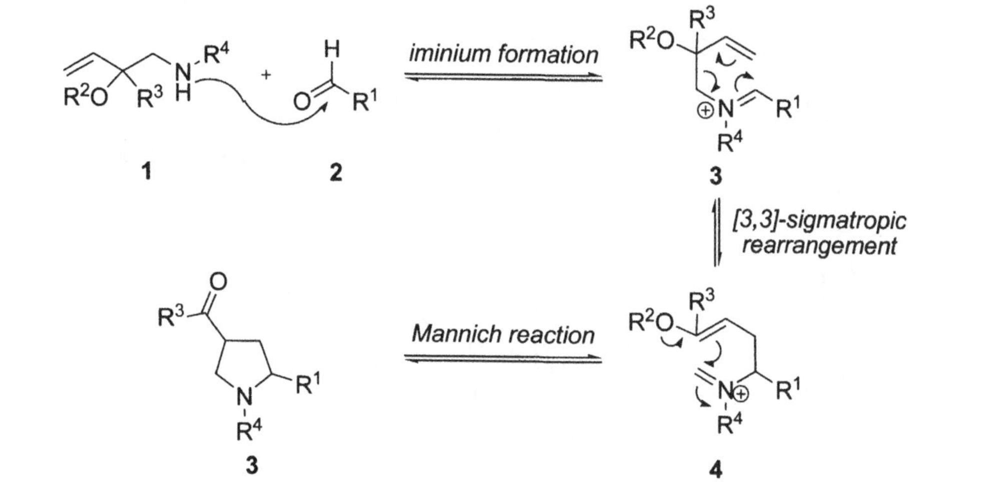

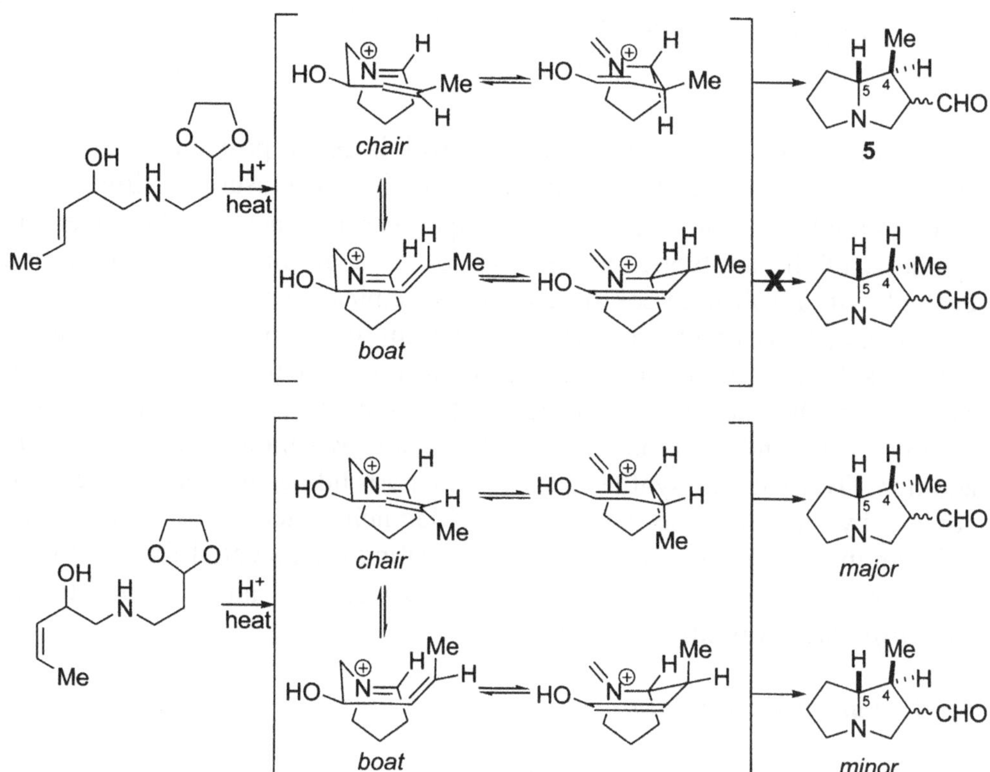

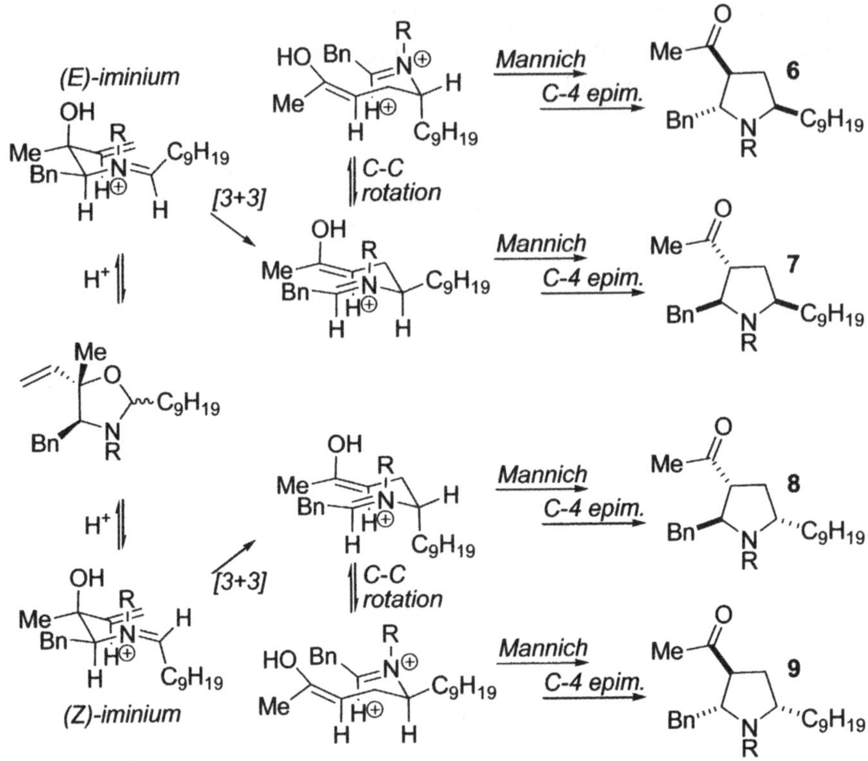

Enantioselective synthesis of pyrrolidines is possible with chiral starting amines, but racemization can occur if ring closure is slower than C–C σ-bond rotation after the aza-Cope rearrangement. The Overman synthesis of the antibiotic preussin serves as a prominent and interesting example of a substrate-controlled stereochemical outcome.[8] A large *N*-substituent (R) was found to favor (*Z*)-iminium formation upon condensation with an aldehyde, contradicting earlier studies with smaller *N*-groups (R = H, Me). After the aza-Cope rearrangement, Mannich addition occurred more rapidly than bond rotation and led to major pyrrolidine product **8**. In addition, Overman has synthesized bicyclic pyrrolidines with stereochemical control by utilizing the transannular Mannich reaction of macrocyclic iminium intermediates, and therefore eliminating the potential for rotation of a C–C σ-bond, in his formal synthesis of *d,l*-crinine[9] and *en route* to the *Aspidosperma* alkaloids (*vide infra*).[10]

2.3.4 Variations and Improvements

The Overman pyrrolidine synthesis, or aza-Cope–Mannich reaction, was developed while troubleshooting a stereochemical challenge encountered by the Overman group as they pursued the amphibian alkaloid, perhydrogephyrotoxin.[1] This elegant method relied on the inherent reactivity associated with charged iminium intermediates of type **12**, and as Overman states: "exploits the facility of this charged sigmatropic equilibrium and directs the course of the rearrangement by capturing the 'product' sigmatropic isomer **13** by an exothermic Mannich cyclization."[5b] Original reaction conditions were quite simple and required only that the tetrafluoroborate salt form of secondary homoallylic amine **10** be treated with aldehyde in refluxing benzene.[1] If one desired to use the free-base form of amine **10**, substoichiometric CSA could be used to promote the cascade. A wide scope of acylpyrrolidines **14** was synthesized using this method that contained aliphatic, aromatic, and heteroaromatic substituents.

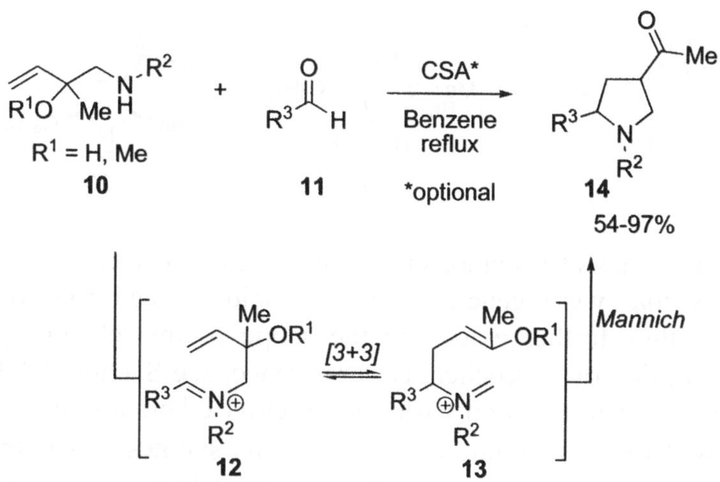

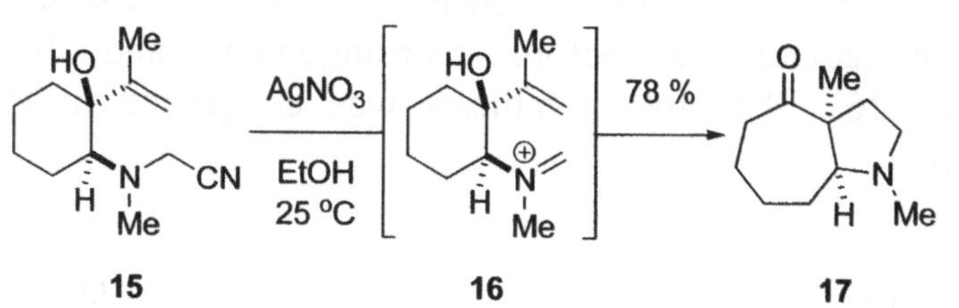

The titled reaction sequence has also been triggered by cyanide loss from appropriately designed cyanoalkylamines.[6a] This silver-mediated process was found useful for aminocyclohexanol substrates with widely varying electronic character. Treatment of compound **15** with silver nitrate ($AgNO_3$, 1.1 equiv) in ethanol at ambient temperature generated intermediate iminium **16**, and after the aza-Cope–Mannich cascade, cleanly afforded 4-oxocycloheptapyrrolidine **17** in racemic form. Within the same manuscript, Overman, *et al.* extended this methodology to an enantiomerically pure hydroindolone **20** from 1-alkenyl-2-aminocyclopentanol **18** via intermediate

19. This example of a ring-enlarging pyrrolidine annulation takes advantage of conformational constraints inherent to a medium-size ring intermediate.

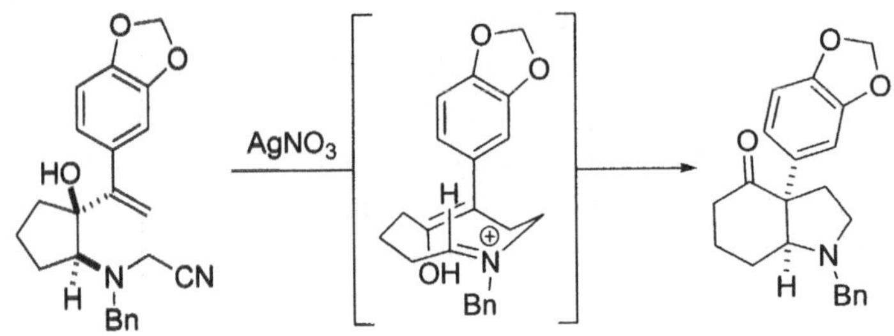

18 19 20

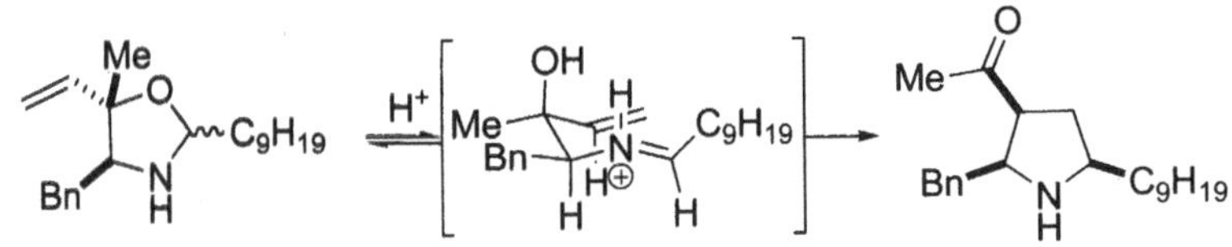

21 22 23

Overman has also demonstrated pyrrolidine synthesis from iminium intermediates that were generated by the acid-induced ring opening of oxazolidines, thus further extending the scope of this chemistry.[8] (For a detailed description of stereochemical preference, see Section 2.3.3.) Lewis acid promoters, such as boron trifluoride diethyl etherate ($BF_3{\cdot}OEt_2$), have also been used for oxazolidine opening. In his synthesis of both (–)- and racemic pancracine,[3] a member of the *Amaryllidaceae* alkaloids first discovered in 1955,[11] Overman used $BF_3{\cdot}OEt_2$ to open oxazolidine **24** to the iminium intermediate **25** before rearrangement and formation of hydroindolone **26**. The total synthesis was completed in 7% over 17 chemical steps.

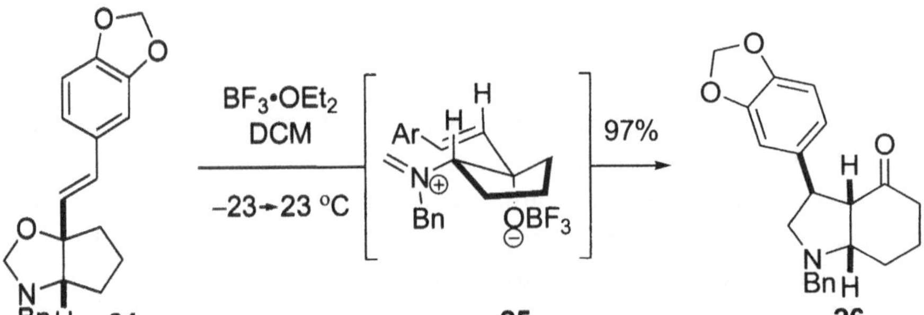

24 25 26

In a recent expansion of the Overman pyrrolidine synthesis, Carballo, *et al.* were successful in the iron(III)-promoted formation of 3-

formylpyrrolidines from 2-hydroxy homoallyltosyl amine (**27**) and both alkyl and aryl aldehydes.[12] Formation of 3-formylpyrrolidines had proven difficult under conventional reaction conditions; however, this method afforded a variety of formyl products, albeit the reduced alcohols were isolated due to direct product instability. Treatment of 2-hydroxy homoallyltosyl amine (**27**) and butyraldehyde (**28**) at ambient temperature in dichloromethane with stoichiometric iron (III) chloride ($FeCl_3$) and trimethylsilyl chloride (TMSCl) afforded pyrrolidine **29**, which was isolated as the primary alcohol **30** in a 99:1 *trans*:*cis* ratio after sodium borohydride ($NaBH_4$) reduction.

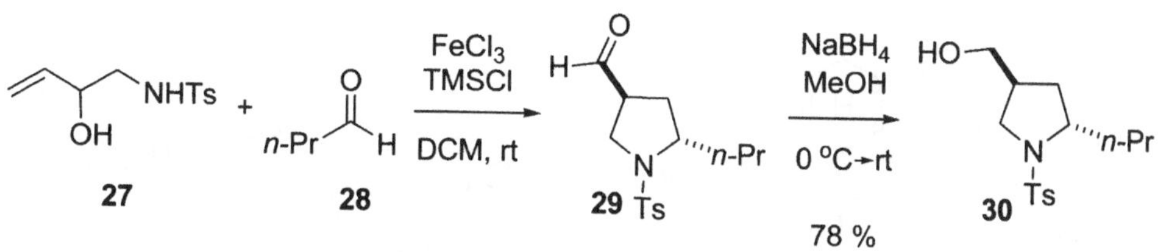

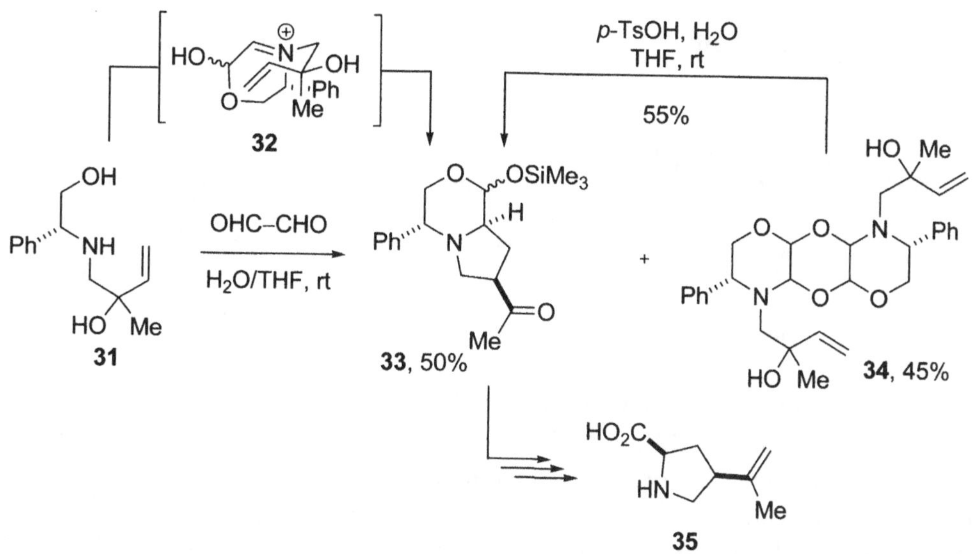

Agami and co-workers developed an interesting and asymmetric approach to proline derivatives in which they harnessed the stereochemical bias inherent to chiral starting materials.[13] An (*R*)-phenyl glycinol-derived chiral auxiliary was built into the amine substrate using standard chemistry. Treatment of homoallylic amine **31** with glyoxal (1.5 equiv) in a water/THF mixture gave fused bicycle **33**. A major tricyclic side product **34** was formed under the reaction conditions but could be transformed into **33** with acidic treatment. The morpholine ring was opened with vinyl chloroformate, and subsequent Jones oxidation produced proline **35**. The stereochemical outcome of this cascade sequence is a direct consequence of cyclization from the less hindered iminium face of intermediate **32**, in addition to a chair-like transition state directing synclinal enol attack for the Mannich step.

2.3.5 Synthetic Utility

The synthetic utility of the Overman pyrrolidine synthesis has been demonstrated in various approaches to a range of alkaloids: the *Amaryllidaceae* alkaloids,[3] *Aspidosperma* alkaloids,[10] *Strychnos* alkaloids,[2] and *Melodinus* alkaloids[14] are several types.

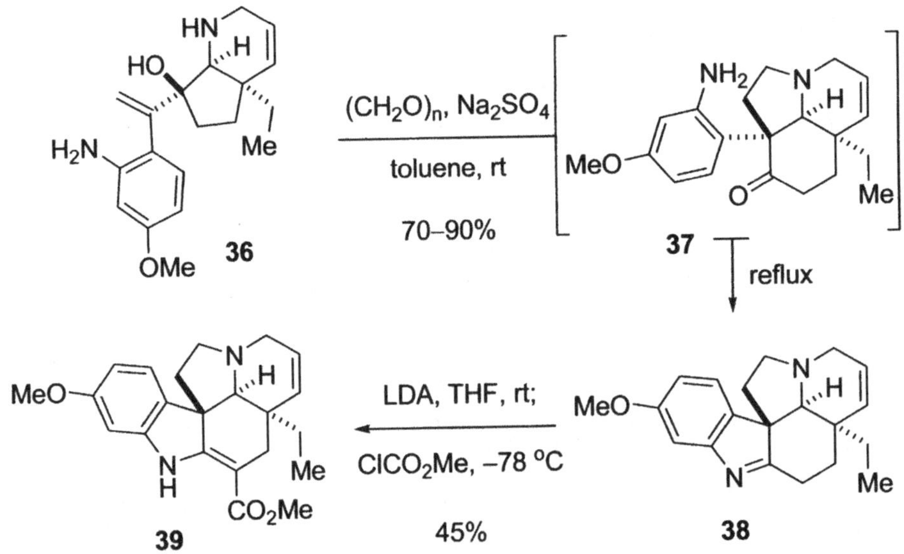

A report by Overman and co-workers on the total synthesis of (±)-16-methoxytabersonine (**39**), a *Aspidosperma* alkaloid that contains an interesting pentacyclic core, serves as an early example of this methodology used in total synthesis.[10] Treatment of aniline **36** with paraformaldehyde in the presence of sodium sulfate (Na_2SO_4) formed an iminium intermediate that spontaneously underwent [3,3]-sigmatropic rearrangement and Mannich addition to form fused tricycle **37**. Subsequent dehydration afforded penultimate pentacycle **38**. Trace formic acid, present in paraformaldehyde, was found sufficient to catalyze the aza-Cope–Mannich cascade and 1,2,6,7-tetradehydro-aspidospermidine-forming dehydration. Stereochemical control was derived from a single "chair-like" rearrangement conformation available to the iminium intermediate. This member of the *Aspidosperma* family of alkaloids, and useful precursor to vindoline, was rapidly accessed in 6% overall yield over 11 steps with stereochemical control.

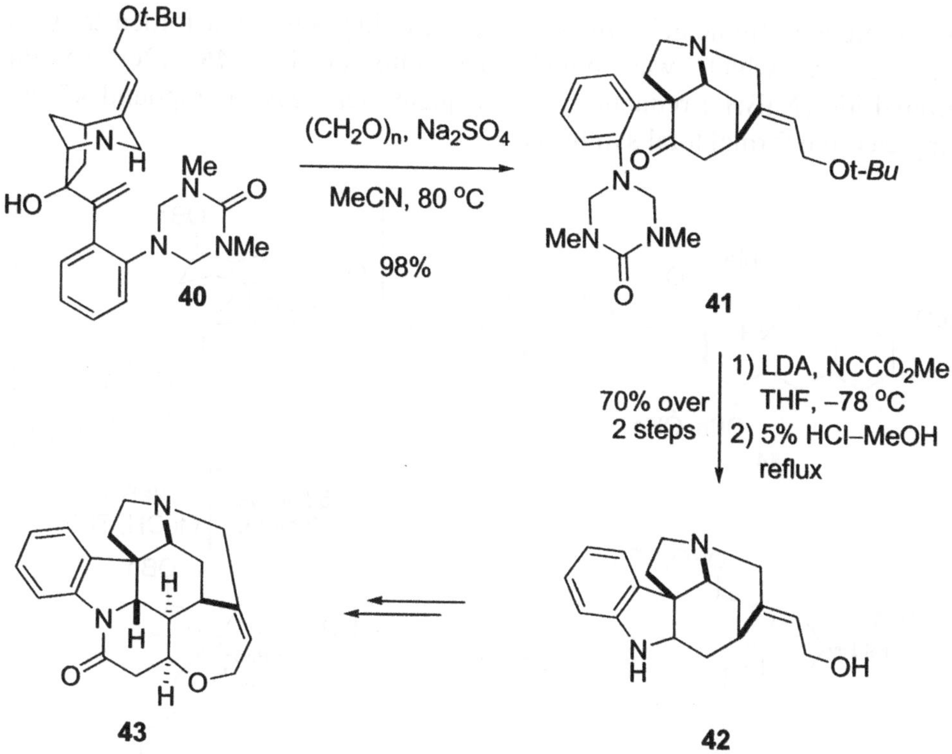

The enantioselective synthesis of both (–)- and (+)-strychnine (**43**) serves as another impressive example from the Overman group.[2] After isolation from *Strychnos ignatii* in 1818[15] and structural elucidation in 1946,[16] only several racemic total syntheses,[17] including a pioneering synthesis by Woodward,[18] had been reported before the first asymmetric route published by Knight, Overman, and Pairaudeau. Their work used the cationic aza-Cope-Mannich reaction to assemble fused tricycle **41** by treatment of azabicyclooctane **40** with paraformaldehyde and Na_2SO_4. Subsequent carbomethoxylation, acid-induced cleavage of both the triazone and *t*-butyl protecting groups, and dehydrative cyclization afforded the pentacyclic strychnan core. Overman's synthesis furnished enantiomerically pure (–)-strychnine (**43**) in 3% overall yield over 20 steps and "provides an important benchmark of the power of the aza-Cope rearrangement–Mannich reaction to solve formidable problems in alkaloid construction."

Other academic groups have used the Overman pyrrolidine synthesis en route to complex alkaloids and other medicinally relevant compounds. The Brummond group at the University of Pittsburgh has reported the formal synthesis of (–)-FR901483 (**48**), an immunosuppressant isolated by researchers at Fujisawa Pharmaceuticals.[19] In this report, Brummond incorporated a tandem cationic aza-Cope rearrangement/Mannich cyclization that proceeds via a bridgehead iminium ion. Treatment of intermediate **44** with *p*-toluenesulfonic acid (*p*-TsOH) in refluxing benzene afforded tricyclic aldehyde **45**, which was transformed into alcohol **46** after sodium borohydride ($NaBH_4$) reduction. Subsequent steps gave compound **47** and completed the formal total synthesis.

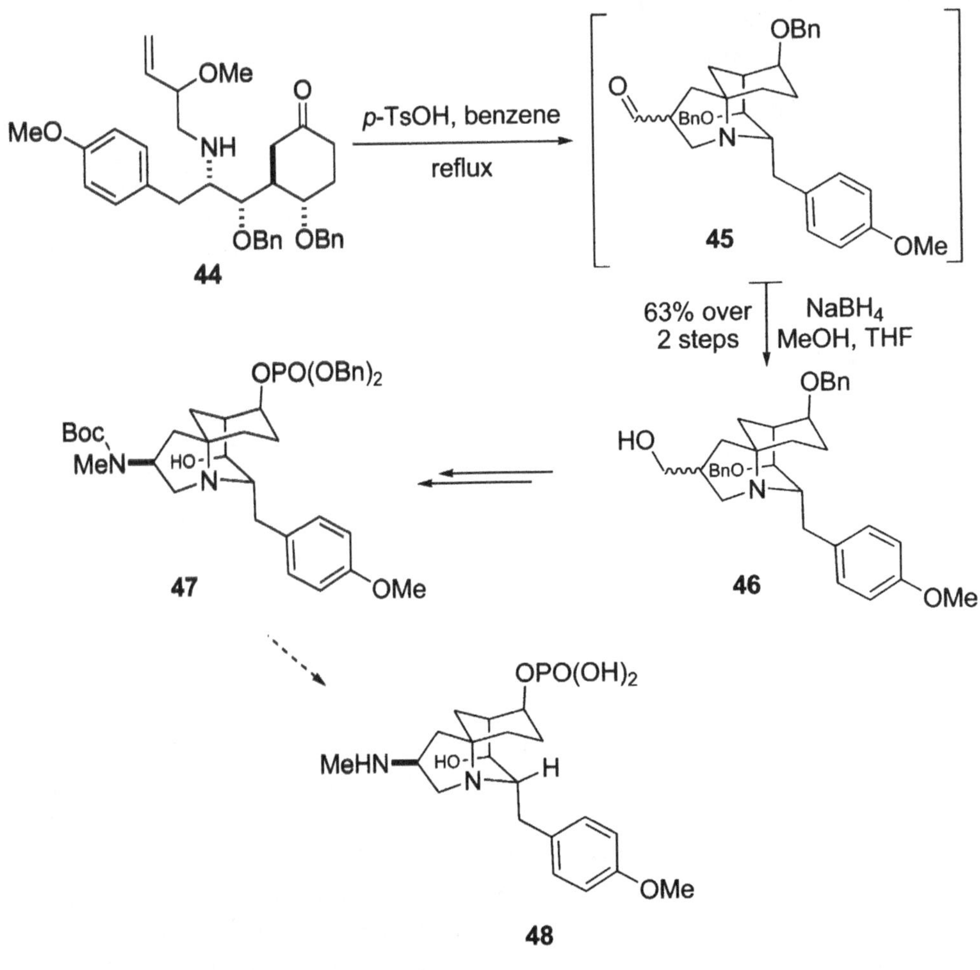

Sonnenschein was first isolated gelsemine (**55**) from Carolina jasmine (*Gelsemium sempervirens*),[20] a plant indigenous to southeastern United States of America, in 1876, but the structure remained unsolved until 80 years later.[21] Since then, several academic groups have reported syntheses of this member of the *Gelsemium* alkaloids; the most provocative of which has come out of the Overman group.[4] Overman's approach consists of an anionic aza-Cope rearrangement and sequential Mannich cyclization to afford the azatricyclo[4.4.0.0^{2,8}]decane core, followed by an intramolecular Heck reaction, and subsequent base-promoted skeletal rearrangement. The key-steps in forming the azatricyclic core represent a stepwise variant of the Overman Pyrrolidine Synthesis and allowed for installation of bromine that was critical for later steps. In this protocol, potassium hydride-promoted generation of imine **49** and aza-Cope rearrangement, followed by quench with methyl chloroformate gave bicycle **52**. After bromination, treatment with trifluoroacetic acid (TFA) led to Mannich cyclization and formation of key tricycle **54**. Overman completed the total synthesis of (±)-gelsemine (**55**) in 1.1% overall yield by way of 26 isolated intermediates.

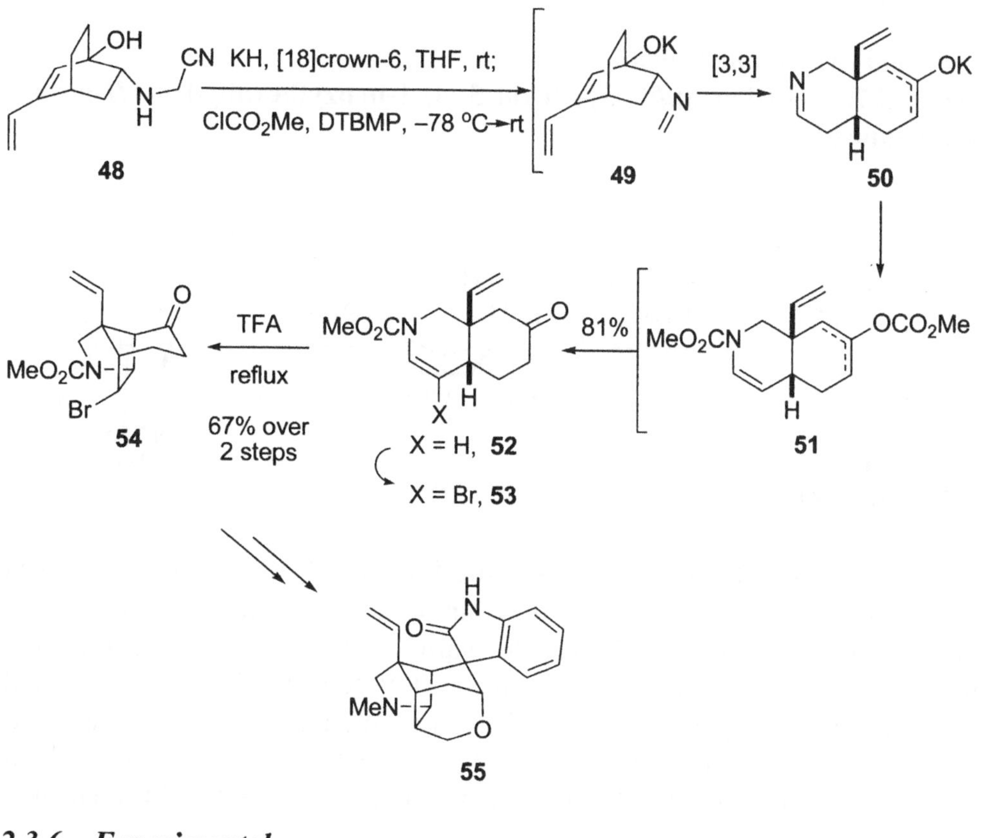

2.3.6 Experimental

Overman Pyrrolidine Synthesis:

1-(5-Phenyl-1-propylpyrrolidin-3-yl)ethanone (58)[1]

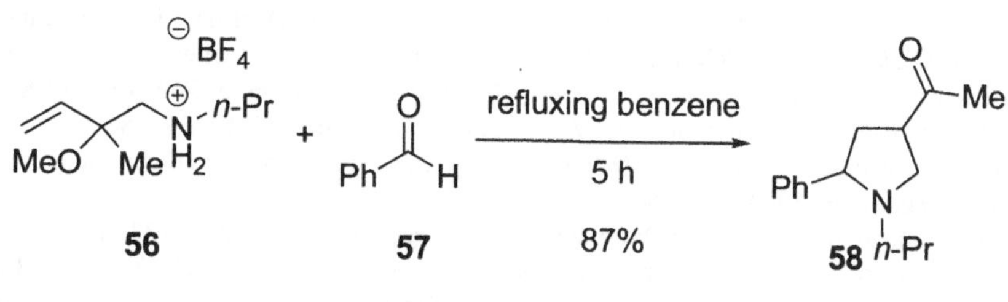

A mixture of benzaldehyde (**57**, 3.3 mmol) and 2-methoxy-2-methyl-*N*-propyl-3-butenammonium tetrafluoroborate (**56**) was heated for 5 h at reflux in 5 mL of benzene. After the mixture cooled to room temperature, 3 mL of 1 N NaOH was added, and the amine product was isolated by ether extraction and dried (Na_2SO_4). Distillation (bulb to bulb; bath temperature 95 °C; 0.01 mm) afforded 3-acetyl-5-phenyl-1-propylpyrrolidine (**58**, a 1:1 mixture of acetyl epimers) in 87% yield.

Overman Pyrrolidine Synthesis:
(3*R*,3a*S*,7a*S*)-3-(Benzo[*d*][1,3]-dioxol-5-yl)-1-benzylhexahydro-1*H*-indol-4(2*H*)-one (26)[3]

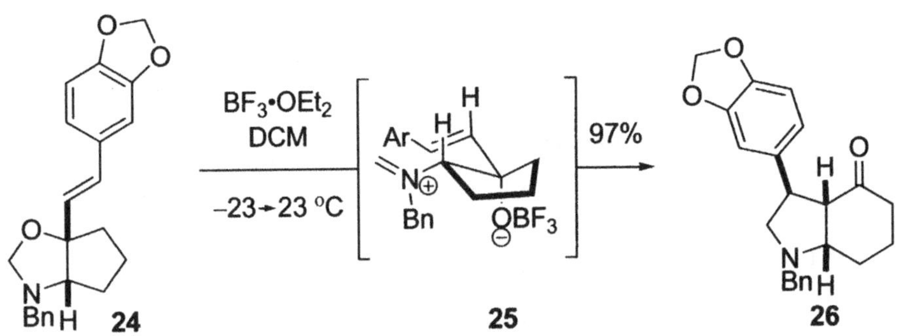

A solution of ***rac*-24** (120 mg, 0.34 mmol) in CH_2Cl_2 (8.2 mL) was allowed to react with $BF_3{\cdot}OEt_2$ (0.11 mL, 0.81 mmol) at −20 °C for 30 min and then the reaction solution was allowed to warm to 23 °C. After 15 min, the resulting solution was quenched with a 1.0 N NaOH solution (4 mL) and extracted with CH_2Cl_2 (3 × 10 mL). The organic portions were dried (K_2CO_3) and concentrated to give 116 mg (97%) of a light yellow oil, which crystallized upon standing. The product was homogeneous by TLC analysis.

2.3.7 References

1. Overman, L. E.; Kakimoto, M.-A. *J. Am. Chem. Soc.* **1979,** *101*, 1310–1312.
2. (a) Fevig, J. M.; Marquis, Jr., R. W.; Overman, L. E. *J. Am. Chem. Soc.* **1991,** *113*, 5085–5086. (b) Angle, S. R.; Fevig, J. M.; Knight, S. D.; Marquis, Jr., R. W.; Overman, L. E. *J. Am. Chem. Soc.* **1993,** *115*, 3966–3976. (c) Knight, S. D.; Overman, L. E.; Pairaudeau, G. *J. Am. Chem. Soc.* **1993,** *115*, 9293–9294. (d) Knight, S. D.; Miledi, R.; Nguyen, Q-T.; Overman, L. E.; Pairaudeau, G. *Bioorg. Med. Chem. Lett.* **1995,** *5*, 749–752. (e) Knight, S. D.; Overman, L. E.; Pairaudeau, G. *J. Am. Chem. Soc.* **1995,** *117*, 5776–5788.
3. (a) Overman, L. E.; Shim, J. *J. Org. Chem.* **1991,** *56*, 5005–5007. (b) Overman, L. E.; Shim, J. *J. Org. Chem.* **1993,** *158*, 4662–4672.
4. (a) Abelman, M. M.; Oh, T.; Overman, L. E. *J. Org. Chem.* **1987,** *52*, 4133–4135. (b) Earley, W. G.; Jacobsen, E. J.; Meier, G. P.; Oh, T.; Overman, L. E. *Tetrahedron Lett.* **1988,** *29*, 3781–3784. (c) Earley, W. G.; Oh, T.; Overman, L. E. *Tetrahedron Lett.* **1988,** *29*, 3785–3788. (d) Overman, L. E.; Sharp, M. J. *J. Org. Chem.* **1992,** *57*, 1035–1038. (e) Madin, A.; O'Donnell, C. J.; Oh, T.; Old, D. W.; Overman, L. E.; Sharp, M. J. *Angew. Chem., Int.*

Ed. **1999,** *38*, 2934–2936. (f) Earley, G. W.; Jacobsen, J. E.; Madin, A.; Meier, G. P.; O'Donnell, C. J.; Oh, T.; Old, D. W.; Overman, L. E.; Sharp, M. J. *J. Am. Chem. Soc.* **2005,** *127*, 18046–18053. (g) Madin, A.; O'Donnell, C. J.; Oh, T.; Old, D. W.; Overman, L. E.; Sharp, M. J. *J. Am. Chem. Soc.* **2005,** *127*, 18054–18065.

5. For reviews, please see [R] (a) Overman, L. E. *Acc. Chem. Res.* **1992,** *25*, 352–359. [R] (b) Overman, L. E. *Aldrichimica Acta* **1995,** *28*, 107–120. [R] (c) Royer, J.; Bonin, M.; Micouin, L. *Chem. Rev.* **2004**, *104*, 2311–2352. [R] (d) Padwa, A.; Bur, S. K. *Tetrahedron* **2007,** *63*, 5341–5378.
6. For discussions on the mechanism of the Overman Pyrrolidine Synthesis, see (a) Jaobsen, E. J.; Levin, J.; Overman, L. E. *J. Am. Chem. Soc.* **1988,** *110*, 4329–4336. (b) Eguchi, T.; Koudate, T.; Kakinuma, K. *Tetrahedron* **1993,** *49*, 4527–4540. (c) Overman, L. E.; Trenkle, W. C. *Isr. J. Chem.* **1997,** *37*, 23–30. (d) Lukowski, M.; Jacobs, K.; Hsueh, P.; Lindsay, H. A.; Milletti, M. C. *Tetrahedron* **2009,** *65*, 10311–10316.
7. Doedens, R. J.; Meier, G. P.; Overman, L. E. *J. Org. Chem.* **1988,** *53*, 685–690.
8. Deng, W.; Overman, L. E. *J. Am. Chem. Soc.* **1994,** *116*, 11241–11250.
9. (a) Overman, L. E.; Mendelson, L. T. *J. Am. Chem. Soc.* **1981,** *103*, 5579–5581. (b) Overman, L. E.; Jacobsen, E. J. *Tetrahedron Lett.* **1982,** *23*, 2741–2744. (c) Overman, L. E.; Mendelson, L. T.; Jacobsen, E. J. *J. Am. Chem. Soc.* **1983,** *105*, 6629–6637. (d) Overman, L. E.; Sugai, S. *Helv. Chem. Acta* **1985,** *68*, 745–749.
10. (a) Overman, L. E.; Sworin, M.; Bass, L. S.; Clardy, J. *Tetrahedron* **1981,** *37*, 4041–4045. (b) Overman, L. E.; Sworin, M.; Burk, R. M. *J. Org. Chem.* **1983,** *48*, 2685–2690. (c) Overman, L. E.; Robertson, G. M.; Robichaud, A. J. *J. Am. Chem. Soc.* **1991,** *113*, 2598–2610.
11. Wildman, W. C.; Kaufman, C. J. *J. Am. Chem. Soc.* **1955,** *77*, 1248–1252.
12. Carballo, R. M.; Purino, M.; Ramírez, M. A.; Martín, V. S.; Padŕon, J. I. *Org. Lett.* **2010,** *12*, 5334–5337.
13. Agami, C.; Couty, F.; Lin, J.; Mikaeloff, A.; Poursoulis, M. *Tetrahedron* **1993,** *49*, 7239–7250.
14. (a) Overman, L. E.; Robertson, G. M.; Robichaud, A. J. *J. Org. Chem.* **1989,** *54*, 1236–1238. (b) Overman, L. E.; Robertson, G. M.; Robichaud, A. J. *J. Am. Chem. Soc.* **1991,** *113*, 2598–2610.
15. Pelletier, P. J.; Caventou, J. B. *Ann. Chim. Phys.* **1818,** *8*, 323.
16. Briggs, L. H.; Openshaw, H. T.; Robinson, R. *J. Chem. Soc.* **1946,** 903–908.
17. (a) Magnus, P.; Giles, M.; Bonnert, R.; Kim, C. S.; McQuire, L.; Merritt, A.; Vicker, N. *J. Am. Chem. Soc.* **1992,** *114*, 4403–4405. (b) Kuehne, M. E.; Xu, F. *J. Org. Chem.* **1993,** *58*, 7490–7497.
18. Woodward, R. B.; Cava, M. P.; Ollis, W. D.; Hunger, A.; Daeniker, H. U.; Schenker, K. *J. Am. Chem. Soc.* **1954,** *76*, 4749–4751.
19. (a) Brummond, K. M.; Lu, J. *Org. Lett.* **2001,** *3*, 1347–1349. (b) Brummond, K. M.; Hong, S.-P. *J. Org. Chem.* **2005,** *70*, 907–916.
20. Sonnenschein, F. L. *Ber. Dtsch. Chem. Ges.* **1876,** *9*, 1182–1186.
21. (a) Lovell, F. M.; Pepinsky, R.; Wilson, A. J. C. *Tetrahedron Lett.* **1959,** *1*, 1–5. (b) Conroy, H.; Chakrabarti, J. K. *Tetrahedron Lett.* **1959,** *1*, 6–13.

2.4 Trofimov Pyrrole Synthesis

Gerald J. Tanoury

2.4.1 Description

The Trofimov pyrrole synthesis (the Trofimov Reaction) refers to the conversion of ketones to pyrroles in the presence of NH_2OH, KOH and DMSO at elevated temperatures. The reaction was shown to occur via hydroxylamine and *O*-vinylhydroxylamine intermediates.

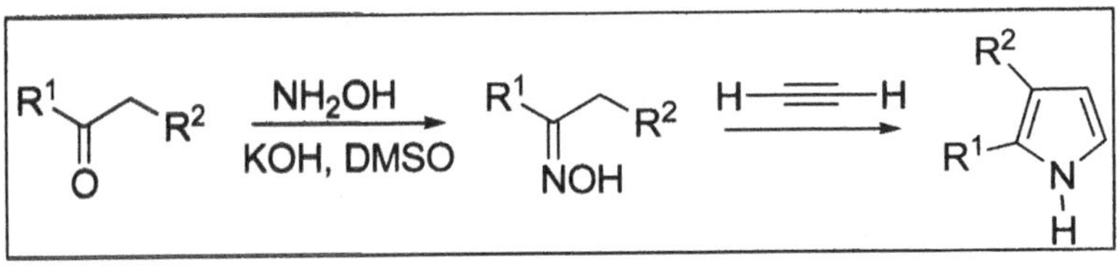

2.4.2 Historical Perspective

In the 1980s, Boris A. Trofimov published several manuscripts describing the synthesis of pyrroles and vinylpyrroles from ketones, hydroxylamine and acetylene.[1] The reaction was believed to occur via hydroxylimine formation, followed by *O*-vinylation, rearrangement, then ring closure. The Trofimov reaction has been studied in limited detail but has been shown to have application to the synthesis of a variety of pyrroles.

2.4.3 Mechanistic Considerations

The Trofimov reaction is thought to occur via conversion of the ketone **1** to the corresponding ketoximine **2**, which underwent vinylation with acetylene under strongly basic condition (KOH/DMSO) to give vinyl ether **3**. After isomerization of the imine to the enamine **4**, a [3,3]-sigmatropic shift generated imine aldehyde **5**. Intramolecular cyclization with concomitant loss of water provided pyrrole **6**. Under the reaction conditions, **6** can react with acetylene to give the vinylpyrrole product **7**. The reaction is typically conducted at elevate temperatures (100–150 °C) and at acetylene pressures ranging from atmospheric pressure to 20 atm or more. The Trofimov reaction has been applied to several keto and alkynyl systems bearing aliphatic, aromatic, and heterocyclic substituents.

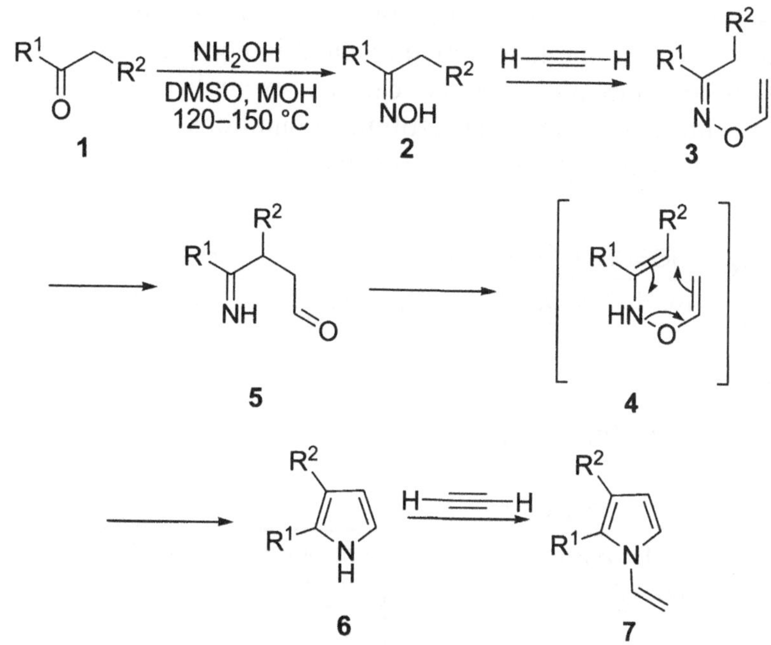

A review of the Trofimov reaction describes the work done up to 1994.[1] The review covers aliphatic and aromatic ketoximines, heterocyclic ketoximines. Typical reaction yields ranged from 40–70%, some examples reported yields > 90%, and several reported single-digit yields. This report will cover the literature published post-1994.[1]

2.4.4 *Variations and Improvements*

Variations of this reaction concern preformation of the hydroxylimine and/or formation of various salts of the hydroxylimine (Cs, K, Na, *etc.*). In addition, the range of Group 1 metals (Li–Cs) have been examined. These variations and their effect on reaction yields are discussed within the context of their synthetic utility (Section 2.4.5).

2.4.5 *Synthetic Utility*

Arylpyrroles from Ketoximines and Alkynes

Arylpyrroles have been generated from the reaction of arylketoximines with acetylene.[2] Under standard Trofimov conditions (KOH/DMSO, acetylene), several aryl pyrroles can be prepared in respectable to good yields. Formation of the nonvinylated product was observed in good yields, and similarly for the N-vinyl pyrroles. The effect of the KOH loading on the reaction yield showed an optimum at 30 wt%, and the effect of temperature on the reaction yield reached an optimum at 100 °C (Table 1). The presence of a sulfide moiety in the ketoximine resulted in complex product mixtures, making this tranformation impractical, as exemplified by **11**. When subjecting oxime sulfide **15** to the Trofimov conditions, a mixture of the free, monovinyl and divinyl pyrroles were obtained.[3]

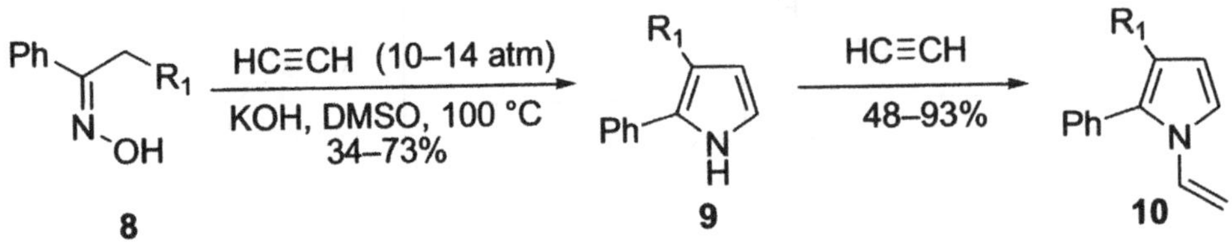

R_1 = H, Me, Et, Pr, *i*-Pr, Bu, $C_{5-9}H_{11-19}$, Ph

Table 1.

KOH/oxime (wt%)*	yield (%)	temperature (°C)**	yield (%)
5	7	50	10
10	23	80	69
15	39	100	76
20	50	120	68
30	76	140	68
100	68	150	65

*100 °C, 3 h, 12 atm; **30 wt% KOH, 12 atm

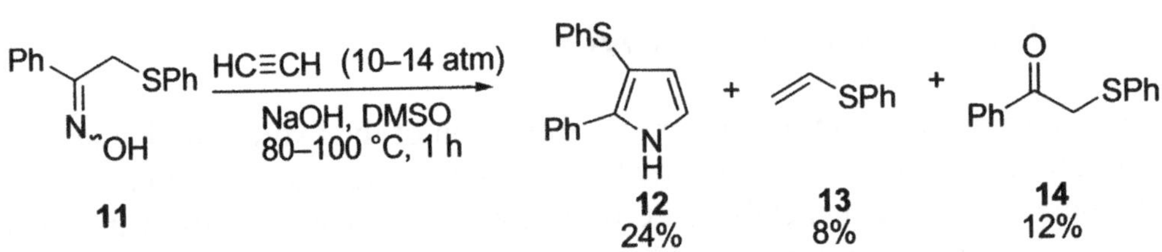

One of the weaknesses of the Trofimov reaction is the requirement for using acetylene under high pressure. To obviate this requirement, acetylene analogs (vinyl halides and 1,2-dihaloethanes) were examined for the utility as components in the Trofimov reaction. The results are promising and indicated that further development of this variation in the Trofimov reaction could lead to a more practical and attractive method for pyrrole synthesis.

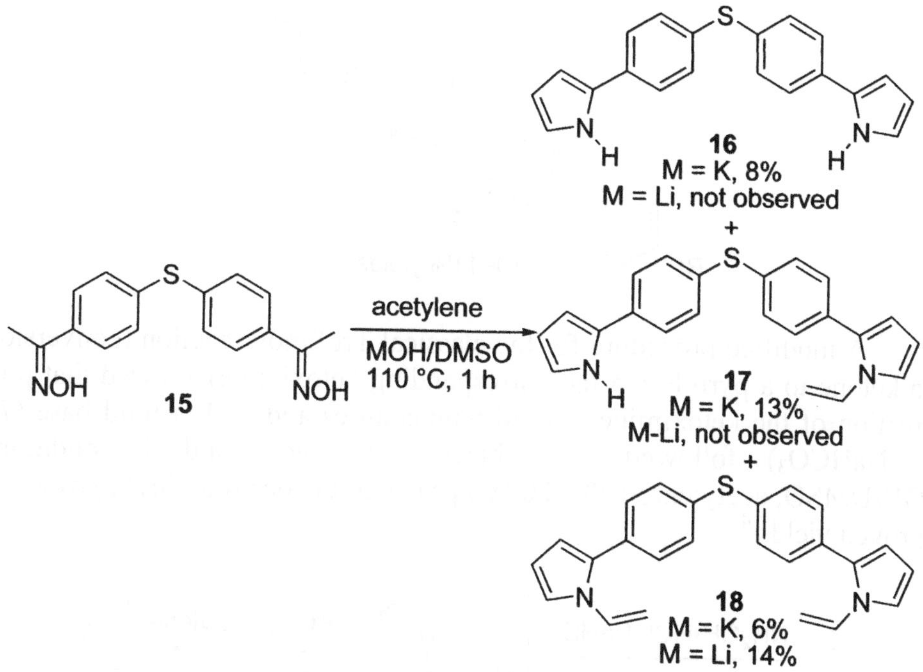

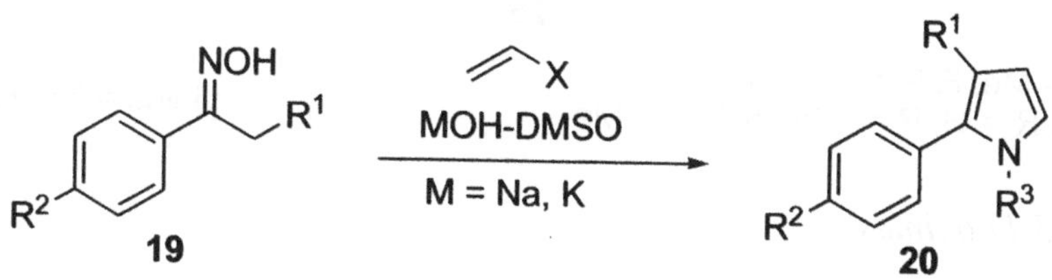

R^1 = H, Me, Et, *i*-Pr, *t*-Bu, EtS, PhS, PrS, *i*-PrS,*t*-BuS, *i*-BuS
R^2 = H, Me

R^3 = H; yields = 19–53%
R^3 = vinyl; yields = 6–19%

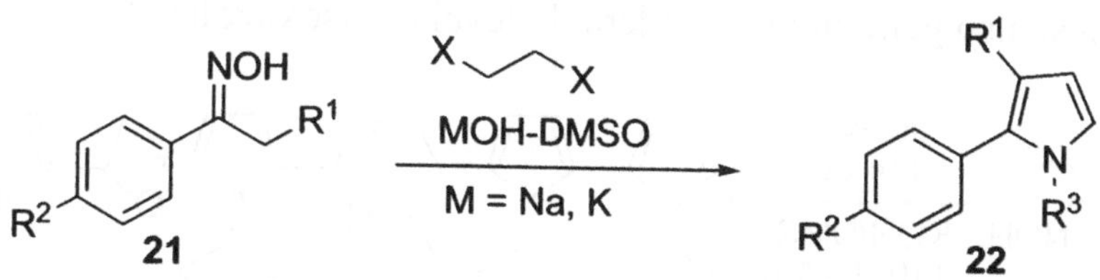

R^1 = H, Me, Et, *i*-Pr, *t*-Bu, Et, PhS, PrS, *i*-PrS, *t*-BuS, *i*-BuS
R^2 = H, Me

R^3 = H; yields = 19–53%
R^3 = vinyl; yields = 6–19%

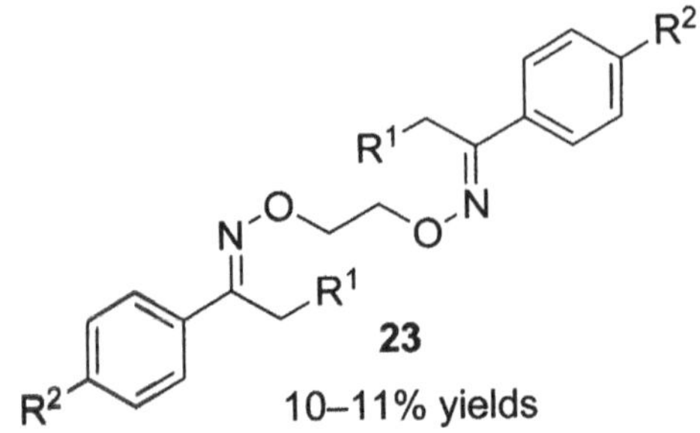

A modified procedure for the classical Trofimov reaction (conversion of a ketone to a pyrrole via the corresponding ketoximine) showed that preformation of the ketoximine at mild temperatures and with a mild base (70 °C, $NaHCO_3$) followed by subjection to the standard conditions (KOH/DMSO, acetylene, 100–120 °C) provided the desired vinyl pyrroles in improved yields.[4]

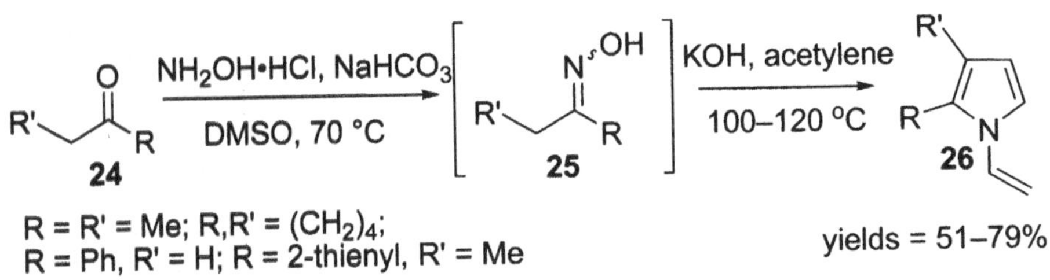

1,2- and 1,3-Dioximes

1,2- and 1,3-dioximines can be converted to the corresponding pyrrole, but complex mixtures are obtained. 1,2-dioximines were successfully converted to bipyrroles, but 2-pyrrolopyridines, 2-ketopyrroles, and *O*-vinylated dioximes were generated. 1,3-Dioximines, although converted to monopyrroles, also generated considerable levels of isoxazoles.[5,6]

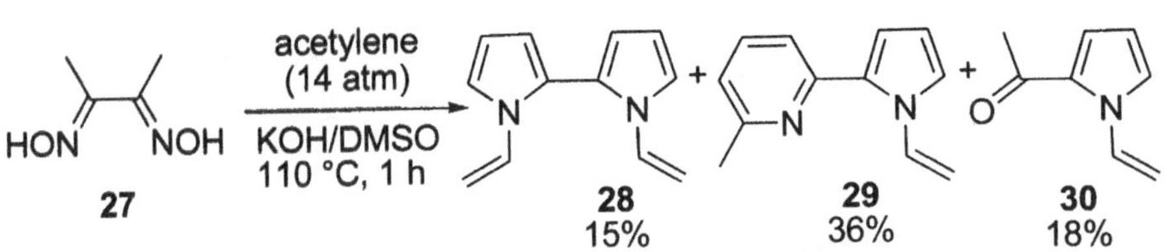

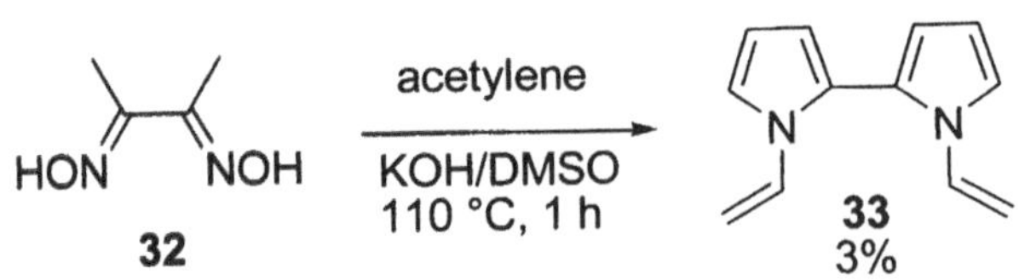

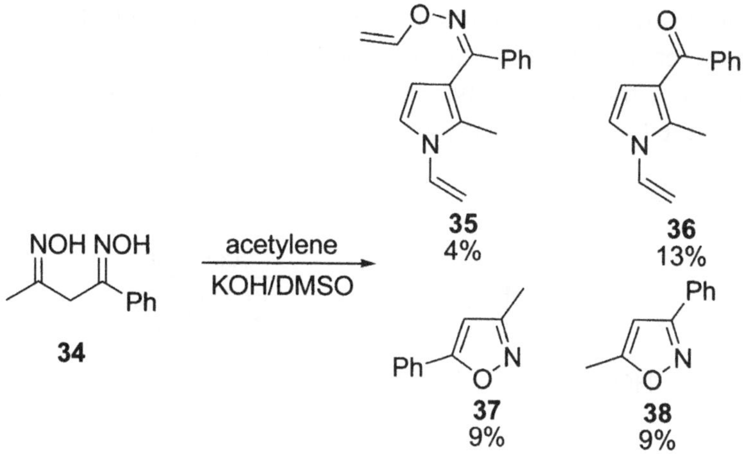

Steroidal Pyrroles

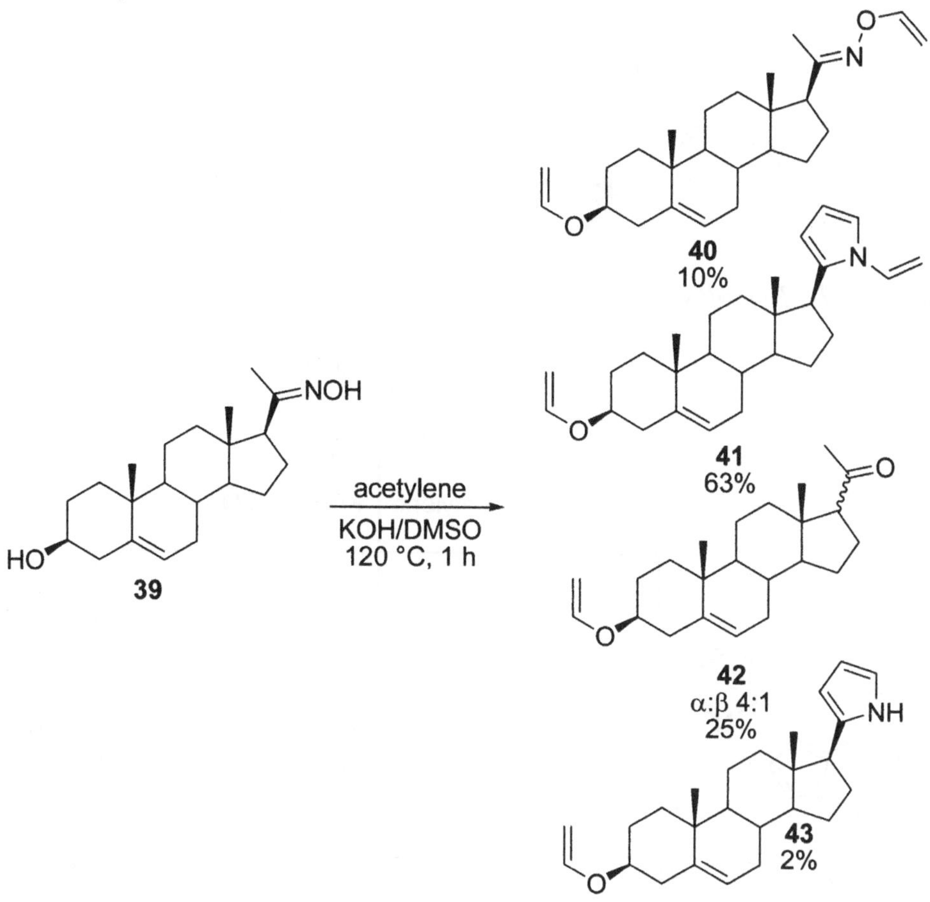

A nice application of the Trofimov reaction to the synthesis of steroidal pyrroles is shown in the reaction above. Reaction of pregnenolone ketoximine **39** with acetylene in KOH/DMSO gave four vinyl ether products

40–41. The desired nonvinylated pyrrole was the minor product.[7] Applying the Trofimov reaction to pregesterone diketoximine **44** gave a mixture of the conjugated pervinylated bispyrrole **45** and ketone **46**.[8] Cholesterol ketoximine gave the corresponding *N*-vinyl pyrrole in 25% yield.[9]

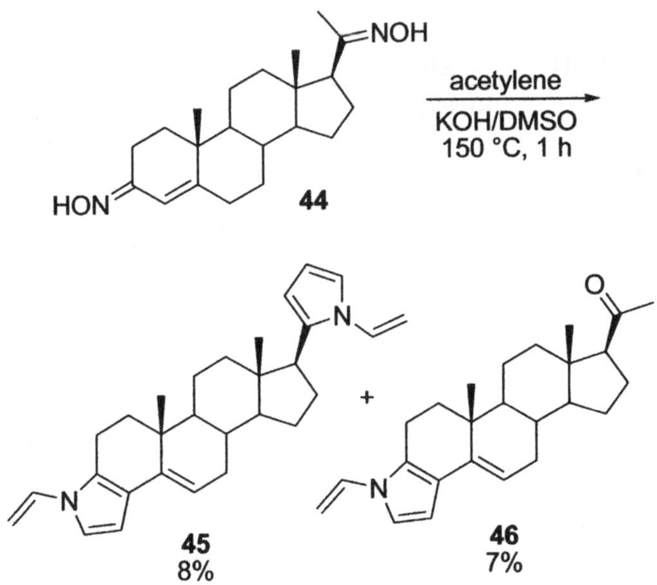

Heterocyclic Pyrroles

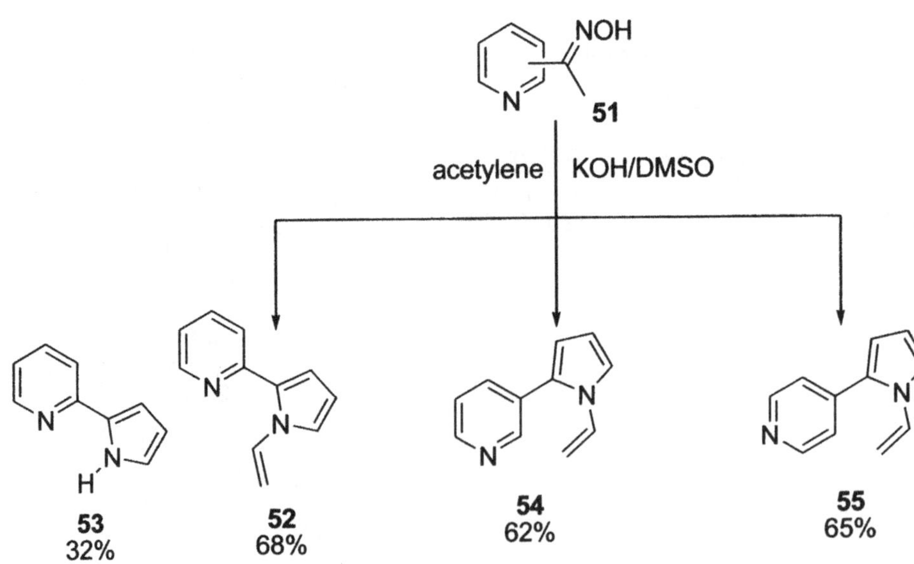

It is not surprising that heterocyclic pyrroles can be prepared from the corresponding heterocyclic ketoximines. Pyrrolopyridines, fused pyrrolopiperidines, and other heterocyclic systems have been prepared.[4,10–12]

Pyrrolothiophenes have been prepared. Compounds **57** and **58**, although unstable and nonisolable, were identified by LCMS analysis.[13,14]

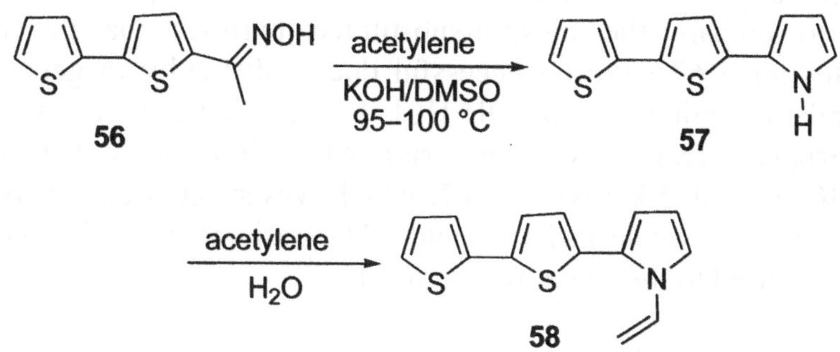

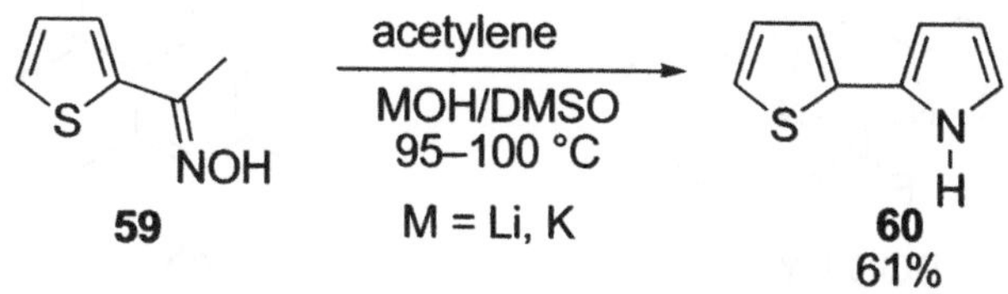

Optoelectronics

Interesting applications of the Trofimov reaction in optoelectronic materials have been reported. The preparation of pyrrolo[2,2]-paracyclophanes is one example. Starting from the acetyl-substituted compound **61**, conversion to the ketoximine followed by reaction with acetylene and KOH/DMSO, in one pot, gave the two pyrroloparacyclophanes in low yield. However, formation of the *O*-vinyl ketoximine [$Cs(OH)_2$, acetylene, 70 °C, 78% yield], followed by conversion to the pyrrole in two separate steps gave **64** in 53% yield.[15,16]

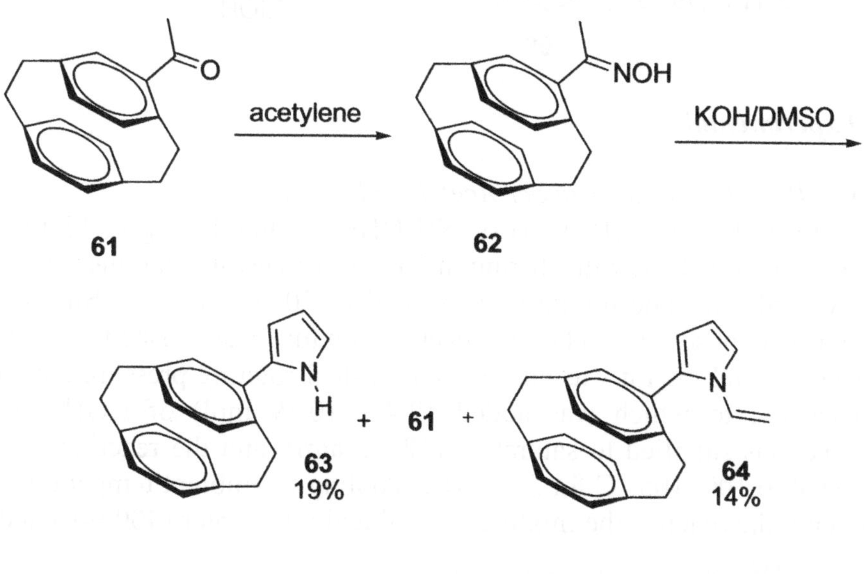

A second area of optoelectronic materials are the BODIPY (boron-dipyrromethene) fluorescent dyes. These materials are prepared by condensation of pyrroles with aldehydes, followed by complexation of boron. Attempts to prepare the mesityl-substituted pyrroles via the hindered mesitylketoximine **67** were unsuccessful due to difficulty in generating **67** under standard conditions ($NH_2OH \cdot HCl$, heat). However, reaction of the mesityl Grignard reagent with proprionitrile followed by reaction with $NH_2OH \cdot HCl$ provided ketoximine **67**, which was successfully converted to the corresponding pyrrole in 23% yield. The cesium salt of ketoximine **67** was also converted to the same pyrrole in 41%.[17]

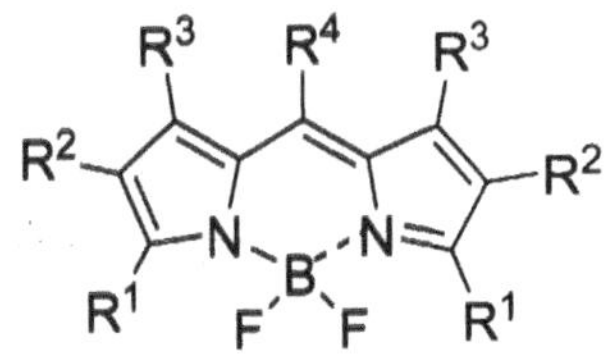

BODIPY

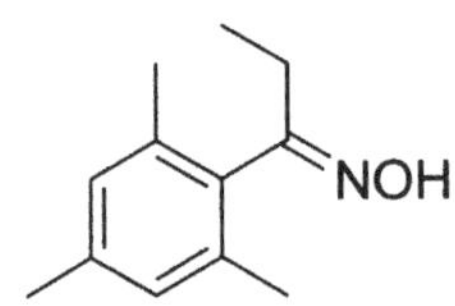

65

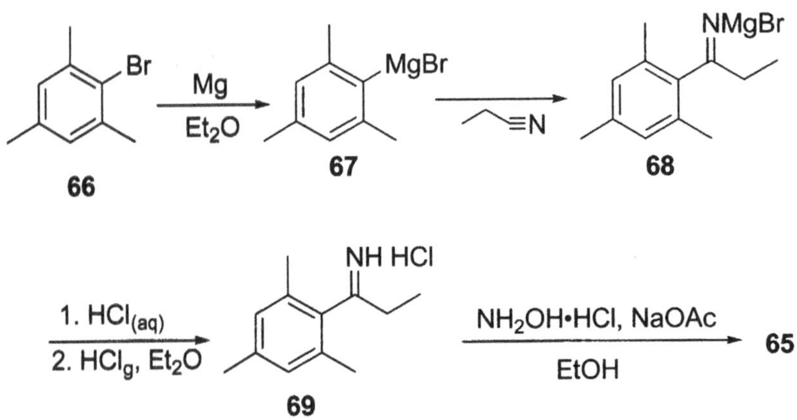

2.4.6 Experimental

Trofimov Pyrrole Synthesis under Acetylene Pressure[4]

A mixture of 10.4 g (0.15 mol) of $NH_2OH \cdot HCl$ and 12.6 g (0.15 mol) of $NaHCO_3$ was dissolved with stirring in 200 mL of DMSO, and then 0.15 mol of **70** was added. The mixture was heated to 70 °C for 3 h. Strong CO_2 liberation was observed. The hot reaction mixture was flushed with Ag to remove CO_2, and then it was charged into a steel rotating pressure reactor of 1 L capacity, to which was added 12.4 g (0.19 mol) of $KOH \cdot 0.5H_2O$. Acetylene was supplied to saturation (12–15 atm), and the reaction mixture was heated to 100–105 °C for 3 h. After cooling to ambient temperature and discharging the reactor, the mixture was diluted with water (400 mL) and the product extracted with ether (6 × 100 mL). The ether extract was washed with water (3 × 100 mL) and dried over K_2CO_3. The ether was removed *in vacuo* and the residue was purified by chromatography (Al_2O_3, eluent hexane) to give the desired product **72** in 79% yield.

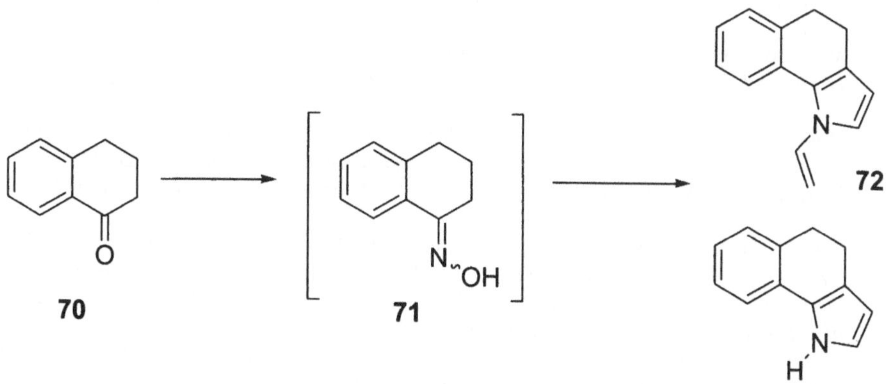

Trofimov Pyrrole Synthesis under a Flow of Acetylene Gas[19]
$NH_2OH \cdot HCl$ (5 mmol) was dissolved in DMSO (10 mL) in a 25-mL flask equipped with a stirrer, and then $NaHCO_3$ (5 mmol) and **73** (5 mmol) were added. An intense CO_2 evolution was observed. The mixture was allowed to stand for 3–4 h at room temperature until the reaction completed. The mixture was heated to 100 °C, and acetylene was fed into the mixture with stirring for 0.5 h. After addition of $KOH \cdot H_2O$ (7.5 mmol), acetylene feeding was continued at the same temperature for 5 h at a rate of ~15 cm^3/min. The mixture was cooled, diluted with water to ~ 30 mL, and extracted with ether (5 × 5 mL). The extracts were washed with water (3 × 5 mL) and dried over K_2CO_3. After removal of the ether, the residue was purified by chromatography (Al_2O_3) eluting with hexane then with 3 : 1 hexane:ether. The *N*-vinyl pyrrole eluted with hexane, and the *NH*-pyrrole eluted with the 3 : 1 hexane:ether mixture to give a 72% yield of **73** and 28% yield of **72**.

2.4.7 References

1. See Trofimov, B. A.; Mikhaleva, A. I. *Heterocycles* **1994,** 1193–1232, and references therein.
2. Korostova, S. E.; Mikhaleva, A. I.; Vasil'tsov, A. M.; Trofimov, B. A. *Russ. J. Org. Chem. (Translation of Zhurnal Organicheskoi Khimii)* **1998,** *34*, 911–948.
3. Vasil'tsov, A. M.; Schmidt, E. Y.; Mikhaleva, A. B. I.; Zorina, N. V.; Zaitsev, A. B.; Petrova, O. V.; Krivdin, L. B.; Petrushenko, K. B.; Ushakov, I. A.; Pozo-Gonzalo, C.; Pomposo, J. A.; Grande, H.-J. *Tetrahedron* **2005,** *61*, 7756–7762.
4. Mikhaleva, A. I.; Shmidt, E. Y.; Ivanov, A. V.; Vasil'tsov, A. M.; Senotrusova, E. Y.; Protsuk, N. I. *Russ. J. Org. Chem.* **2007,** *43*, 228–230.
5. Vasil'tsov, A. M.; Zaitsev, A. B.; Schmidt, E. Y.; Mikhaleva, A. B. I.; Afonin, A. V. *Mendeleev Commun.* **2001,** 74–75.

6. Zaitsev, A. B.; Schmidt, E. Y.; Mikhaleva, A. M.; Afonin, A. V.; Ushakov, I. A. *Chem. Heterocycl. Compd.* **2005,** *41*, 722–729.
7. Vasil'tsov, A. M.; Shmidt, E. Y.; Mikhaleva, A. I.; Afonin, A. V.; Zaitsev, A. B. *Chem. Heterocycl. Compd. (Translation of Khimiya Geterotsiklicheskikh Soedinenii)* **2001,** *37*, 1488–1491.
8. Zaitsev, A. B.; Vasil'tsov, A. M.; Shmidt, E. Y.; Mikhaleva, A. I.; Afonin, A. V.; Il'icheva, L. N., *Russ. J. Org. Chem. (Translation of Zhurnal Organicheskoi Khimii)* **2003,** *39*, 1406–1411.
9. Vasil'tsov, A. M.; Zaitsev, A. B.; Mikhaleva, A. I.; Shmidt, E. Y.; Afonin, A. V. *Chem. Heterocycl. Compd. (Translation of Khimiya Geterotsiklicheskikh Soedinenii)* **2002,** *38*, 60–64.
10. Voskressensky, L. G.; Borisova, T. N.; Varlamov, A. V. *Chem. Heterocycl. Compd. (New York) (Translation of Khimiya Geterotsiklicheskikh Soedinenii)* **2004,** *40*, 326-333.
11. Petrova, O. V.; Mikhaleva, A. I.; Sobenina, L. N.; Schmidt, E. Y.; Kositsyna, E. I. *Mendeleev Commun.* **1997,** 162–163.
12. Petrova, O. V.; Mikhaleva, A. I.; Sobenina, L. N.; Shmidt, E. Y.; Kositsina, E. I. *Russ. J. Org. Chem. (Translation of Zhurnal Organicheskoi Khimii)* **1997,** *33*, 1007–1009
13. Deryagina, E. N.; Russavakaya, N. V.; Vvedenskii, V. Y. *Russ. J. Org. Chem. (Translation of Zhurnal Organicheskoi Khimii)* **1999,** *35*, 1225–1227.
14. Cristina, P.-G.; Pomposo, J. A.; Alduncin, J. A.; Salsamendi, M.; Mikhaleva, A. B. I.; Krivdin, L. B.; Trofimov, B. A. *Electrochim. Acta* **2007,** *52*, 4784–4791.
15. Schmidt, E. Y.; Zorina, N. V.; Zaitsev, A. B.; Mikhaleva, A. b. I.; Vasil'tsov, A. M.; Audebert, P.; Clavier, G.; Meallet-Renault, R.; Pansu, R. B. *Tetrahedron Lett.* **2004,** *45*, 5489–5491.
16. Zaitsev, A. B.; Meallet-Renault, R.; Schmidt, E. Y.; Mikhaleva, A. B. I.; Badre, S.; Dumas, C.; Vasil'tsov, A. M.; Zorina, N. V.; Pansu, R. B. *Tetrahedron* **2005,** *61*, 2683–2688.
17. Meallet-Renault, R.; Clavier, G.; Dumas-Verdes, C.; Badre, S.; Shmidt, E. Y.; Mikhaleva, A. I.; Laprent, C.; Pansu, R.; Audebert, P.; Trofimov, B. A. *Russ. J. Gen. Chem.* **2008,** *78*, 2247–225
18. Schmidt, E. Y.; Mikhaleva, A. I.; Vasil'tsov, A. M.; Zaitsev, A. B.; Zorina, N. V. *ARKIVOC* **2005,** *(7),* 11–17.

Chapter 3 Indoles 83

3.1 Bischler–Möhlau Indole Synthesis

Ji Zhang

3.1.1 Description

The Bischler–Möhlau indole synthesis[1] is a classical chemical reaction that forms substituted indoles. The reaction involves heating excess anilines **1** with α-haloketones **2** (or α-haloacetals), followed by acid-catalyzed (aniline•HX salt, X = Br or Cl, and Lewis acid) cyclization of the resulting intermediate 2-arylaminoketone.

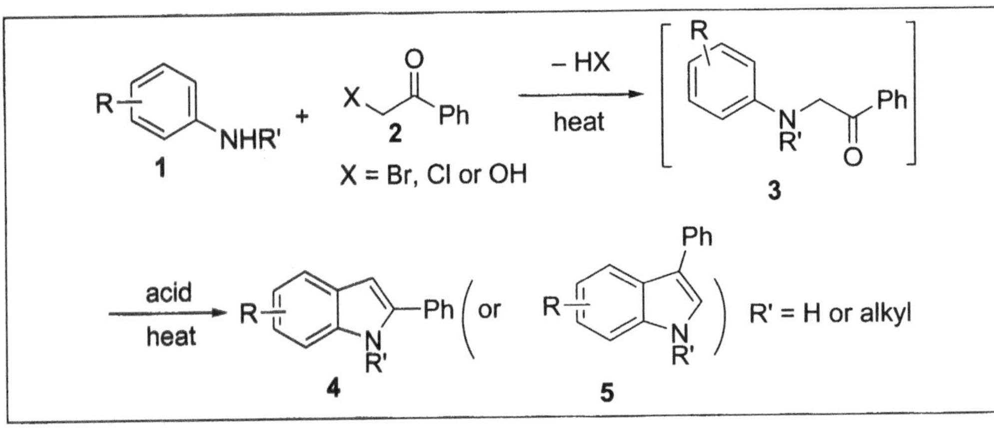

3.1.2 Historical Perspective

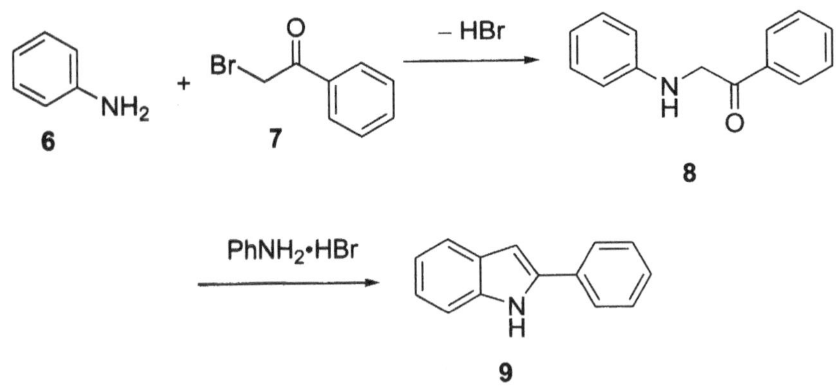

In 1881, Möhlau first disclosed this reaction,[2] which was followed by further study carried out by Bischler in 1892–1893.[3] Typically, this indole synthesis of heating excess anilines with α-haloketones is referred to as the Bischler reaction, however naming the transformation the Bischler–Möhlau indole synthesis is more appropriate. Despite its long history, this classical reaction has received little attention compared to other methods for indole synthesis, perhaps owing to the harsh reaction conditions (usually > 110 °C) and the generation of a mixture of regioisomers (if 3-substituted anilines are used as starting material). Recently, milder conditions have been used, including the use of LiBr as a catalyst[4] and the use of a microwave-assisted solvent-free procedure[5] the solvent-free condition, however, is not very useful on a larger larger scale, especially when both pieces are solids, and the microwaves usefulness depends on the internal reaction temperature.

3.1.3 Mechanism

The mechanism of Bischler–Möhlau indole reaction is quite complicated, and several different mechanisms have been proposed for the rearrangement of α-arylaminoketones to indoles.[5] As shown in the scheme below, the formation of diamine intermediate **10** from the reaction of aniline with α-haloketones. The high temperature promotes the formation of epoxy intermediate **11**, followed by rearrangement through cyclic intermediates **13** and **14**. Finally, the dehydration of **14** generates the desired indole **9**.[6]

via aminoepoxide intermediate:

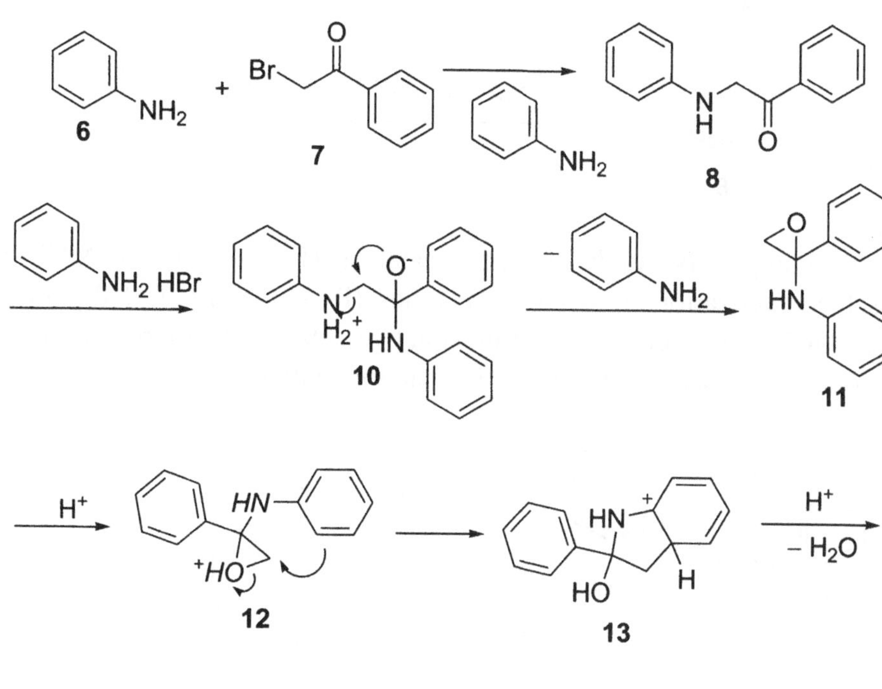

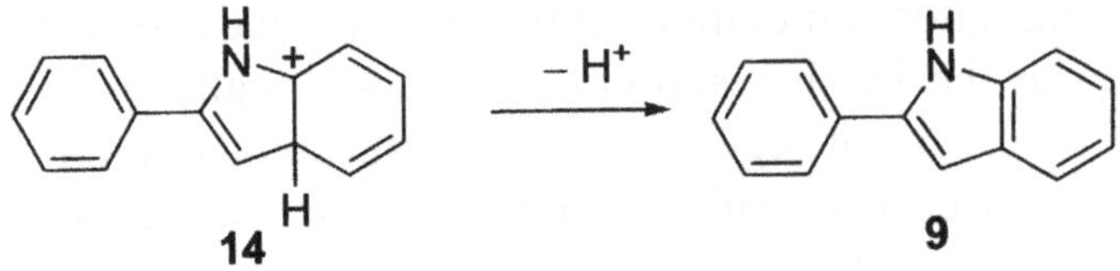

or via imine intermediate:

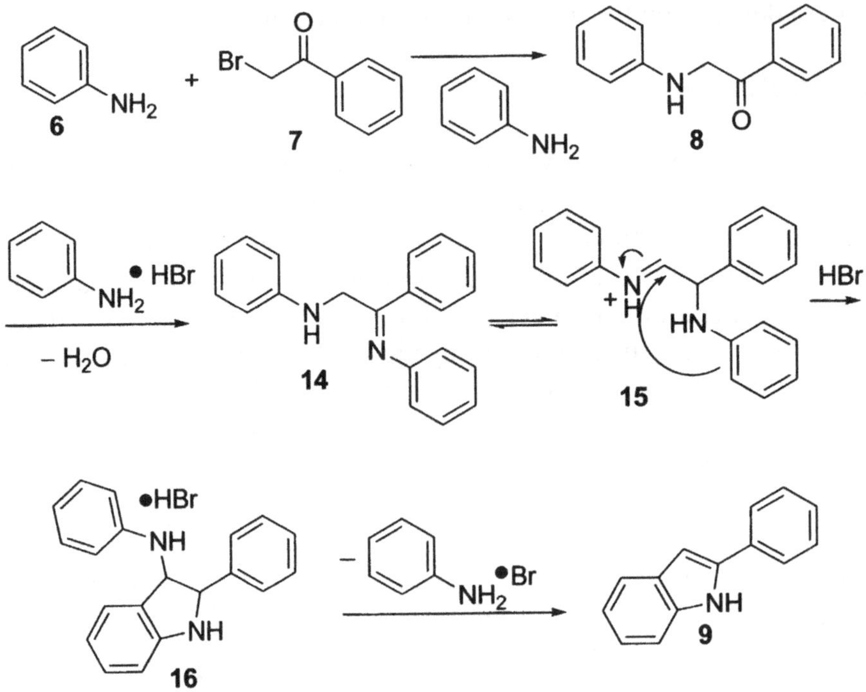

In a comparative study of various catalysts (HCl, $ZnCl_2$ and $AlCl_3$) in the Bischler cyclization of α-amino ketones (R = Me, Et, CH_2Ph) to β-indoles,[7] $AlCl_3$ was found to be the most active catalyst. Isomerization of β- to α-indoles was observed by raising the reaction temperature.

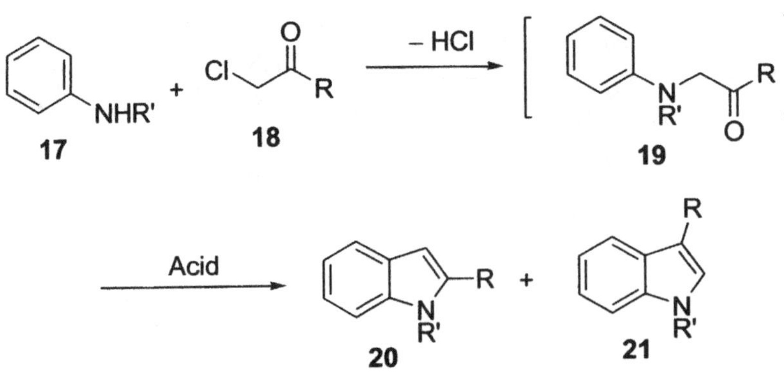

3.1.4 Variations and Improvements

In 1981, Nordlander demonstrated that acetals **22** can be used as reactants in the Bischler–Möhlau indole synthesis, providing 2,3-unsubstituted indoles in good yield.[8] Subsequent modification was made by Sundberg in 1984.[9] From 1998 to 2002, Moody and co-workers developed a modified Bischler indole synthesis by using rhodium(II) acetate to catalyze the reaction of *N*-methylanilines with α-diazo-β-ketoesters via an N–H insertion reaction of a rhodium carbenoid. The resulting α-(*N*-arylamino)ketones cyclize to give indoles upon treatment with BF_3 or an acidic ion exchange resin.[10]

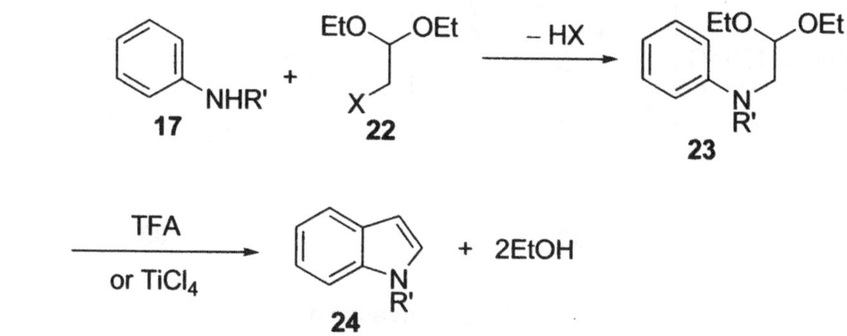

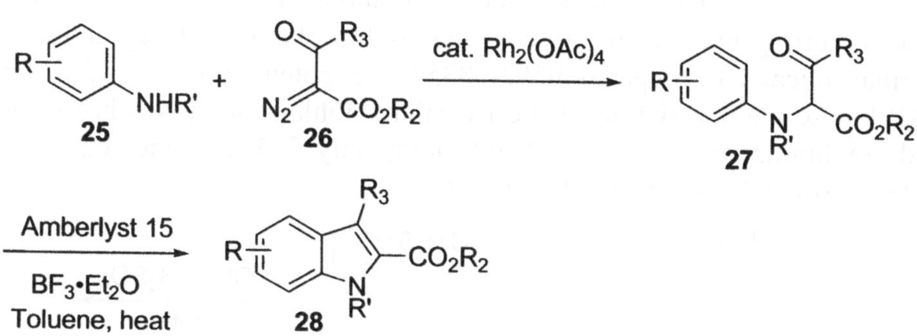

3.1.5 Synthetic Utility

The Bischler–Möhlau indole synthesis was applied in the synthesis of fluvastatin sodium (Lescol) to assemble its indole core. As shown below, reaction of α-chloroketone with *N*-*i*-Pr-aniline at elevated temperature generated a tertiary amine. The resulting *N*-*i*-Pr-aniline tertiary amine underwent a $ZnCl_2$-mediated Bischler–Möhlau indole synthesis at elevated temperature affording the indole core structure of fluvastatin.[11]

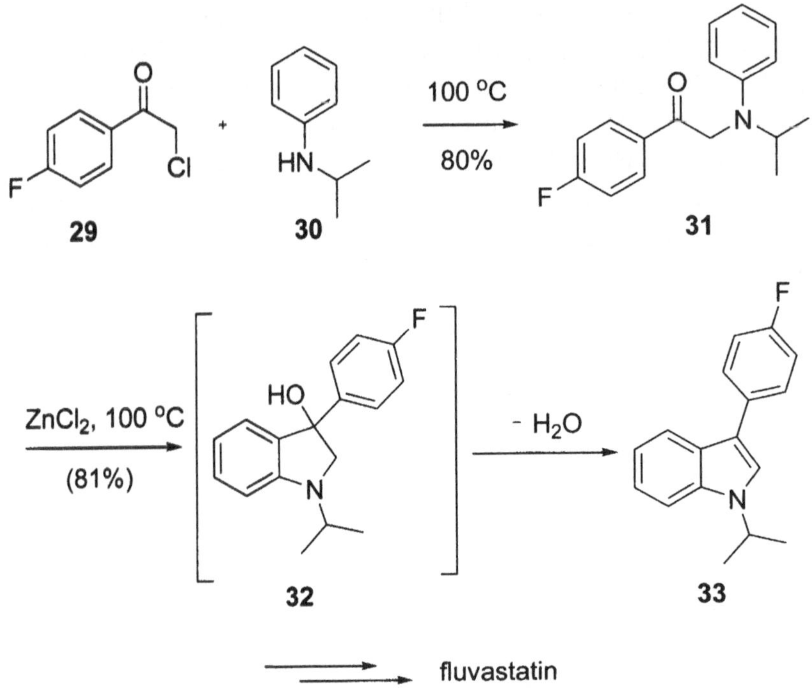

Instead of using α-haloketones, an unsymmetrical benzoin can be used as starting material in place of the α-chloroketone. Henry at Johnson Pharmaceutical has prepared RWJ-68354,[12] a potent inhibitor of the p38 MAP kinase via a variation of the Bischler–Möhlau indole synthesis under mild conditions with a 55% isolated yield; only 2–3% of the regioisomer could be isolated from the mother liquor.

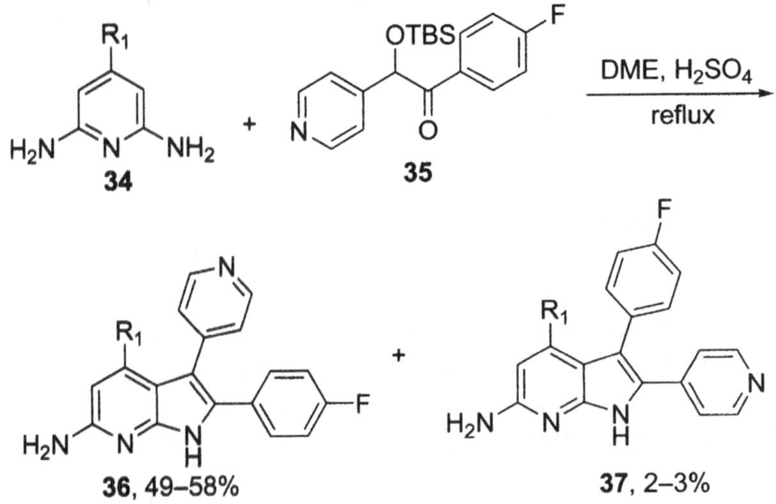

A Bischler–Möhlau indole synthesis was also used in constructing the indole ring of bazedoxifene acetate (Viviant), a novel and highly selective indole estrogen. It is a selective estrogen receptor modulator (SERM) for the treatment of and prevention of osteoporosis. Condensation between α-bromopropiophenone and 4-(benzyloxy)-aniline hydrochloride generated the 3-methyl indole core.[13]

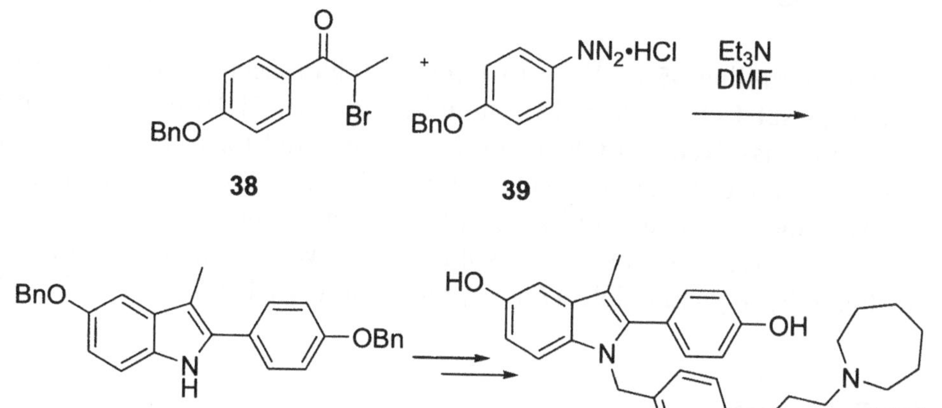

38 39 40 41

bazedoxifene (Viviant)

Steroids bearing heterocycles fused to the A-ring of the steroid nucleus have been of pharmaceutical interest. Zhang at Johnson & Johnson has synthesized novel compounds **43** and **44** using a modified Bischler–Möhlau method (PPA, toluene, reflux, 1 h).[14]

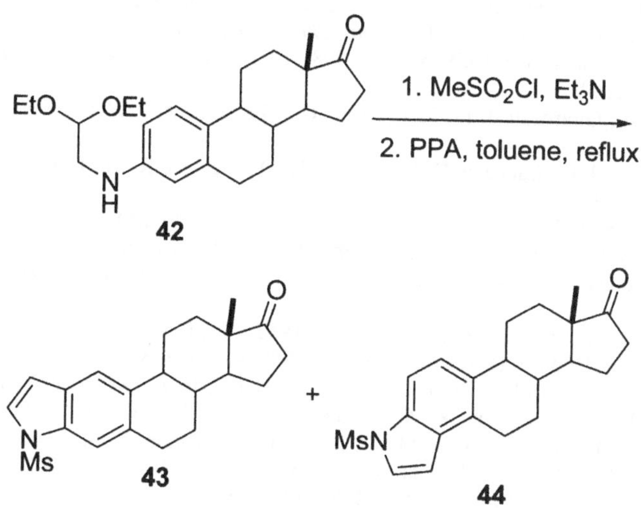

42 43 44

3.1.6 *Experimental*[4]

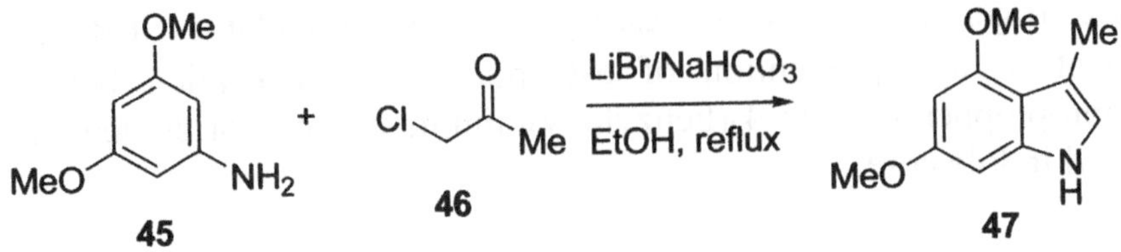

3,5-Dimethoxyaniline **45** (2.00 g, 13.1 mmol), chloroacetone **46** (1.03 mL, 13.1 mmol), $NaHCO_3$ (1.09 g, 13.1 mmol), and LiBr (1.10 g, 13.1 mmol) were partially dissolved in EtOH (36 mL) and refluxed for 6 h. The solvent was evaporated and the crude residue extracted with CH_2Cl_2 (40 mL). The extract was then washed with water (3 × 20 mL), dried ($MgSO_4$), and evaporated to give a yellow-green solid, which was purified by column chromatography to yield indole **47** as a yellow solid (1.84 g, 74%). mp 72–74 °C.

3.1.7 *References*

1. [R] (a) Humphrey, G. R.; Kuethe, J. T. *Chem. Rev.* **2006,** *106*, 2875–2911, (b) Sundberg, R. J. *The Chemistry of Indoles*, Academic Press, New York, 1970, (c) Sundberg, R. J. *Indoles*, Academic Press, London, 1996.
2. (a) Möhlau, R. *Chem. Ber.* **1881,** *14*, 171. (b) Möhlau, R. *Chem. Ber.* **1882,** *15*, 2480.
3. (a) Bischler, A.; Brion, H. *Chem. Ber.* **1892,** 25, 2860. (b) Bischler, A.; Firemann, P. *Chem. Ber.* **1893,** *26*, 1336.
4. Pchalek, K.; Jones, A. W.; Wekking, M. M. T.; Black, D. S. C. *Tetrahedron* **2005,** *61*, 77–82.
5. Sridharan, V.; Perumal, S.; Avendaño, S.; Menéndez, J. C. *Synlett* **2006,** *1*, 91–95.
6. (a) Nelson, K. L.; Seefeld, R. L. *J. Am. Chem. Soc.* **1958,** *80*, 5957–5959. (b) Nelson, K. L.; Robertson, J. C.; Duvall, J. J. *J. Am. Chem. Soc.* **1964,** *86*, 684–687. (c) Weygand, F.; Richter, E. *Chem. Ber.* **1955,** *88*, 499–508. (d) Teuber, H. J.; Schnee, K. *Chem. Ber.* **1958,** *91*, 2089–2094.
7. Galons, H.; Girardeau, J. F.; Farnoux, C. C.; Miocque, M. *J. Heterocyclic Chem.* **1981,** *18*, 561–563.
8. Nordlander, J. E.; Catalane, D. B.; Kotian, K. D.; Stevens, R. M.; Haky, J. E. *J. Org. Chem.* **1981,** *46*, 778–782.
9. Sundberg, R. J.; Laurino, J. P. *J. Org. Chem.* **1984,** *49*, 249–254.
10. (a) Moody, C. J.; Swann, E. *Synlett* **1998,** 135–136. (b) Bashford, K. E.; Cooper, A. L.; Kane, P. D.; Moody, C. J.; Muthusamy, S.; Swann, E. *J. Chem. Soc., Perkin Trans. 1* **2002,** 1672–1687.
11. Walkup, R. E.; Linder, J. *Tetrahedron Lett.* **1985,** *26*, 2155–2158.
12. Henry, J. R.; Dodd, J. H. *Tetrahedron Lett.* **1998,** *48*, 8763–8764.
13. (a) Miller, C. P.; Collini, M. D.; Tran, B. D.; Harris, H. A.; Kharode, Y. P.; Marzolf, J. T.; Moran, R. A.; Henderson, R. A.; Bender, R. H. W.; Unwalla, R. J.; Greenberger, L. M.; Yardley, J. P.; Abou-Gharbia, M. A.; Lyttle, C. R.; Komm, B. S. *J. Med. Chem.* **2001,** *44*, 1654–1657. (b) Miller, C. P.; Harris, H. A.; Komm, B. S. *Drug Fut.* **2002,** *44*, 1654–1657.
14. Zhang, X.; Sui, Z. *Tetrahedron Lett.* **2003,** *44*, 3071–3073.

3.2 Borsche–Drechsel Cyclization

Micheal W. Fultz

3.2.1 Description

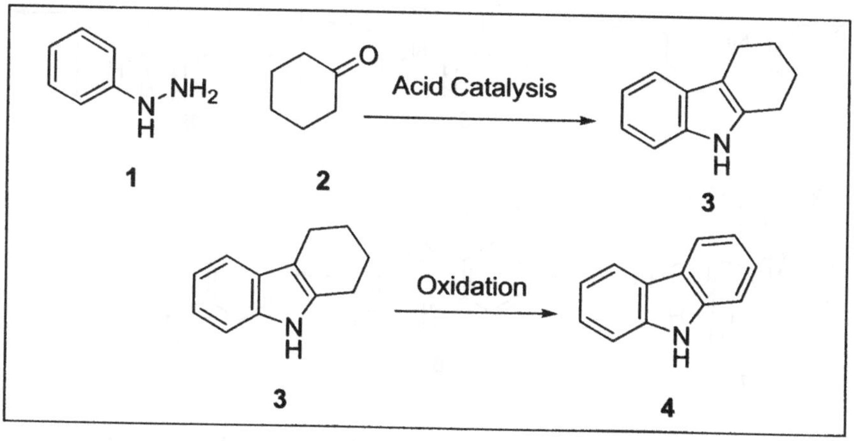

The Borsche–Drechsel cyclization, also known as Borsche carbazole synthesis, is the two-step conversion of phenyl hydrazine and cyclohexanone derivatives to the corresponding carbazole.[1,2] The first step, which is analgous to the Fischer Indole synthesis, converts the phenyl hydrazine **1** and cyclohexanone **2** to the tetrahydrocarbazole **3**. The second step is the oxidation of the terahydrocarbazole to the corresponding carbazole **4**.

3.2.2 Historical Perspective

Drechsel and Borsche discovered the titled reaction and independently worked to optimize the reaction over a span of 50 years.[1–3] This transformation has been extensively exploited and optimized, which has allowed for biological assays of many carbazole derivatives.[4–6]

3.2.3 Mechanism

The first step of the Borsche–Drechsel cyclization proceeds through the same mechanism as the Fischer indole synthesis.[7,8] The aryl hydrazine and cyclohexanone condense to form the iminium **5** and water. The imine nitrogen then coordinates to the Lewis acid, which causes a tautomerization of the hydrazone to the ene–hydrazine **6**. This ene–hydrazine undergoes a [3,3]-sigmatropic rearrangement to form the iminium ion **7**. Rearomitization through deprotonation of the benzylic proton releases the nucleophilic electrons that approach the iminium carbon to form the five-member ring through the favored 5-*exo-trig* cyclization of **8**. Subsequent elimination of ammonia provides the tetrahydrocarbazole **3**.

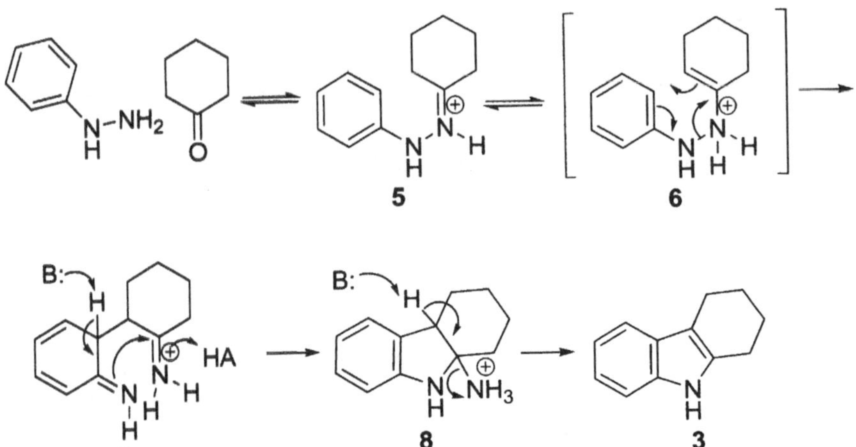

5 6 7 8 3

The second step is the oxidation of the tetrahydrocarbazole to the desired carbazole. This oxidation usually proceeds at high temperature with the assistance of a metal salt such as palladium,[9] mercury,[10] or lead.[11] However, none of these produced completely satisfactory results. It was later determined by Barclay that chloranil provided the carbazole with the highest yields.[12]

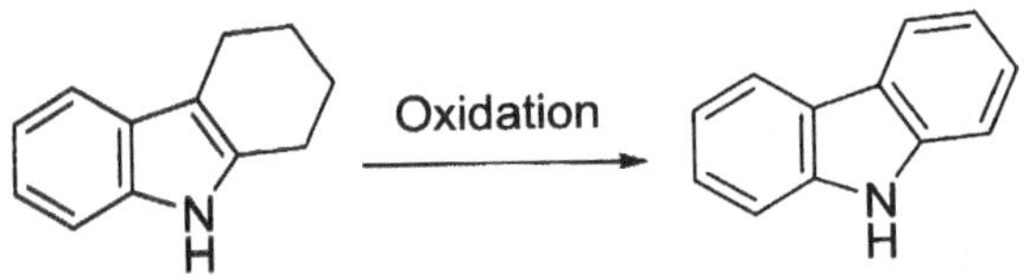

3.2.4 *Variations and Improvements*

Initially, sulfuric acid was used to initate the Borsche–Drechsel cyclization. However, Perkin found that acetic acid provided cleaner products.[10] Recently, other reagents—such as cerium ammonium nitrate[13] (CAN) and the more environmentally friendly acidic ionic liquids[14]—have been used to catalyze the ring formation. It should be noted that people have found it difficult to promote the cyclization when there are substituents at the *ortho*-position of the phenylhydrazine.[15] This difficulty becomes more pronounced when both the hydrazine and the cyclohexanone have α-substitution This difficulty usually leads to low yields compared with other methods of forming the carbazole.

For people who lack the access to designer ionic liquids mentioned previously, Khadilkar[16] was able to achieve the cyclization to form the tetrahydrocarbazole using a chloroaluminate Lewis acid in an easily synthesized ionic liquid to obtain the tetrahydrocarbazole in high yields. This reaction was general enough to withstand both halogens (**9**) and alkyl groups (**10**) on the phenyl hydrazine. The work was general enough that many noncyclic ketones were used in the cyclization process.

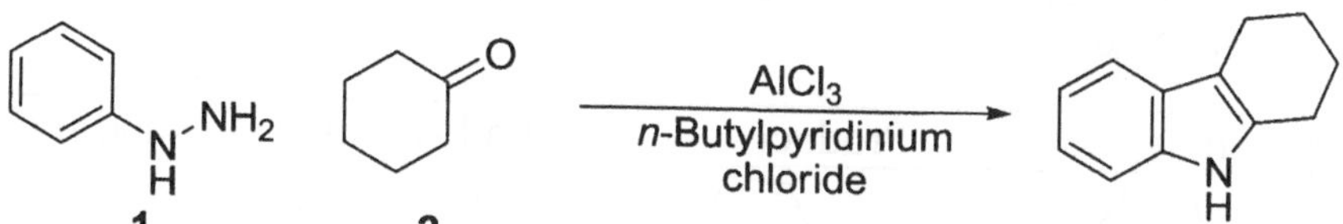

1 **2** **3**

hydrazine	ketone	tetrahydrocarbazole	Time (min)	Yield (%)
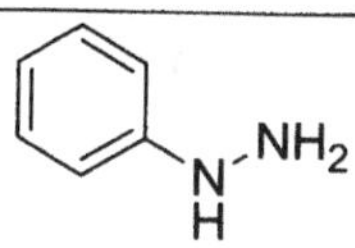	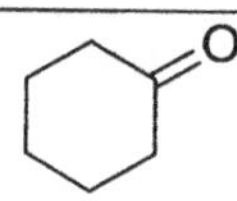	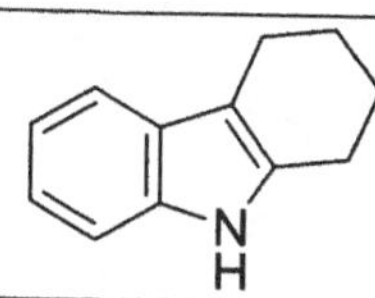	35	92
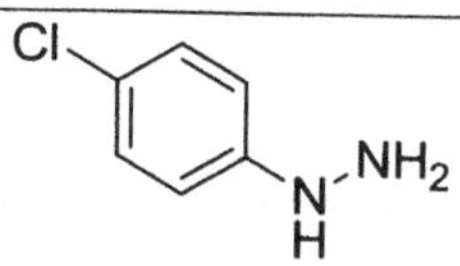 **9**	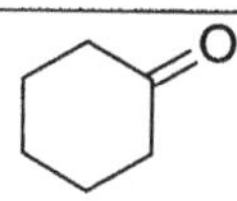	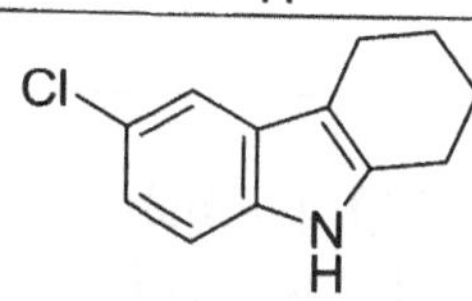	30	90
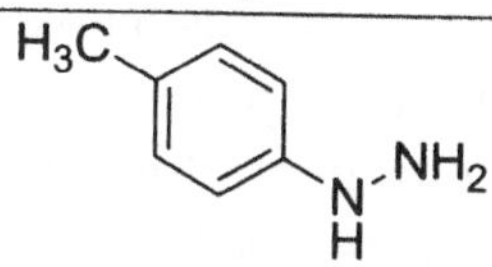 **10**	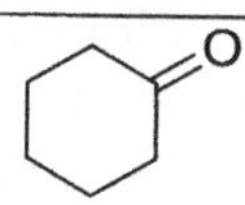	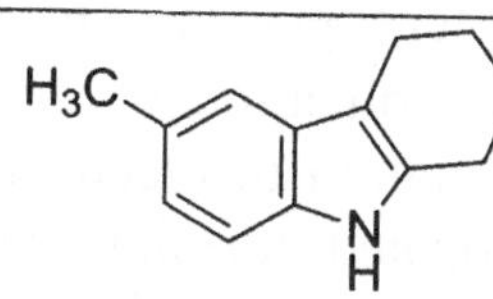	35	92

There has been a great deal of interest in optimizing the second step of the Borsche–Drechsel cyclization. Lead oxide,[11] mercurous acetate,[10] and sulfur[10] were the first oxidizing agents used by investigators. These methods were generally low yielding, and in some cases substituents were cleaved from the ring system, providing the unsubstituted carbazole as the major product.[12] However, in 1945 Barclay and Campbell found that chloranil provided the carbazole in higher yields.[12] Since that time, there have been many improvements and variations. These variations include 2,3-dichloro-5,6-dicyanobenzoquinone (DDQ)[17] and microwave-assisted palladium acetate dehydrogenation[18] as well as the examples that follow.

Hayashi and co-workers[19] compared two methods of oxidations in the conversion of tetrahydrocarbazole to carbazole. Both conditions provided the desired product in more mild conditions and higher yields than previous conditions.[20] Method A used palladium on carbon in acetonitrile under ethylene atmosphere. This oxidation transferred hydrogen from the tetrahydrocarbazole to the ethene to form ethane, which was detected via gas chromatography. Method B involves activated carbon heated in xylenes for 24 ho under an oxygen atmosphere. It is believed this reaction proceeded through a radical dehydration rather than a dehydrogenation.

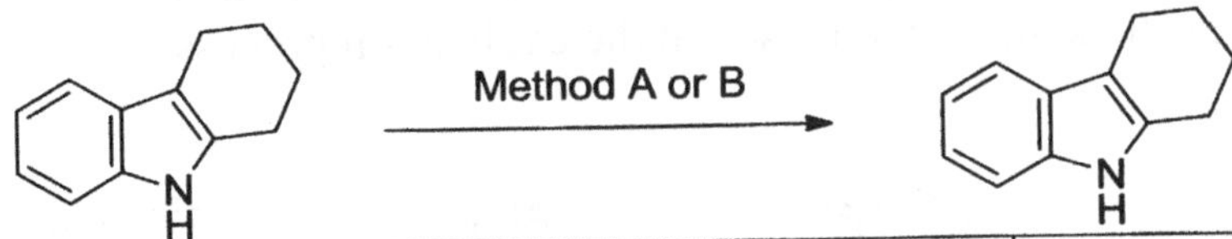

Substrate	Product	Method A (%)	Method B (%)
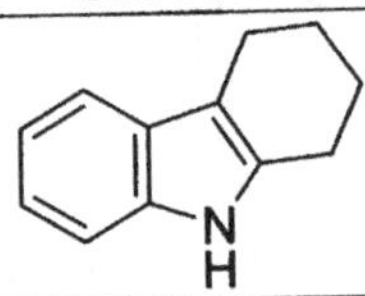	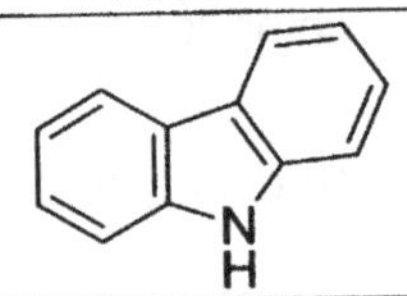	92	77
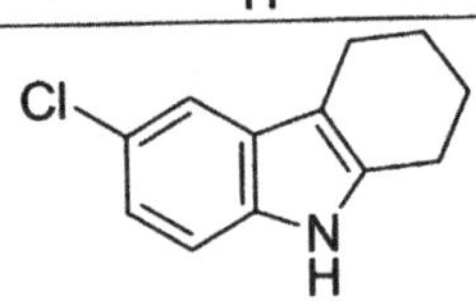	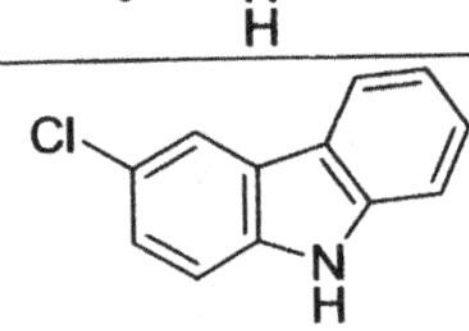	69	53

Bisagni[21] in 1980 published a procedure of converting 2,3,4,9-tetrahydro-1H-carbazol-1-one (**11**) derivatives to the corresponding substituted carbazoles with pyridinium hydrochloride (**12**). The mechanism proceeds through the enolization of the unsaturated ketone to provide the extended enol **13**. Donation of electrons by the indole nitrogen allows the enol alkene to be protonated, forming the secondary alcohol **14**. Dehydration of this alcohol proceeds through an E1 mechanism to provide **15**. Aromatization of the cyclohexadiene proceeds upon deprotonation of the doubly allylic proton and resonance of the ring.

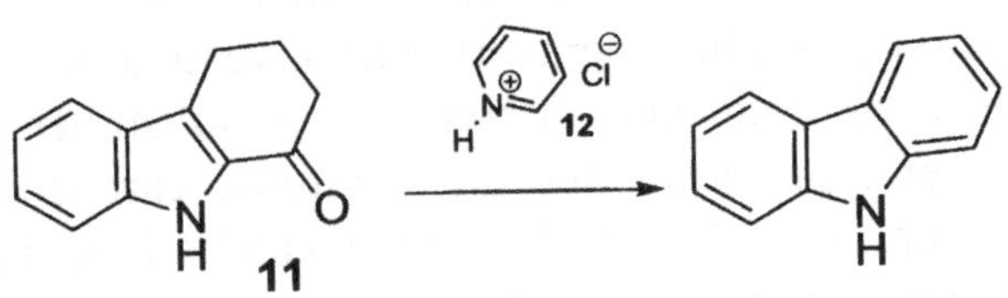

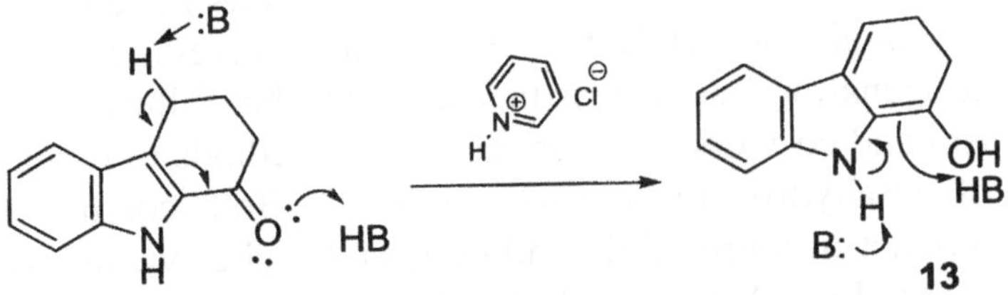

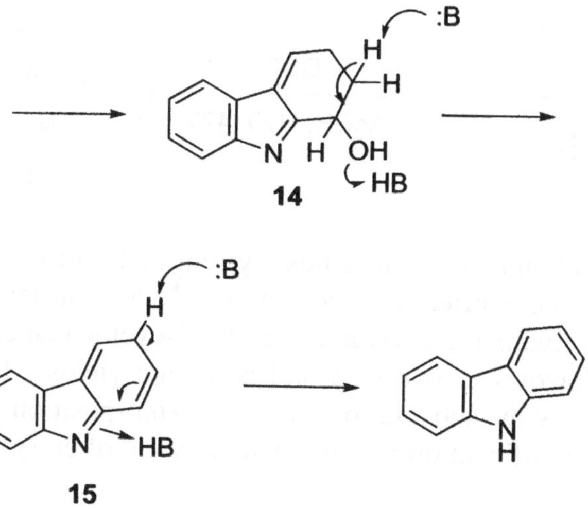

3.2.5 *Synthetic Utility*

General Utility

Kirsh and Dufour[22] used the cyclization in their work to synthesize carbazole alkaloids for biological study. Cyclization of phenyl hydrazine with α-tetralone (**16**) provided the dihydrocarbazole **17** in outstanding yields. This was oxidized to the tetracyclic carbazole **18**. Next a regioselective reduction of a phenyl ring was accomplished in high yields using sodium metal in isoamyl alcohol to provide **19**. Benzylic oxidation of the ring provides the functionalized carbazole **20** in moderate yields with greater than 10:1 regioselectivity of the carbonyl placement.

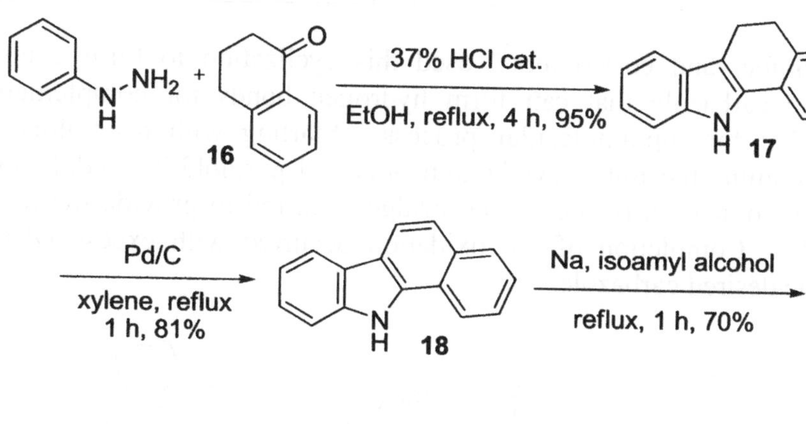

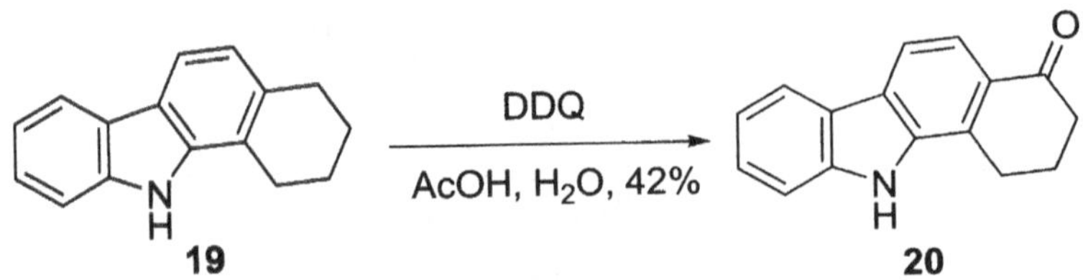

In their attempts to build a host system that will recognize and sense anions in solution, Curiel and co-workers[23] used a tandem cyclization between phenyl hydrazine derivatives and 1,2-cyclohaxanedione to provide pentacyclic carbazoles in one step and moderate yields. It is important to note that even with substitution at the *ortho*-position Curiel did not experience any significant drop in the percent yield of entry 4 as compared to entry 1 or 3.

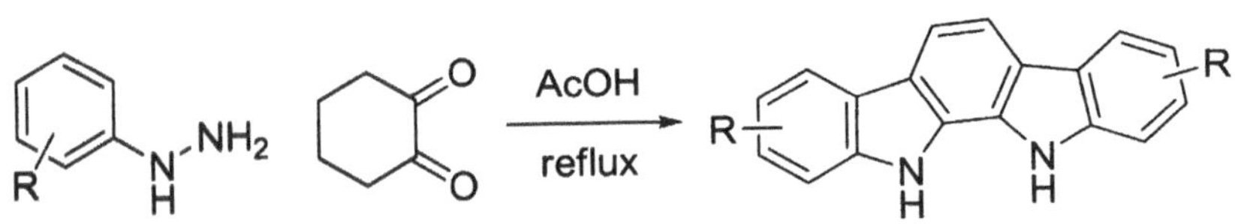

Entry	R =	% Yield	Product
1	H	40	indolo[2,3-a]carbazole
2	4-methyl	47	3,8-dimethylindolo[2,3-a]carbazole
3	4-bromo	41	3,8-dibromoindolo[2,3-a]carbazole
4	2-bromo	42	1,10-dibromoindolo[2,3-a]carbazole

Höpfner and co-workers[24] used this cyclization to form a unique pentacyclic carbazole that can form hydrogen bonds for complementary binding vital for supramolecular process. Starting with α-tetralone and quinolin-8-amine the initial cyclization provided pyrido[3,2-g]indole **21**, in which some oxidation of the ethano bridge occurred to provide the desired product **22**. Completion of the oxidation occurred with excess DDQ to provide the desired carbazole.

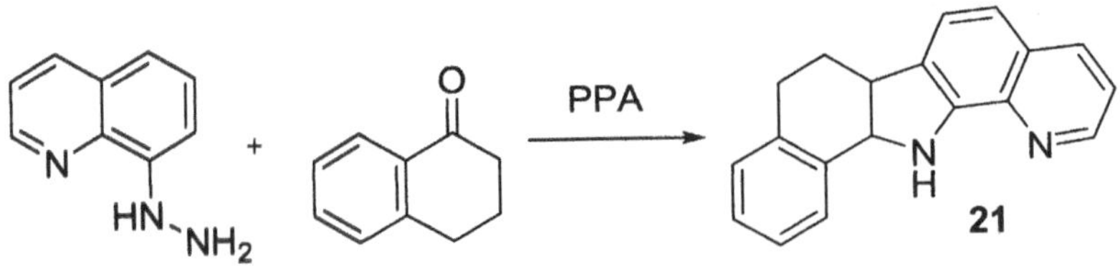

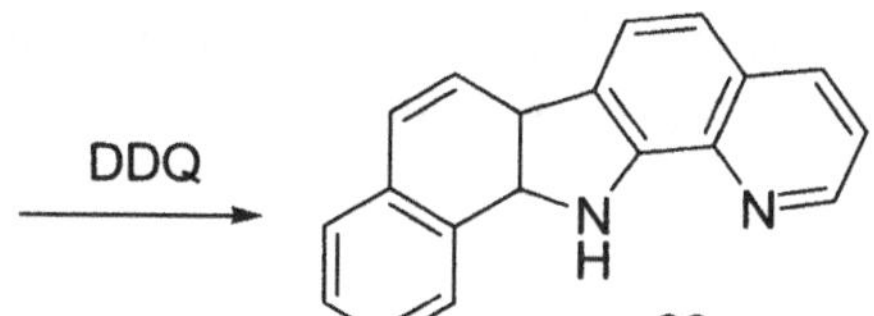

22

Applications in the Total Synthesis of Natural Products

Carbazoles have a long history of examination for biological activity.[4–6] Notably, in 1992 Mauffret and co-workers[25] completed the synthesis of isomeric ellipticine and several other derivatives to test their antitumor properties. Starting with (2,5-dimethyl-3-nitrophenyl)hydrazine **23** was heated in a solution of sodium acetate and hydrochloric acid with cyclohexanone or 4-methoxycyclohexanone to provide the tetrahydrocarbazoles **24a** and **24b** in moderate yields. Refluxing the intermediate in xylene and 2 equiv of DDQ for 3 h provided the advanced intermediates **25a** and **25b**, which were carried on to the targeted derivatives in short order.

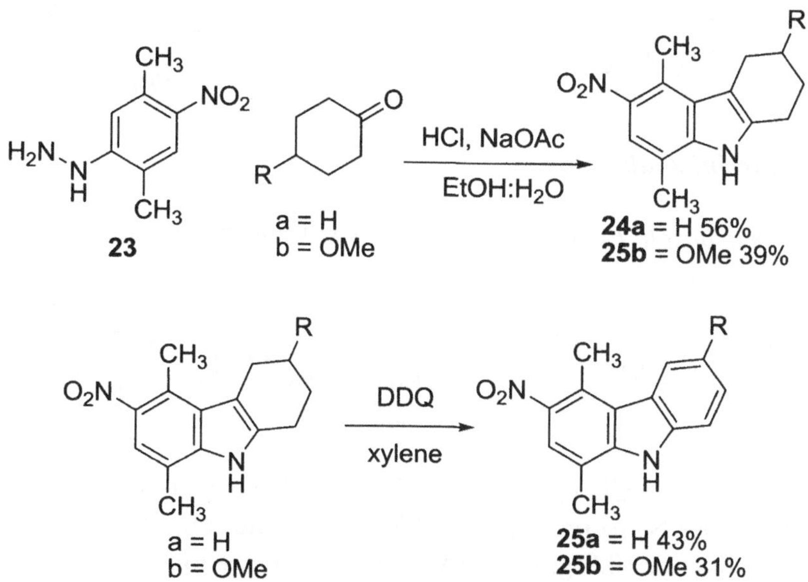

23

24a = H 56%
25b = OMe 39%

25a = H 43%
25b = OMe 31%

Janot and co-workers[26] used the cyclization to achieve the total synthesis of ellipticine and some important derivatives. The synthesis of the ellipticine derivatives began with phenyl hydrazine **26** and 2,5-dimethylcyclohexanone **27** that were heated with hydrochloric acid in ethanol to achieve both the tetrahydrocarbazole **28** and a minor component of indole **29**, which differ only in the regioselectivity of the cyclization. The tetrahydrocarbazole was isolated and then oxidized by treatment with chloranil in refluxing THF to obtain the carbazole **30** in moderate yields. This completed the carbazole core was then taken on to finish the target derivative.

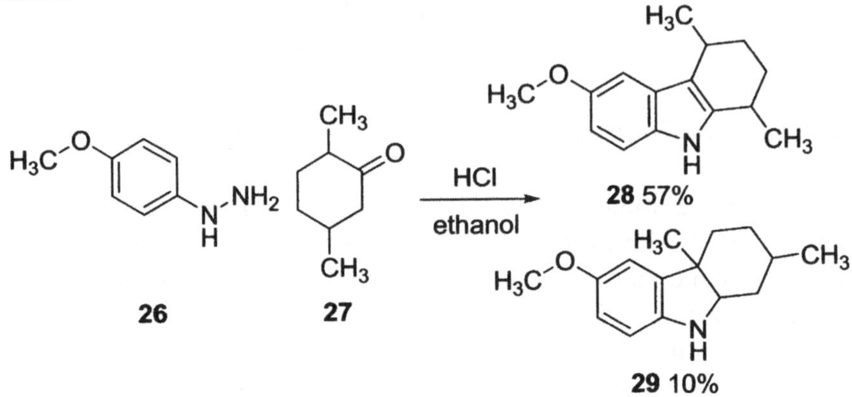

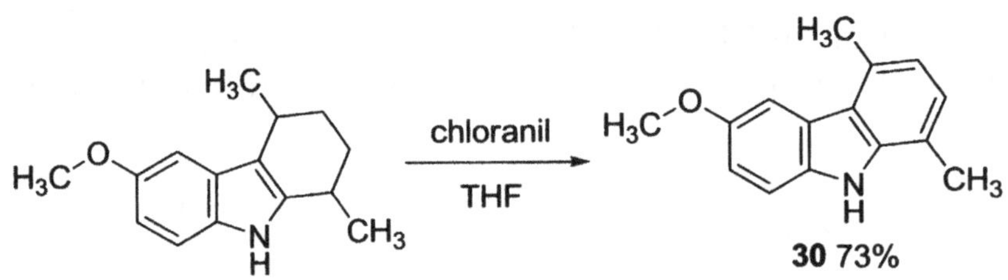

3.2.6 *Experimentals*

Tetrahydrocarbazole Formation Using Lewis Acids in Ionic Liquids[13]

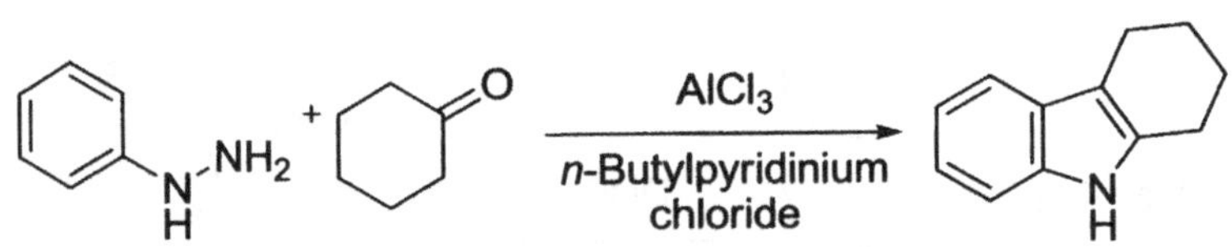

Freshly sublimed $AlCl_3$ (134 g, 1 mol) was added to *n*-BuPyCl (50 g, 0.5 mol) at r.t. under N_2 to provide the chloroaluminate ionic liquid. A mixture of phenyl hydrazine (0.9 g, 8.3 mmol) and cyclohexanone (0.8 g, 8.3 mmol) was added dropwise to ionic liquid (1.5 mL, 6.5 mmol), at 180–185 °C for 3 h under efficient stirring under N_2. After the reaction was complete (monitored by TLC), it was quenched with water/conc. HCl, extracted Et_2O, and the Et_2O solution was passed through anhyd. Na_2SO_4. The Et_2O was removed to obtain the desired product, which was purified by recrystallization (EtOH) (92% yield).

*Borsche–Drechsel Cyclization Reaction of Cyclohexanone with (2,5-Dimethyl-4-Nitrophenyl)Hydrazine to Provide 1,4-Dimethyl-3-Nitro-9*H*-Carbazole*[25]

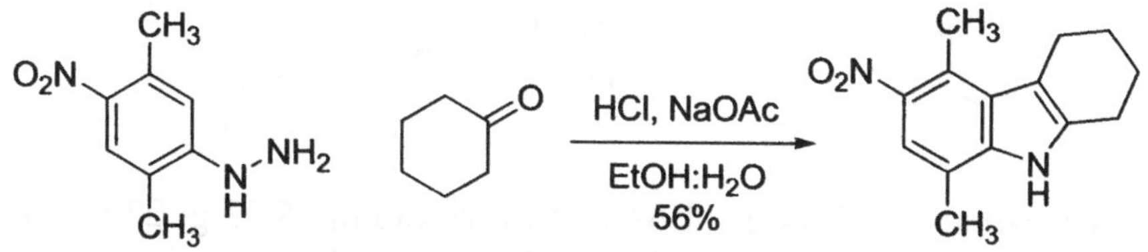

To a suspension of the hydrochloride salt of 15 (1 g, 4.6 mmol) in ethanol (10 mL), a solution of sodium acetate (1 g, 12 mmol) in water (10 mL) was added and then heated until complete dissolution. Cyclohexanone (0.62 g, 6.3 mmol) was added to the reaction mixture, which was then refluxed for 2 h and subsequently evaporated to dryness. The resulting solid was taken up with acetic acid (10 mL) saturated with hydrogen chloride, and refluxed for 10 min. The reaction mixture was poured onto ice and the resulting yellow precipitate was extracted with dichloromethane (× 3). The combined extracts were washed with 30% aqueous sodium hydrogen carbonate until neutral and then with water, dried (Na_2SO_4) and evaporated under reduced pressure. The residue was purified on a silica gel chromatographic column with dichloromethane as eluent to give orange microcrystals (0.6 g, 56%), which were recrystallized from ethanol; mp 208 °C.

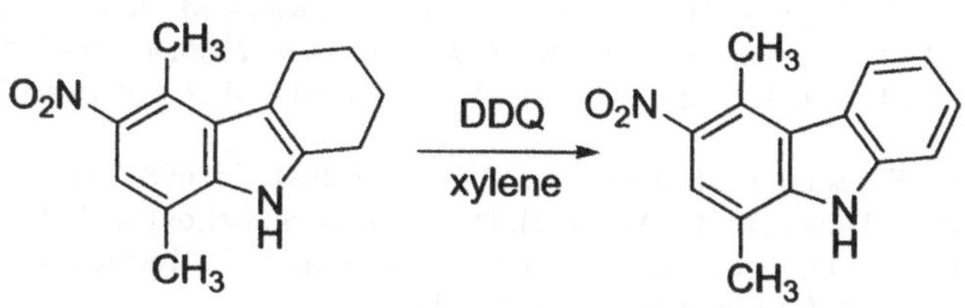

Compound **24a** (1 g, 4 mmol) and DDQ (2 g, 8 mmol) were refluxed in xylene (40 mL) for 3 h. The reaction mixture was filtered while still hot, and the brown precipitate was washed with hot xylene. The filtrate and washings were evaporated under reduced pressure, and the residue was purified by silica gel column chromatography using chloroform:methanol (95:5) as eluent to afford yellow microcrystals (0.43 g, 43%).

*Borsche–Drechsel Cyclization Reaction of 2,3,4,9-Tetrahydro-1*H*-Carbazole to Provide Carbazole*[9]

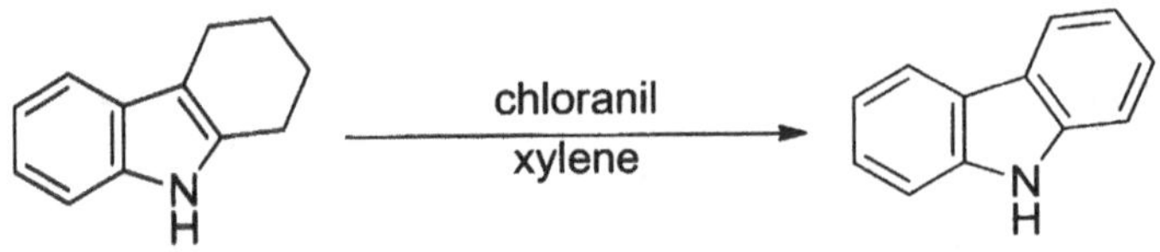

Tetrahydrocarbazole (2.00 g, 11.68 mmol), chloranil (5.77 g, 23.68 mmol), and the minimum volume of boiling, sulfur-free xylene required to form a clear solution were refluxed until a few drops of the solution gave no red color when heated with sodium hydroxide (24 h). The solution was then cooled, separated from tetrachloroquinol, diluted with ether, shaken first with sodium hydroxide and then with water, and finally dried (sodium sulfate). On evaporation of the solvents, the impure crystalline compound separated and was purified by crystallization from methyl alcohol, benzene, or xylene; when this method failed, purification was effected by chromatographic adsorption.

3.2.7 *References*

1. Drechsel, E. *J. Prakt. Chem.* **1858,** *38,* 69.
2. Borsche, W.; Feise, M. *Ber. Dtsch. Chem. Ges.* **1907,** *40,* 378–386.
3. [R] Campbell, N.; Barclay, B. M. *Chem. Rev.* **1947,** *40,* 359–380.
4. Napper, A. D.; Hixon, J.; McDonagh, T.; Keavey, K.; Pons, J.-F.; Barker, J.; You, W. T.; Amouzegh, P.; Flegg, A.; Hamelin, E.; Thomas, R. J.; Kates, M.; Jones, S.; Navia, M. A.; Saunders, J. O.; Distefano, P. S.; Curtis, R. *J. Med. Chem.* **2005,** *48,* 8045–5054.
5. Shintani, A.; Toume, K.; Rifai, Y.; Arai, M. A.; Ishibashi, M. *J. Nat. Prod.* **2010,** *73,* 1711–1713.
6. Takada, K.; Kajiwara, H.; Inamura, N. *J. Nat. Prod.* **2010,** *73,* 698–701.
7. Martin, M. J.; Trudell, M. L.; Aratho, H. D.; Allen, M. S.; LaLoggia, A. J.; Deng, L.; Schultz, C. A.; Tan, Y.-C.; Bi, Y.; Narayanan, K.; Dorn, L. J.; Koehler, K. F.; Skolnick, P.; Cook, J. M. *J. Med. Chem.* **1992,** *35,* 4105–4117.
8. Robinson, G. M.; Robinson, R. *J. Chem. Soc.* **1924,** *125,* 827–840.
9. Cooke G. W.; Gulland, J. M. *J. Chem. Soc.* **1939,** 872–873.
10. Perkin, W. H. Jr; Plants, G. P. *J. Chem. Soc.* **1921,** *119,* 1825–1839.
11. Borsche, W.; Witte, A.; Bothe, W. *Ann.* **1908,** *359,* 52.
12. Barclay, B. M.; Campbell, N. *J. Chem. Soc.* **1945,** 530–533.
13. Varma, P. P.; Sherigara, B. S.; Mahadevan, K. M.; Hulikal, V. *Synth. Commun.* **2009,** *39,* 158–165.
14. Xu, D.-Q.; Wu. J.; Luo, S.-P.; Zhang, J.-X.Wu, J.-Y.; Du. H.-X. Zu, Z.-Y. *Green Chem.* **2009,** *11,* 1239–1246.
15. Kuroki, M.; Tsunashina, Y. *J. Heterocyclic Chem.* **1981,** *18,* 709–714.
16. Rebeiro, G. L.; Khadilkar, B. M. *Synthesis* **2001,** *3,* 370–372.
17. Park, I.-K.; Suh, S.-E.; Lim, B.-Y.; Cho, C.-G. *Org. Lett.* **2009,** *11,* 5455–5456.
18. Sridharan, V.; Martin, M. A.; Menéndez, J. C. *Eur. J. Org. Chem.* **2009,** 4614–4621.
19. Tanaka, T.; Okunaga, K.-I.; Hayashi, M. *Tetrahedron Lett.* **2010,** *51,* 4633–4635.
20. Wolthuis, E. *J. Chem. Educ.* **1979,** *56,* 343–344.
21. Bisagni, E.; Ducrocq, C.; Hung, N. C. *Tetrahedron* **1980,** *36,* 1327–1330.
22. Dufour, F.; Kirsch, G. *Synlett* **2006,** *7,* 1021–1022.

23. Curiel, D.; Cowley, A. Beer, P. D. *Chem. Commun.* **2005,** 236–238.
24. Hung, C.; Höpfner, T.; Thummel, R. P. *J. Am. Chem. Soc.* **1993,** *115,* 12601–12602.
25. Alunni-Bistocchi, G.; Orvietani, P.; Mauffret, O.; Antri, S. E.; Deroussent, A.; Jacquignon, P. C.; Fermandjian, S.; Ricci, A.; Lescot, E. *J. Chem. Soc., Perkin Trans. 1.* **1992,** 2935–2941.
26. Rousselle, D.; Gilbert, J.; Viel, C.; Janot, M. M.-M. C. *R. Acad. Sci. Paris* **1977,** 377–380.

3.3 Buchwald–Hartwig Indole Synthesis

Brian Goess

3.3.1 Description

Indoles can be synthesized via a palladium-mediated intramolecular amination reaction of an appropriately substituted haloarene. This is a specific application of the Buchwald–Hartwig amination reaction[1] that has found use in the synthesis of biologically active natural products.

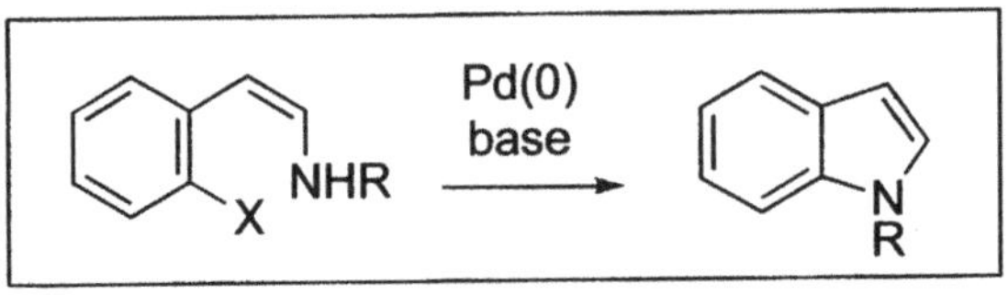

The Buchwald–Hartwig amination reaction has also been applied to the synthesis of *N*-arylhydrazones, which can then be transformed into indoles via the Fisher indole synthesis. Since this reaction sequence does not use a Buchwald–Hartwig amination reaction directly in the synthesis of indoles, it will not be a focus of this chapter.

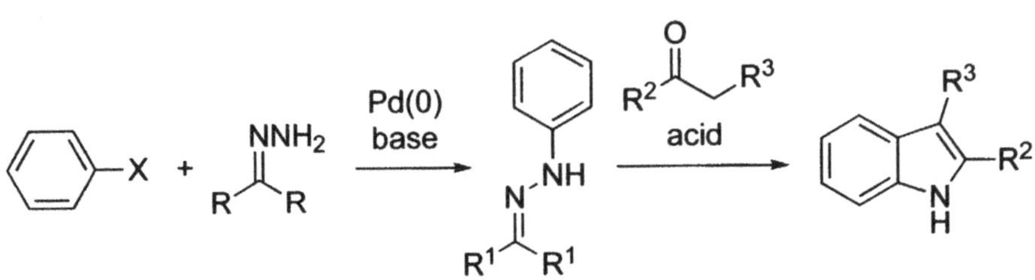

3.3.2 Historical Perspective

The prevalence of the aniline functional group in alkaloid natural products of medicinal interest has inspired a number of metal-catalyzed strategies for the preparation of substituted anilines. One classical method, a variation of the Ullman aryl ether synthesis, features copper catalysis.[2] This method generally suffers from harsh reaction conditions and limited substrate scope. Nonetheless, the simplicity of the transformation and the inexpensiveness of copper catalysis has prompted additional development of this strategy. For instance, cyclohexylphenylamine is formed in excellent yield from the copper(I) iodide-catalyzed amination of iodobenzene.[3]

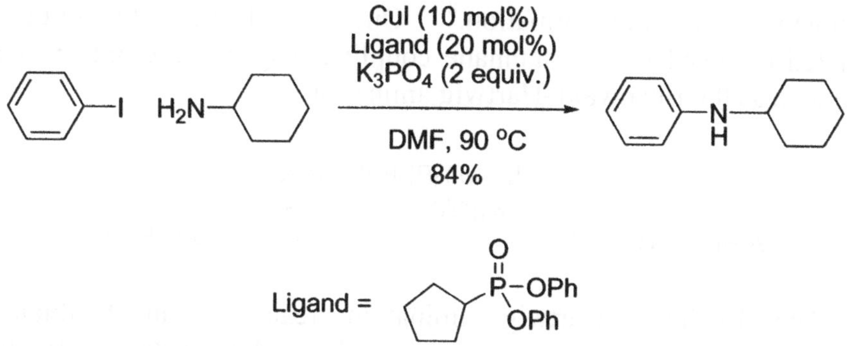

Nickel-catalyzed variations of this process have appeared more recently.[4] A representative transformation is the substitution of *m*-chloroaniline with pyrrolidine in the presence of $Ni(acac)_2$ and an *N*-heterocyclic carbene (NHC) ligand.[5] This reaction gives high yields with both aromatic amines and dialkylamines, though the operational utility of the transformation is diminished by the air-sensitivity of the catalyst system.

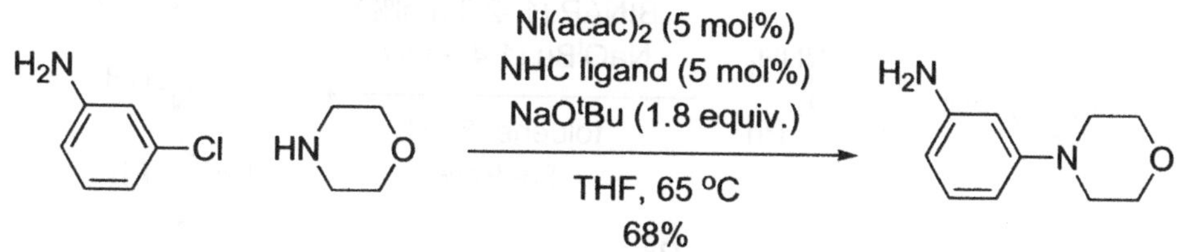

The first demonstration of palladium catalysis in an aryl amination reaction was reported in 1983 by Kosugi and co-workers.[6] In this reaction, an aminostannane is coupled to an aryl halide in a transformation analogous to a Stille cross-coupling.

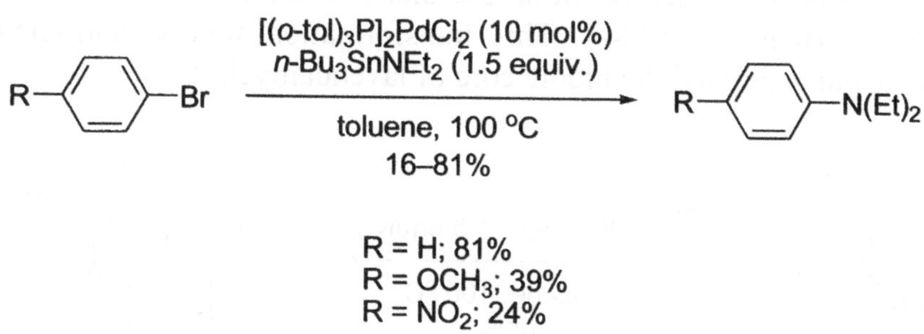

In 1994, Buchwald and co-workers improved this original procedure by developing a method for *in situ* generation of the aminostannane.[7] Simultaneously, Hartwig and co-workers obtained an X-ray crystal structure of a catalytically active Sn-Pd complex, furthering our understanding of the mechanism of the transformation.[8] One year later, both groups had eliminated the need for the stannane coupling reagent, resulting in what is now known as the Buchwald–Hartwig amination.[9]

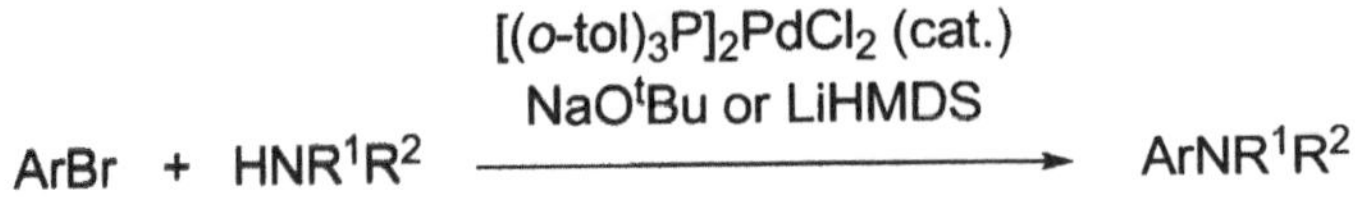

The Buchwald–Hartwig amination reaction can facilitate the synthesis of nitrogen-containing heterocycles. For instance, Buchwald demonstrated that the amination reaction could be used to prepare *N*-arylhydrazones from hydrazones and aryl halides. Acid-catalyzed condensation of the hydrazone with a ketone then yielded the desired indoles.[10] Hartwig described a similar arylhydrazone formation with DPPF as a ligand and Cs_2CO_3 as a base.[11]

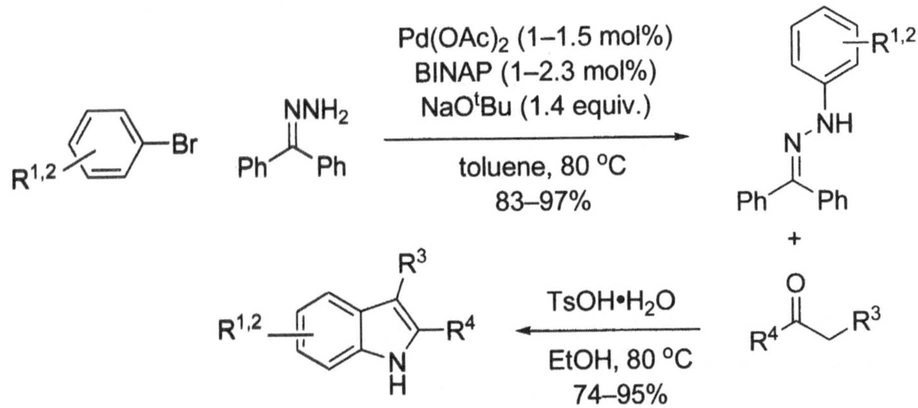

The first Pd-catalyzed intramolecular amination of an aryl halide was reported by Boger in 1984. This transformation was stoichiometric in palladium and generated the indole core of lavendamycin.[12]

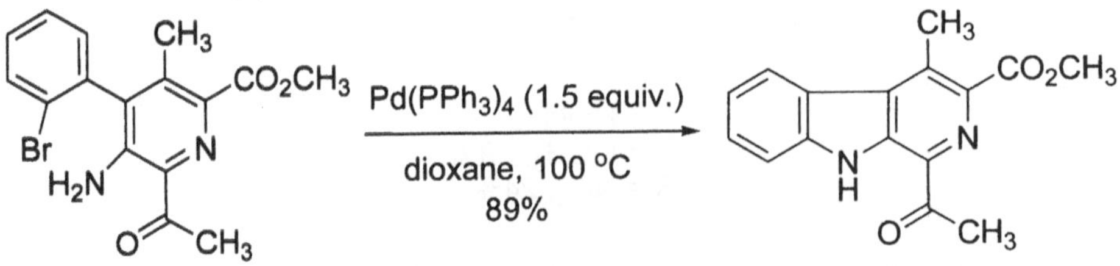

The first related synthesis that was catalytic in palladium was reported by Buchwald and co-workers.[13] Nitrogen heterocycles of various ring sizes could be prepared with this methodology.

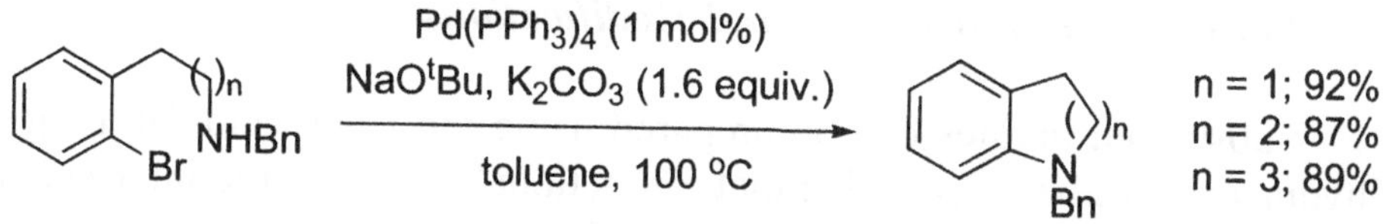

Buchwald quickly parlayed this transformation into a versatile indole synthesis via oxidation of intermediate indolines prepared using the above intramolecular Buchwald–Hartwig amination.[14] Variations leading directly to indoles and related nitrogen-containing heterocycles quickly followed and are described below.

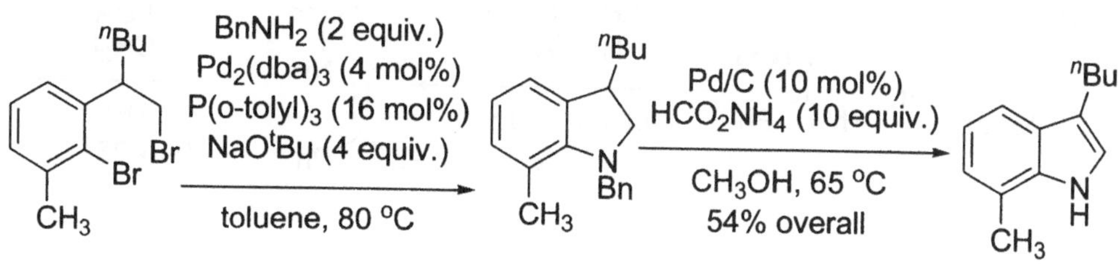

3.3.3 *Mechanism*

The mechanism of the direct intramolecular Buchwald–Hartwig indole synthesis is that of a traditional palladium-catalyzed cross-coupling reaction and begins with loss of a ligand on palladium. Oxidative addition of an appropriately substituted *Z*-vinylhaloarene (**1**) generates intermediate **2**. Deprotonation of **2** and displacement of a halide ligand sets up a reductive elimination on **3** to yield indole **4**.

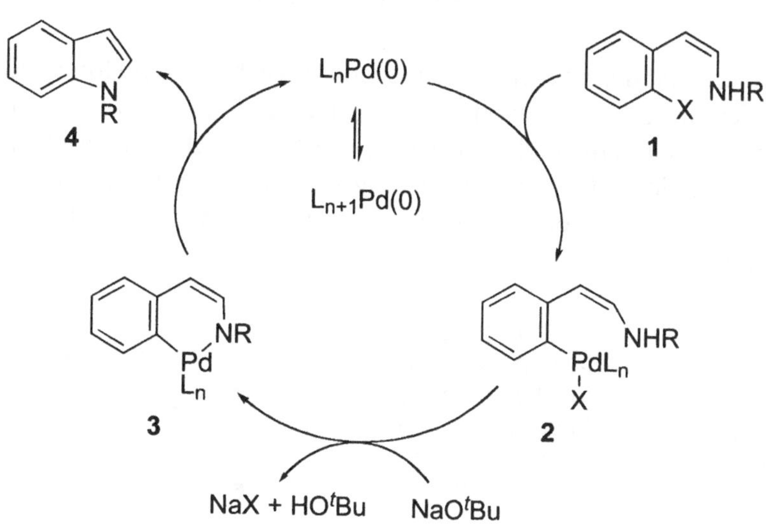

3.3.4 Variations, Improvements and Modifications

Substituted indoles can be prepared using the Buchwald–Hartwig *N*-arylhydrazone synthesis provided that a functionalization of the intermediate *N*-arylhydrazone is carried out before the Fisher indole synthesis synthesis step. Both *N*-alkyl and *N*-aryl indoles may be prepared in this manner.[15] For instance, an *N*-methylindole was prepared as a mixture of regioisomers by first methylating an arylhydrazone.

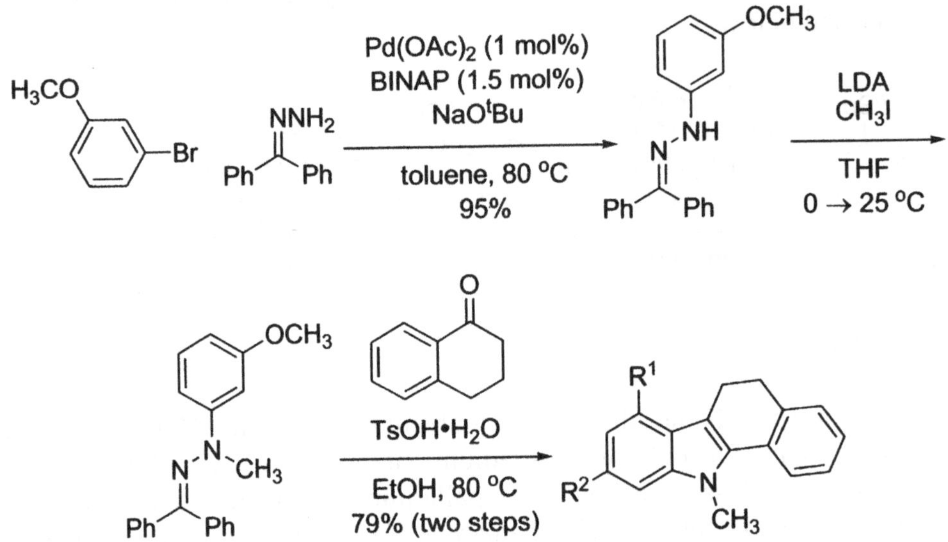

R^1 = H, R^2 = OCH_3; 59%
R^1 = OCH_3, R^2 = H; 20 %

Several tandem reaction sequences have been developed for indole synthesis featuring a Buchwald–Hartwig aryl amination as one component. For instance, a tandem double *N*-arylation of substrate **5** leads to indoles **6** through a cascade of C–N bond forming reactions. A range of structurally and electronically diverse amines can be used successfully in this transformation.[16]

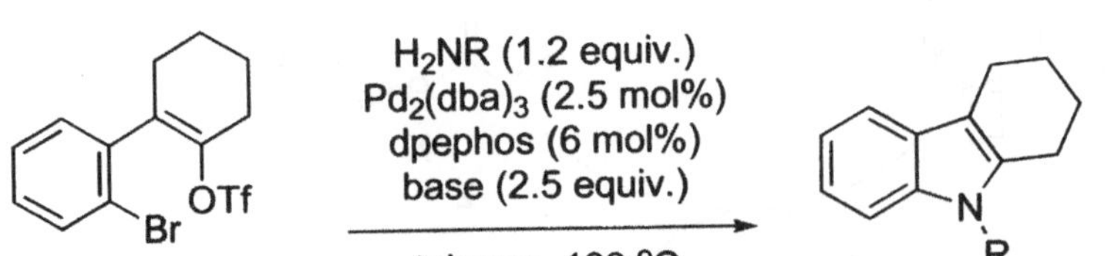

5 **6**

R = Ph; 83%
R = ^{n}Bu; 62%
R = Bn; 70%

A tandem amination/arylation reaction can convert appropriately substituted gem-dibromovinyl substituted anilines into indoles.[17]

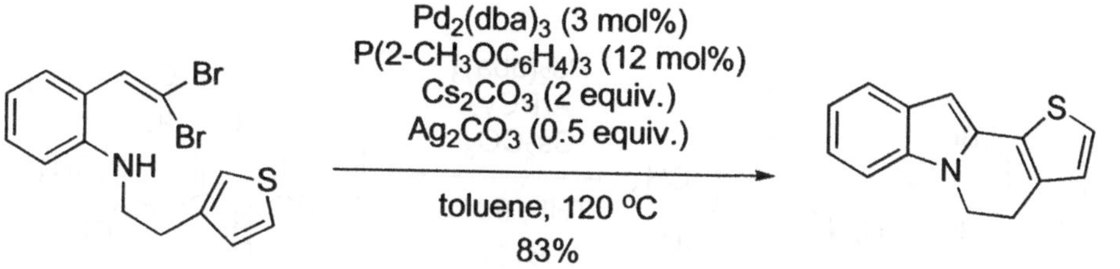

Indoles can also be prepared directly from *o*-chlorophenyl-2-alkyl alkynes and primary amines via a tandem alkyne hydroamination/Buchwald–Hartwig *N*-arylation sequence, as illustrated for the conversion of **7** to **8** via intermediate enamine **9**.[18]

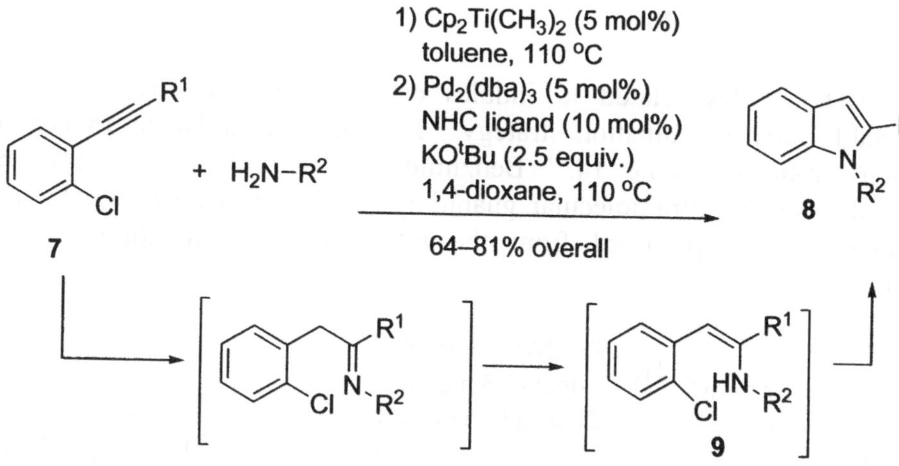

A tandem *ortho*-alkylation/amination reaction between aryl halides and alkyl halides generates indolines in moderate yield.[19] Such indolines can be easily oxidized to indoles.

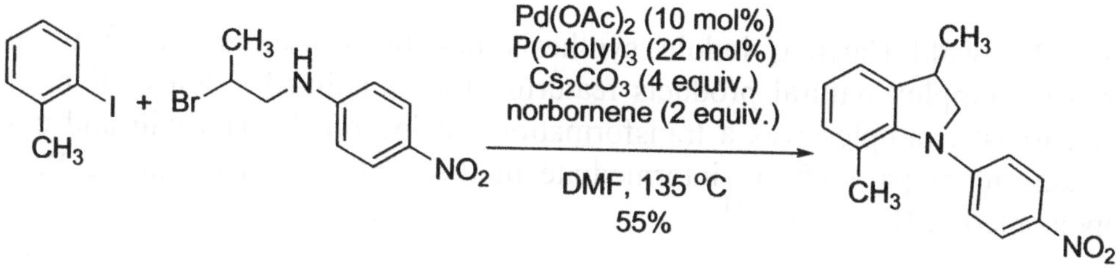

Finally, a tandem *N*-arylation/Heck coupling has also been developed for indole synthesis. Vinologous amide **11** couples with aryldibromide **10** to yield indole **13** via intermediate **12**.[20]

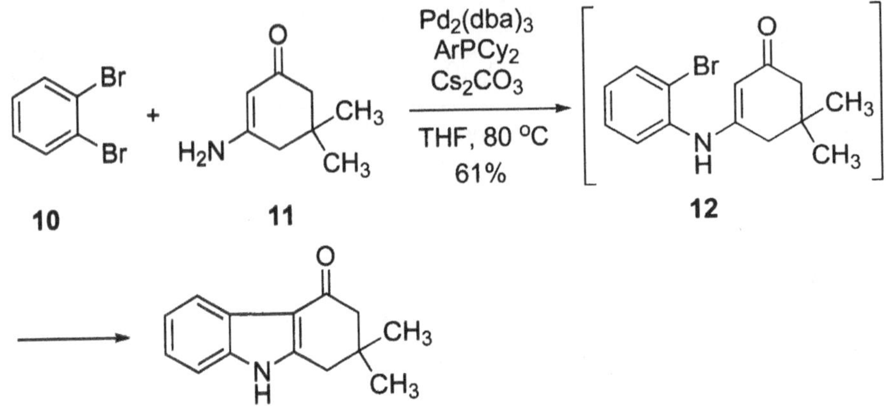

10 **11** **12** **13**

Heterocycles related to indoles can also be prepared using the Buchwald–Hartwig amination strategy. For instance, indazole **15** can be prepared from hydrazine **14**.[21] Benzimidazole derivatives may also be prepared from the intramolecular guanidinylation of aryl bromides.[22] And oxindoles can be prepared from the intramolecular amidation of aryl bromides.[23]

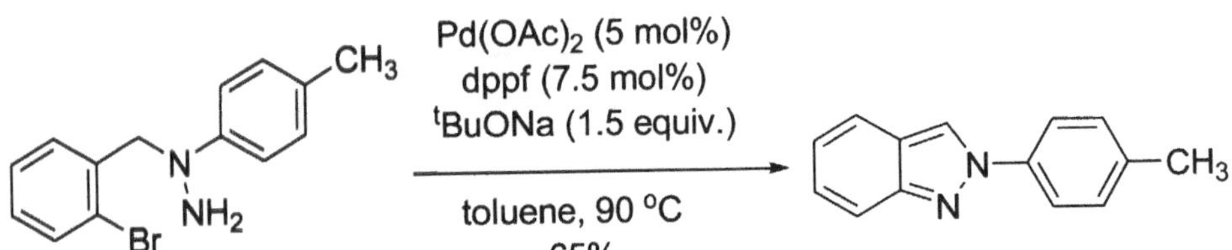

14 **15**

3.3.5 *Synthetic Utility*

The Buchwald–Hartwig indole synthesis has been used successfully to prepare complex natural products featuring functionalized indoles. One of the earliest examples was a transformation carried out by Hartwig and co-workers to prepare **18**, an intermediate in the synthesis of damirone B, a topoisomerase II inhibitor.[24]

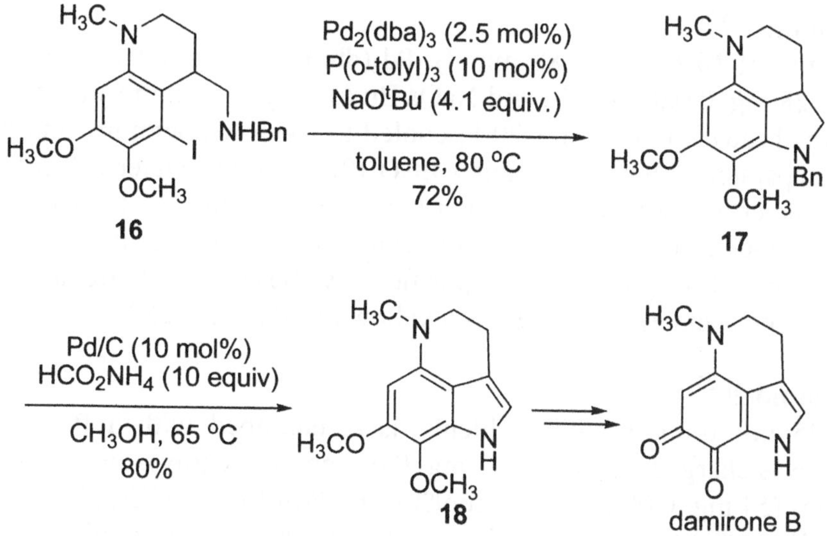

A more recent example illustrates the continuing applicability of this transformation in natural product synthesis, in this case toward **20**, which bears a core resembling that of insecticidal indole diterpene nodulisporic acid A.[25]

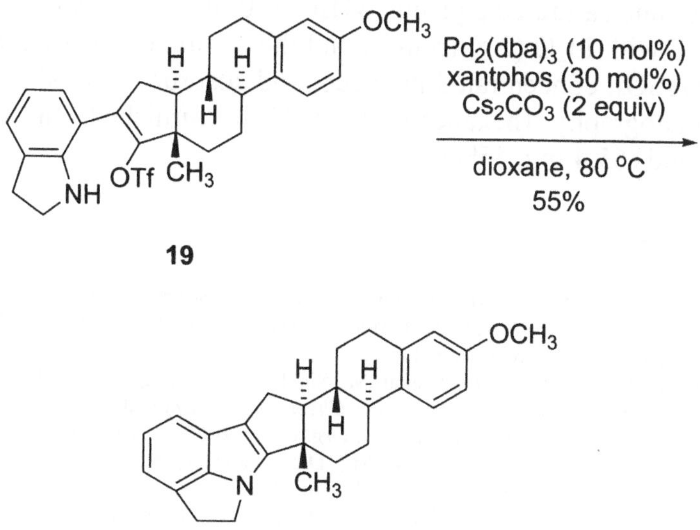

20

3.3.6 Experimental

The examples presented illustrate two of the common ways Buchwald–Hartwig indolizations are run. Both are from the applications to natural product synthesis described above.

Compound 17[24]

A mixture of **16** (0.4 g, 0.88 mmol), $Pd_2(dba)_3$ (20 mg, 0.022 mmol), P(*o*-tolyl)$_3$ (28 mg, 0.088 mmol), and NaO*t*-Bu (0.34 g, 3.52 mmol) in toluene (5 mL) was heated to 80 °C for 20 h, cooled to room temperature, and poured into a separatory funnel containing Et_2O (20 mL) and water (20 mL). The organic layer was washed with water (15 mL) and brine (15 mL), dried over $MgSO_4$, and filtered, and the solvents were removed using a rotary evaporator. The product was purified by flash chromatography (10:1 hexane/ethyl acetate) to yield **17** as a yellow oil (0.21 g, 72%).

Compound 20[25]

To a 25 mL round-bottom Schlenk flask, equipped with a PTFE-coated stirbar, was charged Cs_2CO_3 (202 mg, 0.62 mmol, 2 equiv), dioxane (8 mL), xantphos (54 mg, 0.09 mmol, 30 mol%), and $Pd_2(dba)_3$•$CHCl_3$ (32 mg, 0.031 mmol, 10 mol%) to give a deep brownish red solution. The resulting solution was stirred for 5 min, then treated with a solution of the enol triflate **19** (165 mg, 0.309 mmol, 1 equiv) in THF (3 mL). Next, the Schlenk flask was fitted with a reflux condenser, and the reaction mixture was brought to reflux, yielding a dark brownish yellow solution. After 50 min, the reaction was cooled to room temperature and poured into a vigorously stirred mixture of Et_2O (20 mL) and saturated aqueous NH_4Cl (20 mL). The aqueous layer was then extracted with Et_2O (2 × 15 mL), and the combined organic layers were dried over $MgSO_4$, concentrated in vacuo, and adsorbed onto silica gel (2 g). Flash chromatography (hexanes/EtOAc) 12:1) furnished **20** as a white crystalline solid (65.4 mg, 55%).

3.3.7 References

1. For a comprehensive review of this reaction, found earlier in this series, see [R] Janey, J. M. *Name Reactions for Functional Group Transformations*, Wiley, Hoboken, NJ, **2007**, 564–609, and references for reviews cited therein.
2. [R] (a) Das, P.; Sharma, D.; Kumar, M.; Singh, B. *Curr. Org. Chem.* **2010,** *14,* 754–783. [R] (b) Ley. S. V.; Thomas, A. W. *Angew. Chem., Int. Ed. Engl.* **2003,** *42*, 5400–5449.
3. Rao, H.; Fu, H.; Jiang, Y.; Zhao, Y. *J. Org. Chem.* **2005,** *70*, 8107–8109.
4. (a) Kelly, R. A. III; Scott, N. M.; Díez-Gonzáles, S.; Stevens, E. D.; Nolan, S. P. *Organomettalics* **2005,** *24*, 3442–3447. (b) Chen, C.; Yang, L.-M. *Org. Lett.* **2005,** *7*, 2209–2211. (c) Omar-Amrani, R.; Thomas, A.; Brenner, E.; Schneider, R.; Fort, Y. *Org. Lett.* **2003,** *5*, 2311–2314.
5. Desmarets, C.; Schneider, R.; Fort, Y. *J. Org. Chem.* **2002,** *67*, 3029–3036.
6. Kosugi, M.; Kameyama, M.; Migita, T. *Chem. Lett.* **1983,** *12*, 927–928.
7. Guram, A. S.; Buchwald, S. L. *J. Am. Chem. Soc.* **1994,** *116,* 7901–7902.
8. Paul, F.; Patt, J.; Hartwig, J. F. *J. Am. Chem. Soc.* **1994,** *116,* 5969–5970.
9. (a) Guram, A. S.; Rennels, R. A.; Buchwald, S. L. *Angew. Chem., Int. Ed. Engl.* **1995,** *34,* 1348–1350. (b) Louie, J.; Hartwig, J. F. *Tetrahedron Lett.* **1995,** *36*, 3609–3612.
10. Wagaw, S.; Yang, B. H.; Buchwald, S. L. *J. Am. Chem.* Soc. **1998,** *120,* 6621–6622.
11. [R] Hartwig, J. F. *Angew. Chem., Int. Ed. Engl.* **1998,** *37*, 2090–2093.

12. (a) Boger, D. L.; Panek, J. S. *Tetrahedron Lett.* **1984,** *25*, 3175–3178. (b) Boger, D. L.; Duff, S. R.; Panek, J. S.; Yasuda, M. *J. Org. Chem.* **1985,** *50*, 5782–5789. (c) Boger, D. L.; Duff, S. R.; Panek, J. S.; Yasuda, M. *J. Org. Chem.* **1985,** *50*, 5790–5795.
13. Wolfe, J. P.; Rennels, R. A.; Buchwald, S. L. *Tetrahedron Lett.* **1996,** *52*, 7525–7546.
14. Aoki, K.; Peat, A. J.; Buchwald, S. L. *J. Am. Chem. Soc.* **1998,** *120*, 3068–3073.
15. Wagaw, S.; Yang, B. H.; Buchwald, S. L. *J. Am. Chem.* Soc. **1999,** *121,* 10251–10263.
16. Willis, M. C.; Brace, G. N.; Holmes, I. P. *Angew. Chem., Int. Ed.* **2005,** *44*, 403–406.
17. Bryan, C. S.; Lautens, M. *Org. Lett.* **2008,** *10*, 4633–4636.
18. Siebeneicher, H.; Bytschkov, I.; Doye, S. *Angew. Chem., Int. Ed.* **2003**, *42*, 3042–3044.
19. Rudolph, A.; Rackelmann, N.; Turcotte-Savard, M.-O.; Lautens, M. *J. Org. Chem.* **2009,** *74*, 289–297.
20. Edmondson, S. D.; Mastracchio, A.; Parmee, E. M. *Org. Lett.* **2000,** *2*, 1109–1112.
21. Song, J. J.; Yee, N. K. *Org. Lett.* **2000,** *2*, 519–521.
22. Evindar, G.; Batey, R. A. *Org. Lett.* **2003,** *5*, 133–136.
23. Yang, B. H.; Buchwald, S. L. *Org. Lett.* **1999,** *1*, 35–37.
24. Peat, A. J.; Buchwald, S. L. *J. Am. Chem. Soc.* **1996,** *118,* 1028–1030.
25. Smith, A. B., III; Kürti, L.; Davulcu, A. H.; Cho, Y. S.; Ohmoto, K. *J. Org. Chem.* **2007,** *72,* 4611–4620.

3.4 Cadogan–Sundberg Indole Synthesis

Jie Jack Li

3.4.1 *Description*

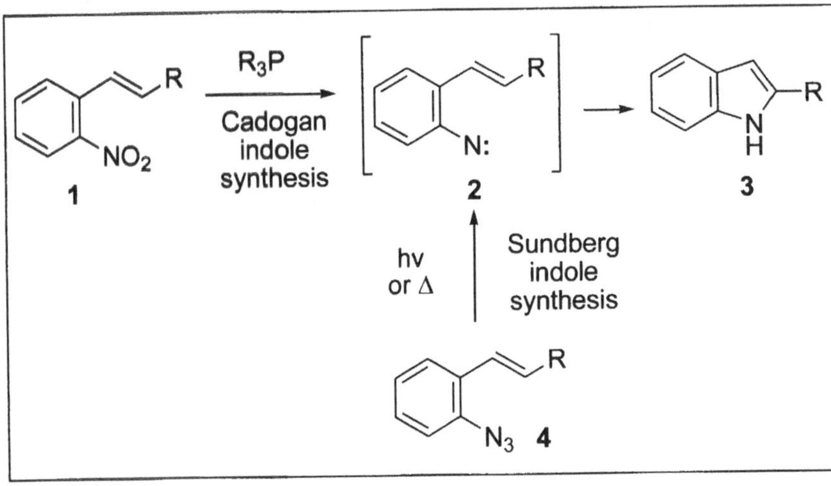

The Cadogan reaction refers to the deoxygenation of *o*-nitrostyrenes **1** or *o*-nitrostilbenes with trialkyl phosphite or trialkylphosphine and subsequent cyclization of the resulting intermediate nitrene **2** to form indoles **3**.[1] The reductive cyclization protocol has also been exploited to prepare a variety of *N*-containing heterocyclic compounds including carbazoles, indazoles, benzimidazole, benzotriazoles, anthranils, phenazines, phenothiazines, quinolines, and related compounds.

On the other hand, the Sundberg indole synthesis refers to the synthesis of indoles **3** via either thermolysis or irradiation of *o*-azidostyrene **4** via the intermediacy of nitrene **2**.[2] Since the Sundberg indole synthesis and the Cadogan reaction share the common intermdiate nitrene, they are sometimes grouped together as the Cadogan–Sundberg indole synthesis.

3.4.2 *Historical Perspective*

In 1962, Prof. John I. G. Cadogan at University of St. Andrews observed that aromatic *C*-nitroso compounds were readily deoxygenated by triethyl phosphite. Carbazoles were prepared from 2-*nitroso*biaryls in this fashion albeit in low yields. Later on that year, he published an article titled "Reduction of Nitro Compounds by Triethyl Phosphite: A New Cyclization Reaction" in the journal *Proceedings of the Chemical Society*.[3] In the paper, Cadogan described that 2-*nitro*biaryls were also readily deoxygenated by triethyl phosphite to provide the corresponding carbazoles with higher yields than the corresponding 2-*nitroso*biaryls. For instance, refluxing 2-nitrophenyl-pyridine (**5**) in triethyl phosphite gave pyrido[1,2-*b*]indazole (**7**) essentially quantitatively. Cadogan also presciently proposed nitrene **6** as the purported intermediate.

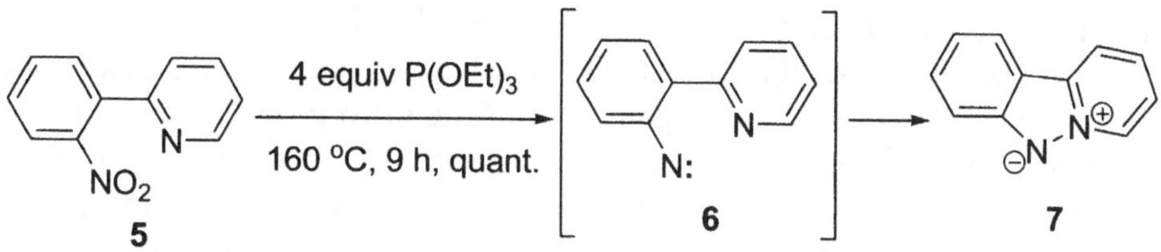

Cadogan further extended the methodology to make carbazoles, indoles, indazoles, and thiazoles.[4–6] Meanwhile in 1965, at the beginning of his independent career, Sundberg at the University of Virginia, also explored the Cadogan reaction to synthesize additional indoles.[7–9] At first, he prepared β-alkyl-*o*-nitrostyrenes and β-acyl-*o*-nitrostyrenes. While deoxygenation using triethyl phosphite in reflux gave β-alkyl-indoles in 51–71% yields; β-acyl-indoles were obtained in lower yields (16–19%).[7] In 1966, Sundberg painstakingly isolated and characterized all the major products and by-products, whose formations could be reconciled with the involvement of the nitrene intermediate. In 1967, he carried out deoxygenation of β,β-disubstitued-*o*-nitrostyrenes to prepare α,β-disubstitued indoles via 1,2-shift of the substituents.[9] Although formation of the products and by-products could be explained by the nitrene intermediacy, he also suggested the intermediacy of *N*-hydroxyindole **8** as shown below. Unfortunately, his proposal did not gain traction in future publications.

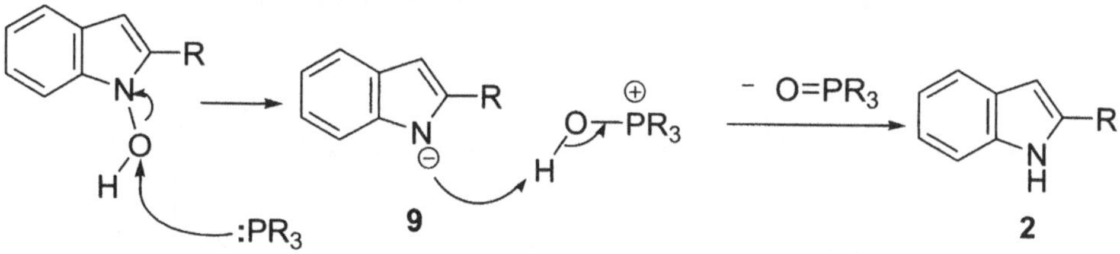

N-hydroxyindole, **8**

The Cadogan–Sundberg indole synthesis seemed to be domant for nearly half a century until the 1990s. It is now experiencing a renaissance and has been employed in medicinal chemistry, total synthesis of natural products, and polymers.

3.4.3 *Mechanism*

Although Cadogan's conviction of the nitrene intermediacy wavered somewhat in 1966,[5] he quickly reembraced the nitrene intermediacy after more experimental evidence emerged.[6] The formation of the phosphorus-imidate was one of the many impetuses that swayed his shift. This nitrene mechanism has now been generally accepted by the chemistry community.[10]

As shown below, the first step of the deoxygenation takes place when trialkyl phosphite attacks the oxygen atom of the nitro group. The adduct **10** then expels a molecule of trialkyl phosphate to give rise to the key intermediate nitroso compound **11**. Addition of the second molecule of trialkyl phosphite to nitroso **11** then gives adduct **12**, which expels another molecule of trialkyl phosphate to provide the pivatol intermediate nitrene **13**. Cyclization of nitrene **13** to indoline **14** was followed by re-aromatization to deliver indole **2**.

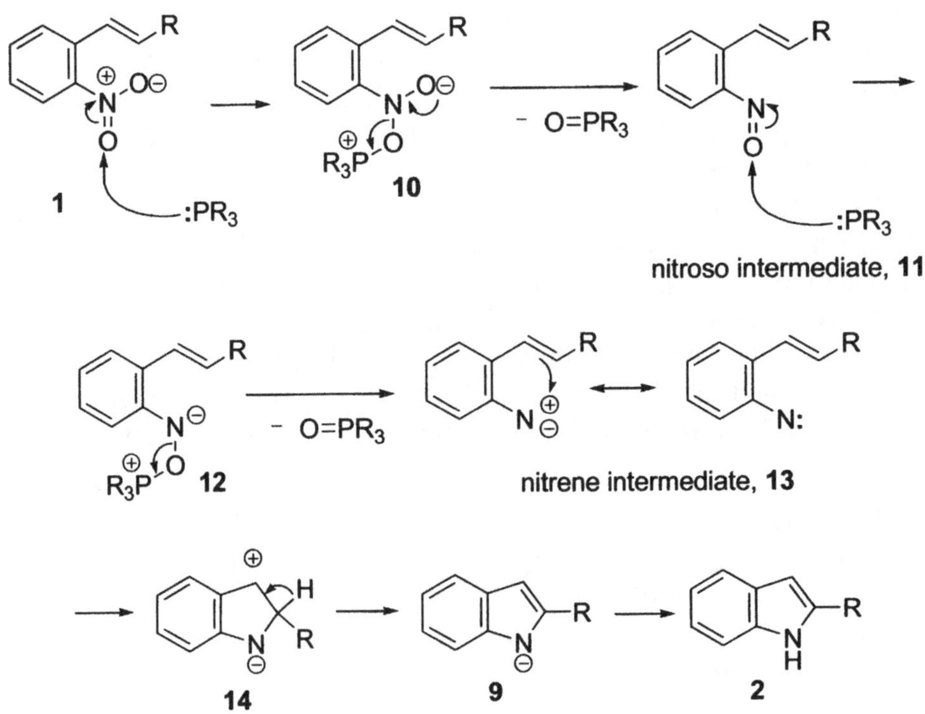

1 10 nitroso intermediate, 11

12 nitrene intermediate, 13

14 9 2

3.4.4 Variations and Improvements

Variations of Reducing Agents

Cadogan determined the order of reactivities of the phosphorus reagents in 1969:[6] $(EtO)_2PMe$ >> $(Et_2N)_3P$ ~ $EtOP(NEt_2)_2$ > $(EtO)_3P$ ~ $(i\text{-}PrO)_3P$ >>> PCl_3 (inactive).

Söderberg pioneered the combination of palladium and CO as the replacement of triethyl phosphite as the reducing agent for the Cadogan–Sundberg indole synthesis.[11–14] Carbon monoxide is a useful reducing agent. The construction of the indole skeleton *via* reductive deoxygenation of 2-nitro substituted arenes with carbon monoxide as a reducing agent has emerged as a versatile method for the synthesis of indoles. Various approaches have been reported for this transformation using transition metal compounds containing palladium,[11–14] iron,[15] ruthenium,[16] or rhodium[16] as the catalyst.

In one example of Söderberg's modification, 1,2-dihydro-4(3*H*)-carbazolone (**16**) was prepared by reductive cyclization of nitrostyrene **15** using $Pd(dba)_2$ and CO.[13] Carbazolone **15** is an important intermediate to ondansetron (**17**), a potent $5HT_3$ receptor agonist used for the prevention of nausea caused by chemotherapy and radiation treatment of cancer patients.

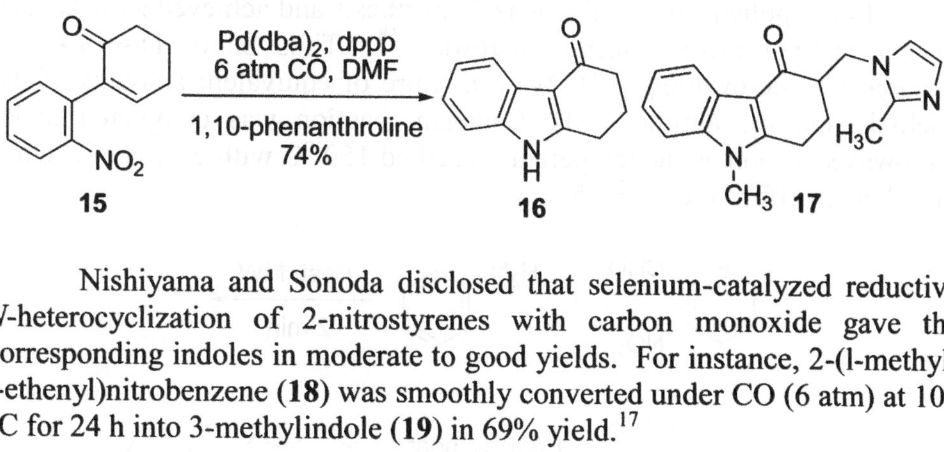

Nishiyama and Sonoda disclosed that selenium-catalyzed reductive *N*-heterocyclization of 2-nitrostyrenes with carbon monoxide gave the corresponding indoles in moderate to good yields. For instance, 2-(l-methyl-l-ethenyl)nitrobenzene (**18**) was smoothly converted under CO (6 atm) at 100 °C for 24 h into 3-methylindole (**19**) in 69% yield.[17]

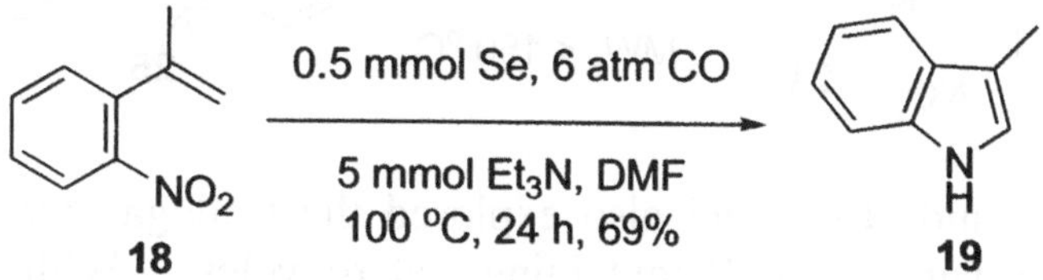

Variations of Substrates and Cylization Products

Cadogan himself further extended the scope of the Cadogan reaction. As early as in 1965, he already prepared carbazoles, indoles, indazoles, thiazoles and related compounds.[4] In 1966, he synthesized phenothiazine (**21**) and anthranil **23**, respectively, employing the reductive cyclization of nitro-compounds by triethyl phosphite.[5]

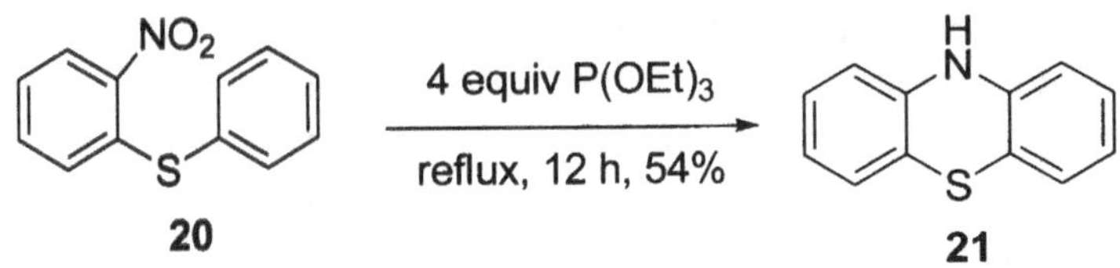

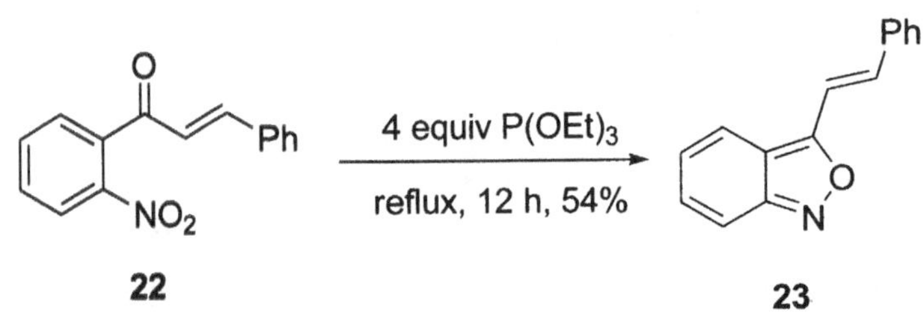

Bose applied Cadogan's indazole synthesis and achieved an enhanced greener one-pot process using microwave.[18] Thus Schiff base **24** was prepared by microwaving (400 W) a mixture of equivalent amounts of the aldehyde and the aniline. The Cadogan reaction was completed under microwave as soon as the temperature reached 150 °C with excess of triethyl phosphate to afford indazole **25**.

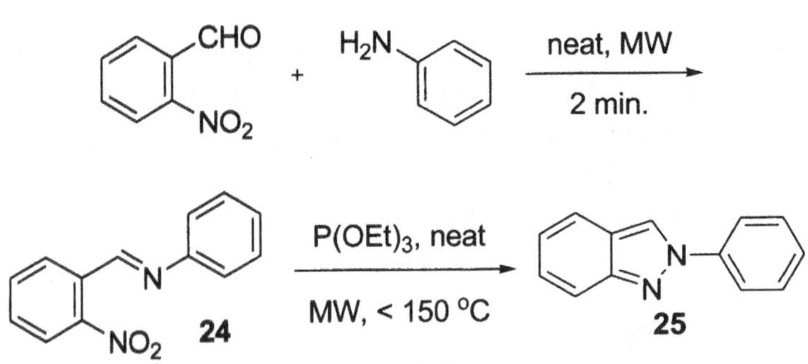

Creenia and Horaguchi also explored the Cadogan reaction for the synthesis of a variety of *N*-containing heterocycles. With the aide of microwave, a less reactive and less toxic triphenylphosphine could be used in place of triethyl phosphite.[19] For example, a one-pot, two-step synthesis of benzimidazole **26** was accomplished by microwaving (200 W) a mixture of equivalent amounts of the aldehyde and the aniline in the presence of 4 equiv of PPh_3. Meanwhile, 2-nitrodiphenylamine (**27**) was converted to phenazine (**28**) under similar conditions.

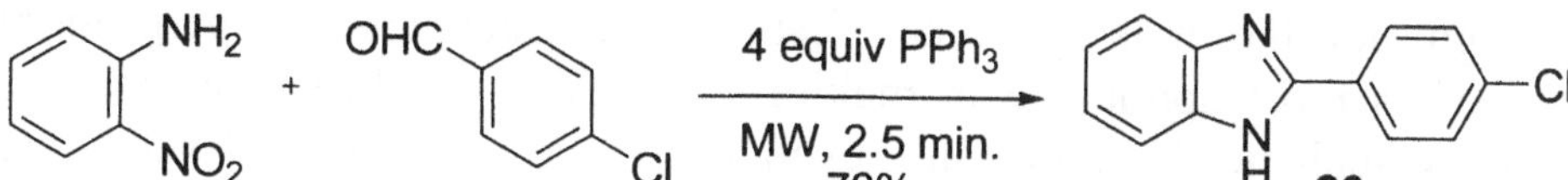

26

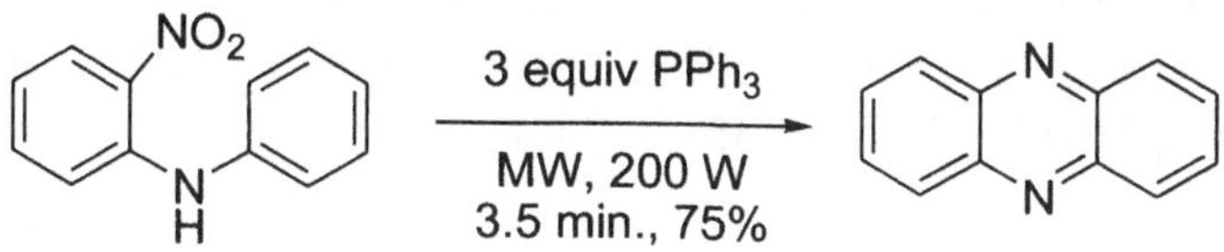

27 **28**

The Sundberg Indole Synthesis

Gribble used the Sundberg indole synthesis to make 2-nitroindole (**30**) by simply refluxing β-nitro-*o*-azidonitrostyrene (**29**) in xylene.[20] This was the first synthesis of 2-nitroindole (**30**).

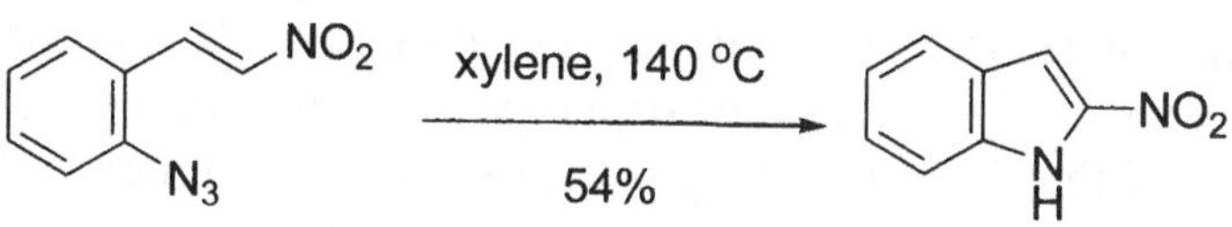

29 **30**

The Cadogan protocol did not work well for making anti-HIV natural product siamenol **32**, giving rise to a mixture of three isomers.[21] However, the Sundberg indole synthesis of azido-phenol **31** worked better, giving siamenol (**32**) and its regioisomer **33** in 78% yield and in almost 1 : 1 ratio.

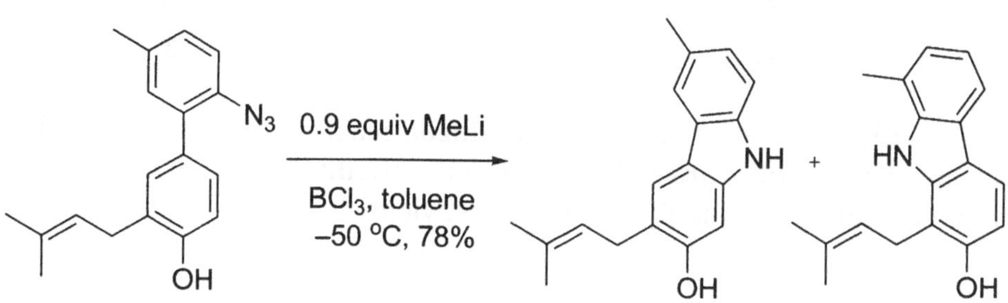

31 **32** (siamenol) **33**

3.4.5 Synthetic Utility

Carbazoles

Although not the most practical feature, the Cadogan–Sundberg indole synthesis is most applicable in making carbazoles. Dehaen carried out a microwave-enhanced Cadogan cyclization to prepare 2-substituted carbazoles and other fused heterocyclic systems.[22] Substrate **34**, easily assembled via a Suzuki coupling, was treated with triethyl phosphite under microwave conditions with a ceiling temperature of 210 °C to give 1,9-dihydropyrrolo[2,3-*b*]carbazole (**35**) in 73% yield.

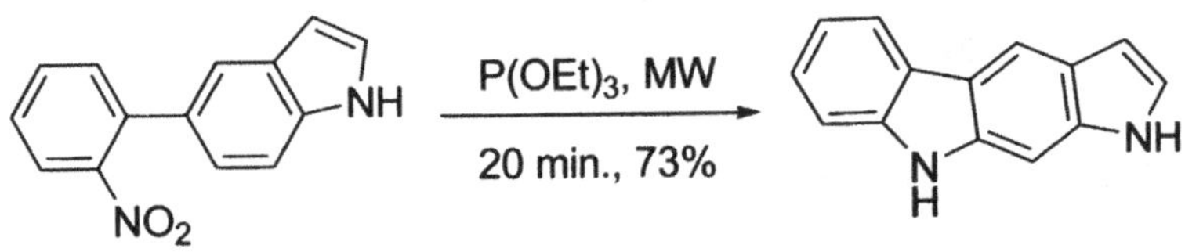

34 **35**

Freeman and co-workers also advocated the use of triphenylphosphine in place of toxic triethyl phosphite.[23] Cyclization of various 2-nitrobiphenyls was achieved via reductive deoxygenation of the nitro group using a slight excess of triphenylphosphine in a suitable solvent. They discovered that higher boiling solvents afforded higher yields across a range of substrates. For example, nitrobiphenyl **36** was converted to carbazole **37** in good yield when *o*-dichlorobenzene (*o*-DCB) was used as the solvent. The aldehyde functionality was preserved during the reaction.

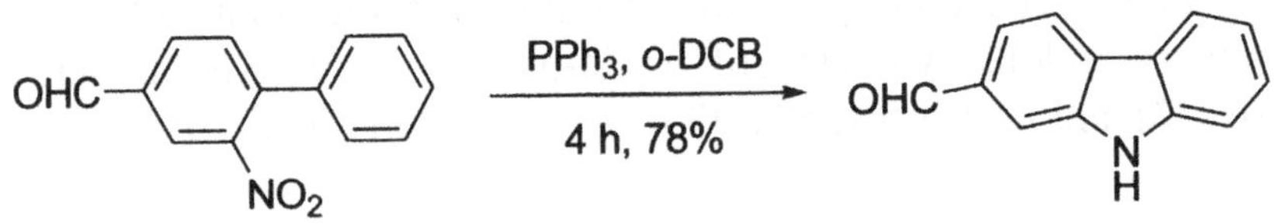

36 **37**

Merlic took advantage of the Cadogan reaction to make bisindoles.[24] Thus cyclization of *trans*-stilbene **38** with triethyl phosphite resulted in unsymmetrical 2,2′-bisindolyl **39** in modest yield.

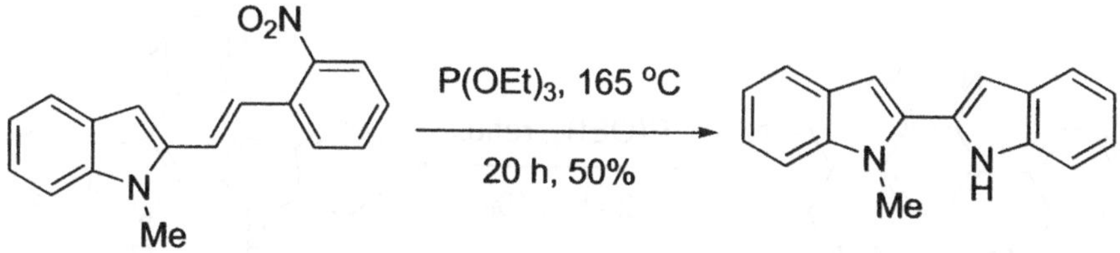

38 39

Oligomers and Polymers

Poly(2,7-carbazole)s find utility in electronic and optical devices such as polymeric light-emitting diodes (PLEDs). Leclerc engineered a synthesis of diiodolecarbazole as the monomer for polymerization using the Cadogan reaction.[25] The reductive cyclization of bis-nitro compound **40** was carried out in refluxing triethyl phosphite to give a mixture of two isomers: **41** and **42**.

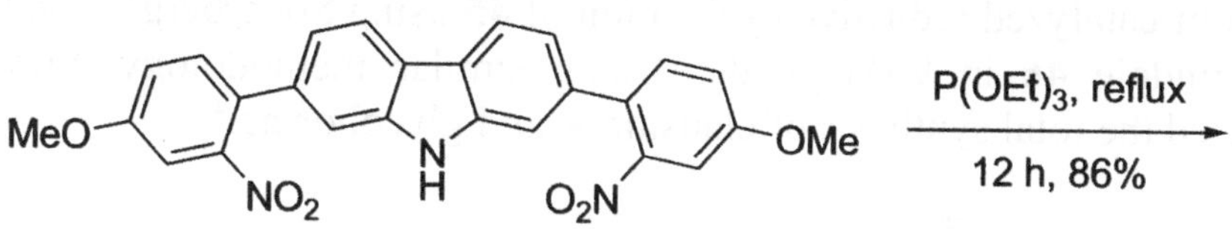

40

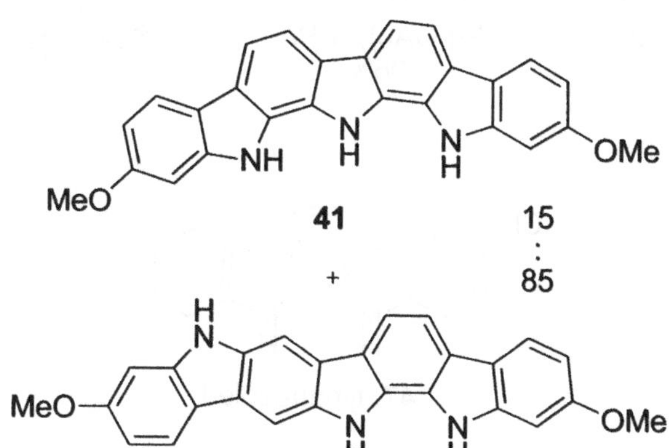

41 15 : 85

+

42

Applying Leclerc's protocol,[25] Valiyaveettil prepared thiadiazole-fused indolo[2,3-*a*]carbazole **44** from bis-nitro compound **43**.[26] Monomer **44**, in turn, was electropolymerized to yield a stable polymer.

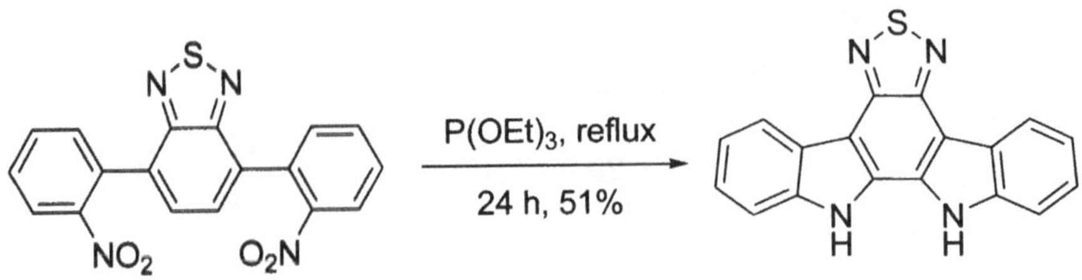

43 44

Natural Products and Medicinal Chemistry

The aglycon tjipanazoles and the glycosidic tjipanazoles are natural products. These aglycons have been shown to be potent and selective inhibitors of human cytomegalovirus. Kuethe *et al.* employed a Cadogan–Sundberg indole synthesis to achieve the total synthesis of tjipanazoles B, D, E, and I (**47**).[27] Reductive cyclization of *trans*-nitrostyrene **45** under the classic Cadogan–Sundberg conditions gave biindole **46** in 73% yield. Alternatively, palladium-catalyzed reductive cyclization of **45** using Söderberg's conditions gave biindole **46** in 96% yield. Using similar methodology, they also completed the total synthesis of clausine V and glycoborine.[28]

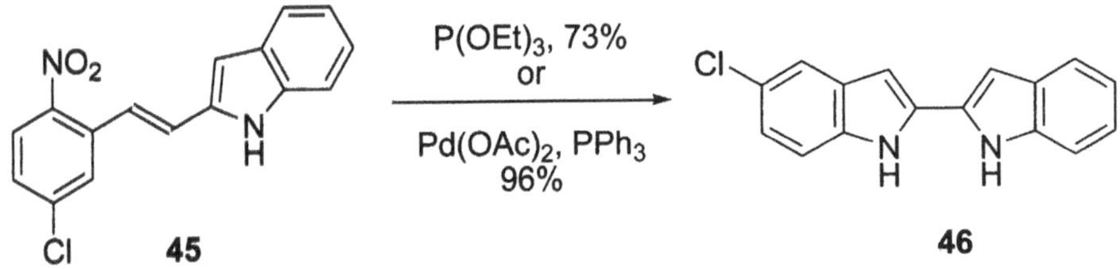

45 46

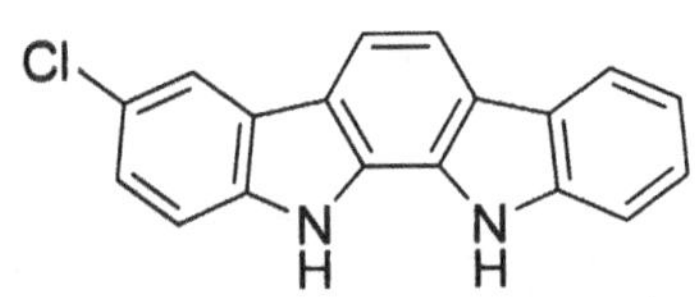

47, tjipanazole I

Kuethe and co-workers also employed the Cadogan–Sundberg indole synthesis to prepare a novel class of kinase domain receptor (KDR) inhibitors.[29] The KDR kinase is a tyrosine kinase that has a high affinity for vascular endothelial growth factor (VEGF) and is believed to be a primary mediator of tumor induced angiogenesis. Compounds that inhibit, modulate, or regulate the KDR receptor are useful for the prevention and treatment of tumor-induced angiogenesis. Reductive cyclization of nitrostyrene **48** with $P(OEt)_3$ afforded **49**. Reaction of **48** under the optimized conditions for reductive cyclization (0.1 mol % of $Pd(TFA)_2$, 0.7 mol% of 2,2,6,6-

tetramethyl-4-piperidone (TMP), 15 psi of CO, DMF, 80 °C) for a similar substrate led to low conversions. The deprotection of the masked quinolin-2-one moiety of chloroquinoline **49** was accomplished in a straightforward manner under acidic conditions. Therefore, hydrolysis of chloroquinoline **49** in a 1 : 1 mixture of AcOH/H_2O gave freebase **50** in 93% yield.

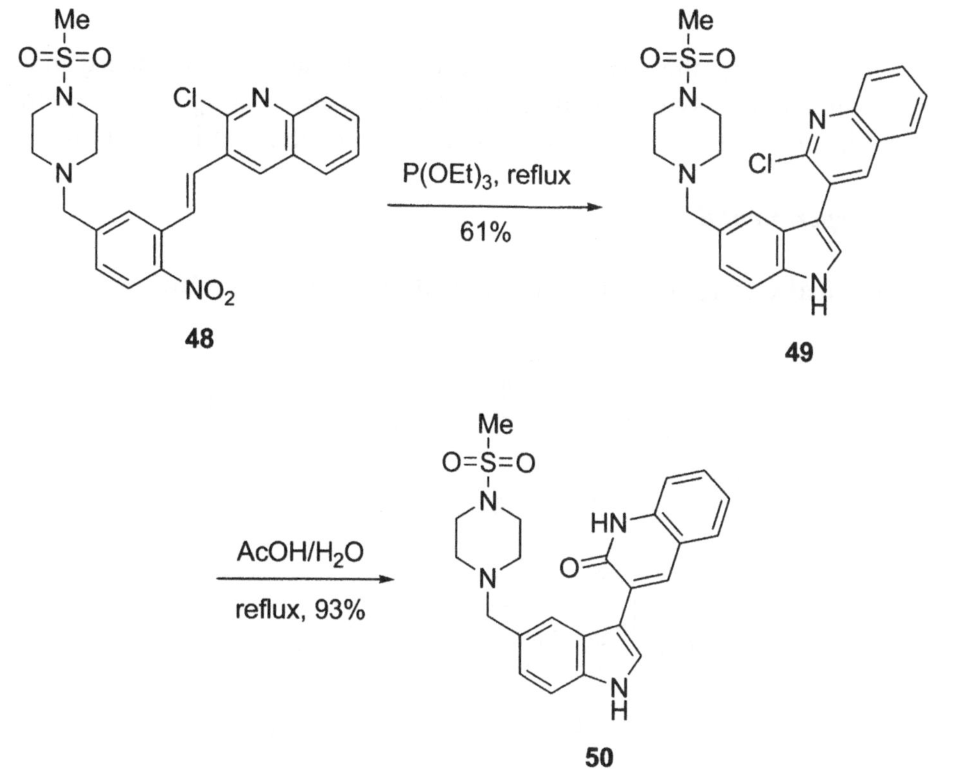

Li and colleagues took advantage of the Cadogan–Sundberg indole synthesis to prepare botulinum neurotoxin A light chain (BoNT/A LC) endopeptidase inhibitors.[30] Botulinum neurotoxins secreted by strains of the anaerobic spore-forming bacterial species *Clostridium botulinum* are the most potent neurotoxins known and are categorized as category A (highest priority) bioterrorist agents by the Centers for Disease Control and Prevention (CDC). In Li's synthesis, stilbene **51** was refluxed in neat triethyl phosphite for 1 h to deliver the key intermediate **52**, which was converted to botulinum neurotoxin A inhibitors in several additional steps.

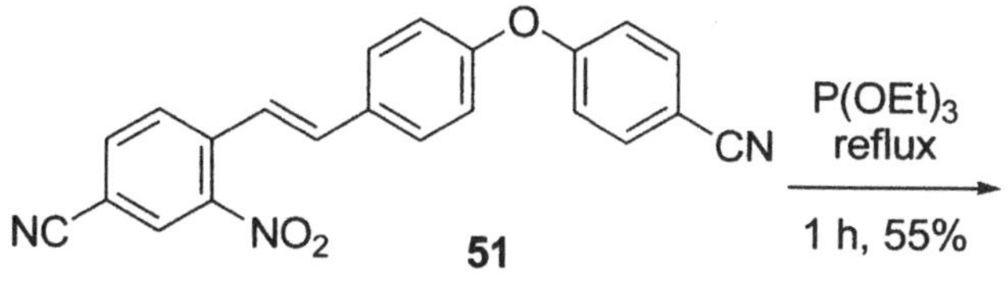

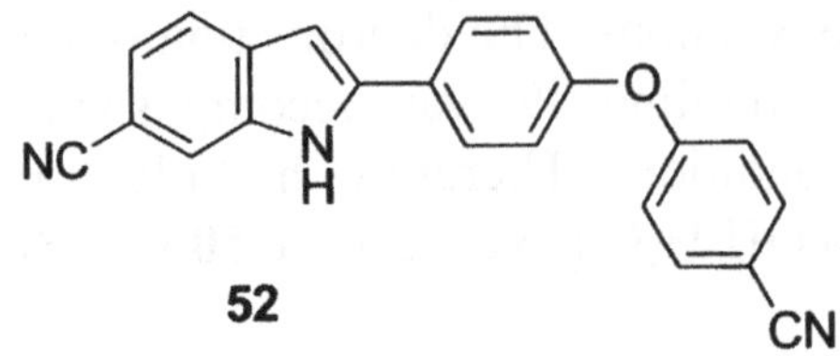

52

In summary, the exciting applications of Cadogan–Sundberg indole synthesis in the total synthesis of natural products and in drug discovery are a good indication that the reaction may receive more attention in the future.

3.4.6 Experimental

The Classical Cadogan–Sundberg Conditions to Prepare Indole 52[30]

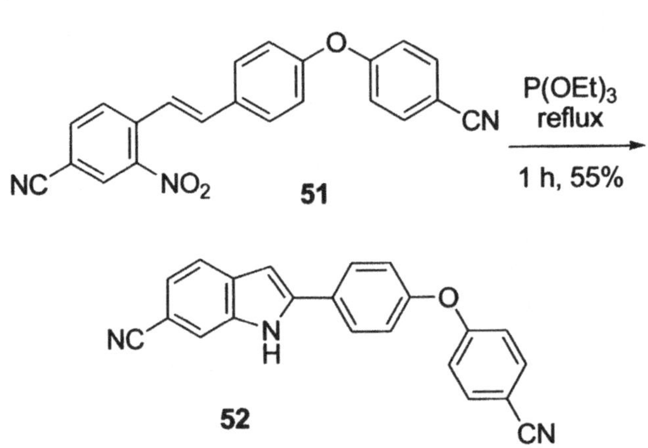

51 → 52

Compound **51** (5.0 g, 13.6 mmol) was suspended in triethyl phosphite (30 mL) and heated to reflux using a heating mantle for 1 h. The solution was cooled to room temperature, and excess triethyl phosphite was removed by distillation under vacuum. The residue was purified by flash chromatography using 25% ethyl acetate in hexane to yield a light yellow solid **52** (2.5 g, 55% yield).

Söderberg's Modified Conditions to Prepare Indole 55[13]

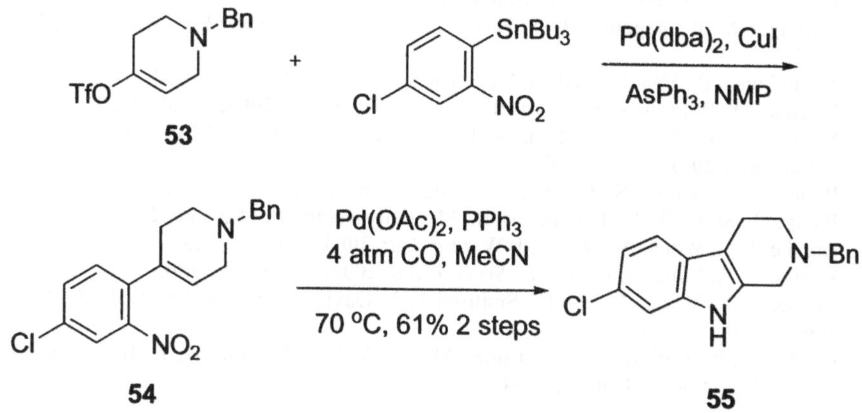

To an oven-dried, threaded ACE glass pressure tube was added **54** (469 mg, 1.43 mmol), $Pd(OAc)_2$ (19.2 mg, 0.085 mmol), PPh_3 (90 mg, 0.34 mmol), and MeCN (5 mL). The tube was fitted with a pressure head, and the solution was saturated with CO (four cycles of 4 atm of CO). The reaction mixture was heated at 70 °C (oil-bath temperature) under CO (4 atm) until all starting material was consumed (15 h), as judged by TLC. The solvent was evaporated under reduced pressure, and the crude product was purified by chromatography (hexanes/EtOAc, 8 : 2) to give **55** (380 mg, 1.28 mmol) as a white solid. The yield in two steps starting from **53** was 61%.

3.4.7 References

1 [R] Gribble, G. W. *Perkin 1* **2000,** 1045–1075.
2 Sundberg, R. J.; Russell, H. F.; Ligon, W. V.; Lin, L.-S. *J. Org. Chem.* **1972,** *37,* 719–724.
3 Cadogan, J. I. G.; Cameron-Wood, M. *Proc. Chem. Soc.* **1962,** 361.
4 Cadogan, J. I. G.; Cameron-Wood, M.; Mackie, R. K.; Searle, R. J. G. *J. Chem. Soc.* **1965,** 4831–4837.
5 Cadogan, J. I. G.; Mackie, R. K.; Todd, M. J. *Chem. Commun.* **1966,** 491a.
6 Cadogan, J. I. G.; Todd, M. J. *J. Chem. Soc. C* **1969,** 2808–2813.
7 Sundberg, R. J. *J. Org. Chem.* **1965,** *30,* 3604–3610.
8 Sundberg, R. J. *J. Am. Chem. Soc.* **1966,** *88,* 3781–3789.
9 Sundberg, R. J.; Yamazaki, T. *J. Org. Chem.* **1967,** *32,* 290–294.
10 Tsunashima, Y.; Kuroki, M. *J. Heterocyclic Chem.* **1981,** *18,* 315–318.
11 Söderberg, B. C.; Shriver, J. A. *J. Org. Chem.* **1997,** *62,* 5838–5845.
12 Söderberg, B. C.; Rector, S. R.; O'Neil, S. N. *Tetrahedron Lett.* **1999,** *40,* 3657–3660.
13 Scott, T. L.; Söderberg, B. C. G. *Tetrahedron Lett.* **2002,** *43,* 1621–1624.
14 Dantale, S. W.; Söderberg, B. C. G. *Tetrahedron* **2003,** *59,* 5507–5514.
15 Crotti, C.; Canini, S.; Rindone, B.; Tollari, S.; Demartin, E. *J. Chem. Soc., Chem. Commun.* **1986,** 784–786.
16 Ragaini, F.; Sportiello, P.; Cenini, S. *J. Organomet. Chem.* **1999,** *577,* 283–291.
17 Nishiyama, Y.; Maema, R.; Ohno, K.; Hirose, M.; Sonoda, N. *Tetrahedron Lett.* **1999,** *40,* 5717–5720.
18 Varughese, D. J.; Manhas, M. S.; Bose, A. K. *Tetrahedron Lett.* **2006,** *47,* 6795–6797.

19 Creenia, E. C.; Kosaka, M.; Muramatsu, T.; Kobayashi, M.; Iizuka, T.; Horaguchi, T. *J. Heterocyclic Chem.* **2009,** *46,* 1309–1317.

20 Pelkey, E. T.; Gribble, G. W. *Tetrahedron Lett.* **1997,** *38,* 5603–5606.

21 Naffziger, M. R.; Asgburn, B. O.; Perkins, J. R.; Carter, R. G. *J. Org. Chem.* **2007,** *72,* 9857–9865.

22 Appukkutan, P.; Van der Eycken, E.; Dehaen, W. *Synlett* **2005,** 127–133.

23 Freeman, A. W.; Urvoy, M.; Criswell, M. E. *J. Org. Chem.* **2005,** *70,* 5014–5019.

24 Merlic, C. A.; You, Y.; McInnes, D. M.; Zechman, A. L.; Miller, M. M.; Deng, Q. *Tetrahedron* **2001,** *57,* 5199–5212.

25 Bouchard, J.; Wakim, S.; Leclerc, M. *J. Org. Chem.* **2004,** *69,* 5705–5711.

26 Balaji, G.; Shim, W. L.; Parameswaran, M.; Valiyaveettil, S. *Org. Lett.* **2009,** *11,* 4450–4453.

27 Kuethe, J. T.; Wong, A.; Davies, I. W. *Org. Lett.* **2003,** *5,* 3721–3723.

28 Kuethe, J. T.; Childers, K. G. *Adv. Synth. Catal.* **2008,** *350,* 1577–1586.

29 Kuethe, J. T.; Wong, A.; Qu, C.; Smitrovich, J.; Davies, I. W.; Hughes, D. L. *J. Org. Chem.* **2005,** *70,* 2555–2567.

30 Li, B.; Pai, R.; Cardinale, S. C.; Butler, M. M.; Peet, N. P.; Moir, D. T.; Bavari, S.; Bowlin, T. L. *J. Med. Chem.* **2010,** *53,* 2264–2276.

3.5 Fukuyama Indole Synthesis

Ji Zhang

3.5.1 Description

The Fukuyama indole synthesis[1,2] is a novel tin-mediated chemical transformation of *o*-isocyanostyrene derivatives **1**. Conversion of the α-stannoimidoyl radical **2** results in the formation of 3-substituted indoles **3a** or 2,3-disubstituted indoles **3b,c**. Alternatively, 2,3-disubstituted indoles were formed from 2-alkenylthioanilides **4** *via* imidoyl radical species **5** and followed by radical cyclization to form indole **6**.

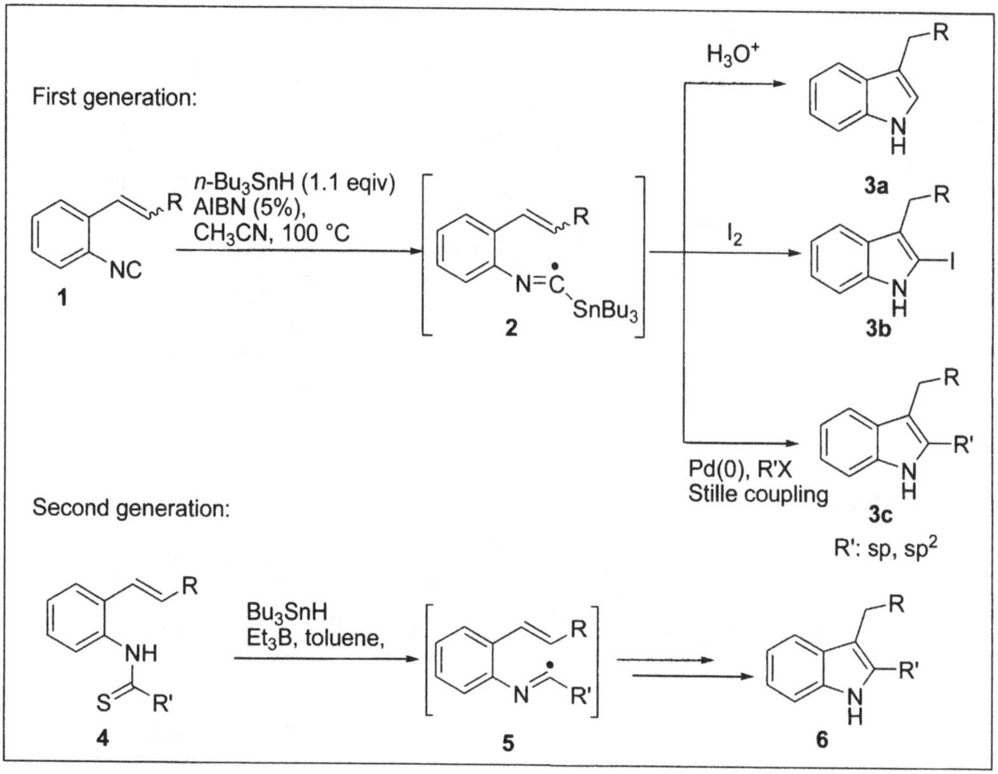

The *in situ* formation of stannylindole **7** from α-stannoimidoyl radical **2** can be used under the Stille conditions to afford a variety of 2,3-disubstituted indoles **8**[1] and was used in the synthesis of indolocarbazoles.[3] 2-Alkenylthioanilides **4a** and **4b** can be prepared by one of three general methods from quinolines **9** (Method A or B) or 2-iodoaniline **12** (Method C).[2]

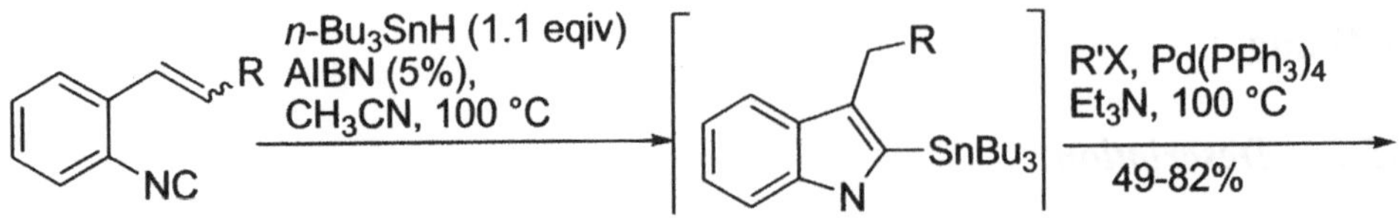

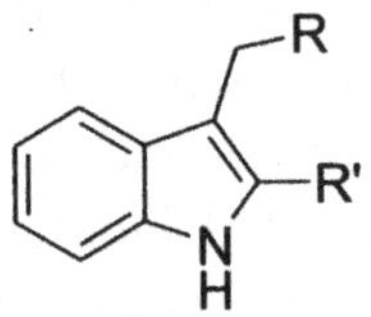

8

R= CO_2Me, CH_2OTHP, n-Bu
R' = Ph, 4-Ac-Ph, Bn, 1-cyclohexenyl, 1-hexenyl
X= Br, I, OTf

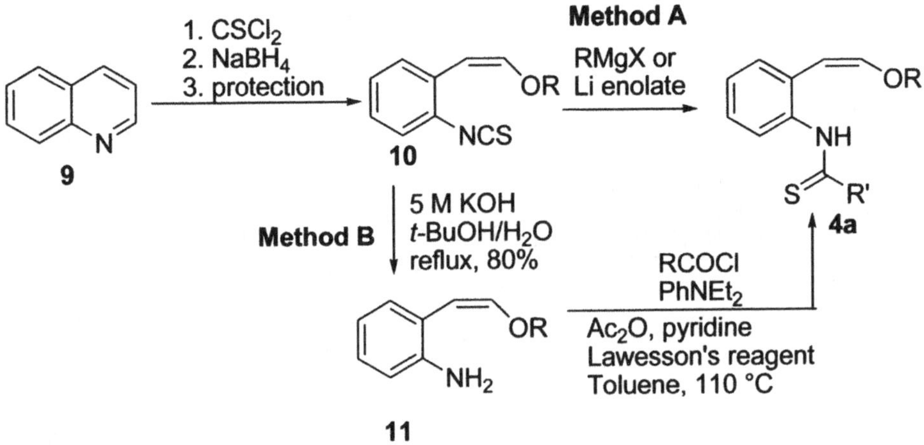

Method C

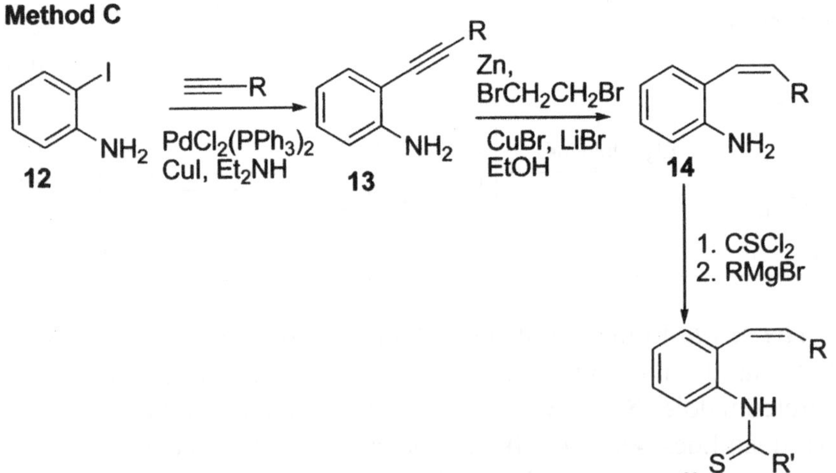

3.5.2 *Historical Perspective*

In 1994, Fukuyama disclosed the first-generation transformation,[1] followed by the further modification in 1999 using 2-alkenylthioanilide **4** for the radical cyclization as the second-generation method.[2] The later process (**4** to **6**) is complementary to the previously reported protocol (**1** to **3**) in which introduction of sp[3]-hybridized substituent at the 2-position was not easy. Although the attachment of sp[2]-hybridized substituents, such as phenyl groups to the 2-position could be accomplished using palladium-catalyzed methods like the Stille cross-coupling reaction. Furthermore, this procedure provided a mild radical cyclization reaction, at room temperature by using Et_3B to replace the original initiator, AIBN Today, as the one of the simplest methods for synthesizing substituted indoles, the Fukuyama indole reaction has been used in numerous natural product syntheses, such as (+)-vinblastine[4] and (–)-strychnine.[5]

3.5.3 *Mechanism*

A proposal for the mechanism of the Fukuyama indole synthesis is proposed[1,2] as shown below. Treatment of the isonitriles **1** with tributyltin hydride and a catalytic amount of AIBN, affords α-stannoimidoyl radical **2**, followed by cyclization, to give radical **16**. It was found that the substrates bearing radical-stabilizing groups at the β-position gave indoles **3** in excellent yield after tautomerization, and acidic workup. Similarly, when thioamide derivatives such as 2-alkenylthioanilide **18** are subjected to radical-initiating conditions, radical **5** or imidoyl radical species are formed, which then undergo radical cyclization to furnish 2,3-disubstiuted indoles **6**.

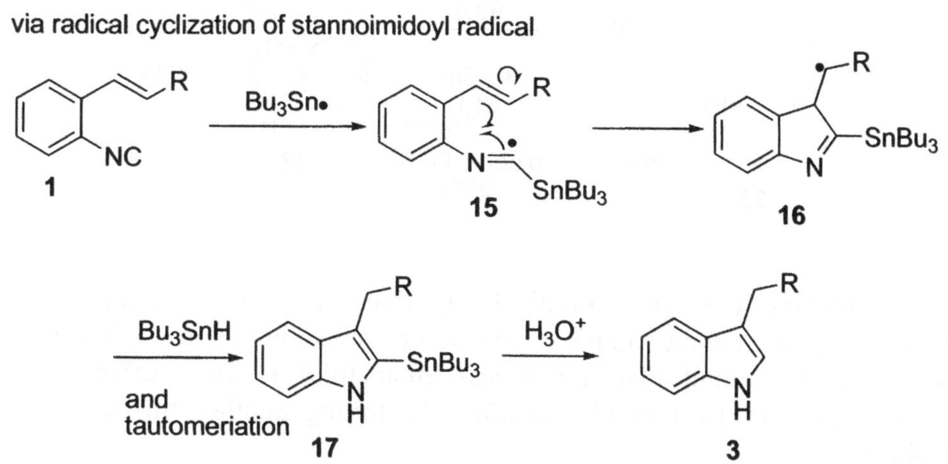

via imidoyl radical species

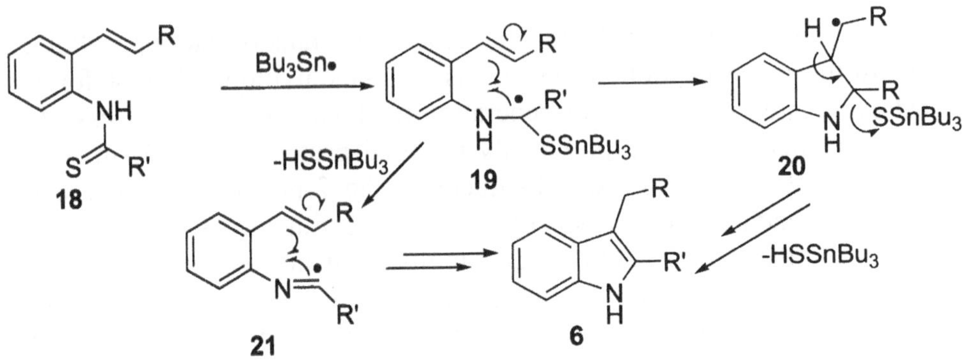

3.5.4 *Variations and Improvements*

Et_3B/O_2 replacement of AIBN as the radical initiator allowed the reaction to be carried out at room temperature, instead of at 80 °C, allowing milder conditions to complete the transformation. Fukuyama also found that hypophosphorous acid could be employed as an alternative radical reducing agent to tin hydride, thus, developing a metal-free process. For example, after heating a mixture of the substrate, hypophosphorous acid, triethylamine and AIBN in *n*-PrOH at 100 °C for 20 min, 2-cyclohexylindole was prepared in 71% yield.[2]

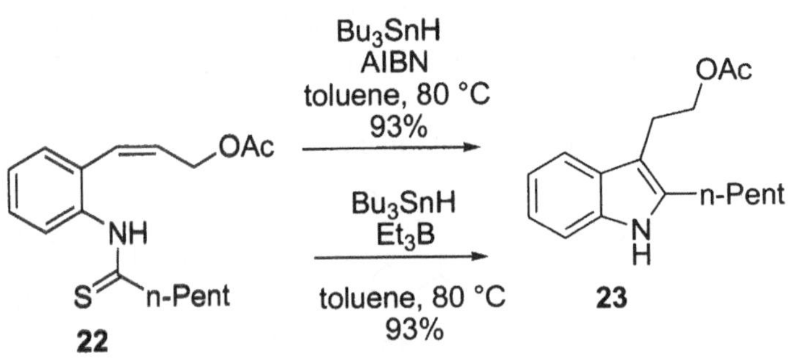

Instead of the standard radical conditions using tri-*n*-butyltin hydride, Fukuyama examined the radical cyclization by screening using various thiols and eventually found that the excess ethanethiol is quite effective for promoting the cyclization (**24** to **25**),[6] illustrating another tin-free indole synthesis.

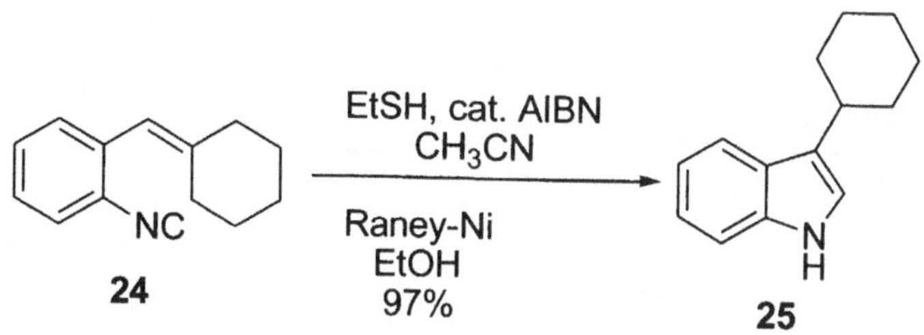

24 25

3.5.5 *Synthetic Utility*

The Fukuyama indole synthesis was used during the total synthesis of aspidophytine **28**. Tin-mediated indole formation was followed by the treatment of the 2-stannyl indole intermediate with iodine and gave the 2-iodoindole derivatives **27** in 85% yield.[7]

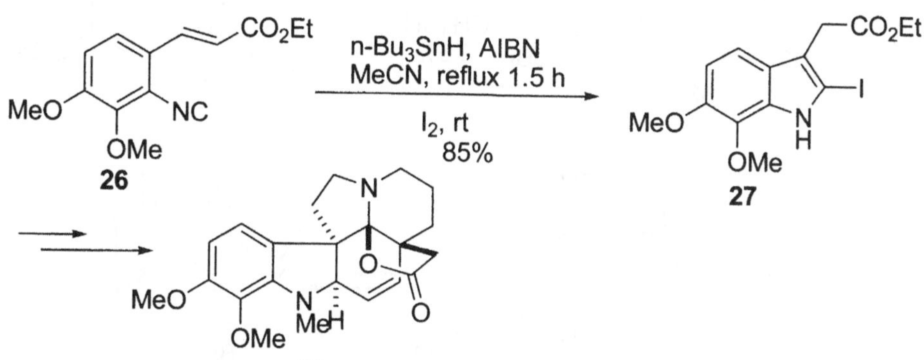

26 27 28

Aspidophytine

Fukuyama also employed this radical cyclization for the total synthesis of gelsemine **32**, where the spiroindolinone was assembled in three steps from bromo-aniline.[8]

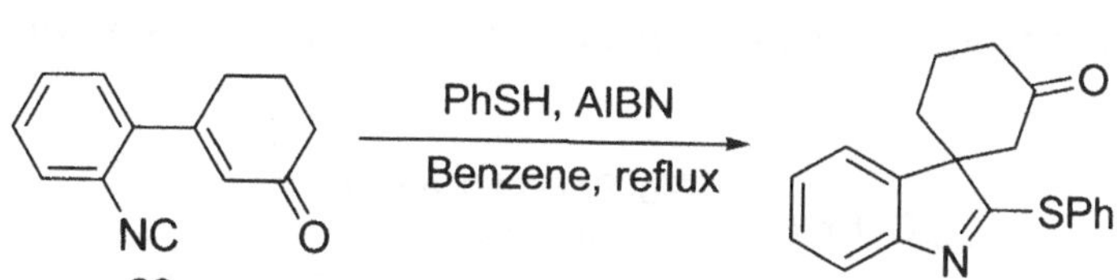

29 30

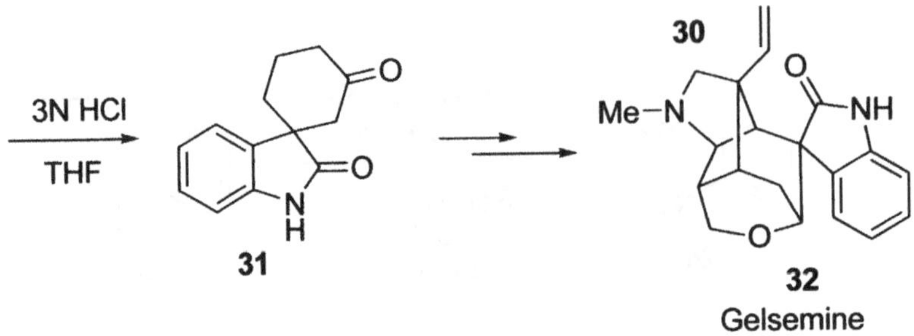

In the total synthesis of discorhabdin A, the key intermediate **34** was prepared in three steps *via* the tin-mediated radical cyclization in 76% yield by Fukuyama and co-workers.[9]

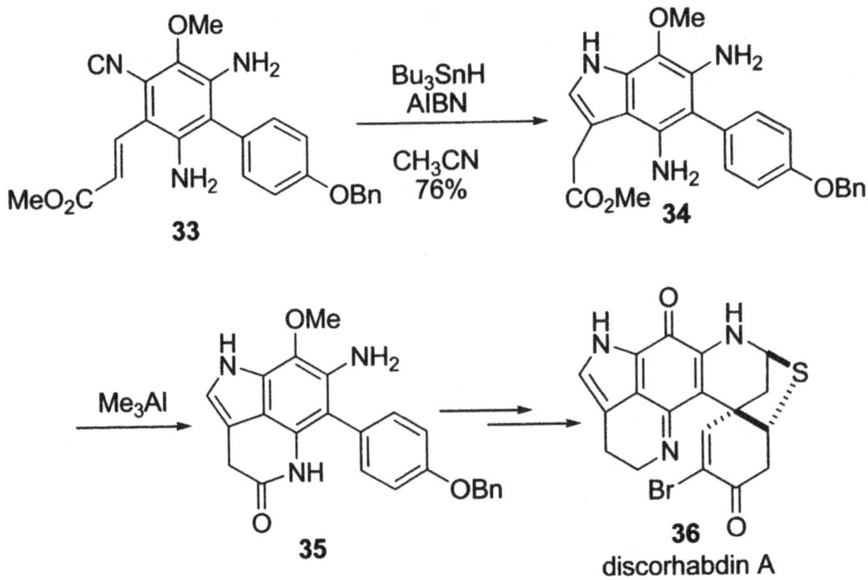

Recently, other groups have used this facile and mild tin-mediated Fukuyama indole synthesis. For example, a new route to the general core of the paullones **40**, potential cyclin-dependent kinase (CDK) inhibitors that are particularly efficient against three disease-relevant kinases was reported.[10]

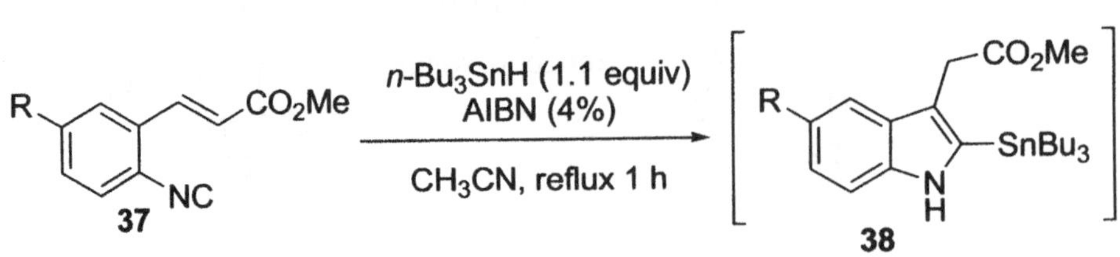

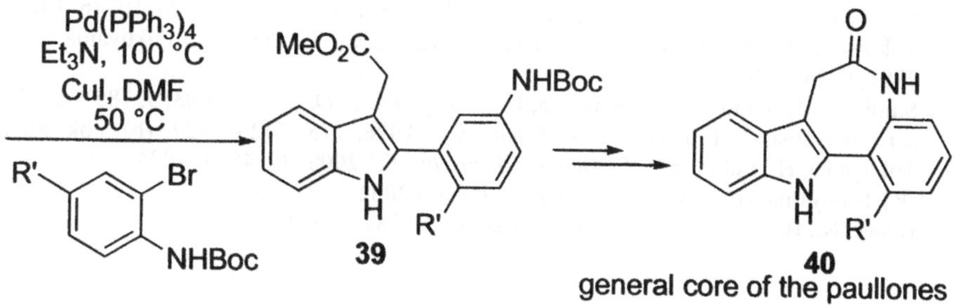

3.5.6 *Experimental*

General Procedure (first generation, combined with Stille cross-coupling)[1]

A solution of the isonitrile (0.85 mmol), *n*-Bu_3SnH (0.93 mmol), and AIBN (0.04 mol) in dry acetonitrile (5 mL) was heated to 100 °C for 1 h under argon atmosphere. The reaction mixture was cooled to room temperature, and triethylamine (1.02 mol), $Pd(PPh_3)_4$ and bromobenzene (1.02 mmol) were added. The mixture was heated for additional 5 h under argon atmosphere. The reaction mixture was portioned between hexane and CH_3CN. Ether was added to the combined hexane layer and the organic layer was washed with a 1:1 mixture of 3 N HCl and brine. The extracts were washed with brine, dried over Na_2SO_4, and evaporated to dryness. The crude product was purified by flash silica gel chromatography to give desired indole.

General Procedure (second generation)[2]

To a stirred solution of 2-alkenylthioanilide (0.025 M in toluene) and *n*-Bu_3SnH (2.0 equiv) was added Et_3B (0.10 equiv, 1.0 M in hexane) at room temperature under an argon atmosphere. After TLC analysis showed that the starting material had been consumed, the reaction mixture was diluted with EtOAc, washed with saturated aqueous KF and brine, and dried over $MgSO_4$. Filtration and concentration afford the crude product. The desired indole was purified by flash chromatography on silica gel.

3.5.7 *References*

1. Fukuyama, T.; Chen, X.; Peng, G. *J. Am. Chem. Soc.* **1994,** *116*, 3127–3128.
2. Tokuyama, H.; Yamashita, T.; Reding, M. T.; Kaburagi, Y.; Fukuyama, T. *J. Am. Chem. Soc.* **1999,** *121*, 3791–3792.
4. (a) Kobayashi, Y.; Fukuyama, T. *J. Heterocyclic Chem.* **1998,** *35*, 1043–1055. (b) Kobayashi, Y.; Peng, G. *Tetrahedron Lett.* **1999,** *40*, 1519–1522. [R] Knolker, H.-J.; Reddy, K. R. *Chem. Rev.* **2002,** *102*, 4303–4427.
4. (a) Yokoshima, S.; Ueda, T.; Kobayashi, S.; Sato, A.; Kuboyama, T.; Tokuyama, H.; Fukuyama, T. *J. Am. Chem. Soc.* **2002,** *124*, 2137–2139. (b) Yokoshima, S.; Ueda, T.; Kobayashi, S.; Sato, A.; Kuboyama, T.; Tokuyama, H.; Fukuyama, T. *Pure Appl. Chem.* **2003,** *75*, 29–38.

5. Kaburagi, Y.; Tokuyama, H.; Fukuyama, T. *J. Am. Chem. Soc.* **2004,** *126*, 10246–10247.
6. Tokuyama, H.; Watanabe, M.; Hayashi, Y.; Kurokawa, T.; Peng, G.; Fukuyama, T. *Synlett* **2001,** *9*, 1403–1406.
7. Sumi, S.; Matsumoto, K.; Tokuyama, H.; Fukuyama, T. *Org. Lett.* **2003,** *5*, 1891–1893.
8. (a) Fukuyama, T. Liu, G. *J. Am. Chem. Soc.* **1996,** *118*, 7426–7427. (b) Yokoshima, S.; Tokuyama, H.; Fukuyama, T. *Angew. Chem., Int. Ed.* **2000,** *39*, 4073–4075.
15. [R] Tokuyama, H.; Fukuyama, T. *Chem. Record* **2002,** *2*, 37–45.
16. Henry, N.; Běněteau, J. B. V. *Synthesis* **2006,** *22*, 3895–3901.

3.6 Gassman Oxindole Synthesis

Daniel P. Christen

3.6.1 Description

In the early 1970s, Professor Gassman and his group described a general method for the alkylation of anilines.[1] This method, which proceeded through intermediary azasulfonium salts that underwent a Sommelet–Hauser-type rearrangement, was subsequently modified to produce indole and oxindole derivatives.[2–4] The Gassman oxindole synthesis (**1**→**3**), an important variation of the Gassman indole synthesis (**4**→**5**), was originally reported in 1973. The reaction, a one-pot, multistep process, converts an aniline derivative into the corresponding oxindole by sequential treatment of an aniline (**1**) with *tert*-butyl hypochlorite, ethyl (methylthio)acetate, triethylamine and finally hydrochloric acid. Subsequent reduction of the resulting thiol intermediate **2** leads to the final oxindole derivative **3**.

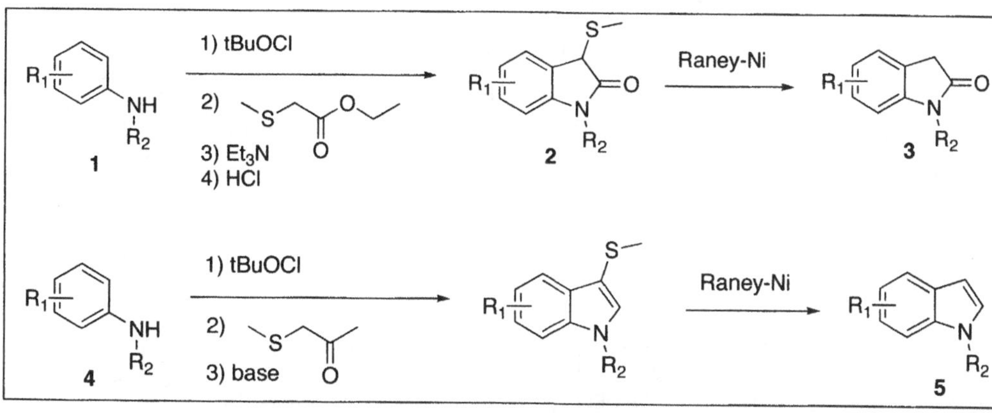

Generally, the reaction works well for electron-deficient anilines. A modified approach using *in situ* prepared chlorosulfonium salts is preferred for more electron-rich anilines.[5] Substituted thiol derivatives can be used to produce 3-oxindoles, which can be further reduced to 3-indole derivatives that are not available via the Gassman indole synthesis. The reaction is equally useful for the preparation of isatin derivatives (*vide infra*).[6] Historically, the Gassman oxindole synthesis has found applications in the preparation of heterocycles with medical or insecticidal properties. The reaction also has potential applications for the synthesis of oxindole containing natural products.[7]

3.6.2 Mechanism

The proposed mechanisms for the two main variations of the Gassman oxindole synthesis are shown below. The *N*-chloroaniline starting material **6** is prepared from an aniline and *tert*-butyl hypochlorite.

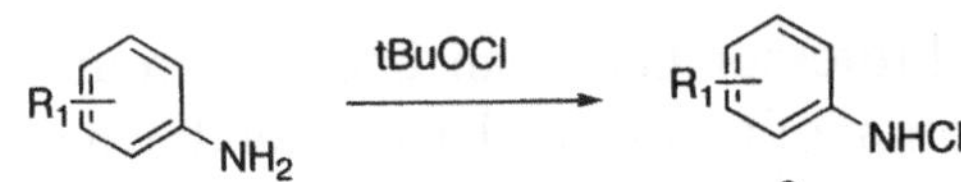

6

Further reaction of *N*-chloroaniline **6** with ethyl (methyl)thioacetate yields azasulfonium ion **7**, which forms ylide **8** upon deprotonation with triethylamine. After the ylide undergoes a [2,3]-Sommelet–Hauser-like rearrangement, a proton transfer and subsequent rearomatization produces the intermediate aniline derivative **9**, which upon cyclization forms a transient tetrahedral anion. Collapse of the anion with loss of ethoxide produces the stable oxindole derivative **2**. Removal of the methylthiol functionality with Raney nickel produces the final product **3**.

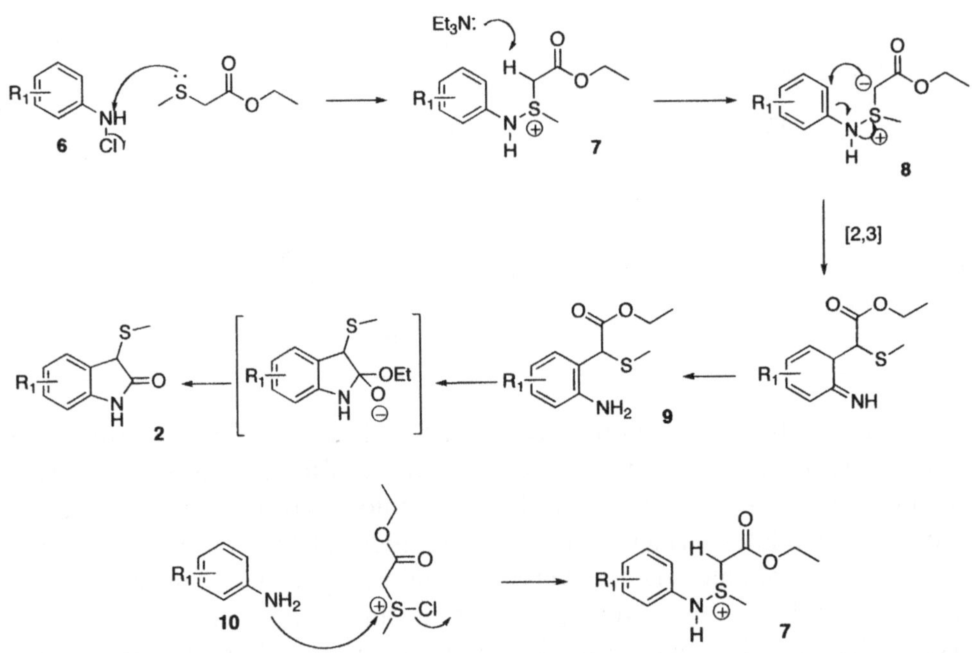

In the second variation of the Gassman oxindole synthesis, the nucleophile–electrophile pairing is reversed and the aniline **10** attacks a positively charged chlorosulfonium ion to produce sulfonium ion **7**. After that, the reaction proceeds along the same mechanistic route as the one proposed for the original Gassman oxindole synthesis.

3.6.3 *Variations and Improvements*

The Gassman oxindole synthesis has been modified by treacting the aniline with chloro sulfonium ions instead of preparing *N*-chloroaniline derivatives with *tert*-butyl hypochlorite as in the original Gassman approach. One common, but experimentally more difficult approach, uses chlorine to prepare highly reactive chlorosulfonium salts that are then added to the desired aniline.[8] This procedure, also developed in Professor Gassman's group, has been surpassed by newer methods that form the intermediate sulfonium ions by reacting ethyl (methylthio)acetate with sulfuryl chloride.[9]

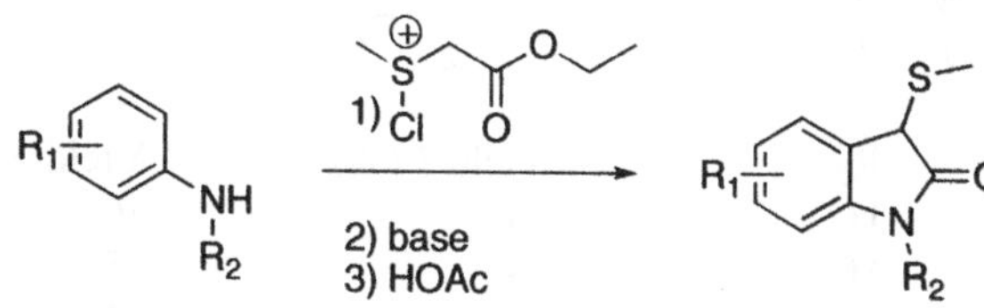

More recently, Wright *et al.* have shown that sulfoxides (**11**) can be converted into the desired chlorosulfonium intermediates by using oxalyl chloride.[10] This operationally more simple approach can be used make a variety of oxindoles. Additional examples using these approaches for the *in situ* preparation of the chlorosulfonium salts will be shown below.

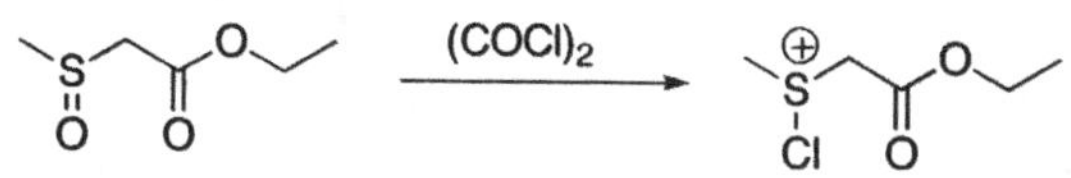

11

3.6.4 *Synthetic Utility*

The original Gassman oxindole synthesis (**1**→**3**) produces **2** in yields ranging from 30 to 80 %.[11] Selected examples from the original Gassman paper are shown below. The reaction tolerates mildly electron-donating substituents (*e.g.*, CH_3) and strongly electron-withdrawing substituents (*e.g.*, NO_2). As expected, regioselectivity for *ortho*- and *para*-substituted aniline precursors is not an issue. For *meta*-substituted anilines, however, two regioisomers can form during the reaction; experimentally, the ratio of the two possible isomers is highly dependent on the nature of the aniline starting material.

From an experimental perspective, the reaction proceeds with the addition of a stoichiometric amount of *tert*-butyl hypochlorite (dissolved in CH_2Cl_2) to a stirred solution of the aniline in dichloromethane at –65 °C. After a few minutes, a solution of ethyl methylthioacetate (stoichiometric amount, dissolved in CH_2Cl_2) is added. The sulfonium salt usually does not precipitate out of solution. After an hour at –65 °C, a stoichiometric amount of triethylamine is added. The solution is allowed to warm to room temperature and the reaction mixture is then worked up. After dissolving the organic concentrate in diethyl ether, 2 N HCl is added and the reaction is stirred overnight at room temperature. During this time, the oxindole product **2** precipitates out of solution. The calculated yields are based on the isolated precipitate. Additional product can be obtained by concentration of the ether layer. No additional purification was carried out.

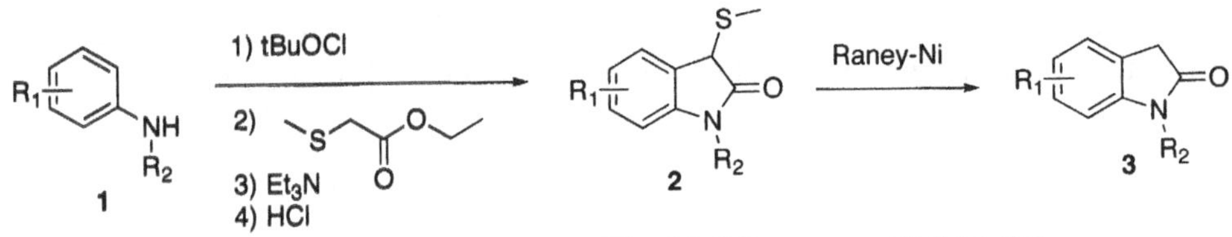

R_1	R_2	Yield **2** (%)	Yield **3** (%)
H	H	84	76
p-CH_3	H	34	55
o-CH_3	H	67	72
H	CH_3	46	77
p-$CO_2C_2H_5$, *o*-CH_3	H	66	67
p-NO_2	H	51	ND (not determined)

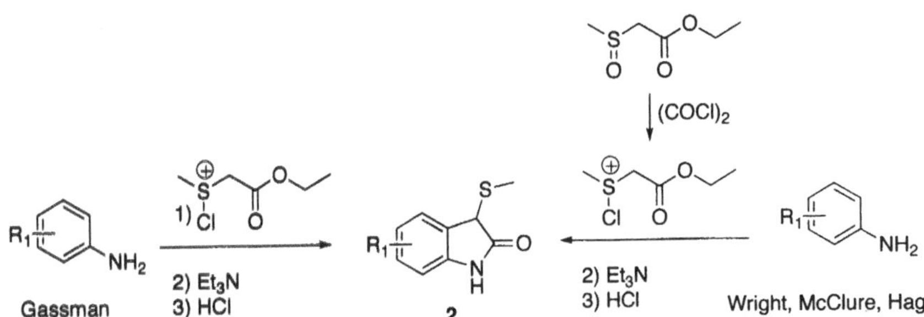

Gassman varaition

Wright, McClure, Hageman variation

R_1	Yield **2** (%) Gassman	Yield **2** (%) Wright
H	65	63
p-OCH_3	53	50
o-CH_3	62	82
p-NO_2	12	8
3,4,5-$(CH_3O)_3$	ND (not determined)	44
3-CH_3S	ND	50 (1:1 mixture isomers)

While the above work was being carried out in the Gasssman group, an alternative procedure to make oxindoles was developed as well.[12] In this approach, ethyl methylthioacetate is first treated with chlorine gas to make an intermediate chlorosulfonium ion that is then treacted with the aniline to form the oxindoles. A few examples from Gassman's lab are shown. Data from Wright, McClure, and Hageman are shown for comparison purposes.[13] Wright's group converted the sulfoxide analogs **11** of ethyl methylthioacetate into the corresponding chlorosulfonium salts by using oxalyl chloride. For smaller scale reactions, this approach is operationally easier than trying to prepare chlorine gas solutions. Generally, the results for both approaches are comparable.

The experimental yields from these reactions are consistent with those obtained from the original Gassman approach. The reaction of the *p*-nitro-aniline gave a much poorer yield.

As mentioned above, the outcome of the Gassman oxindole reaction of 3-substituted anilines is highly dependent on the starting aniline. Reaction of 3-nitroanline **12** using the original Gassman protocol gave exclusively isomer **13** (67 % yield). Gassman *et al.* claim that switching to an electron-donating group leads to the isolation of the other regioisomer.[14] In more complex cases, however, mixtures of the possible isomers are obtained. Thus preparation of a series of tetrahydrophthalimide substituted indoline-2(3*H*)-ones by Karp and Condon led to the production of mixtures containing both possible isomers.[15] Reaction of **14** with chlorosulfonium salt **15** (prepared from ethyl methylthioacetate and chlorine) gave the two regioisomers **16** and **17** in isolated yields of 39 and 23%, respectively. Compound **17** crystallized from the reaction mixture and was recrystallized from methylene chloride/hexanes to give a white solid. The major product was obtained after concentration of the reaction mixture and filtering off the solid material after addition of toluene.

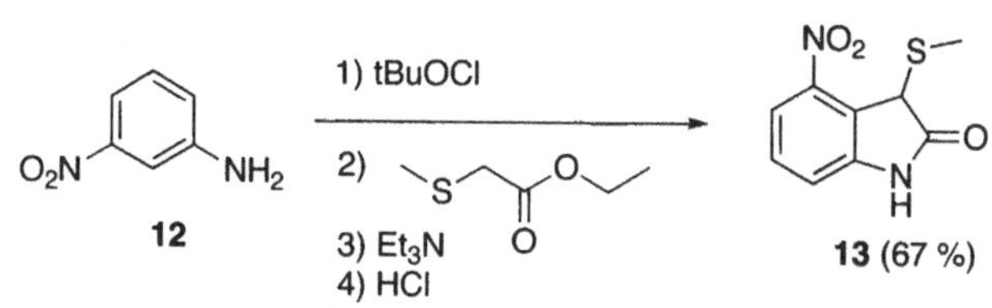

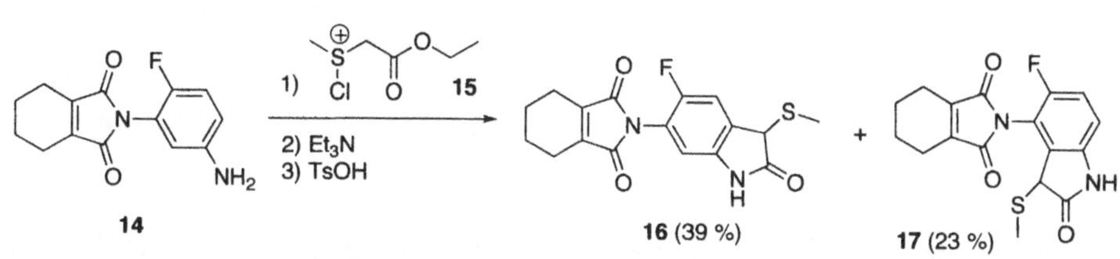

The preparation of 3-substituted oxindoles can be readily achieved by using alkylated derivatives of ethyl methylthioacetate.[4] For example, reaction of aniline with compound **18** gave oxindole derivative **19** in 64% yield. Reduction of **19** with Raney nickel gave oxindole **20** in 70% yield, while reduction with $LiAlH_4$ gave the corresponding indole **21** in 76% yield. The two-step route to indoles complements the Gassman indole synthesis.

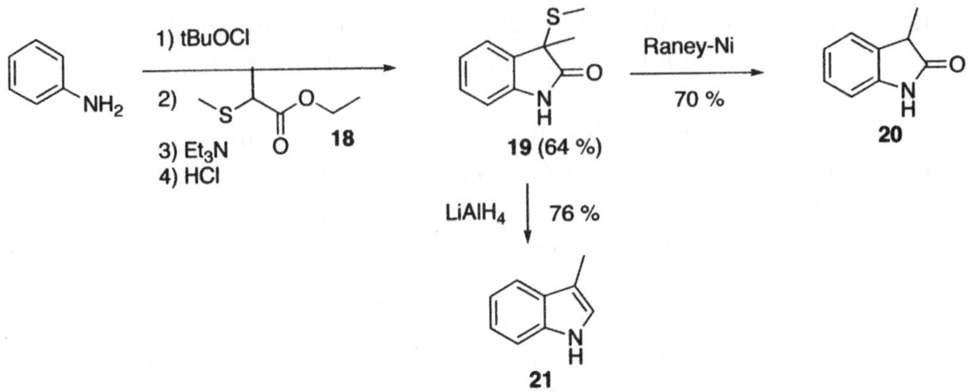

Larger-scale applications of the Gassman oxindole synthesis are possible. Reactions in the 100 g plus range have been reported for all of the common synthetic approaches. For example, Walsh and co-workers reported the synthesis of oxindole **22** in 92 g yield using the original Gassman procedure.[16]

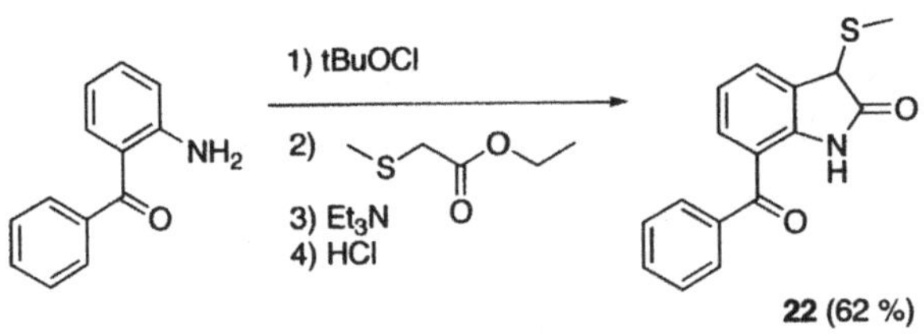

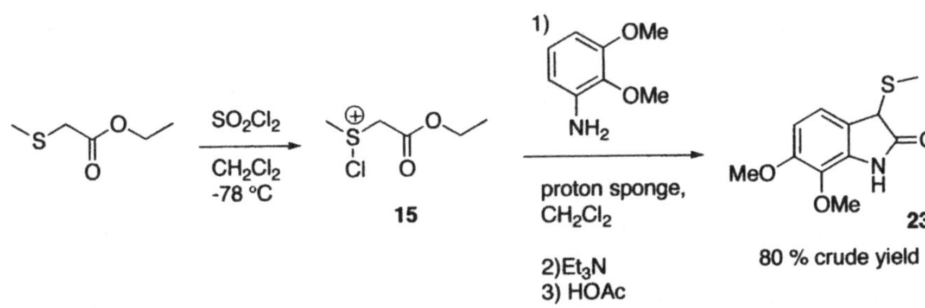

Applications of the Gassman oxindole synthesis in total synthesis are uncommon. Savall and McWhorter prepared a 6,7-dihydroxyindole derivative, part of the potent antihelmintic compound paraherquamide A, by using a chlorosulfonium ion (**15**) obtained from ethyl methylthioacetate and sulfuryl chloride.[17] The intermediate oxindole **23** was obtained in 80% crude yield. The starting aniline was obtained in good yield from 2,3-dimethoxybenzoic acid via a modified Curtius rearrangement. Removal of the thiomethyl functionality with Raney nickel gave the final product in 62% yield.

More recently, Danishefsky *et al.* used a Gassman oxindole synthesis to complete a total synthesis of the alkaloid phalarine.[18] The oxindole synthesis was used to prepare an indole ring precursor in the natural product after a more direct approach using the Fischer indole synthesis had failed. The ethyl ester of thiomethylacetic acid was converted into an activated chlorosulfonium ion by using sulfuryl chloride. The complex aniline derivative **24** and proton sponge were then added at –78 °C. After further treatment of the reaction mixture with triethylamine and acetic acid, oxindole **25** was obtained in 66% yield. The oxindole was converted into the corresponding indole **26** in 90% yield by using borane and Raney nickel reductions to remove the carbonyl and thiomethyl functionalities.

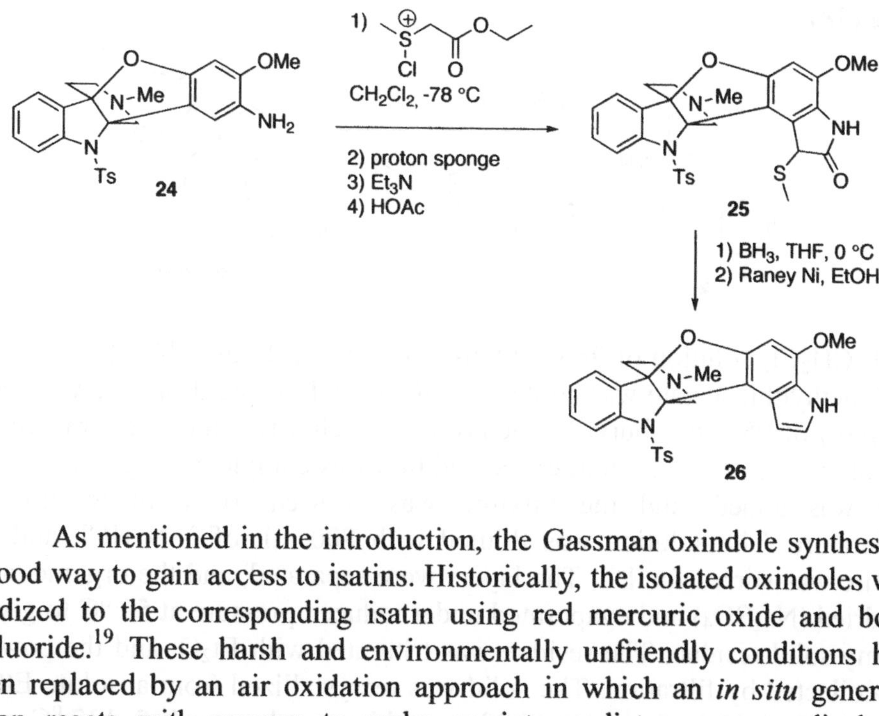

As mentioned in the introduction, the Gassman oxindole synthesis is a good way to gain access to isatins. Historically, the isolated oxindoles were oxidized to the corresponding isatin using red mercuric oxide and boron trifluoride.[19] These harsh and environmentally unfriendly conditions have been replaced by an air oxidation approach in which an *in situ* generated anion reacts with oxygen to make an intermediate peroxo radical that decomposes into the isatin.[20] Both approaches are shown in the scheme below.

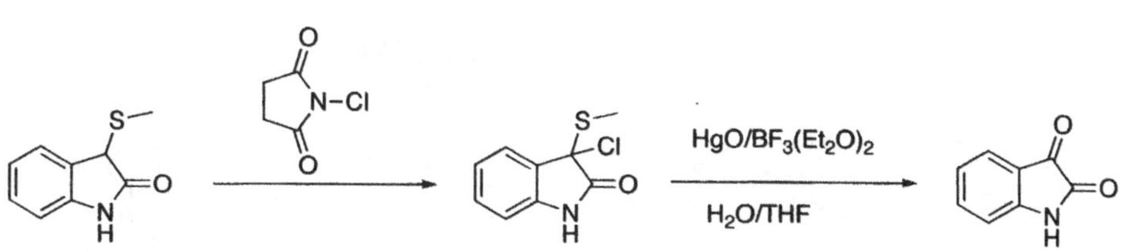

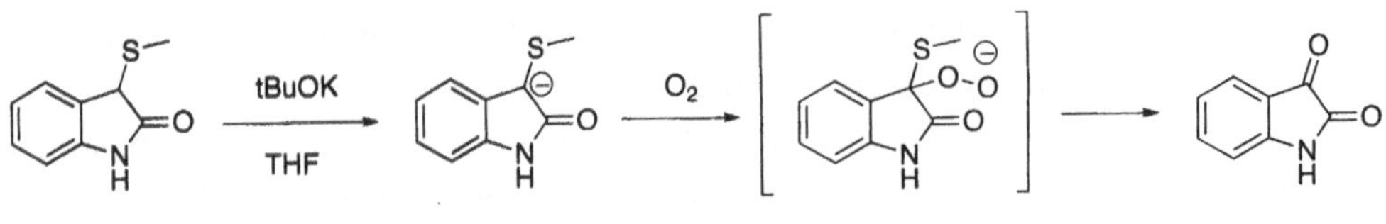

Typical yields for the air oxidation reaction were in the range of 30 to 60%. These yields are lower than those typically reported for the NCS/HgO route (40–80% yield).

3.6.5 *Experimental*

Preparation of 7-Benzoyl-1,3-dihydro-3-methyl-3-(thiomethyl)-2*H*-indol-2-one (28)[21]

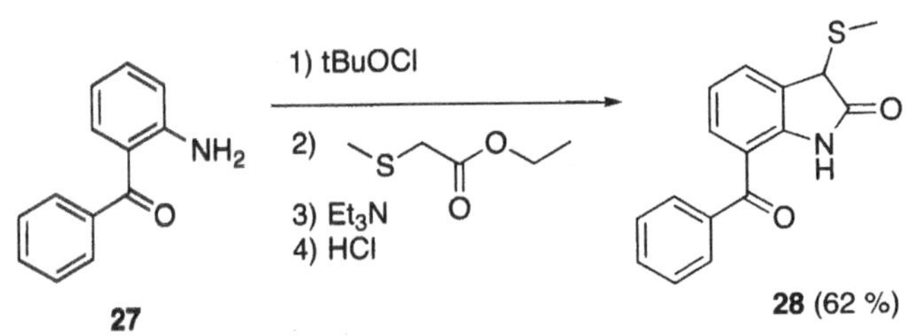

A 2 L CH_2Cl_2 solution of 98 g (0.5 mol) of aniline **27** and 74 g (0.5 mol) of ethyl methylthioacetate was cooled to –70 °C and treated dropwise with 56 g (0.5 mol) of 95% *tert*-butyl hypochlorite at such a rate that the temperature did not exceed –65 °C; 1 h after the addition was complete, 51 g (0.5 mol) of Et_3N was added, and the mixture was allowed to warm to ambient temperature. The solution was diluted with 800 mL of 3 N HCl, and the mixture was stirred for 1 h. The layers were separated, and the organic layer was dried (Na_2SO_4) and evaporated under reduced pressure at 50 °C to give a gummy solid residue. The residue was triturated with Et_2O, and the powder was collected by filtration. The solid was recrystallized from absolute EtOH to yield 92 g (62%) of compound **28** as a white powder, mp 135–137 °C.

Preparation of 3-(Methylthio)-6,7-dimethoxyoxindole (30)[22]

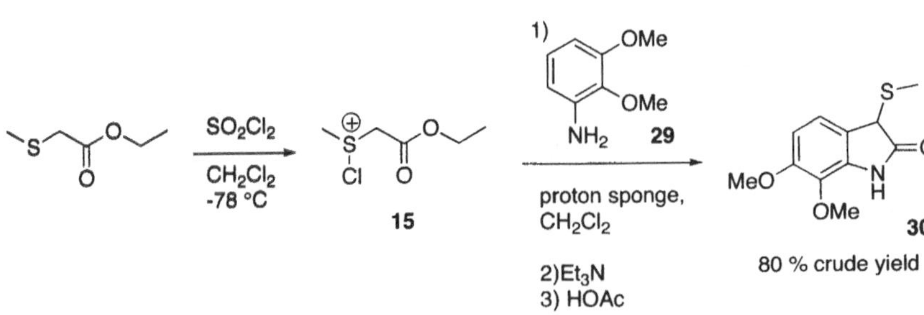

Sulfuryl chloride (2.47 mL, 30.7 mmol) was added to a cold (–78 °C) solution of ethyl (methylthio) acetate (3.95 mL, 30.7 mmol, dissolved in 200 mL anhydrous CH_2Cl_2). After 15 min, a solution of **29** (4.7 g, 31 mmol) and 1,8-bis(dimethylamino)naphthalene (6.58 g, 30.7 mmol) in CH_2Cl_2 (100 mL) was added over 1 h. After the mixture was stirred for 2 h, a solution of triethylamine (4.28 mL, 30.7 mmol, in 10 mL CH_2Cl_2) was added dropwise, and the reaction was allowed to warm to room temperature. This mixture was washed with water (3 × 100 mL). The combined aqueous layers were back-extracted with CH_2Cl_2 (100 mL). The combined organic layers were washed with brine, dried over $MgSO_4$, filtered, and concentrated to yield ethyl 2-amino-3,4-dimethoxy-α-(methylthio)benzeneacetic acid as a brown oil. The crude material was taken up in glacial acetic acid (200 mL) and stirred under N_2 for 3 h. The acetic acid was removed by azeotropic rotary evaporation using toluene to yield a brown tacky solid that was suspended in Et_2O (100 mL), stirred for 30 min., filtered, and washed with cold Et_2O. The crude product (5.6 g) was obtained in 80% yield. An analytically pure sample was obtained by recrystallization from toluene to yield orange crystals: mp 168–170 °C dec.

3.6.6 References

1 (a) Gassman, P. G.; Gruetzmacher, G. D.; Smith, R. H. *Tetrahedron Lett.* **1972,** 497. (b) Gassman, P. G.; Gruetzmacher, G. D. *J. Am. Chem. Soc.* **1973,** *95*, 588–560. (c) Gassman, P. G.; Gruetzmacher, G. D. *J. Am. Chem. Soc.* **1974,** *96*, 5487–5495.

2 (a) [R] Gribble, G. W. *Perkin 1* **2000,** 1045–1075. (b) Gassman, P. G.; Gruetzmacher, G. D. *J. Am. Chem. Soc.* **1974,** *96*, 5487–5495.

3 (a) Gassman, P. G.; van Bergen, T. J.; Gilbert, D. P.; Cue, Jr., B. W. *J. Am. Chem. Soc.* **1974,** *96,* 5495–5508. (b) Gassman, P. G.; van Bergen, T. J. *Org. Synth. Coll. Vol. 6.* **1988,** *6*, 601.

4 (a) Gassman, P. G.; van Bergen, T. J. *J. Am. Chem. Soc.* **1973,** *95*, 2718. Gassman, P. G.; van Bergen, T. J. *J. Am. Chem. Soc.* **1974,** *96*, 5508–5512.

5 Gassman, P. G.; Gruetzmacher, G. D.; van Bergen, T. J. *J. Am. Chem. Soc.* **1974,** *96*, 5512–5517.

6 (a) Gassman, P. G.; Cue, B. W. Jr.; Luh, T.-Y. *J. Org. Chem.* **1977,** *42*, 1344–1348. (b) Gassman, P. G.; Halweg, K. M. *J. Org. Chem.* **1979,** *44*, 628–629.

7 Peddibothla, S. *Cur. Bioactive Compds.* **2009,** *5*, 20–38.

8 (a) Karp, G. M. *J. Org. Chem.* **1992,** *57*, 4765–4772. (b) See also reference 5.

9 (a) Wierenga, W.; Griffin, J.; Warpehoski, M. A. *Tetrahedron Lett.* **1983,** *24*, 2437–2440. (b) Bramson, H. N.; Corona, J.; Davis, S. T.; Dickerson, S. H.; Edelstein, M.; Frye, S. V.; Gampe, R. T. Jr.; Harris, P. A.; Hassell, A.; Holmes, W. D.; Hunter, R. N.; Lackey, K. E.; Lovejoy, B.; Luzzio, M. J.; Montana, V.; Rocque, W. J.; Rusnak, D.; Shewchuk, L.; Veal, J. M.; Walker, D. H.; Kuyper, L. F. *J. Med. Chem.* **2001,** *44*, 4339–4358.

10 Wright, S. W.; McClure, L. D.; Hageman, D. L. *Tetrahedron Lett.* **1996,** *37*, 4631–4634.

11 (a) Gassman, P. G.; van Bergen, T. J. *J. Am. Chem. Soc.* **1973,** *95*, 2718. (b) Gassman, P. G.; van Bergen, T. J. *J. Am. Chem. Soc.* **1974,** *96*, 5508–5512.

12 Gassman, P. G.; Gruetzmacher, G. D.; van Bergen, T. J. *J. Am. Chem. Soc.* **1974,** *96*, 5512–5517.

13 Wright, S. W.; McClure, L. D.; Hageman, D. L. *Tetrahedron Lett.* **1996,** *37*, 4631–4634.

14 (a) Gassman, P. G.; Gruetzmacher, G. D.; van Bergen, T. J. *J. Am. Chem. Soc.* **1974,** *96*, 5512–5517. (b) Also see reference 4.

15 Karp, G. M.; Condon, M. E. *J. Heterocyclic Chem.* **1994,** *31*, 1513–1520.

16 Walsh, D. A.; Shamblee, D.A.; Welstead, W. A. Jr.; Sancilio, L. F. *J. Med. Chem.* **1982,** *25*, 446–451.

17 Savall, B. M.; McWhorter, W. N. *J. Org. Chem.* **1996,** *61*, 8696–8697.

18 Li, C.; Chan, C.; Heimann, A. C.; Danishefsky, S. J. *Angew. Chem., Int. Ed.* **2007,** *46*, 1448–1450.

19 Gassman, P. G.; Cue, B. W., Jr.; Luh, T.-Y. *J. Org. Chem.* **1977,** *42*, 1344–1348.

20 Gassman, P. G.; Halweg, K. M. *J. Org. Chem.* **1979,** *44*, 628–629.

21 Walsh, D. A.; Shamblee, D. A.; Welstead, W. A. Jr.; Sancilio, L. F. *J. Med. Chem.* **1982,** *25*, 446–451.

22 Savall, B. M.; McWhorter, W. N. *J. Org. Chem.* **1996,** *61*, 8696–8697.

3.7 Larock Indole Synthesis

Jennifer Xiaoxin Qiao

3.7.1 Description

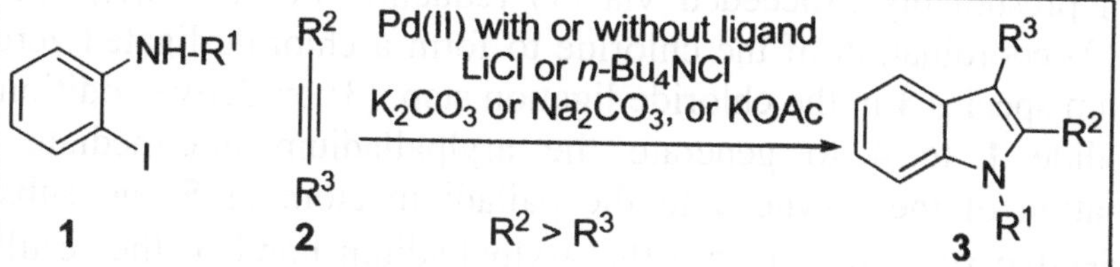

Larock indole synthesis, also known as Larock heteroannulation, is the one-pot palladium-catalyzed heteroannulation of *ortho*-iodoaniline and its derivatives **1** and internal alkynes **2** for the synthesis of 2,3-disubstituted indoles **3**.[1–14]

A wide variety of disubstituted alkynes **2** can be used as coupling parters and the substituted pattern of R^2 or R^3 groups does not have a significant effect on the yield of the reaction. The nitrogen atom on the *o*-iodoanilines **1** can be diversely substituted (R^1). The best results were obtained treating *o*-iodoaniline or the corresponding *N*-methyl, *N*-acetyl, and *N*-tosyl derivatives with an excess of the internal alkyne (usually 1.5–2 equiv, for volatile alkynes, 5 equiv), an excess amount of sodium or potassium carbonate base (usually 5 equiv), 1 equiv of LiCl or n-Bu_4NCl and occasionally adding 5 mol% of PPh_3 at 100 °C in DMF. Under these conditions, 2,3-disubstituted indoles were isolated in good to excellent yields.[2] With unsymmetrical alkynes the process is usually highly regioselective: the larger alkyne substituent (R^2) almost always locates at the 2-position of the indole ring. In particular, when R^2 is a silyl group, the reaction gives almost exclusively 2-silylindoles, which can be further protodesilylated, halogenated, or coupled with alkenes via a Pd-catalyzed reaction.

3.7.2 Historical Perspective

Professor Richard C. Larock reported the Larock indole synthesis in 1991.[1] Since then, the reaction has been studied by both his group and other researchers. The reaction is versatile and practical for the construction of indole ring in a single operation. The reaction has been widely employed for the preparation of indole derivatives in both industrial (on a multikilogram scale) and academic settings. The broad range of products synthesized according to the references cited herein demonstrates the scope and utility of this reaction.

3.7.3 *Mechanism*

Larock proposed the mechanism of the Larock indole synthesis.[2] The reaction presumably proceeded via (1) reduction of the Pd(II) species to Pd(0), (2) coordination of the chloride to form a chloride-ligated zerovalent palladium species **4** in the chloride ligation step, (3) oxidative addition of the aryl iodide **1** to **4** to generate the arylpalladium intermediate **5**, (4) coordination of the alkyne **2** to the palladium atom of **5** and subsequent regioselective *syn*-insertion into the arylpalladium bond of the resulting π-alkyne-σ-organopalladium intermediate **6** to generate the resulting vinylic palladium intermediate **7** in the carbopalladation step, (5) nitrogen displacement of the halide in **7** to form a six-membered, heteroatom-containing palladacycle **8**, and (6) reductive elimination of **8** to form the indole **3** and regenerate Pd(0).

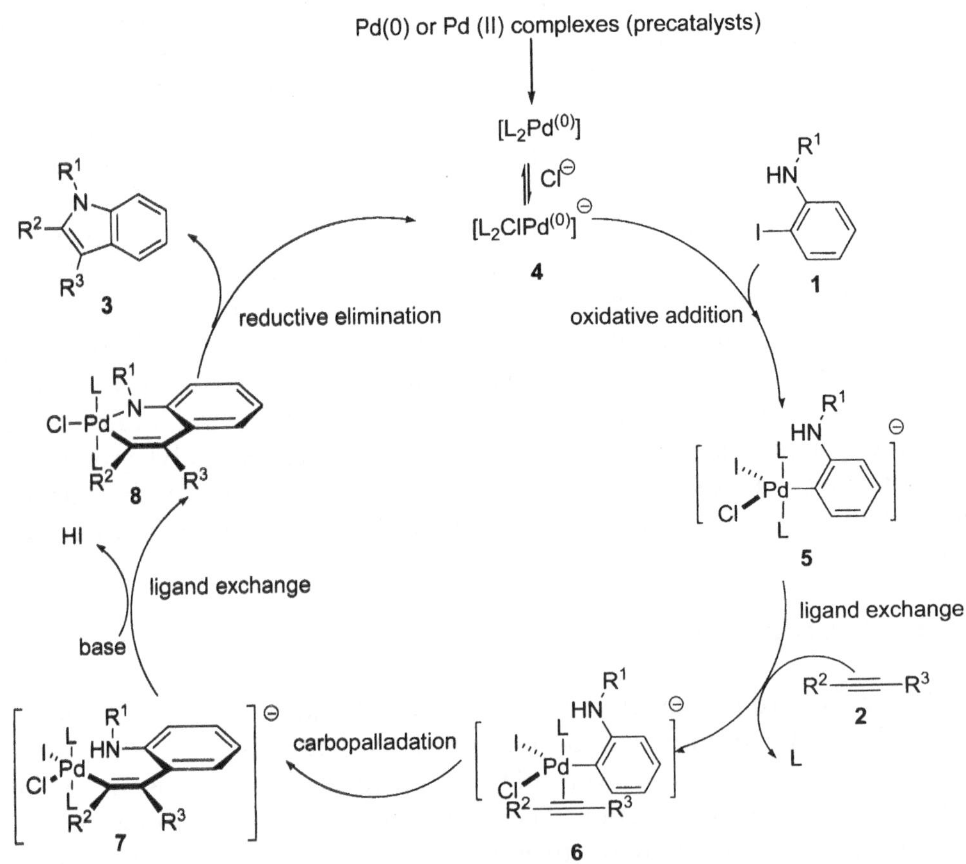

The carbopalladation step determines the regiochemical outcome of the reaction with unsymmetrical alkynes. Steric effect controls the conversion of intermediate **6** into the carbopalladation adduct **7** so as to direct the organic residue preferentially to the less hindered end of the carbon-carbon triple bond and the palladium moiety to the more hindered end. On the other hand, coordinating effect of the neighboring alcohol of HO-substituted alkyne influences the formation of **7** so that the added palladium ends up close to the coordinating group during ligand exchange step with the alkyne.

3.7.4 Variations and Improvements

Under typical Larock heteroannulation conditions, the cheaper and/or more readily accessible *o*-bromoanilines or *o*-chloroanilines failed to react with the internal alkynes. Researchers from Boehringer–Ingelheim extended the scope and utility of Larock indole synthesis, and first reported that a variety of *o*-bromo- or *o*-chloroanilines **9** or **10** was reacted with internal alkynes **11** in the presence of $Pd(OAc)_2$, K_2CO_3, and utilizing 1,1'-bis(di-*tert*-butylphosphino)ferrocene **12** as the ligand in NMP at 110–130 °C to afford 2,3-disubstituted indoles **13** in 60 to 99% yields and with excellent regioselectivity.[15] The major by-products observed under these conditions were dihydrophenazines **14**, which were minimized at lower reaction concentrations. A patent regarding the palladium-catalyzed indolization of *o*-bromo/chloroanilines was published later by the same group of researchers.[16]

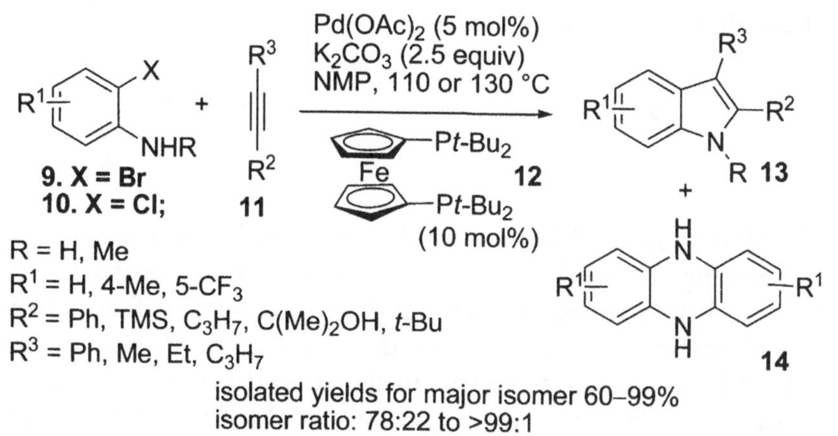

As mentioned early in the chapter, one of the general features of Larock indole synthesis is that the reaction gave exclusively 2-silyl indoles when a silyl-substituted alkyne was present, and the corresponding 2-silyl indoles can be further functionalized. In this respect, Denmark and Baird recently reported a sequential Larock heteroannulation and silicon-based cross-coupling reaction for the synthesis of 2,3-disubstituted indoles **15** from *o*-iodoanilines **16** and silyl indoles **17**, followed by cross-coupling of the sodium salt of the resulting annulation product **18** with aryl halides **19**.[17] The silyl group in **17** played a dual role, serving both as a directing group in the Larock annulation step and as an activating group for the subsequent silicon-based cross-coupling reaction. Slight modifications of the Larock indole synthesis was achieved by adding 3 equiv of H_2O and rapid stirring of the reaction mixture. In addition, *t*-butoxylsilyl ether is crucial for the hydrolysis and then cross-coupling steps. And catalysts such as allyl palladium chloride (APC) and **20** are important for the cross-coupling step.

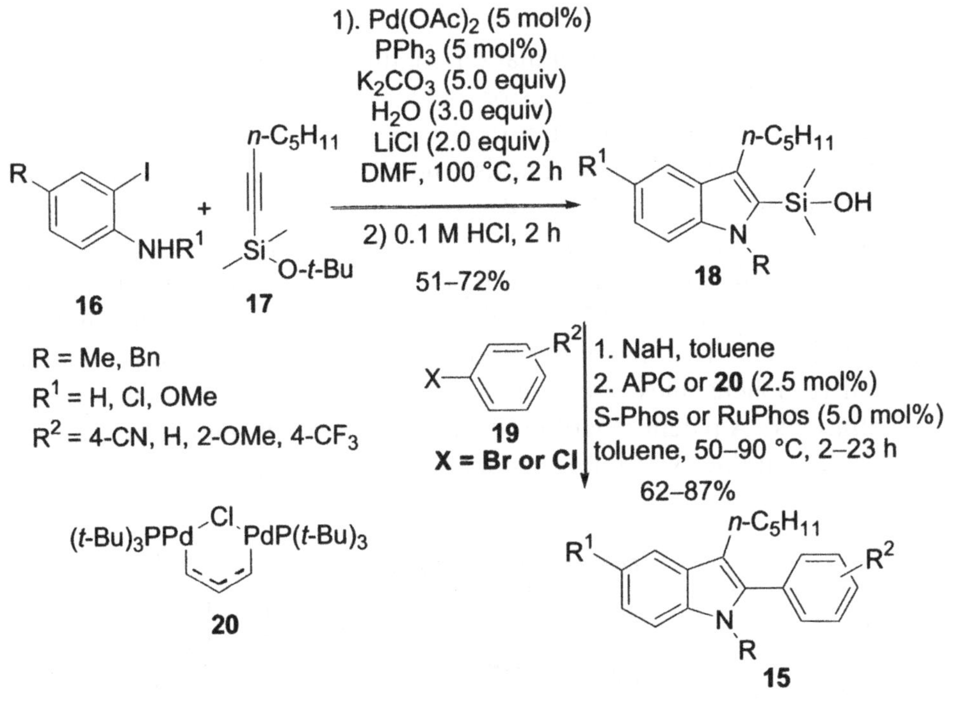

2-Iodobenzoic acid **21** was used as the coupling partner in a one-pot Curtius rearrangent/palladium-catalyzed indolization process for the synthesis of 2,3-disubstituted indoles **22**.[18] A synergistic effect between the two steps of the process was observed, with the by-product of the first reaction (NaCl) serving as a reagent in the second synthetic step. With no lithium chloride added, but adding 3 equiv of Na_2CO_3 provided the best yields.

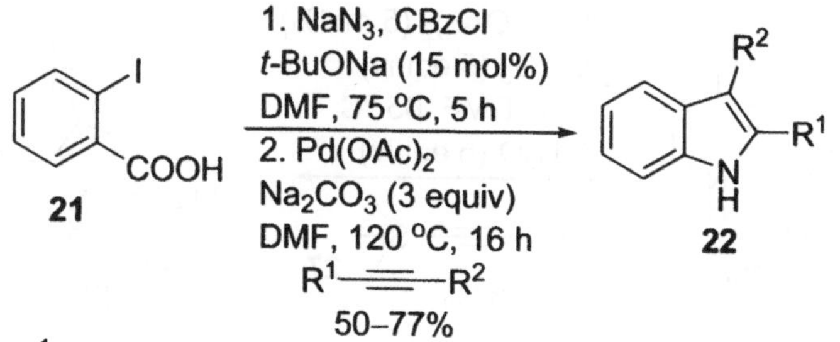

R^1 = Ph, Me, TMS,
R^2 = Ph, *t*-Bu, CH_2CH_2OBn
or $R^1R^2 = (CH_2)_4$

In addition, the above multicomponent process was used to prepare the indole *N*-carboxamide derivatives **23** via Larock heteroannulation reaction.

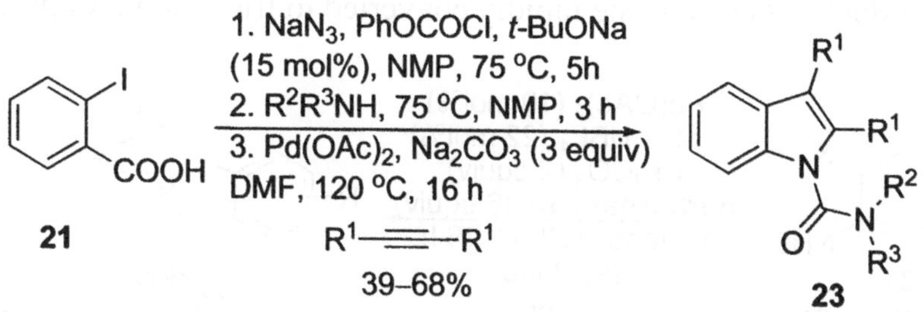

R^1 = Ph, Pr
NHR^2R^3 = morpholine, piperidine, pyrrolidine, (*S*)-$NH_2CH(Me)Ph$, $NH_2CH_2CH_2Ph$

Researchers at Wyeth adapted the Larock indole synthesis approach and developed a synthetic methodology for highly functionalized pyrroles from 2-amino-3-iodoacrylates **24** and internal alkynes in moderate to excellent yields and with good regioselectivity.[19] For example, when a significantly differentiated trimethylsilyl-propargyl alcohol **25** was reacted, a single pyrrole regioisomer **26** formed accompanied by desilylation and deacylation under reaction conditions. When 1-phenyl-2-trimethylsilyl-acetylene **27** was used, a complete reversal of expected regiochemistry occurred to give a single isomer **28** in 81% yield.

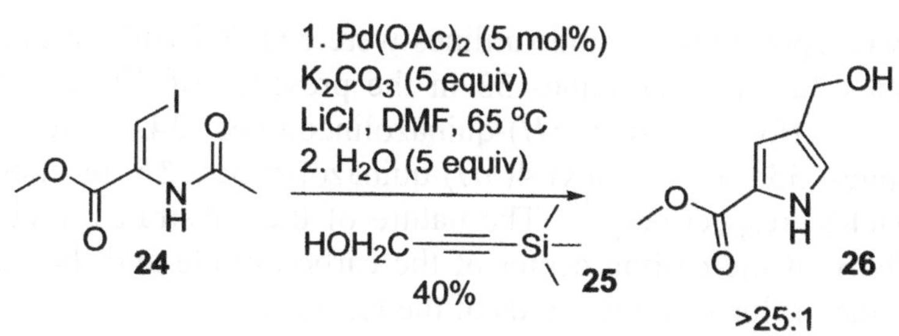

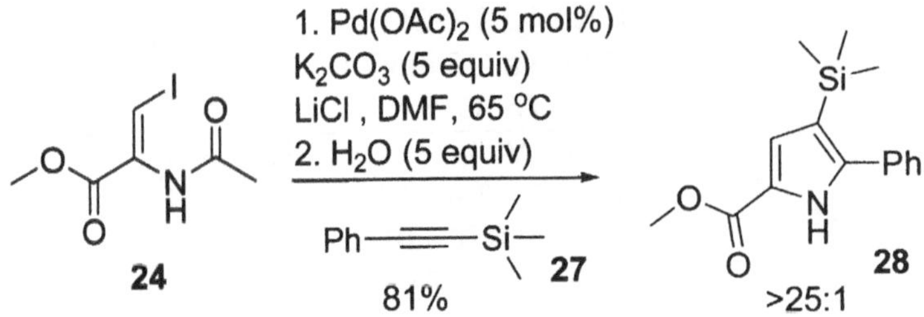

Norbornadiene as an acetylene synthon was used for the synthesis of functionalized indolines **29** and indoles **30** from *o*-bromo or *o*-chloroanilines **31**.[20] The annulation step tolerated a variety of functional groups, including electron-donating and electron-withdrawing groups at several positions. The subsequent retro Diels–Alder reaction to generate related indoles was developed under two conditions: (A) heating in ethylene glycol at 170 °C, and (B) treatment with silica gel in xylenes at 170 °C. The annulated haloaniline products **29** can also be rapidly converted to tricyclic indolines.

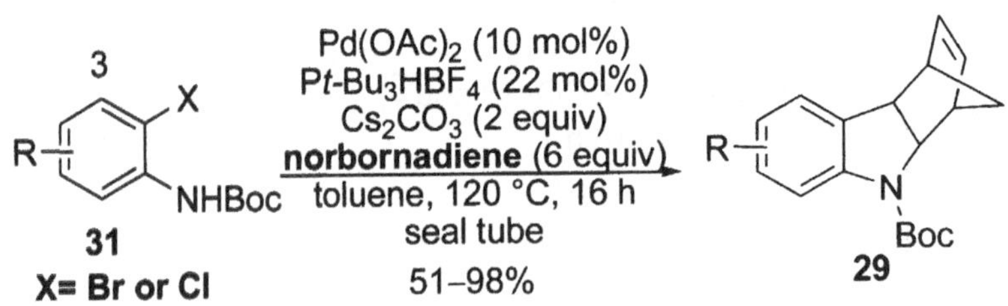

R = H, 4-Me, 3-Me, 4-Cl, 5-CF_3, 3-COOMe, 5-NO_2, 6-Cl

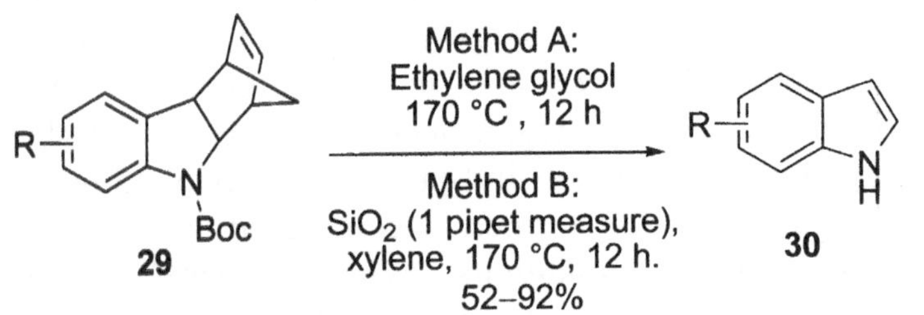

R = 5-Me, 5-F, 5-Cl, 5-OMe, 5-NO_2, 7-F

It was reported that *o*-iodoanilines were coupled with isocyanates **31**, carbodiimides **32**, or ketenimines **33** in the presence of 300 psi of carbon monoxide to afford 2,4-(1*H*,3*H*)-quinazolinediones **34**, 2-amino-4(3*H*)-quinazolinones **35**, and 2-alkyl-4(3*H*)-quinazolinones **36** in moderate to excellent yields, respectively.[21] The nature of the substrates, including the electrophilicity of the carbon center of the carbodiimide and the stability of the ketenimine, influenced the yields of the reaction.

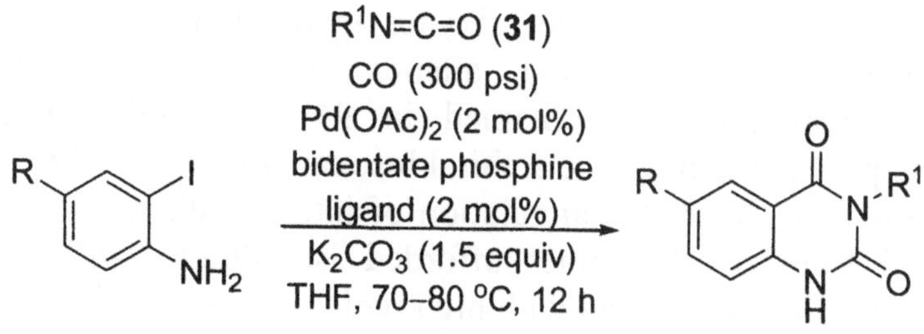

34

R = H, Cl, Me, OH
R^1 = *p*-Cl-Ph, *p*-Br-Ph, *p*-MeO-Ph, *p*-Me-Ph

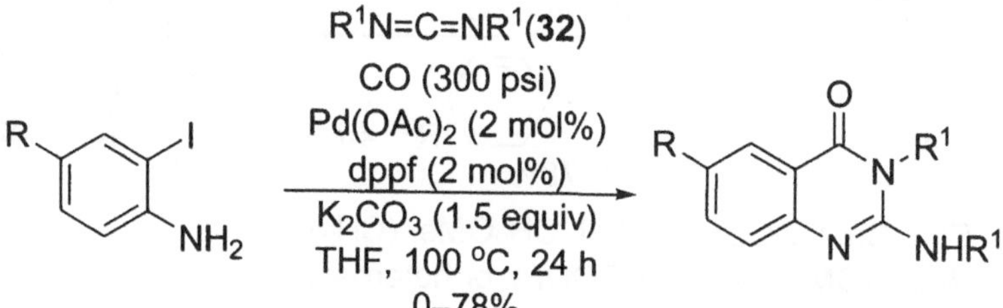

35

R = H, Cl, Me, OH, CN
R^1 = *p*-Cl-Ph, *p*-Br-Ph, Ph, *p*-Me-Ph, C_6H_{11}, *i*-C_3H_7

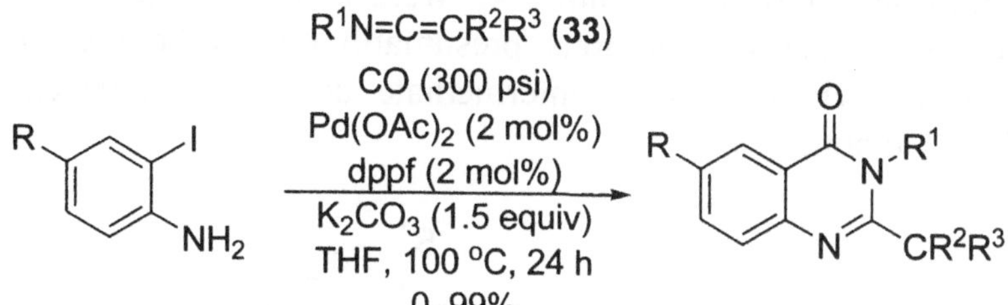

36

R = H, Cl, Me, OH, CN
R^1 = Ph, *p*-Me-Ph, *n*-C_4H_9
CR^2R^3 = C(Me)(COOEt), C(Me)(COPh), $C(COOEt)_2$, $C(Ph)_2$

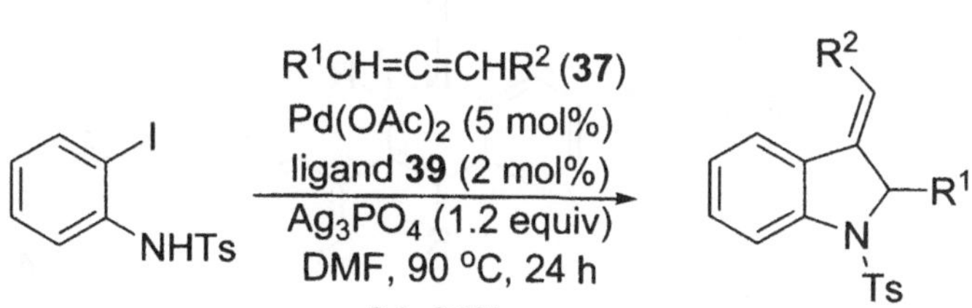

38

R^1 = *n*-C_8H_{17}, *n*-C_3H_7
R^2 = H, *n*-C_3H_7
R^1R^2 = $(CH_2)_{10}$

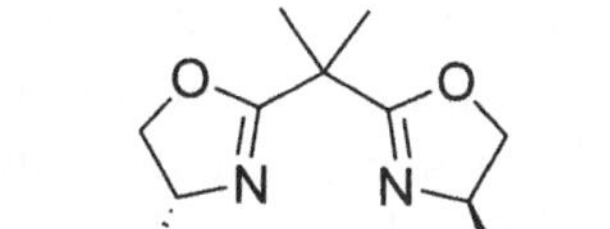

39

When the coupling partner alkynes in the Larock indole synthesis were replaced with substituted allenes **37**, the corresponding 3-methyleneindolines **38** were obtained in 94–95% yields and 80–82% enantiomeric pure in the presence a chiral bisoxazoline ligand **39**.[22,23]

On the other hand, the annulation of internal alkynes **40** and imines **41** generated from *o*-iodoanilines afforded isoindolo[2,1-*a*]indoles **42** in moderate to excellent yields and with high regioselectivity.[24,25]

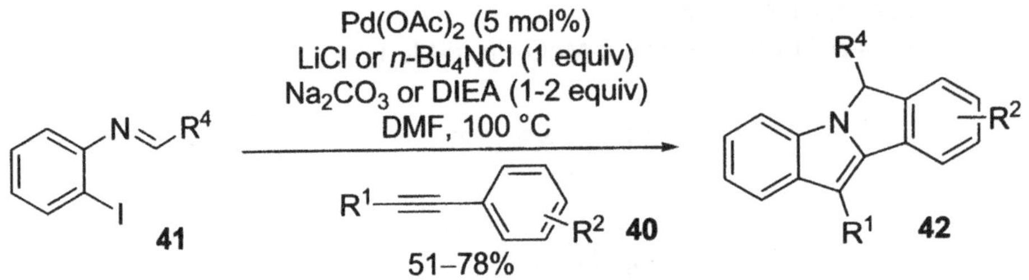

R^1 = Ph, Et, *n*-Bu, COOEt, CH_2OH, CH_2OMe, $(CH_2)_4OH$
R^2 = H, 3-Me, 3-OMe, 3-CF_3, 3-COOEt, or heteroaryl
R^4 = alkyl, aryl

Unexpectedly, the highly substituted quinoline derivatives **43** were not observed, and tetracyclic indoles **43** were isolated as the only products under the above reaction conditions, presumably due to the exclusive 5-*exo*-addition of the vinylpalladium intermediate across the adjacent carbon-nitrogen double bond.

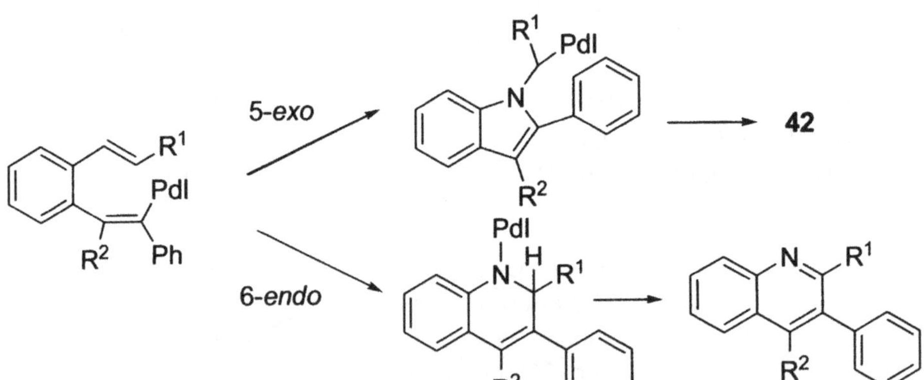

A series of substituted β-carbolines **44** and γ-carbolines **45** were synthesized by the palladium-catalyzed iminoannulation of internal alkynes **46** with the *tert*-butylimines generated from *N*-substituted 3-iodoindole-2-carboxaldehydes **47** and 2-haloindole-3-carboxaldehydes **48**, respectively.[26] The chemistry is effective for a wide range of alkynes, including aryl-, alkyl-, hydroxymethyl-, ethoxycarbonyl-, and trimethylsilyl-substituted alkynes. When an unsymmetrical internal alkyne was employed, this method generally gave two regioisomers.

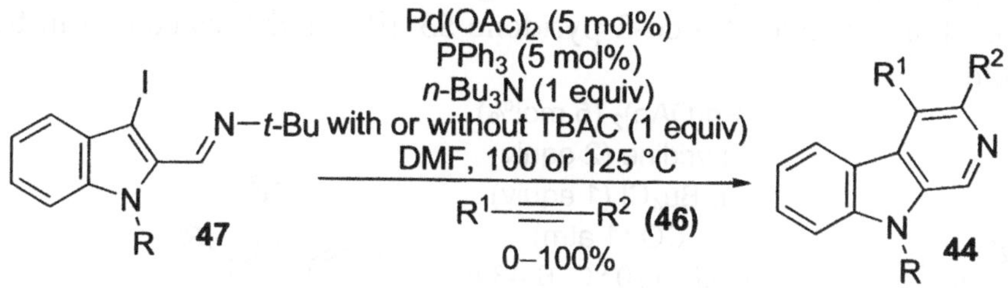

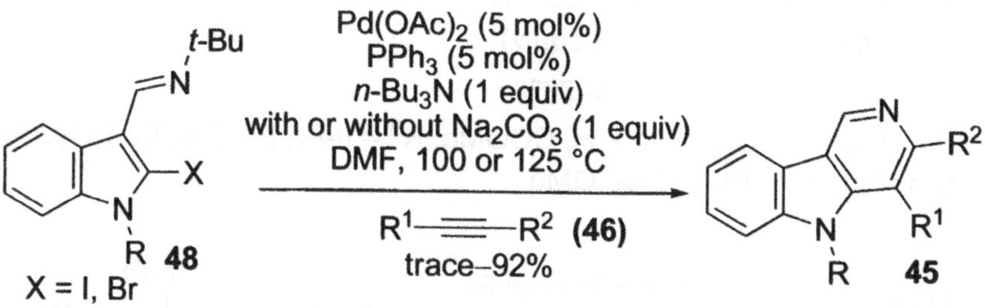

X = I, Br
R = H, Me, MOM, Ts, Bu
R^1 = Ph, Me, *n*-Pr, CH_2OH, COOEt
R^2 = Ph, TMS, *t*-Bu, *n*-Pr, COOEt, CH_2OH

On the other hand, the *γ*-carbolines **49** were prepared by intramolecular iminoannulation of the corresponding *N*-alkynyl-2-bromo-1*H*-indole-3-*tert*-butylimines derived from the aldehyde **50**.[27]

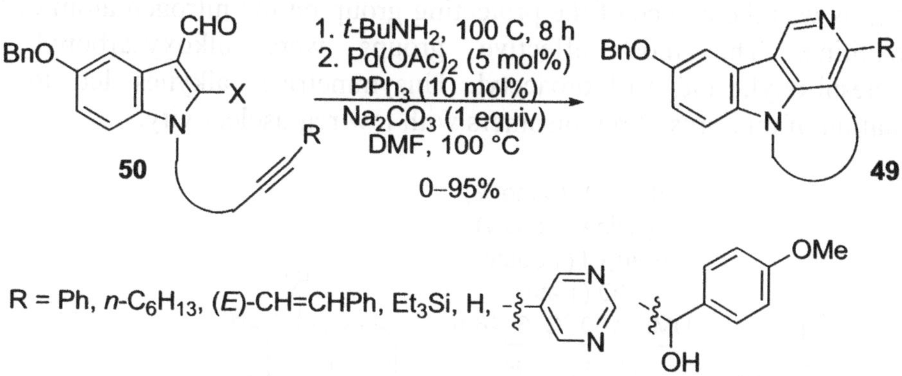

When *o*-iodoanilines were replaced with *o*-iodophenols, a variety of substituted coumarins **51** were synthesized in good yields by the palladium-catalyzed coupling of *o*-iodophenols **52** with internal alkynes and 1 atm of carbon monoxide.[28] For unsymmetrical internal alkynes two regioisomers were usually observed. The regioselectivity is governed by steric factors with the larger substituent on the triple bond ending up at the 3-position of the coumarin, which is consistent with regioselectivity observed in the Larock indole synthesis. Heterocyclic analogues of *o*-iodophenol were also effective. For example, 3-iodo-2-pyridone **53** affords the azacoumarin **54** in 70% yield.

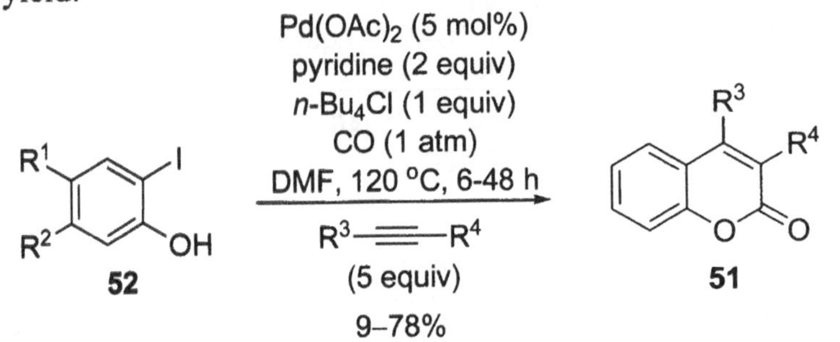

52 **51**

R^1 = H, CO_2Et, OMe; R^2 = H, COOMe; R^3 = *n*-Pr, Ph, Me, Et, CH_2OBn;
R_4 = *t*-Bu, *n*-Pr, Ph, $SiMe_3$, OMe

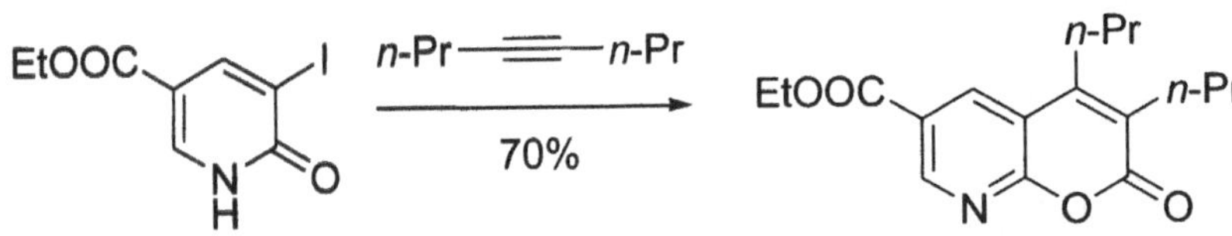

53 **54**

Similarly, 3,4-disubstituted 2-quinolones **55** can be prepared via the palladium-catalyzed annulation of internal alkynes with *N*-substituted *o*-iodoanilines under 1 atm of carbon monoxide.[29] A crucial aspect of the synthesis was the choice of the protecting group on the nitrogen atom of the iodoaniline. The most effective groups were alkoxycarbonyl, *p*-toluenesulfonyl, and trifluoroacetyl. Unsymmetrical alkynes led to the formation of mixtures of regioisomers with low regioselectivity.

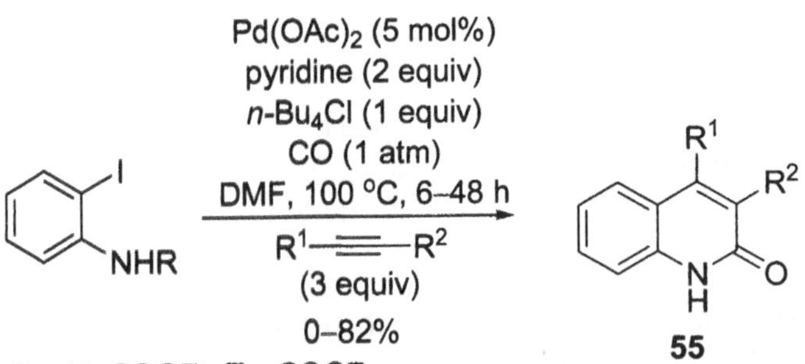

55

R = H, COOEt, Ts, $COCF_3$
R^1 = *n*-Pr, Ph, 2-OMePh, Me, 5-pyrimidinyl, CH_2OH, CH_2OBn, Et
R^2 = *n*-Pr, Ph, Et, *n*-Bu, *t*-Bu, CH_2OH, CH_2OMe, CH_2OBn, $SiMe_3$, COMe

On the other hand, 3,4-disubstituted isocoumarins and polysubstituted α-pyrones **56** were generated in good yields using palladium-catalyzed annulation of internal alkynes with halogen- or triflate-containing aromatic esters and α,α-unsaturated esters **57**, respectively.[30] The methodology provided a simple, convenient, and regioselective route to isocoumarins and α-pyrones containing aryl, silyl, ester, *tert*-alkyl, and other hindered groups.

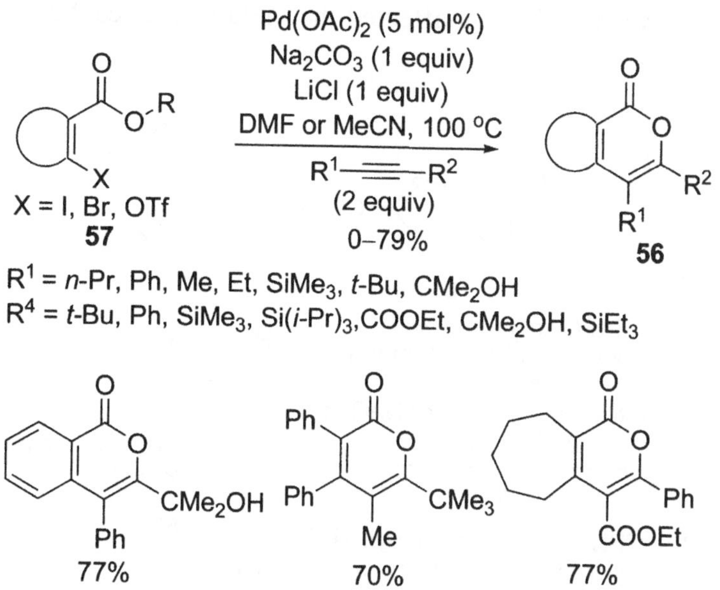

3.7.5 *Synthetic Utility*

General Utility

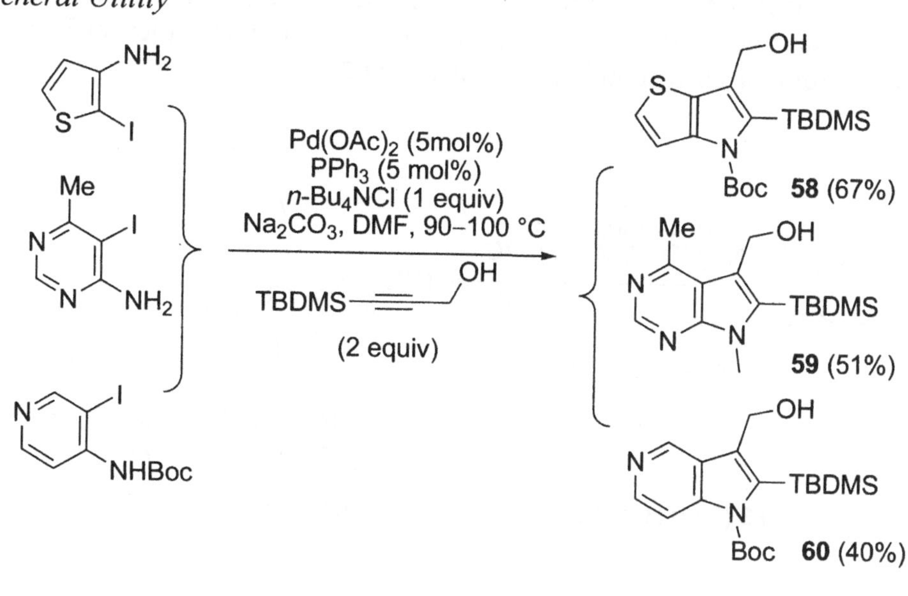

o-Iodoanilines in the Larock indole synthesis can be replaced with vicinal iodo-substituted heterocyclic amines to prepare heterocondensed pyrroles, including thieno[3,2-b]pyrroles **58**, pyrrolo[2,3-*d*]pyrimidine **59**, and azaindoles such as **60**.[31]

Similarly, reaction of 5-amino-4-iodobenzothiophene **61** with internal acetylenes gave 7,8-disubstituted thieno[3,2*e*]indoles **62** under Larock heteroannulation conditions.[32] The choice of the base such as Na_2CO_3, NaOAc, or KOAc significantly affected the yield of the reaction of different alkynes.

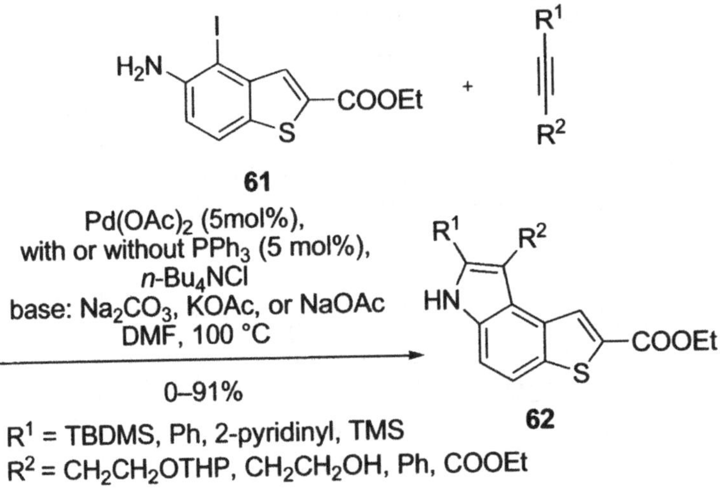

2,3-Diaryl-substituted pyrrolo[3,2-*b*]pyridine-5-carbonitriles **63** exhibiting c-Met receptor tyrosine kinase inhibition activity were reported.[33] The key intermediates **64** were prepared using the Larock indole synthesis. A variety of aryl/heteroaryl substitution in the 2-position was then efficiently explored from the corresponding iodo analogs **65**.

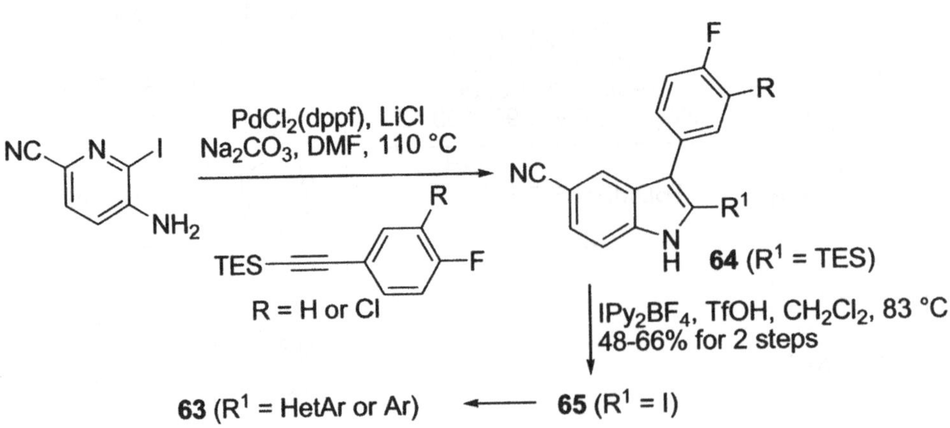

The Larock indole synthesis was employed for the preparation of indole **66**, which was elaborated to give indole **67** for activity screening and safety assessment.[34] The key step involves the coupling of iodoaniline **68** with bis-TES-propargyl alcohol **69** in the presence of $Pd(OAc)_2$ in DMF with 5 equiv of Na_2CO_3 to afford a mixture of **66** and **70** (96:4 ratio). Maxalt (**71**) was also obtained in a similar manner by the same research group.[35]

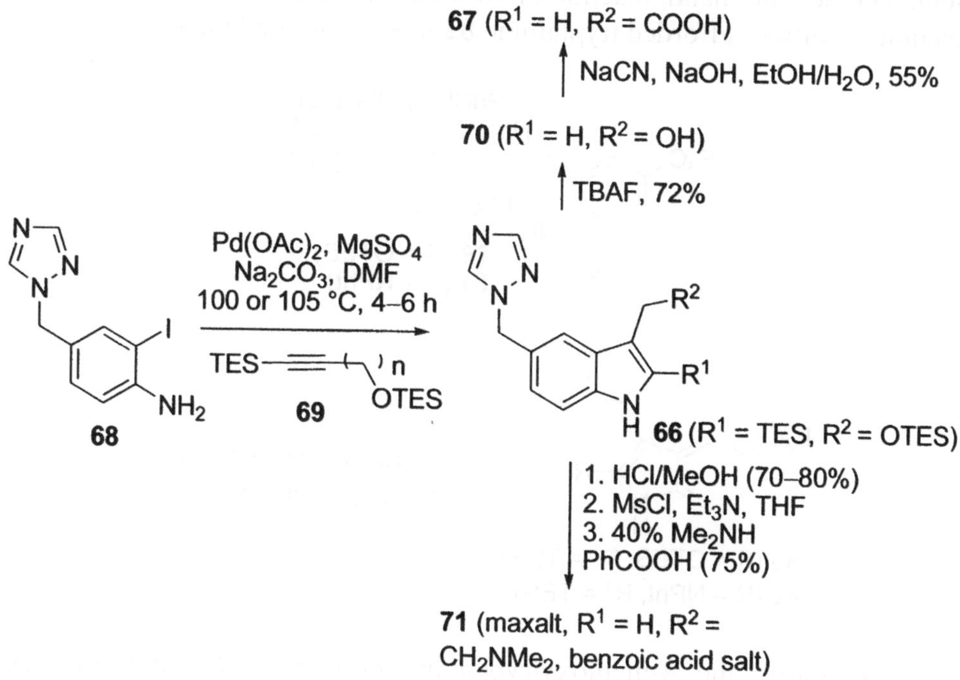

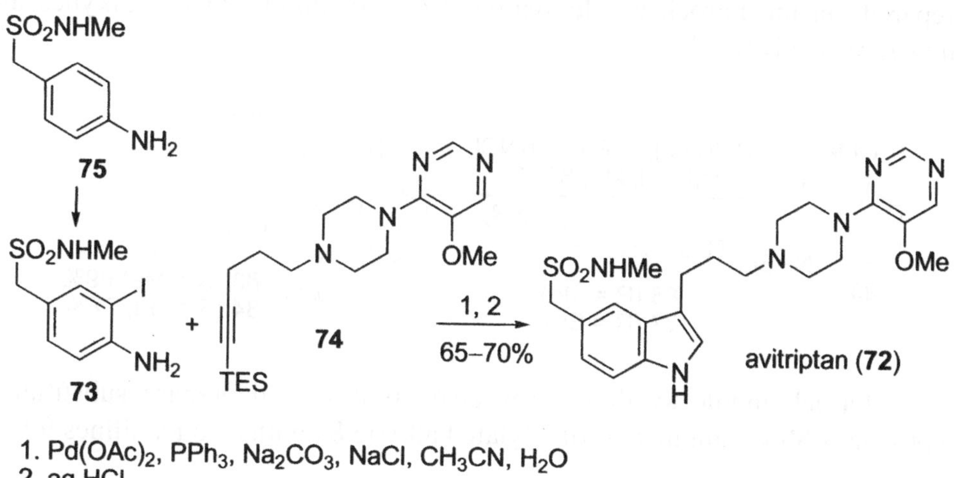

1. $Pd(OAc)_2$, PPh_3, Na_2CO_3, NaCl, CH_3CN, H_2O
2. aq HCl

The Larock heteroannulation was used to prepare multikilogram quantities of avitriptan (**72**).[36] The palladium-catalyzed heteroannulation of iodoaniline **73** and acetylene **74**, followed by acid hydrolysis of the silyl group furnished avitriptan in 65–70% overall yield from aniline **75**.

5-Trifluoromethyl)tryptophol **76** and tryptamine **77** were prepared via the Larock heteroannulation.[37] Reaction of (trifluoromethyl) iodoaniline **78** with (triethylsilyl)acetylene **79** in the presence of $PdCl_2$(dppf), 2 equiv of Na_2CO_3, and 1 equiv of LiCl in DMF at 100 °C gave tryptophol **80** in 44% yield. On the other hand, reaction of **78** with acetylene **81** under the identical reaction conditions afforded tryptamine **82** in 83% isolated yield.

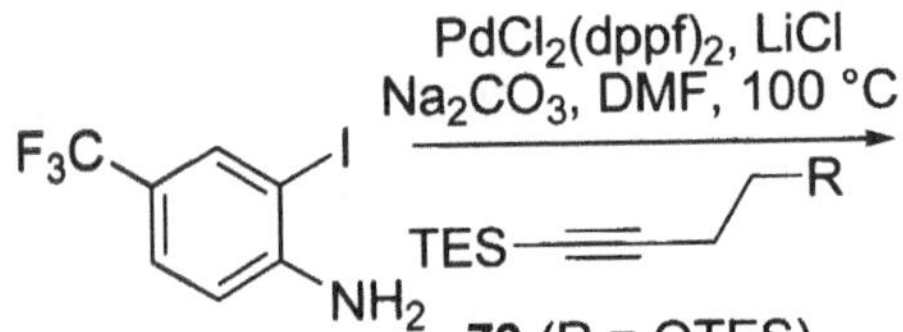

78 **79** (R = OTES)
81 (R = NPht)

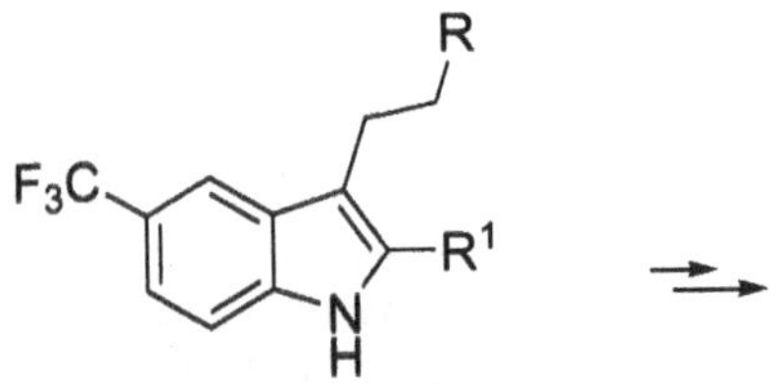

80 (R = OTES, R¹ = TES)
82 (R = NPht, R¹ = TES)

76 (R = OH, R¹ = H)
77 (R = NH₂, R¹ = H)

Similarly, the 4-methoxytryptamine derivatives **83** and **84** were prepared via the Larock indole synthesis of iodoaniline **85** and alkynes **86** and **87**, respectively.[38]

85 **86** (R = Me)
87 (R = Bn)
83 (R = Me) 69%
84 (R = Bn) 77%

Larock indole synthesis was also employed to prepare substituted tryptophans **88** via annulation of silylated alkyne **89** with *o*-iodoanilines **90**.[39]

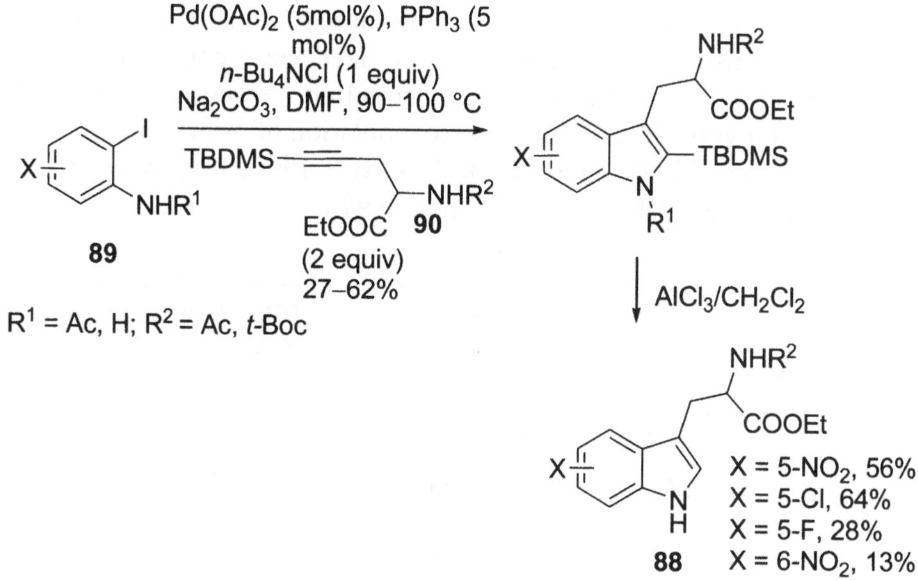

On the other hand, efficient asymmetric synthesis of optically active substituted tryptophans **91** was realized via Larock heteroannulation of *o*-iodoaniline **92** and alkyne **93** derived from the Schöllkopf chiral auxillary to afford the corresponding key intermediate **94** as the major isomeric product.[40]

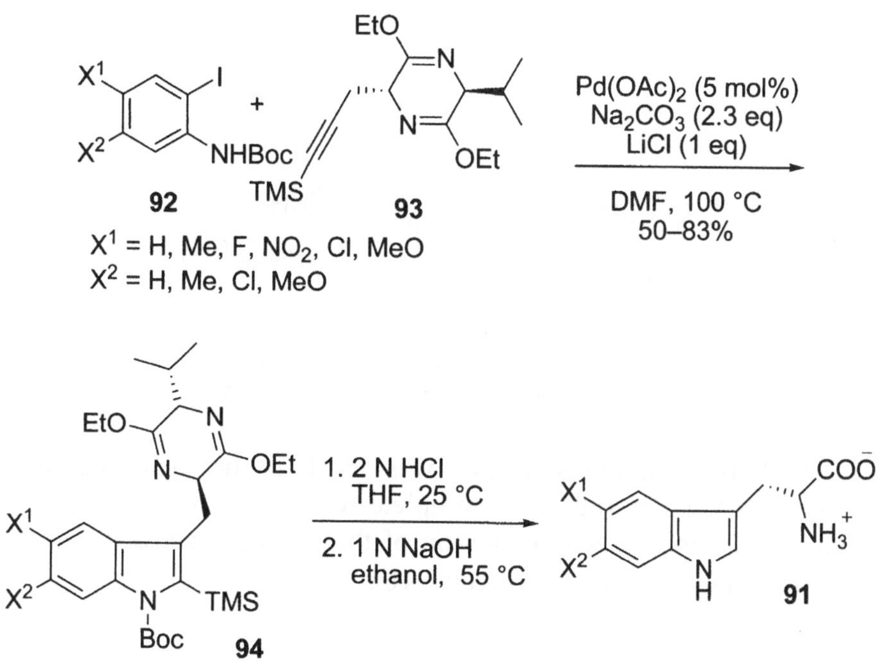

Two (*S*)-β-methyl-2-aryltryptamine-based gonadotropin-releasing hormone antagonists **95** and **96** were synthesized via a consecutive Larock indole synthesis and Suzuki cross-coupling reaction.[41] The key transformation involved the Larock heteroannulation of *o*-iodoaniline **97** with chiral silyl alkyne **98** in the presence of $Pd(OAc)_2$, PPh_3, 1 equiv of LiCl, and 2.5 equiv of K_2CO_3 in DMF at 100 °C to give **99** in 72% yield.

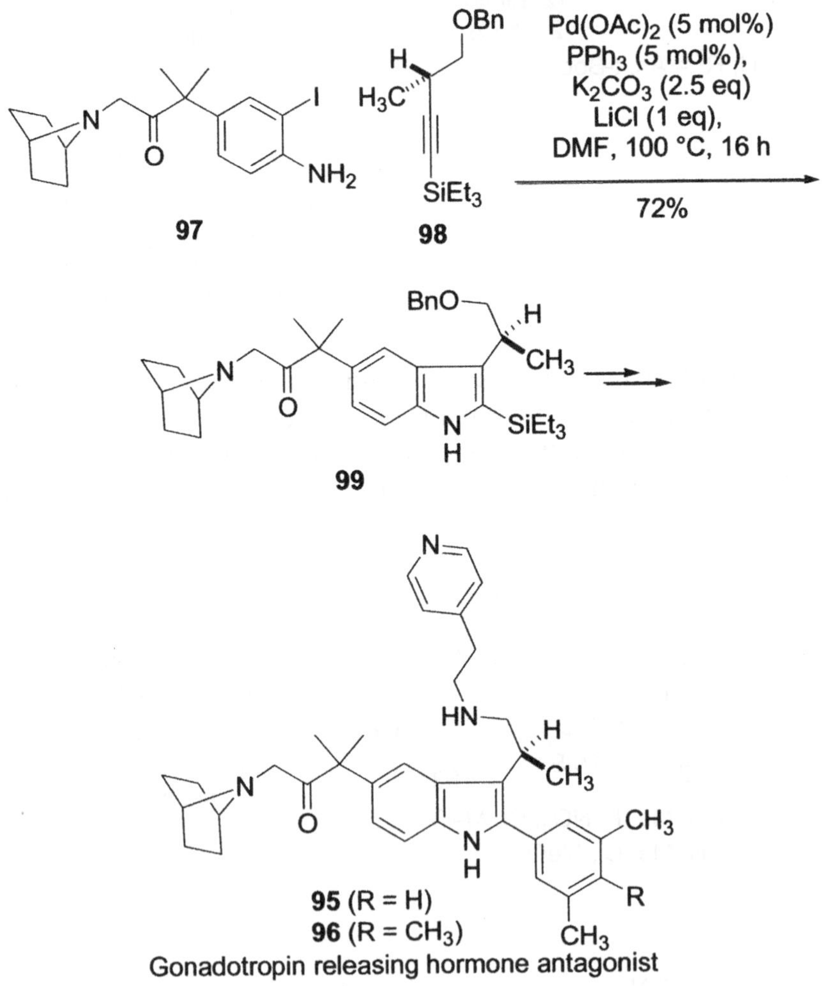

Gonadotropin releasing hormone antagonist

The Larock indole synthesis was used in the synthesis of α-C-glucosyl-*iso*-tryptophan **100**.[42] A complete reversal of regioselectivity was observed in the larock heteroannulation of the *N*-Ts-protected *o*-iodoaniline **101** with α-C-glucosylpropargyl glycine **102**.

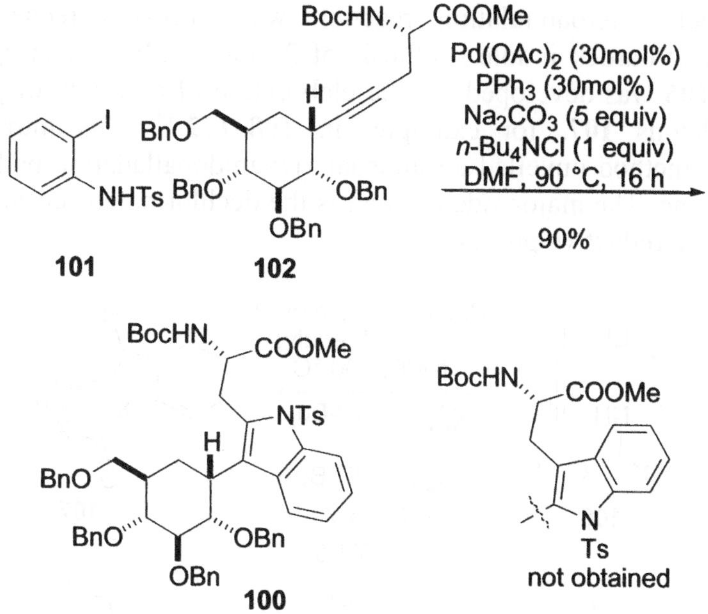

Pd-catalyzed annulation of *meso*-hexynyl Zn(II) porphyrin **103** with 4-amino-3-iodopyridine efficiently provides *meso*-3-(5-azaindolyl)-substituted Zn(II) porphyrin **104** as the major product, which assembles to form a slipped cofacial dimer by the complementary coordination of the pyridine moiety to the Zn(II) center.[43] Isomer **105** was predicted to be the major isomer based on the general trend for Larock heteroannulation. In contrast, the reaction gave only a trace amount of **104**.

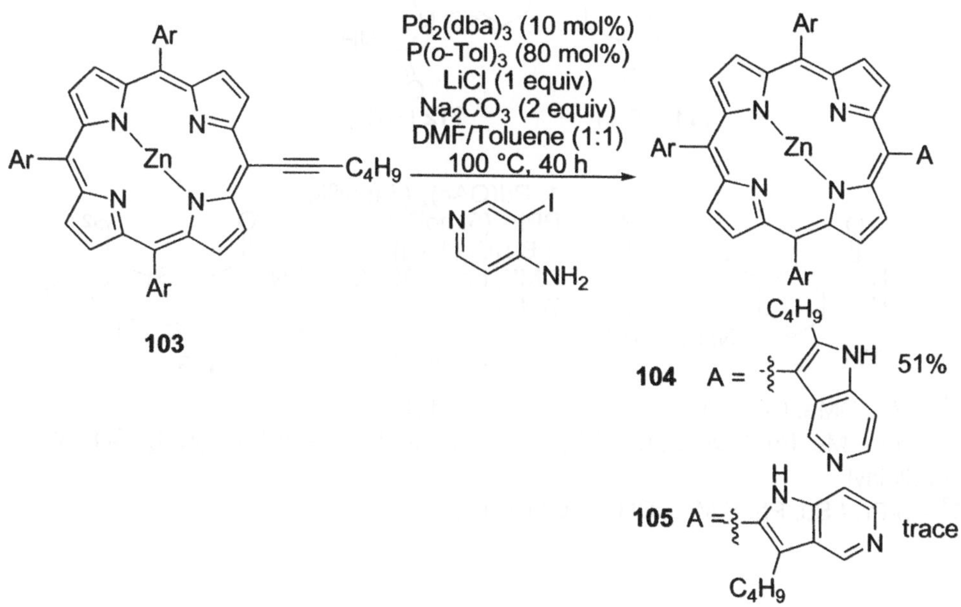

Larock heteroannulation reaction was further extended and an intramolecular Larock indole synthesis of 2-chloroanilines bearing tethered acetylenes **106** was developed for the elaboration of a variety of polycyclic indole skeletons **107**, for example, in **108–112**.[44] This intramolecular indolization method unveiled an unusual *syn*-anidopalladation pathway of a tethered alkyne. The major side product is the dechlorinated starting material as a result of a reductive process.

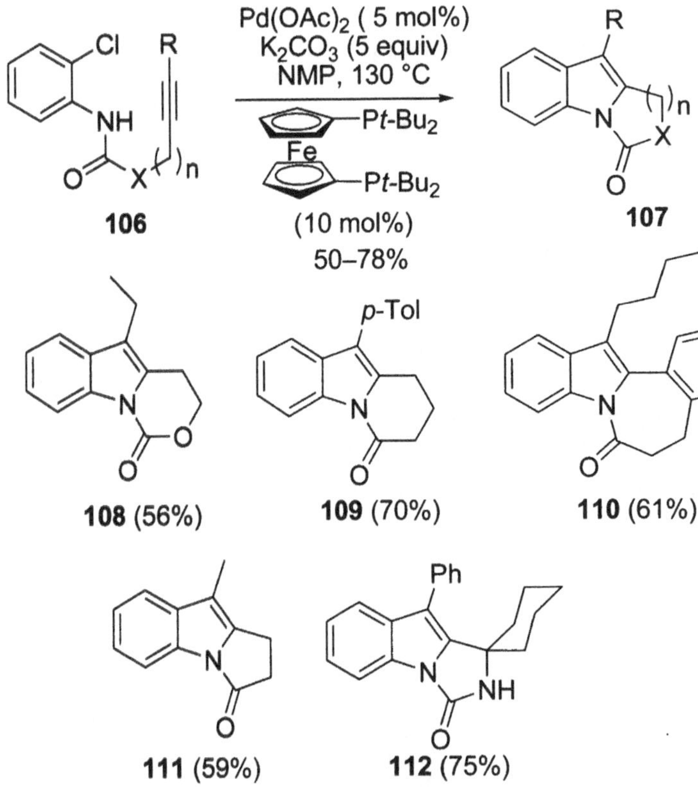

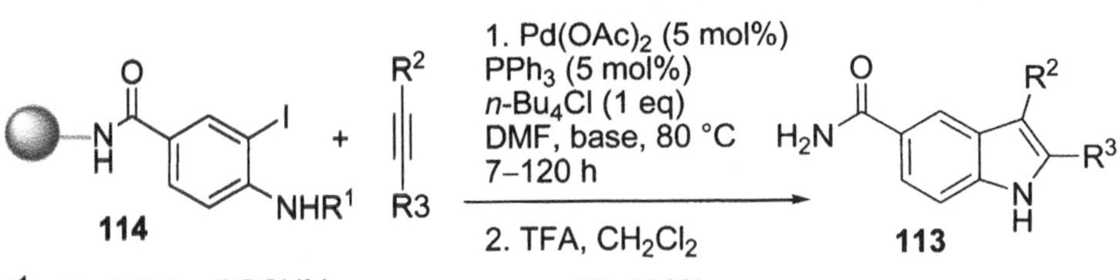

R^1 = H, COMe, $COCHMe_2$

R^2 = *n*-Pr, Me, Ph, COOEt, CH_2CH_2OH, CH_2CH_2Cl, *p*-MeOPh-CH_2CH_2, CH_2-*N*-pyrrolidinyl

R^3 = *n*-Pr, *t*-Bu, Ph, $SiMe_3$, CH_2-*N*-pyrrolidinyl

Larock indole synthesis was applied in a solid-phase synthesis of trisubstituted inodles **113** using an amide group as a linker starting from **114**.[45]

On the other hand, a traceless solid-phase synthesis of 2,3-disubstituted indoles **115** was developed using $PdCl_2(PPh_3)_2$ as the precatalyst, tetramethylgaunidine (TMG) as the base and under double coupling conditions.[46] Solution-phase conditions, in this case, caused incomplete reactions, and large quantities of multiple acetylene insertion side products were observed, presumably due to the poor solubility of the inorganic bases.

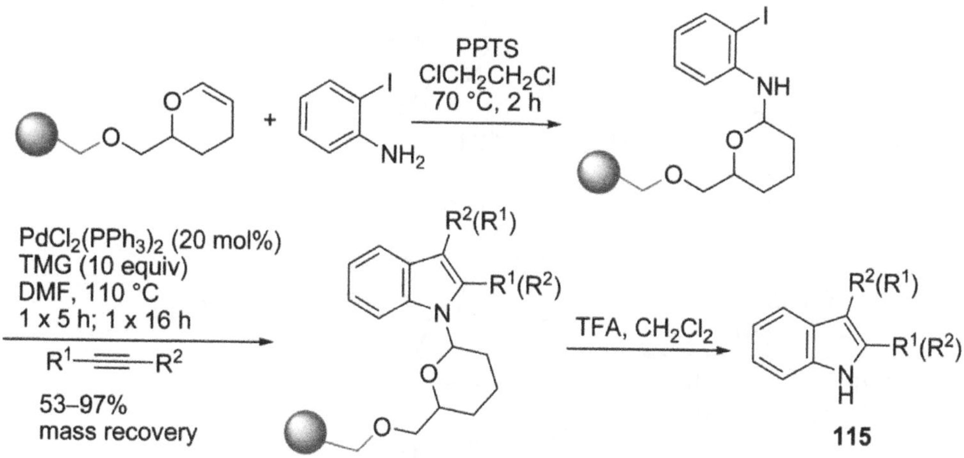

115

R^1 = H, COMe, COCHMe$_2$
R^2 = *n*-Pr, Me, Ph, COOEt, CH_2CH_2OH, CH_2CH_2Cl
R^3 = *n*-Pr, *t*-Bu, Ph, SiMe$_3$

Applications to the Total Synthesis of Natural Products

As mentioned previously, efficient asymmetric synthesis of optically active substituted tryptophans **91** was realized via Larock heteroannulation methodology by Cook and co-workers.[40] This application has been extended to the first asaymmetric synthesis of L-isotryptophan **116** and L-benz[*f*]tryptophan **117** and an improved synthesis of tryprostatin A **118**.[40]

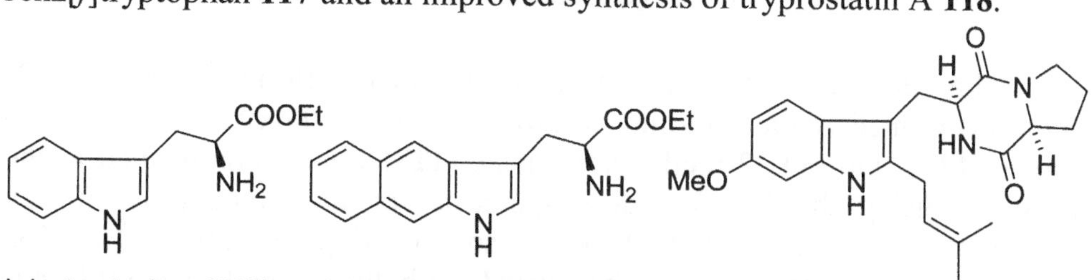

L-isotryptophan (**116**) L-benz[*f*]tryptophan (**117**) Tryprostatin A (**118**)

Similarly, Cook and co-workers used the Larock indole synthesis to prepare the key intermediate 4-methoxytryptophan derived from intermediate **119** in the total syntheses of Mitragynine (**120**), 9-methoxy-N_b-methylgeissoschizol (**121**) and 9-methoxygeissoschizol (**122**).[47,48] Thus Larock heteroannulation of Boc-protected 2-iodo-3-methoxyaniline **123** and a suitable silyl substituted internal alkyne such as **124** regiospecifically afforded the indole derivative **119**, a key intermediate for the total synthesis of mitragynine, 9-methoxy-N_b-methylgeissoschizol, and 9-methoxygeisso-schizol.

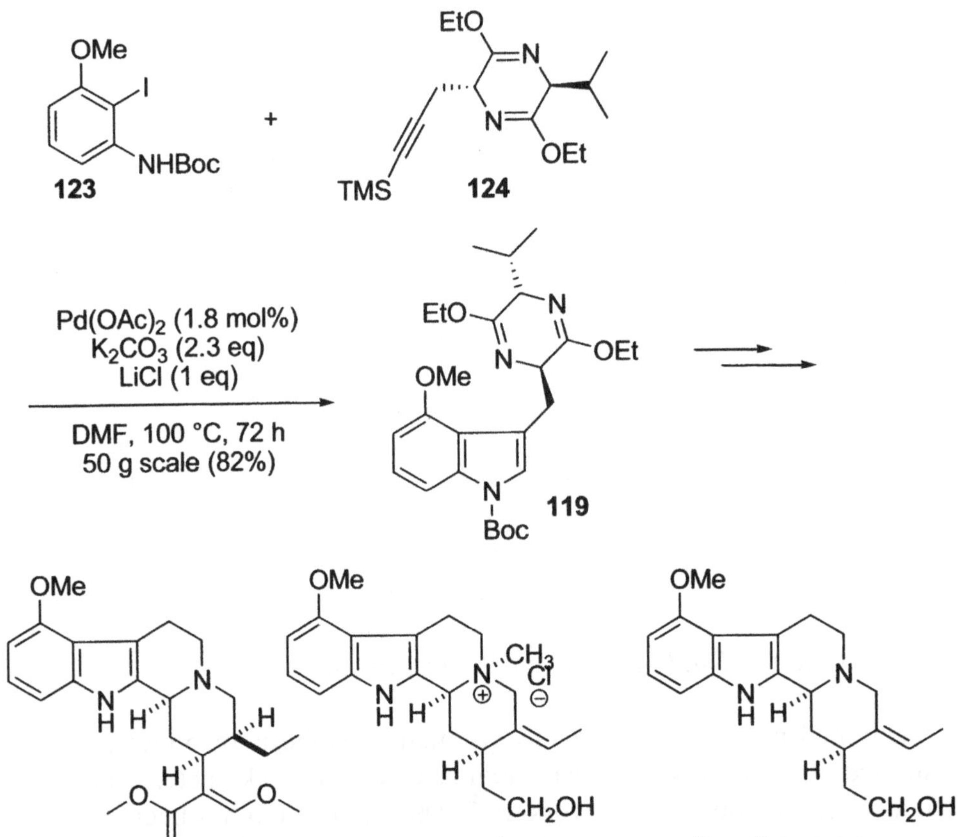

Earlier on, the total syntheses of (–)-fuchsiaefoline (**125**), (–)-12-methoxy-N_b-methylvoachalotine (**126**), (+)-12-methoxy-N_a-methylvellosimine (**127**), and (+)-12-methoxyaffinisine (**128**) were accomplished in the same lab using the larock indole synthesis for the preparation of key precursor 7-methoxy-D-tryptophan in an enantiopure form.[49,50] The annulation process could readily be carried out both on small scale (100 mg)

and on large scale (100 g, 75%). The Batoli indole synthesis has been used to synthesize 7-substiuted indoles, but only in moderate yields.[51]

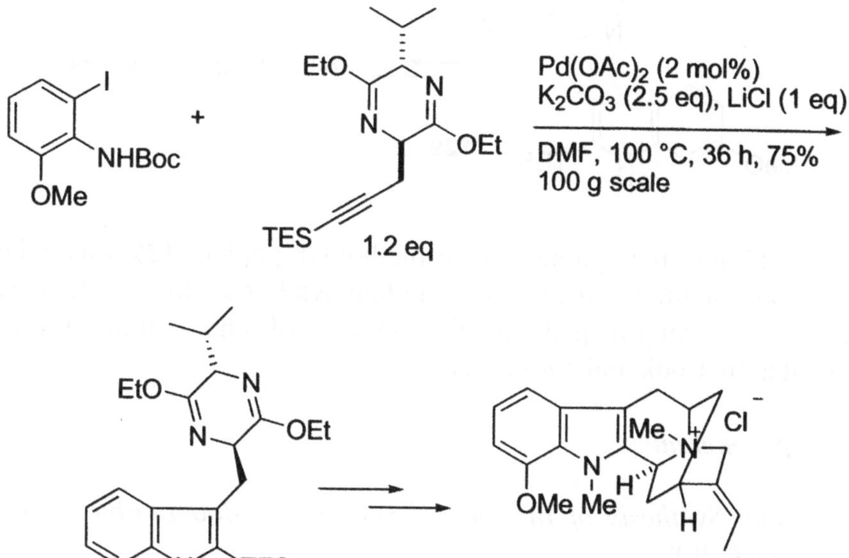

(−)-fuchsiaefoline (**125**)

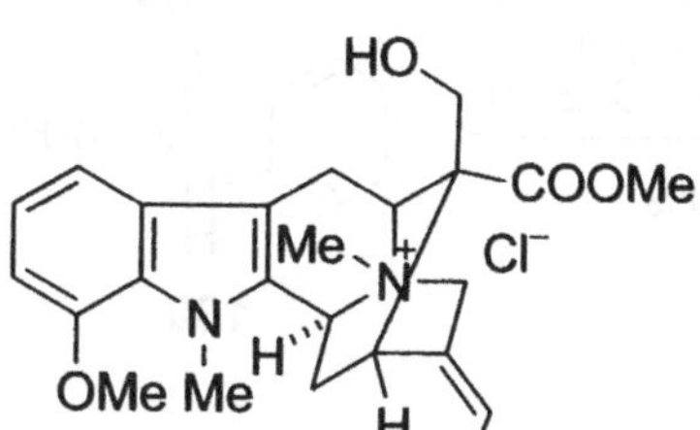

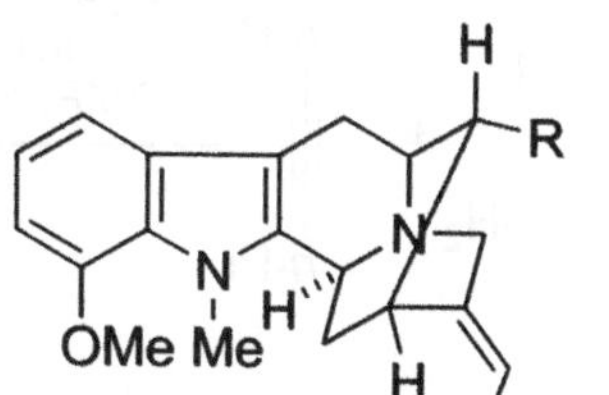

(−)-12-methoxy-N_b-methylvoachalotine (**126**)

R = CHO, (+)-12-methoxy-N_a-methylvellosimine (**127**)
R = CH_2OH, (+)-12-methoxyaffinisine (**128**)

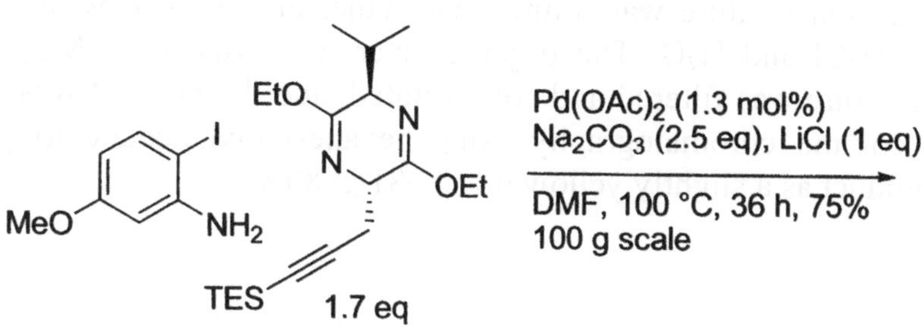

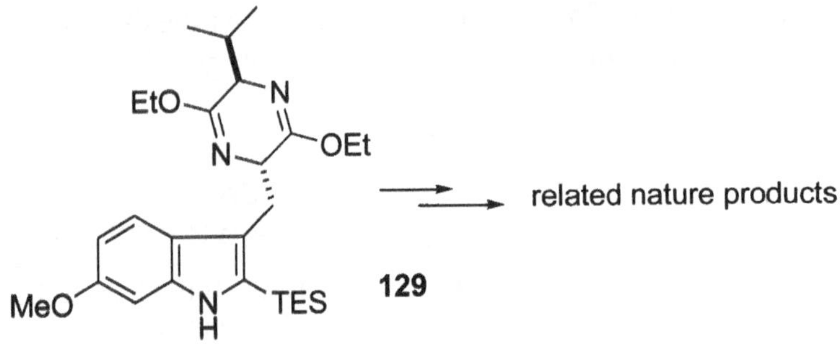

In addition, the synthesis of 6-methoxytrypophan **129** was achieved under similar conditions of the larock indole synthesis. Intermediate **129** is one of the key starting reagents for the synthesis of other natural products as demonstrated by Cook and co-workers.[52]

3.7.6 Experimental

Larock Indole Synthesis of Internal Alkynes with o-Iodoaniline: **2,3-Di-*n*-propylindole** (130)[2]

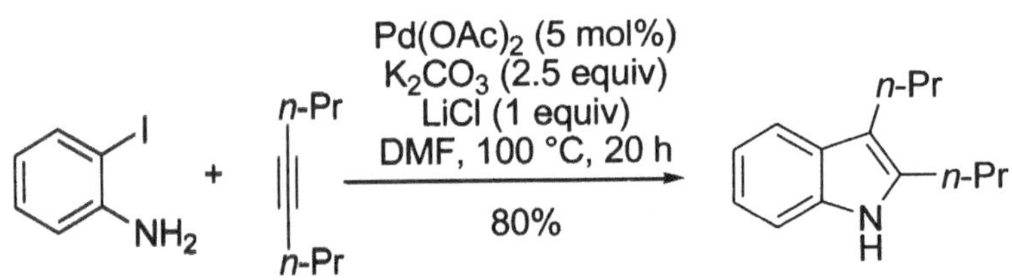

130

$Pd(OAc)_2$ (2.81 mg, 0.0125 mmol), LiCl, (0.021 g, 0.5 mmol), K_2CO_3 (0.35 g, 2.50 mmol), *o*-iodoaniline (0.11 g, 0.50 mmol), oct-4-yne (0.28 g, 2.50 mmol), and DMF (10 mL) were added to a 4-dram vial equipped with a stirring bar and Teflon-lined screw cap. After being heated for 20 h at 100 °C, the reaction mixture was diluted with ether and washed with saturated aqueous NH_4Cl and H_2O. The organic layer was dried over $MgSO_4$. The reaction mixture was filtered and concentrated, and the product was purified by flash column chromatography using hexanes/ethyl acetate to give the desired product as a slightly yellow oil (0.081 g, 80%).

Modified Larock Indole Synthesis of Internal Alkynes With o*-Bromo or* o*-Chloroanilines:* **3-Ethyl-2-isopropenyl-5-methyl-1*H*-indole** (131)[15]

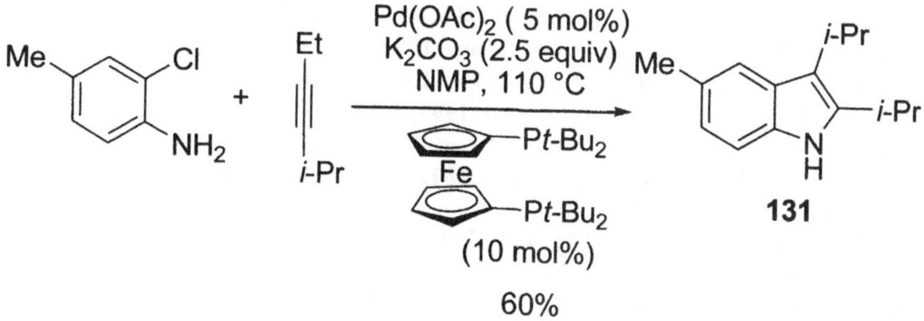

131

A 100 mL reaction flask equipped with a stirring bar and thermocouple were charged with $Pd(OAc)_2$ (56 mg, 0.25 mmol), D*t*-BPF (142 mg, 0.3 mmol), 2-chloro-4-methylaniline (0.71 g, 5 mmol), 2-methyl-1-hexen-3-yne (0.94 g, 10 mmol), K_2CO_3 (1.73 g, 12.5 mmol), and DMF (50 mL). The flask was purged with argon. The reaction was heated to 110 °C and stirred for overnight. The reaction was complete after 20 h at 110 °C. The mixture was filtered through a layer of celite and washed with ethyl acetate (10 mL). The filtrate was diluted with water (50 mL), and extracted with EtOAc (100 mL × 3). The combined organic phase was washed with water (50 mL × 4) and brine, dried over $MgSO_4$, filtered, and concentrated to give a dark brown residue. The product was purified via silica gel (hexane/EtOAc, 50:1). The desired product was obtained as an off-white solid (0.6 g, 60% yield).

3.7.7 *References*

1. Larock, R. C.; Yum, E. K. *J. Am. Chem. Soc.* **1991,** *113*, 6689–6690.
2. Larock, R. C.; Yum, E. K.; Refvik, M. D. *J. Org. Chem.* **1998,** *63*, 7652–7662.
3. [R] Larock, R. C. *Pure Appl. Chem.* **1999,** *71*, 1435–1442.
4. [R] Larock, R. C. *J. Organomet. Chem.* **1999,** *576*, 111–124.
5. [R] Cacchi, S. *J. Organomet. Chem.* **1999,** *576*, 111–124.
6. [R] Gribble, G. W. *Perkin 1* **2000,** 1045–1075.
7. [R] Poli, G.; Giambastiani, G.; Heumann, A. *Tetrahedron* **2000,** *56*, 5959–5989.
8. [R] Battistuzzi, G.; Cacchi, S.; Fabrizi, G. *Eur. J. Org. Chem.* **2002,** 2671–2681.
9. [R] Undheim, K. *Handbook of Organopalladium Chemistry for Organic Synthesis* **2002,** *1*, 409–492.
10. [R] Alonso, F.; Beletskaya, I. P.; Yus, M. *Chem. Rev.* **2004,** *104*, 3079–3159.
11. [R] Kirsch, G.; Hesse, S.; Comel, A. *Cur. Org. Synth.***2004,** *1*, 47–53.
12. [R] Cacchi, S.; Fabrizi, G. *Chem. Rev.* **2005,** *105*, 2873–2920.
13. [R] Humphrey, G. R.; Kuethe, J. T. *Chem. Rev.* **2006,** *106*, 2875–2911.
14. [R] Zeni, G.; Larock, R. C. *Chem. Rev.* **2006,** *106*, 4644–4680.
15. Shen, M.; Li, G.; Lu, B. Z.; Hossain, A.; Roschangar, F.; Farina, V.; Senanayake, C. H. *Org. Lett.* **2004,** *6*, 4129–4132.
16. Li, G.; Liu, J.; Lu, B. Z.; Roschangar, F.; Senanayake, C. H.; Shen, M. US Pat. 2005/0209465.
17. Denmark, S. E.; Baird J. D. *Tetrahedron* **2009,** *65*, 3120–3129.
18. Leogane, O.; Lebel, H. *Angew. Chem., Int. Ed.* **2008,** *47*, 350–352.

19. Crawley, M. L.; Goljer, I.; Jenkins, D. J.; Mehlmann, J. F.; Nogle, L.; Dooley, R.; Mahaney, P. E. *Org. Lett.* **2006,** *8*, 5837–5840.
20. Thansandote, P.; Hulcoop, D. G.; Langer, M.; Lautens, M. *J. Org. Chem.* **2009,** *74*, 1673–1678.
21. Larksarp, C.; Alper, H. *J. Org. Chem.* **2000,** *65,* 2773–2777.
22. Zenner, J. M.; Larock, R. C. *J. Org. Chem.* **1995,** *60*, 482–483.
23. Zenner, J. M.; Larock, R. C. *J. Org. Chem.* **1999,** *64,* 7312–7322.
24. Roesch, K. R.; Larock, R. C. *Org. Lett.* **1999,** *1,* 1551–1553.
25. Roesch, K. R.; Larock, R. C. *J. Org. Chem.* **2001,** *66,* 412–420.
26. (a) Zhang, H.; Larock, R. C. *Org. Lett.* **2001,** *3*, 3083–3086. (b) Zhang, H.; Larock, R. C. *J. Org. Chem.* **2002,** *67*, 9318–9330.
27. (a) Zhang, H.; Larock, R. C. *Org. Lett.* **2002,** *4*, 3035–3038. (b) Zhang, H.; Larock, R. C. *J. Org. Chem.* **2003,** *68*, 5132–5138.
28. (a) Kadnikov, D. V.; Larock, R. C. *Org. Lett.* **2000,** *2*, 3643–3646. (b) Kadnikov, D. V.; Larock, R. C. *J. Org. Chem.* **2003,** *68*, 9423–9432.
29. Kadnikov, D. V.; Larock, R. C. *J. Org. Chem.* **2004,** *69*, 6772–6780.
30. Larock, R. C.; Doty, M. J.; Han, X. *J. Org. Chem.* **1999,** *64*, 8770–8779.
31. Wensbo, D.; Erlksson, A.; Jeschke, T.; Annby, U.; Gronowitz, S.; Lams, T. *Tetrahedron Lett.* **1993,** *34*, 2823–2826.
32. Van Snick, W.; Dehaen, W. *Tetrahedron* **2009,** *65*, 8497–8501.
33. Koolman, H.; Timo Heinrich, T.; Böttcher, H.; Rautenberg, W.; Reggelin, M. *Bioorg. Med. Chem. Lett.* **2009,** *19*, 1879–1882.
34. Chen, C.-Y.; Lieberman, D. R.; Street, L. J.; Guiblin, A. R.; Larsen, R. D.; Verhoeven, T. R. *Synth. Commun.* **1996,** *26*, 1977–1984.
35. Chen, C.-Y.; Lieberman, D. R.; Larsen, R. D.; Reamer, R. A.; Verhoeven, T. R.; Reider, P. J. *Tetrahedron Lett.* **1994,** *35*, 6981–6984.
36. Humora, M. J.; Modi, S. P.; Srivastava, S. K.; Williams, A. D. U.S. Pat. 5,550,239, 1996.
37. Parmentier, J.-G.; Poissonnet, G.; Goldstein, S. *Heterocycles* **2002,** *56*, 465–476.
38. Gathergood, N.; Scammells, P. J. *Org. Lett.* **2003,** *5*, 921–923.
39. Jeschke, T.; Wensbo, D.; Annby, U.; Gronowitz, S.; *Tetrahedron Lett.* **1993,** *34*, 6471–6474.
40. (a) Ma, C.; Liu, X.; Yu, S.; Zhao, S.; Cook, J. M. *Tetrahedron Lett.* **1999,** *40*, 657–660. (b) Ma, C.; Liu, X.; Li, X.; Flippen-Anderson, J.; Yu, S.; Cook, J. M. *J. Org. Chem.* **2001,** *66*, 4525–4542.
41. Walsh, T. F.; Toupence, R. B.; Ujjainwalla, F.; Young, J. R.; Goulet, M. T. *Tetrahedron* **2001,** *57*, 5233–5241.
42. Nishikawa, T.; Wada, K.; Isobe, M. *Biosci. Biotechnol. Biochem.* **2002,** *66*, 2273–2278.
43. Maeda, C.; Shinokubo, H.; Osuka, A. *Org. Lett.* **2007,** *9*, 2493–2496.
44. Liu, J.; Shen, M.; Zhang, Y.; Li, G.; Khodabocus, A.; Rodriguez, S.; Qu, B.; Farina, V.; Senanayake, C. H.; Lu, B. Z. *Org. Lett.* **2006,** *8*, 3573–3575.
45. Zhang, H.-C.; Brumfield, K. K.; Maryanoff, B. E. *Tetrahedron Lett.* **1997,** *38*, 2439–2442.
46. Smith, A. L.; Stevenson, G. I.; Swain, C. J.; Castro, J. L. *Tetrahedron Lett.* **1998,** *39*, 8317–8320.
47. Ma, J.; Yin, W.; Zhou, H.; Cook, J. M. *Org. Lett.* **2007,** *9*, 3491–3494
48. Ma, J.; Yin, W.; Zhou, H.; Liao, X.; Cook, J. M. *J. Org. Chem.* **2009,** *74*, 264–273.
49. Zhou, H.; Liao, X.; Cook, J. M. *Org. Lett.* **2004,** *6*, 249–252.
50. Zhou, H.; Liao, X.; Yin, W.; Ma, J.; Cook, J. M. *J. Org. Chem.* **2006,** *71*, 251–259.
51. (a) Bartoli, G.; Palmieri, G.; Bosco, M.; Dalpozzo, R. *Tetrahedron Lett.* **1989,** *30*, 2129–2132. (b) Dobbs, A. *J. Org. Chem.* **2001,** *66*, 638–641. (c) Dobson, D.; Todd, A.; Gilmore, J. *Synth. Commun.* **1991,** *21*, 611–617. (d) Dobson, D.; Gilmore, J.; Long, D. A. *Synlett* **1992,** 79–80.
52. (a) Liu, X.; Deschamp, J. R.; Cook, J. M. *Org. Lett.* **2002,** *4*, 3339–3342. (b) Yu, J.; Wearing, X. Z.; Cook, J. M. *J. Org. Chem.* **2005,** *70*, 3963–3979.

3.8 Martinet Dioxindole Reaction

Martin E. Hayes

3.8.1 Description

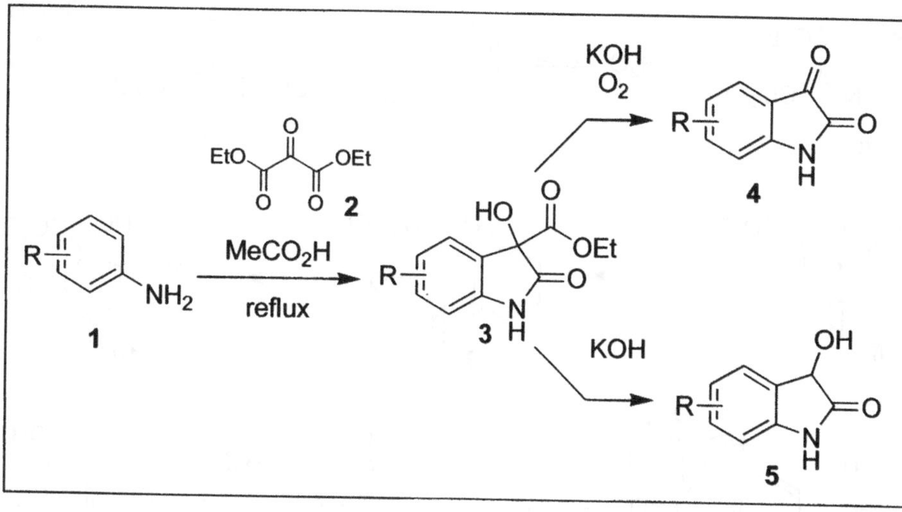

The Martinet reaction is the conversion of primary or secondary anilines in combination with a mesoxalic ester to give an isolable 3-hydroxy-3-carboxy-oxindole (**3**) which, depending on the reaction conditions, can be converted to either an isatin (**4**) or the corresponding 3-hydroxyoxindole (**5**) with a base such as potassium hydroxide.[1]

3.8.2 Historical Perspective

Joseph Martinet first reported[2] the conversion of anilines to dioxindoles upon combination with diethylmesoxalic ester in 1913 while earning a Ph.D. under the direction of A. Haller.[3] He later went on the develop syntheses of indigols as well as develop practical methods for preparing diazo-indole dyes, which have been used industrially.[4]

3.8.3 Mechanism

While no detailed mechanistic studies have been reported, a reasonable mechanism has been proposed previously.[5] Under acid conditions, the aniline and mesoxalic esters likely exist in equilibrium with a Schiff's base such as **6**. However, an anilide (**8**) also can form irreversibly, which then undergoes an intramolecular Friedel–Craft-type alkylation to give an isolable 3-hydroxy-3-carboalkoxy-oxindole such as **10** at elevated temperatures. The

3-carboxyoxindole (**10**) can be further functionalized under basic conditions to give either dioxindole such as **13** or the corresponding isatin (**4**). In the presence of oxygen[2] and a base such as sodium or potassium hydroxide, the isatin is isolated as the major product, however if oxygen is excluded the dioxindole (**13**) is the observed product.

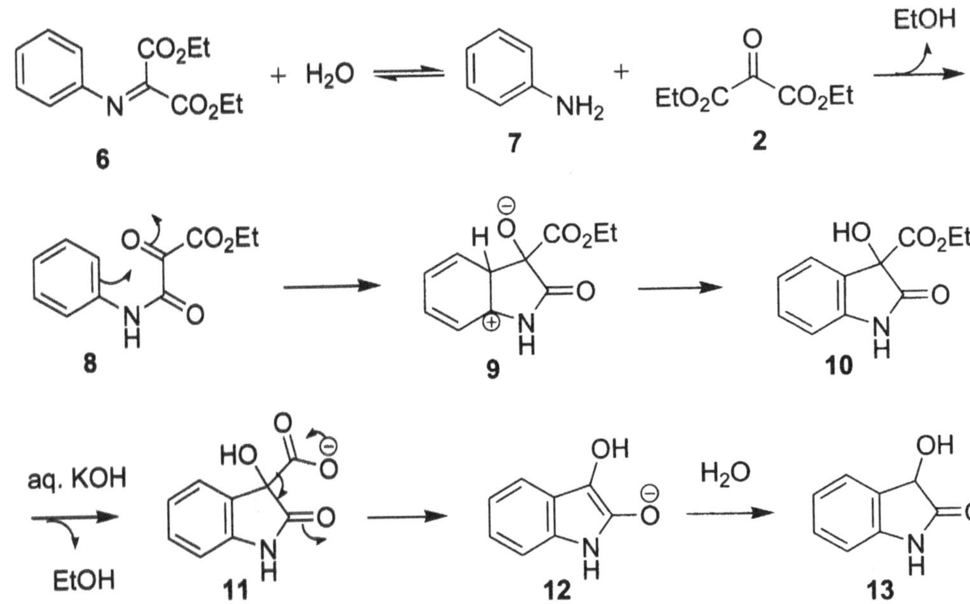

3.8.4 *Variations and Improvements*

The reaction conditions initially reported by Martinet[2] remain the most widely employed for the preparation of dioxindoles. Thermal condensations of anilines with diethylmesoxalic esters in acetic acid at reflux temperatures provide 3-hydroxy-3-carboethoxyoxindoles such as **10**. The reaction mixture is heated for only a short time, typically 10–20 min, and then allowed to sit at room temperature for several hours, allowing the newly formed oxindole to precipitate from solution. A basic workup can be used; however, neutralization is often omitted in favor of washing the solid with a solvent such as petroleum ether. The 3-hydroxy-3-carboxyoxindole can be conveniently converted to the corresponding isatin via aeration of the intermediate oxindoles such as **10** in a basic solution at elevated temperatures. Again, heating is most often applied for a short period of time, 10–20 min., though protocols with extended heating and aeration steps have been reported.[6] To obtain the decarboxylated oxindoles such as **13**, treatment with potassium or sodium hydroxide under inert atmosphere is required. The majority of synthetically useful applications of the Martinet reaction involve conversion to the isatin derivatives. There are no reports of metal-catalyzed

Martinet reactions, though treatment of preformed *N*-phenylglyoxamides such as **14** with Cu(II) catalysts have been reported to give either dioxindole **15** or isatin **16** in good yields.[7]

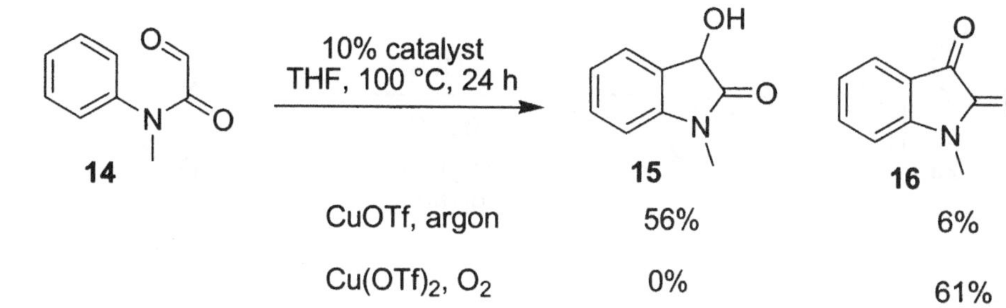

The regiochemical outcome of the Martinet reaction is influenced by steric considerations. For example, 3-phenylaniline (**17**) under the standard Martinet conditions provides 6-phenylisatin (**18**) as the only product.[6] Substitution on nitrogen is tolerated with limited examples of *N*-alkyl reactants such as *p*-anisole (**19**)[8] and *N*-phenyl-*N*-ethyl amine (**21**).[2]

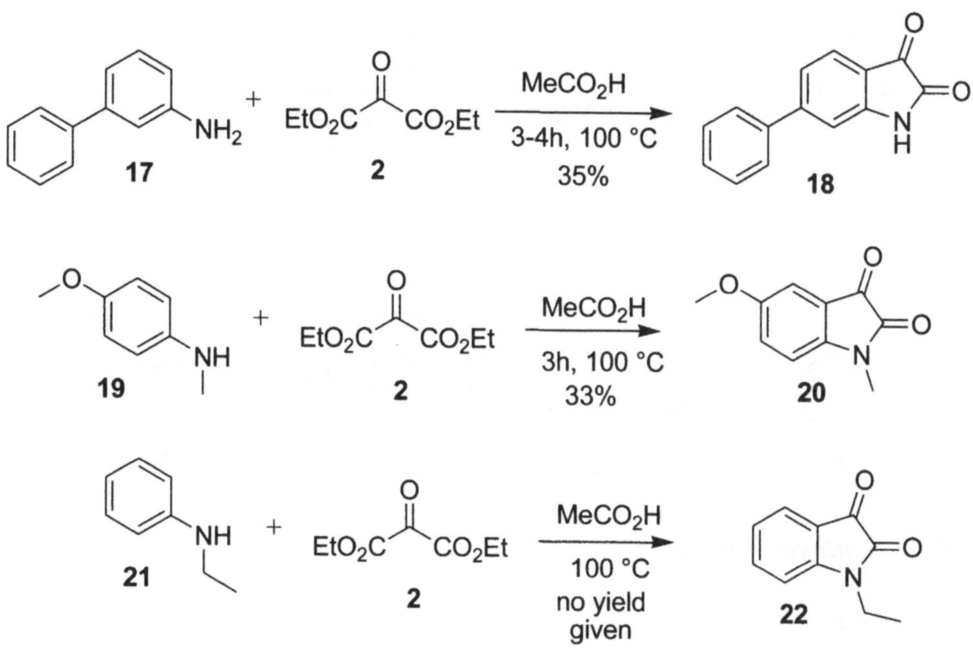

Tetrahydroquinolines are also reported to react in a Martinet fashion with mesoxalic esters to give condensed tricycles such as **24**.[9] α- and β-Amino-napthalenes have been reported to participate in the reaction as well, giving the linear tricycles such as **26**.[10] The reaction is tolerant of *ortho*-

substitution on the aniline as shown in the conversion[6] of **27** to **28** as well as electron-withdrawing substituents on the aniline (**30**).[11]

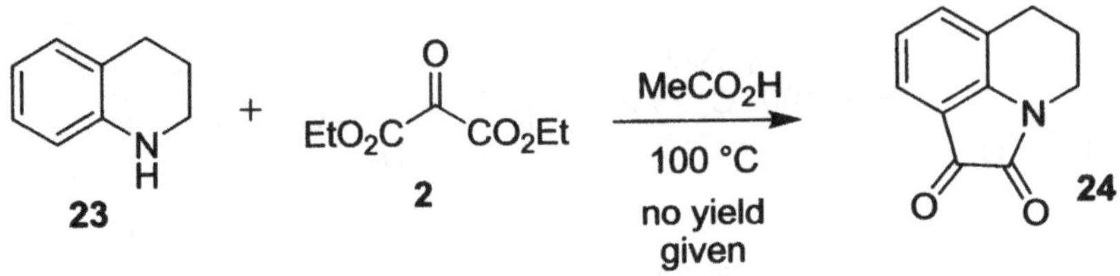

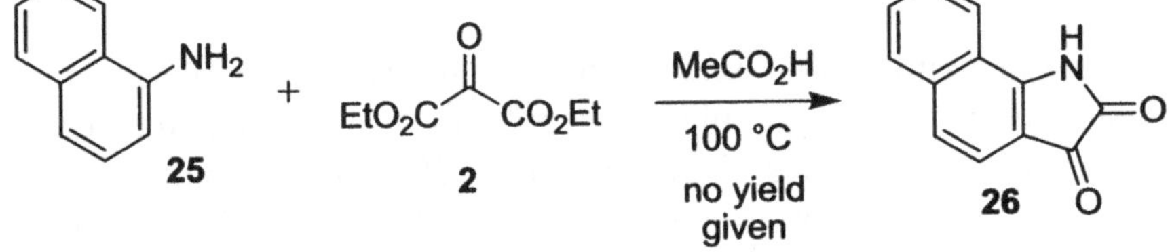

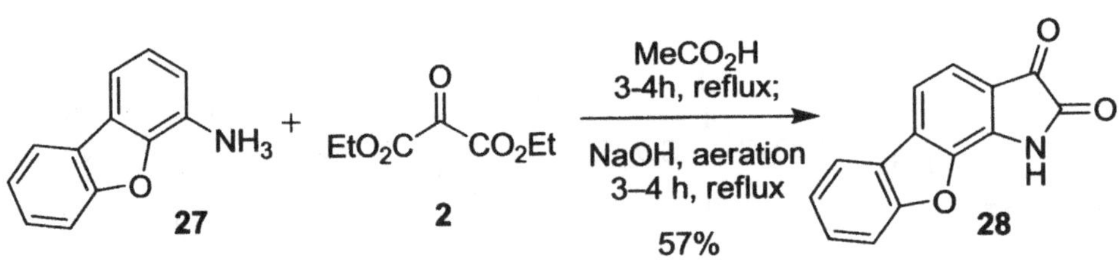

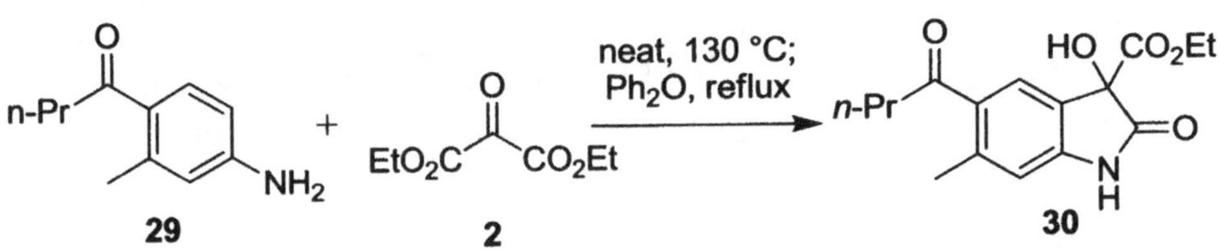

3.8.5 Synthetic Utility

General Utility

Anilines with electron-donating groups at the *meta*-position are reported to provide good yields in the Martinet reaction, as would be expected from the proposed mechanism. For example, 3-5-diemethoxyaniline (**31**) reacts to give an 88% yield of the 3-hydroxy-3-carboethoxyoxindole **32**.[12]

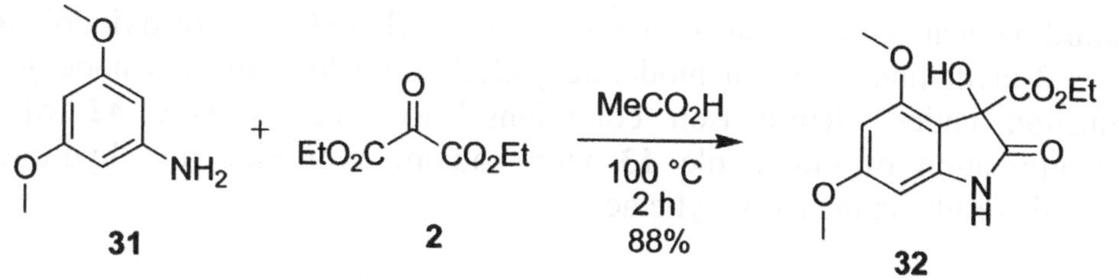

Characterization of the products arising from regioisomeric dimethoxyanilines reveal unexpected byproducts such as **35**[13] and **38**.[14] These byproducts presumably arise from competitive side-reactions of the electron-rich anilines and not from intermediates of the Martinet reaction.

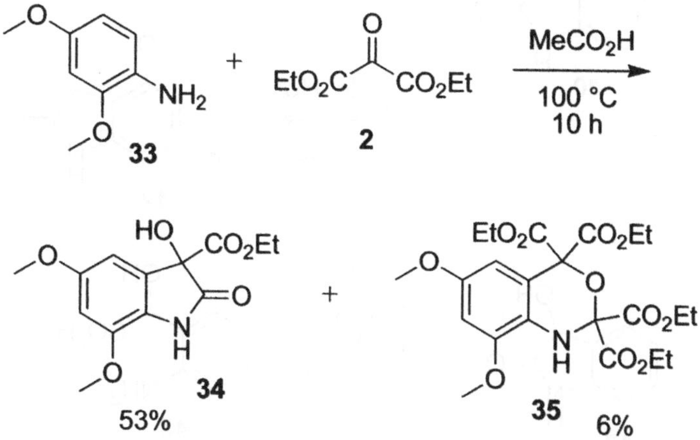

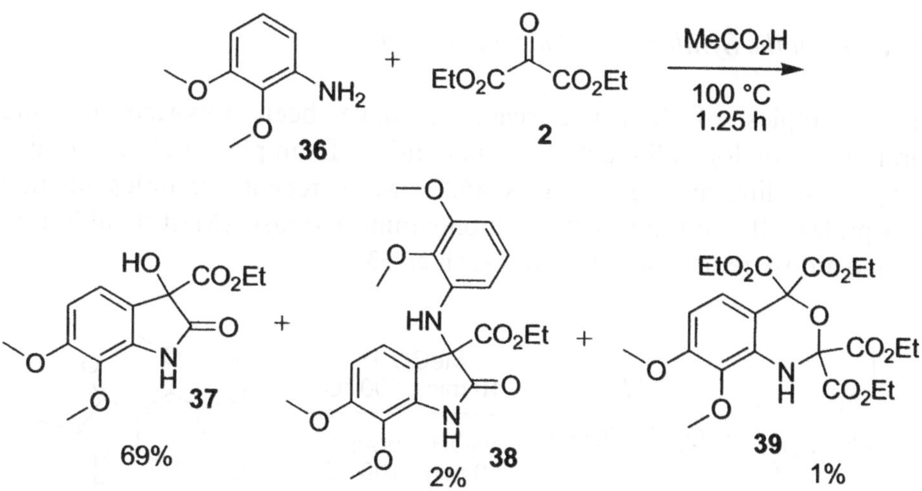

The majority of Martinet[15] reactions in the literature report the preparation of substituted isatins.[16] There are some examples, though, of preparations of the 3-hydroxy-3-carboethoxyoxindoles as intermediates toward the synthesis of related heterocycles. LAH reduction of oxindole **40** gives 3-methylindole **41** in moderate yield.[17] Oxindole **40** also undergoes reduction under Clemmenson conditions[18] to give oxindole **42** while saponification of dioxindole **13** with sodium hydroxide and hydrogen peroxide results in the carboxylic acid **43**.[19]

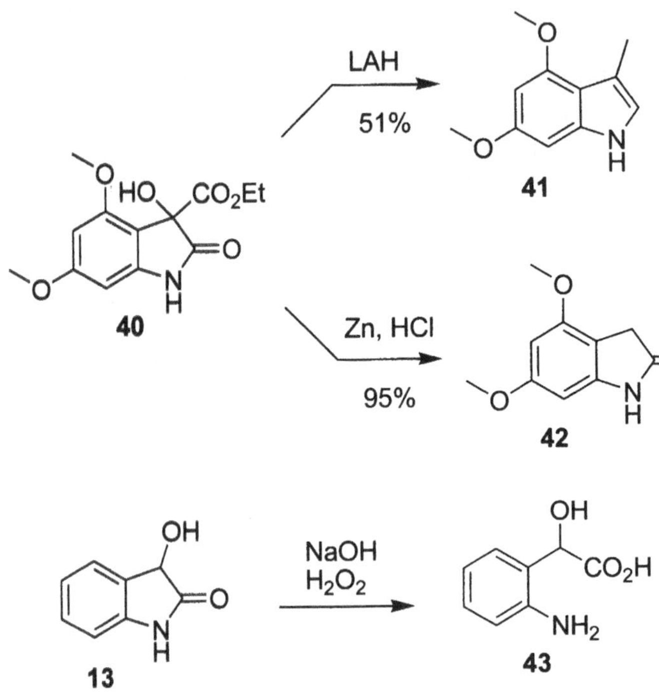

Applications in the Synthesis of Pharmaceuticals

Several examples of Martinet reactions have been reported for the preparation of biologically active compounds. Examples include studies targeting Mescaline analogs such as **45**.[20] More recent examples include aminopeptidase II inhibitors (**48**),[21] monoamine oxidase (MAO) inhibitors (**50**),[22] and aldose reductase inhibitors such as **53**.[23]

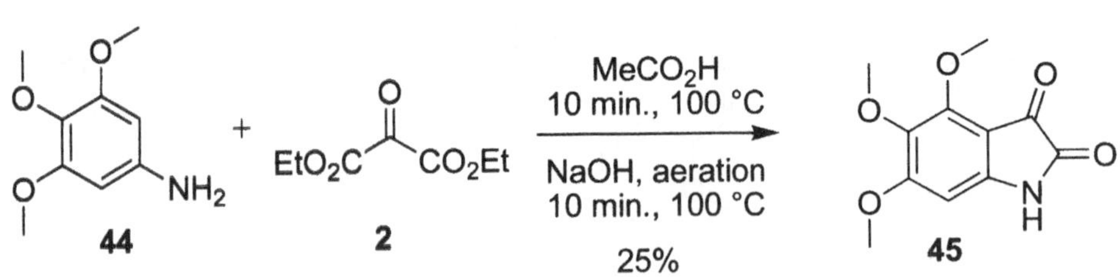

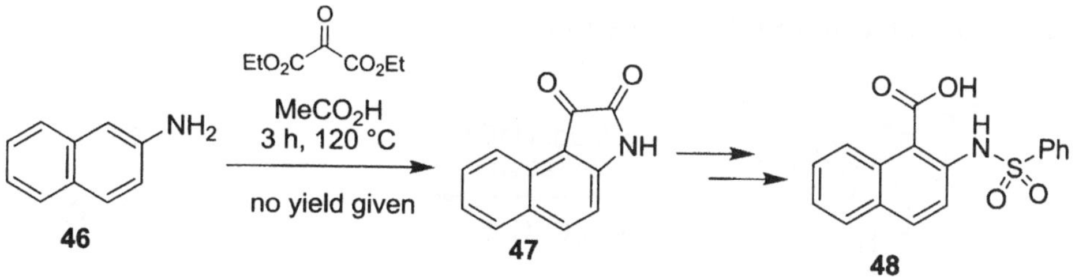

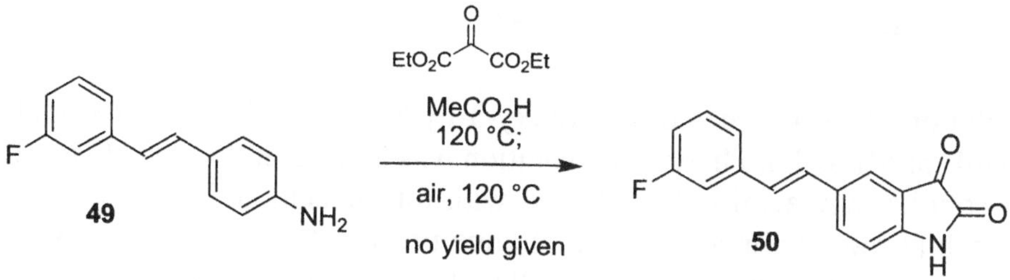

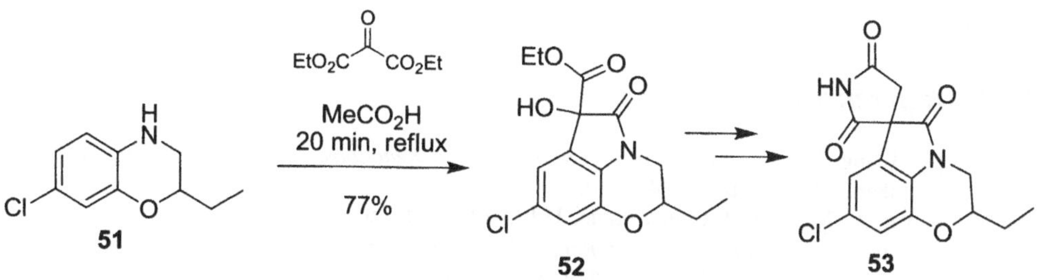

3.8.6 *Experimental*

Preparation of 4,6-Dimethoxy-3-hydroxy-3-carboethoxyoxindole (32)[12]

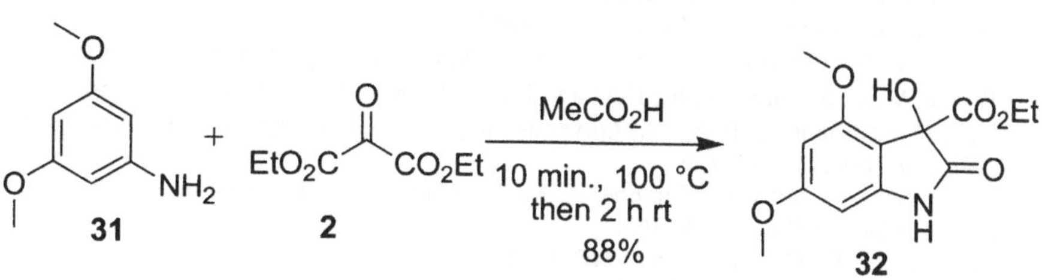

Diethyl mesoxalate and 50 mL of acetic acid was heated on a steam bath for 10 min, then allowed to stand for 2 h at room temperature. The resulting white precipitate was washed with petroleum ether and dried to afford the title compound (20 g, 88%): mp 188.5–190.5 °C. An analytical sample (mp 191.5–194 °C) was obtained by recrystallization from acetic acid.

Preparation of 4,6-Dimethoxyisatin (54)[12]

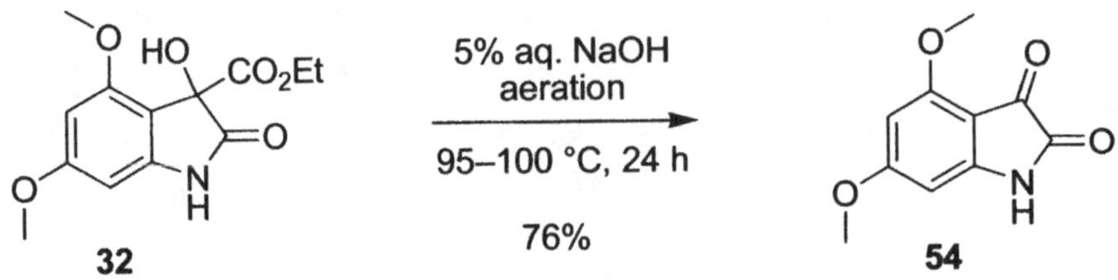

A solution of 19.4 g of 4,6-dimethoxy-3-hydroxy-3-carboethoxyoxindole in 200 mL of 5% sodium hydroxide solution was aerated at 95–100 °C for 24 h. The solution was acidified to pH = 4 with formic acid and allowed to stand for 15 h at room temperature. The orange precipitate was collected, washed with water, and dried to give the title compound (10.9 g, 76%). Recrystallization from 2-methoxyethanol afforded an analytical sample: mp 300 °C.

3.8.7 References

1. [R] Wang, Z. in *Comprehensive Organic Name Reactions and Reagents, vol. 2*; Wiley: Hoboken, NJ, 2009, pp. 1838–1840.
2. Guyot, A.; Martinet, J. *Compt. Rend.* **1913**, *166*, 1625–1628.
3. Guignard, M. M.; *et al. Compt. Rend.* **1918**, *166*, 1310.
4. Martinet, J. *Bull. Soc. Chim. Fr.* **1929**, *45*, 101–109.
5. Taylor, A. *J. Chem. Res., Miniprint* **1980**, *10*, 4154–4171.
6. Langenbeck, V. W.; Rühlmann, K.; Reif, H. H.; Stolze, F. *J. Prakt. Chem.* **1956**, *4*, 136–146.
7. Tang, B.-X.; Song, R.-J.; Wu, C.-Y.; Liu, Y.; Zhou, M.-B.; Wei, W. T.; Deng, G. B.; Yin, D.-L.; Li, J.-H. *J. Am. Chem. Soc.* **2010**, *132*, 8900–8902.
8. Brand, K.; Völcker, E. *Arch. Pharm.* **1934**, *272*, 257–273.
9. Martinet, J. *Compt. Rend.* **1918**, *166*, 998–1000.
10. Martinet, J. *Compt. Rend.* **1918**, *166*, 851–853.
11. Goschke, R.; Ferrini, P. G.; Sallmann, S. A. US Pat. Appl. 4476132 A1, 1984.
12. Brown, V. H.; Skinner, W. A.; DeGraw, J. I. *J. Heterocl. Chem.* **1969**, *6*, 539–543.
13. Hall, R. J.; Shannon, P. V. R.; Oliveira-Campos, A.-M. F.; Qu Joao R. P. *J. Chem. Res. Miniprint*, **1992**, *1*, 114–156.
14. Dharmasens, P.; Oliveira-Campos, A.-M. R.; Queiroz, M.-J.; Shannon, P. V. R. *J. Chem. Res., Miniprint* **1996**, *1*, 316–363.
15. [R] Sumpter, W. C. *Chem. Rev.* **1945**, *37*, 443–479.
16. [R] Sumpter, W. C. *Chem. Rev.* **1944**, *34*, 393–434.
17. Black, D. StC.; Rothnie, N. E.; Wong, L. C. H. *Tetrahedron Lett.* **1980**, *21*, 1883–1886.
18. Black, D. StC.; Ivory, A. J.; Kumar, N. *Tetrahedron* **1996**, *52*, 4697–4708.
19. Martinet, J. *Compt. Rend.* **1918**, *166*, 952–955.
20. Benington F.; Morin, R. D.; Clark, L. C. Jr. *J. Org. Chem.* **1955**, *20*, 1455–1456.
21. Kawai, M.; BaMaung, N. Y.; Findanze, S. D.; Erickson, S. A.; Tedrow, J. S.; Sanders, W. J.; Vasudevan, A.; Park, C.; Hutchins, C.; Comess, K. M.; Kalvin, D.; Wang, J.; Zhang, Q.; Lou, P.; Tucker-Garcia, L.; Bouska, J.; Bell, R. L.; Lesniewski, R.; Henkin, J.; Sheppard, G. S. *Bioorg. Med. Chem. Lett.* **2006**, *16*, 3574–3577.
22. Van der Walt, E. M.; Milczek, E. M.; Malan, S.F.; Edmondson, D. E.; Castagnoli, N. Jr.; Bergh, J. J.; Petzer, J. P. *Bioorg. Med. Chem. Lett.* **2009**, *19*, 2509–2513.
23. Masuzawa, K.; Okamura, K.; Fujimori, S.; Kinoshita, S.; Matsukubo, H. Eur Pat. Appl. 254149, 1988.

3.9 Mori–Ban Indole Synthesis

Ying Han

3.9.1 *Description*

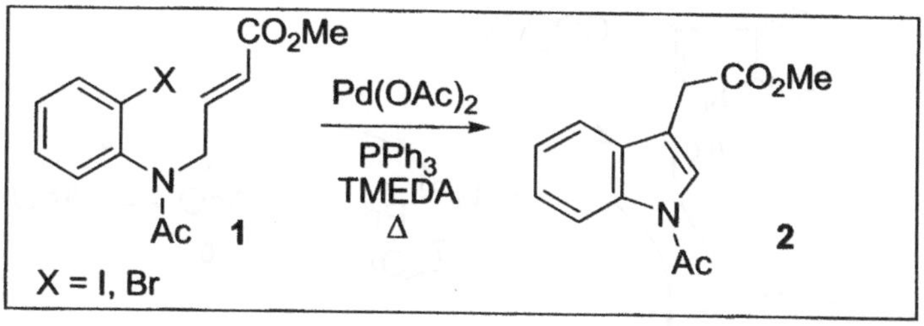

Mori–Ban reaction refers to a synthesis of indole derivatives by an intramolecular Heck reaction of *o*-halo-aniline with pendant olefin catalyzed by a low-valent metal complex Pd and Ni.[1–3] The preferred *ortho*-halogen is bromine or iodine (chlorine has very low reactivity under this case). More often, the *o*-iodo-*N*-allylaniline is more reactive than corresponding *o*-bromo- and *o*-chloro-substrate. Also, it has been found that the catalyst can be deactivated under the reaction. Therefore, a periodic provision of fresh catalyst normally gives higher overall yields than that using the same total amount of catalyst with one addition. In the past three decades, the Mori–Ban reaction has been improved and applied to variety of organic synthesis.

3.9.2 *Historical Perspective*

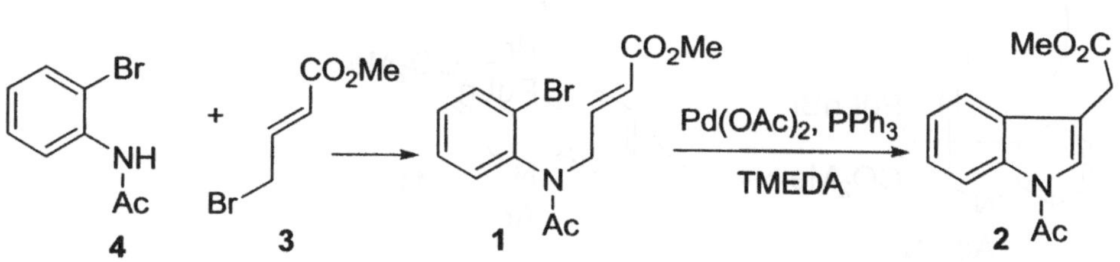

Starting in 1977, Mori and Ban[1–3] at Hokkaido University discovered that Pd can catalyze the intramolecular reaction of *o*-halo-*N*-allylanilines to indoles under Heck reaction conditions.[4] It was the first intramolecuar Heck reaction to the synthesis of heterocycles. For example, the compound **1**, which was prepared from methyl α-bromocrotonate (**3**) and 2-bromo-*N*-acetylaniline **4**, was adopted as a starting material. It was treated with $Pd(OAc)_2$ (2 mol%) and PPh_3 (4 mol%) in the presence of tetramethylethylenediamine (TMEDA, 200 mol%) under a stream of nitrogen at 125 °C for 5.5 h; methyl 1-acetyl-3-indolyl acetate was obtained in a yield of 43% via an intramolecular Heck reaction followed by olefin isomerization to afford the fully aromatic product. Although yields from the initial report were moderate, they have been greatly improved over the years.

3.9.3 *Mechanism*

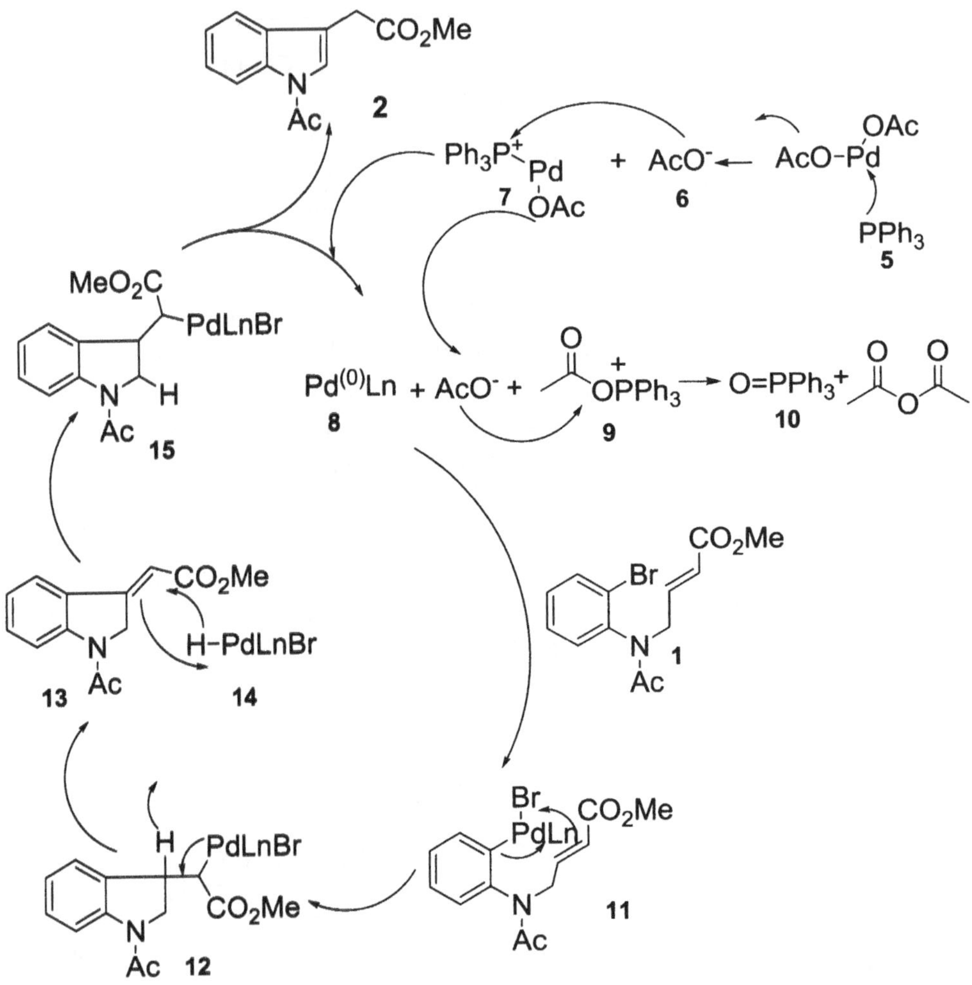

When $Pd(OAc)_2$ was mixed with triphenylphosphine, the reaction provides acetoxy(triphenylphosphonio)palladium intermediate **7**, which is reduced to Pd(0)Ln, a low-valent active palladium species for starting the catalytical cycle. Next, $Pd^{(0)}Ln$ with **1** undergoes an oxidative addition to form intermediate **11**, followed by an intramolecular insertion generating a cyclic intermediate **12**. β-H reductive elimination of **12** gives **13** and **14**. Finally, Pd-H is added to the C=C bond, and a β-H reductive elimination yields the target product **2**, regenerating the catalytically active species **8** to restart the next catalytical cycle.

3.9.4 *Variations and Improvements*

There are many variations and improvements on the application of Heck cyclization to the synthesis of indoles (the Mori–Ban indole synthesis) in the past three decades. The cyclization of *o*-halo-*N*-allylanilines to indoles is a general and efficient methodology that has been improved by Larock.[5] These improvements include cyclozations of *o*-halo-*N*-allylanilines and *o*-halo-*N*-acryloylanilines into indoles and oxindoles. For example, the conversion of **16** to **17** can be performed at lower temperature, shorter reaction time, and with less catalyst to give 3-methylindole **17** in 87% yield.

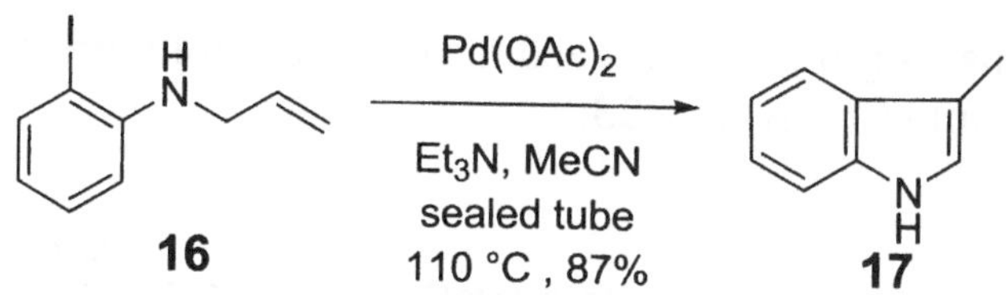

16 **17**

Larock's improved conditions, which have been widely adopted, are catalytic (2 mol%) $Pd(OAc)_2$, *n*-Bu_4NCl, DMF, base (usually Na_2CO_3), 25 °C, 24 h (also known as the Jeffrey's conditions). Larock extended his work in several ways, particularly in regard to Pd-catalyzed cross-coupling of *o*-allylic and *o*-vinylic anilides with vinyl halides and triflates to produce 2-vinylindoles.[6–9]

3.9.5 *Synthetic Utility*

As the first application of Heck reaction in making heterocyclic compounds —namely indole derivatives, the Mori–Ban reaction has been widely used in variety synthesis of indoles. In a program to synthesize CC-1065 analogs, Sundberg prepared indole **19** from *o*-bromo-*N*-allylaniline **18** in an excellent yield using the Jeffrey's conditions.[10] Silver carbonate and sodium carbonate were less effective than triethylamine.

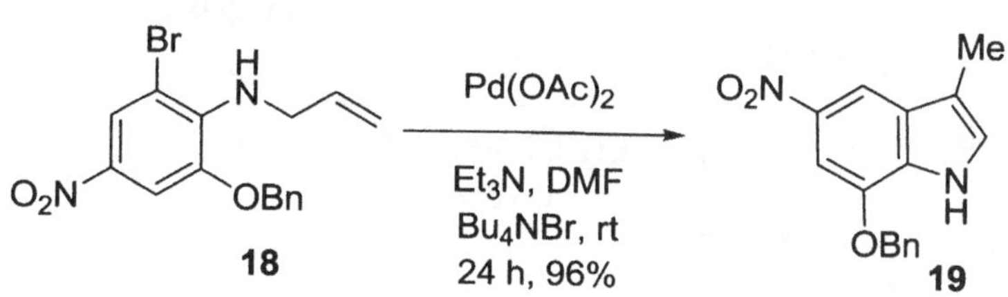

18 **19**

Similarly, Sakamoto has prepared the CC-1065/duocarmycin pharmacophore **22** starting with **20**.[11] Silver carbonate prevented unwanted isomerization of the exocyclic double bond in **21**. Tietze and co-workers applied this method to the synthesis of the A-unit of CC-1065 and analogs.[12]

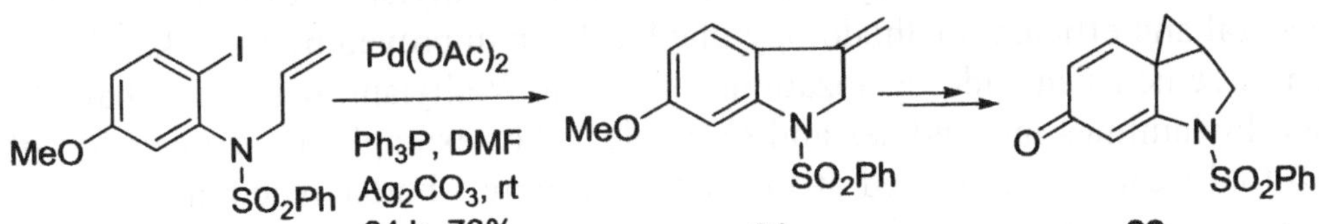

20 **21** **22**

Hoffman's group synthesized the desoxyeserolin precursor **24** from *N*-pyrrolidone aniline derivative **23**.[13]

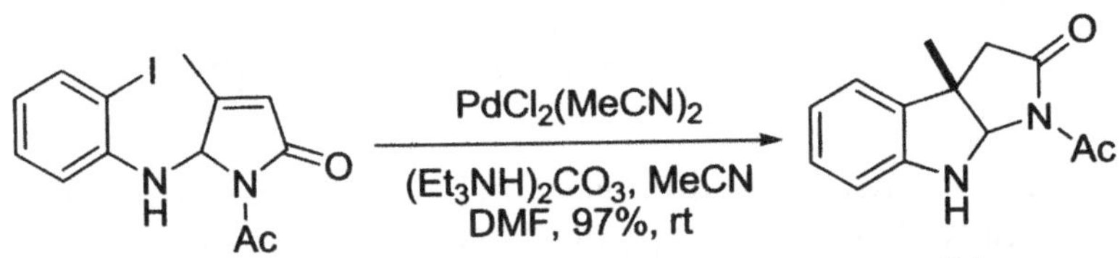

23 **24**

Macor and co-workers explored the Mori–Ban reaction to synthesize several antimigraine analogs of Sumatriptan and homo-tryptamines as potent and selective serotonin reuptake inhibitors.[14,15] The second bromine on substrate **25** was not significantly deleterious to the reaction although a small amount of the 7-bromoindole **26** might be sacrificed at the end of the reaction to consume the active palladium catalyst. This approach to 7-bromoindole **26** can provide a general method accessing 7-bromoindoles, which then could be further manipulated for the synthesis of more complex 7-substituted indoles.

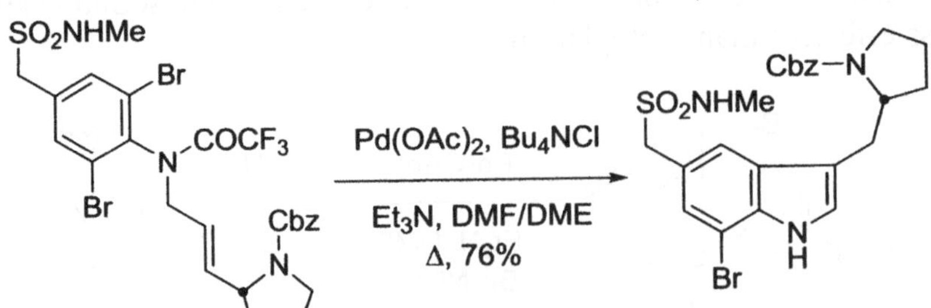

25 **26**

A solid phase synthesis of indoles using this technology has been successfully adapted.[15] An array of 1-acyl-3-alkyl-6-hydroxyindoles is readily assembled from acid chlorides, allylic bromides, and 4-bromo-3-nitroanisole. The Rink amide resin **27** was used to synthesize *N*-benylindole-3-acetamides and related indoles **28** via Heck cyclization by Zhang and Maryanoff.[16]

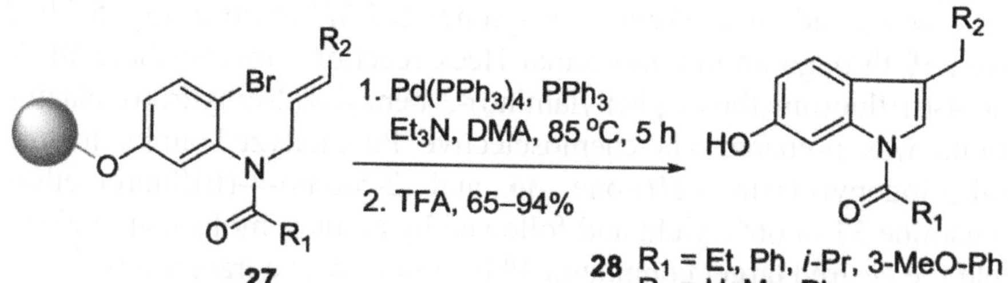

27

28 R_1 = Et, Ph, *i*-Pr, 3-MeO-Ph; R_2 = H, Me, Ph

A first tandem Heck reaction from o-ally-*N*-acryloylanilines leading to tricyclic pyrrolo[1,2-*a*]indoles or pyridino[1,2-*a*]indoles was discovered by Hegedus and Danishefsky.[17,18] Grigg and his colleagues parlayed their Pd-catalyzed tandem polycyclization-anion capture sequence into a treasure trove of syntheses starting with *N*-allyl-o-haloanilines.[19] Diels–Alder and olefin metathesis reaction can be interwoven into the sequence or can serve as the culmination step, a wide variety of nucleophiles. An example of the transformation of **29** to **30** is shown below, in which indole is the terminating nucleophile.[19d]

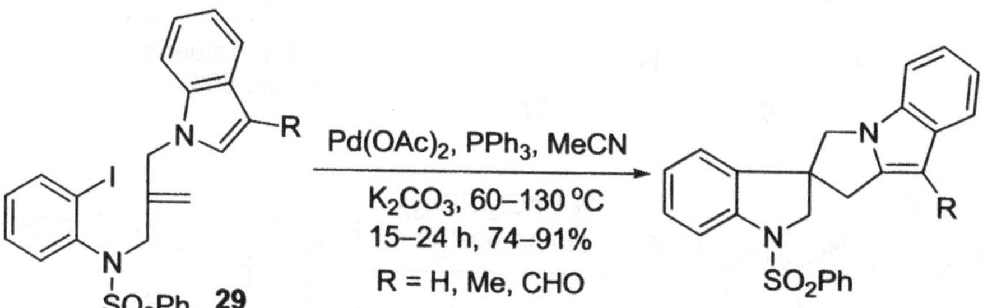

29 **30**

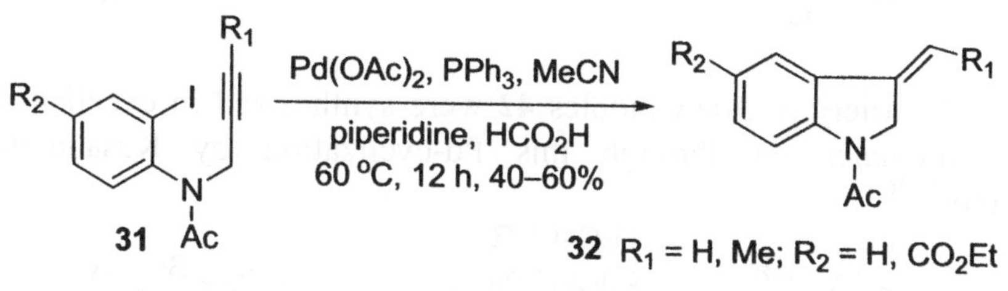

31 **32** R_1 = H, Me; R_2 = H, CO_2Et

Grigg's group also discovered that a 5-*exo-dig* Pd-catalyzed cyclization of *N*-acetylenic-*o*-haloanilines **31** to yield 3-*exo*-alkylidene indolines **32** occurs in the presence of a hydride source, such as formic acid.[20]

Grigg continued his alkyne cyclization to trapping with stannnaes to give 3-*exo*-dienes,[21] alkynes to afford tetracycles[19b, 22] and alkenes leading to cyclopropanes, an example of which is illustrated.[23]

Maes and co-workers have synthesized 5*H*-pyridazino[4,5-*b*]indole skeleton **39** through an intramolecular Heck reaction. Intermediate **38**, 5-(2-bromo-4-(trifluoromethoxy)phenylamino)-2-benzyl-1,2-dihydropyridazin-3-(6*H*)-one was prepared via chemoselective Pd-catalyzed amination of 2-benzyl-5-iodopyridazin-3(2*H*)-one **36** and 2-bromo-4-(trifluoromethoxy)-benzenamine **37** in 66% yield and followed by an intra molecular cyclozition to yield the desired target compound **39** by a microwave iradiations.[24]

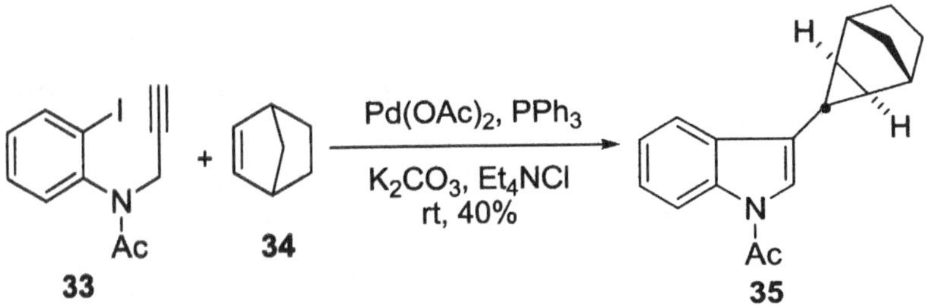

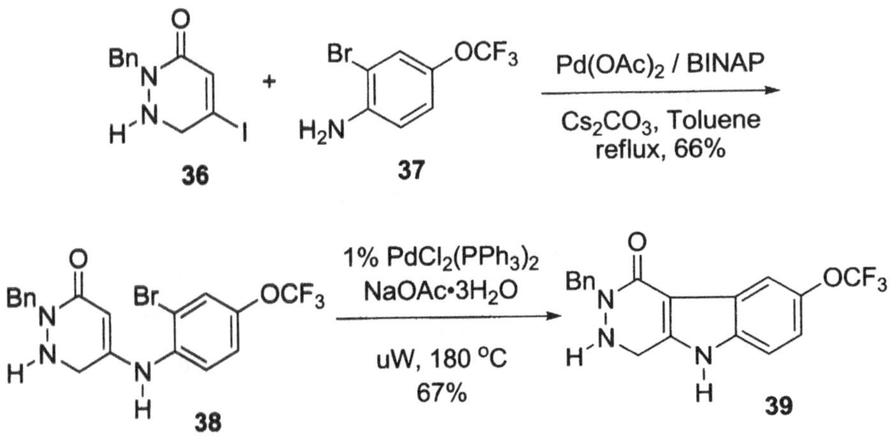

A variety of 3-acylindoles **42** were synthesized in excellent yields via intermediates **41** through this Pd-cyclization by Kasaharan and Sakamoto.[25,26]

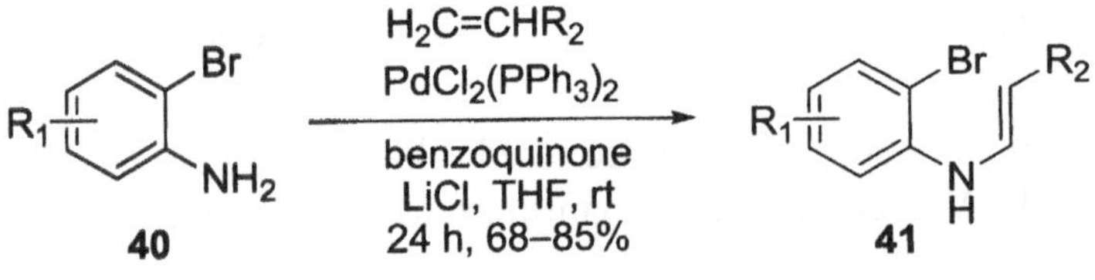

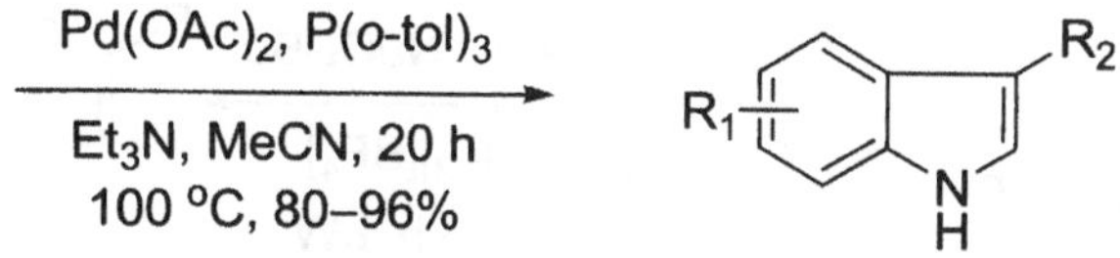

R_1 = H, 6-OMe, 4-, 5-, 6-CO_2Me **42**
R_2 = Ac, CO_2Et

The Mori–Ban reaction also has been expanded to synthesize 3-carboethoxy-2-trifluoromethylindole,[27] 2-carbobenzyloxy-4-hydroxymethyl-3-methylindoles which are present in the antibiotic nosiheptide. A nice variation utilizes the *in situ* synthesis of 2-iodoanilino enamines and subsequent cyclization as shown for the preparation of indoles **45**.[28]

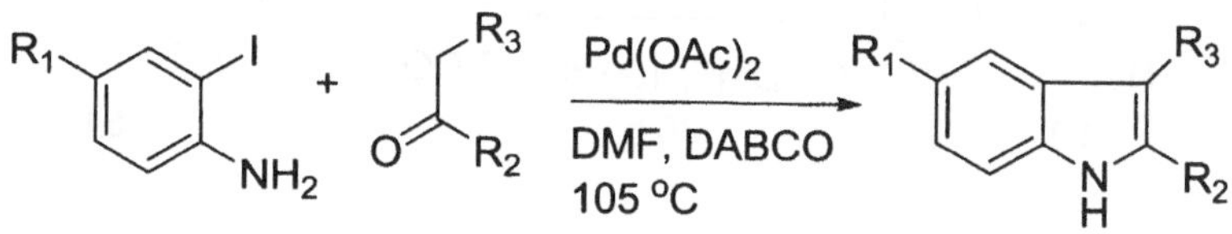

43 **44** 53–82%

R_1 = H, CN, CH_2 -imidazolyl
R_2 = CO_2H, TMS
R_3 = H **45**
R_2, R_3 = cycloalkyl

Bedford and his co-workers have carrid out a successful method for synthesis of flurocarbazoles via a Mori–Ban reaction.[29]

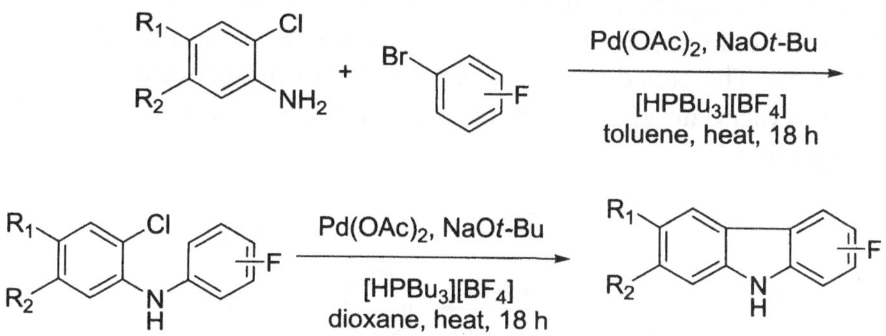

46 68–86% 42–69%

47
R_1 = Me; R_2 = OMe; F = 1-F
R_1 = H; R_2 = H; F = 3-F
R_1 = F; R_2 = H; F = 3-F

Due to the more convenient nature of the one-step, Bedford's group also has explored one-step synthesis of fluorocarbazoles under microwave conditions in reseanable yield by his previous optimized reaction conditions.[29,30]

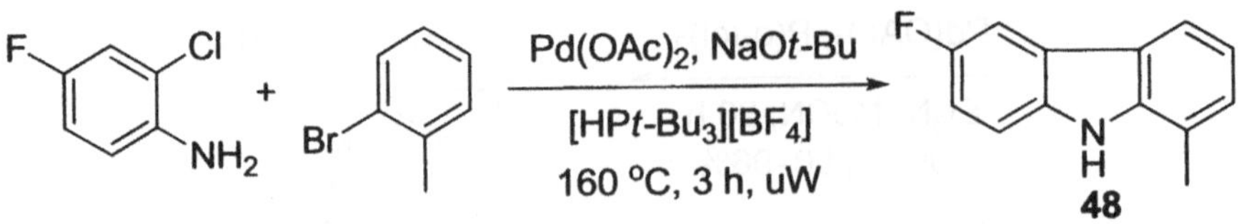

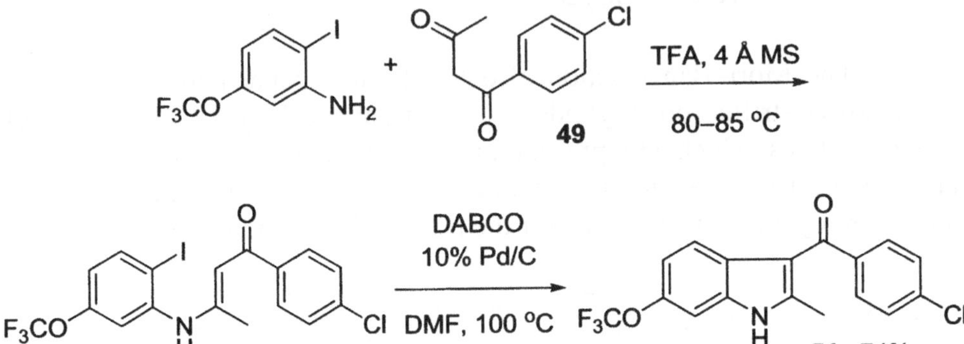

50 81% **51** 74%

Maligres and co-workers noted that the enamine **50** was formed by TFA catalyzed reaction of 2-iodo-5-(trifluoromethoxy)aniline with **49**. The intramolecular Heck reaction of **50** was successfully performed with DABCO or $NaHCO_3$ in the presence of $Pd(OAc)_2$ or Pd on charcoal in DMF at 90 °C to provide the benzoyl indole **51** directly, in 70–74% yields, after crystallization from heptane/EtOAc.[31]

Palladium catalysts have been applied to prepare *o*-vinylaniline derivatives for subsequent (non-Pd) cyclization. Although Pd may not be involved in the indole ring-forming step, these reactions are still of interest to organic chemists. The palladium-catalyzed reaction of *o*-iodoacetanilide with ethyl α-methoxyacrylate gives *o*-vinylacetanilide **52**. Acid treatment affords 2-caroethoxyindole **53**.[32]

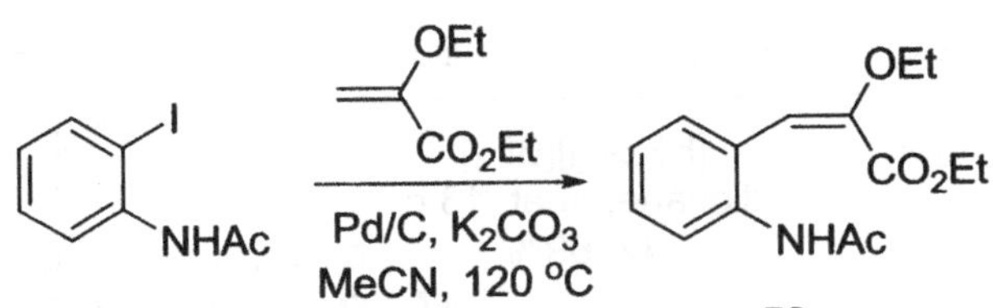

52

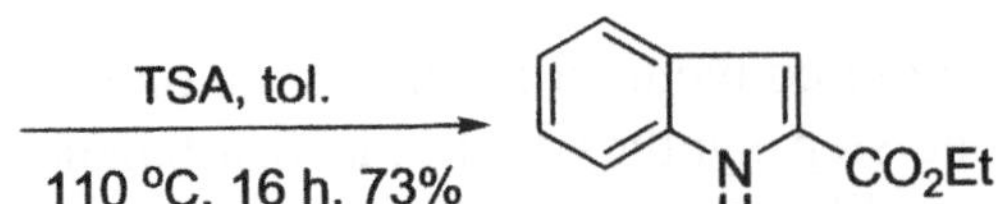

53

The reaction of substituted *o*-bromoacetanilides with styrenes followed by selenium-induced cyclization of the resulting *o*-styrylacetanilides affords 2-arlindoles.[33] Substituted *o*-bromonitrobenzenes react with ethyl vinyl ether under the influene of $Pd(OAc)_2$ to give the corresponding *o*-ethoxyethenylnitrobenzenes. Zinc reduction then yields indoles.[34] The one-step Pd-catalyzed conversion of *o*-bromoanilines to indoles **54** with enamines (or with *N*-vinyl-2-pyrrolidone) has been reported.[35]

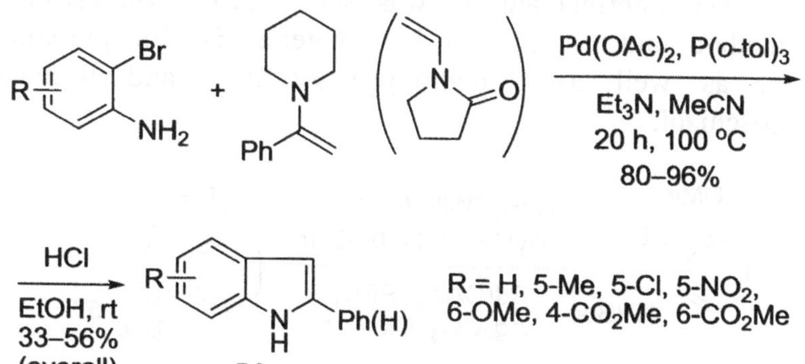

54

Ogasawara applied a Heck reaction of *o*-iodoaniline derivatives with dihydrodimethoxyfuran and vinylene carbonate to give intermediates that are readily cyclized to indole **55** with acids.[36]

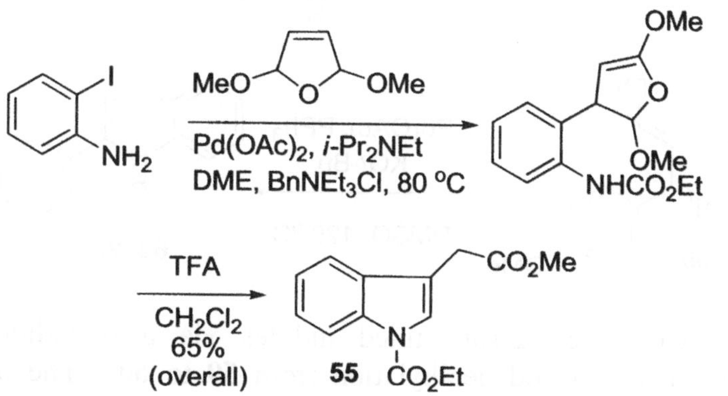

55

In the total synthesis of clavicipitic acid and aurantioclavine, Jia and co-workers have successfully prepared the key intermediate **56** via a Mori–Ban reaction by direct coupling of 3-nitro-2-iodoaniline with (*S*)-2-*N*,*N*-di-*tert*-butoxycarbonyl-5-oxopentane under standard Pd-catalyzed indole synthesis conditions, yielding the optically pure 4-nitrotryptophan derivative **56** in 80% yield.[37]

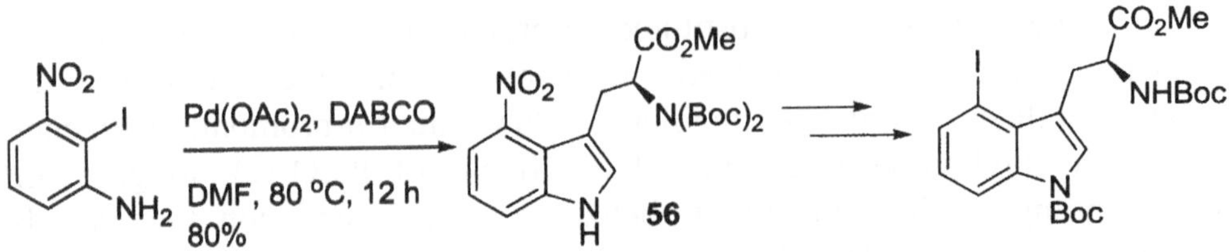

Cook and colleagues improved Mori–Ban reaction and developed the general approach to the total synthesis of 9-methoxy-substituted indole alkaloids.[38] The intermediate **57** was successfully synthesized via the modified method and provided a key material **58** for preparations of mitragynine, as well as 9-methoxygeissoschizol and 9-methoxy-N_b-methylgeissoschizol.

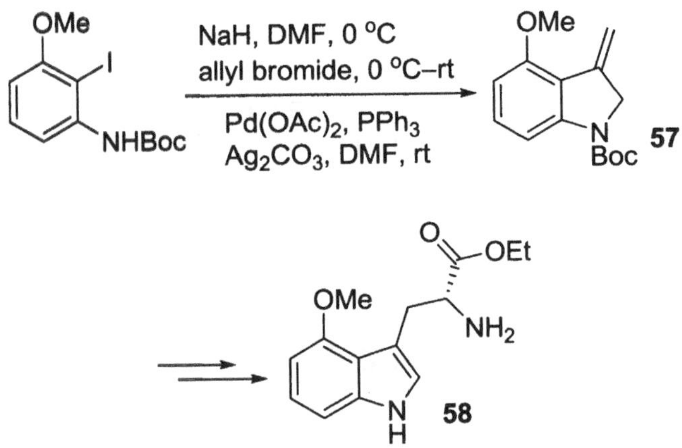

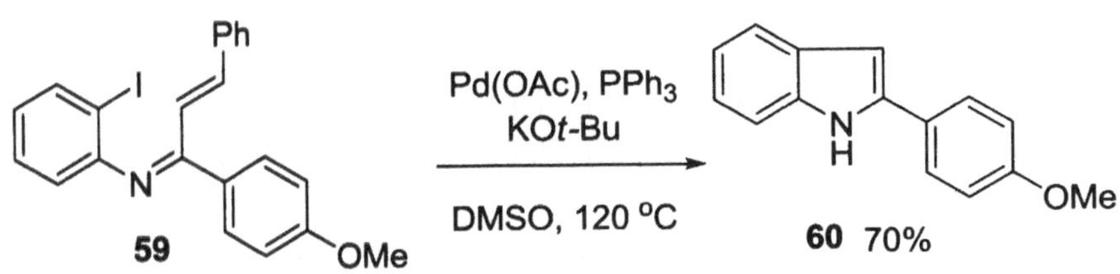

Sun synthesized 2-substituted indoles via a palladium-catalyzed domino Heck reaction and dealkylation from **59** to **60**. The method was optimized by applying varietity of catalysts and solvents, resulting in $Pd(OAc)_2$ and DMSO at 110–130 °C with high yields.[39]

3.9.6 Experimental

The Mori–Ban Conditions to Prepare Indole[10]

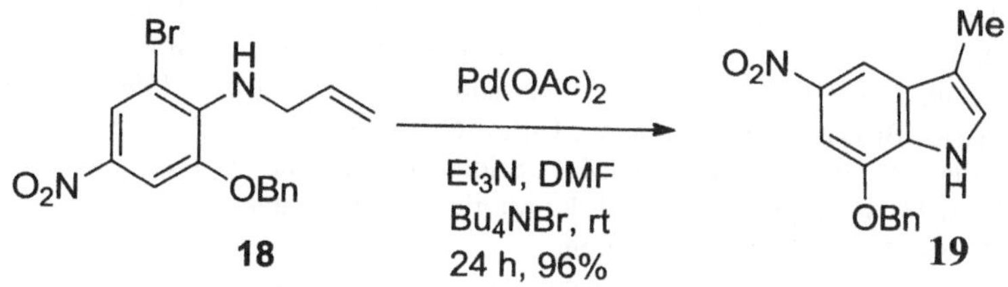

2-(Benzyloxy)-6-bromo-4-nitro-*N*-(2-propenyl)aniline **18** (5.82 g, 16 mmol), tetra-*n*-butylammonium bromide (5.16 g, 16 mmol), and triethylamine (4.05 g, 40 mmol) were dissolved in 15 mL of DMF. Palladium acetate (72 mg, 0.32 mmol) was added, and the reaction mixture was stirred for 24 h. Then, the reaction mixture was diluted with ethyl acetate; filtered through Celite, washed with H_2O, 5% HC1, and saturated aqueous NaCl; dried; and evaporated to give a brown-black solid. The crude product in CH_2C1_2 was filtered through a silica gel column to remove colloidal palladium. Evaporation of the eluate yielded 4.32 g (96%) of **19** as a light orange solid that was pure by NMR and TLC. Recrystallization from ethyl acetate produced orange-brown needles: mp 191–192 °C.

3.9.7 References

1. Mori, M.; Chiba, K; Ban, Y. *Tetrahedron Lett.* **1977,** 1037–1040.
2. Mori, M.; Ban, Y. *Tetrahedron Lett.* **1979,** 1133–1136.
3. Mori, M.; Kanda, N.; Oda, I.; Ban, Y. *Tetrahedron* **1985,** 5465–5474.
4. Terpko, M. O.; Heck, R. F. *J. Am. Chem. Soc.* **1979,** *101*, 5281–5283.
5. Larock, R. C.; Babu, S. *Tetrahedron Lett.* **1987,** *28*, 5291–5294.
6. Larock, R. C.; Hightower, T. R.; Hasvold. L. A.; Peterson, K. P. *J. Org. Chem.* **1996,** *61*, 3584–3585.
7. Larock, R. C.; Yang, H.; Pace, P.; Cacchi, S.; Fabrizi, G. *Tetrahedron Lett.* **1998,** *39*, 18851–888.
8. Larock, R. C.; Pace, P.; Yang, H. *Tetrahedron Lett.* **1998,** *39*, 2515–2518.
9. Larock, R. C.; Pace, P.; Yang, H.; Russell, C. E.; Cacchi, S.; Fabrizi, G. *Tetrahedron* **1998,** *54*, 9961–9980.
10. Sundberg, R. J.; Pitts, W. J. *J. Org. Chem.* **1991,** *56*, 3048–3054.
11. Sakamoto, T.; Kondo, Y.; Uchiyama, M.; Yamanaka, H. *J. Chem. Soc., Perkin Trans. 1* **1993,** 1941–1942.
12. (a) Tietze, L. F.; Grote, T. *J. Org. Chem.* **1994,** *59*,192–196. (b) Tietze, L. F.; Buhr, W. *Angew. Chem., Int. Ed. Engl.* **1995,** *34*, 1366–1368. (c) Tietze, L. F.; Hannemann, R.; Buhr, W.; Lögers, M.; Menningen, P.; Lieb, M.; Starck, D.; Grote, T.; Döring, A.; Schuberth, I. *Angew. Chem., Int. Ed. Engl.* **1996,** *35*, 2674–2677.
13. Hoffmann, H. M. R.; Schmidt, B.; Wolff, S. *Tetrahedron* **1989,** *45*, 6113–6126.
14. (a) Macor, J. E.; Black, D. H.; Post, R. J.; Ryan, K. *Tetrahedron Lett.* **1992,** *33*, 8011–8014. (b) Macor, J. E.; Ogilvie, R. J.; Wythes, M. J. *Tetrahedron Lett.* **1996,** *37*, 4289–4292.
15. Yun, W.; Mohan, R. *Tetrahedron Lett.* **1996,** *37*, 7189–7192.

16. Zhang, H.-C.; Maryanoff, B. E. *J. Org. Chem.* **1997,** *62*, 1804–809
17. Hegedus, L. S.; Allen, G. F.; Olsen, D. J. *J. Am. Soc. Chem.* **1980,** *102*, 3583–3587.
18. Danishefsky, S.; Taniyama, E. *Tetrahedron Lett.* **1983,** *24*, 15–18.
19. (a) Grigg R. *J. Heterocyclic Chem.* **1994,** *31*, 631–639. (b) Grigg, R.; Dorrity, M. J.; Malone, J. F.; Sridharan, V.; Sukirthalingam, S. *Tetrahedron Lett.* **1990,** *31*, 1343–1346. (c) Burns, B.; Grigg, R.; Santhakumar, V; Sridharan, V.; Stevenson, P.; Workun, T. *Tetrahedron Lett.* **1992,** *48*, 7297–7320. (d) Grigg, R.; Fretwell, P.; Meerholtz, C.; Sridharan, V. *Tetrahedron* **1994,** *50*, 359–370. (e) Grigg, R.; Sansano, J. M. *Tetrahedron* **1996,** *52*, 13441–13454. Grigg, R.; Brown, S.; Sridharan, V.; Uttley, M. D. *Tetrahedron Lett.* **1998,** *39*, 3247–3250. (f) Evans, P.; Grigg, R.; Ramzan, M. I.; Sridharan, V.; York, M. *Tetrahedron Lett.* **1999,** *40*, 3021–3024. (g) Grigg, R.; Major, J. P.; Martin, F. M.; Whittaker, M. *Tetrahedron Lett.* **1999,** *40*, 7709–7711; and references therein.
20. Burns, B.; Grigg, R.; Sridharan, V.; Worakun, T. *Tetrahedon Lett.* **1988,** *29*, 4325–4328.
21. Burns, B.; Grigg, R.; Ratananukul, P.; Sridharan, V.; Stevenson, P.; Sukirthalingam, S.; Worakun, T. *Tetrahedron Lett.* **1988,** *29*, 5565–5568.
22. Grigg, R.; Loganathan, V.; Sridharan, V. *Tetrahedron Lett.* **1996,** *37*, 3399–3402.
23. Brown, D.; Grigg, R.; Sridharan, V.; Tambyrajah, V.; Thornton-Pett, M. *Tetrahedron* **1998,** *54*, 2595–2606.
24. Franck, P.; Hostyn, S; Dajka-Hala'sz, B.; Polonka-Ba'lint, A.; Monsieurs, K.; Ma'tyus, P.; Maes, B. U. W. *Tetrahedron* **2008,** *64*, 6030–6037.
25. Kasaharan, A.; Izumi, T.; Murakami, S.; Yanai, H.; Takatori, M. *Bull. Chem. Soc. Jpn.* **1986,** *59*, 927–928.
26. Sakamoto, T.; Nagano, T.; Konda, Y.; Yamanaka, H. *Synthesis* **1990,** 215–218.
27. (a) Latham, E. J.; Stanforth, S. P. *Chem. Commun.* **1996,** 2253–2254. (b) Latham, E. J.; Stanforth, S. P. *J. Chem. Soc., Perkin Trans. 1* **1997,** 2059–2063.
28. Chen, C.; Liebermann, D. R.; Laren, R. D.; Verhoeven, T. R.; Reider, P. J. *J. Org. Chem.* **1997,** *62,* 2676–2677.
29. Bedford, R. B.; Betham, M.; Charmant, J. P. C; Weeks, A. L. *Tetrahedron* **2008,** *64*, 6038–6050.
30. Bedford, R. B.; Betham, M. *J. Org. Chem.* **2006,** *71*, 9403–9410.
31. Maligres, P. E.; Humphrey, G. R.; Marcoux, J. F.; Hillier, M. C.; Zhao, D.; Krska, S. Grabowski, E. J. J. *Org. Process Res. Dev.* **2009,** *13,* 525–534.
32. Sakamoto, T.; Kondo, Y.; Yamanaka, H. *Heterocycles* **1998,** 1312–1314.
33. Izumi, T.; Sugano, M. Konno, T. *J. Heterocyclic Chem.* **1992,** *29*, 899–904.
34. Kasahara, A.; Izumi, T.; Li; X.-P. *Chem. Ind.* **1998,** 50–51.
35. Kasahara, A.; Izumi, T.; Kikuchi, T.; Li; X-P. *J. Heterocyclic Chem.* **1987,** *24*, 1555–1556.
36. (a) Samizu, K.; Ogasawara, K. *Synlett* **1994,** 499–500. (b) Samizu, K.; Ogasawara, K. *Heterocycles* **1995,** *41*, 1627–1629. (c) Sakagami, H.; Ogasawara, K. *Heterocycles* **1995,** *51*, 1131–1135.
37. Xu, Z.; Hu, W.; Liu, Q.; Zhang, L; Jia, Y. *J. Org. Chem.* **2010,** *75*, 7626–7635.
38. Ma, J.; Yin, W.; Zhou, H.; Liao, X.; Cook; J. M. *J. Org. Chem.* **2009,** *74*, 264–273.
39. Mao, H.; Wan, J.-P.; Pan, Y.; Sun, C. *Tetrahedron Lett.* **2010,** *51,* 1844–1846.

3.10 Sandmeyer Isatin Synthesis

Ian S. Young

3.10.1 Description

The reactivity of the isatin scaffold[1–3] allows it to serve as a starting material for diverse applications from medicinal chemistry to total synthesis. At the turn of the 20th century Traugott Sandmeyer reported two methods for the synthesis of this essential building block.

The Sandmeyer diphenylurea isatin synthesis[4] uses symmetrical diphenylthiourea **1** as a substrate for generation of cyanoformamidine **2**, via treatment of **1** with potassium cyanide in the presence of lead carbonate. Reaction of formamidine **2** with ammonium sulfide leads to thioamide **3**, which can undergo acid-induced cyclization and hydrolysis to produce isatin (**4**). Examples of the diphenylurea isatin synthesis in the literature are scarce. This is due to Sandmeyer developing the more practical isonitrosoacetanilide isatin synthesis 16 years later.

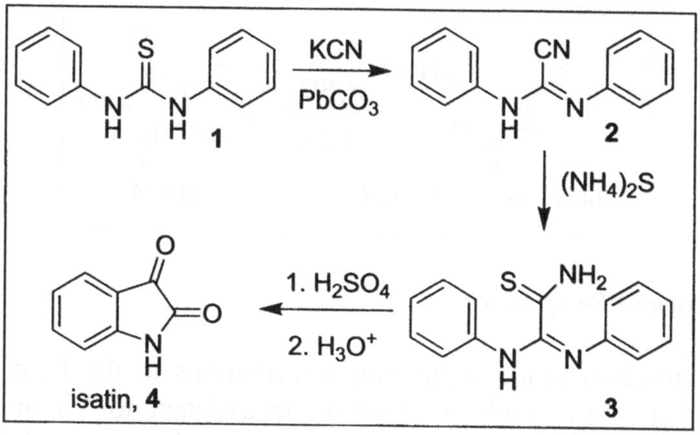

Sandmeyer Isonitrosoacetanilide Isatin Synthesis

The Sandmeyer isonitrosoacetanilide synthesis[5,6] has found regular use for the synthesis of substituted isatins. The two-step procedure begins with condensation of aniline (**5**), chloral hydrate,[7] and hydroxylamine (hydrochloride or sulfate) in water to yield the intermediate isonitrosoacetanilide **6**. In later modifications, sodium sulfate was included in the reaction mixture. The exact role of the sodium sulfate is not known, although it serves a role more complicated than salting out the product.[8] The isonitrosoacetanilide **6** is smoothly converted to isatin (**4**) through heating with acid, sulfuric being most commonly used. Polyphosphoric acid[9] and even anhydrous hydrogen fluoride[10] (although the substrate scope is not general) have been used as alternative acid catalysts for cylization to the isatin. Since the second step involves nucleophilic attack by the aromatic ring, electron-withdrawing substituents (especially in the *meta*-position) lead to decreased reactivity. There are potential issues with the Sandmeyer isatin synthesis that can cause limitations. 3-substituted anilines produce a mixture of 4- and 6-substituted isatins. *N*-Alkyl anilines generally do not perform well in the sequence, although there are exceptions.[11] The harsh cyclization conditions (heating in concentrated sulfuric acid) generally limit the functionality that can be included on the starting aniline. Despite the potential drawbacks detailed above, the Sandmeyer isonitrosoacetanilide isatin synthesis has found common use for the preparation of these heterocycles.

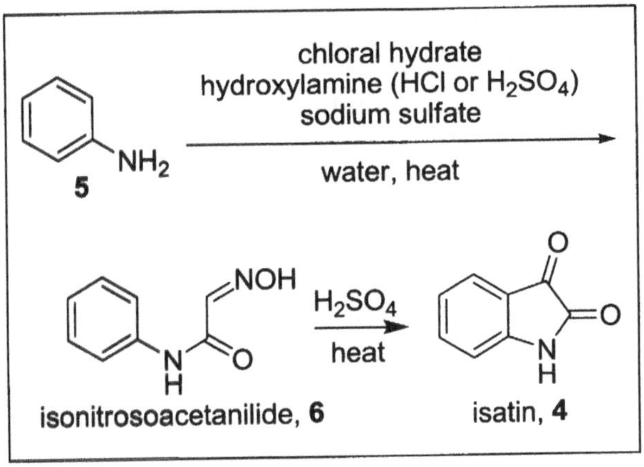

3.10.2 Historical Perspective

Traugott Sandmeyer made significant contributions to the field of organic synthesis.[12] In 1884, while working as an assistant to Victor Meyer, he disclosed that diazotized anilines can be converted to aryl halides,[13,14] the reaction most closely associated with his name. Sandmeyer started a position with the J. R. Geigy Dye Stuff manufacturing company in 1888 and was instrumental in elevating the company to the leading dye manufacturer in Europe. In addition to discovering two isatin syntheses (diphenylthiourea in 1903[4] and isonitrosoacetanilide in 1919[5]), he also reported a synthesis of indigo from thiocarbanilide in 1899,[15] work that likely was the catalyst for his first reported isatin synthesis.

3.10.3 Mechanism

There has been limited mechanistic work regarding the two steps of the Sandmeyer isatin synthesis, although Wang offered a proposal in an earlier summary of the reaction.[6]

Formation of the Isonitrosoacetanilide

Reaction of chloral (**7**, or hydrate) with hydroxylamine (**8**) produces chloral oxime (**10**) after dehydration. Aniline then displaces one of the chlorides to form the dichloroamine **11**, which undergoes hydrolysis to yield the desired isonitrosoacetanilide **6**.

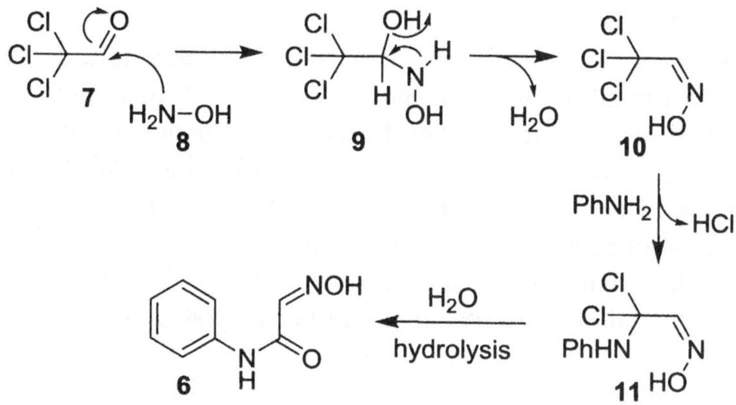

Conversion of the Isonitrosoacetanilide to Isatin

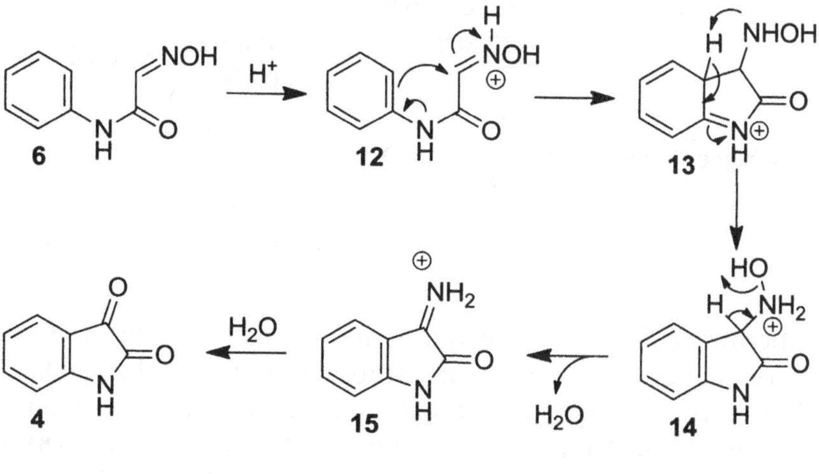

Activation of the oxime of **6** through protonation induces a Friedel–Crafts type acylation. After re-aromatization to **14**, loss of water yields imine **15**, which hydrolyzes upon workup to yield isatin (**4**).

3.10.4 Variations and Improvements

4,6-Dibromoisatin (**18**) is a key starting material for the total synthesis of members of the convolutamydine family.[16–18] 4,6-Dibromoisatin is produced in only 10% overall yield from 3,5-dibromoaniline (**16**)[19] using the standard Sandmeyer procedure, due to the isonitrosoacetanilide **17** being formed in 11%. The origin of the low yield is the limited water solubility of the aniline. Under the typical reaction temperatures employed (90–100 °C) for isonitrosoacetanilide synthesis, formation of a black resinous material hinders product isolation.

Garden demonstarted that lowering the temperature (60–80 °C) eliminated the formation of the resinous material, but the aniline existed as a viscous oil and did not mix efficiently with the aqueous reaction contents.[17] To overcome this problem, addition of one-third volume of ethanol led to a fine dispersion of the insoluble aniline throughout the aqueous layer. Using this modified procedure, the isonitrosoacetanilide could be isolated in 82–86% yield. This ethanol/lower temperature strategy offered an improvement in isonitrosoacetanilide yield for other aniline substrates as well.

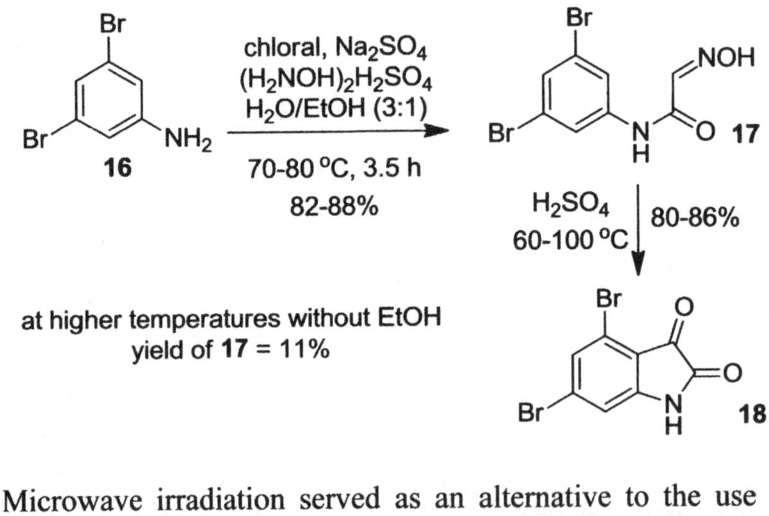

Microwave irradiation served as an alternative to the use of lower temperatures and ethanol as a co-solvent for the control of the black resinous material formed during isonitrosoacetanilide synthesis. Jnaneshwera prepared isonitrosoacetanilide **17** en route to 4,6-dibromoisatin (**18**) in 80% yield.[18] This increase in yield was also realized for other problematic isatins when microwave irradiation was used.

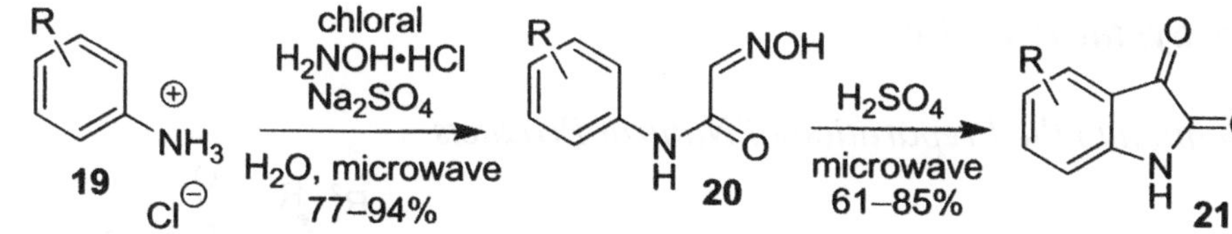

An alternative preparation of isonitrosoacetanilide **24** was developed by Rewcastle, and proceeds through the intermediate acetal **23** via treatment of aniline **22** with a chloral alternative $(AcO)_2CHCOCl$.[20] This procedure is advantageous over the traditional procedure in situations where the aniline substrate used is either water insoluble, or contains an electron donating group at the two position. Drawbacks are that an additional step is required and the chloral alternative has to be prepared.

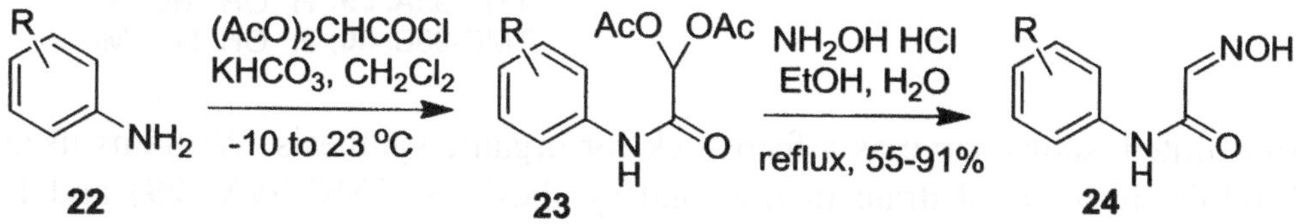

Pinto and Neto demonstrated that if the isatin formation was run using an imidazolium based ionic liquid **27** as the solvent, the reaction could be stoped at isatin oxime **26**. The choice of acid and ionic liquid counter anion had a substantial influence on this reaction efficiency.[21]

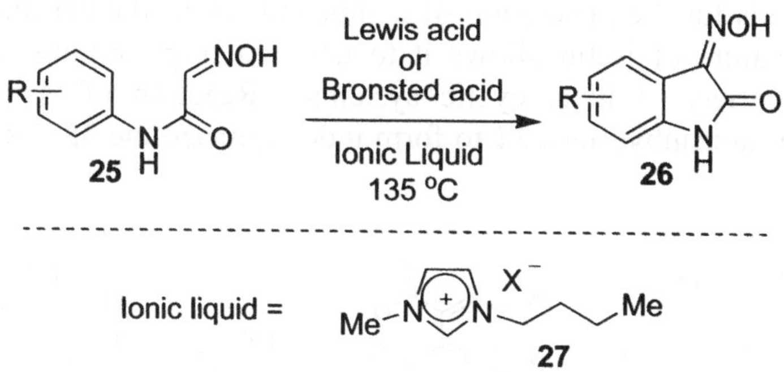

3.10.5 Synthetic Utility

Application to the Preparation of Natural Products

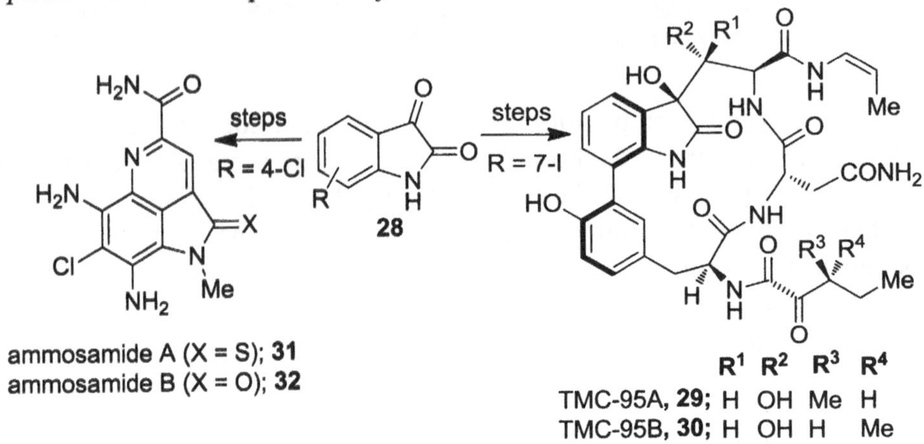

Substituted isatins serve as a feedstock for organic synthesis. Williams used 7-iodoisatin as a substrate in the total synthesis of TMC-95A (**29**) and B (**30**).[22] Fennical constructed ammosamides A (**31**) and B (**32**) around a 4-chloroisatin nucleus.[23]

Application to Medicinal Chemistry

Perhaps the greatest value of the Sandmeyer synthesis is the preparation of isatin scaffolds for the generation of compounds of medicinal interest. The ambiphilic nature of isatin allows it to undergo ring forming reactions to produce a variety of heterocyclic systems. Reaction of isatin **33** with substituted isatoicanhydrides **34** to form indoloquinazoline **35**,[24] serves as an example.

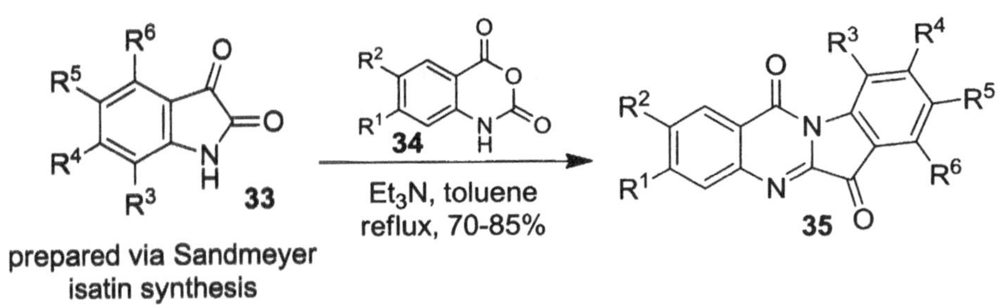

The enhanced electrophilicity of the isatin 3-position carbonyl allows for the regioselective preparation of analogs. Reaction of isatin **36** with arylthiohydrazide **37** prepared analog **38** to study for anti-osteoarthritis properties.[25]

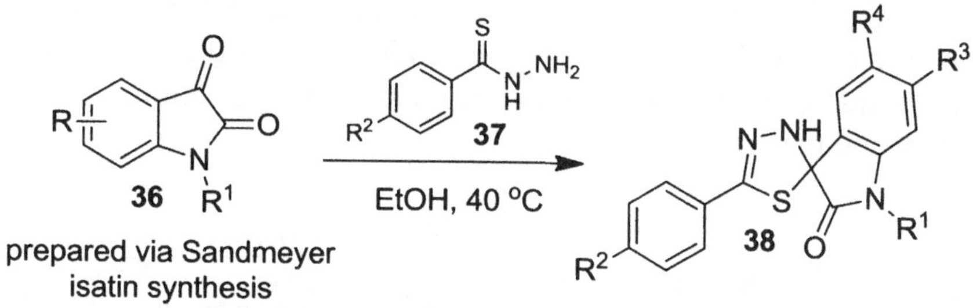

Although many of the anilines used in the Sandmeyer isatin synthesis have minimal functionality, examples exist where potentially sensitive motifs can survive the harsh reaction conditions. The 6-azauracil substituted aniline **39** could be converted to isatin **40** en route to the polyheterocyclic compound **41**.[26] A trifluoroacetyl protected amine could also be carried through the two-step Sandmeyer sequence to yield the isatin of interest.[27]

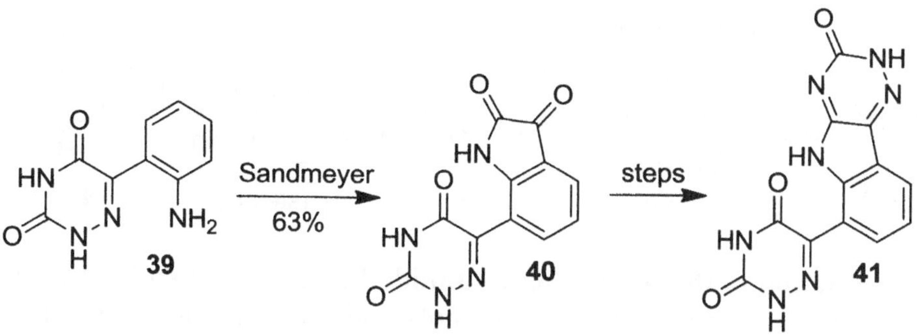

The isatin products from the Sandmeyer synthesis can serve as a source of *ortho*-arylaminoacids through degradation. Hydrolysis of isatin **42**, yielded aminoacid **43**, which was converted to the anticancer compound DMXAA (**44**).[28]

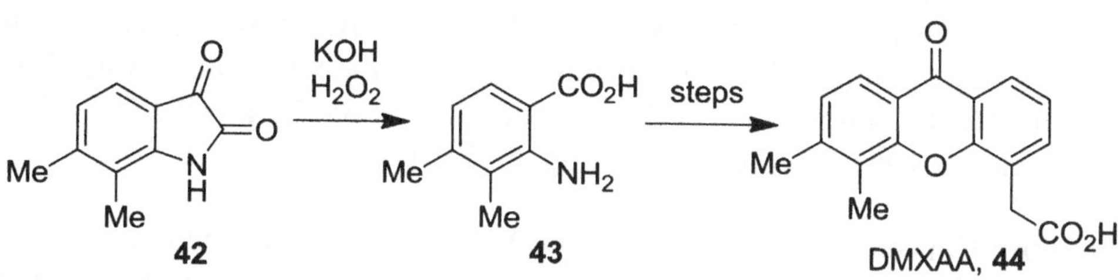

Application to Biological Chemistry

To study the conformational changes of tryptophan synthase by ^{15}N-NMR, indole enriched in *N*-15 was required. Phillips used a Sandmeyer isatin synthesis starting with *N*-15-labeled aniline (**45**) to prepare **46** in good yield over two steps.[29] Reduction with lithium aluminium hydride then produced the labeled indole (**47**). This strategy was chosen over the Fischer or Japp–Klingeman indole syntheses based on efficiency and availability of the *N*-15-labeled starting materials.

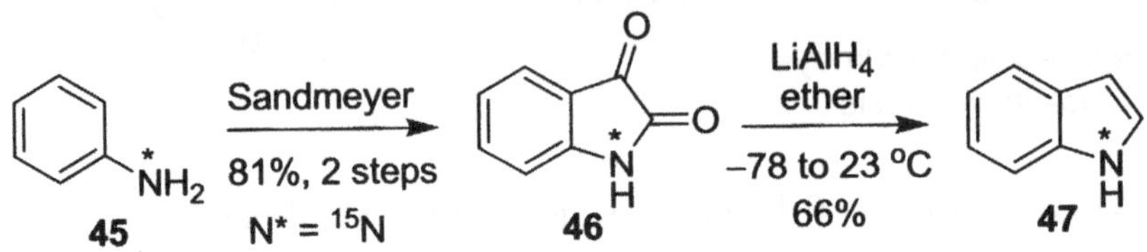

3.10.6 Experimental

Preparation of Isatin[8]

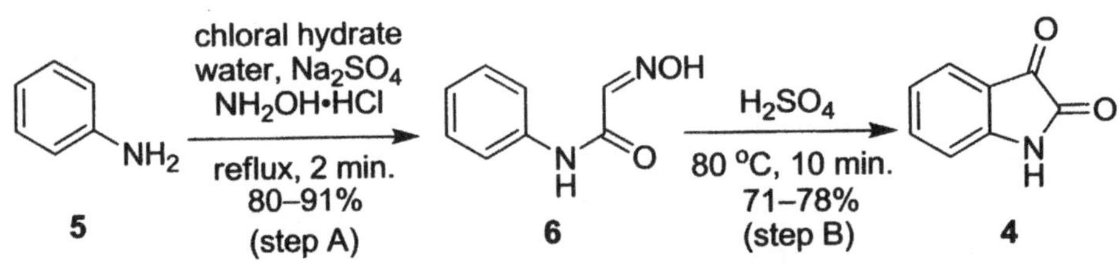

(A) Formation of the isonitrosoacetanilide In a 5-L round-bottom flask are placed chloral hydrate (90 g, 0.54 mol) and water (1200 mL). To this solution are then added, in order: crystallized sodium sulfate (1300 g), a solution of aniline (**5**, 46.5 g, 0.5 mol) in water (300 mL) to which concentrated hydrochloric acid (51.2 g, 43 mL, 0.52 mol) has been added to dissolve the amine, and finally a solution of hydroxylamine hydrochloride (110 g, 1.58 mol) in water (500 mL). The flask is heated to boiling over 40–45 min, and after 1–2 min of vigorous boiling the reaction is complete. During the heating period, some crystals of the isonitrosoacetanilide (**6**) separate. On cooling the solution in running water the remainder crystallizes, is filtered with suction, and air dried. The yield is 65–75 g (80–91% of the theoretical amount) of a product melting at 175 °C.

(B) Isatin Concentrated sulfuric acid (600 g, 326 mL) is warmed to 50 °C in a 1-L round-bottom flask fitted with an efficient mechanical stirrer, and to this, dry isonitrosoacetanilide (**6**, 75 g, 0.46 mol) is added as such a rate as to keep the temperature between 60 and 70 °C but not higher. External cooling should be applied at this stage so that the reaction can be carried out more rapidly. After the addition of the isonitroso compound is finished, the solution is heated to 80 °C and kept at this temperature for about 10 min to complete the reaction. Then the reaction mixture is cooled to room temperature and poured upon 10–12 times its volume of cracked ice. After standing for about 30 min, the isatin is filtered with suction, washed several times with cold water to remove the sulfuric acid, and then dried in air. The yield of crude isatin (**4**), which melts at 189–192 °C, is 47–52 g (71–78% of the theoretical amount). This product is pure enough for many purposes.

For purification, the crude product (200 g) is suspended in 1 L of hot water and treated with sodium hydroxide (88 g) in water (200 mL). The slurry is stirred mechanically, and the isatin passes into solution. Dilute hydrochloric acid (290–300 mL, prepared by diluting one volume of concentrated hydrochloric with two volumes of water) is then added, with stirring, until a slight precipitate appears. The mixture is then filtered at once, the precipitate is rejected, and the filtrate is made acidic to Congo red paper with hydrochloric acid. The solution is then cooled rapidly, and the isatin that separates is filtered with suction and air dried. The pure product thus obtained weighs 150–170 g and melts at 197–200 °C.

3.10.7 References

1. [R] Sumpter, W. C. *Chem. Rev.* **1944,** *34*, 407–434.
2. [R] Popp, F. D. *Adv. Heterocycl. Chem.* **1975,** *18*, 1–58.
3. [R] da Silva, J. F. M.; Garden, S. J.; Pinto, A. C. *J. Braz. Chem. Soc.* **2001,** *12*, 273–324.
4. Sandmeyer, T. *Z. Fab. Textile Chem.* **1903,** *2*, 129–137.
5. Sandmeyer, T. *Helv. Chim. Acta.* **1919,** *2*, 234–241.
6. Wang, Z. "*Sandmeyer Isatin Synthesis*" In *Comprehensive Organic Name Reactions and Reagents*, *3V set*, Wiley: Hoboken, NJ, pp. 2467–2470, **2009**.
7. [R] Luknitskii, F. I. *Chem Rev.* **1975,** *75*, 259–289.
8. Marvel, C. S.; Hiers, G. S. *Org. Synth.* **1941,** *Coll. Vol. 1*, 327.
9. Gainor, J. A.; Weinreb, S. M. *J. Org. Chem.* **1981,** *46*, 4317–4319.
10. Karnes, H. A.; Kybett, B. D.; Wilson, M. H.; Margrave, J. L.; Newman, M. S. *J. Am. Chem. Soc.* **1965,** *87*, 5554–5558.
11. Kurkin, A. V.; Bernovskaya, A. A.; Yurovskaya, M. A. *Tetrahedron: Asymmetry* **2009,** *20*, 1500–1505.
12. Sandmeyer, T. *Nature* **1922,** *109*, 720–721.
13. Sandmeyer, T. *Ber.* **1894,** *17*, 1633–1635.
14. Sandmeyer, T. *Ber.* **1894,** *17*, 2650–2653.
15. [R] Fierz-David, H. E. *The Fundamental Processes of Dye Chemistry,* D. Van Norstrand Co., New York, **1921,** pp. 161–167.
16. Nakamura, T.; Shirokawa, S.-i.; Hosokawa, S.; Nakazaki, A.; Kobayashi, S. *Org. Lett.* **2006,** *8*, 677–679.
17. Garden, S. J.; Torres, J. C.; Ferreira, A. A.; Silva, R. B.; Pinto, A. C. *Tetrahedron Lett.* **1997,** *38*, 1501–1504.
18. Jnaneshwara, G. K.; Bedekar, A. V.; Deshpande, V. H. *Synth. Commun.s* **1999,** *29*, 3627–3633.
19. Baker, B. R.; Schaub, R. E.; Joseph, J. P.; McEvoy, F. J.; Williams, J. H. *J. Org. Chem.* **1952,** *17*, 149–156.
20. Rewcastle, G. W.; Sutherland, H. S.; Weir, C. A.; Blackburn, A. G.; Denny, W. A. *Tetrahedron Lett.* **2005,** *46*, 8719–8721.
21. Pinto, A. C.; Moreira Lapis, A. A.; Vasconcellos da Silva, B.; Bastos, R. S.; Dupont, J.; Neto, B. A. D. *Tetrahedron Lett.* **2008,** *49*, 5639–5641.
22. Albrecht, B. K.; Williams, R. M. *PNAS* **2004,** *101*, 11949–11954.
23. Hughes, C. C.; Fenical, W. *J. Am. Chem. Soc.* **2010,** *132*, 2528–2529.

24. Sharma, M. V.; *et al. Bioorg. Med. Chem. Lett.* **2002,** *12*, 2303–2307.
25. Bursavich, M. G.; Gilbert, A. M.; Lombardi, S.; Georgiadis, K. E.; Reifenberg, E.; Flannery, C. R.; Morris, E. A. *Bioorg. Med. Chem. Lett.* **2007,** *17*, 5630–5633.
26. Hlavac J.; Slouka, J. *J. Heterocyclic Chem.* **1997,** *34*, 917–919.
27. Huffman, W. F.; Hall, R. F.; Grant, J. A.; Wilson, J. W.; Heible, J. P.; Hahn, R. A. *J. Med. Chem.* **1983,** *26*, 933–935.
28. Rewcastle, G. W.; Atwell, G. J.; Zhuang, L.; Baguley, B. C.; Denny, W. A. *J. Med. Chem.* **1991,** *34*, 217–222.
29. Osborne, A.; Teng, Q.; Wilson, M., E.; Phillips, R. S. *J. Biol. Chem.* **2003,** *278*, 44083–44090.

3.11 Sommelet–Hauser Rearrangement

Alexandros L. Zografos

3.11.1 Description

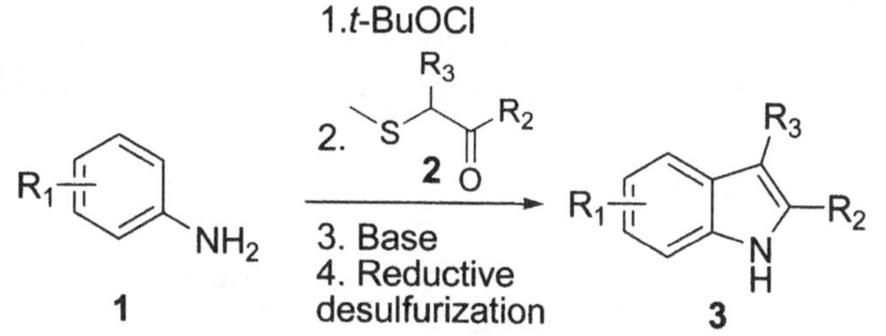

Sommelet-Hauser rearrangement in indole synthesis
Gassman indole synthesis

Typical Sommelet–Hauser rearrangement involves a [2,3]-Wittig rearrangement of benzylic quaternary ammonium salts on treatment with alkali metal amides *via* the formation of ammonium ylide intermediates.[1] Expansion of the Sommelet–Hauser rearrangement in indole synthesis by using azasulfonium ylide intremediates was first reported by Gassman in the so-called Gassman indole synthesis. According to that report, a hypohalite, a *β*-carbonyl sulfide **2** derivative, and a base are sequentially added to aniline or a substituted aniline derivative **1** to provide 3-thioalkoxyindole, which under reductive desulfurization produces the corresponding indole **3**.

3.11.2 Historical Perspective

The Sommelet–Hauser rearrangement was introduced by M. Sommelet in 1937 when he first faced the unique transformation by keeping benzhydryltrimethylammonium hydroxide in desiccator over P_2O_5, exposed to sunlight and he found that it was rearranged to give (*o*-benzylbenzyl)dimethylamine in modest yield.[2] The same result occurred when the substrate was heated to 145 °C instead of irradiating it, suggesting that sunlight provided only the necessary heat for the transformation. During the next decade several research groups reported similar rearrangements following the well-known, at this period, Stevens rearrangement of quaternary ammonium salts;[3] however, it was C. R. Hauser who investigated the new rearrangement extensively. Hauser and co-workers observed that when benzyltrimethylammonium iodide was treated with $NaNH_2$ in liquid ammonia, dimethyl-(2-methylbenzyl)-amine was isolated as the sole product in excellent yield.[4]

In 1974, Paul Gassman and co-workers recognized the potential of the described rearrangement and reported a one-pot, general method for the synthesis of indoles based on the rearrangement of the nonisolable aza-sulfonium ylides.[5–7] In the very first example of this method, aniline **4** was treated sequentially with *t*-BuOCl, methylthio-2-propanone **5** and triethylamine to provide methylthioindole **6** in 69% yield. Raney-nickel-mediated desulfuriza-tion afforded 2-methylindole **7** in 79% yield.

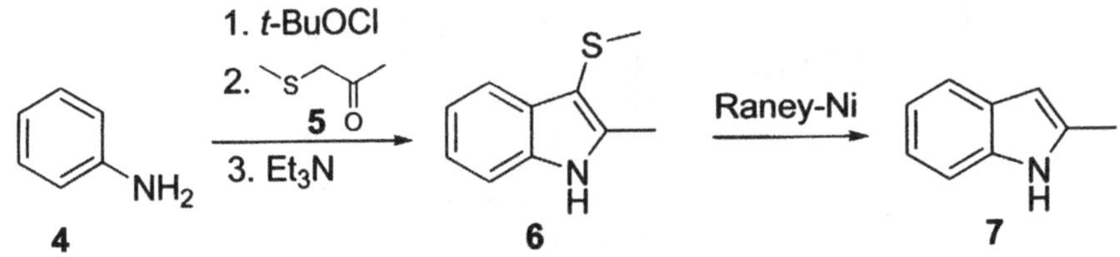

3.11.3 Mechanism

The mechanism of the [2,3]-migration in a quaternary ammonium salt when treated with strong bases in the classic Sommelet–Hauser rearrangement is well studied, usually in contrast with the related [1,2]-migration observed in the same systems as a consequence of the Stevens rearrangement.[8] Its mechanism was easily clarified by intermediate isolation and labeling experiments.[9,10]

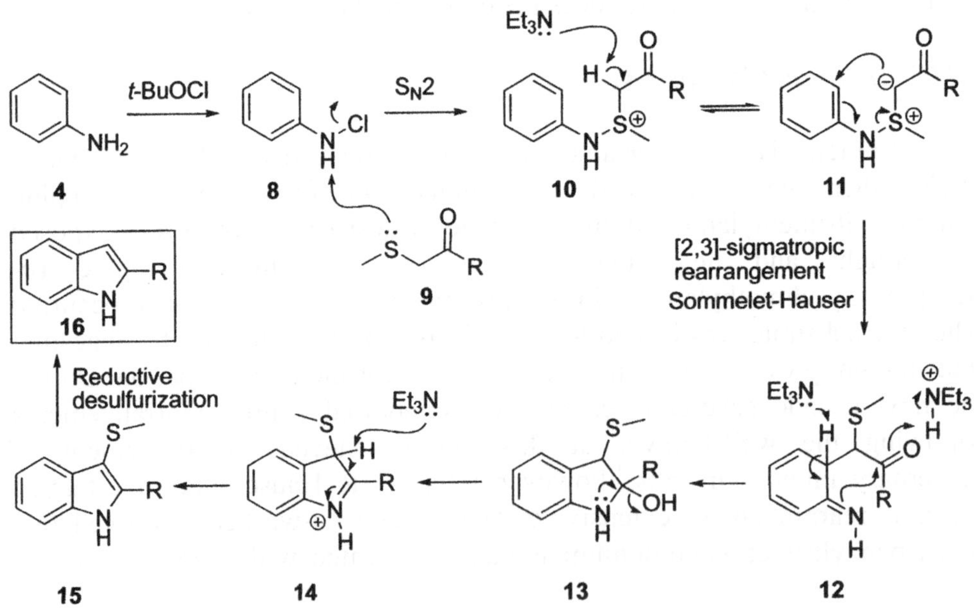

Years later when Gassman introduced the Sommelet–Hauser rearrange-ment in indole synthesis, he proposed an expanded mechanism, including the [2,3]-migration of azasulfonium salts for the indolization of aniline with methylthioketones, as is depicted in the scheme above. In accordance to the original reference,[5] anilines react readily with *t*-butyl hypochlorite, or with variety of other hypohalites, to produce *N*-chloroanilines **8**. These *N*-chloroanilines react with good nucleophiles such as sulfides **9**, providing azasulfonium salts **10**. Azasulfonium salts are then deprotonated with the appropriate base (in the described case, triethylamine) to trigger an intramolecular Sommelet–Hauser rearrangement producing compound **12**. Proton transfer and rearomatization leads to aromatic amino-ketone, which under intramolecular addition of the free amine to the carbonyl moiety is expected to provide intermediate *α*-aminoalcohol **13**. Dehydration finishes the reaction sequence affording the polyfunctional thioindole derivative **15**. Finally, quench of the reaction mixture with a desulfurization reagent, such as Raney-nickel, yields the corresponding indole derivative **16**.

3.11.4 Variations and Improvements

Several variations exist for the classic Sommelet–Hauser rearrangement. The most important concern the use of base-free conditions for the [2,3]-sigmatropic rearrangement based on the fluoride anion induced desilylation of substituted benzyldialkyl-[(trimethylsilyl)methyl] ammonium halides[11] and the asymmetric versions of Sommelet–Hauser rearrangement that appeared recently in the literature.[12] None of the above has ever been used in a modification of the Gassman indole synthesis.

The only modification reported in Gassman indole synthesis came from Gassman himself in a sequel of the original publication.[7] The modification overcomes the only serious limitation of the described methodology, concerning the inability of substituted anilines with cation stabilizing groups, as *p*-methoxy, in *ortho*- or *para*-positions of the aromatic ring, to react with methylthiosulfides, due to the instability of the formed *N*-chloroanisidines. Gassman found that when ketosulfides were treated with chlorine or bromine, a complex is formed that is readily reacted with *p*- or *o*-substituted electron-rich anilines to provide the aza-sulfonium salts. Treatment under the standard basic conditions affords the highly important, in the natural product synthesis, 5- and 7-methoxyindoles.

3.11.5 Synthetic Utility

General Utility[5–7]

Gassman indole synthesis provides a single regioisomer of 3-thioindole derivative **17** when *ortho/para* substituted anilines **1** are employed, the yields of which are moderate to good as is depicted in the scheme below. The reaction provides some advantages on the preparation of 7-substituted indoles compared to other methods (*e.g.*, Fisher–indole synthesis due to the sluggishness of the reaction and the low reported yields).

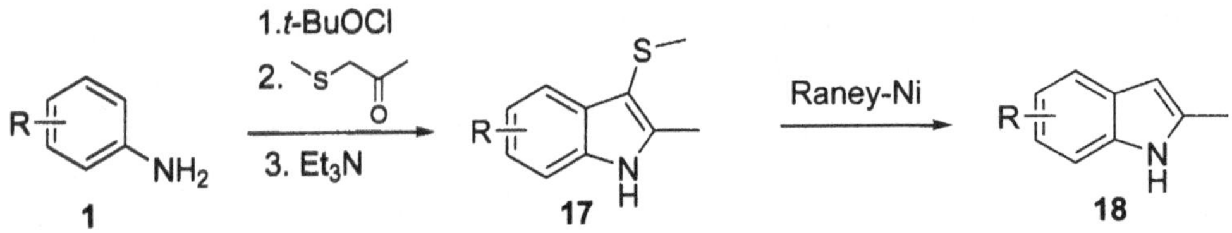

1	17	18
R= *p*-OC(O)CH_3	68%	72%
R= *p*-CH_3	60%	80%
R= *p*-Cl	72%	74%
R= *p*-CO_2Et	58%	83%
R= *o*-CH_3	72%	73%

The *meta*-substituted anilines **19** provide one or two isomeric products (**20** and/or **21**), depending on the nature of the substituents.

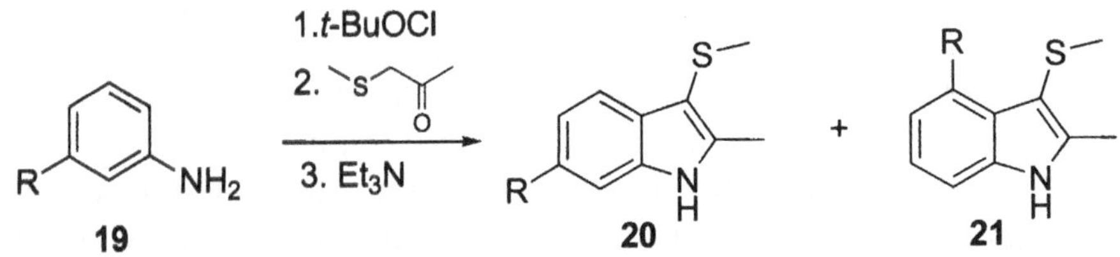

19	20	21
R= CH_3	35%	23%
R= NO_2	0%	82%

In general the described method is advantageous based on the use of inexpensive readily available starting materials, mild conditions, as all steps can be run below zero and for its avoidance of either acidic or strong basic conditions enabling the elaboration of sensitive indole derivatives to elevated temperatures, acids or strong bases.

Based on the described generality, a variety of 2-substituted indoles **16** can be prepared by using the appropriate starting acylsulfides **9**, as in the examples described below.

It should be noted that higher yield of the unsubstituted indole (**15**, R = H) is obtained by using methylthioacetaldehyde dimethyl acetal (57% yield) instead of methylthioacetaldehyde (30%).

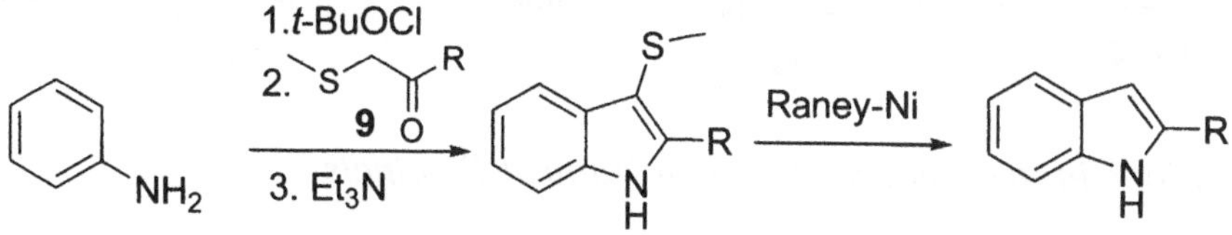

4 **15** **16**

R = H 30%
R = Ph 81% 74%

When α-substituted-β-ketosulfides **2** are used instead of methylthioketosulfides **9**, intermediate **22** is formed and dehydrated to provide indolenine **23**. Reductive elimination under various conditions offers 2,3-disubstituted indoles **3**.

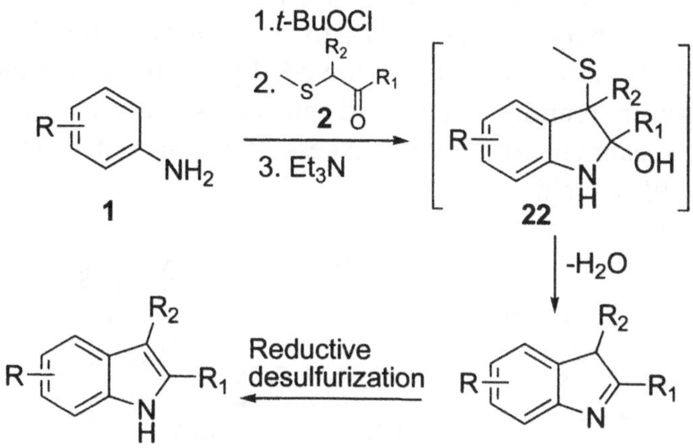

1 **22** **3** **23**

Based on this modification a series of highly substituted carbocyclic indoles can be prepared as the five-fused **26** or six-fused indoles **29** presented below.

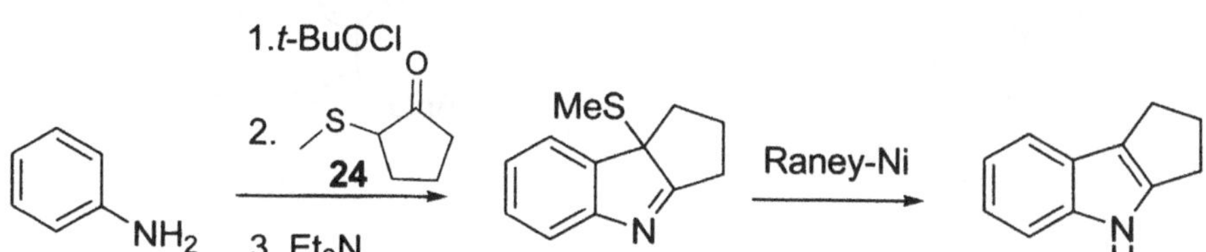

4 **25** **26**

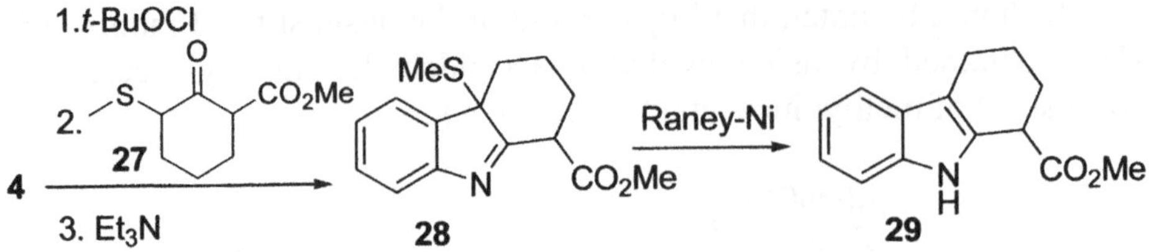

Applications in the Total Synthesis of Natural Products

Although nitrogen ylides have been extensively used in organic transformations, especially those concerning the formation of nitrogen heterocycles,[13] not many references are available for the use aza-sulfonium ylides and the Sommelet–Hauser rearrangement in the synthesis of indole derivatives.

In 1981, Wierenga first introduced a modification of Gassman oxindole synthesis for the synthesis of indole derivative **34**, part of the left-hand segment of the antitumor agent CC-1065.[14] In accordance to that, addition of derivative **30** in a equimolar amount of the hindered base, [1,8-bis(dimethylamino)naphthylene (R_3N), to the chloride complex of ethyl-*α*-(mercaptomethyl)propionate **31**, followed by a triethylamine catalyzed Sommelet–Hauser rearrangement and an acid-induced cyclization gave oxindole **32**. Treatment of **32** with excess $BH_3{\cdot}SMe_2$ at room temperature afforded **33** in 95% yield.

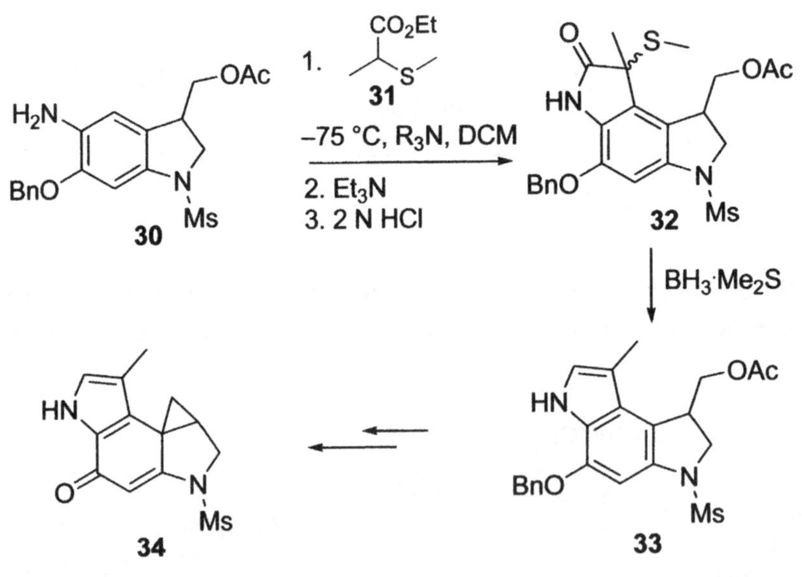

In 1992, Smith and his group described the total synthesis of indole diterpenes (+)-paspalicine and (+)-paspalinine based on the use of Sommelet–Hauser reaction for the formation of the polycyclic indole framework.[15,16] Thus treatment of thiomethylketone **36**, prepared from ketone **35** and dimethylsulfide using LDA in THF/HMPA, with *N*-chloroaniline and exposure of the resultant azasulfonium salt to triethylamine gave the aniline derivative **37** in 93% yield. Desulfurization with Raney-nickel in ethanol at room temperature provided the corresponding aniline in 82% yield. No spontaneous cyclization was observed and treatment with *p*-toluenesulfonic acid was essential for the formation of indole **38** in 81% yield, which represents the basic core of the natural compounds.

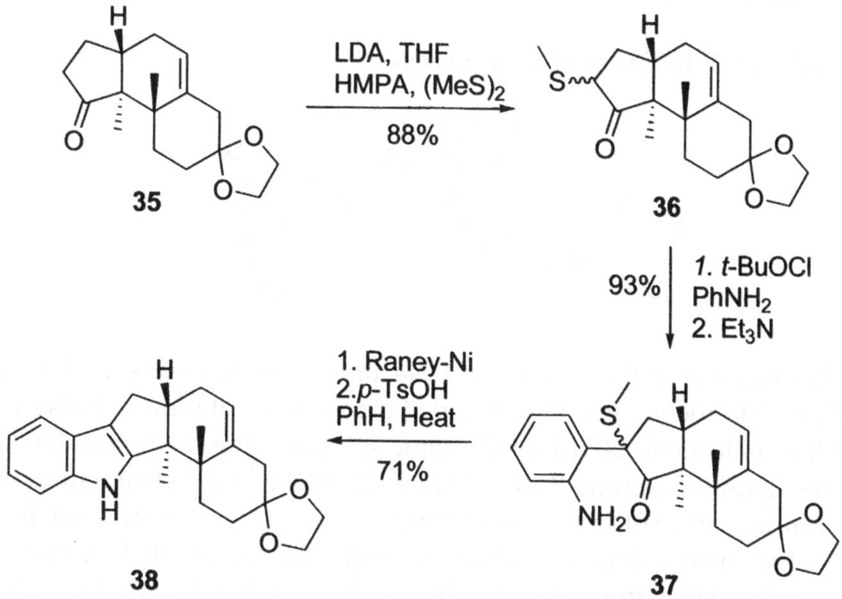

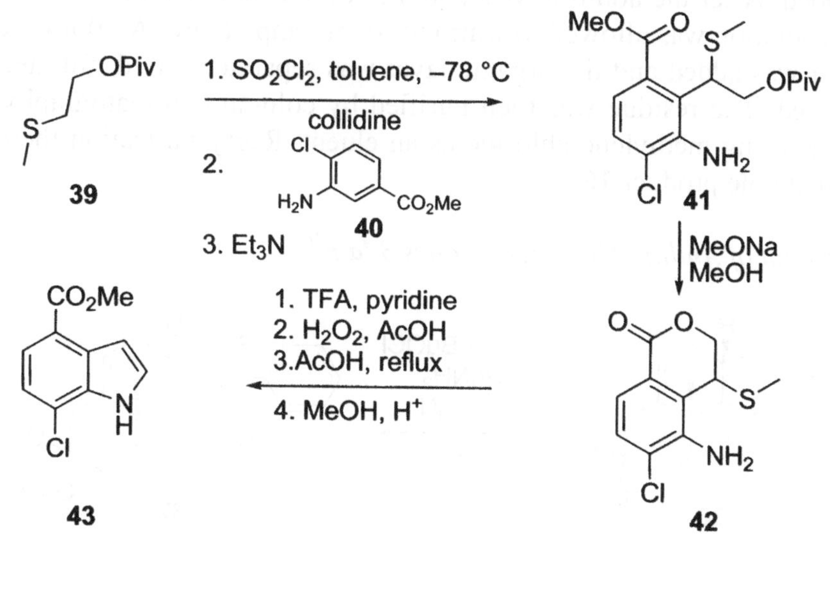

In 2003, Alper and co-workers, based on an earlier, reported the synthesis of substituted tosyl-indoles from substituted anilines and isopropylsulfides,[17,18] introduced a practical way for the elaboration of multigram scale of the otherwise difficult accessible methyl 7-chloroindole-4-carboxylate, based on the desulfurization and cyclization of a prior Sommelet–Hauser rearranged product **41**, derived from the reaction of methylsulfide **39** with *o*-chloroaniline derivative **40**.[19]

3.11.6 Experimental

Preparation of 2-Methyl-3-alkylthioindole[5]

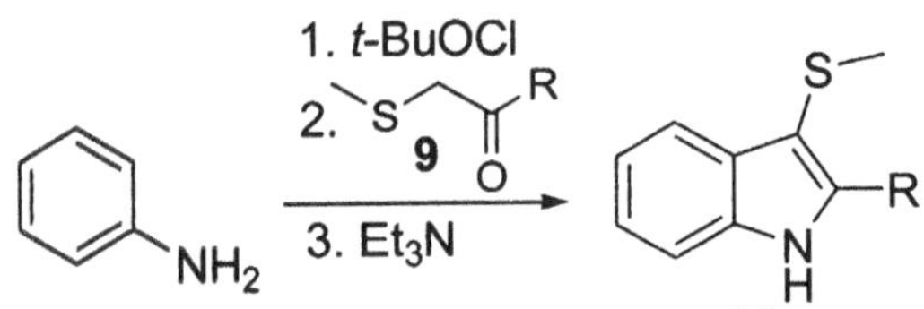

4 **16**

To a vigorously stirred solution of 0.044 mol of the aniline **4** in 150 mL of methylene chloride at –65 °C was added dropwise a solution of 0.044 mol of *t*-butyl hypochlorite in 20 mL of the same solvent. After 5–10 min, 0.044 mol of the methylthio-2-ketone **9** dissolved in 20 mL of methylene chloride was added causing an exotherm and stirring at –65 °C was continued for 1 h. Usually, the intermediate azasulfonium salt had precipitated at this stage. Subsequently, 0.044 mol of triethylamine in 20 mL of methylene chloride was added. After the addition was completed, the cooling bath was removed and the solution was allowed to warm to room temperature. A 50 mL portion of water was added and the organic layer was separated, dried, filtrated, and evaporated. The residue was then purified by column chromatography over silica gel using methylene chloride as an eluent. Recrystallization then gave the pure indole product **16**.

Preparation of Anilino Cyclopentanones ***37a,b***[16]

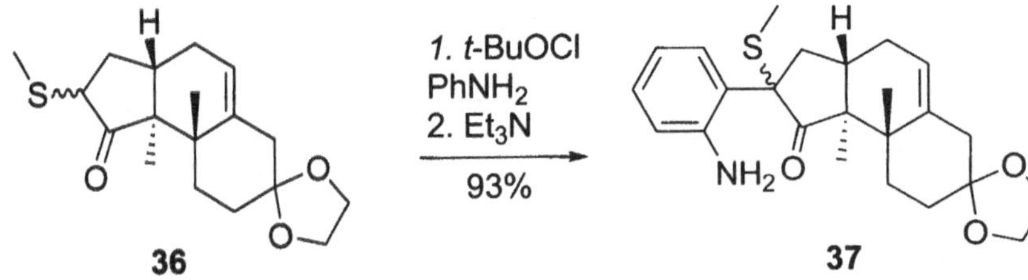

36 **37**

A solution of aniline (0.65 mL, 7.19 mmol) in 30 mL of dichloromethane was cooled to –78 °C and treated with *t*-butyl hypochlorite (0.86 mL, 7.19 mmol) dissolved in 30 mL of dichloromethane. After 15 min, a solution of **36a,b** (2.21 g, 6.84 mmol) in 30 mL of dichloromethane was introduced dropwise, and the mixture was stirred 1 h at –78 °C. Triethylamine (2.1 mL, 15.05 mmol) was added, and the reaction mixture was warmed to room temperature and poured into water. The aqueous layer was extracted with dichloromethane, and the combined organic layers were washed with brine, dried over $MgSO_4$, and evaporated *in vacuo*. Flash chromatography (hexane/ethyl acetate, 6:1) afforded 2.64 g (93% yield) of ketones **37a,b** as a mixture of diastereomers.

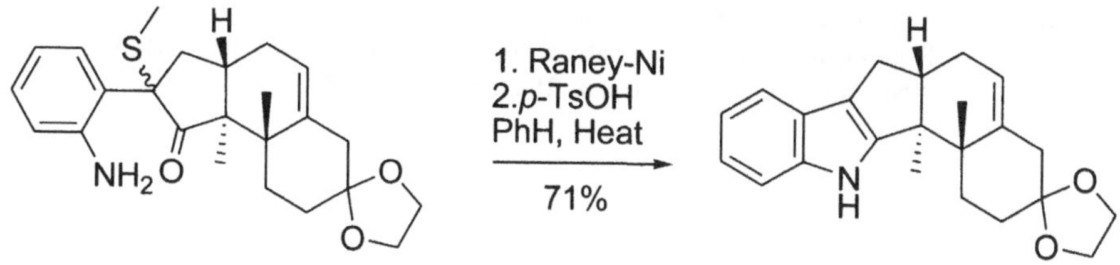

37 **38**

Indole Ketal 38
A solution of **37a,b** (2.59 g, 6.26 mmol) in 200 mL of absolute ethanol was stirred with an excess of W-2 Raney-Nickel (Aldrich) for 2 h at room temperature. The supernatant was decanted and filtered through a pad of Florisil. The catalyst was washed thoroughly with ethyl acetate, and the washings were then filtered. Concentration *in vacuo* and flash chromatography (hexane/ethyl acetate, 5:2) furnished 1.89 g (82% yield) of the desulfurized compound as a colorless solid.

A solution of *p*-toluenesulfonic acid (44 mg, 0.23 mmol) in benzene (40 mL) was heated at reflux under a Dean–Stark trap for 15 min. A solution of the desulfurized ketone from the step described above, (1.69 g, 4.59 mmol) in 30 mL of benzene was then added and heated at reflux for 15 min. After cooling to room temperature, the mixture was quenched with saturated aqueous sodium bicarbonate and extracted with ethyl acetate. The organic layers were washed with brine, dried over $MgSO_4$, and evaporated *in vacuo*. Flash chromatography (hexane/ethyl acetate, 5:1) yielded 1.39 g (87% yield) of indole **38** as a colorless solid: mp 159–160 °C.

3.11.7 References

1. [R] Pine, S. H. *Org. React.* **1970,** *18*, 403–464.
2. Sommelet, M. A. *Compt. Rend.* **1937,** *205*, 56.
3. [R] Marko, I. E. In *Comp. Org. Synth. 3,* 913–974, Pergamon, Oxford, UK, **1991**.
4. Kantor, S. W.; Hauser, C. R. *J. Am. Chem. Soc.* **1951,** *73*, 4122–4131.

5. Gassman, P. G.; van Bergen, T. J.; Gilbert, D. P.; Cue, B. W. Jr. *J. Am. Chem. Soc.* **1974,** *96,* 5495–5508.
6. Gassman, P. G.; van Bergen, T. J. *J. Am. Chem. Soc.* **1974,** *96,* 5508–5512.
7. Gassman, P. G.; Gruetzmacher, G.; van Bergen, T. J. *J. Am. Chem. Soc.* **1974,** *96,* 5512–5517.
8. Ghigo, G.; Cagnina, S.; Maranzana, A.; Tonachini, G. *J. Org. Chem.* **2010,** *75,* 3608–3617.
9. Hauser, C. R.; Van Eenam, D. N. *J. Am. Chem. Soc.* **1957,** *79,* 5512–5530.
10. Jones, F.; Hauser, C. R. *J. Org. Chem.* **1961,** *26,* 2979–2981.
11. Shirai, N; Sato, Y. *J. Org. Chem.* **1988,** *53,* 194–196.
12. Tayama, E.; Kimura, H. *Angew. Chem., Int. Ed.* **2007,** *46,* 8869–8871.
13. [R] Beall, L. S.; Padwa, A. *Adv. Nitrogen Heterocycles* **1998,** *3,* 117–158.
14. Wierenga, W. *J. Am. Chem. Soc.* **1981,** *103,* 5621–5623.
15. Smith, A. B., III; Sunazuka, T.; Leenay, T. L.; Kingery-Wood, J. *J. Am. Chem. Soc.* **1990,** *112,* 8197–8198.
16. Smith, A. B., III; Kingery-Wood, J.; Leenay, T. L.; Nolen, E. G.; Sunazuka, T. *J. Am. Chem. Soc.* **1992,** *114,* 1438–1449.
17. Murai, Y.; Masuda, G.; Inoue, S.; Sato, K. *Heterocycles* **1991,** *32,* 1377–1386.
18. Murai, Y.; Kobayashi, S..; Inoue, S.; Sato, K. *Heterocycles* **1992,** *34,* 1017–1029.
19. Alper, P. B.; Nguyen, K. T. *J. Org. Chem.* **2003,** *68,* 2051–2053.

3.12 Stollé Oxindole Synthesis

Gerald J. Tanoury

3.12.1 Description

The Stollé Oxindole Synthesis (the Stollé Reaction)[1–4] refers to the conversion of α-chloroacyl chlorides (**1**) to oxindoles (**3**) in the presence of a Lewis acid. The reaction occurs via amide formation and subsequent annulation via an intramolecular Friedel–Crafts alkylation.

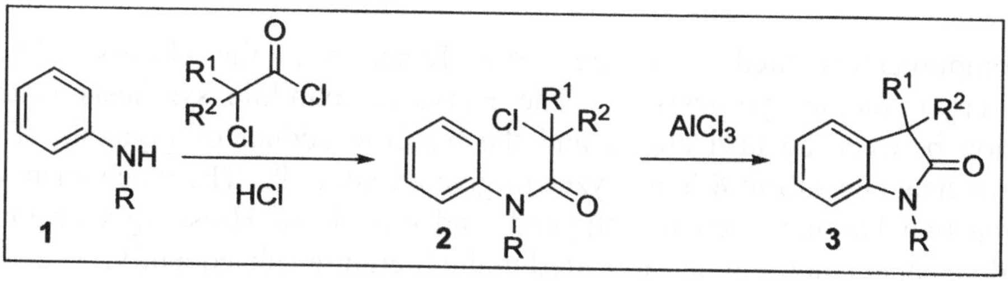

3.12.2 Historical Perspective

From 1913 to 1930, R. Stollé published several manuscripts describing the synthesis of oxindoles from anilines and α-chloroacyl chlorides. His typical substrates were chloroacetanilides or the corresponding *N,N*-diphenylamide. The manuscripts also described the synthesis of isatines (2,3-dioxindoles) from arylamines and oxalyl chloride. The reactions require high temperatures (typically > 150 °C) to complete the cyclization step.

3.12.3 Mechanism

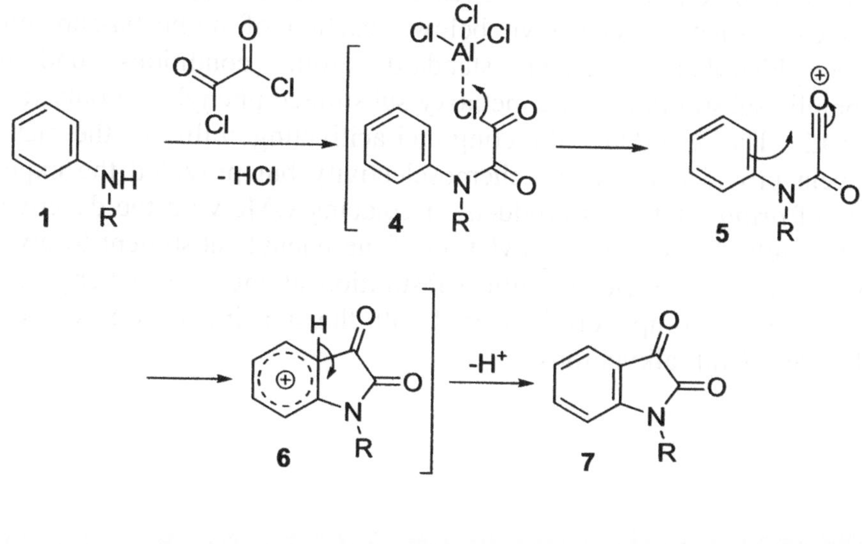

The Stollé reaction is thought to occur via a typical mechanism for amide formation from an amine and acid chloride, followed by Friedel–Crafts alkylation or acylation. No definitive mechanistic work has been performed on this reaction, but incorporating the mechnistic understandings of two steps provides a firm basis for understanding the mechanism of this reaction. Formation of the mono-amide from oxalyl chloride and aniline provides intermediate **4**, which in the presence of $AlCl_3$ undergoes intramolecular electrophilic aromatic substitution to the desired 2,3-dioxindole (isatin) **7** via intermediates **5** and **6**.

3.12.4 Variations and Improvements

A complimentary method to the Stollé Reaction is the closely related Hinsberg oxindole synthesis.[5–7] The Hinsberg oxindole synthesis is the reaction between an aryl amine and the bisulfite adduct of glyoxal. The intermediate ene sulfite **8** hydrolyzes to give oxindole **9**. The mechanism of the reaction has not been investigated, and outside of Hinsberg's original oxindole papers, nothing has appeared in the literature subsequently.

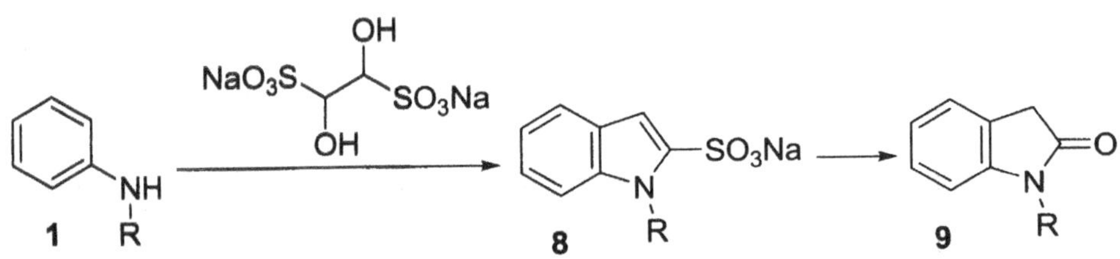

3.12.5 Synthetic Utility

One of the weaknesses of the Stollé reaction is regiochemical selectivity with substituted analines. As shown below, reaction of mono-(*m*-substituted) chloroacetyldiarylamine under standard Stollé conditions underwent electrophilic substitution at the methoxy-substituted phenyl ring only to give **11** and **12**. The *ortho*/*para*-directing and acitivating ability of the methoxy substituent provided exclusive chemoselectivity but provided the expected mixture of *ortho*- and *para*-products. Replacing OMe with the deactivating chloride resulted in selective alkylation of the phenyl substituent to give **14**, completely excluding electrophilic substitution at the chlorophenyl group. These results are completely in accord with the substituent effects observed with Friedels–Craft alkylations.[8,9]

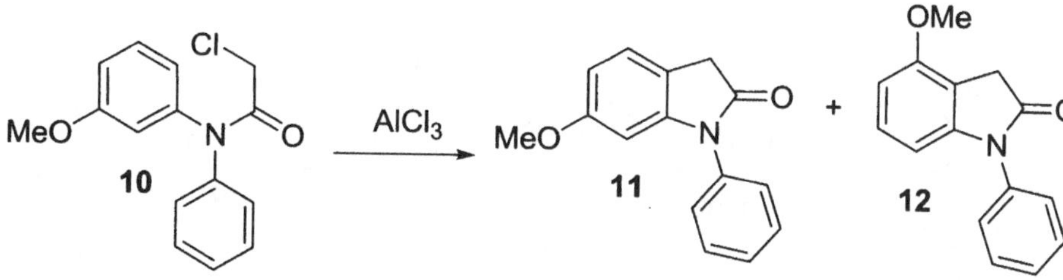

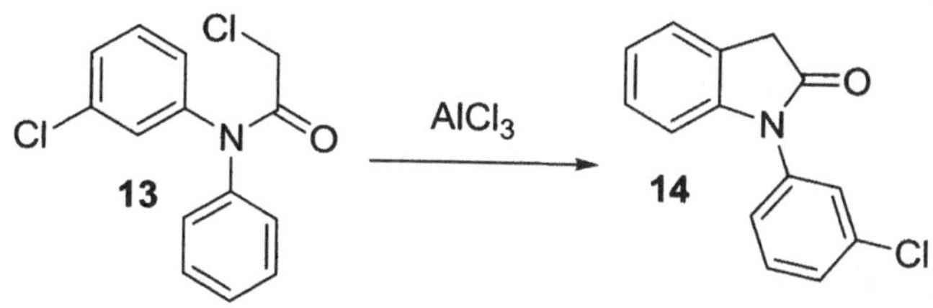

Another interesting chemoselectivity event is the Stollé reaction the chloroacetyl derivative of 2-aminobiphenyl **15**. In the annulation step of the reaction, cyclization occurred exclusively on the nonsubstitued phenyl moiety to give azepinone **17** as opposed to the expected oxindole product **16**. Again, substiutent effects are the reason for the unexpected chemoselectivity: the amide functionality deactivates the *N*-phenyl ring relative to the 2-phenyl substituent.[10]

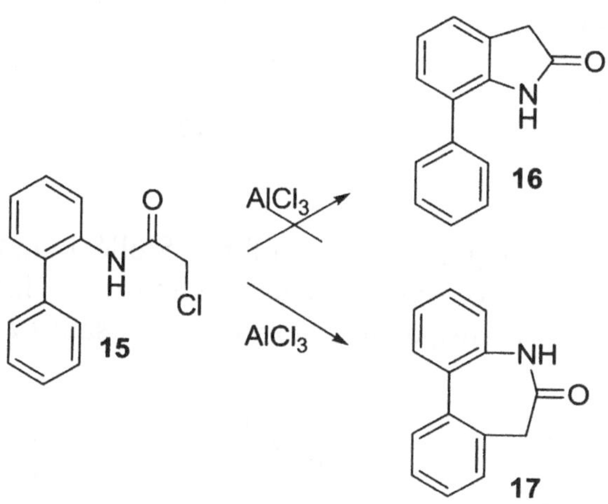

The Stollé reaction of arylamines with oxalyl chloride has provided interesting avenues into dioxoindoles. As shown below, reaction of 4,5-dihydro-[1,2]dithiolo[3,4-*c*]quinoline-1-thiones **18** with oxalyl chloride in refluxing toluene gave the corresponding quinoline-4,5-diones **20**. The reaction did not require Lewis acid catalysis, as the HCl generated in the acylation step provided sufficient conditions for the cyclization step.[11,12]

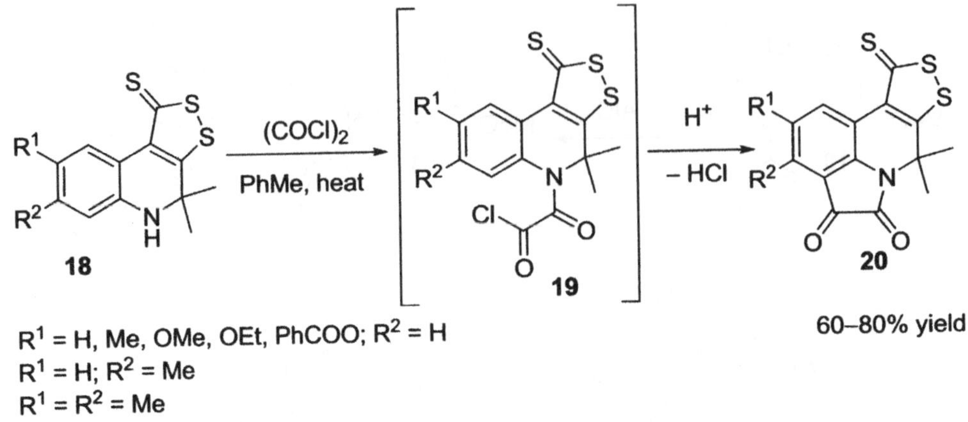

The reaction of *N*-phenyl-2-naphthylamine with oxalyl chloride in diethyl ether at ambient temperature for 3 days gave the phenylbenzo[*e*]isatin **21** shown below in 60% yield, again without the use of a Lewis acid. Analogously, reaction of *N*-methyl-2-phenoxatinylamine and oxalyl chloride in the presence of $AlCl_3$ gave the corresponding isatin product **22** in 40% yield.[13,14]

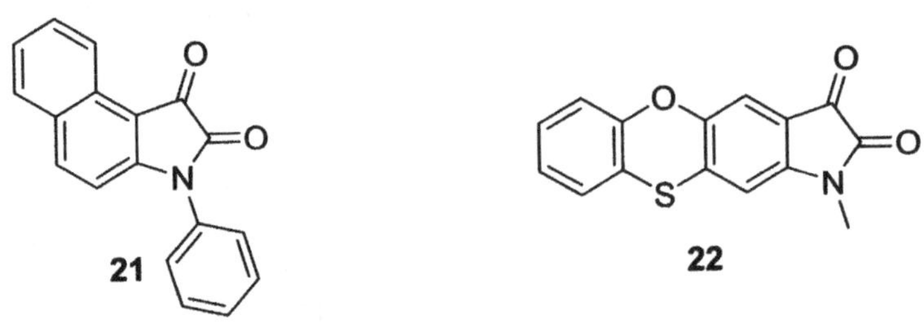

In a very interesting application of the Stollé oxindole synthesis, reaction of *O*-acylated-*N*-glycosylphenylamine with oxalyl chloride followed by deprotection gave the corresponding hexopyranosylisatins **23–25**.[15]

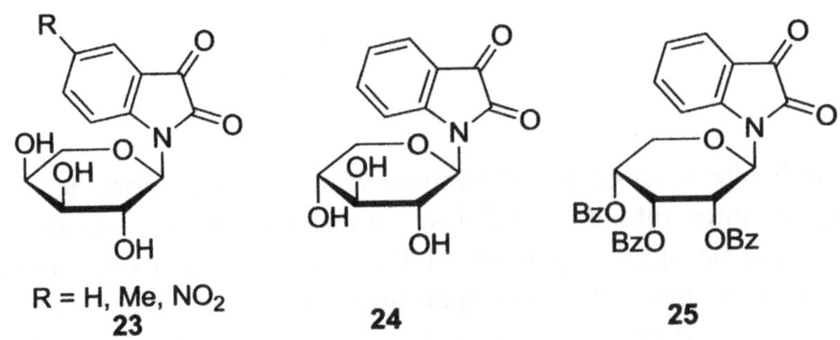

3.12.6 Experimental

Synthesis of N-*Phenyloxindole via Lewis Acid-Mediated Stollé Reaction*[9]

A solution of 84.6 g of diphenylamine (0.5 mol) and 80 mL of chloroacetyl chloride (1.0 mol) in 500 mL of toluene was refluxed for 1.5–2.0 h under N_2, cooled, and concentrated in vacuo. The residue was recrystallized from EtOH to give 103 g (84%) of *N,N*-diphenyl-2-chloroacetamide as tan crystals, mp 113–116 °C.

A mixture of 32.5 g of the amide (0.132 mol) and 41 g of $AlCl_3$ (0.307 mol) was mixed while being heated in an open beaker until an internal temperature of 180–190 °C was attained (***Note:* copious evolution of HCl**) and then heated an additional 10 min. The molten mass was allowed to cool to approximately 70 °C and then was treated with crushed ice and 100 mL of 1 N HCl. The crude product that solidified was filtered, washed with H_2O, and air dried. Recrystallization from absolute EtOH gave the desired oxindole **3** ($R = Ph$; $R^1 = R^2 = H$) as light golden colored crystals, 22.5 g (81.5%), mp 117–119 °C.

Lewis Acid-Free Stollé Reaction[13]

To a solution of 11 g (50 mmol) of *N*-phenyl-2-naphthylamine in 200 mL of dry ether was added dropwise 6.3 g (50 mmol) of oxalyl chloride. The mixture was stirred at ice-bath temperature and allowed to warm slowly to room temperature. The solution was stirred at room temperature for 3 days during which time 8.2 g (60%) of the deep red 1-phenylbenzo[*e*]isatin (**21**), mp 230–230.5 °C, precipitated from the ether.

3.12.7 References

1. Stollé, R. *Ber.* **1913,** *46*, 3915–3916.
2. Stollé, R. *Ber.* **1914,** *47*, 2120–2122.
3. Stollé, R.; Bergdoll, M.; Luther, M.; Auerhahn, A.; Wacker, W. *J. Prakt. Chem.* **1922,** *105*, 137–148.
4. Stollé, R.; Bergdoll, M.; Luther, M.; Auerhahn, A.; Wacker, W. *J. Prakt. Chem.* **1930,** *128*, 1–43.
5. Hinsberg, O. *Ber. Dtsch. Chem. Ges.* **1888,** *21*, 110–118.
6. Hinsberg, O, Simeoff, A. *Ber. Dtsch. Chem. Ges.* **1898,** *24*, 250–254.
7. Hinsberg, O. *Ber. Dtsch. Chem. Ges.* **1908,** *27*, 1367–1373
8. Przheval'skii, N. M.; Grandberg, I. I. *Chem. Heterocycl. Compd. (Translation of Khimiya Geterotsiklicheskikh Soedinenii)* **1982,** *18*, 716–719.
9. Sarges, R. Howard, H. R.; Koe, B. K. Weissman, A. *J. Med. Chem.* **1989,** *32*, 437–444.
10. Brown, R. F. C.; Butcher, M. *Tetrahedron. Lett.* **1971,** 667–670.
11. Shikhaliev, Kh. S.; Leshcheva, E. V.; Medvedeva, S. M. *Chem. Heterocycl. Compd. (Translation of Khimiya Geterotsiklicheskikh Soedinenii)* **2002,** *38*, 755–756.
12. Medvedeva, S. M.; Leshcheva, E. V.; Shikhaliev, Kh. S.; Solov'ev, A. S. *Chem. Heterocycl. Compd. (Translation of Khimiya Geterotsiklicheskikh Soedinenii)* **2006,** *42*, 534–539.

13. Shirley, D. A.; Jackson, Thomas G. *J. Organomet. Chem.* **1964,** *2*, 188–190.
14. Loloiu, G.; Maior, O. *Rev. Roum. Chim.* **1997,** *42*, 67–69.
15. Yartseva, I. V.; Ektova, L. V.; Sakharova, V. I.; Dobrynin, Y. V.; Yavorskaya, N. P.; Nikolaeva, T. G.; Sof'ina, Z. P.; Preobrazhenskaya, M. N. *Zh. Org. Khim.* **1977,** *13*, 1743–1749.

4.1 Nierenstein Reaction

Narendra B. Ambhaikar

4.1.1 Description

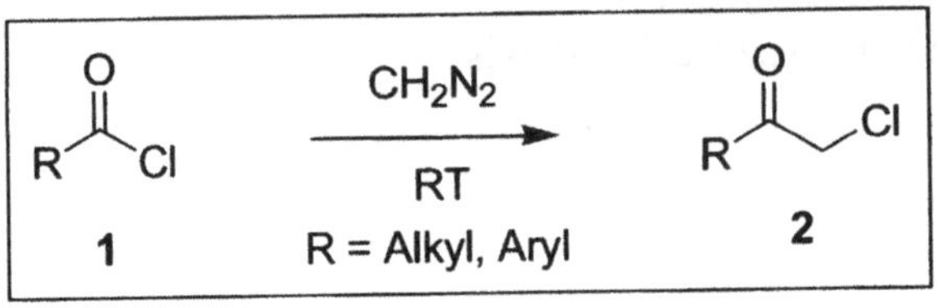

Nierenstein reaction is the reaction of an acid chloride **1** with diazomethane at room temperature to yield a α-chloroketone **2**.[1a] It differs from the Arndt–Eistert[1b] reaction in that the latter specifically forms diazoketone, which is used to form the higher homolog of the substrate acid chloride.

4.1.2 Historical Perspective

In 1915, Maximilian Nierenstein and Douglas Arthur Clibbens at the University of Bristol reported that when an ethereal solution of 1 equivalent of freshly diazomethane is added to an acid chloride at "laboratory temperature" and the mixture is stirred, the corresponding chloromethylketone **2** is obtained.[1] The sequence of addition and the temperature appeared to be rather specific. Nierenstein proposed that this tendency of acid chlorides to form the chloroketones was analogous to the reaction of aldehydes with diazomethane to yield ketones **3** reported by Schlotterbeck.[2]

Schlotterbeck Reaction

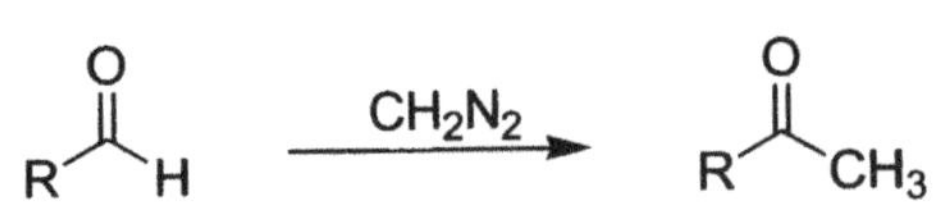

3

1

However, the first report by Nierenstein appears to have provoked considerable debate at the time. Notably, R. Robinson, then at the University of Manchester, responded to this report by saying Nierenstein's procedure could not be reproduced in their laboratories and hydrogen chloride had to be added to drive the reaction to the chloromethyl ketone **2**.[3] According to Robinson, the corresponding diazoketone **4** was observed to form in the absence of an external source of hydrogen chloride. However, Nierenstein maintained that the interaction of diazomethane with aromatic acyl chloride proceeded on the same lines as the Schlotterbeck reaction and required no special interpretation as was suggested by Robinson.[4]

It is interesting that the conditions used by Robinson and co-workers were different from the ones described by Nierenstein, and they involved the addition of *acid chloride to excess diazomethane* rather than stoichiometric in the reverse order, with the reaction mixture cooled with ice and the ethereal solution of acid chloride being added to diazomethane rather than the reverse![5] Thus, unlike Nierenstein, Robinson and co-workers had added the acid chloride to the diazomethane solution whereby the tendency for the nonformation of the chloroketone might be dominant.

4.1.3 Mechanism

Nierenstein reaction takes place when 1 equivalent of diazomethane is added to the acid chloride solution at 35 °C.[6] Nierenstein reaction is now accepted to take place through the *in situ* generation of HCl during the formation of diazoketone **4**, which reacts with it further to form the chloromethyl ketone **2**. Thus diazoketone is an intermediate in the reaction.

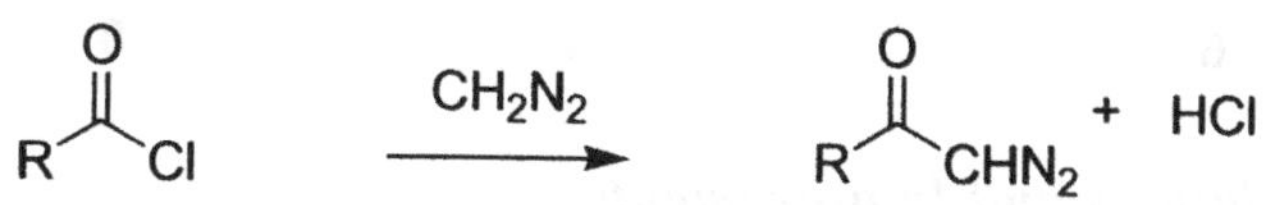

1 **4**

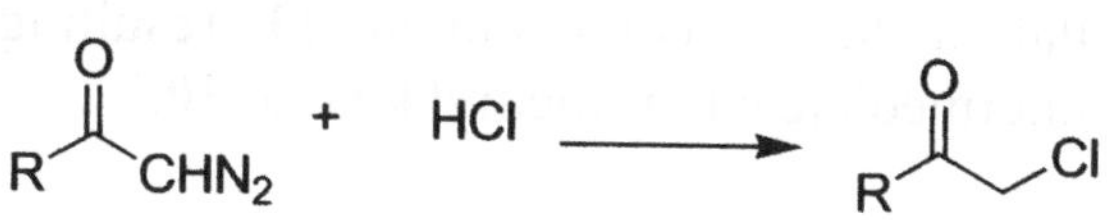

4 **2**

With excess diazomethane, methyl chloride is generated.

$$CH_2N_2 + HCl \longrightarrow CH_3Cl + N_2$$

5

Observations from both Nierenstein and Robinson have enabled the understanding of this reaction. It is also accepted that excess diazomethane at lower temperatures leads to diazoketones because hydrogen chloride liberated gets consumed by its preferential reaction with excess diazomethane to form methyl chloride **5**.[7] This explains why hydrogen chloride gas added through an external source to a solution of diazoketone can drive the Nierenstein to completion. Further evidence to support this theory was offered by McPhee and Klingsberg[8] in the procedure for the synthesis of benzyl chloromethyl ketone from phenylacetyl chloride, by which the diazoketone is separately treated with hydrogen chloride at 0 °C and also by Bhatt *et al.* in their synthesis of α-chloromethyl-3,4,5-triacetoxy acetophenone in which diazoketone is the key intermediate.[9]

One vital observation made by Nierenstein was that there was a considerable reluctance of acyl bromides to form the corresponding bromomethyl ketones.[6] Acid bromides on reaction with diazomethane were also found to form substituted 1,4-dioxanes or specifically 3,6-dibromo-3,6-diphenyl-1,4-dioxane **8**, resulting from the dimerization of adducts. The dimerization according to him could be attributed to the poor mobility of the bromide compared to the chloride group.

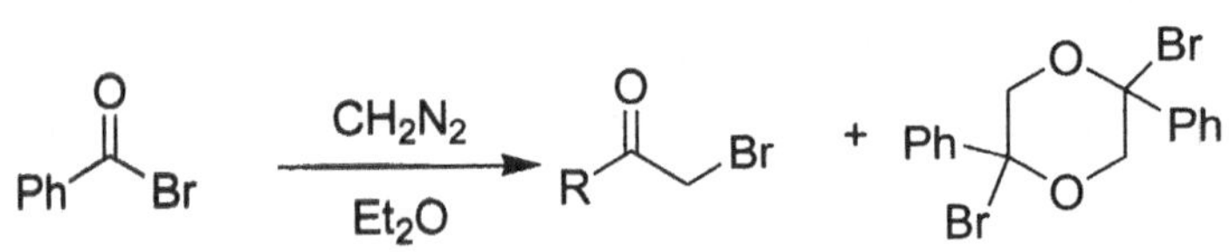

6 **7** **8**

4.1.4 *Variations and Improvements*

Nierenstein reported that when *ortho*-acetoxyacyl chlorides, for example **9**, were used, distillation yielded coumaranone **11**, resulting from the self-reaction of the intermediate chloromethyl ketone **10**.[1]

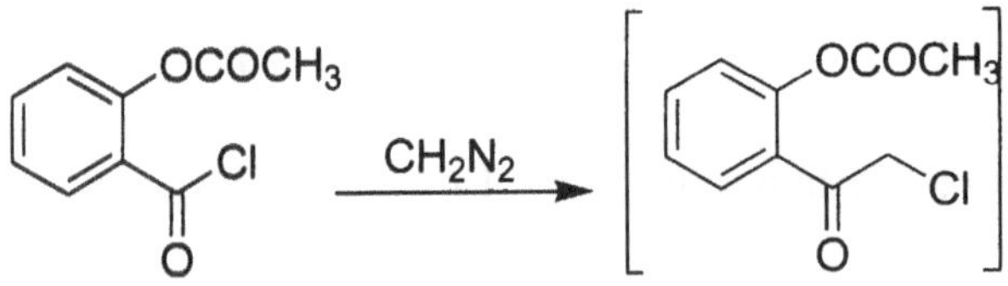

9 **10**

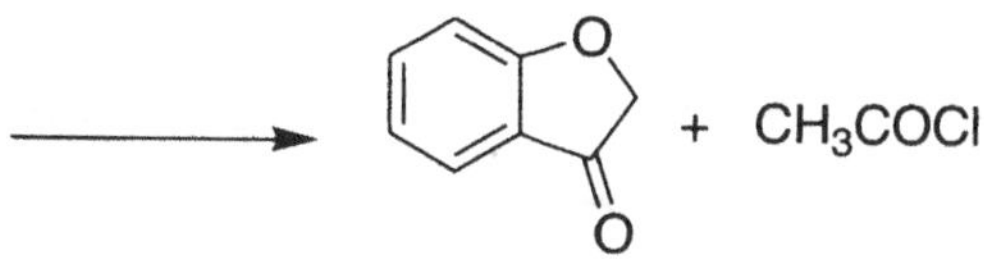

11, coumararone

4.1.5 *Synthetic Utility*

Synthesis of Fisetol

In efforts to apply the reaction to total synthesis, Nierenstein reported the synthesis of fisetol **15**, a product of disintegration of the flavonol fisetin. It was reported during his studies on the action of diazomethane on aromatic acyl chlorides.[10] Some other flavonols are quercetin, datiscetin, myricetin, quercetagetin, and gosspyetin. The synthesized molecule was observed to show the same properties as that prepared by Sonn and Falkenheim by a different method, thus confirming its structure.

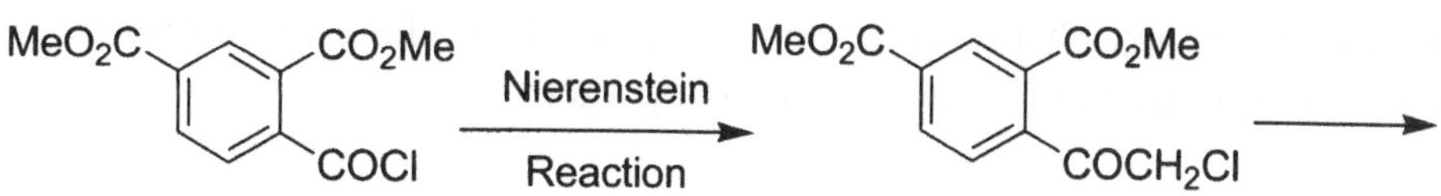

12 **13**

14 **15**, fisetol

Synthesis of 2,5-Diacylthiophenes

An interesting application of the Nierenstein reaction in the synthesis of heterocycles came from Miyahara at Kyushu University in Japan. Using the Nierenstein reaction to prepare chloromethyl ketones **17**, Miyahara reported a methodology for a facile synthesis of 2,5-diacylthiophenes **20** bearing a variety of substitutents from readily available nonthiophenic precursors.[11] The methodology involved a base-catalyzed condensation of diketo sulfides **18** with glyoxal to yield **20** under mild conditions.

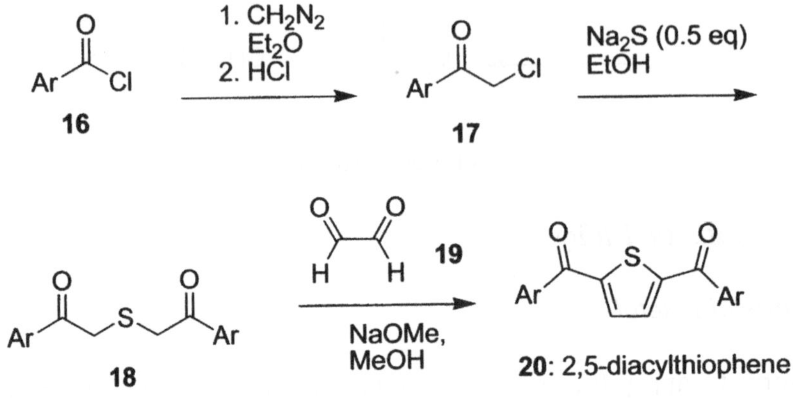

Synthesis of Aminothiazole

The Nierenstein reaction has been applied in the synthesis of 4-(β-chloromethyl)-2-aminothiazole by Carroll and Smith in 1933.[13] The α′,β-dichloromethylethyl ketone **22** was prepared *via* the Nierenstein reaction of the corresponding chloropropionyl chloride **21**. Reaction with thiourea yielded the cyclized product **23**. The authors additionally prepared the same ketone using an alternative approach involving the addition of ethylene to chloroacetyl chloride with aluminium chloride as a condensing agent.

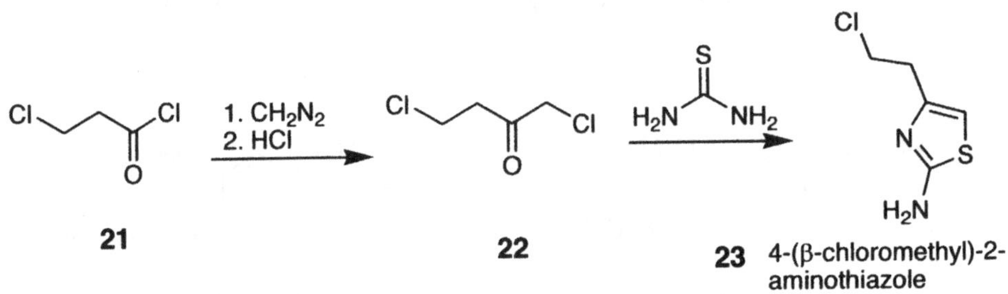

4.1.6 Experimental

Synthesis of 3,5-Dimethoxy-α-chloroacetophenone[12]

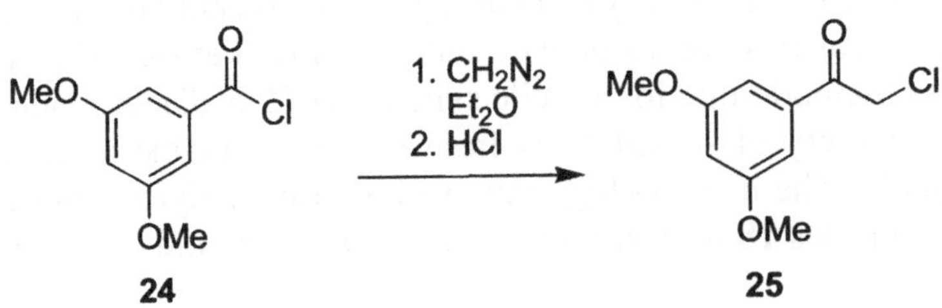

To a solution of 3,5-dimethoxybenzoyl chloride **24** (10.0 g, 0.05 mol) in 50 mL of ether, was added a solution of diazomethane, prepared from *N*-methyl-*N*-nitroso-*p*-toluenesulfonamide (32 g), portion-wise with shaking. After standing overnight at room temperature, 50 mL of concentrated hydrochloride acid were gradually added to the reaction mixture cooled in an ice bath with shaking. After the evolution of nitrogen had ceased, the lower layer was discarded and washed well with water. After evaporation of ether the crystals were collected and washed with methanol giving 8.6 g of the product **25**, 80.4% yield. Recrystallization from methanol gave colorless needless, mp 77.5–78 °C.

Synthesis of α′,β-Dichloromethylethyl Ketone (1,4-dichloro-butanone-2) through the Nierenstein Reaction[13]

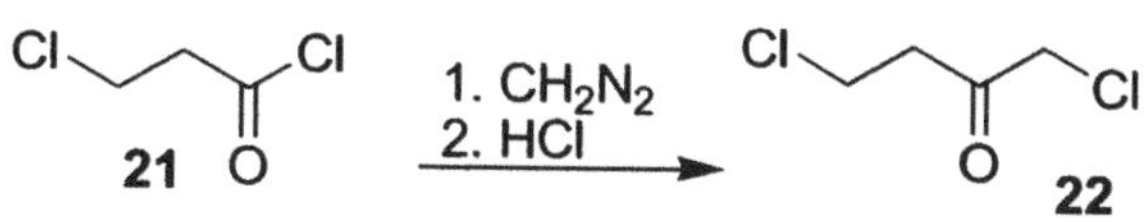

A total of 4.75 g of β-chloropropionyl chloride **21** was dissolved in 25 g of absolute ether, and the solution was cooled to –5 °C; 1.8 g of diazomethane was dissolved in 75 g of absolute ether contained in a 500-mL suction flask, cooled to –5 °C, and protected from moisture by a calcium chloride tube. This solution was treated with the ethereal solution of β-chloropropionyl chloride, taking about 5 min for the addition. The temperature was maintained at –5 °C and the solution was allowed to stand for 1 h, at the end of which period a second portion of 1.8 g of diazomethane was added to the solution, nitrogen being again evolved. The solution was now placed in an ice-box and allowed to stand for 36 h. The yellow-colored solution was treated with dry hydrogen chloride until the color was pale yellow and the evolution of nitrogen had ceased. During the treatment with hydrogen chloride, the temperature was held below 10 °C. The ether was evaporated by a current of dry air, and the residue was fractionated at reduced pressure. The portion distilling at 65 °C under a pressure of 3 mm was collected: 3.2 g.

4.1.7 References

1 (a) Clibbens, D. A.; Nierenstein, M. *J. Chem. Soc.* **1915,** 1491. [R] (b) Bachmann, W. E.; Struve, W. S. *Org. React.* **1942,** *1*, 38.
2 Schlotterbeck, F. *Ber.* **1907,** *40,* 479.
3 Bradley, W.; Robinson, R. *Nature* **1928,** 130–131.
4 Malkin, T.; Nierenstein, M. *J. Am. Chem. Soc.* **1930,** *52,* 1504–1508.

5 Bradley, W.; Robinson, R. *J. Am. Chem. Soc.* **1930,** *52,* 1558–1565.
6 Lewis, H. H.; Nierenstein, M.; Rich, E. M. *J. Am. Chem. Soc.* **1925,** *47,* 1728–1730.
7 [R] Sonnetag, N. O. V. *Chem. Rev.* **1953,** *52,* 237–416.
8 McPhee, W. D.; Klingsberg, E. K. *Org. Synth. Coll. V 3*, **1955,** 119.
9 Bhatt, M. V.; Iyer, B. H.; Guha, P. C. *Cur. Sci.* **1948,** *9*, 264.
10 Nierenstein, M.; Wang, D. G.; Warr, J. C. *J. Am. Chem. Soc.* **1924,** *46,* 2551–2555.
11 Sonn; Falkenheim, *Ber.* **1922,** *55,* 2975.
12 Miyahara, Y. *J. Heterocyclic Chem.* **1979,** 1147–1151.
13 Carroll, R. H.; Smith, G. B. L. *J. Am. Chem. Soc.* **1933,** *55*, 370–373.

4.2 Davidson Oxazole Synthesis

Nadia M. Ahmad

4.2.1 Description

The Davidson oxazole synthesis describes the preparation of substituted oxazoles **2** from the condensation of *β*-keto-esters **1** with ammonia or ammonium acetate. A general reaction scheme is depicted below.

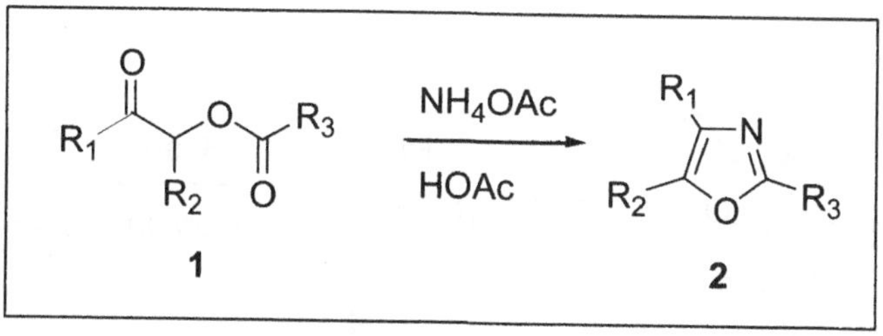

4.2.2 Historical Perspective

Davidson, Weiss and Jelling found that the reaction of acyl derivatives of benzoin with ammonium acetate in boiling glacial acetic acid afforded excellent yields of 2-substituted-4,5-diphenyloxazole.[1] This reaction had previously been mentioned in the literature by Japp and co-workers;[2] however, it was Davidson who examined the mechanistic detail of the synthesis in further depth.

This synthesis of oxazoles was much studied by Davidson and co-workers, who investigated both the mechanism of formation of oxazoles and also a general method to prepare substituted oxazoles. It was eventually found that the reaction gave better results when synthesizing 2,4,5-tri-substituted oxazoles with an aromatic substituent at C-5 and mediocre yields for the formation of 2,4-disubstituted or mono-substituted oxazoles.

4.2.3 Mechanism

A mechanism is shown illustrating the reaction of the *β*-keto ester with ammonium formate. An intra-molecular reaction of the intermediate results in imine **3**. A further intra-molecular self-condensation is now possible, which proceeds to give the protonated species **4**. A basic work-up then affords oxazole **2**.

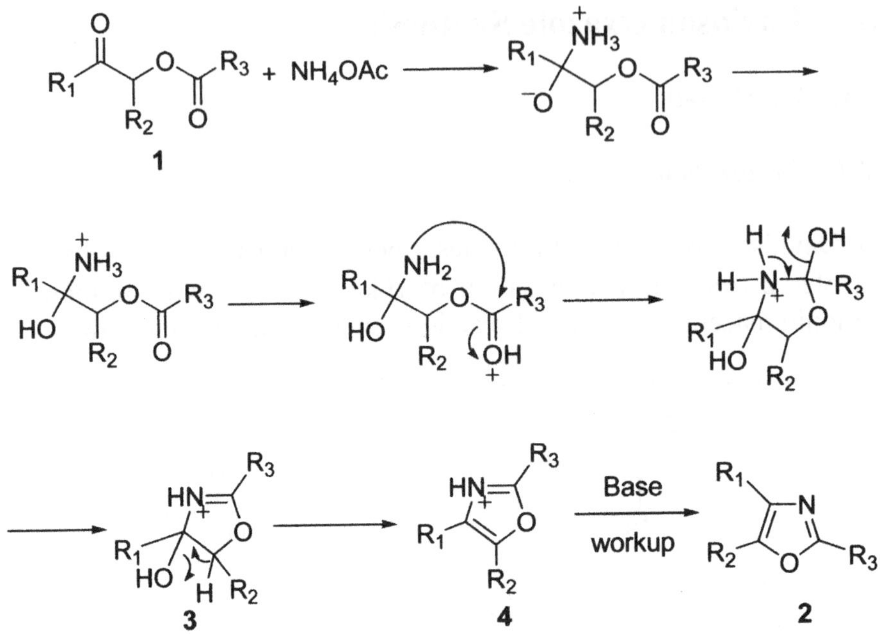

It has also been suggested that the reaction mechanism proceeds *via* an enamine intermediate **7**; these have in fact been synthesized by other methods and converted to oxazoles by boiling in glacial acetic acid.[4,5]

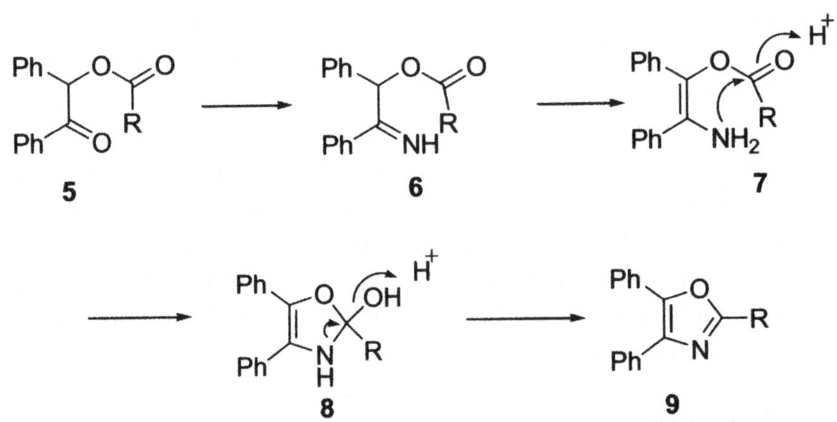

4.2.4 *Variations, Improvements or Modifications*

The acyloin ester can be prepared from treatment of the corresponding α-bromo ketone **10** with the sodium or potassium salt of the desired carboxylic acid **11**. This reaction can be carried out neat in the carbocylic acid used or in ethanol. Using this method, the ester may not necessarily need to be isolated and the addition of ammonium acetate to the mixture, or passing through ammonia through it, results in the desired oxazole **13** in a one-pot procedure.[6] An example is illustrated below.[7]

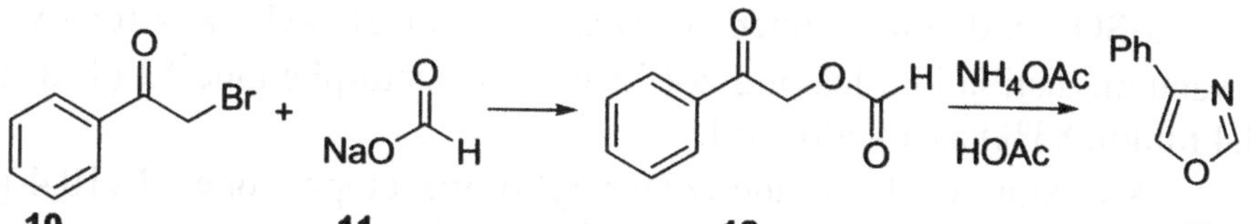

10 11 12 13

4.2.5 *Synthetic Utility*

Jefferys has modified the Davidson oxazole synthesis by using phenacyl bromides and ammonium acetate to yield 2-methyl-4-(or 4,5-di-)-substituted oxazoles.[8] He was also able to extend the syntheses outlined by Japp and Davidson to obtain 4,5-dialkyl and 4-aryl oxazoles. These oxazoles together with 2-methyloxazole were then quaternized and transformed into cyanine and merocyanine dyes.

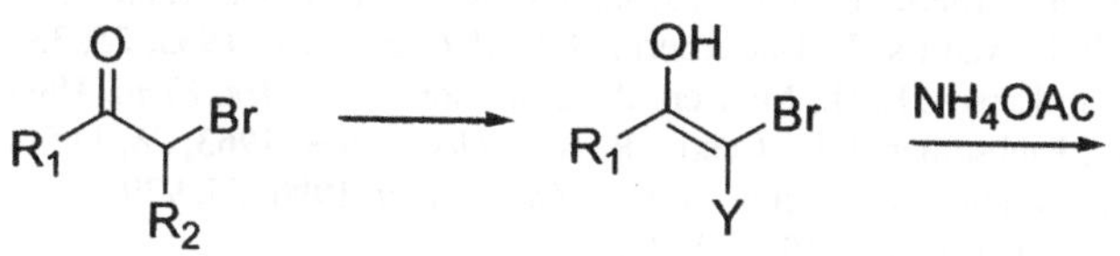

14 15

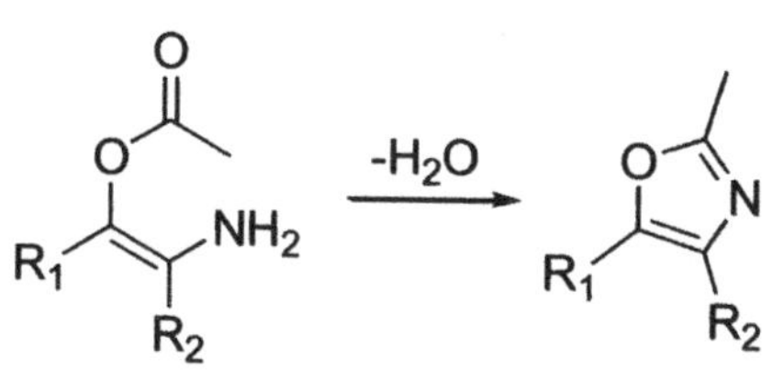

16 17

4.2.6 *Experimental*

2-Phenyl oxazole 13[6]

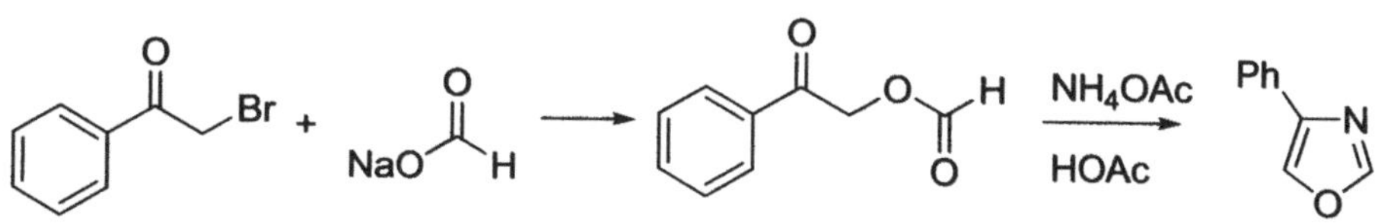

10 11 12 13

A mixture of bromoacetophenone **10** (0.100 mol) and sodium formate **11** (0.13 mol) in DMF (200 mL) was stirred for 2 h. The solution was then poured into water (800 mL) and extracted with CH_2Cl_2 until the aqueous layer was colorless. The organic phase was washed with brine, dried over Na_2SO_4, and concentrated *in vacuo*. Residual DMF was removed by vacuum distillation, leaving α-(formyloxy)aceto-phenone **12** (13.6 g, 63.4 mmol, 83%) as a yellow oil.

A mixture of this crude α-(formyl oxy)acetophenone **12** (10.4 g, 63.4 mmol) and ammonium acetate (24.3 g, 315 mmol) in acetic acid (125 mL) was heated under reflux for 90 min. After cooling, it was added to water (500 mL) and extracted with CH_2Cl_2. The organic phase was neutralized by careful addition (foaming) of saturated $NaHCO_3$ solution, washed with brine, dried over Na_2SO_4, concentrated *in vacuo*, and distilled to yield 2-phenyl-oxazole **13** (4.52 g, 49.5%).

4.2.7 References

1. Davidson, D.; Weiss, M.; Jelling, M. *J. Org. Chem.* **1937,** *2*, 319.
2. Davidson, D.; Weiss, M.; Jelling, M. *J. Org. Chem.* **1937,** *2*, 328.
3. (a) Japp, F. R. *Ber.* **1883,** *16*, 2636. (b) Japp, F. R. *J. Chem. Soc.* **1883,** *43*, 9.
4. Strzybny, P. P. E.; van Es, T.; Backeberg, O. G. *J. Org. Chem.* **1963,** *28*, 3381.
5. Jakobsen, H. J.; Larsen, E. H.; Madsen, P.; Lawesson, S.-O. *Ark. Kemi.* **1965,** 519.
6. Aldous, D. L.; Riebsomer, J. L.; Castle, R. N. *J. Org. Chem.* **1963,** *28*, 1151.
7. Whitney, S. E.; Winter, M.; Rickborn, B. *J. Org. Chem.* **1990,** *55*, 929.
8. Jeffreys, R. A. *J. Chem. Soc.* **1952,** 4823.

4.3 Fischer Oxazole Synthesis

Noha Maklad

4.3.1 Description

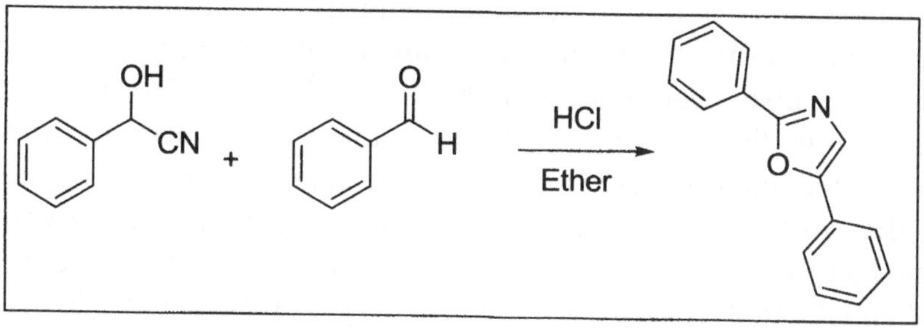

Fischer oxazole synthesis, reported in 1896 by Emil Fischer, is one of the earliest methods reported for the synthesis of 2,5-disubstituted oxazoles. It entails the reaction of cyanohydrin with aldehyde under dry acidic conditions to form the corresponding oxazole.[1]

4.3.2 Historical perspective

Emil Fischer was born in Euskirchen, Germany on October 9, 1852. He died in Berlin on July 15, 1919. A 1902 Nobel Laureate, Fischer is considered a great genius of organic chemistry to the present day.[2,3] He was one of Adolf von Baeyer's most prominent students; under Baeyer's supervision he received his doctorate in Strasbourg in 1874. His scientific achievements and contributions to the field are of a great importance and influence. Fischer's accomplishments range from phenylhydrazine synthesis (his first major discovery) to various contributions in natural products synthesis, including but not limited to carbohydrates, nucleic acids, and peptides synthesis, all of which have shaped a great deal of the organic chemistry field to this day.

Fischer's name is associated with a great number of important organic transformations, such as the Fischer indole synthesis, Fischer esterification, Kiliani–Fischer synthesis, and the Fischer oxazole synthesis. The Fischer oxazole synthesis was one of his first contributions in his early years in the Berlin University (currently Humboldt University). In 1896 he published a new synthesis of 2,5-diaryloxazoles using an acid-catalyzed condensation of cyanohydrins with aldehydes.

4.3.3 *Mechanism*

Fisher's oxazole formation occurs by the condensation of cyanohydrins with aldehydes in the presence of HCl in dry ethereal solution.[1] The cyanohydrins and aldehydes used are usually aromatic in nature, although there arc aliphatic examples that have been reported. There has been little study of the reaction mechanism such as that shown by Ingham and Cornforth in the early nineteenth century; the reaction details have also been published by others.[4–6] The first step of the mechanism is the addition of HCl to the cyanohydrin to from an iminochloride intermediate **2**. This intermediate then reacts with the aldehyde, which is followed by water loss to give a chloro-oxazoline intermediate **4**. Isomerization of two protons occurs, followed by the loss of an HCl molecule to form the 2,5-diaryloxazole end product.

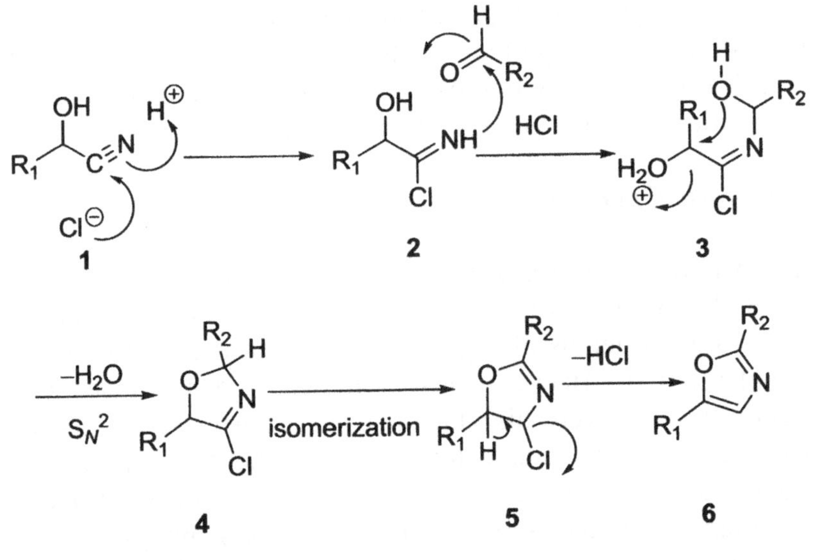

In the presence of moisture, the chloro-oxazoline **4** is converted to oxazolidone of structure **7**, which can form the 2,5-disubstituted oxazole end product if warmed up with phosphoryl chloride ($POCl_3$), where it is believed to reform intermediate **4**, and then to proceed to the oxazole end-product formation.[5]

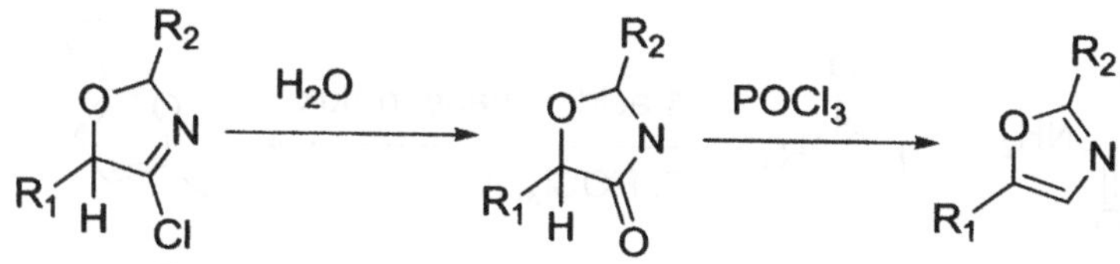

4 7 6

4.3.4 *Standard Method, Variations, and Improvements*

The standard method as discussed above is simply the condensation of the aldehyde and the cyanohydrins in a moisture-free HCl/ether solution.

The cyclization in the presence of moisture was studied and discussed by Cornforth *et al.* Under such conditions, oxazolidone **7** is formed, and the reaction halts, although the final oxazole product can still be formed by using either $POCl_3$ or thionyl chloride.[5] Cope and Hancock showed that compounds of structure **7** can undergo ring-chain tautomerism to give **8** (Schiff base of aldehyde R_2CHO and α-hydroxy amide $R_1C(OH)CONH_2$).[5,7] This realization has prompted a variation to the synthesis using the corresponding α-hydroxy amide with the corresponding aldehyde to give not only the 2,5-diaryl oxazole but also to allow the incorporation of alkyl substituents as well, which was not possible using Fischer's original conditions. An example of such variation is the synthesis of oxazoles **11a** and **11b.** In toluene, cyclization is acid catalyzed in the presence of *p*-toluene sulfonic acid (PTSA) or acetic acid. For example, lactamide (**9**) reacts with benzaldehyde to give the corresponding oxazolidone intermediate in a ~40% yield, this is then heated in $POCl_3$, which proceeds to give 2-phenyl-5-methyloxazole (**11a**) in ~12% yield (for the two steps). Similarly, lactamide and heptaldehyde form the corresponding oxazolidone in acetic acid, which then gives 2-hexyl-5-methyl oxazole (**11b**). The yields for the latter cannot be clearly derived from Cornforth's reported work.

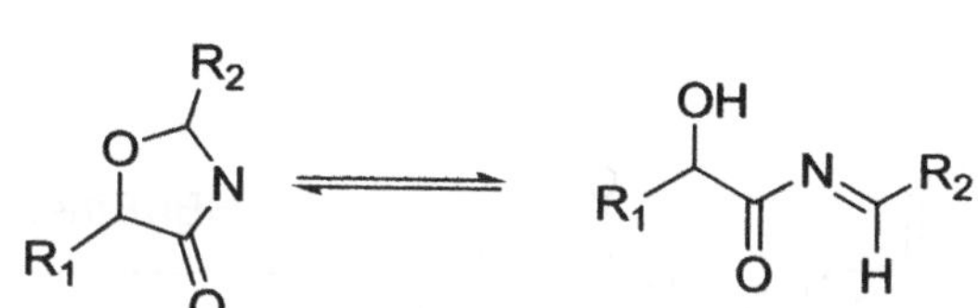

7 8

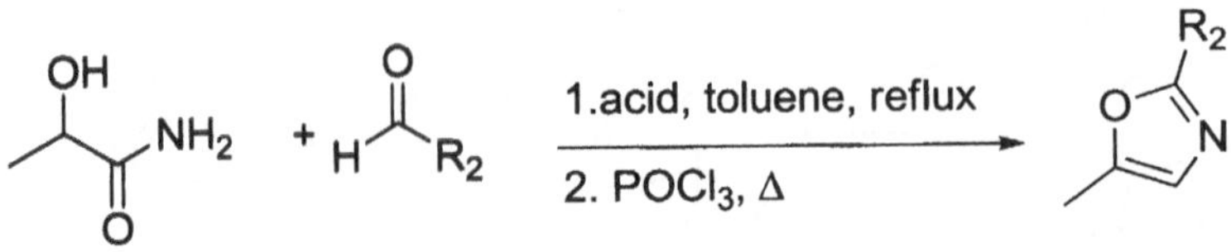

9 | 10a R_2 = phenyl, 10b R_2 = heptyl | 11a R_2 = phenyl, 11b R_2 = hexyl

In 1977 Davis, Ternai *et al.* introduced an interesting variation to the synthesis.[8] In their efforts to form substituted 2,5-diaryl oxazoles, a simple modification of the Fischer oxazole synthesis rendered a good yielding process. Acyl cyanide replaces cyanohydrins, which reacts with aldehyde to form 2,5-diaryl-4-chloro- (or bromo-) oxazoles. An example for such variation is the synthesis of the chloro-oxazoles **14**. Benzoyl cyanide (**12**) reacts with aldehyde **13a** at 0 °C in the presence of HCl, and the mixture is stirred overnight to give **14a** in a 75% yield. Similarly, benzoyl cyanide reacts with aldehyde **13b** to give **14b** in a 65% yield. 4-Bromo-oxazoles can also be formed using such variation. An example is the synthesis of **16** by the reaction of *p*-nitrobenzoyl cyanide (**15**) with benzaldehyde in the presence of HBr; the 4-bromo adduct is formed in a 68% yield.

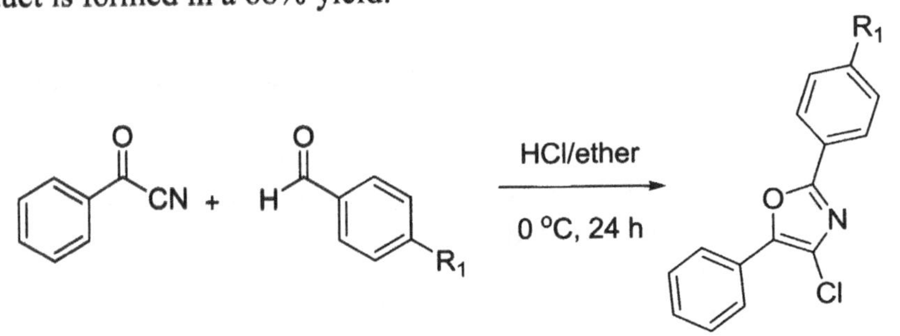

12 | **13a–b** | **a**, $R_1 = NO_2$; **b**, R_1 = OMe | **14a**, 75%; **14b**, 65%

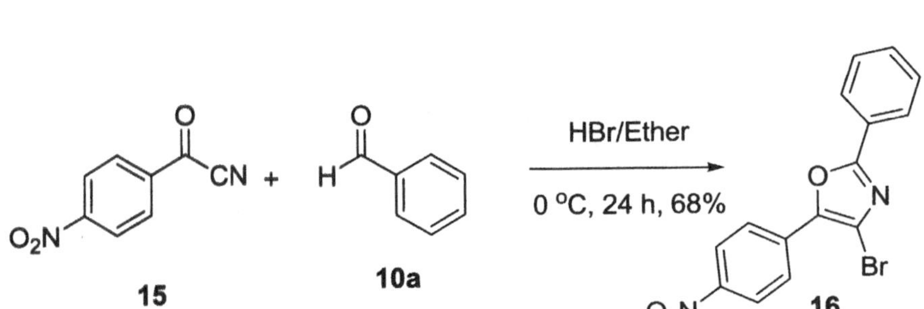

15 | **10a** | **16**

The mechanism for this variation is believed to go through an acylimidoyl chloride/bromide intermediate **18**; the intermediate then attacks the aldehyde to give 4-halo-oxazoles of structure **20**.

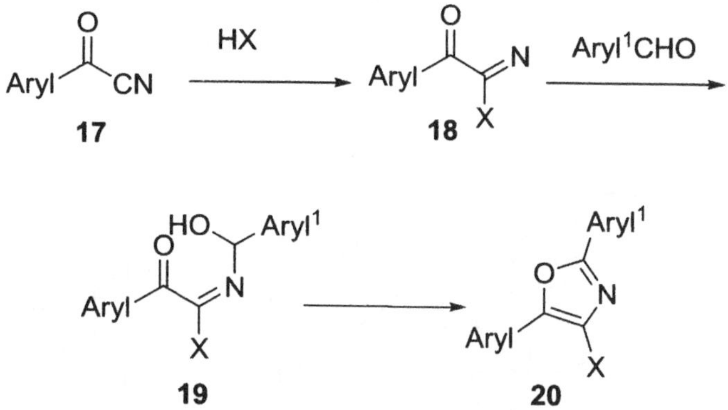

4.3.5 *Synthetic Utility*

Natural Products

Fischer oxazole synthesis and its variations are used generally in the synthesis of 2,5-disubstituted oxazoles. An example of its utility is the one pot two-step synthesis of halfordinol, a parent compound for *Rutaceae* alkaloids. The synthesis follows Fisher's original conditions,[9] although the acid-catalyzed cyclization occurs in two steps rather than one. This minor twist is adopted to ensure the formation of the di-chloro intermediate **22**. This measure prevents the formation of the unwanted (opposite) regio-isomer. This is generally seen under standard conditions and occurs by the hydrolysis of the starting cyanohydrin to the parent aldehyde, followed by the transformation of the reacting aldehyde to its corresponding cyanohydrin.[1,8,9] While sustaining dry conditions the di-chloro adduct (formed in the presence of $SOCl_2$) then undergoes cyclization in the presence of nicotinaldehyde (**23**). Halfordinol is recrystallized from methanol in a 16.5% yield.

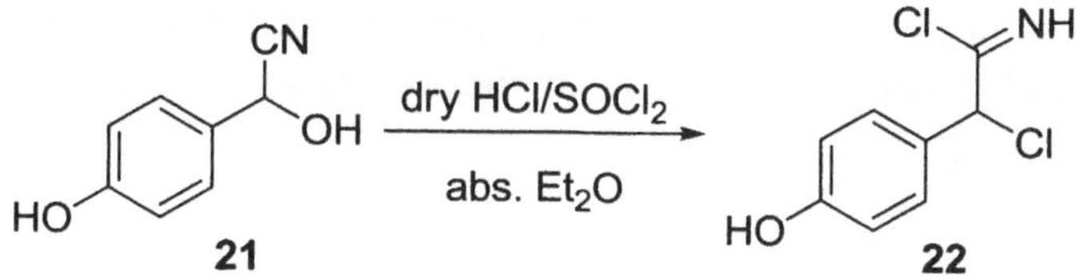

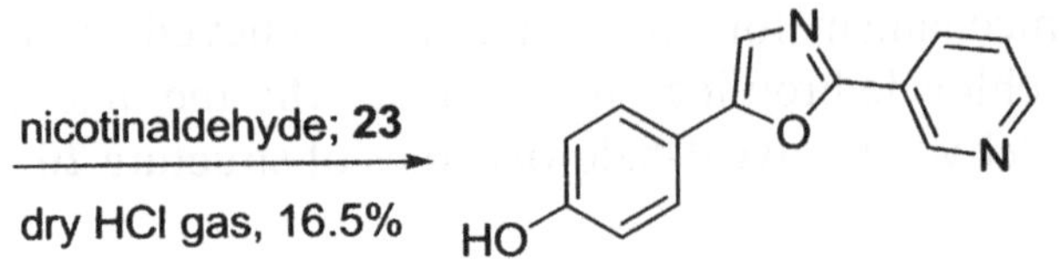

halfordinol, 24

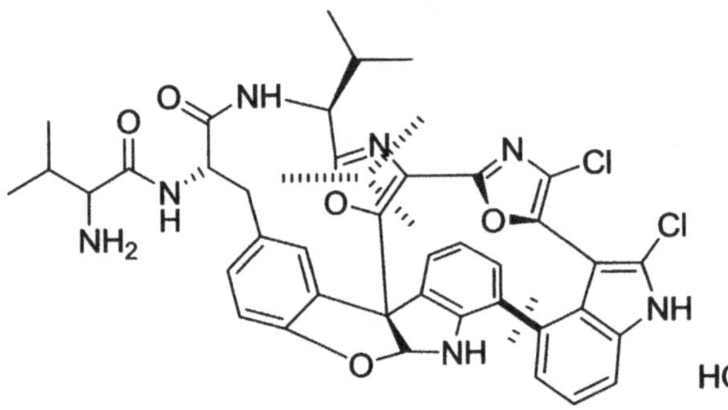

diazonamide A (revised structure); 25

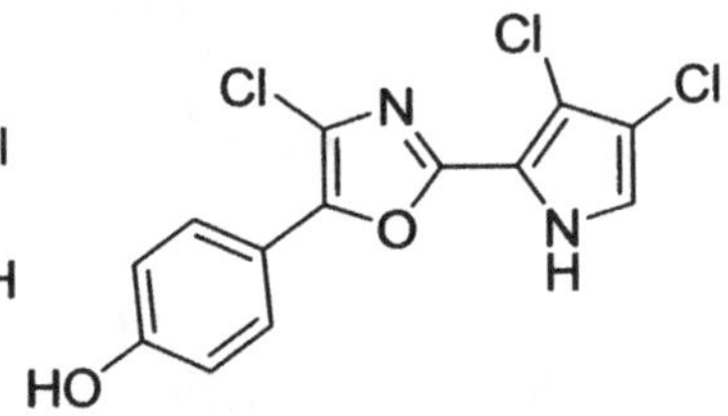

phorbazole A; 26

One interesting example that pertains to the use of the Davis variation is the synthesis of analogues of diazonamide A, cytotoxic ingredients of *Diazona chinensis*—and analogs of phorbazole A (a marine natural product) as reported by Liebscher *et al.*[10] This report adds to the variation the ability to place non-aromatic substituents at C-1 (position 2) of the oxazole. The cyclization proceeds in the presence of 1.2 mol of $BF_3{\cdot}Et_2O$, possibly used for the activation of the nitrile group to form the acylimidoyl chloride intermediate.

For diazonamide A analog such as compounds of structure **29a–c**, synthesis proceeds by the reaction of trichloroethoxycarbonyl (Troc-group) protected indole-3-ylcarbonyl cyanide **27** with aldehydes **28** to give 4-chloro-oxazole adducts of structure **29**. Compound **29a** (R_1 = Me) is formed in a 77% yield, **29b** (R_1 = CH_2Br) in a 43% yield, and **29c** (R_1 = CH_2CH) in a 74% yield. One example of phorbazole A analogs' synthesis is the reaction of 4-Troc-*O*-benzoyl cyanide **30** with aldehyde **31** to give the chloro-oxazole **32** in a 21% yield. The chloro-oxazoles **29** and **32** thus formed undergo deprotection and other transformations to form the analog intended for use in the synthesis of the above natural products' analog.

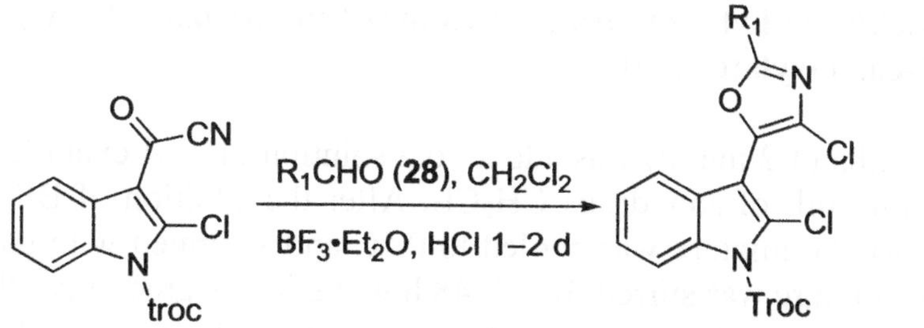

27

29a R_1 = Me, 77%
29b R_1 = CH_2Br, 43%
29c R_1 = CH_2CH, 74%

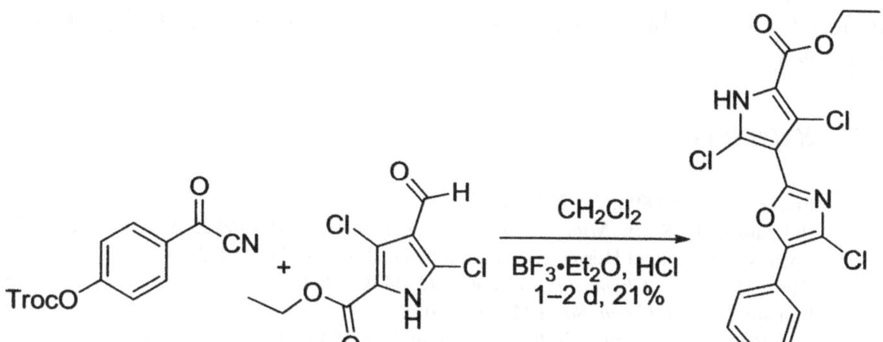

30 31 32

4.3.6 Experimental

Halfordinol (24)

Dry HC1 gas was passed and saturated in an ice-cold solution of freshly prepared crystals of 0.94 g of *p*-hydroxymandelonitrile (**21**) in 45 mL of anhydrous ether. After an addition of 1.12 g of $SOC1_2$, the reaction mixture was stirred for 10 min with external cooling. Then an addition of 0.75 g of nicotinaldehyde was followed and the reaction mixture was saturated with dry HCl once again. After standing at room temperature for 2 days, the reaction mixture was poured into water, and the separated organic layer was further extracted with aq. HCl. Neutralization of the combined aqueous layers with Na_2CO_3 resulted in the precipitation of halfordinol (**24**), which was collected on a filter and recrystallized from methanol to fine cream needles of mp 254–255 °C yield, 248 mg (16.5%).

2,2,2-Trichloroethyl-2-chloro-3-(4-chloro-2-methyloxazol-5-yl)-1*H*-indole-1-carboxylate (29a).

Aldehyde **28a** (1.2 mmol) was added to a solution of acyl cyanide **27** (1 mmol) in 40 mL of anhydrous CH_2Cl_2. After the addition of $BF_3 \cdot Et_2O$ (1.2 mmol), the mixture was cooled to 0 °C and saturated with gaseous HCl. The mixture was stirred for 24–48 h while it was gradually allowed to warm up to room temperature. It was then poured into 100 ml of ice water and was extracted with (3 × 30 ml) CH_2Cl_2. The combined organic layers were washed with sat. aq. $NaHCO_3$ solution, H_2O, and brine and dried ($MgSO_4$). After stripping off the solvent, the remainder was purified by flash chromatography (hexane–CH_2Cl_2 or CH_2Cl_2) to give **29a** (77%).

4.3.7 *References*

1. Fischer, E. *Ber.* **1896,** *29*, 205–214.
2. [R] James, L. K. In *Nobel Laureates in Chemistry 1901–1992*, American Chemical Society & Chemical Heritage Foundation, Washington, DC, 1993.
3. Kunz, H. *Angew. Chem., Int. Ed.* **2002,** *41*, 4439–4451.
4. Ingham, B. H. *J. Chem. Soc.* **1927,** 692–700.
5. Cornforth, J. W.; Cornforth, R. H. *J. Chem. Soc.* **1949,** 1028–1030.
6. [R] Brooks, D. A. In *Name Reactions in Heterocyclic Chemistry*; Wiley: Hoboken, NJ, **2005**. (b) [R] Cornforth, J. W. In *Heterocyclic Compounds,* 5; Wiley: New York, **1957**. (c) [R] Wiley, R. H. *Chem. Rev.* **1945,** *37*, 401–442.
7. (a) Cope, A. C.; Hancock, E. M. *J. Am. Chem. Soc.* **1942,** *64*, 1503–1506. (b) Cope, A. C.; Hancock, E. M. *J. Am. Chem. Soc.* **1944,** 66, 1448. (c) Bishop, D. C.; Bowman, R. E.; Campbell, A.; Jones, W. A. *J. Chem. Soc.* **1963,** 2381–2385.
8. Davis, M.; Lakhan, R.; Ternai, B. *J. Heterocyclic Chem.* **1977,** *14*, 317–318.
9. (a) Onaka, T. *Tetrahedron Lett.* **1971,** *46*, 4393–4394. (b) Brossi, A.; Wenis, E. *J. Heterocyclic Chem.* **1965,** *2,* 310–312.
10. (a) Radspieler, A.; Liebscher, J. *Synthesis* **2001,** 745–750. (b) [R] Yeh, V. S. C. *Tetrahedron* **2004,** *60*, 11995–12042. (c) Rudi, A.; Stein, Z.; Green, S.; Goldberg, I.; Kashman, Y.; Benayahu, Y.; Schleyer, M. *Tetrahedron Lett.* **1994,** *35*, 2589–2592.

4.4 Japp Oxazole Synthesis

Xiaojun Han

4.4.1 Description

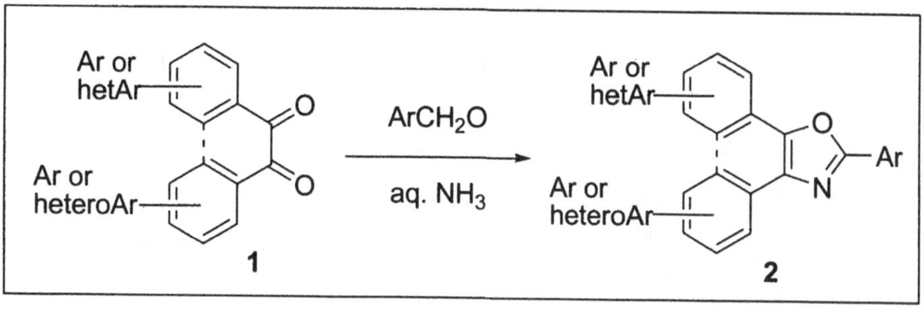

The Japp oxazole synthesis is the reaction of 1,2-aromatic diketones **1** and aromatic aldedhyes in the presence of ammonia to form oxazoles **2**.[1,2]

4.4.2 Historical Perspective

Professor Japp reported the reaction of phenanthraquinone **3** with benzadehyde and aqueous ammonia to form oxazole **4** in 1880.[1]

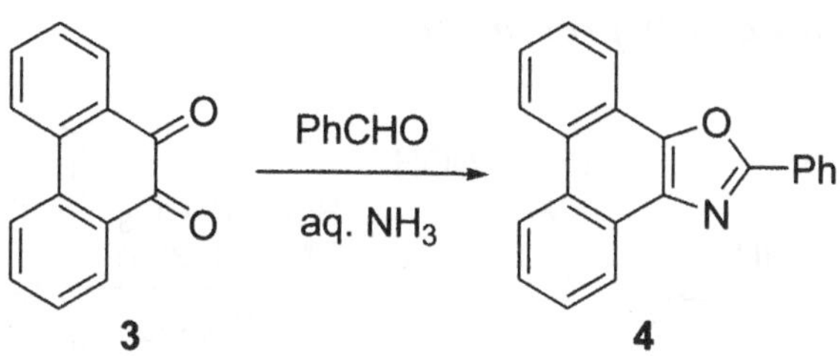

4.4.3 Mechanism

Day proposed the following mechanism in 1940, after he made a more thorough study of intermediates of Japp's original reaction.[2]

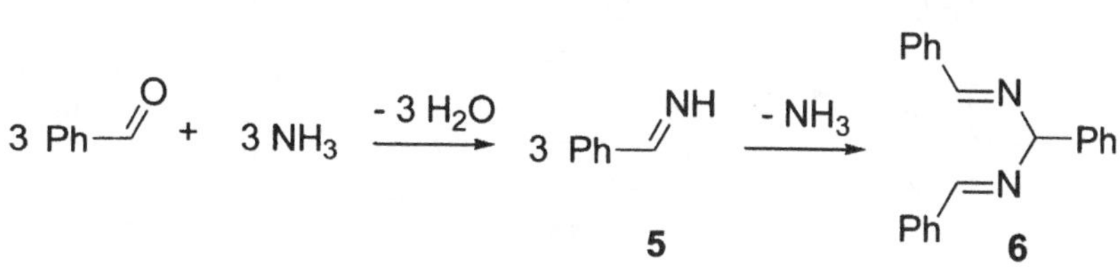

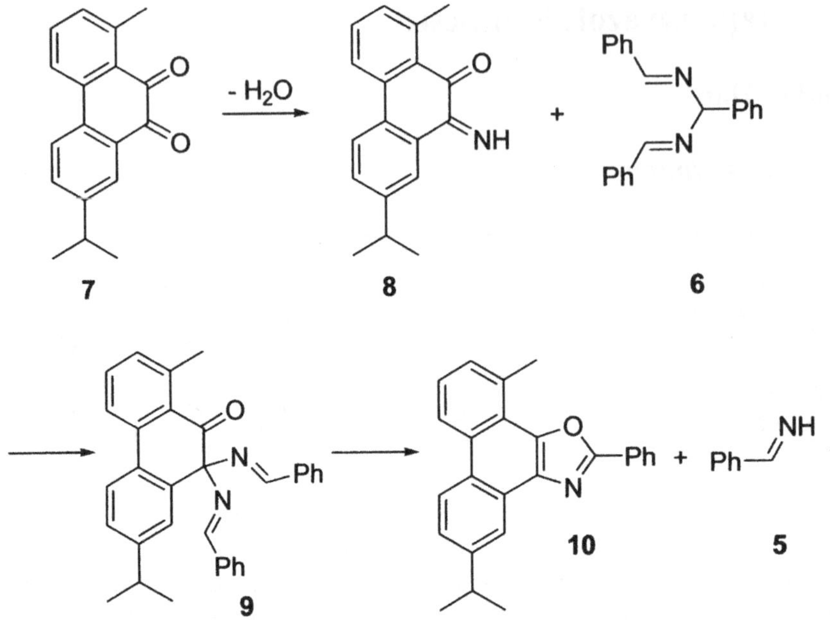

Benzaldehyde reacted with ammonia to form imine **5**, which then self-condensed to form intermediate **6**. In the meantime, imine **8**, formed by the reaction of diketone **7** and ammonia reacted with intermediate **6** to form intermediate **9** and imine **5**. After loss of one molecule of imine **5**, oxazole **10** was formed from intermediate **9**.

4.4.4 Variations and Improvements

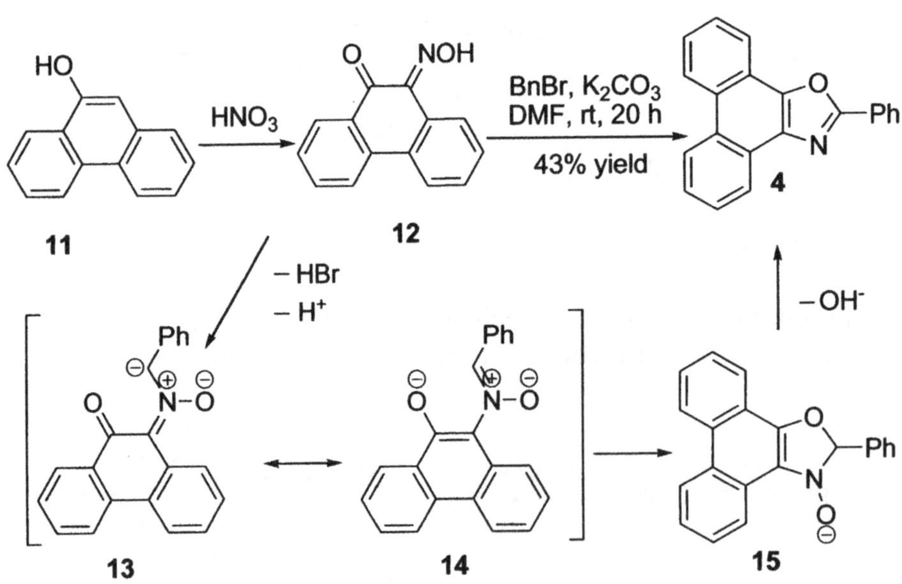

Katritzky[3] reported that oxime **12**, formed by nitrosation of enol **11**, reacted with benzyl bromide in the presence of K_2CO_3 in DMF at room temperature to form oxazole **4** in 43% yield. Since benzyloxime **16** formed by the reaction of diketone **3** and *o*-benzylhydroxyamine didn't form oxazole **4** in the presence of K_2CO_3 in DMF, even under forcing conditions (heating at 60 °C), he proposed the following for its formation. The reaction of oxime **12** with benzyl bromide formed nitrone **13**, which could also exist as resonance structure **14**, followed by cyclization and loss of HO^-, afforded oxazole **4**. Alkyl halides could also used as electrophiles to react with 1,2-diketone nitrones to form oxazoles (**17** → **18**).

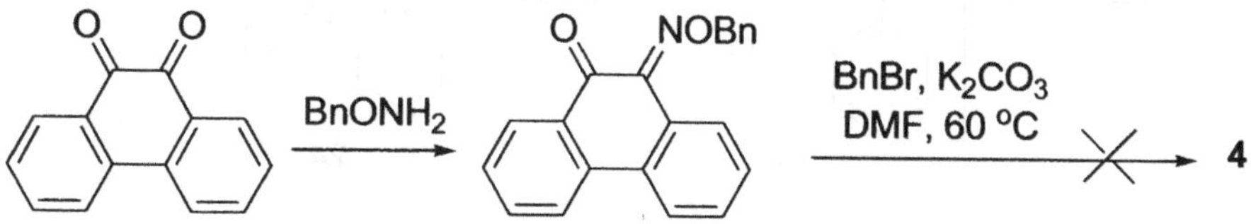

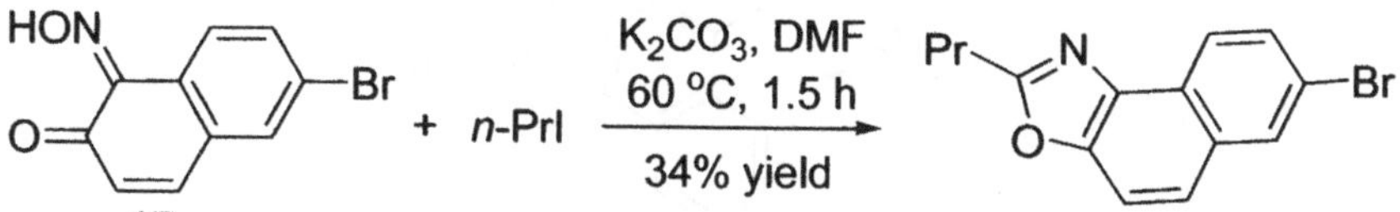

Nicolaides[4] observed that the reflux of oxime **19** in PhMe formed oxazole **4** in 37% yield, and proposed its formation by a sequence of reactions as follows. A homolytic reaction between oxime **19** and PhMe led originally to the formation of radicals **20** and **21** from oxime **19** and PhMe, respectively; then the combination of the two radicals formed intermediate **22**. After loss of one molecule of MeOH from **22**, *o*-quinone-imine **23** was formed. *o*-Quinone-imine **23** tautomerized to hydroxyl-imine **24**, which after cyclization and loss of of one molecule of H_2 formed oxazole **4** in 37% yield. Oxime **19** could also react with other methyl containing aromatics and heteroaromatics to form oxazoles (**19** → **25**).

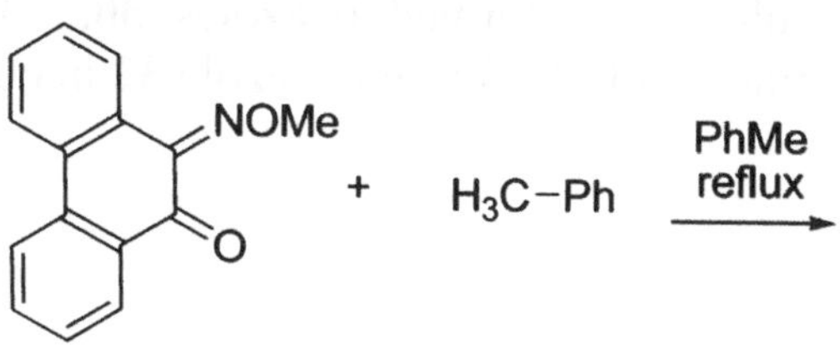

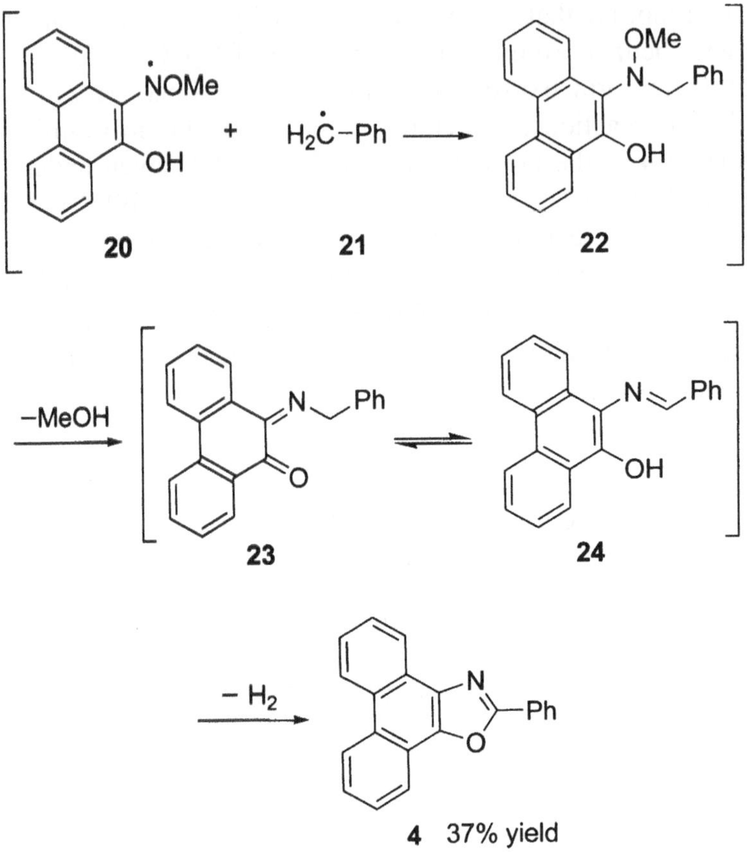

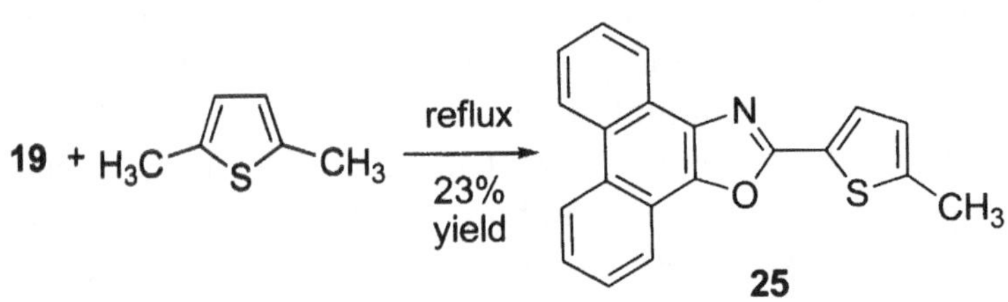

Vanderesse[5] described that the reaction of 3,5-di-*tert*-butyl-1,2-quinone **26** and primary amines **27** to form oxazoles **30**. Presumably, the reaction of diketone **26** and primary amines **27** formed keto-imine **28**, which isomerize to hydroxy-imine **29**, its subsequent cyclization and loss of one molecule of H_2 formed oxazoles **30**. The reaction of diketone **26** and amino amide **31** formed oxazole **32** in 60% yield.

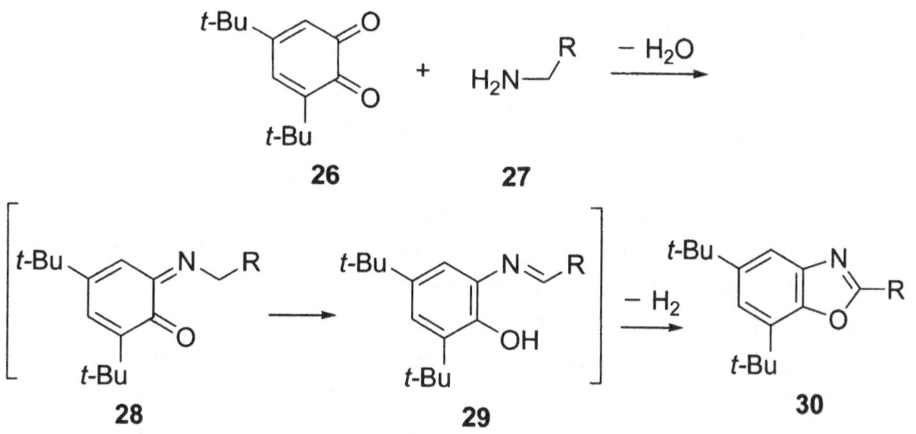

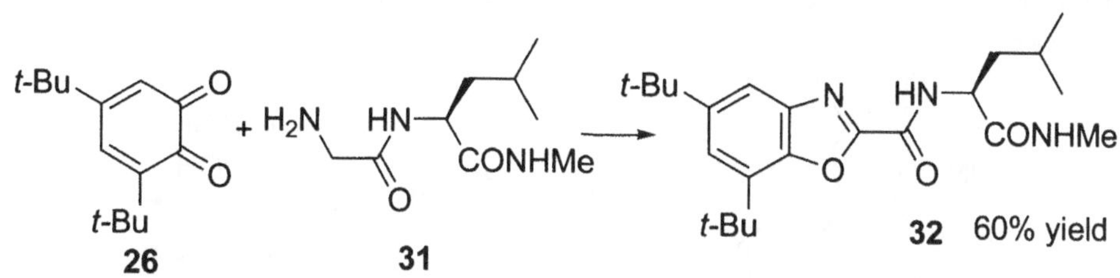

4.4.5 *Synthetic Utility*

Oxazole **35** was studied for sensing water in organic solvents by photo-induced electron transfer since it exhibited a weak emission in organic solvents but demonstrated a drastic enhancement in the fluorescence intensity with increasing water content in organic solvents. Oxazole **35** was formed in 32% yield by the reaction of diketone **33** and aldehydes **34** in the presence of NH_4OAc in AcOH at 90 °C for 8 h.[6]

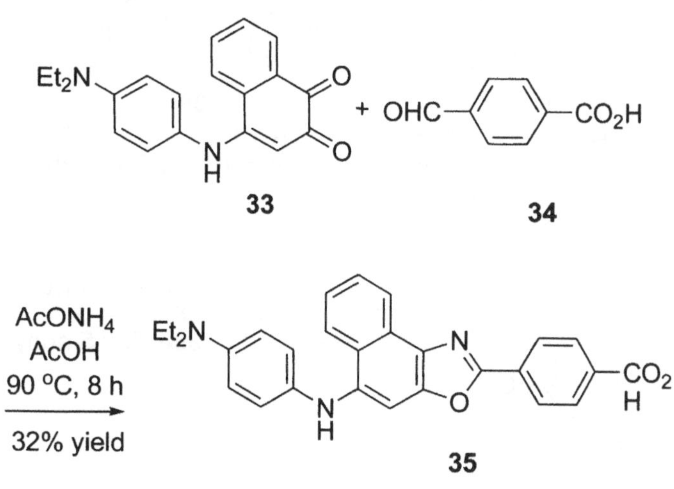

Harima[7] synthesized oxazole **38** (55% yield) and **39** (16% yield) by the reaction of diketone **36** and aldehyde **37** in the presence of NH_4OAc in AcOH at 90 °C for 2 h. They found that in solution **38a** exhibited much stronger absorption and fluorescence intensity than regioisomers **39a,** and *N*-substituted analogs **38b** and **39b** showed strong solid-state fluorescence properties. These compounds were studied for application in dye-sensitized solar cells. They also found that compounds **38c–e** possessed mechano-fluorochromism, which is a change in fluorescent color induced by mechanical stress, being accompanied with a reversion to the original fluorescent color by heating, recrystallization, or exposure to solvent vapor. These mechanofluorochromic fluorescent dyes could be a promising class of organic dyes for rewritable photo-imaging and electroluminescence devices.[8]

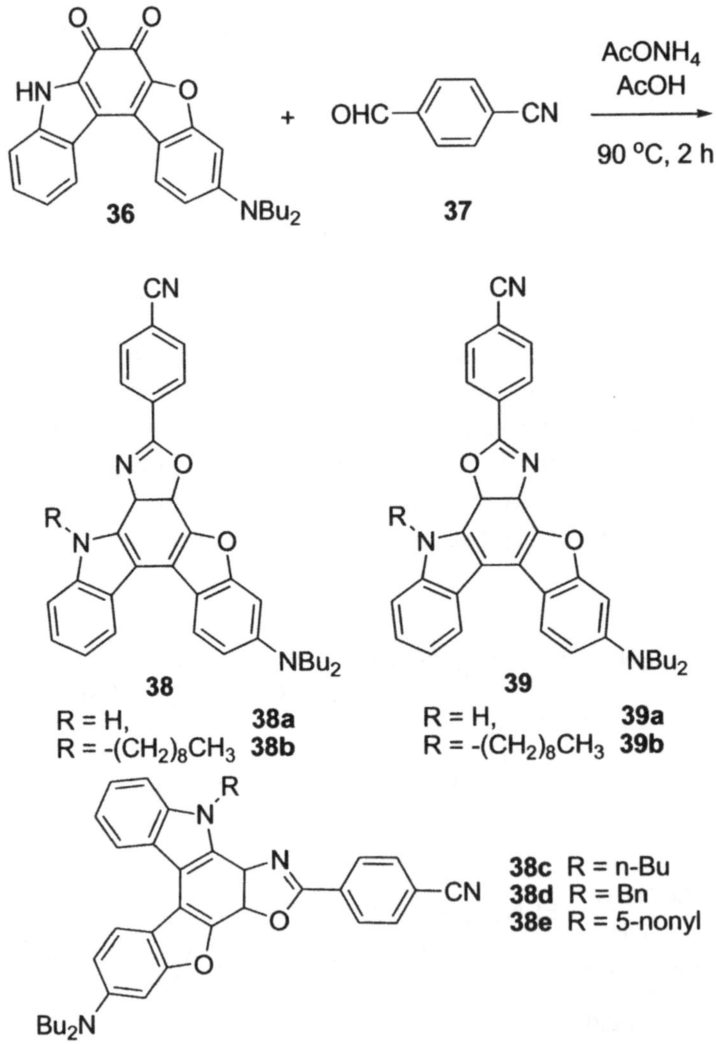

Sécheresse reported size-controlled formation of silver nanparticles (**43**) by direct bonding of ruthenium complex **42** and silver nanoparticles. Oxazole **42** was formed by the reaction of diketone **40** and aldehydes **41** under the influence of NH_4OAc. These metallic nanoparticles may find applications in DNA sequencing, catalysis, optics, nanoscale electronics and antimicrocrobials.[9]

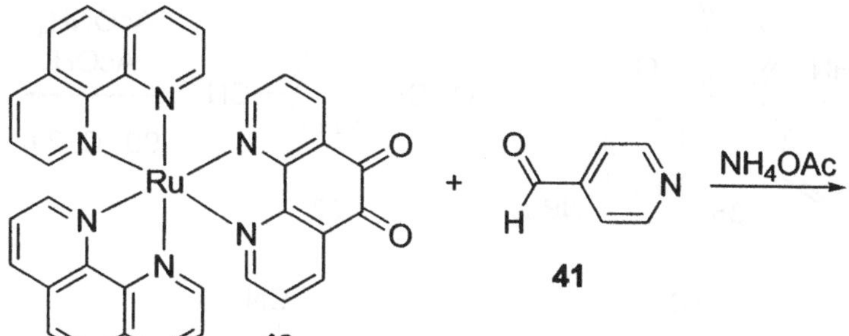

40 **41**

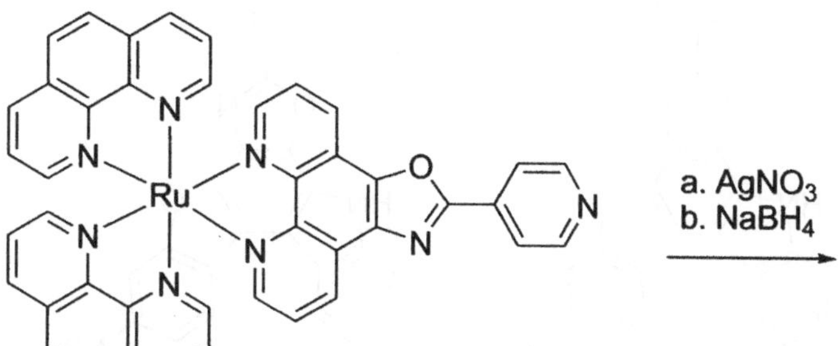

42

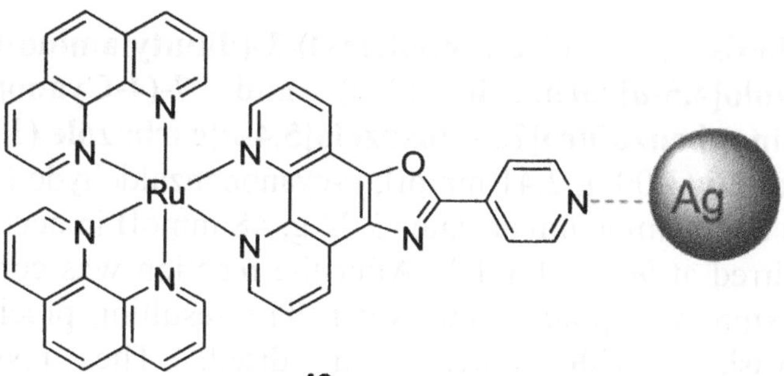

43

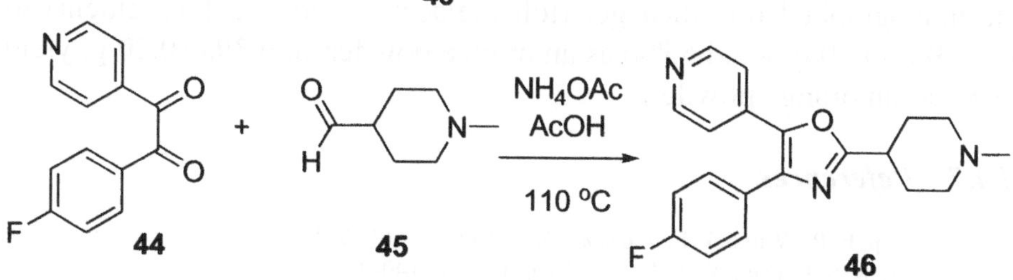

44 **45** **46**

Biftu reported that oxazole **46**, synthesized form diketone **44** and aldehydes **45** as inhibitors of parasite *c*GMP-dependent protein kinase as novel anticoccidial agents.[10]

4.4.6 Experimental

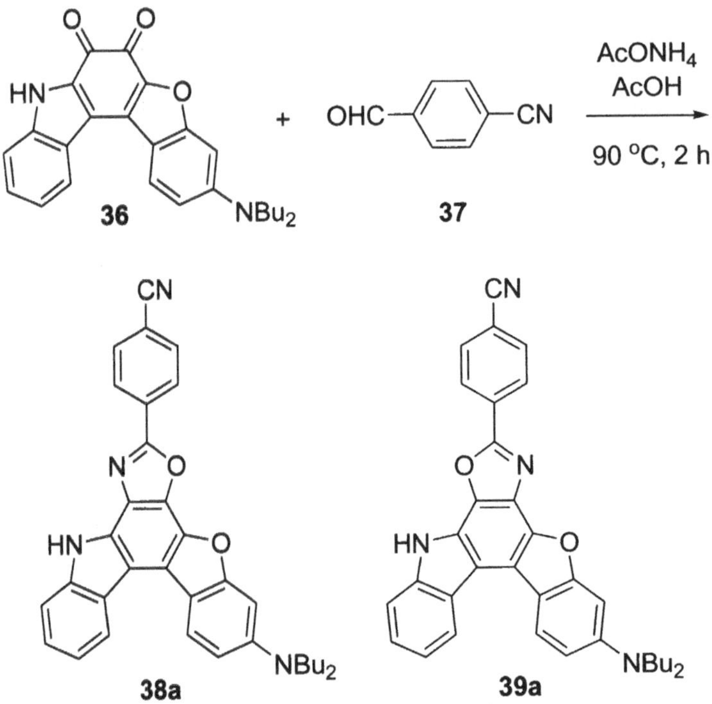

The Synthesis of 7-(4-Cyanophenyl)-3-(dibutylamino)benzofuro [2,3-*c*]-oxazolo[4,5-*a*]carbazole (38a) **and 7-(4-Cyanophenyl)-3-(dibutylamino) benzofuro[2,3-*c*]oxazolo[5,4-*a*]carbazole** (39a)[7]

A solution of **36** (1.00 g, 2.41 mmol), *p*-cyanobenzaldehyde (**37**, 0.32 g, 2.41 mmol) and ammonium acetate (3.72 g, 48 mmol) in acetic acid (30 mL) was stirred at 90 °C for 1 h. After the reaction was complete, the reaction mixture was poured into water. The resulting precipitate was filtered, washed with water, and dried. The residue was chromatographed on silica gel (toluene/acetic acid = 5:1 as eluent) to give **38a** (0.70 g, yield 55%) as an orange powder, and **39a** (0.20 g, yield 16%) as an orange powder.

4.4.7 References

1. Japp, F. R.; Wilcock, E. *J. Chem. Soc.* **1880,** *37,* 661–672.
2. Kreps, S. I.; Day, A. R. *J. Org. Chem.* **1941,** *6,* 140–156.
3. Katritzky, A. R.; Wang, Z.; Hall, C. D.; Akjmedov, N. G.; Shestopalov, A. A.; Steel, P. J. *J. Org. Chem.* **2003,** *68,* 9093–9099.

4. Nicolaides, D. N.; Papageorgiou, G. K.; Stephanidou-Stephanatou, J. *Tetrahedron* **1989,** *45,* 4585–4592.
5. Lourak, M.; Vanderesse, R.; Vicherat, A.; Jamal-Eddine, Marraud, M. *Tetrahedron Lett.* **2000,** *41,* 8773–8776.
6. Ooyama, Y.; Egawa, H.; Yoshida, K. *Eur. J. Org. Chem.* **2008,** 5239–5243.
7. Ooyama, Y.; Kagawa, Y.; Harima, Y. *Eur. J. Org. Chem.* **2007,** 3613–3621.
8. Ooyama, Y.; Kagawa, Y.; Fukuoka, H.; Ito, G.; Harima, Y. *Eur. J. Org. Chem.* **2009,** 5321–5326.
9. Mayer, C. R.; Dumas, E.; Secheresse, F. *Chem. Commun.* **2005,** 345–347.
10. Biftu, T.; Feng, D.; Ponpipom, M.; Girotra, N.; Liang, G.-B.; Qian, X.; Bugianesi, R.; Simeone, J.; Chang, L. *Bioorg. Med. Chem. Lett.* **2005,** *15*, 3296–3301.

4.5 Schöllkopf Oxazole Synthesis

Brian A. Lanman and Xiaojun Han

4.5.1 Description

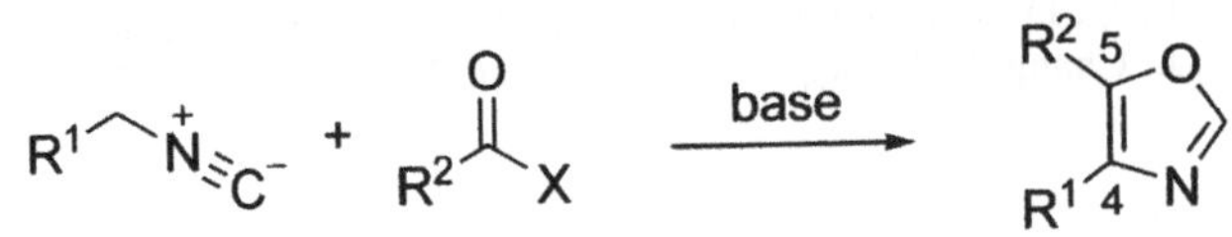

1 2 3

The Schöllkopf oxazole synthesis, also known as the Schöllkopf reaction, is the base-promoted condensation of an alkyl isocyanide (**1**) and an acylating agent (**2**) to produce an oxazole substituted at either (or both) the 4- or 5-position (**3**).[1–5]

4.5.2 Historical Perspective

Ulrich Schöllkopf (1927–1998)[6] and Rolf Schröder of the University of Göttingen first reported the preparation of 4,5-disubstituted oxazoles (**3**) from the condensation of α-metalated alkyl isocyanides[7] and a range of acid chlorides, esters, and amides in 1971.[8]

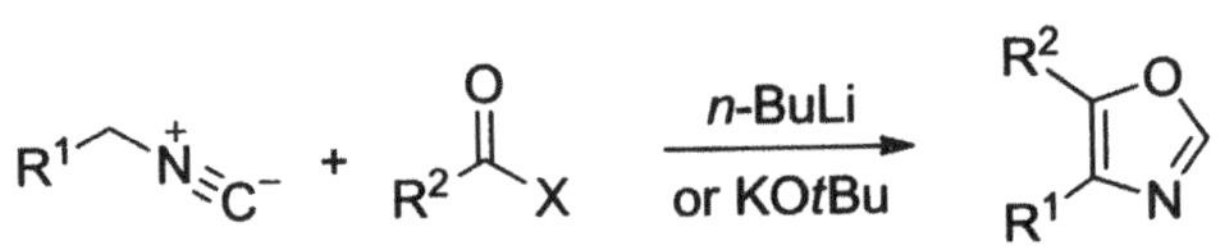

1 2 41–83% 3

X = Cl, OMe, OEt, NMe_2, N(Me)Ph
R^1 = H, Ph, CO_2Et, $CH{=}CMe_2$
R^2 = H, Me, Et, *i*-Pr, Ph, Bn, OEt, 1,2-epoxy-2-methylpropyl

Subsequent research has expanded the range isocyanides, bases, and acylating agents compatible with this transformation and has led to considerable use of the Schöllkopf reaction in both the academic and industrial synthetic communities (*vide infra*).

4.5.3 Mechanism

In 1968, Schöllkopf and Gerhart reported that alkyl isocyanides (**1**) could be metalated at the α-carbon by deprotonation with *n*-butyllithium.[9] Schöllkopf and Schröder[8] subsequently found that treatment of the resulting α-lithio isocyanides (**4**) with acylating agents (**2**) led to α-isocyano carbonyl compounds (**5**),[10] which were readily deprotonated under the basic reaction conditions to afford α-isocyanoenolates (**6**). Enolates **6** exist in a dynamic equilibrium[11] with their corresponding 2-metallo-oxazole isomers (**7**); both isomers are capable of reacting with external electrophiles, depending on the identity of the electrophile.[12–14] In the case of the Schöllkopf reaction, acidification of the reaction mixture upon workup leads (via either isomer) to the formation of 4,5-disubstituted oxazoles (**3**).[15]

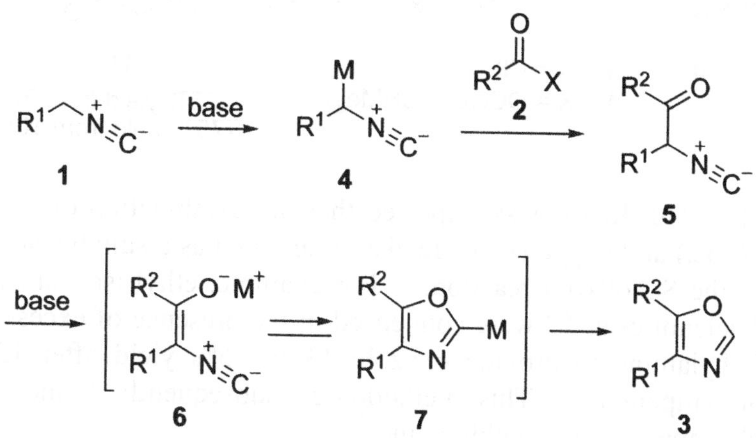

Due to the acidity of the α-isocyano carbonyl compounds (**5**), two equivalents of **4** (relative to acylating agent **2**) are typically employed when the Schöllkopf reaction is conducted using preformed metallo-isocyanides: one equivalent to react with acylating agent **2** and one equivalent to deprotonate the resulting α-isocyano carbonyl compound **5**.[12] The need for excess metallo-isocyanide is relieved when *N,N*-dialkylamides (**2**; X = NR_2) are employed as acylating agents, however, as the dialkylamide liberated during the formation of **5** is sufficiently basic to subsequently deprotonate **5** and generate enolate **6**.[8,12,16]

4.5.4 *Variations and Improvements*

Although Schöllkopf and Schröder initially reported only the use of acid chlorides, chloroformates, carboxylic esters, and dimethylamides as acylating agents in the Schöllkopf reaction,[8] subsequent investigations have revealed additional acylating agents, such as anhydrides,[17] *N*-acyl imidazoles,[18,19] and benzo[*d*][1,3]oxazin-4-ones,[20] to also be compatible with this transformation. In the case of aliphatic acylating agents, acid anhydrides sometimes provide superior yields of oxazole product as compared with the corresponding acid chlorides (*cf.*, **8** + **10** (*vs.* **9**) → **11**).[17]

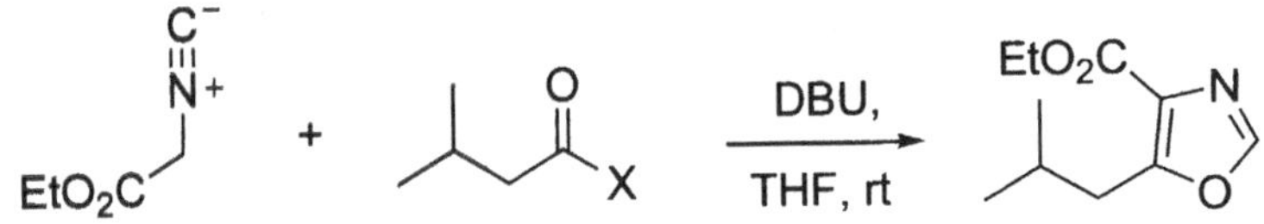

8 **9** (X = Cl) **10** (X = OCOCH$_2$CHMe$_2$) **11** (25% yield from **9**) (75% yield from **10**)

In 1980, Kozikowski reported that the combination of a selenol ester (*cf.*, **12**) and copper(I) oxide also functioned as a suitable acylating agent in the Schöllkopf reaction.[21] For example, ethyl isocyanoacetate (**8**) and selenol ester **12** were condensed in the presence of excess Cu_2O and triethylamine to provide oxazole **13** in 85% yield after 12 h at ambient temperature. This methodology subsequently found use in Weinreb's approach to amphimedine.[22]

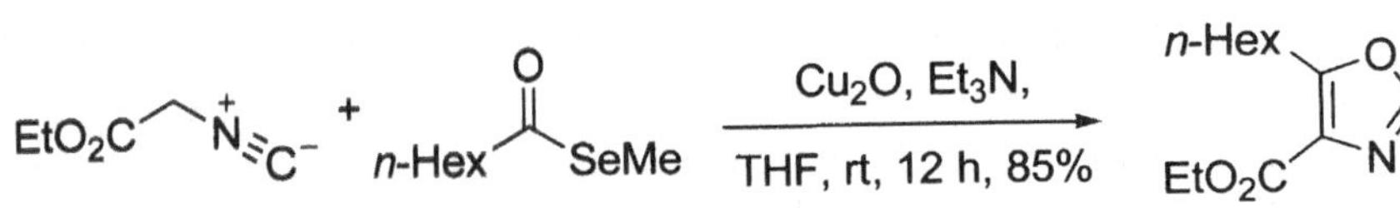

8 **12** **13**

In 1982, Hamada and Shioiri reported that carboxylic acids could be directly employed as acyl donors in the Schöllkopf reaction via *in situ* activation (presumably via a mixed anhydride or azidocarbonyl species) using diphenyl phosphorazidate (DPPA).[23] Thus reaction of a mixture of carboxylic acid, isocyanide, DPPA, and potassium carbonate in DMF afforded the anticipated oxazole product in good yield (*cf.*, **14** + **15** → **16**). Notably, this reaction was compatible with *N*-protected α-amino acid starting materials and proceeded with minimal epimerization. Although diethyl phosphorocyanidate (DEPC) has found similar use in the in situ activation of carboxylic acids for modified Schöllkopf reactions, attempts to employ this methodology with optically active α-amino acid substrates have proven inferior to DPPA-mediated activation, affording either racemized products (using triethylamine as a base) or poor yields of oxazole products (using sodium hydride as a base).[24,25]

Hamada and Shioiri have subsequently demonstrated the utility of this methodology through the synthesis of a range of natural products[25,26] including prumycin,[27] L-daunosamine (*vide infra*),[28] mugineic acid,[29] and the hydroxyamino acid moiety of AI-77-B.[30]

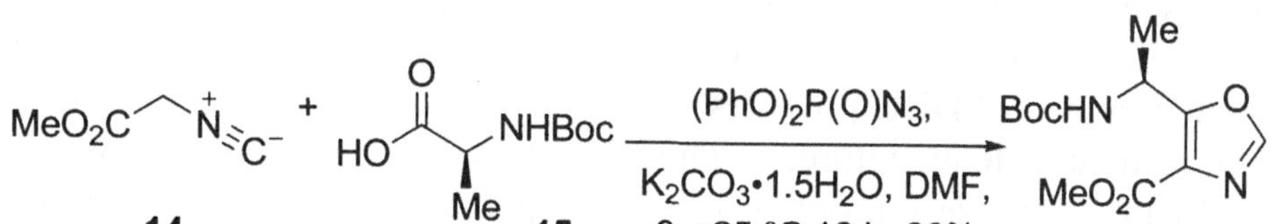

14 15 16

Although Schöllkopf and Schröder initially reported using only *n*-butyllithium and potassium *tert*-butoxide in their generation of metallo-isocyanides (**4**) from alkyl isocyanides (**1**),[8] lithium diisoproylamide (LDA) has subsequently proven a useful complement to these bases, allowing for the metalation of ethyl isocyanide and other alkyl isocyanides that are not effectively metalated with *n*-butyllithium.[31]

Choice of base can play a crucial role in the success of the Schöllkopf reaction, as the reactivity of α-metalated isocyanides varies as a function of the metal counterion. For example, although lithium ethyl isocyanoacetate (**8**) condenses readily with a wide range of acid chlorides, lithio-**8** fails to react with less reactive ester electrophiles and affords only isocyanide dimerization products.[12] Likewise, only α-metalated alkyl isocyanides lacking electron-withdrawing groups at the α-position are sufficiently nucleophilic to react with amide electrophiles in the Schöllkopf reaction.[8,12]

In 1972, Matsumoto and co-workers demonstrated that the need for excess metalated isocyanide (**4**) in Schöllkopf reactions employing acid halide and anhydride electrophiles could be overcome by using an excess of a mild organic base (*e.g.*, triethylamine or DBU) to deprotonate isocyanide **1** in situ (*cf.*, **14** + **17** → **18**).[32] Subsequent investigations revealed a number of additional bases, such as potassium carbonate,[23] sodium hydroxide,[33] and sodium hydride,[28,32,34] to also be compatible with the Schöllkopf reaction—particularly in the case of isocyanide starting materials (**1**) bearing acidifying electron-withdrawing groups (*e.g.*, ester or sulfone) on the isocyanide α-carbon.

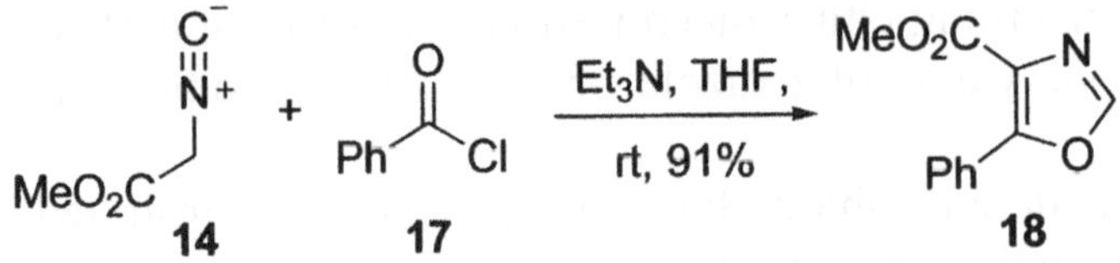

One drawback of the use of weak bases as *in situ* promoters for the Schöllkopf oxazole synthesis is the long time typically required to drive such reactions to completion. For example, the condensation of methyl isocyanoacetate (**14**) and 3,4,5-trimethoxybenzoyl chloride (**19**) in the presence of DBU provides only an 8% yield of oxazole **21** after 2 h at ambient temperature. This limitation can be overcome by employing Verkade's prophosphatrane superbase (**20**),[35] which, in the case of the reaction of **14** and **19**, results in near-quantitative formation of **21** after 2 h at ambient temperature.[36] This procedure additionally benefits from the limited solubility of the resulting phosphonium salt in weakly polar organic solvents, which allows for its ready separation from the reaction product by filtration.

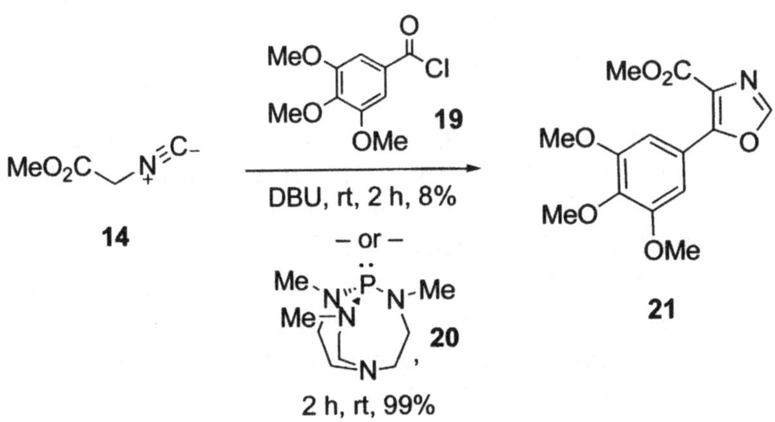

Soon after the initial discovery of the Schöllkopf reaction, van Leusen[37] and researchers at Merck[38,39] reported that this protocol could be extended to the preparation of thiazoles through the use of carboxymethyldithioates[37] or thionoesters (*cf.*, **22**)[38,39] as acylating agents. Both groups found that the identities of the base and acylating agent were crucial to the success of these transformations. Thus, although van Leusen failed to observe thiazole formation using monothiocarboxylate- or dithiocarboxylate acylating agents in potassium hydroxide-promoted Schöllkopf reactions,[37] Hartman and Weinstock found that they could readily obtain thiazoles (*e.g.*, **23**) from the condensation of thionoesters (*e.g.*, **22**) and isocyanides bearing α-electron-withdrawing groups (*e.g.*, **8**) using sodium cyanide as a base. (Carboxylic esters fail to provide the corresponding oxazole products under the sodium cyanide-promoted reaction conditions, however).[38]

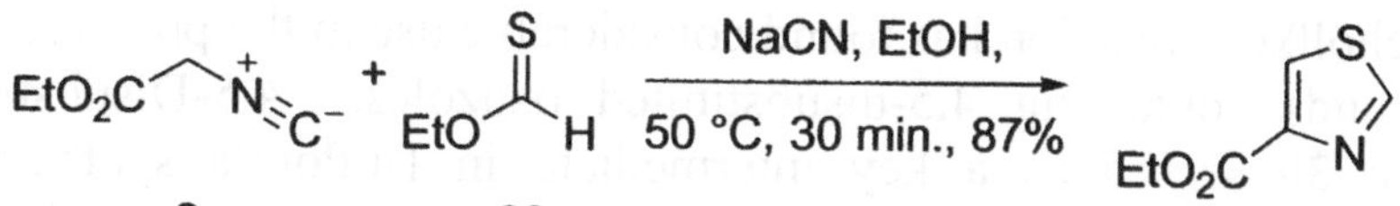

8 **22** **23**

Carbon disulfide[40,41] and isothiocyanates[42,43] have also proven suitable electrophiles for the preparation of thiazoles via the Schöllkopf reaction. Isocyanides bearing electron-withdrawing α-substituents are insufficiently nucleophilic to react with nitriles to afford the corresponding imidazoles,[44] however *N*-alkyl 4-tosylimidazoles are accessible through the sodium hydride-mediated condensation of toluenesulfonylmethyl isocyanide and imidoyl chlorides.[45] Furthermore, the more electron-rich *p*-tolylthiomethyl isocyanide reacts smoothly with carbodiimides and nitriles to afford the corresponding imidazoles.[44]

In related studies on the transformations of tosylmethyl isocyanide (**24**; TosMIC), van Leusen and co-workers found that the based-mediated condensation of **24** and a range of aldehydes yielded the corresponding 5-substituted oxazoles (*cf.*, **28**). This methodology thus provided an alternative to the Schöllkopf reaction in the preparation of these compounds,[46] and has subsequently become known as the *van Leusen oxazole synthesis*.

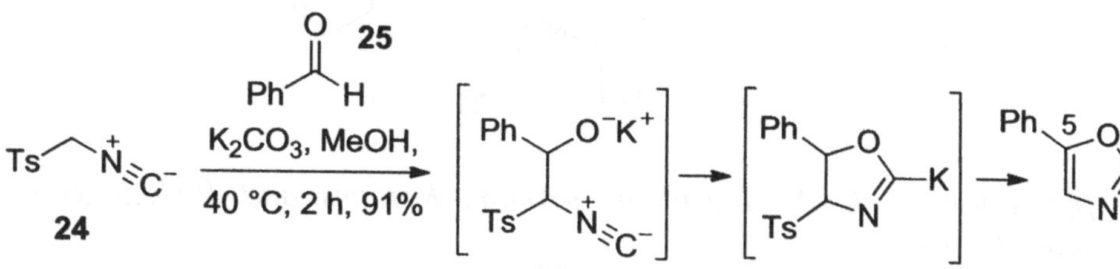

24 **26** **27** **28**

4.5.5 *Synthetic Utility*

Scope and Utility

Application to the Synthesis of 4,5-Disubstituted Oxazoles

The Schöllkopf reaction has found considerable use in the preparation of compounds containing 4,5-disubstituted oxazoles. 4,5-Disubstituted oxazole **30** served as a key intermediate in DuPont's synthesis of benzamidine **31**, a Factor Xa inhibitor.[47] The Schöllkopf reaction has similarly found use in the preparation of 4,5-disubstituted oxazoles for β_3-adrenergic receptor agonist,[48] Fe(II)-form-selective *E. coli* methionine aminopeptidase inhibitor,[49] and prostacyclin receptor antagonist[50] programs. The robustness of this synthetic methodology has also led to the use of the Schöllkopf oxazole synthesis as a test reaction in the evaluation of a number of flow-reactor systems.[51–53]

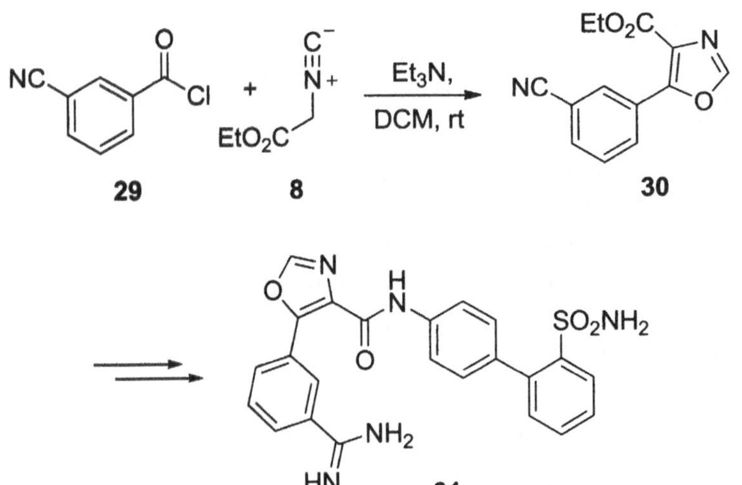

Application to the Synthesis of α-Amino Ketones and β-Hydroxy-α-Amino Esters

The Schöllkopf reaction additionally serves as a useful method for the preparation of α-amino ketones (*cf.*, **14** → **32** & **33**).[17,32] Acid-catalyzed hydrolysis of the oxazole products of the Schöllkopf reaction (*e.g.*, **18**) leads initially to α-amino ketone intermediates (*cf.*, **32**), which can either be isolated or—in the case of intermediates derived from oxazoles bearing C-4 ester substituents—de-alkylated and decarboxylated to provide the related α-amino ketones (*cf.*, **33**).[32]

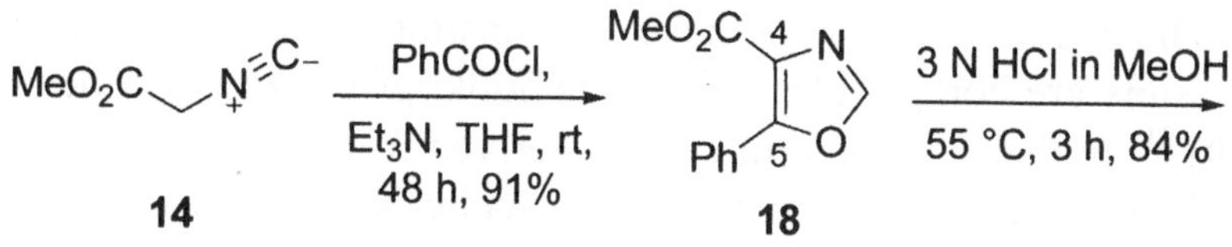

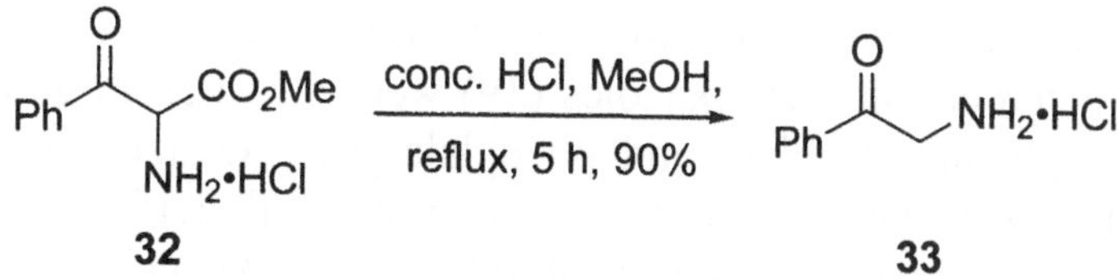

α-Amino ketone intermediates (*cf.*, **34**) obtained from the hydrolysis of oxazoles bearing aliphatic substituents at C-5 can additionally serve as precursors to *anti*-amino alcohols (*e.g.*, **35**) via *anti*-selective hydrogenation/dynamic kinetic resolution using a ruthenium–BINAP catalyst.[54] The corresponding *syn*-diastereomers are also accessible from **34** via hydrogenation of the corresponding benzoyl-protected substrates with $RuCl_2$[(*S*)-BINAP].[54,55]

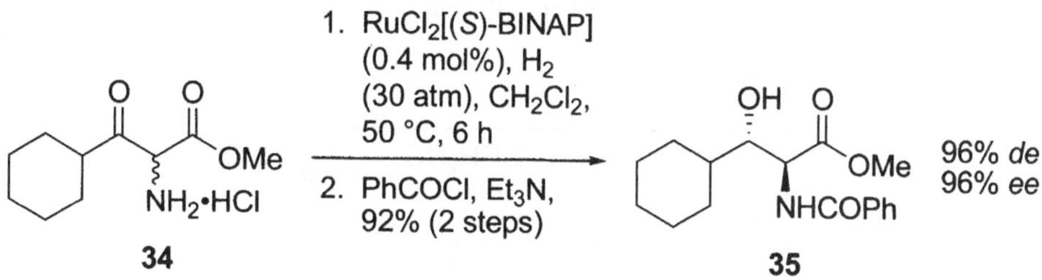

In Schöllkopf reactions proceeding via isocyanide intermediates bearing multiple carbonyl-containing α-substituents (*e.g.*, **36**), enolization is observed to occur toward the more electron-deficient carbonyl group, affording the oxazole product with the more electron-rich carbonyl group in the 4-position.[12] Additional electron-withdrawing groups attached to the isocyanide α-carbon are reported to also facilitate cyclization of the α-isocyano carbonyl intermediate (*e.g.*, **36**) to the oxazole product.[12]

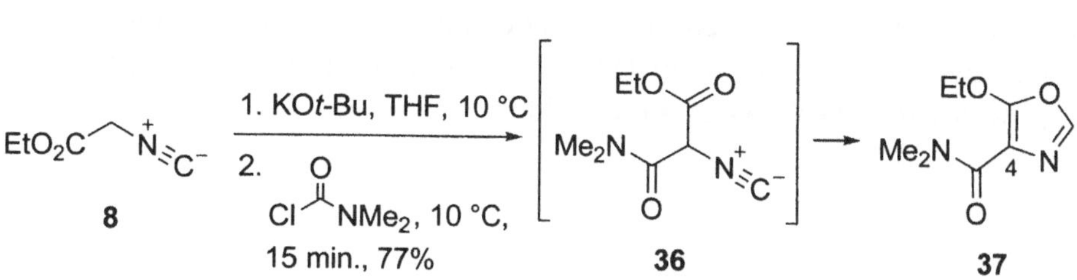

Ohba and co-workers have demonstrated that *N*-protected α-amino esters are compatible with the Schöllkopf oxazole synthesis (*cf.*, **38**→**39**).[31] In the case of amino esters derived from natural amino acids (*e.g.*, **38**), the presence of an additional acidic N–H bond in the *N*-Boc ester substrate necessitated the use of an added excess of metalated isocyanide (2.5 equiv was found to be optimal) to obtain maximal yields. Under optimized conditions, oxazoles (**39**) were obtained in good yield from *N*-Boc glycine, alanine, and phenylalanine. Oxazole formation from *N*-Boc serine (which possesses an additional acidic site in its hydroxylic side chain) proceeded in good yield (66%) using 3.5 equiv lithiated methyl isocyanide. Notably, no epimerization was detected in the reaction of *N*-Boc alanine methyl ester with lithiated methyl or ethyl isocyanide under these conditions. Minor epimerization was observed (91–92% *ee* product) with substrates that lacked a carbamate NH hydrogen (*e.g.*, *N*-Boc proline methyl ester), however.[31]

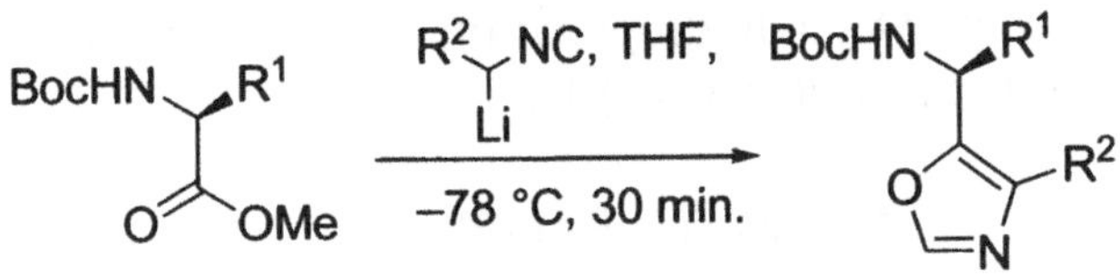

38 **39**

R^1	R^2	$LiCH_2NC$ (equiv.)	Yield
H	H	2.5	73%
Me	H	"	76%
CH_2Ph	H	"	74%
CH_2OH	H	"	41%
CH_2OH	H	3.5	66%
Me	Me	2.5	91%
Me	Ph	"	47%

Lactones have also been found to be competent substrates for the Schöllkopf oxazole synthesis. For example, treatment of aldonic acid lactone **40** with ethyl α-isocyanoacetate (**8**) and 1,5-diazabicyclo[4.3.0]non-5-ene (DBN) at ambient temperature provided acyclic sugar oxazole **41** in 49% yield.[56]

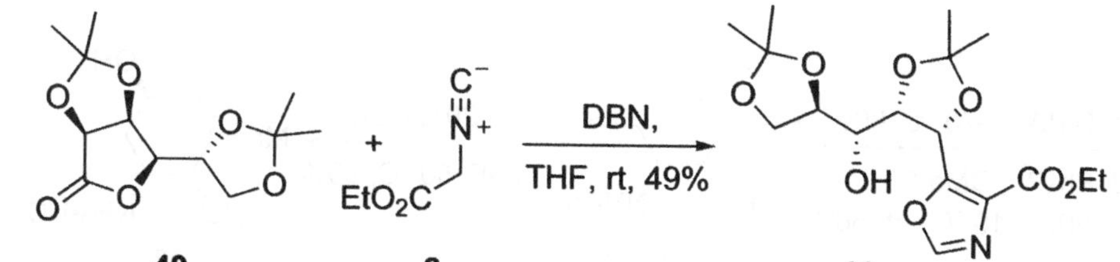

40 8 41

Applications in the Total Synthesis of Natural Products

Synthesis of L-Daunosamine (47)[28]

Hamada and Shioiri have employed their modification of the Schöllkopf reaction (*i.e.*, DPPA-mediated carboxylic acid activation)[23] in the synthesis of L-daunosamine, the carbohydrate component of several important anthracycline antibiotics, including adriamycin and daunomycin.[28] In their synthesis, lithium L-methoxymethyllactate (**42**), prepared in two steps from commercially available L-lactic acid, was treated with diphenylphosphoryl azide (DPPA), and sodium methyl isocyanoacetate was added to the resulting acyl azide to afford oxazole **43** in 70% yield. NMR analysis of **43** in the presence of the chiral shift reagent Eu(facam)$_3$ revealed **43** to be formed without detectible racemization. Treatment of **43** with methanolic HCl afforded lactone **44** in quantitative yield. Protection of the primary amine followed by catalytic hydrogenation over 5% rhodium on alumina subsequently provided **45** in 91% yield. Compound **45** was then treated with DIBAL at –70 °C, and the resulting lactol was incubated with (methoxymethylene)triphenylphosphorane to afford enol ether **46**. Heating of **46** in 20% HCl–tetrahydrofuran at 50 °C for 10 h subsequently afforded L-daunosamine (**47**) in 90% yield (26% yield over eight steps).

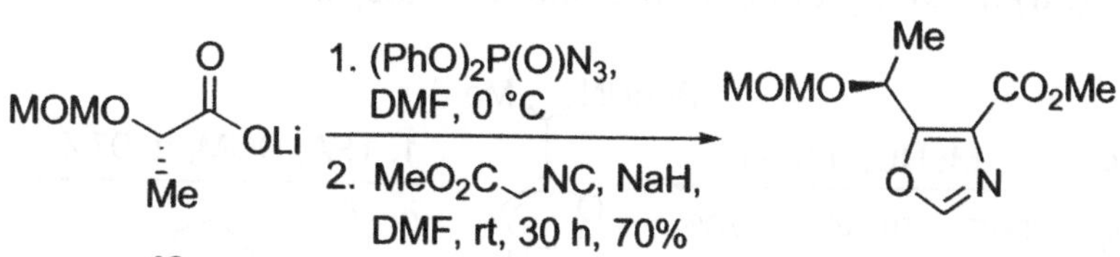

42 43

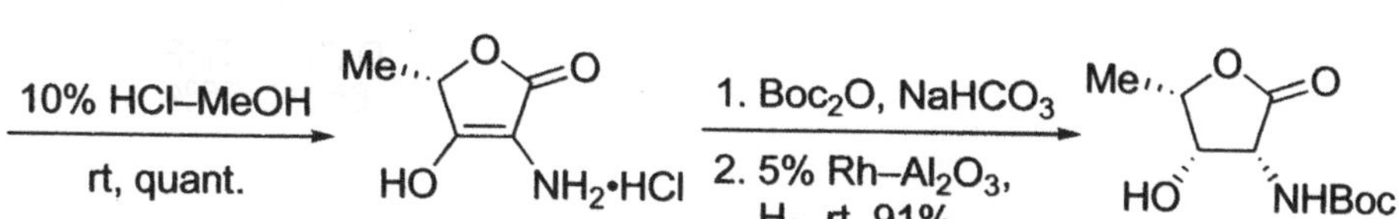

44 45

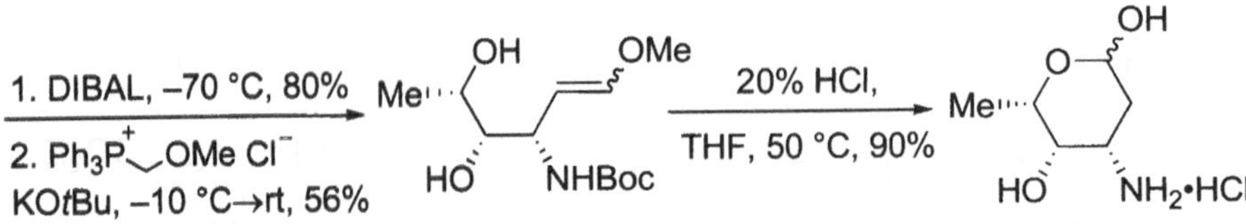

46 **47**

*Synthesis of (–)-Normalindine (**56**)*[57]

Ohba's observation that *N*-Boc amino acid esters could participate in the Schöllkopf reaction without α-carbon epimerization led to the use of this transformation as a key step in the synthesis of (–)-normalindine (**56**).[57] In this synthesis, *N*-Boc methyl alanine (**48**) was treated with α-lithiated methyl isocyanide to provide oxazole **49** without detectible racemization. Deprotection of **49** followed by alkylation with (bromoethyl)indole **50** then supplied (aminoethyl)indole **51**. Subsequent acylation with monoethyl malonate followed by Bischler–Napieralski cyclization and reduction of the resulting iminium salt afforded a 3:1 mixture of tetrahydrocarboline isomers (**52**), favoring the desired β-epimer. Reduction of the diastereomeric esters with DIBAL furnished the corresponding aldehydes, which were then treated with ethyl (triphenylphosphoranylidene)acetate to provide α,β-unsaturated ester **53** as a mixture of four diastereomers. Heating the resulting ethyl enoate diastereomers in boiling toluene for one day selectively cyclized the tetrahydrocarboline β-epimers to provide epimeric ester intermediates **54** in 49% yield. Heating **54** in 5:1 xylenes–AcOH led to dehydration, affording pyridine **55** in moderate yield. The ester moiety of **55** was subsequently saponified, and the resulting acid converted to the corresponding amine by Curtius rearrangement. The resulting 3-aminopyridine was then de-aminated using butyl nitrite to afford (–)-normalindine (**56**) in 26% yield over the final four steps.

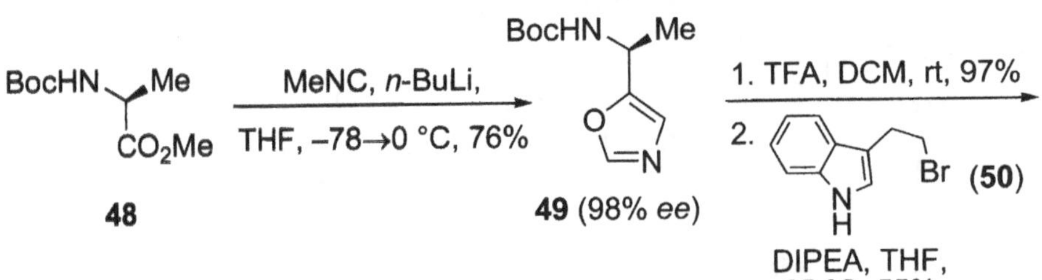

48 **49** (98% *ee*)

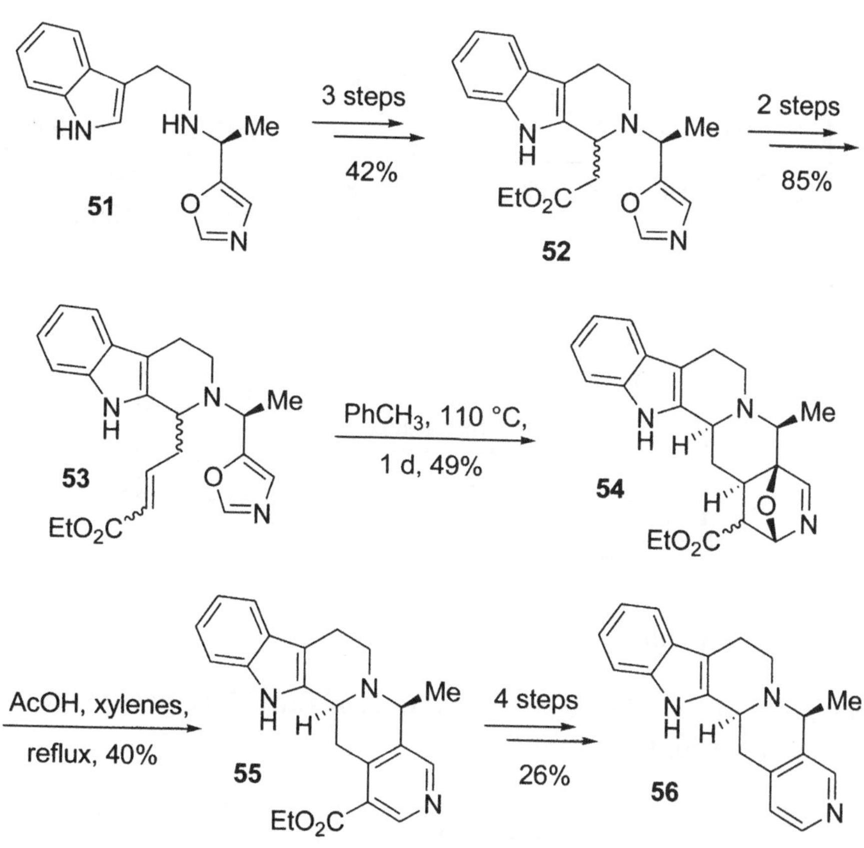

*Synthesis of Ligularone (**60**)*[58]

The Schöllkopf oxazole synthesis also played a key role in the Jacobi lab's synthesis of the furanoterpene ligularone (**60**).[58] Treatment of bicyclic lactone **57** (prepared by Bäyer–Villiger oxidation of the corresponding perhydroindanone)[59] with (isocyanomethyl)lithium afforded oxazole **58** in 51% yield. The use of DMF as a solvent and the avoidance of high substrate concentrations were crucial to the success of this reaction, serving to suppress intermolecular decomposition pathways.[60] Swern oxidation of alcohol **58** and subsequent addition of lithiopropyne afforded a 55:45 mixture of propargylic alcohols, which was then oxidized to provide ynone **59** (67% yield, three steps). Heating ynone **59** in refluxing ethylbenzene for 26 h subsequently effected a sequential Diels–Alder/retro-Diels–Alder sequence, furnishing the terminal furan ring of ligularone (**60**).

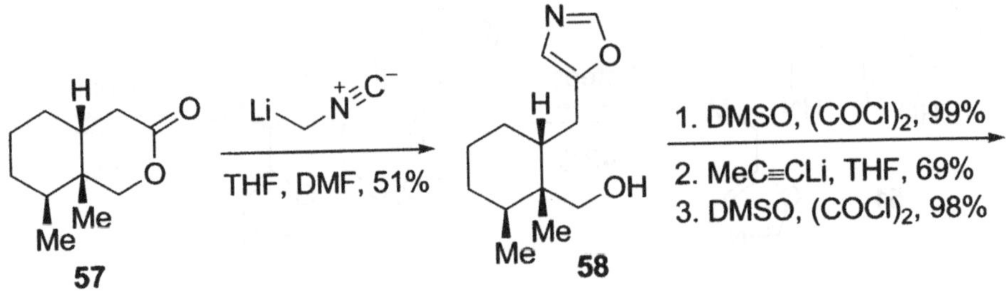

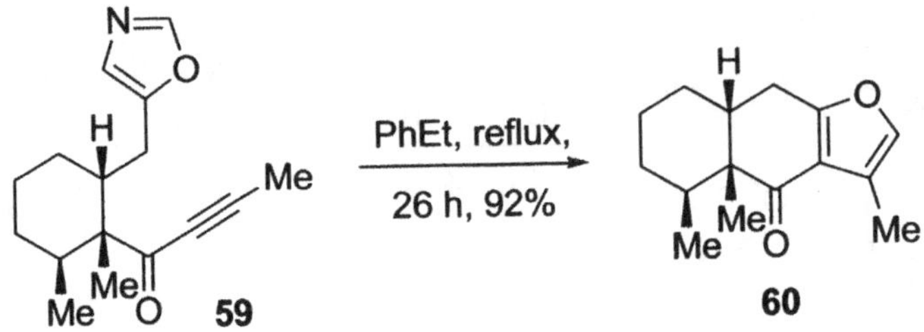

4.5.6 *Experimental*

Original Schöllkopf Procedure: **Synthesis of Ethyl 5-isopropyloxazole-4-carboxylate (62)**[12]

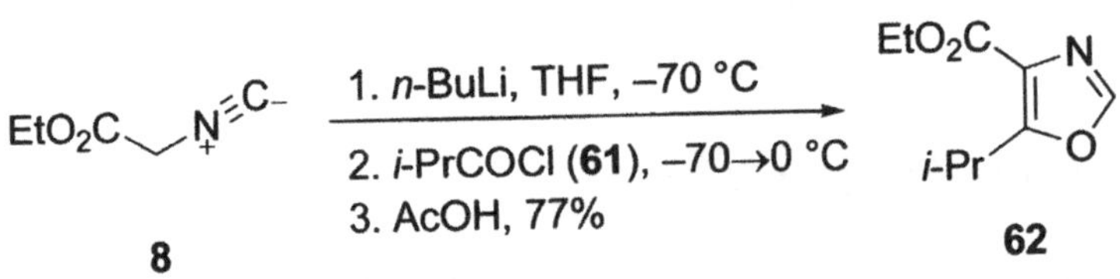

n-Butyllithium (2.5 M in pentane; 16 mL, 40 mmol) was added drop-wise to a stirred solution of ethyl isocyanoacetate (**8**; 4.37 mL, 40 mmol) in tetrahydrofuran (60 mL) at –70 °C over 15 min such that the reaction temperature did not exceed –60 °C. Isobutyryl chloride (**61**; 2.09 mL, 20 mmol) was then added, keeping the reaction temperature below –50 °C. The reaction mixture was then allowed to warm to 0 °C; glacial acetic acid (1.14 mL, 20 mmol) was added, and the reaction mixture was concentrated in vacuo at 25 °C. The residue was partitioned between ethyl ether and water; the ether layer was dried over magnesium sulfate, filtered, and concentrated *in vacuo*. Excess **8** was hydrolyzed by shaking the residue between 2 N aqueous HCl and ethyl ether for 2 min. The ether layer was then separated, dried over magnesium sulfate, and concentrated *in vacuo*. Vacuum distillation of the residue provided ethyl 5-isopropyloxazole-4-carboxylate (**62**; 2.81 g, 15.3 mmol, 77% yield).

In situ *Isocyanide Deprotonation:* **Synthesis of Methyl 5-phenyloxazole-4-carboxylate** (18)[32]

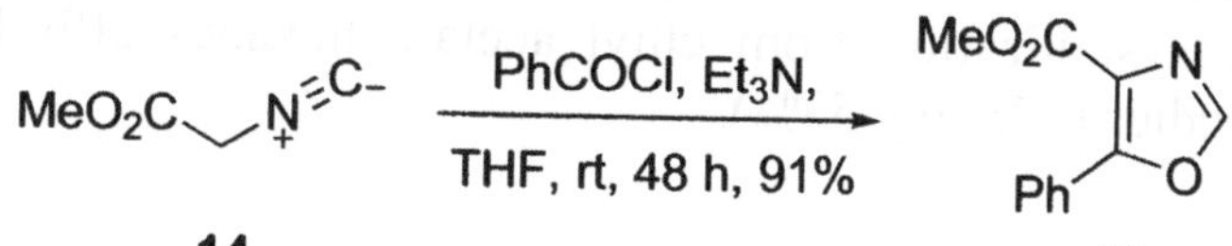

A mixture of methyl isocyanoacetate (**14**; 1.84 mL, 20 mmol), benzoyl chloride (2.15 mL, 20 mmol), and triethylamine (8.4 mL, 60 mmol) in THF (30 mL) was stirred at ambient temperature for 48 h. The reaction mixture was then concentrated *in vacuo*, and the residue was partitioned between water and ethyl acetate. The organic layer was separated, dried over magnesium sulfate, filtered, and concentrated *in vacuo*. The concentrate was triturated with *n*-hexane and collected by vacuum filtration. Recrystallization from methanol subsequently furnished methyl 5-phenyloxazole-4-carboxylate (**18**; 3.70 g, 18.2 mmol, 91% yield).

DPPA-Mediated Carboxylic Acid Activation: **Synthesis of (*S*)-Methyl 5-(2-(4-(benzyloxy)phenyl)-1-(*tert*-butoxycarbonylamino)ethyl) Oxazole-4-carboxylate** (64)[23]

Methyl isocyanoacetate (**14**; 0.728 mL, 8.00 mmol) in DMF (3 mL) was added to a stirred suspension of (*S*)-3-(4-(benzyloxy)phenyl)-2-((*tert*-butoxycarbonyl)amino)propanoic acid (**63**; 743 mg, 2.00 mmol) and potassium carbonate sesquihydrate (661 mg, 4.00 mmol) in DMF (4 mL) at ambient temperature and the resulting mixture was stirred at ambient temperature for 5 min, then cooled to 0 °C. Diphenyl phosphorazidate (DPPA; 0.474 mL, 2.2 mmol) in DMF (3 mL) was added, and the resulting mixture was stirred at 0 °C for 2 h, then at ambient temperature for 14 h. The reaction mixture was subsequently diluted with 1:1 benzene/ethyl acetate (150 mL) and sequentially washed with 30-mL portions of water, 10% aqueous citric acid, water, and saturated aqueous sodium bicarbonate. The organic layer was then dried over magnesium sulfate, filtered, and concentrated *in vacuo*. Purification of the residue by column chromatography (silica gel (40 g), 4:1 benzene/ethyl acetate) afforded (*S*)-methyl 5-(2-(4-(benzyloxy)phenyl)-1-((*tert*-butoxycarbonyl)amino)ethyl)-oxazole-4-carboxylate (**64**; 633 mg, 70%) as colorless crystals. Recrystallization from ethyl acetate–hexanes afforded **64** as colorless needles (478 mg, 53%).

4.5.7 References

1. [R] Wang, Z. *Schöllkopf Oxazole Synthesis*; in *Comprehensive Organic Name Reactions and Reagents, vol. 3*; Wiley: Hoboken, NJ, 2009; pp 2529–2532.
2. [R] Hoppe, D. *Angew. Chem., Int. Ed. Engl.* **1974,** *13*, 789–804.
3. [R] Schöllkopf, U. *Angew. Chem., Int. Ed. Engl.* **1977,** *16*, 339–348.
4. [R] Schöllkopf, U. *Pure Appl. Chem.* **1979,** *51*, 1347–1355.
5. [R] Turchi, I. J. *Ind. Eng. Chem. Prod. Res. Dev.* **1981,** *20*, 32–76.
6. Ulrich Schöllkopf received his Ph.D. under the guidance of Georg Wittig and was one of the earliest students to work on the Wittig reaction. (a) Wittig, G. *Science* **1980,** *210*, 600–604. (b) Wittig, G.; Schöllkopf, U. *Chem. Ber.* **1954,** *87*, 1318–1330.
7. Readily prepared by dehydration of the corresponding *N*-formyl amines: (a) Hertler, W. R.; Corey, E. J. *J. Org. Chem.* **1958,** *23*, 1221–1222, (b) Ugi, I.; Meyr, R. *Chem. Ber.* **1960,** *93*, 239–248, (c) Ugi, I.; Fetzer, U.; Eholzer, U.; Knupfer, H.; Offermann, K. *Angew. Chem.* **1965,** *77*, 492–504.
8. Schöllkopf, U.; Schröder, R. *Angew. Chem., Int. Ed. Engl.* **1971,** *10*, 333.
9. Schöllkopf, U.; Gerhart, F. *Angew. Chem., Int. Ed.* **1968,** *7*, 805–806.
10. These intermediates are typically unstable, undergoing further deprotonation and cyclization to the corresponding oxazoles; however, in unusual cases, such intermediates have proven isolable: Larionov, O. V.; de Meijere, A. *Angew. Chem., Int. Ed.* **2005,** *44*, 5664–5667.
11. The position of this equilibrium is strongly influenced by the identity of the oxazole substituents: Jacobi, P. A.; Ueng, S.; Carr, D. *J. Org. Chem.* **1979,** *44*, 2042–2044. Solvent effects likewise appear to influence the position of this equilibrium: Li, B.; Buzon, R. A.; Zhang, Z. *Org. Synth.* **2010,** *87*, 16–25.
12. Schröder, R.; Schöllkopf, U.; Blume, E.; Hoppe, I. *Liebigs Ann. Chem.* **1975,** 533–546.
13. Pridgen, L. N.; Shilcrat, S. C. *Synthesis* **1984,** 1048–1050.
14. Dondoni, A.; Fantin, G.; Fogagnolo, M.; Medici, A.; Pedrini, P. *J. Org. Chem.* **1987,** *52*, 3413–3420.
15. The acid-catalyzed cyclization of β-keto isocyanides to oxazoles has previously been reported in Hagedorn, I.; Etling, H. *Angew. Chem.* **1961,** *73*, 26–27.
16. Schregenberger and Seebach have alternatively suggested that the tetrahedral intermediate initially obtained from the reaction of metalated isocyanide and amide is stable to the reaction conditions and decomposes only on workup to afford the oxazole product: Schregenberger, C.; Seebach, D. *Liebigs Ann. Chem.* **1986,** 2081–2103.
17. Suzuki, M.; Iwasaki, T.; Miyoshi, M.; Okumura, K.; Matsumoto, K. *J. Org. Chem.* **1973,** *38*, 3571–3575.
18. Henneke, K.-W.; Schöllkopf, U.; Neudecker, T. *Liebigs Ann. Chem.* **1979,** 1370–1387.
19. Rachoń, J.; Schöllkopf, U. *Liebigs Ann. Chem.* **1981,** 1186–1189.
20. Matsumoto, K.; Suzuki, M; Yoneda, N.; Miyoshi, M. *Synthesis* **1976,** 805–807.
21. Kozikowski, A. P.; Ames, A. *J. Am. Chem. Soc.* **1980,** *102*, 860–862.
22. Subramanyam, C.; Noguchi, M.; Weinreb, S. M. *J. Org. Chem.* **1989,** *54*, 5580–5585.
23. Hamada, Y.; Shioiri, T. *Tetrahedron Lett.* **1982,** *23*, 235–236.
24. Hamada, Y.; Seiji, M.; Shioiri, T. *Heterocycles* **1982,** *17*, 321–324.
25. Shioiri, T.; Hamada, Y. *Heterocycles* **1988,** *27*, 1035–1050.
26. Hamada, Y.; Kawai, A.; Matsui, T.; Hara, O.; Shioiri, T. *Tetrahedron* **1990,** *46*, 4823–4846.
27. Hamada, Y.; Shioiri, T. *Tetrahedron Lett.* **1982**, *23*, 1193–1196.

28. Hamada, Y.; Kawai, A.; Shioiri, T. *Tetrahedron Lett.* **1984,** *25*, 5409–5412.
29. Hamada, Y.; Shioiri, T. *J. Org. Chem.* **1986,** *51*, 5489–5490.
30. Kawai, A.; Hara, O.; Hamada, Y.; Shioiri, T. *Tetrahedron Lett.* **1988,** *29*, 6331–6334.
31. Ohba, M.; Kubo, H.; Seto, S.; Fugii, T.; Ishibashi, H. *Chem. Pharm. Bull.* **1998,** *46*, 860–862.
32. Suzuki, M.; Iwasaki, T.; Matsumoto, K.; Okumura, K. *Synth. Commun.* **1972,** *2*, 237–242.
33. Zhang, L.; Xu, X.; Tan, J.; Pan, L.; Xia, W.; Liu, Q. *Chem. Commun.* **2010,** *46*, 3357–3359.
34. Tormyshev, V. M.; Mikhalina, T. V.; Rogozhnikova, O. Y.; Troitskaya, T. I.; Trukhin, D. V. *Russ. J. Org. Chem.* **2006,** *42*, 1031–1035.
35. Lensink, C.; Xi, S. K.; Daniels, L. M.; Verkade, J. G. *J. Am. Chem. Soc.* **1989,** *111*, 3478–3479.
36. Tang, J.; Verkade, J. G. *J. Org. Chem.* **1994,** *59*, 7793–7802.
37. Oldenziel, O. H.; van Leusen, A. M. *Tetrahedron Lett.* **1972,** *27*, 2777–2778,
38. Hartman, G. D.; Weinstock, L. M. *Synthesis* **1976,** 681–682.
39. Hartman, G. D.; Weinstock, L. M. *Org. Synth.* **1979,** *59*, 183–189.
40. Schöllkopf, U.; Porsch, P.-H.; Blume, E. *Liebigs Ann. Chem.* **1976,** 2122–2125.
41. van Leusen, A. M.; Wildeman, J. *Synthesis* **1977,** 501–502.
42. Suzuki, M.; Moriya, T.; Matsumoto, K.; Miyoshi, M. *Synthesis* **1982,** 874–875.
43. Solomon, D. M.; Rizvi, R. K.; Kaminski, J. J. *Heterocycles* **1987,** *26*, 651–674.
44. van Leusen, A. M.; Schut, J. *Tetrahedron Lett.* **1976,** *32*, 285–288.
45. van Leusen, A. M.; Oldenziel, O. H. *Tetrahedron Lett.* **1972,** *23*, 2373–2374.
46. van Leusen, A. M.; Hoogenboom, B. E.; Siderius, H. *Tetrahedron Lett.* **1972,** *23*, 2369–2372.
47. Fevig, J. M.; Pinto, D. J.; Han, Q.; Quan, M. L.; Pruitt, J. R.; Jacobson, I. C.; Galemmo, R. A., Jr.; Wang, S.; Orwat, M. J.; Bostrom, L. L.; Knabb, R. M.; Wong, P. C.; Lam, P. Y. S.; Wexler, R. R. *Bioorg. Med. Chem. Lett.* **2001,** *11*, 641–645.
48. Shearer, B. G.; Chao, E. Y.; Uehling, D. E.; Deaton, D. N.; Cowan, C.; Sherman, B. W.; Milliken, T.; Faison, W.; Brown, K.; Adkison, K. K.; Lee, F. *Bioorg. Med. Chem. Lett.* **2007,** *17*, 4670–4677.
49. Wang, W.-L.; Chai, S. C.; Ye, Q.-Z. *Bioorg. Med. Chem. Lett.* **2009,** *19*, 1080–1083.
50. Brescia, M.-R.; Rokosz, L. L.; Cole, A. G.; Stauffer, T. M.; Lehrach, J. M.; Auld, D. S.; Henderson, I.; Webb, M. L. *Bioorg. Med. Chem. Lett.* **2007,** *17*, 1211–1215.
51. Baumann, M.; Baxendale, I. R.; Ley, S. V.; Smith, C. D.; Tranmer, G. K. *Org. Lett.* **2006,** *8*, 5231–5234.
52. Hornung, C. H.; Mackley, M. R.; Baxendale, I. R.; Ley, S. V. *Org. Process Res. Dev.* **2007,** *11*, 399–405.
53. Carter, C. F.; Lange, H.; Ley, S. V.; Baxendale, I. R.; Wittkamp, B.; Goode, J. G.; Gaunt, N. L. *Org. Process Res. Dev.* **2010,** *14*, 393–404.
54. Makino, K.; Goto, T.; Hiroki, Y.; Hamada, Y. *Tetrahedron: Asymmetry* **2008,** *19*, 2816–2828.
55. Makino, K.; Goto, T.; Ohtaka, J.; Hamada, Y. *Heterocycles* **2009,** *77*, 629–634.
56. Eitelman, S. J.; Hall, R. H.; Jordaan, A. *J. Chem. Soc., Chem. Commun.* **1976,** 923–924.
57. Ohba, M.; Kubo, H.; Fujii, T.; Ishibashi, H.; Sargent, M. V.; Arbain, D. *Tetrahedron Lett.* **1997,** *38*, 6697–6700.
58. Jacobi, P. A.; Walker, D. G. *J. Am. Chem. Soc.* **1981,** *103*, 4611–4613.
59. Evans, D. A.; Sims, C. L.; Andrews, G. C. *J. Am. Chem. Soc.* **1977,** *99*, 5453–5461.
60. Jacobi, P. A.; Walker, D. G.; Odeh, I. M. A. *J. Org. Chem.* **1981,** *46*, 2065–2069.

Chapter 5 Other Five-Membered Heterocycles **259**

5.1 Bamberger Imidazole Cleavage

Narendra B. Ambhaikar

5.1.1 Description

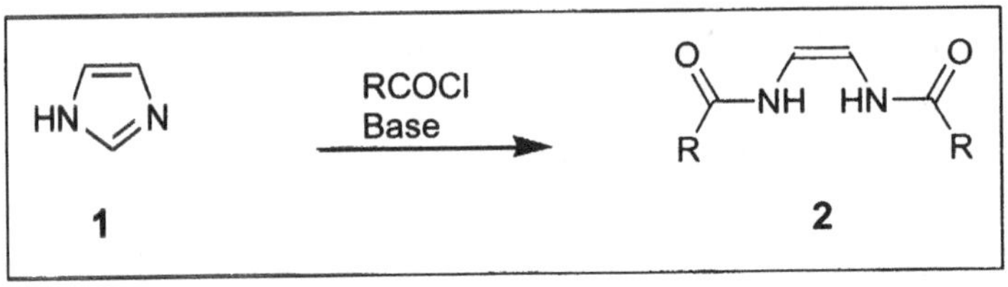

The reaction of imidazole with an acid chloride or an acylating agent in an inert solvent to form a bis-*N*-substituted derivative in aqueous alkali forces the intermediate to undergo hydrolytic ring cleavage and give bis-*N*-substituted *cis*-olefins is known as the Bamberger imidazole cleavage.[1] The reaction is general for unsubstituted as well as some substituted imidazoles. It is possible to achieve the cleavage of imidazole rings even under the mild conditions of protein modification. The reaction is typically run as a two-step process, involving first the isolation of *N*-acyl imidazole followed by further acylation of the second N and then the reaction with water, resulting in the breakdown to the product.

5.1.2 Historical Perspective

In 1893, E. Bamberger at the University of Heidelberg reported that an imidazole in the presence of an acylating agent such as an acid chloride in aqueous alkaline medium, underwent fission to yield *N,N*-diaminoacyl-substituted alkene.[1] Over the years since its discovery, the reaction has found application in the synthesis of natural and nonnatural products involving the preparation of 1,2-diamino vicinal alkenes. It has also been used as part of a detection technique for histidine residues and imidazole components.

5.1.3 Mechanism

Much later after the discovery of the Bamberger Imidazole Cleavage, R. F. Pratt and co-workers[2] at Wesleyan University systematically investigated its mechanism in 1980. Pratt as well as Vliegenthart[3] and Avaeva[4] showed that it was also possible to achieve the cleavage of imidazole rings under mild conditions of protein modification using diethyl pyrocarbonate (DEP) as the acylating agent. DEP is a reagent that is also commonly used in the specific modification of histidine residues in proteins.[5] Pratt additionally noted that at low diethyl pyrocarbonate concentrations (< 10 mM), in phosphate buffer at pH 6, further carbethoxylation (or acylation) of the *N*-carbethoxyimidazole **i** was rate determining to the cleavage, while at higher diethyl pyrocarbonate concentrations, the rate-determining step was the breakdown of a tetrahedral intermediate 2-hydroxy-4-imidazoline **iii**. The stepwise mechanism is shown below.[2]

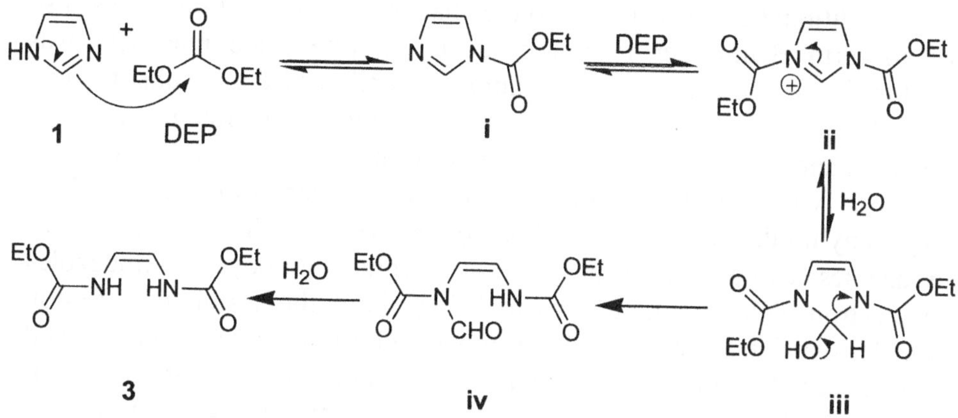

Pratt also noted that at pH 6 and with short reaction times recovery of intermediate **iii** was almost quantitative but at higher pH and longer reaction times a mixture of **iv** and **3** or just **3** was obtained. Pure **iv** could also be quantitatively transformed into **3** under the same conditions.

5.1.4 Variations and Improvements

Bamberger Imidazole Cleavage Shown by Benzimidazoles and Adenosines

In 1968 Ben-Ishai *et al.* at the Israel Institute of Technology extended the scope of this reaction by carrying out benzoylation and carbobenzoxylation of benzimidazoles and substituted adenines such as compound **4**. Cleavage of the imidazole ring of adenine **4** was accomplished in ethyl acetate with aqueous base to yield Bamberger fission product compound **5** in 74% yield.[6] They successfully applied this approach to 9-benzyladenine, 6-benzylamino-9-benzyladenine and 2′3′-*O*-isopropylidene-5′-trityladeno-sine. It is interesting that under similar experimental conditions adenine itself afforded two mono-acyl derivatives rather than undergoing the Bamberger imidazole cleavage.

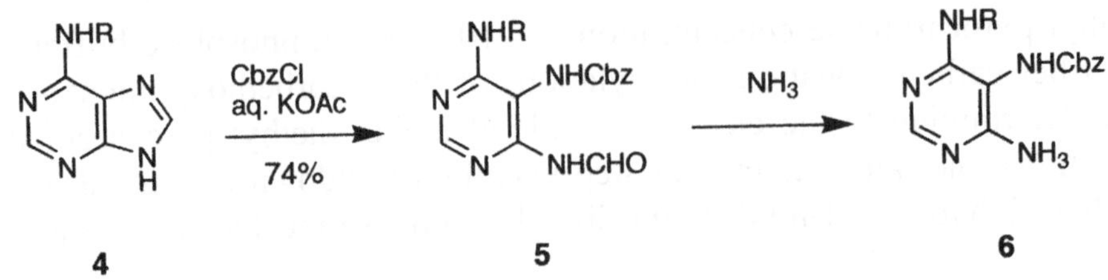

In additional studies it was found that *N*-substituted imidazoles also reacted to form Bamberger fission products; however, the reactivity was observed to depend on the electron-withdrawing ability of the *N*-substituents.[7] For example, imidazoles **7** with electron-withdrawing groups like nitrophenyl, dinitrophenyl and monocarbobenzyloxy underwent reaction to yield product **9** while those with electron donors such as *N*-benzylimidazole, *N*-*p*-methoxyphenylimidazole and *N*-phenylimidazole did not. Thus benzimidazoles exhibit superior reactivity while monocyclic imidazoles require electron-withdrawing groups to bring about the Bamberger fission reaction. Ben-Ishai *et al.* explained that monocyclic imidazole system proved more stable than the imidazole ring than in the condensed benzimidazole or purine system and therefore required an electron withdrawing group to reduce the resonance stability of the imidazole to promote the reaction.

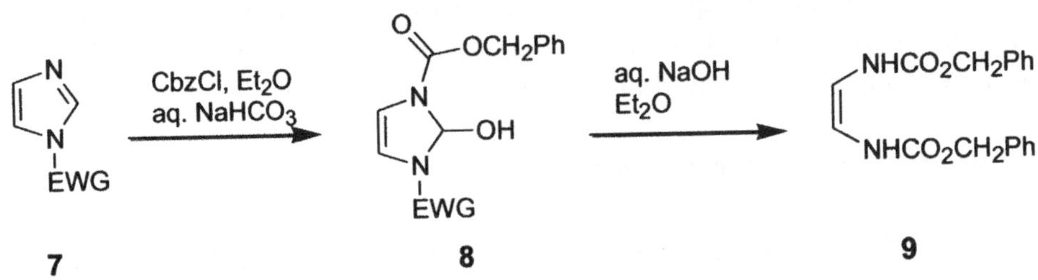

EWG = electron withdrawing group

Synthesis of 6-Aryl-4,5-dibenzamido-1,2,3,6-tetrahydropyridines

In their efforts to reinvestigate a 1928 synthesis by van der Merwe[8] using histamine and *p*-anisaldehyde as starting materials, Stocker and Evans[9] in 1990 reported an application of the Bamberger imidazole cleavage in the general synthesis of compounds such as **13**. They proved that **12** was actually the result of cyclization of the corresponding Schiff's base rather than just the Schiff's base that van der Merwe had earlier suggested.[10]

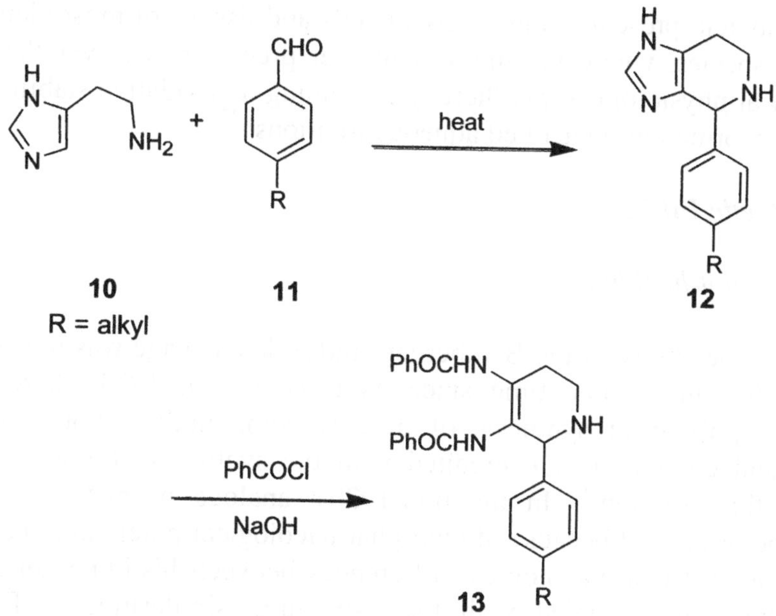

10 R = alkyl 11 12 13

Synthesis of Imidazole-2-thione from Imidazoles

In an effort to find an efficient approach for the synthesis of L-ergothioneine, Yadan[11] developed an elegant methodology to prepare functionalized imida-2-thiones from the corresponding imidazoles. It involved an ANRORC[12] process consisting of a sequential cleavage and reformation of the thiourea ring with phenyl chlorothionoformate *via* a Bamberger type intermediate **15**.

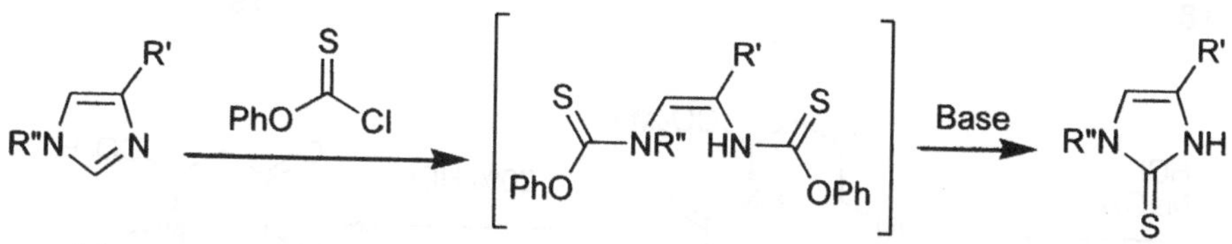

14 15 16

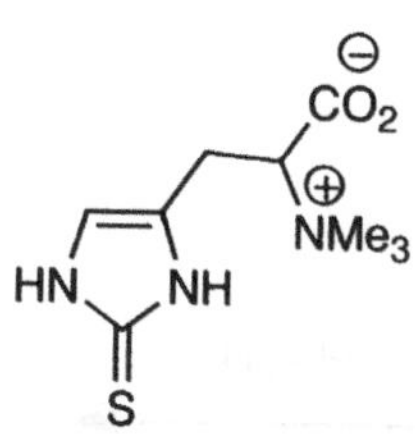

17: L-(+)-ergothioneine

Yadan *et al.* successfully applied the methodology toward a novel synthesis of enantiopure L-(+)-ergothioneine (**17**). L-(+)-Ergothioneine is a natural amino acid present ubiquitously in cells and tissues of most plant and mammalian species, where the thione tautomer predominates over the thiol form largely at physiological pH thereby explaining its oxidative stability and lack of dimer formation in aerated aqueous solutions.[13]

5.1.5 *Synthetic Utility*

Synthesis of Thiol-histidine

Although the specificity of the Bamberger imidazole cleavage was not clearly understood for quite some time since its discovery in 1893, there were attempts to apply it in synthesis of organic compounds.[14] For example, Harington and co-workers are credited with the synthesis of thiolhistidine (**22**) using this reaction.[15] In the mid-1920s, analogs of histamine were thought to be of interest because of their pharmacological potential in part due to the pronounced pharmacological differences between histidine, an amino acid, and histamine coupled with their structural similarities. Despite workers efforts, this work went without much success.

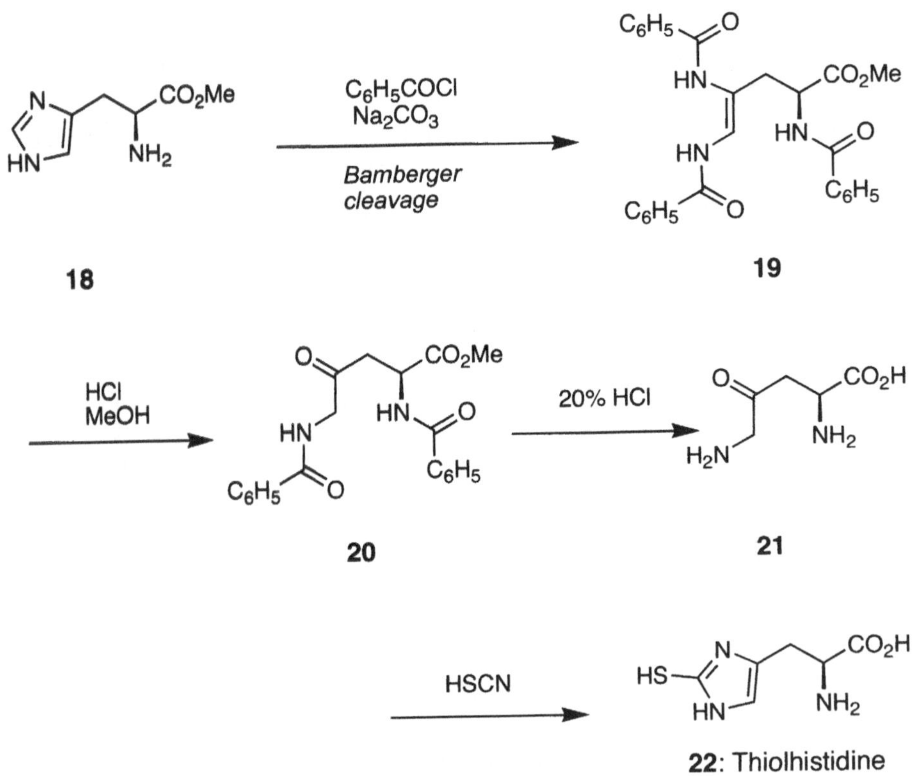

Application to the Synthesis of Caissarone Hydrochloride

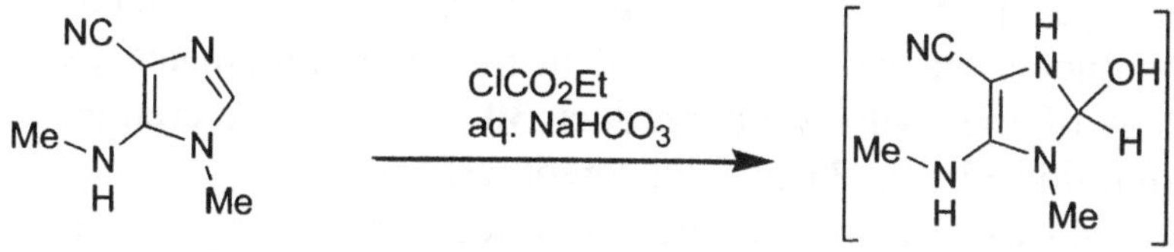

23 24

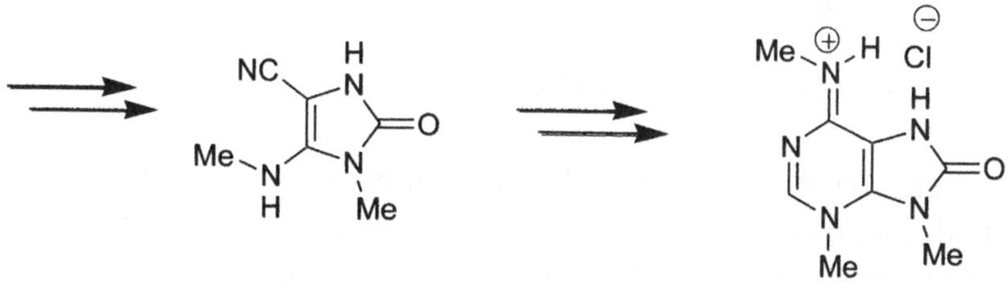

25 26, caissarone HCl

Itaya *et al.*[16] have applied the Bamberger imidazole cleavage reaction and subsequent reclosure of 1-alkyl-5-(alkylamino)imidazole-4-carbonitriles to transform imidazoles to their corresponding 2-*oxo* derivatives during their efforts to find an alternative entry into the synthesis of caissarone hydrochloride (**26**), a 3,6,9-trialkylpurine.

Synthesis of Chiral Diamines

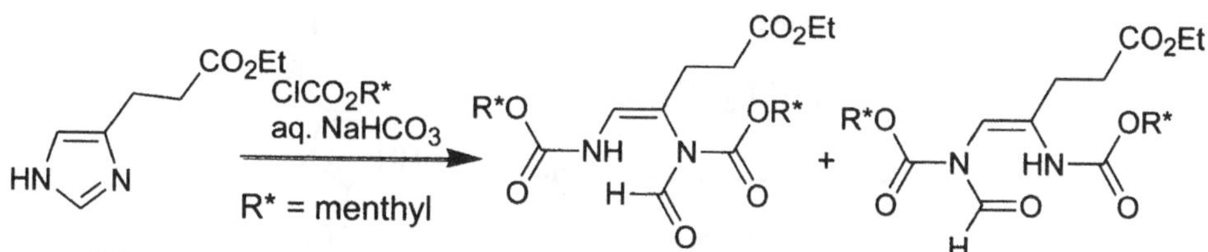

27 28 29

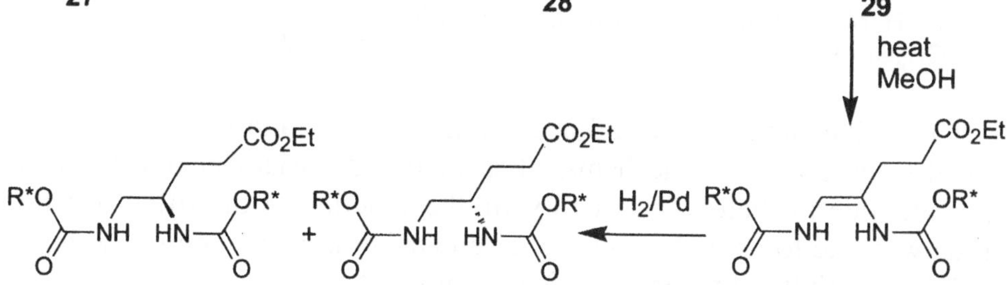

32 31 30

32:31 5:1

Altman *et al.*[17] have used the Bamberger imidazole cleavage reaction of substituted imidazoles in producing enantio-enriched vicinal di-acylamines. Specifically ring cleavage of **25** with (–)-menthyl chloroformate introduces chiral carbamate substituents on the double bond in menthyl carbamates **28** and **29**. Hydrogenation of the hydrolyzed product **30** leads to preferential formation of vicinal diamide **32** over **31** in a diastereomeric ratio of 5:1.

Altman's result shows that the Bamberger imidazole cleavage using asymmetric carbamates provided entry into vicinal diamines albeit with modest facial selectivity via hydrogenation.

Synthesis of Trifluoromethyl Histamine and Histidine

Cohen and co-workers have reported a facile synthesis of 2-(trifluoromethyl)histamine and 2-(trifluoromethyl)-L-histidine using histamine and L-histidine as starting materials, repectively.[18] The desire to use histamine and histidine was mainly due to ready availability, which would lead to rapid synthetic sequences and a direct entry into the fluorinated L-amino acid series. *cis*-1,2-Dibenzamido-1-alkenes (**34**) prepared by the Bamberger imidazole cleavage of corresponding imidazoles (**33**) were condensed with trifluoroacetic anhydride at reflux (40 °C) to yield 2-trifluoromethyl substituted imidazoles (**35**) in about 70% yield.

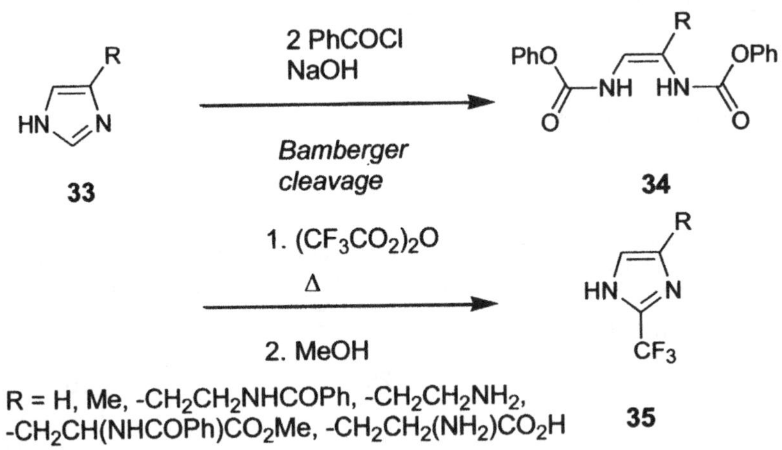

The conversion of **34** to **35** according to Cohen and co-workers was speculated to go through a mechanism that involved a triacylated species as the requisite intermediate. The facility of the reaction with trifluoroacetic anhydride was due to its high reactivity as an acylating agent and/or the high electrophilic reactivity of the trifluoroacetyl carbonyl group.

5.1.6 Experimental

The Classical Bamberger Imidazole Conditions to Prepare N,N'- **Dicarbethoxy-1,2-diaminoethene (compound 30)**[17]

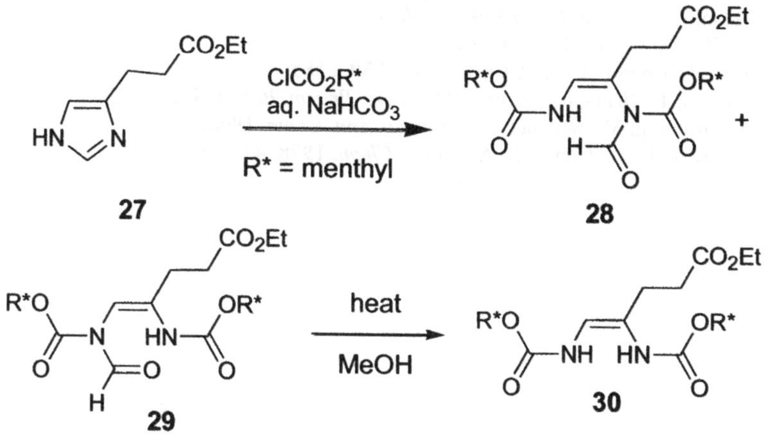

(–)-Menthyl chloroformate (3.5 g, 0.0016 mol) in EtOAc (10 mL) and $NaHCO_3$ (1.35 g, 0.016 mol) in water (15 mL) were simultaneously added from two separate funnels over 20 min into a solution of the ester **27** (1.008 g, 6 mmol) in EtOAc (50 mL) while being cooled in an ice bath and stirred. Stirring was continued for 1 h with cooling and for 12 h at room temperature, keeping the pH above 8. The organic layer was separated, washed with water, dried (Na_2SO_4), and concentrated. The residue was applied to a silica column prepared in hexane. Menthol and dimenthyl carbonate were eluted with 5% EtOAc/hexanes. Elution with 10% EtOAc/hexane yielded **28** and **29** as an oil (2.4 g, 73%).

Deformylation

The mixture of **28** and **29** (550 mg, 1 mmol) was heated in absolute methanol (15 mL) for 6 h at 60 °C. Deformylation was monitored by TLC (elution with 25% EtOAc/hexane) or by ^1H NMR until the absorption at $\delta = 9.19$ and 9.24 ppm disappeared. The crude mixture containing **30** was immediately subjected to hydrogenation.

5.1.7 References

1. Bamberger, E. *Liebigs Ann. Chem.* **1893,** *273*, 342.
2. Grace, M. E.; Loosemore, M. J.; Semmel, M. L.; Pratt, R. F. *J. Am. Chem. Soc.* **1980,** *102*, 6784.
3. Vliegenthart, J. F. G.; Dorland, L. *J. Biochem.* **1970,** *117*, 319.
4. Avaeva, S. N.; Krasnova, V. T. *Biorg. Khim.* **1975,** *1*, 1600.
5. F. Schneider, *Angew. Chem., Int. Ed. Engl.* **1978,** *17*, 583.
6. Babad, E.; Ben-Ishai, D. *J. Heterocyclic Chem.* **1968,** *5*, 679.
7. Altman, J.; Ben-Ishai, D. *J. Heterocyclic Chem.* **1968,** *6*, 235.
8. Van der Merwe, P. Z. *Physiol. Chem.* **1928,** *177*, 30.
9. Stocker, F. B.; Evans, A. J. *J. Org. Chem.* **1990,** *55*, 3370.

10. Stocker, F. B.; Fordice, M. W.; Larson, J. K.; Thorstenson, J. H. *J. Org. Chem.* **1966,** *31,* 2380.
11. Xu, J.; Yadan, J. C. *Synlett* **1995,** 239.
12. ANRORC: addition of nucleophile, ring opening, ring closure.
13. Xu, J.; Yadan, J. C. *J. Org. Chem.***1995,** *60*, 6295.
14. [R] Fox, S. W. *Chem. Rev.* **1942,** *42*, 47.
15. Ashley, J. N.; Harington, C. R. *J. Chem. Soc.* **1930,** 2585.
16. Itaya, T.; Kanal, T.; Iwata, M.; Azuma, M. *Chem. Pharm. Bull.* **1997,** *45*, 75.
17. Altman, J.; Grinberg, M.; Wilcheck, M. *Liebigs Ann. Chem.* **1990,** 339.
18. Kimoto, H.; Kirk, K. L.; Cohen, L. A. *J. Org. Chem.* **1978,** *43*, 3403.

5.2 Dimroth Triazole Synthesis

Matthew D. Hill

5.2.1 Description

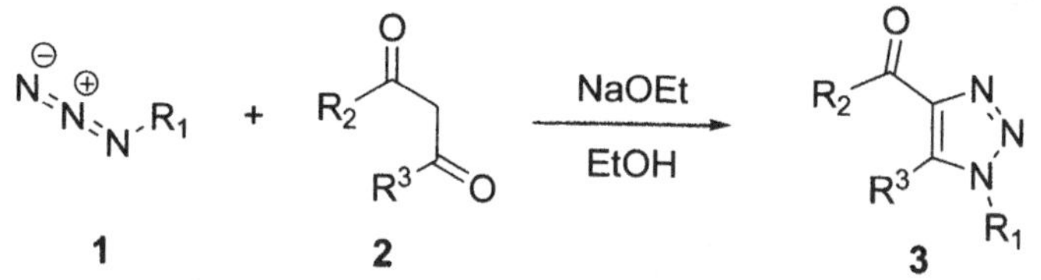

The venerable Dimroth triazole synthesis is a base-catalyzed condensation of an azide **1** with an active methylene compound **2** to provide a 1,2,3-triazole derivative **3**. Commonly, this reaction is run with an alkoxide base in the corresponding alcohol solvent at ambient temperature or reflux.[1,5,9,11–14,16]

5.2.2 Historical Perspective

Otto Dimroth was a professor at the University of Tübingen in Germany when he reported the Dimroth triazole synthesis in 1902[1], and subsequently the Dimroth rearrangement in 1909[2] (see Section 9.4). His name is commonplace in chemistry labs throughout the world for the water condenser that bears his name.

5.2.3 Mechanism

The Dimroth triazole synthesis is a stepwise addition/condensation process that occurs with complete regioselectivity. Deprotonation of an active methylene compound, most commonly achieved using alkoxide bases in alcohol solvents (protic polar), followed by nucleophilic attack on the terminal azide nitrogen leads to cyclization and generation of a dihydro-1,2,3-triazole **6**. Aromatization completes the sequence and affords tri-substituted 1,2,3-triazoles **3**. Consistent with this mechanism, electron-donating substituents on the azide **1** retard the process, while electron-withdrawing groups increase reaction efficiency. More recently, alternative base and solvent combinations have been reported and include common reagents such as potassium *tert*-butoxide,[3,4] potassium hydroxide,[5] and potassium carbonate.[6]

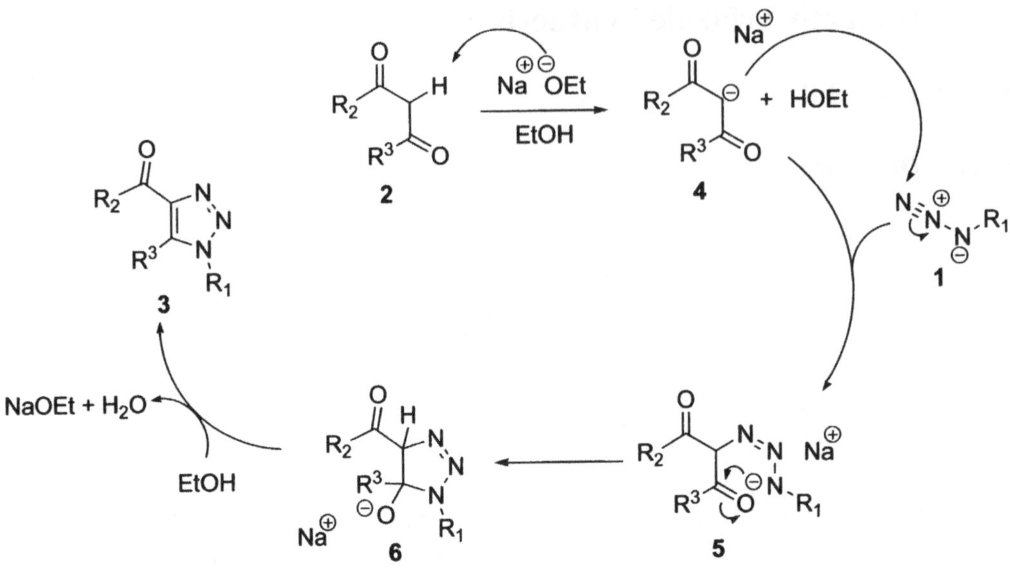

5.2.4 *Variations and Improvements*

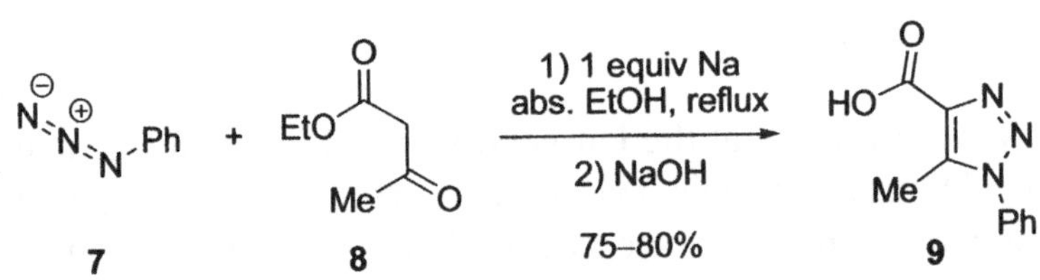

State-of-the-art condensation chemistry has changed little since the turn of the 20th century when Dimroth first reported his novel method for 1,2,3-triazole synthesis.[1] Still, replacement of the alkoxide base and alcohol solvent with several alternatives has extended the scope of the original reaction[3–7] and has made possible a variety of interesting tandem processes[2,14,15] (*vide infra*).

An early modification to the Dimroth procedure used potassium *tert*-butoxide in THF to replace the standard alkoxide base in absolute alcohol. Lieber et al. at DePaul University were able to use *n*-hexyl azide (**10**) and phenylacetonitrile (**11**) to provide access to the *N*-hexyl triazole **12**.[3] This marked the first time an *N*-alkyl azide was used to synthesize the corresponding product in synthetically useful yield.

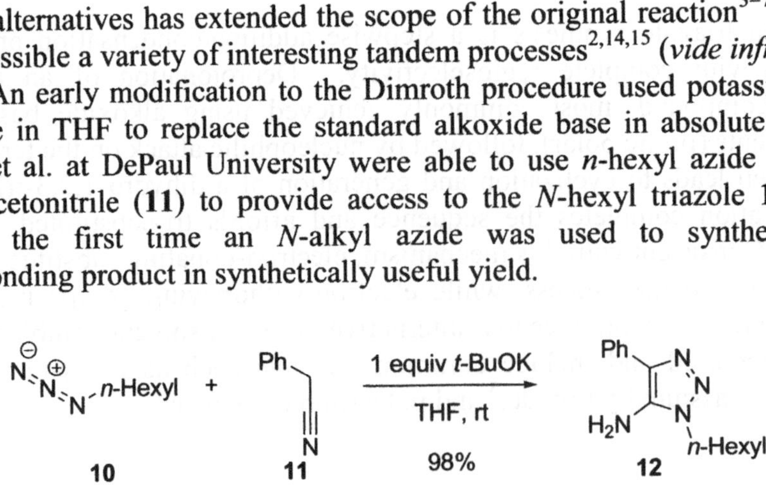

Townsend *et al.* found potassium hydroxide in DMF a suitable reaction system while pursuing nucleoside antibiotics at the University of Michigan.[5] Their use of a sugar-containing azide **13** was a noteworthy extension of the traditional substrate scope.

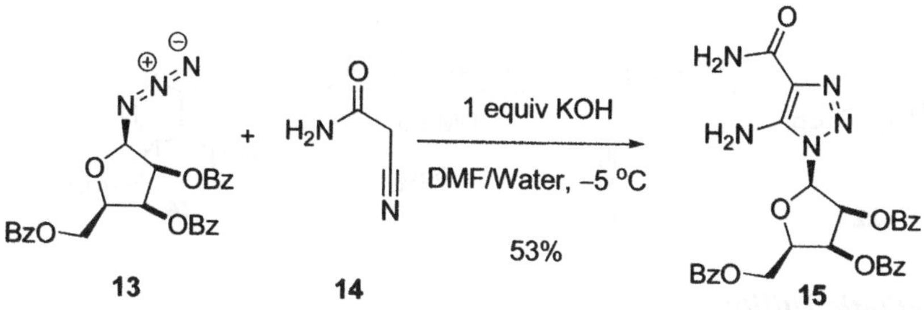

13 **14** **15**

A simple thermal modification was communicated by L'Abbé and Beenaerts at the Univeristy of Leuven.[7] While preparing triazoles to study the thermal ramifications of electron-withdrawing *N*-1 substituents on 1,2,3-triazoles, they discovered that lowering the temperature profile of the Dimroth reaction to 0–20 °C prevented the Dimroth rearrangement. Significant to this report was the first successful isolation of the triazole **18** from the reaction of 4-nitrophenzyl azide (**16**) with ethyl cyanoacetate (**17**).

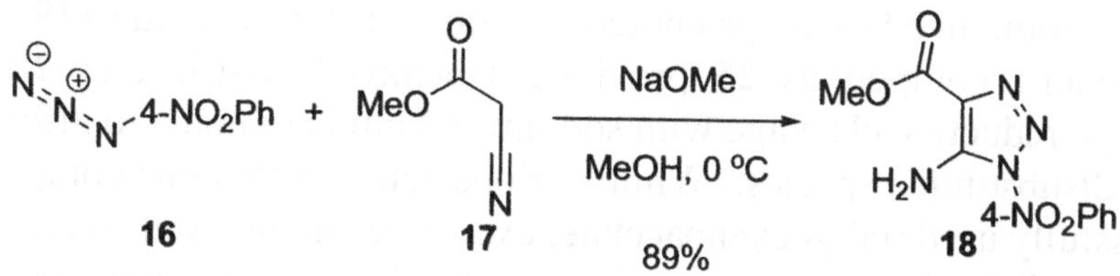

16 **17** **18**

Cottrell *et al.* at Merck Sharp and Dohme Research Laboratories reported a mild procedure that used potassium carbonate in dimethyl sulfoxide.[4] These conditions were compatible with highly functionalized benzyl azides, and extended the substrate scope of active methylene compounds to acetoacetone and benzoylacetone. This extension provided access to acyl-1,2,3-triazoles **21** for the first time *via* the Dimroth triazole synthesis.

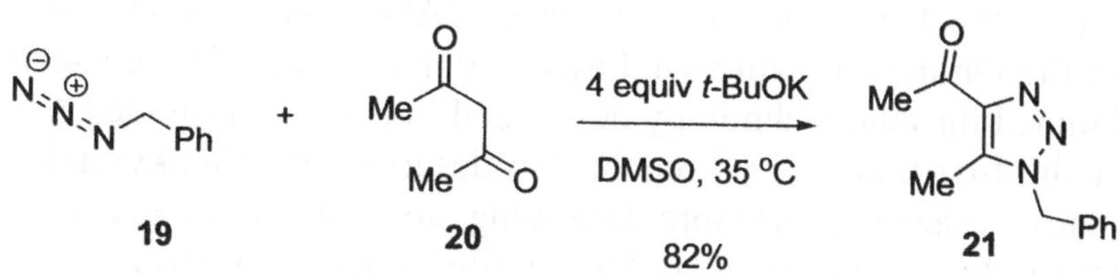

19 **20** **21**

Gibson, Thomas, and Rowley further extended the list of viable bases for the Dimroth synthesis while pursing $GABA_A$ ligands at Merck Sharp & Dohme Research Laboratories.[6] They found phenylthiomethyl azide (**22**) was a suitable electrophile when potassium carbonate in DMSO was used in the reaction.

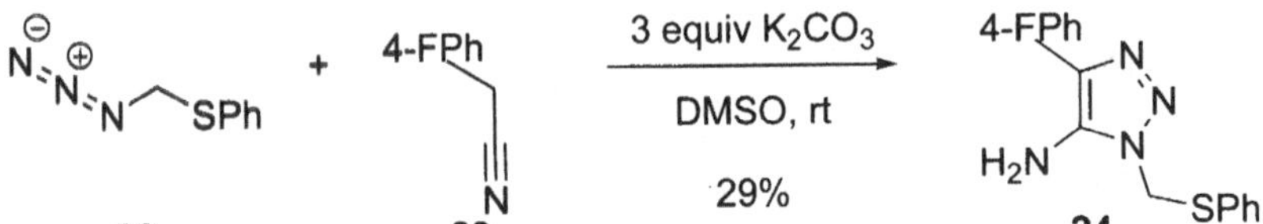

22 23 24

5.2.5 *Synthetic Utility*

General Utility

As reported in 1902, the substrate scope of the Dimroth triazole synthesis was limited to aromatic azides.[1] An early extension of this methodology was reported in 1956 by Hoover and Day at the University of Pennsylvania.[8] 1H-1,2,3-Triazoles were of particular interest at the time as potential modifiers of nucleic acid metabolism. As part of a program directed at cancer chemotherapies, they replaced the azide aromatic moiety with a benzyl substituent. Sodium ethoxide-promoted reaction of benzyl azide (**19**) with active methylene compounds **25** provided 1-benzyl-1,2,3-triazoles **26** that could undergo reductive cleavage with sodium in liquid ammonia to afford the desired 4,5-disubstituted species. While various active methylene compounds were successfully used (ethyl cyanoacetate, cyanoacetamide, cyanoacetic acid, and malononitrile), the yields were low to modest when compared with aromatic substrates.[1]

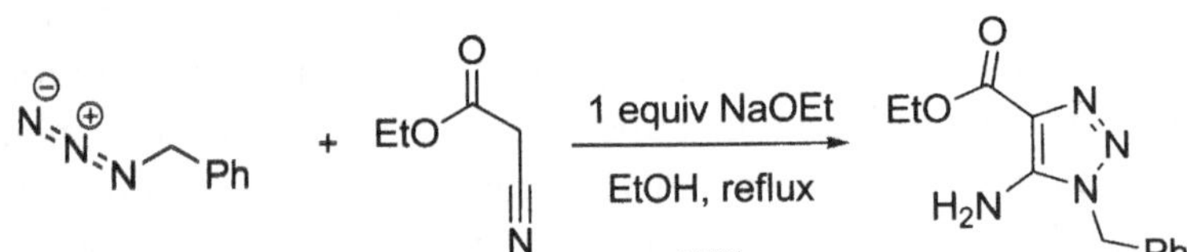

19 25 26

While at the University of Colorado, L'Abbé and Hassner extended the scope of the Dimroth triazole synthesis to vinyl azides.[9] These substrates could be obtained through technology developed in their lab provided *N*-vinyl triazoles for the first time.[10] This report also described β-haloalkyl azides as suitable *N*-vinyl triazole precursors following an elimination event. It is unclear whether elimination occurs before or after triazole formation.

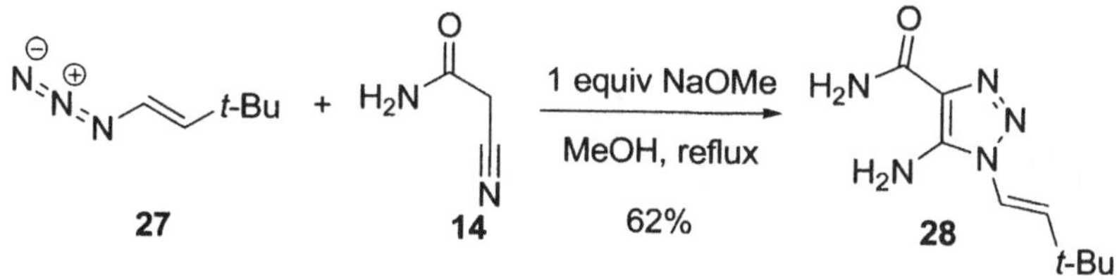

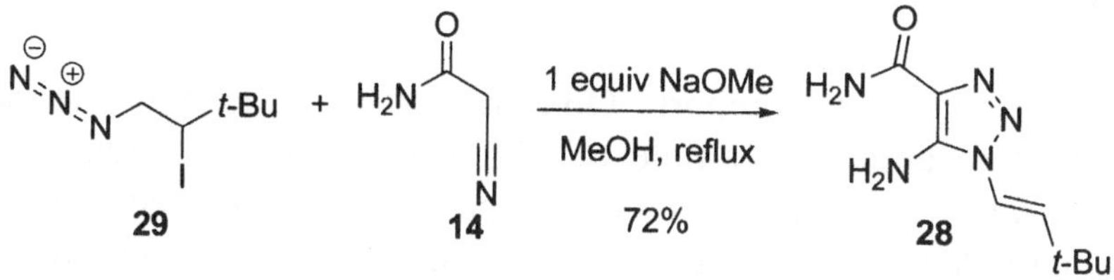

In a report describing the synthesis of D-ribofuranose glycone-containing cyclonucleosides, Ewing *et al.* found potassium hydroxide in DMF the optimal reagent system (*vide supra*).[11] Subsequent to heterocycle formation, the triazole **31** was chemically joined to the pentafuranose system by synthesis of an azole.

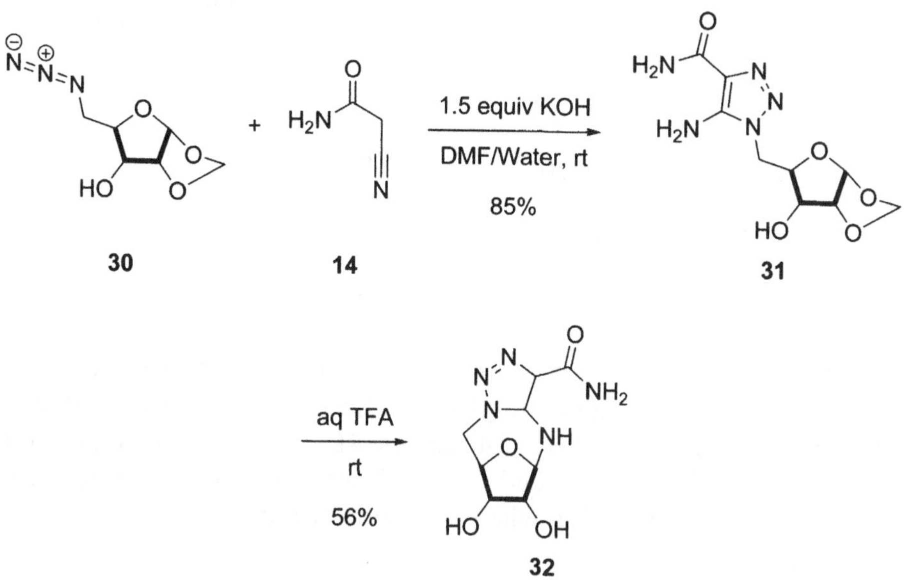

A recent report from the Ukrainian group of Obushak *et al.* takes advantage of the conventional Dimroth triazole synthesis conditions to prepare 4-thiazole and 4-benzothiazole-containing triazoles **35**.[12] 2-Benzothiazolylacetonone, 1,3-benzothiazol-2-ylacetonitrile, and (4-aryl-1,3-thiazol-2-yl)acetonitriles were found to be suitable starting materials and provided the corresponding products in moderate to excellent yield when reacted with aryl azides **33**.

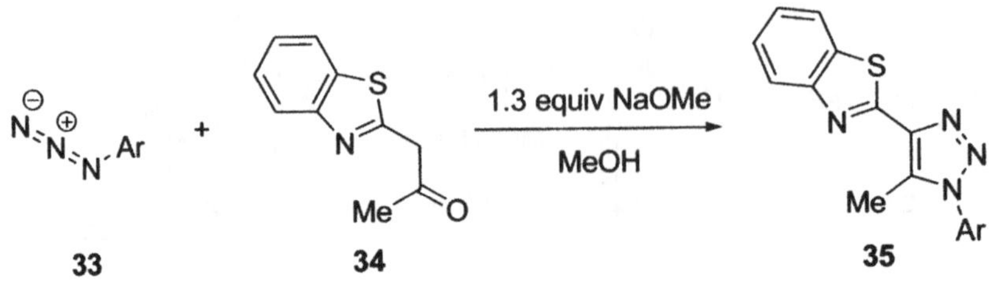

In a closely related report, Obushak *et al.* extended the active methylene scope to another example of a 6,5-fused bicycle.[13] Once again, traditional Dimroth triazole synthesis conditions gave the desired triazole **37** from 1H-benzimidazol-2-ylacetonitrile **36** and aryl azides en route to [1,2,3]triazolo-[4′,5′:4,5]pyrimido[1,6-*a*]benzimidazole **38**.

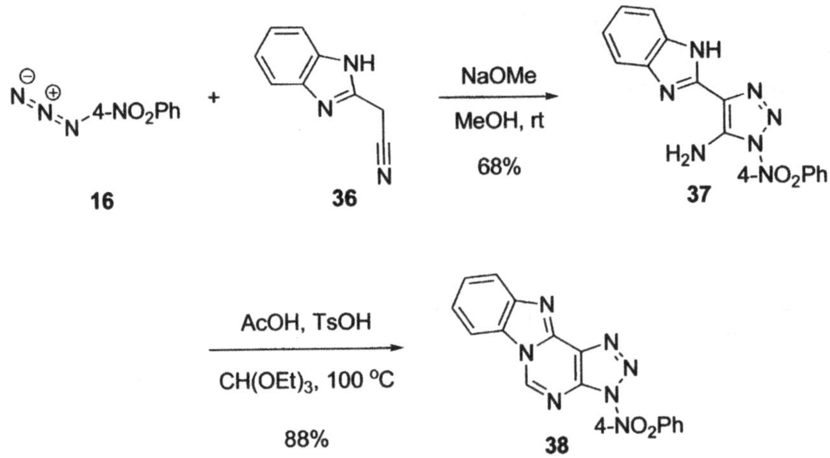

A third report from Obushak, Pokhodylo, and Matiychuk used (arylsulfonyl)acetones **40** and (arylsulfonyl)acetonitriles as activated methylenic building blocks for 1,2,3-triazoles **41**.[14] Conventional Dimroth triazole synthesis conditions provided 4-arylsulfonyl triazoles in moderate to good yield with mild or no heating when aryl azides were used.

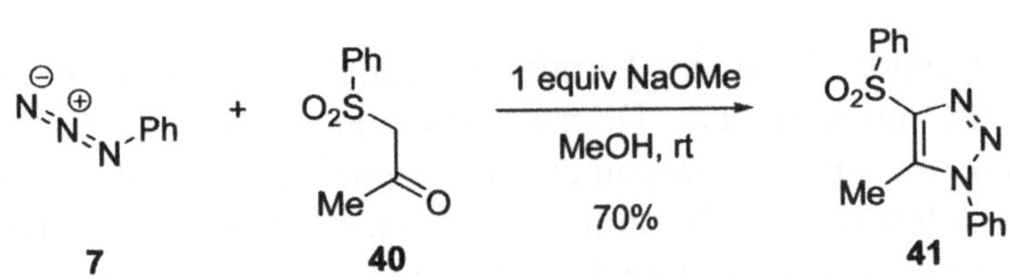

Utility of the Dimroth Triazole Synthesis in Tandem Reaction Sequences

The first observed tandem process accompanied the discovery of the Dimroth triazole synthesis. Otto Dimroth reported on the propensity of his newly formed triazole product to undergo a rearrangement in which a cyclic C–N bond was broken, followed by a C–N bond forming event in which the endocyclic and exocyclic nitrogens had reversed (See Section 9.4).[2]

Rational substrate design, with appropriately functionalized aryl azides, has been shown to afford a variety of fused heteroaromatic polycycles. One report by Smalley et al. from the University of Salford describes exploitation of the Dimroth process to afford 1,2,3-triazolo[1,5-*a*]quinazolines from *o*-azidobenz-aldehyde **(42)**, *o*-azidoacetophenone, and benzonitrile with various active methylene substrates.[15] It is interesting that the authors found an organic base, piperidine, was optimal for some substrates, while either sodium ethoxide or Amberlite resin worked better for others.

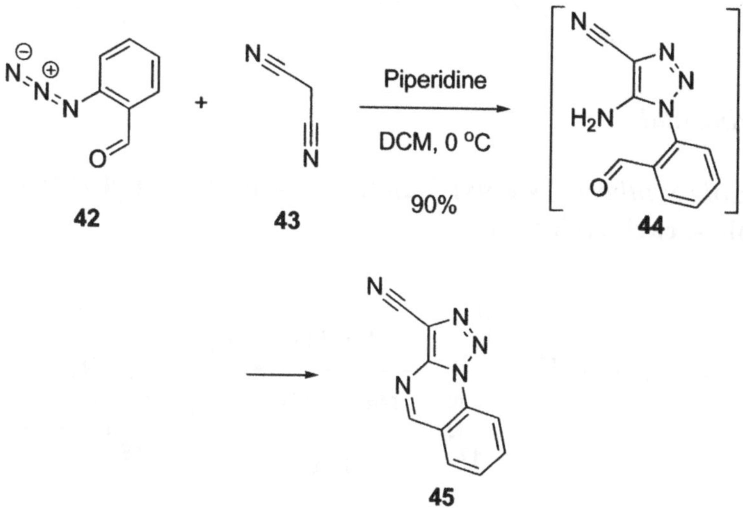

While pursuing a program directed at new ligands for $GABA_A$ receptors, Jones and Chambers developed a rapid approach to C-5 substituted 1,2,3-triazolo-[1,5-*a*]quinazolines.[16] Reaction of 2-azidobenzoic acid **(46)** with an isoxazole acetonitrile **47** under classical Dimroth triazole synthesis conditions gave the desired core skeleton with a chemical handle for rapid derivitization.

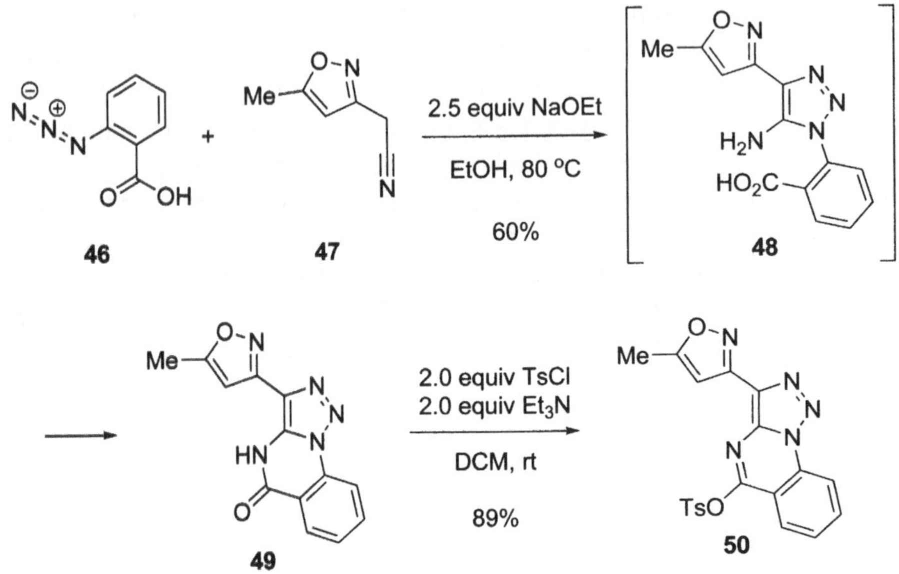

5.2.6 *Experimental*

Dimroth Triazole Synthesis: **Methyl 5-amino-1-(4-nitrophenyl)-1*H*-1,2,3-triazole-4-carboxylate** (18)[5]

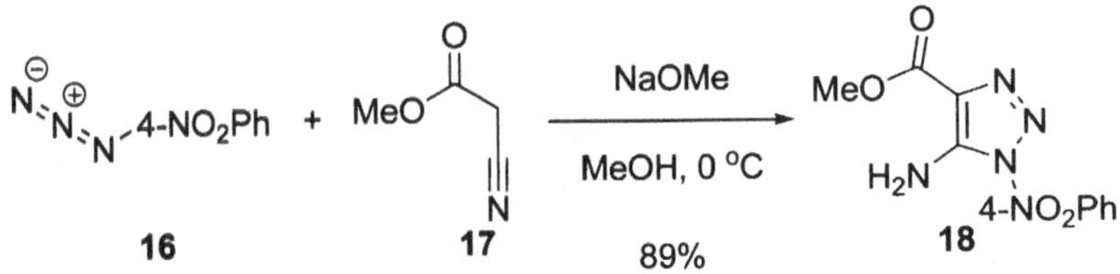

Equimolar amounts (24 mmol) of methyl 2-cyanoacetate (**17**) and 4-nitrophenylazide (**16**) were added dropwise to a methanol solution of sodium methoxide (0.6 g Na in 40 mL of dry methanol). The mixture was allowed to react at 0 °C for 1 h and then at room temperature for 1 day. After removal of the solvent, the residue was treated with water, and the resulting semicrystalline material was filtered off, washed with water and dried in vacuo. This yielded methyl 5-amino-1-(4-nitrophenyl)-1*H*-1,2,3-triazole-4-carboxylate (**18**) in 89% (mp 224 °C).

5.2.7 *References*

1. Dimroth, O. *Ber. Dtsch. Chem. Ges.* **1902,** *35*, 1029–1038.
2. Dimroth, O. *Liebigs Ann. Chem.* **1909,** *364*, 183–226.
3. Lieber, E.; Rao, N. R.; Rajkumar, T. V. *J. Org. Chem.* **1958,** *24*, 134–135.

4. Cottrell, I. F.; Hands, D.; Houghton, P. G.; Humphrey, G. R.; Wright, S. H. B. *J. Heterocyclic Chem.* **1991,** *28*, 301–304.
5. Acevedo, O. L.; Krawczyk, S. H.; Townsend, L. B. *J. Org. Chem.* **1986,** *51*, 1050–1058.
6. Gibson, K. R.; Thomas, S. R.; Rowley, M. *Synlett* **2001,** 712–714.
7. L'Abbé, G.; Beenaerts, L. *Tetrahedron* **1989,** *45*, 749–756.
8. Hoover, J. R. E.; Day, A. R. *J. Am. Chem. Soc.* **1956,** *78*, 5832–5836.
9. L'Abbé, G.; Hassner, A. *J. Heterocyclic Chem.* **1970,** *7*, 361–366.
10. Fowler, F. W.; Hassner, A.; Levy, L. A. *J. Am. Chem. Soc.* **1967,** *89*, 2077–2082.
11. Ewing, D. F.; Goethals, G.; Mackenzie, G.; Martin, P.; Ronco, G.; Vanbaelinghem, L.; Villa, P. *Carbohydr. Res.* **1999,** *321*, 190–196.
12. Pokhodylo, N. T.; Matiychuk, V. S.; Obushak, M. D. *Chem. Heterocycl. Compd.* **2009,** *45*, 483–488.
13. Pokhodylo, N. T.; Matiychuk, V. S.; Obushak, M. D. *Chem. Heterocycl. Compd.* **2009,** *45*, 245–247.
14. Pokhodylo, N. T.; Matiychuk, V. S.; Obushak, M. D. *Synthesis* **2009,** 2321–2323.
15. Porter, T. C.; Smalley, R. K.; Teguiche, M.; Purwono, Bambang. *Synthesis* **1997,** 773–777.
16. Jones, P.; Chambers, M. *Tetrahedron* **2002,** *58*, 9973–9981.

5.3 Finnegan Tetrazole Synthesis

Timothy T. Curran

5.3.1 Description

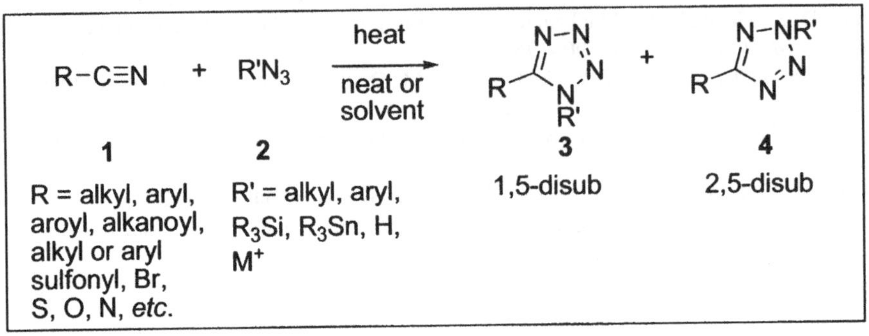

The Finnegan tetrazole synthesis is the reaction of a functionalized nitrile **1** with hydrazoic acid or an alkyl, aryl, or inorganic azide **2** to generate tetrazole **3** and/or **4**. Typically, the 1,5-disubstituted tetrazole or 1*H*-5-substituted tetrazole **3** is the predominant or exclusive product. The reaction can be promoted thermally or by use of mild organic, inorganic, or Lewis acids. The substituents allowed on both the nitrile and azide span a wide variety of functionality; however, electron poor nitriles react more smoothly than electron-rich nitriles using standard protocol.

5.3.2 Historical Perspective

Finnegan and co-workers[1] reported the preparation of tetrazoles using the above method due to a need to improve the preparation of tetrazoles. Methods at the time were not ready for large scale use and suffered due to a variety of reasons. For example, methods lacked readily available starting materials; were multistep; oftentimes resulted in modest yields; used excess reagents; and/or had high concentrations of hydrazoic acid, posing safety hazards. The conditions developed by Finnegan and co-workers used slight excess of azide, in an appropriate polar organic solvent like DMF. The reactions were typically heated at steam-bath temperature to provide the product. As can be seen in the following table, the reaction was quite general and provided good yield of the desired 5-substituted-1*H*-tetrazole products **7**. A variety of azide salts were used. Neither safety information nor incidents were reported with these reactions.

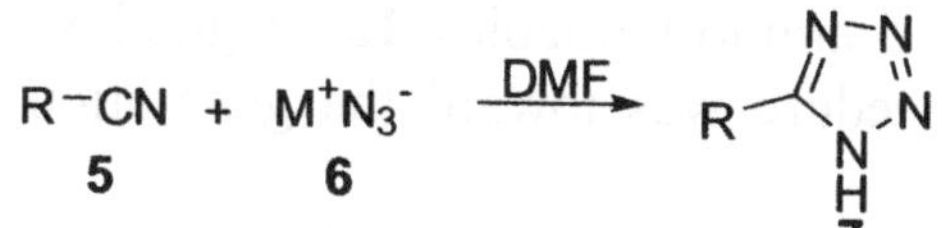

5 **6** **7**

R	M+	Temp °C, Time h	Yield %
$HOCH_2CH_2$–	NH_4	120-125, 24	93
MeS–	NH_4	95, 6	91
C_7F_{15}–	NH_4	95, 4	quant
Ph–	NH_4	125, 7	quant
Ph–	Me_3NH	100, 3	90
Ph–	Na	reflux, 20	51

In Finnegan's initial report, the scope and understanding of the reaction was investigated, but the literature of the time did not show a broad range of use for the tetrazole moiety. Once the tetrazole moiety found application in a variety of areas, the need for improvement in the chemistry spurred ingenuity. For example, the tetrazole functionality has been noted as being a carboxylic acid equivalent, a *cis* peptide bond mimetic, and has been used as such in both the pharmaceutic and agriculture industry.[2] Due to the energetics associated with the tetrazole functional group, this moiety has been employed in the synthesis of many potential propellant or explosive agents. A method reported recently by Sharpless[3] was used to prepare a variety of tetrazoles.[4] Tricyanomethane was treated with sodium azide and zinc bromide ($ZnBr_2$) to provide *tris*-tetrazolylmethane **9** in 38% yield. Tetrazole products were then *N*-alkylated with chloro-ethanol and further reacted with "binders," generating high-energy, oligomeric tetrazoles with the desire to generate the next generation of propellants.

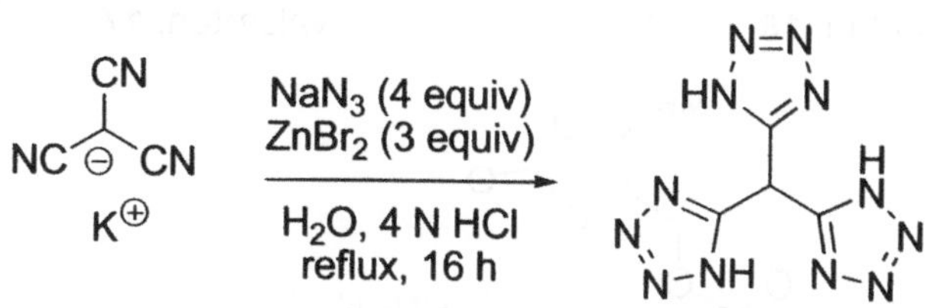

8 38% **9**

In a modification to the Finnegan reaction, preparation of 5-amino-tetrazoles has been accomplished via preparation of intermediate **12** from cyano–azide **11**. Reaction of **11** with a variety of amines **10** at room temperature provided the 5-amino-tetrazoles **13** in good yield. Again, the application using this procedure was toward the generation of high energy materials.[5]

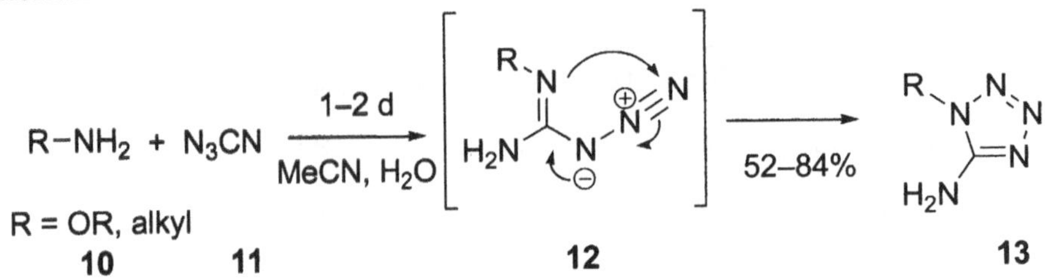

The use of tetrazoles in the pharmaceutical industry has improved quality of life in a key therapeutic area. The tetrazole moiety is a key structural component of compounds known as angiotensin II receptor blockers used in the treatment of hypertension. Materials known as sartans exemplified by **14** through **18** contain the tetrazole moiety. In part, the need for robust chemistry to prepare sartans has encouraged development of modified Finnegan tetrazole syntheses and tetrazole synthesis as a whole.

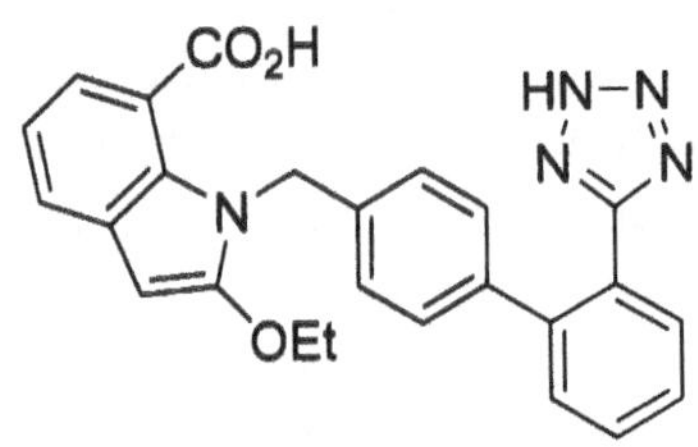

candesartan, **14**

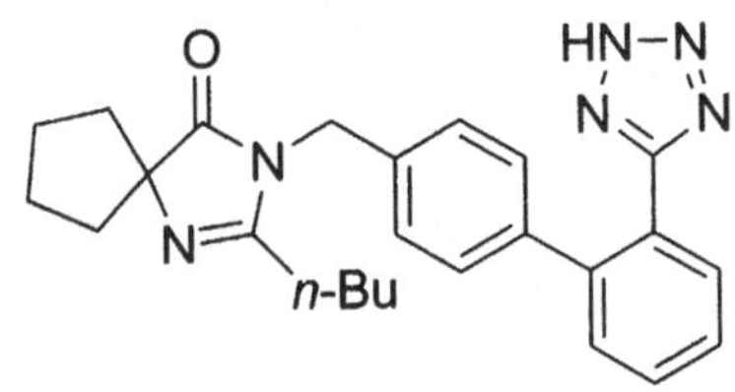

irbesartan, **15**

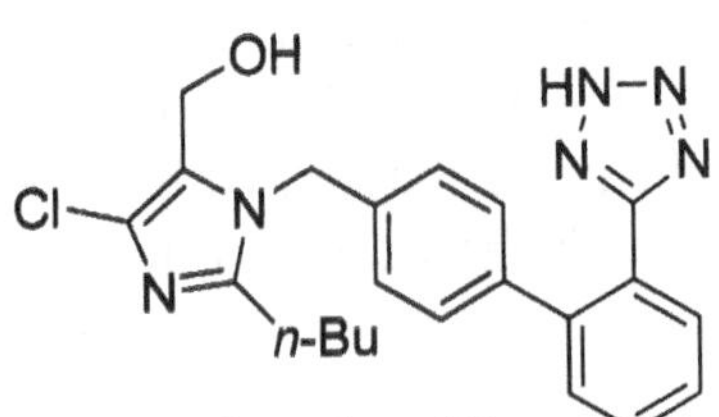

losartan, **16**

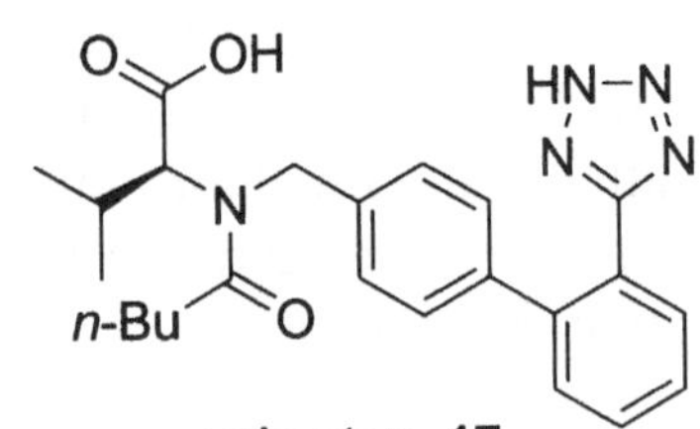

valsartan, **17**

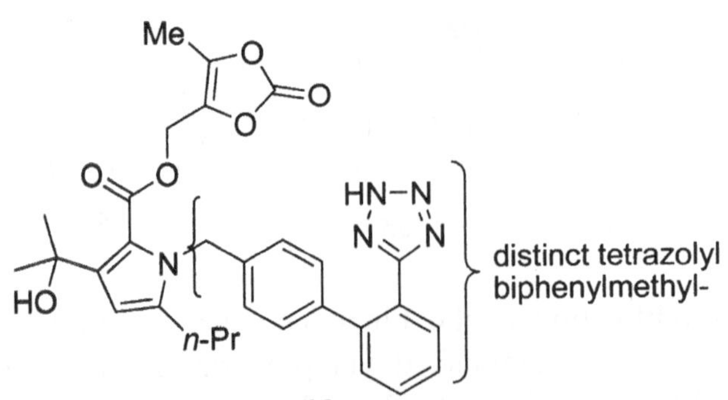

olmesartan, **18**

The key repeating component of these antihypertensives is the tetrazolylbiphenyl methyl moiety, which has been derived from the corresponding nitrile. During the course of development, groups have oftentimes benchmarked against how well their method faired when applied to the synthesis of the corresponding biphenyl nitrile. Two recent reports reviewing methods to prepare tetrazoles have appeared.[6]

5.3.3 Mechanism

The suggested mechanism for the Finnegan tetrazole synthesis as put forth in the original paper was a two step process that started with the addition of an azide **21** to an activated nitrile **20** followed by cyclization to form the tetrazole. The authors noted that hydrazoic acid, amine hydrazides, and Lewis acids doubled the yield of some reactions when these materials were added at the 4–10 mol% level, suggesting that activation of the nitrile may be a critical initial step. The authors actually drew the resonance structure **19** in the lead manuscript. In addition, it was noted that electron-withdrawing groups on the nitrile facilitated the addition of the azide to the nitrile and that the reaction was found to be subject to general acid catalysis with BF_3 being noted as equivalent to NH_4Cl in promoting the reaction.

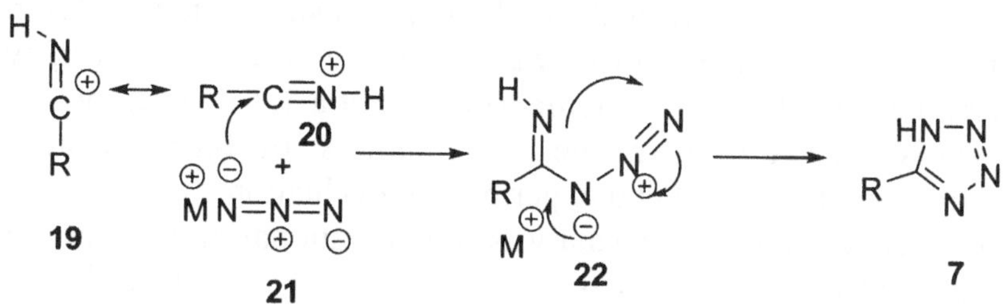

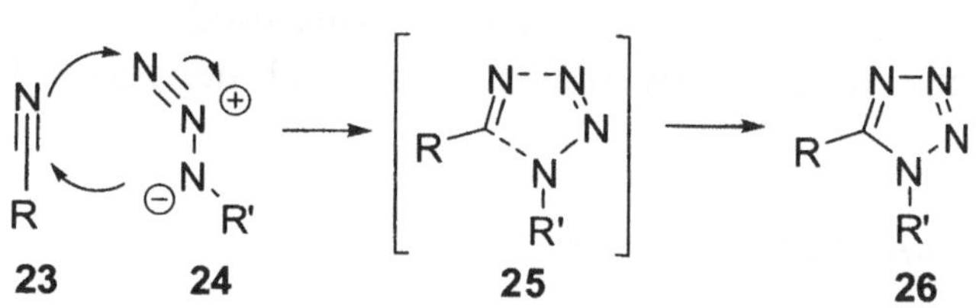

There has been debate over the mechanism through the years, with some contesting that the reaction took place via a concerted [3 + 2]-cycloaddition mechanism (depicted in reaction of **23** and **24**).[7] Koldobskii and co-workers[8] suggested that dialkyl ammonium azide (a practically unionized salt based on electrochemical measurements) underwent Diels–Alder type cycloadditions to yield tetrazoles. In addition, a modest solvent effect was observed, and calculated activation parameters were both consistent with a Diels–Alder cycloaddition process.

Sharpless and co-workers recently reported high yields of the [3 + 2] cycloaddition products when tosyl or acyl nitriles **28** were reacted with several different alkyl or aryl azides **27**.[9] After the tetrazole-forming step, the Ts group can be easily displaced with a variety of nucleophiles and acyl group when appropriately selected R′ groups are employed (*e.g.*, *p*-nitrophenol) were shown to undergo subsequent functionalization. While these reactions do generally provide high yields, the use of acyl cyanides with α-hydrogens were unstable at elevated temperature. A more recent report described using $ZnBr_2$ to promote the reaction of benzyl azide (R = Bn) with acetyl cyanide (R′ = Me) at room temperature and was much cleaner than the thermal reaction.[10]

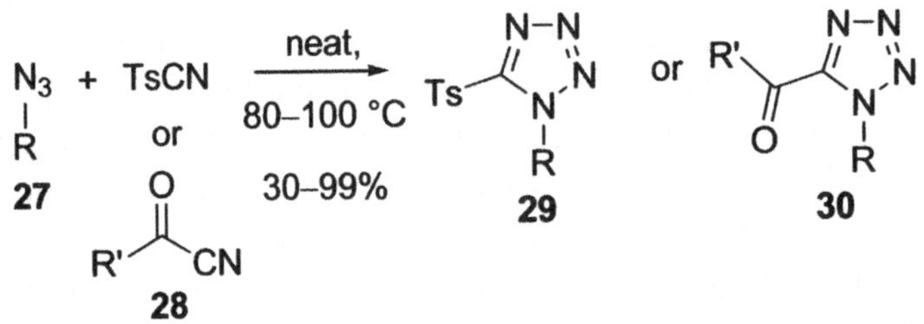

It was recently suggested that for alkyl or aryl substituted azides, most agreed that the concerted mechanism was likely the case. However, when salts of hydrazoic acid are used (as ionic azide is not a 1,3-dipole), or when a strongly EWG is attached to the nitrile, many authors still favor the stepwise process based on computational and experimental evaluation.[11–13]

In the case of using acid catalysts, computational and experimental information currently supports Lewis acid activation of the nitrile.[11,12] Such activation does not necessarily contradict mechanisms in which Al or Zn azides are suggested to form, as azides are typically used in excess and could be involved in the nitrile activation process and delivery of the azide separately.

5.3.4 Variations and Improvements

Variations of Promoters and Azide

A variety of azides have been used for this chemistry. The substituent on the azide could be aryl or alkyl. Several different salts were described in Finnegan's initial report. In addition, trialkyl stannanes and silanes have been employed. Due to sterics, the 2,5-disubstituted isomer was reported to be isolated when trialkyl stannanes were employed. This chemistry was showcased in the preparation of a biphenyl component **32**, a popular component used in the preparation of sartans.[14]

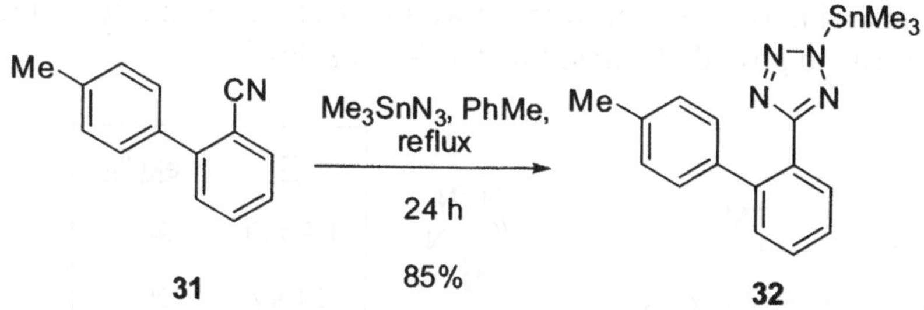

Use of $TMSN_3$ has been shown to be general and the reaction with nitriles was found to be promoted with catalytic dialkyl tin oxide[15] or tetra-*n*-butylammonium fluoride (TBAF).[16] In the former case, good yield of the 1,5-disubstituted material was generated. Wittenberger *et al.* proposed initial generation of intermediate **33**, followed by cycloaddition to form the 2,5-disubstituted tetrazole **34** followed by rearrangement into **35**. While shown as an intramolecular reaction, this rearrangement could also occur intermolecularly. Both alkyl and aryl nitriles were shown to react smoothly yielding the 5-substituted 1*H*-tetrazole **7**.

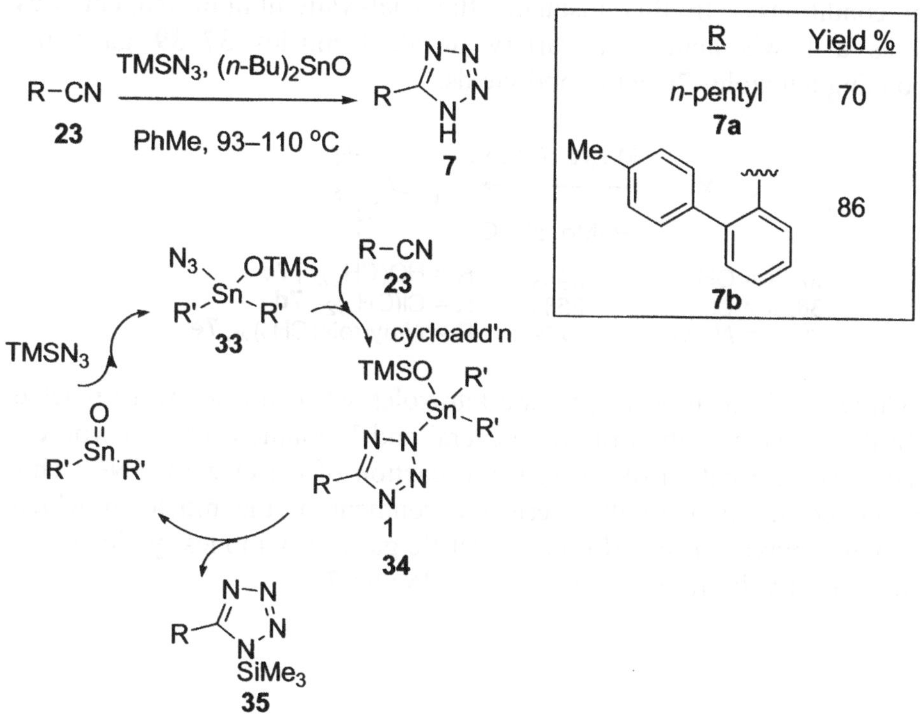

For the TBAF-promoted reaction, it was noted that the reaction does not require a full equivalent of fluoride source; the "carrier" was not identified but could likely be the tetrazole anion. Again, this method was general for both alkyl and aryl nitriles **36** and worked well with sterically hindered nitriles like *t*-butyl nitrile to provide 5-substituted-1*H*-tetrazoles **7**.

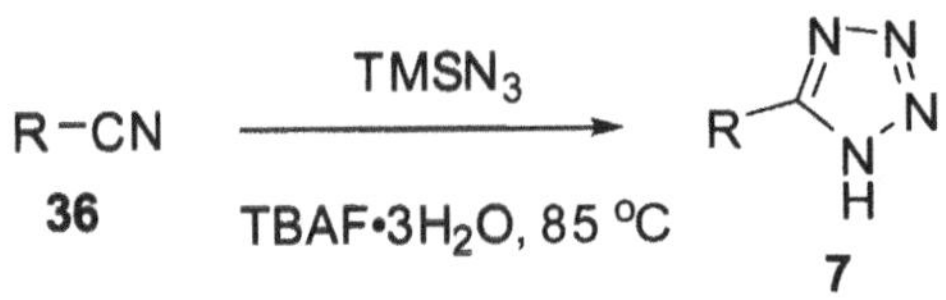

R	Yield %
4-Ac-Ph	94
2-furyl	95
5-indolyl	82
t-Bu	82

Trialkylaluminum reagents have been used in combination with $TMSN_3$ to generate tetrazoles.[17] While at the time of the report, this procedure reported by Huff was an improvement to existing procedures and was shown to be somewhat general. The authors pointed out that the presence of certain functionality (amides) required large excess of reagents, and that esters could be transesterified. Yet hydroxyls and halides proved stable to the reaction conditions. In this instance, the trialkylaluminium reagent was viewed as a Lewis acid. A variety of alkyl nitriles **37–39** gave the corresponding tetrazoles **7c–e** in good yields.

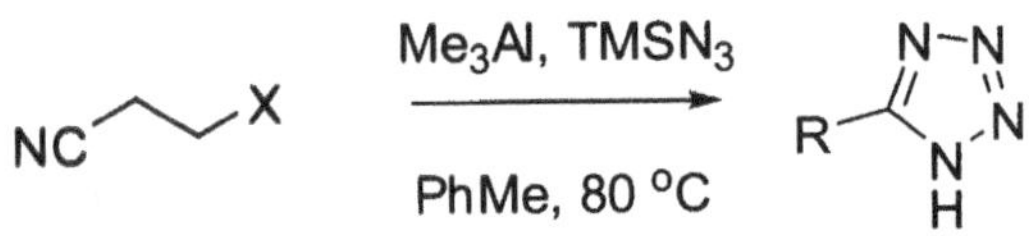

37, X = OH	89%	R = HO(CH2)2-, **7c**
38, X = Cl	65%	R = Cl(CH2)2-, **7d**
39, X = *N*-pyrrol	87%	R = *N*-pyrrole(CH2)2-, **7e**

Cu_2O has been used to provide tetrazoles when nitriles were reacted with $TMSN_3$.[18] This method proved general and Yamamoto and co-workers proposed *in situ* formation of CuN_3 from reaction of Cu_2O with HN_3. This method proved sensitive to the steric environment of the nitrile in which reaction with *t*-butyl nitrile and biphenylnitrile gave only modest yields of 36 and 50%, respectively, for the conversion of **36** into **7**.

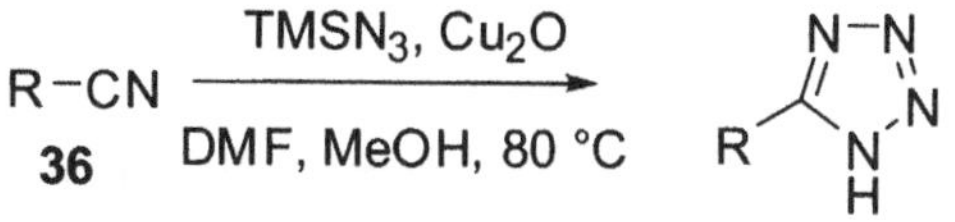

36 → 7

R	Yield %
Ts	77
t-Bu	36
n-Bu	55
p-NO$_2$-Ph	96
p-HO-Ph	87
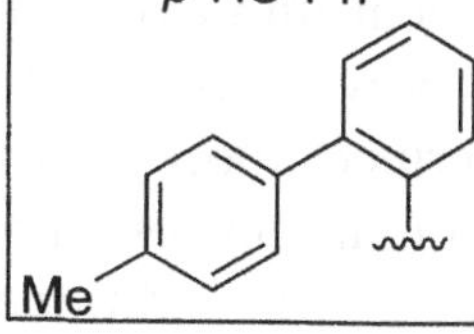	50*

*10 mol% Cu$_2$O used at 120 °C

$Cu_2(OTf)_2$•PhH complex was used to catalyze tetrazole formation from alkyl azides and electron deficient nitriles. In this instance,[19] the sterics seem to modulate the amount of 1,5- and 2,5-disubstituted tetrazoles. Increased Cu loading provided more 2,5-disubstituted tetrazole. Bosch *et al.* reported either stirring at room temperature or using microwave irradiation to promote the reaction. Mild conditions and modest to excellent yields of the 1,5-disubstituted tetrazole **41** were reported using 10 mol% catalyst. However, when more than 50 mol% of $Cu_2(OTf)_2$•PhH complex was used, the 2,5-disubstituted tetrazole predominated. The rationale for the difference in selectivity was suggested to be due to Cu forming a complex, with both the nitrile and the azide and altering the mode of addition.

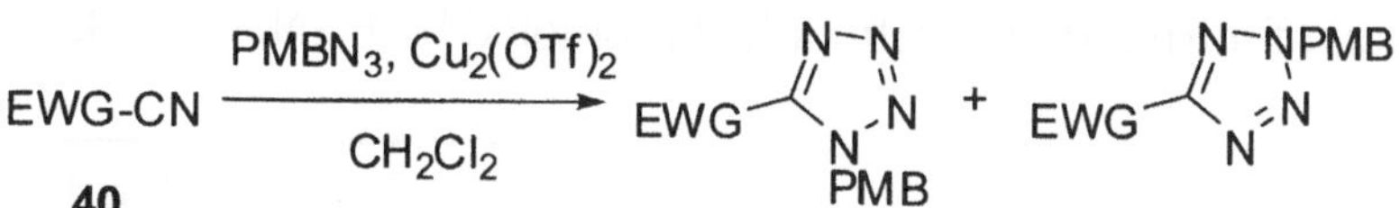

40 → **41** + **42**

R	Temp °C, Time h	Yield %, **41**:**42**
EtOCO	20, 48	95, >99:<1
EtOCO	80, MW, 2	88, >99:<1
MeCO	20, 48	81, 90:10
MeCO	80, MW, 2	37, 90:10
PhCO	20, 48	74, 92:8
PhCO	80, MW, 2	77, 95:5
Ts	80, MW, 2	99, 80:20

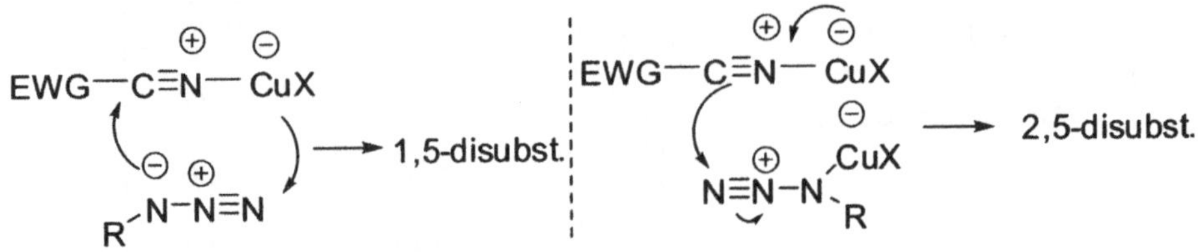

More traditional improvements have come from considering many of the improvements previously mentioned. For example, Koguro and co-workers have described more scale-friendly reaction conditions than the traditional Finnegan reaction conditions using NaN_3 with NH_4Cl in DMF. As previously mentioned, the traditional Finnegan conditions gave rise to an explosive and shock-sensitive substance, NH_4N_3, which formed *in situ*, sublimed, and could find its way into overheads in large reaction vessels and plant plumbing. Koguro and co-workers described the use of NaN_3 in combination with other amine salts in toluene to provide high yield of the 5-substituted 1*H*-tetrazole **7**. The reaction was general, and the workup simple. These authors suggested the formation of $R_3NH^+N_3^-$ salts to facilitate the reaction.[20]

In this work, a variety of amines salts of HCl, HOAc and H_2SO_4 was tested. The stronger acids, HCl and H_2SO_4, were deemed superior, and substituted benzenes such as PhMe, $PhNO_2$ and xylenes were described as solvents that provided successful reaction. Protic solvents were shown to give poor yields under these reaction conditions. Note that this method allows for successful reaction with anilines, phenols, esters, and carbamates as shown for the conversion of nitrile **36** into tetrazole **7** promoted by $Et_3NH^+N_3^-$ **43**.

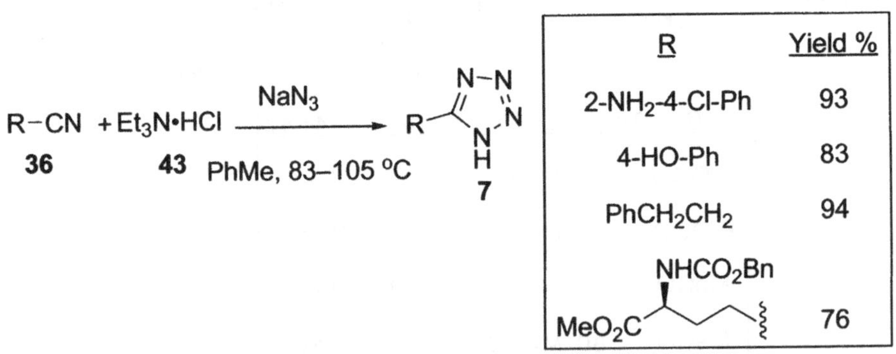

Information reported by Koguro and co-workers was used by Aventis to scale up a leukotriene D_4-receptor antagonist and PPARγ agonist **45**.[21] Bridge *et al.* found that $Et_3NH^+Cl^-$ reduced the amount $Et_3NH^+N_3^-$ sublimate (in comparison to NH_4Cl), which was observed when NMP was used as solvent but the azide sublimate was not eliminated. Switching to $Et_3NH^+OAc^-$ produced an equivalent reaction rate in NMP while further reducing the sublimate. While xylenes, acetophenone, hexan-1-ol, and cyclohexanone could alternatively be used as solvent, best yields were obtained by using anisole at 130 °C for 6 h, and nitrile **44** was converted into tetrazole **45** in 90% yield.

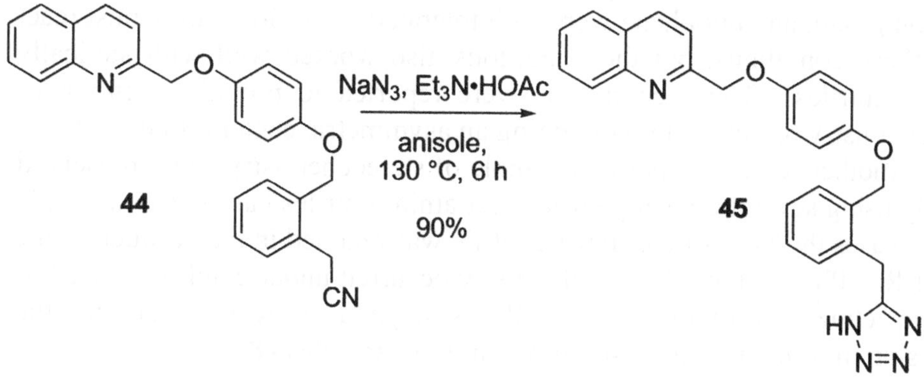

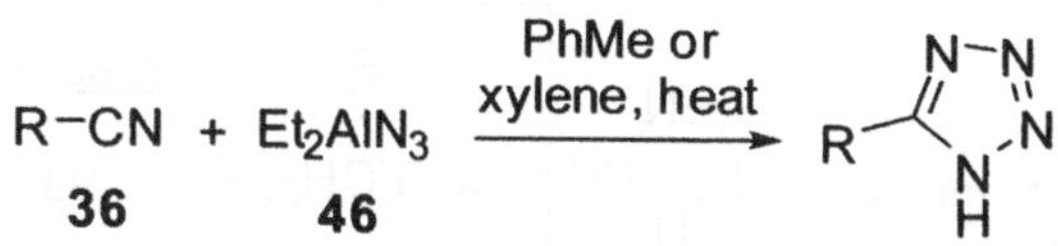

R	Temp °C, Time h	Yield %
BnS−	45, 4.5	46
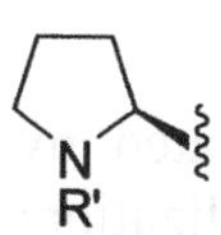 R' = Cbz	50, 9	96
R' = Boc	40, 30	57
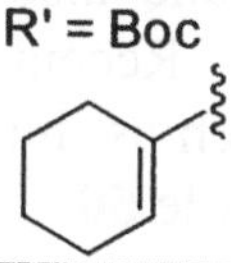	65	55
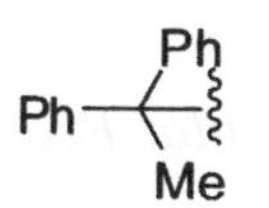	85, 18	45
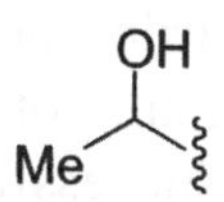	0 to rt, 24	65
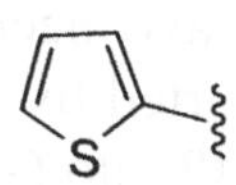	55, 12	81

Conditions similar to those reported by Koguro have been applied successfully with stereogenic centers α to the nitrile with retention of stereochemistry.[22]

Further modification was derived by adding a Lewis acid to the azide. Sedelmeir and Aureggi showed this reaction to be very general and provided good yields of the 5-substituted-tetrazole.[23] The conditions were also conducive to retaining the stereochemistry α to the nitrile. Note that thiocyanates, hydroxyl groups, sterically hindered nitriles, electron-rich heterocycles, and α,β-unsaturated nitriles all worked well under these conditions. This chemistry requires simple modification of NaN_3 with R_2AlCl to create the reactive R_2AlN_3, which promotes the reaction. A drawback for this methodology is that a pyrophoric reagent is used; however, the generality and simplicity of the reaction provides a useful methodology for the organic chemist.

As can be observed by the yields reported, the benzyl carbamate, hydroxyl group, and thiophene were well tolerated. The Boc group must react under these conditions, but the conditions also worked well with sterically hindered nitriles. These conditions were reported to provide retention of stereochemistry on substrates containing an asymmetric center α to the nitrile.

Another recently reported example of stereochemistry being preserved partially using several optically enriched α-amino-nitriles has appeared. While in some cases the % *ee* of the starting nitrile was retained in the product, in the case of R = Ph, erosion of stereochemistry occurred under conditions used to promote tetrazole formation.[24] In all cases, yields were excellent for the conversion of α-amino nitriles **47** into α-amino tetrazoles **48**.

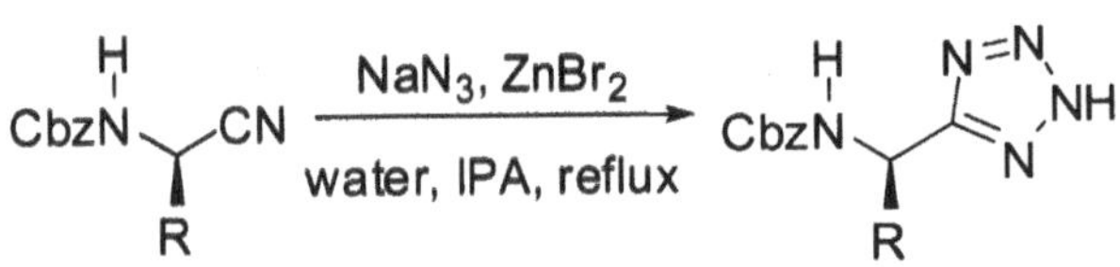

47 **48**

R	Yield% (ee%)
$MeS(CH_2)_2$-	93 (87)
$CbzHN(CH_2)_3$-	90 (92)
Ph	98 (39)

Variations of Substrates: Intramolecular Reactions

Intramolecular reactions between nitriles and azides can be facilitated by proximity effects if steric interactions are not too burdensome for cyclization. There are several examples of the successful thermal cyclization and also instances in which the reactions have been Lewis acid promoted. Recent thermal cyclizations have been reported by Sharpless and Demko such as the conversion of *N*-cyano-alkyl azide **49** into 6-5-5 tricyclic, fused tetrazole **50**.[25]

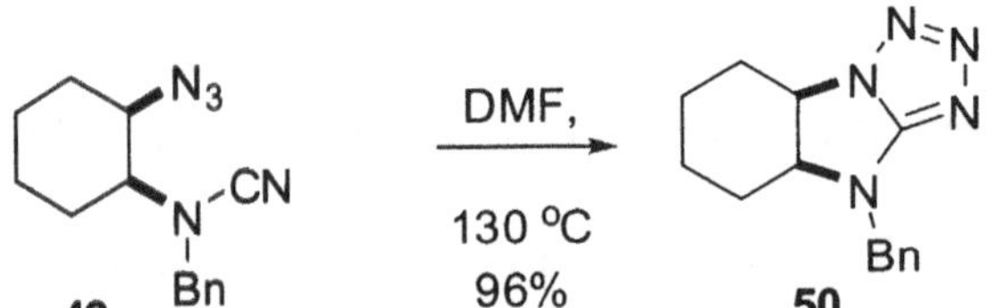

49 **50**

The corresponding *trans*-compound (of **49**) required higher temperature and gave a modest 53% yield. Addition of an EWG to the nitrogen-bearing the CN group led to poor yield due to increased instability under the reaction conditions.

Additional substrates could easily be prepared *in situ* by reaction with NaSCN with the corresponding alkyl halide or appropriately activated alcohol, *e.g.*, **51**. The 6-5 fused ring system could be formed, as exemplified in the conversion of **51** into **52** via displacement followed by cyclization. The 5-5 ring system formed in excellent yield as well while, the 7-5-fused ring system did not react to form the desired bicyclic tetrazole. Other high-boiling, aprotic solvents (*e.g.*, DMSO) have been reported for promoting the intramolecular cyclization[26] as well as running the reaction in a sealed tube.[27]

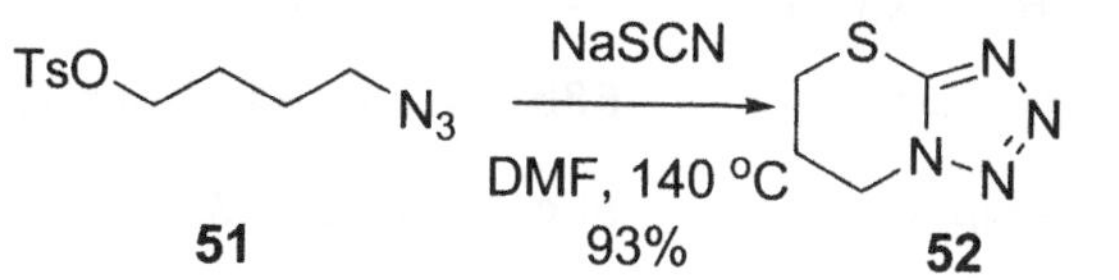

Lewis acid catalysis has worked nicely for promoting intramolecular cyclization. Hannesian reported this modification which generated nitrile **55** using BF_3•OEt_2 by the initial generation of oxonium **54** from acetal **53**. Azido-nitrile **55** subsequently underwent [3 + 2] cyclization to form a mixture of isomers **56** and **57**. The selectivity for this tandem oxonium formation, cyanide addition, cyclization was governed by steric interactions in the transition state. Electron deficient R groups (*e.g.*, *p*-O_2NPh) were reported to provide inferior selectivity. When starting material was enantiopure, enantio-enriched products were obtained.[28]

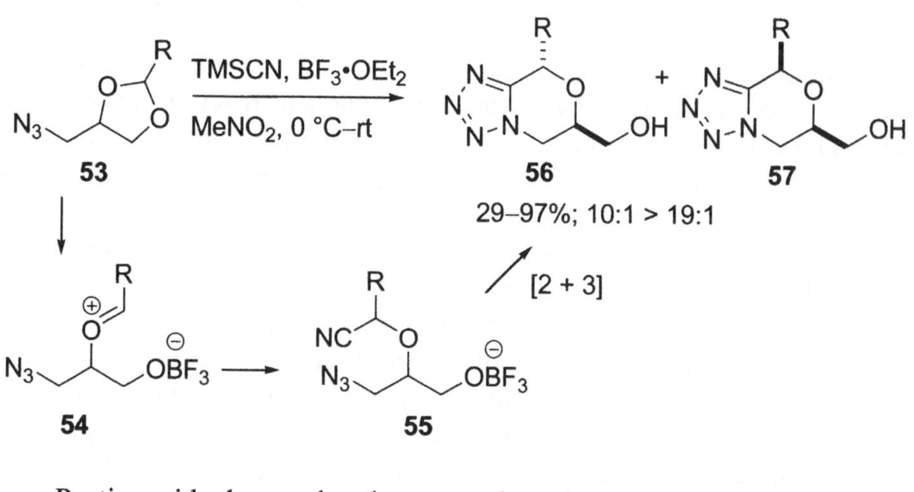

Protic acids have also been used to intramolecularly promote the cyclization of a nitrile with an azide. Chlorosulfonic acid was used to promote the intramolecular cyclization of a few substrates into their tetrazoles. Note that the presence of chlorosulfonic acid along with the intramolecular nature of this substrate allowed this reaction to occur at room temperature. As one might expect, the pure *E*-isomers were reluctant to cyclize, which suggests that for this system, little alkene isomerization occurred. Poor to good yield was obtained for these three azido-α,β-unsaturated nitriles **57a–c.**[29]

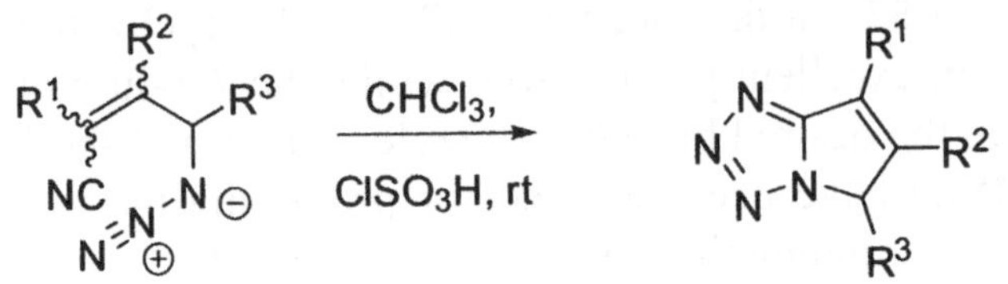

57a, R1 = Me, R2 = R3 = H

57b, R2 = H, R1 = R3 = H

57c, R1 = R2 = Me, R3 = H

58a, R1 = Me, R2 = R3 = H, 57%

58b, R2 = H, R1 = R3 = H, 25%

58c, R1 = R2 = Me, R3 = H, 73%

Nonclassical Promoters of the Finnegan Reaction

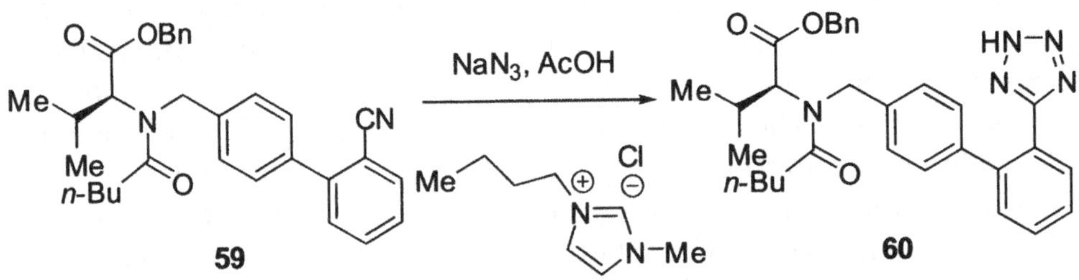

59 **60**

140 °C: 73%
microwave: 88%

A study on the use of ionic liquids in combination with or without microwave irradiation has been used to promote the Finnegan reaction to prepare tetrazoles. The counter ions used as ionic liquid had a large impact, while the cationic portion of the substrates screened had little impact (Me, *n*-Bu, *n*-hexyl, and *n*-octyl-3-methylimidazolium were used). Simple ions like Br^- or Cl^- were superior to PF_6^- or TFO^-.[30] In this work, several nitriles were screened, and microwave irradiation was shown to improve the yield, probably due to shortening the reaction time at elevated temperature. The method proved to promote reactions with electron-rich and -poor nitriles and was showcased by application to a sartan intermediate **60**. General procedures for the use of microwave alone to promote the Finnegan reaction to prepare tetrazoles have appeared using reagents and solvents reported above.[31,32]

Use of mesopourous ZnS nanospheres (MZnSS) has been reported to promote the Finnegan reaction to prepare tetrazoles. The reactions still required high temperature (120 °C) and used catalytic amounts of MZnSS which could be re-used. This method proved fairly general and high-yielding as exemplified by the conversion of the amino-benzonitrile **61** into tetrazole **62**.[33]

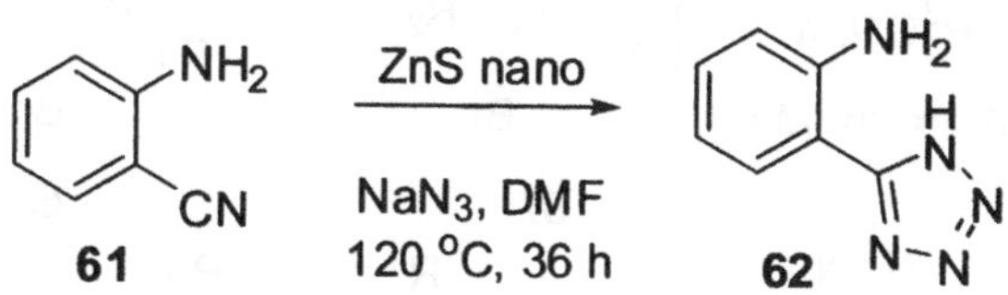

61 **62**

95%

Other solid supports have been reported to promote the Finnegan tetrazole reaction, like catalytic $FeCl_3$–SiO_2.[34] The procedure was shown to be general and the catalyst could be re-used.

The Finnegan tetrazole synthesis has also provided an opportunity to showcase fluorous chemistry.[35] In this instance, a polyfluorinated tin azide **64** was used in excess and reacted with electron-rich nitrile **63** to provide excellent yield of the tetrazole **65**. This procedure was quite general, providing yields of tetrazoles in modest to excellent yield.

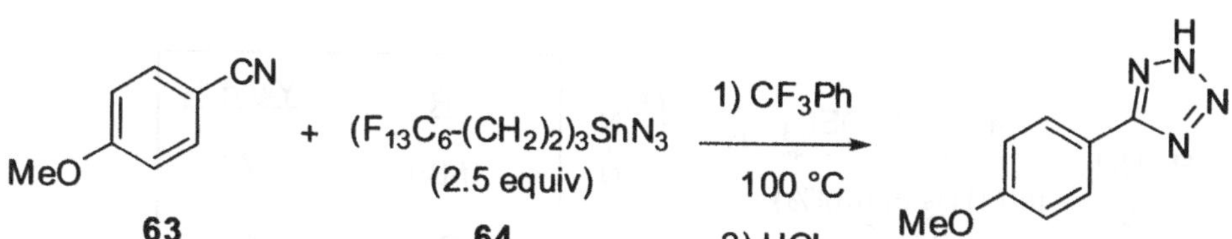

63 **64** **65**

95%

Reactions with in situ Generated Nitrile Equivalents

Activation of a nitrile with an electrophile other than an acid or Lewis acid has also shown applicability for entry into a 1,5-disubstituted tetrazole. A bromonium ion generated from an alkene **66** proved electrophilic enough to trap a nitrile which when treated with an azide source ($TMSN_3$) provided the bromo-alkyl tetrazole **69**. A broad range of alkenyl substrates and nitriles were shown to provide fair to excellent yield of the tetrazole.[36] Of course, this method is limited in that nucleophilic functionality cannot be present or it will likely react. Considering the substituents of the products generated, it would be difficult to cleanly provide these intermediates via direct nucleophilic addition to the tetrazole.

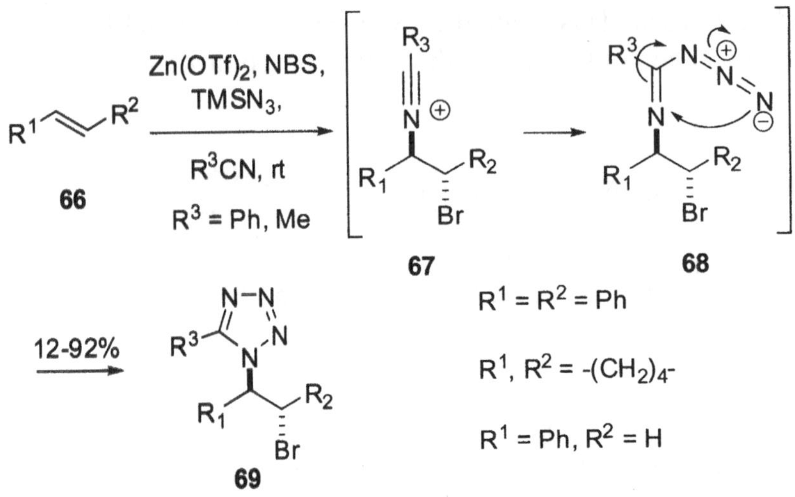

Generation of the nitrile or a nitrile equivalent has been successfully utilized for tetrazole formation. For example, reaction of an amine **70** with an orthoformate in the presence of NaN_3 and $In(OTf)_3$ was proposed to provide the azido-imine (Finnegan intermediate **22**), which cyclized to the tetrazole **71**. These conditions also proved general for a range of amines and provided the 1-substituted-5*H*-tetrazole.[37]

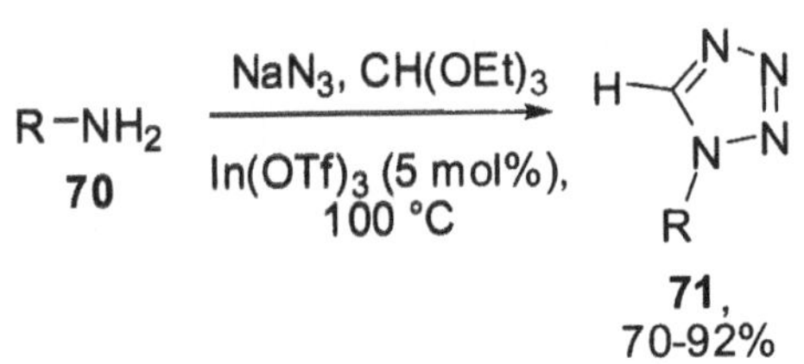

R	Yield %
t-Bu	70
Bn	85
EtO_2C-Ph	80
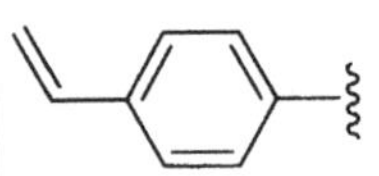	82

Pd-Promoted Selective Preparation of 2,5-Disubstituted Tetrazoles

Yamamoto and co-workers[38] reported the selective preparation of the 2-allyl tetrazole. The method was found to be general and provided moderate to good yields of 2-allyl-5-substituted tetrazoles. The proposed mechanism was formation of a Pd–N_3 bond by the reaction of $TMSN_3$ with methylallylcarbonate and Pd providing intermediate **73**. Subsequent [3 + 2] cycloaddition with nitrile **36** occurred, followed by allylation via reductive elimination at the point of Pd attachment shown as intermediate **74**. The paper mainly described reaction of heteroatom bonded nitriles (heteroatom–CN) and, interestingly, malononitriles. The reaction provided good yields for mono-tetrazole formation in the case of bis-nitriles. Simple alkyl nitriles such as valeronitrile did not react to form the desired tetrazole, and *p*-O_2NPhCN provided a modest 40% yield. A series of *N*-cyano-indoles were described and when EWG were attached to the indole ring, the reactions proved generally faster and yields higher than when electron donating groups were present. A brief screen of catalysts and solvents was described and the best catalyst and solvent were tri-2-furylphosphine ligated Pd and octane, respectively, with about 2 mol% Pd loading. Modest to good yields were obtained of the 2-allyl-5-substituted tetrazole **75**.

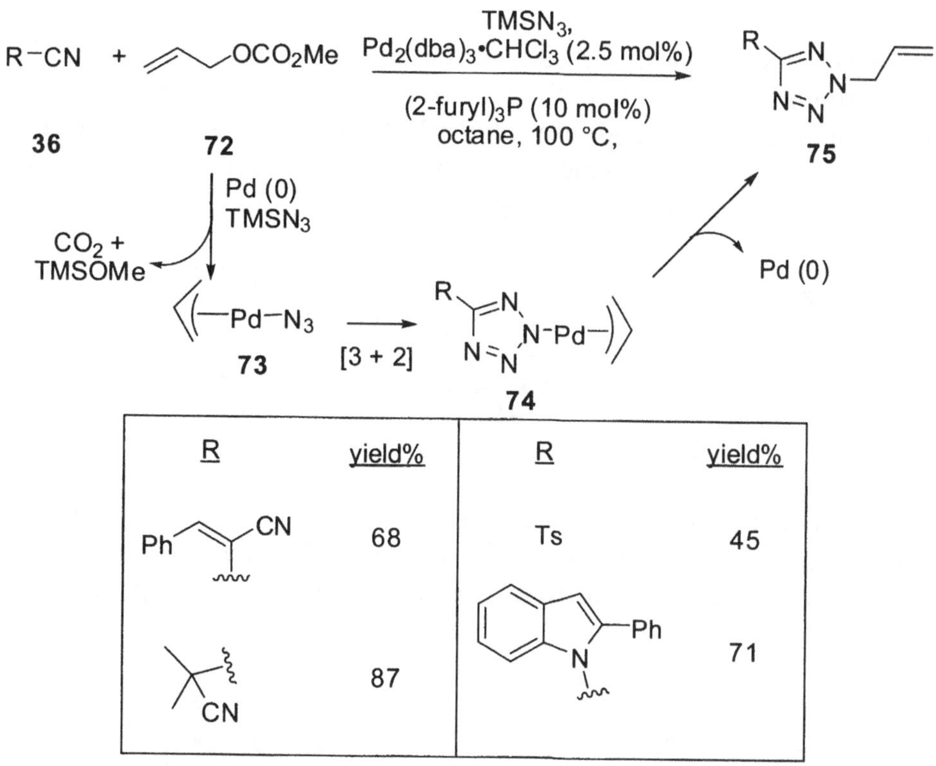

5.3.5 *Synthetic Utility*

Tetrazoles as Potential Therapeutic Agents

The utility of the Finnegan tetrazole synthesis is derived in part due to the tetrazole moiety itself. Tetrazoles have been used to prepare a broad range of potential therapeutic agents due to the claim that they are carboxylic acid surrogates.[2] In this arena, tetrazoles have been used to support medicinal efforts for ACAT inhibitors,[39] as neuroprotective agents,[40] and a broad range of potential therapeutic agents in medicinal chemistry efforts. Specifically for the ACAT inhibitors, the tetrazole was generated either by reaction of nitriles

75 or **76** under the standard Finnegan protocol (NaN_3, NH_4Cl, DMF) or pyridinium trifluoroacetate to provide tetrazoles **77** or **78**. In either case, the tetrazole-containing building blocks were further elaborated into materials for testing.

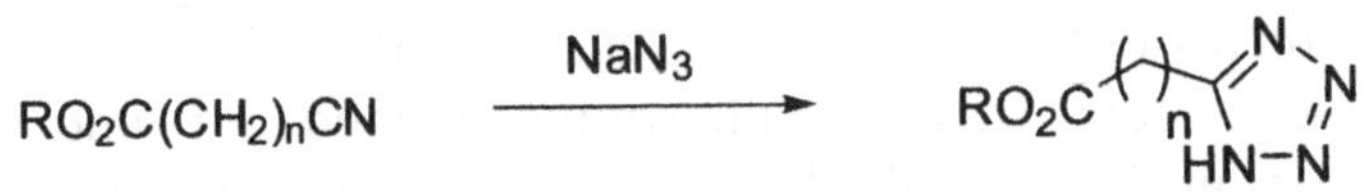

75, R = Et, n = 0 TFA, pyr, 60 °C **77**, R = Et, n = 0

76, R = Me, n = 2 NH_4Cl, DMF, 100 °C **78**, R = Me, n = 2

Toward the synthesis of sartans, Galante described the use of Zn halides to promote the formation of the tetrazole. In this work, the author also described the preparation and use of $Zn(N_3)_2$ bis-pyridine complex and mentioned the successful implementation of other Lewis acids like $MgCl_2$ to give the desired product **80**, all in very good yield.[41]

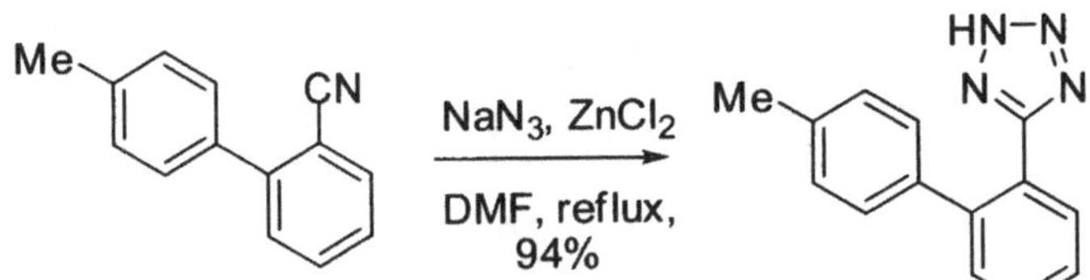

79 **80**

Glycosyltetrazoles were prepared as inhibitors dehydroquinate,[42] which would potentially be used as antibacterial or herbicidal agents. The traditional Finnegan tetrazole method was used and enabled preparation of these substrates.

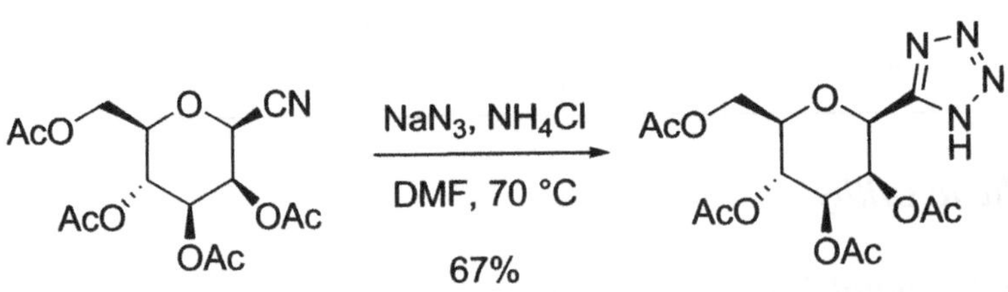

81 **82**

Sharpless's click methodology has led to application in the preparation of glycoclusters of calixarene.[43] With the high-yielding formation of tetrazole **84** due in part to the EWG on the CN group, tosyl substitution with an alkoxide enabled easy attachment with further elaborations to the calixarene.

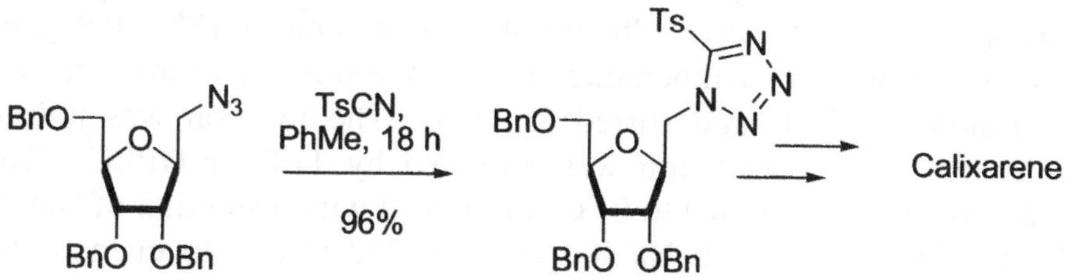

83 84

Safety

While the use of tetrazoles as potential propellants or explosives was mentioned, the energy of the intermediate and concerns of the generation of hydrazoic acid and subsequent issues derived therefrom should be mentioned. Popular Finnegan tetrazole conditions employed NH_4Cl and NaN_3 in DMF at elevated temperature. Under such conditions, ammonium azide is formed, which can sublime. Ammonium azide is described in Bretherick's as being explosive on rapid heating and is also friction and impact sensitive. Other conditions used for the Finnegan tetrazole synthesis specifically considering the workup of some reactions leads to the formation of hydrazoic acid, which can be handled safely as a dilute solution yet is toxic, violently explosive, and varibly sensitive in > 50% purity. Under many of these reaction conditions, hydrazoic acid can be detected in the headspace and proper precautions should be taken to avoid accumulation. Trace amounts of heavy metals can also catalyze the rapid degradation of hydrazoic acid.[44] Other salts of N_3^- have been shown to be less prone to sublimation and less energetic or have an onset temperature that allows safe engineering of the reaction. When planning to run the Finnegan reaction, thoughts on reaction mechanism and safety concerns should be in the forefront.

5.3.6 *Experimental*

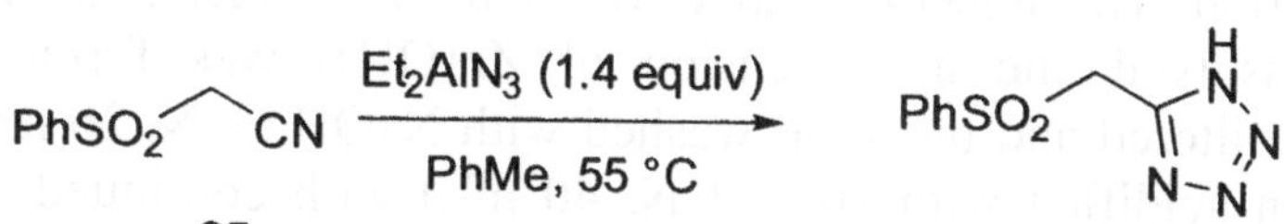

85 86

Preparation of 5-Phenylsulfonylmethyl-1*H*-tetrazole with Diethyl Aluminum Azide[34]

An oven-dried flask was charged with NaN_3 (5.46 g, 84 mmol) and Et_2AlCl (46.7 mL, 84 mmol, 2.7 M in PhMe) at 0 °C under inert conditions and stirred for 15 min. The white heterogeneous mixture was then warmed to room temperature and stirred for 4–6 h. During the formation of the reagent, a suspension of NaCl was formed. Phenylsulfonyl acetonitrile (**85**, 10.9 g, 60 mmol) was added at room temperature in two portions. The mixture was gradually heated to 55 °C and stirred for 2 h. The reaction was judged complete when > 98% conversion was observed by TLC or HPLC. The reaction mixture was then cooled to 0 °C and treated with a solution of NaOH (15%, 67 mL, 252 mmol) with $NaNO_2$ (17.4 g, 252 mmol; to quench any hydrazoic acid in the reaction mixture). The pH of the aqueous phase was adjusted to 1.5 (from pH ~13) using HCl (6 M). The solution was then extracted with EtOAc (3 × 20 mL), which was concentrated to dryness to provide crude product. Crude product was redissolved in EtOAc (30 mL) and extracted with K_2CO_3 (aq. 10%, 3 × 25 mL). The combined aqueous phase was cooled to 0 °C and treated with HCl (6 M) to adjust the pH to ~2.5. Extraction of this acidified solution with EtOAc (3 × 20 mL), drying, and evaporation was followed by recrystallization from EtOAc/PhMe and gave 10.2 g of **86**, 76% yield.

Preparation of 5-(4-Methoxyphenyl)Tetrazole Using Zinc Halide Method in Water[3]

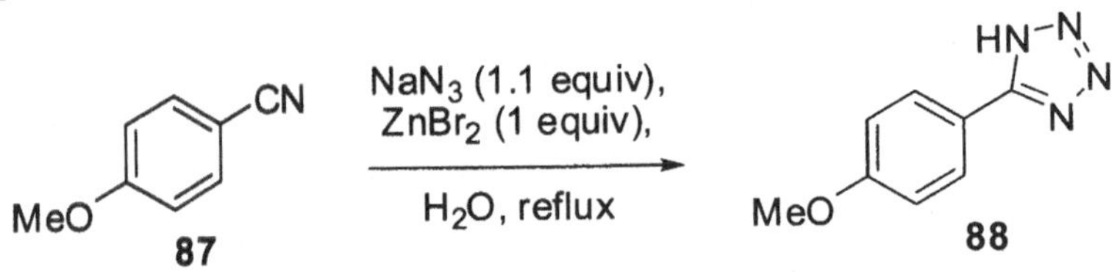

To a 250-mL round-bottom flask was added the nitrile (**87**, 20 mmol), NaN_3 (1.43 g, 22 mmol), $ZnBr_2$ (4.5 g, 20 mmol), and water (40 mL). The reaction mixture was refluxed for 24 h with vigorous stirring. HCl (3 N, 30 mL) was added followed by the addition of EtOAc (100 mL), and vigorous stirring continued until all solids were dissolved and the pH was ~ 1. The organic phases were separated, and the aqueous phase extracted with EtOAc (2 × 100 mL). The combined organic layers were concentrated *in vacuo* and the residue was stirred with aqueous NaOH (0.25 M, 200 mL) until all initial solids were dissolved and a suspension of $Zn(OH)_2$ was formed. The suspension was filtered and the solid washed with NaOH (1 N, 20 mL). The filtrate was then acidified with HCl (3 N, 40 mL) with continued vigorous stirring, which caused the tetrazole **88** to precipitate. The tetrazole **88** was filtered and washed with 3 N HCl (2 × 20 mL) and dried to give a white to off-white solid 3.04 g, 86% yield.

5.3.7 *References*

1 Finnegan, W. G.; Henry, R. A.; Lofquist, R. *J. Am. Chem. Soc.* **1958,** *80*, 3908–3911.
2 [R] (a) Singh, H.; Chawla, A. S.; Kapoor, V. K.; Paul, D.; Malhotra, R. K. *Prog. Med. Chem.* **1980,** *17,* 151–183. (b) Ek, F.; Wistrand, L.-G.; Frejd, T. *Tetrahedron* **2003,** *59*, 6759–6769.

3 Demko, Z. P.; Sharpless, K. B. *J. Org. Chem.* **2001,** *66,* 7945–7950.
4 Chafin, A.; Irvin, D. J.; Mason, M. H.; Mason, S. L. *Tetrahedron Lett.* **2008,** *49*, 3823–3826.
5 Joo, Y.-H.; Shreeve, J. M. *Org. Lett.* **2008,** *10*, 2665–4667.
6 (a) Brigas, A. F. *Product Class 30: Tetrazoles* in *Science of Synthesis*, Category 2, Volume 13, Georg Thieme Verlag, Stuttgart, GE, **2004,** Storr, R. C. and Gilchrist, T. L., Eds. (b) Butler, R. N. in *Comprehensive Heterocyclic Chemistry II*, *vol. 4*, 621–678; Pergamon, New York, **1996.**
7 Huisgen, R. *Angew. Chem., Int. Ed. Engl.* **1963,** *10*, 565–632.
8 [R] Ostrovskii, V. A.; Poplavskii, V. S.; Koldobskii, G. I.; Erusalimskii, G. B. *Chem. Heterocycl. Compds* **1992,** *28*, 1027–1030.
9 Demko, Z. P.; Sharpless, K. B. *Angew. Chem., Int. Ed. Engl.* **2002,** *41*, 2110–2113; *ibid*, 2113–2116.
10 Clémençon, I. F.; Ganem, B. *Tetrahedron* **2007,** *63*, 8665–8669.
11 Himo, F.; Demko, Z. P.; Noodlemen, L.; Sharpless, K. B. *J. Am. Chem. Soc.* **2003,** *125*, 9983–9987.
12 Himo, F.; Demko, Z. P.; Noodlemen, L.; Sharpless, K. B. *J. Am. Chem. Soc.* **2002,** *124*, 12210–12216.
13 Jursic, B. S.; Zdravkovski, Z. *J. Mol. Structure (Theochem)* **1994,** *312*, 11–22.
14 Duncia, J. V.; Pierce, M. E.; Santella, J. B. III *J. Org. Chem.* **1991,** *56*, 2395–2400.
15 Wittenberger, S. J.; Donner, B. G. *J. Org. Chem.* **1993,** *58*, 4139–4141.
16 Amantini, D.; Beleggia, R.; Fringuelli, F.; Pizzo, F.; Vaccaro, L. *J. Org. Chem.* **2004,** *69*, 2896–2898.
17 Huff, B. E.; Staszak, M. A. *Tetrahedron Lett.* **1993,** *34*, 8011–8014.
18 Jin, T.; Kiahara, F.; Kamijo, S.; Yamamoto, Y. *Tetrahedron Lett.* **2008,** *49*, 2824–2827.
19 Bosch, L.; Vilarrasa, J. *Angew. Chem., Int. Ed.* **2007**, *46*, 3926–3930.
20 Koguro, K.; Oga, T.; Mitsui, S.; Orita, R. *Synthesis* **1998,** 910–914.
21 Bridge, A. W.; Jones, R. H.; Kabir, H.; Kee, A A.; Lythgoe, D. J.; Nakach, M.; Pemberton, C.; Wrightman, J. A. *Org. Process Res. Dev.* **2001,** *5*, 9–15.
22 Aureggi, V.; Franckevičius, V.; Kitching, M. O.; Ley, S. V.; Longbottom, D. A.; Oelke, A. J.; Sedelmeier, G. *Org. Synth.* **2008,** *85*, 72–87.
23 Aureggi, V.; Sedelmeier, G. *Angew. Chem., Int. Ed.* **2007,** *46*, 8440–8444.
24 Demko, Z. P.; Sharpless, K. B. *Org. Lett.* **2002,** *4*, 2525–2527.
25 Demko, Z. P.; Sharpless, K. B. *Org. Lett.* **2001,** *3*, 4091–4095.
26 Davis, B.; Brandstetter, T. W.; Smith, C.; Hackett, L.; Winchester, B. G.; Fleet, G. W. J. *Tetrahedron Lett.* **1995,** *36*, 7507–7510.
27 Anegundi, R. I.; Puranik, V. G.; Hotha, S. *Org. Biomol. Chem.* **2008,** *6*, 779–786.
28 Hanessian, S.; Simard, D.; Deschênes-Simard, B.; Chenel, C.; Haak, E. *Org. Lett.* **2008,** *10*, 1381–1384.
29 Dulcere, J.-P.; Tawil, M.; Santelli, M. *J. Org. Chem.* **1990,** *55*, 571–575.
30 Schmidt, B.; Meid, D.; Kieser, D. *Tetrahedron* **2007,** *63*, 492–496.
31 Alterman, M.; Hallberg, A. *J. Org. Chem.* **2000,** *65*, 7984–7989.
32 Roh, J.; Artamonova, T. V.; Vávorová, K.; Koldobskii, G. I.; Hrabálek, A. *Synthesis* **2009,** 2175–2178.
33 Lan, L.; Li, B.; Liu, W.; Jiang, L.; Xu, Z.; Yin, G. *Chem. Commun.* **2010,** *46*, 448–450.
34 Nasrollahzadeh, M.; Bayat, Y.; Habibi, D.; Moshaee, S. *Tetrahedron Lett.* **2009,** *50*, 4435–4438.
35 Curran, D. P.; Hadida, S.; Kim, S.-Y. *Tetrahedron* **1999,** *55*, 8997–9006.
36 Hajra, S.; Sinha, D.; Bhowmick, M. *J. Org. Chem.* **2007,** *72*, 1852–1855.
37 Kundu, D.; Majee, A.; Hajra, A. *Tetrahedron Lett.* **2009,** *50*, 2668–2670.
38 Kamijo, S.; Jin, T.; Yamamoto, Y. *J. Org. Chem.* **2002,** *67*, 7413–7417.
39 O'Brien, P. M.; Sliskovic, D. R.; Picard, J. A.; Lee, H. T.; Purchase, C. F. II, Roth, B. D.; White, A. D.; Anderson, M.; Mueller, S. B.; Bocan, T.; Bousley, R.; Hamelehle, K. L.; Homan, R.; Lee, P.; Krause, B. R.; Reindel, J. F.; Stanfield, R. L.; Turluck, D. *J. Med. Chem.* **1996,** *39*, 2354–2366.
40 Hung, K.-Y.; Harris, P. W. R.; Brimble, M. A. *Synlett* **2009,** 1233–1236.
41 Galante, R. J. US Pat. 5502191A, 1996.
42 Buchanan, J. G.; Clelland, A. P. W.; Johnson, T.; Rennie, R. A. C.; Wightman, R. H. *J. Chem. Soc., Perkin Trans. 1*, **1992,** 2593–2601.

43 Dondoni, A.; Marra, A. *Tetrahedron* **2007,** *63*, 6339–6345.
44 Pitt, M. J. In *Bretherick's Handbook of Reactive Chemical Hazards, 6th ed., vol. 1;* Butterworth Heinemann, Oxford, UK, 1999, pps. 1602–1603, 1685.

5.4 Hantzsch Thiazole Synthesis

James Kempson

5.4.1 Description

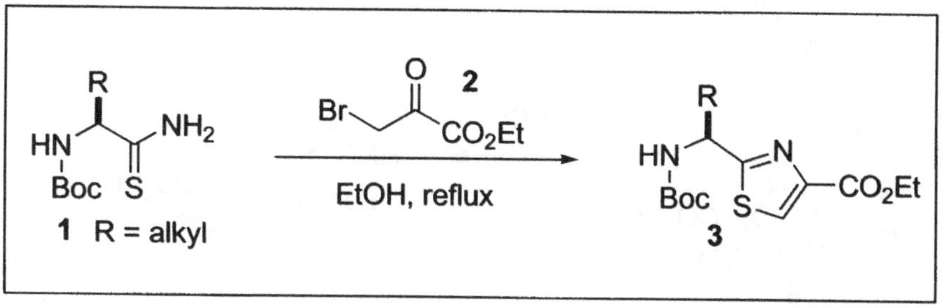

The Hantzsch reaction, discovered in 1889, remains one of the most reliable routes to thiazoles.[1,2] The reaction involves a [3 + 2] atom cyclization between thioamide **1**, and an α-halo carbonyl compound **2** and is one of the most direct routes to thiazoles. The reaction can also be carried out with thioureas, thiosemicarbazides and other compounds containing the —N—C=S structural unit.[3–32]

5.4.2 Mechanism

The reaction proceeds by way of nucleophilic attack by sulfur on the carbon atom bearing the halogen; the acylic intermediate formed in this way has been isolated in a few cases. The thiazoles are normally obtained by treating the components together in ethanol or a similar solvent and lead to excellent yields of the heterocycle. However, since this reaction generates one equivalent of hydrogen bromide, significant loss of optical purity can occur with substrates bearing an α-chiral center.

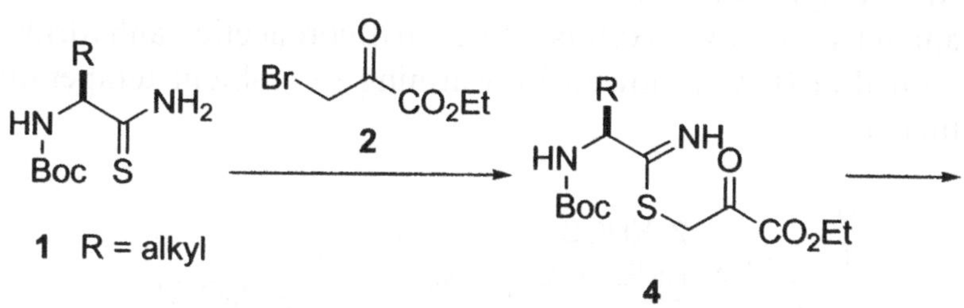

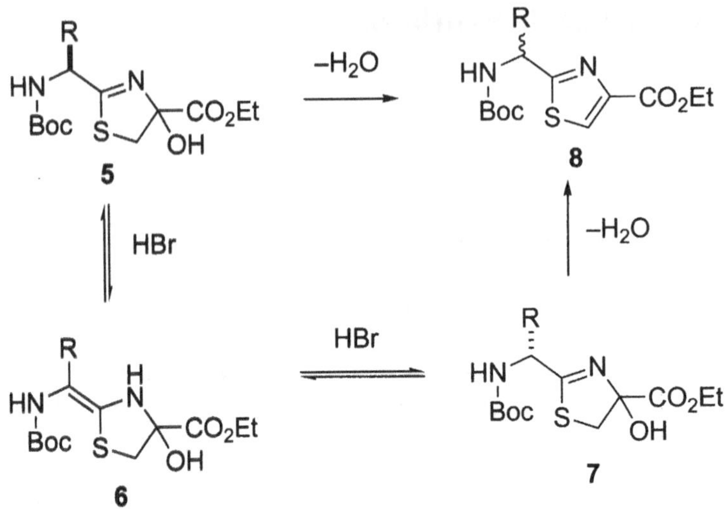

5.4.3 *Variations and Improvements*

Due to the problem of racemization, a number of attempts to adapt the synthesis have been made. Schmidt decided to limit the exposure time of the hydroxythiazoline to the racemizing effect of the hydrobromic acid by adding the dehydrating agent, trifluoroacetic anhydride, to increase the rate of conversion of the intermediate thiazoline to the thiazole.[33] In addition, the reaction was carried out at –10 °C. The combined effect of these modifications, was the production of thiazoles with enantiomeric excesses of up to 90%. Kelly and co-workers published the use of calcium carbonate to remove the unwanted hydrobromic acid and Holzapfel in work published between 1990 and 1992 later combined all of the aforementioned modifications which formed the basis of the widely used modified Hantzsch condensation for the formation of thiazoles.[34,35] In a widely used procedure, sodium hydrogen carbonate or potassium hydrogen carbonate is stirred in an aprotic solvent, such as DME, along with the thioamide, and the solution is then treated with ethyl bromopyruvate at 0 °C. When the reaction contains significant amounts of hydroxythiazoline, trifluoroacetic anhydride and pyridine are added at 0 °C followed by warming to ambient temperature to produce the thiazole.

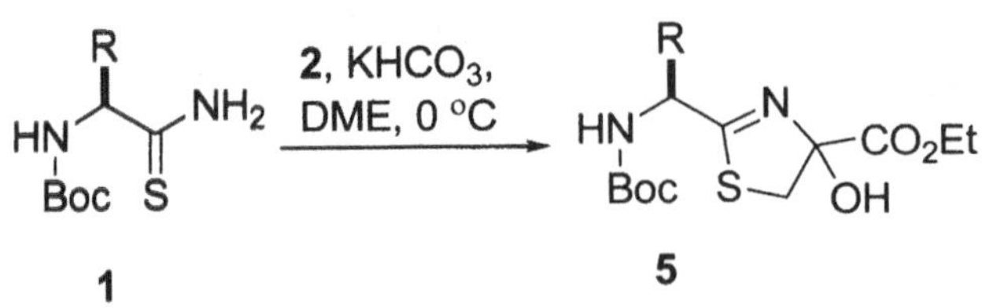

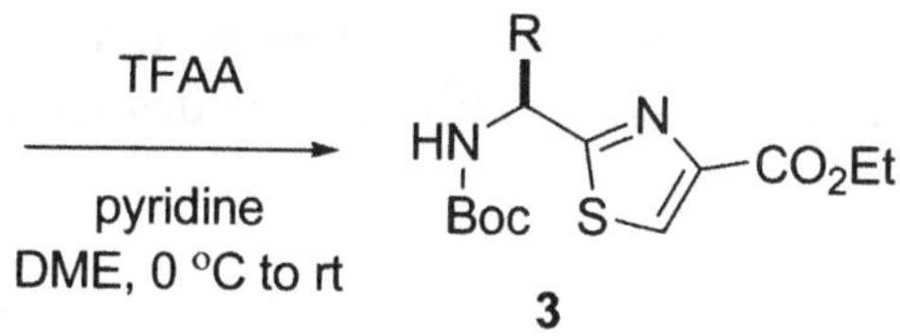

This method was successful in producing a wide range of optically pure thiazoles; however, with some amino acids a small amount of racemization was still observed. Further studies by Meyers and co-workers examined both the initial hydroxythiazoline formation as well as several variations in the dehydration step. Optimal conditions were to conduct the hydroxythiazoline formation step at –15 °C; elimination of the intermediate hydroxythiazoline used a 2,6-dimethylpyridine and TFAA mixture while maintaining the temperature at –15 °C. This method produced the valine derived thiazole with an enantiomeric excess of >99% and Meyers subsequently used this method to complete a total synthesis of bistatramide C.[36,37]

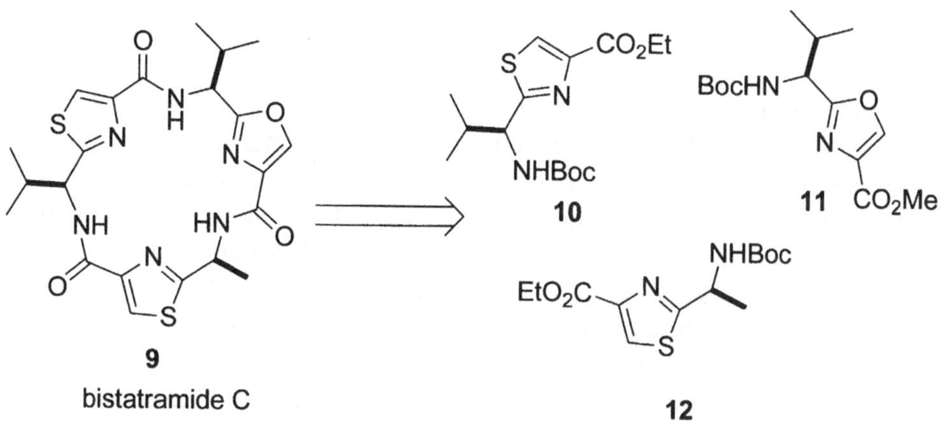

These general conditions have been employed in a wide variety of synthetic campaigns, including the synthesis of several heterocyclic natural products such as nostacyclamide, the promothiocins, the *Lissoclinum* cyclic peptides, and more recently GE2270A.[38–41]

Nicolaou and co-workers revisited these general conditions during their landmark total synthesis of thiostrepton.[42,43] Notable features in ring construction during this synthesis include higher reaction temperatures than those used in the Meyers method and the isolation of the crude hydroxythiazoline before the elimination step, which was then conducted employing TFAA in the presence of pyridine followed by triethylamine. These reaction conditions also led to *N*-trifluoroacetylation of the free –NH group in the target molecule and so subsequent treatment with sodium methoxide in methanol was necessary to liberate the desired product.

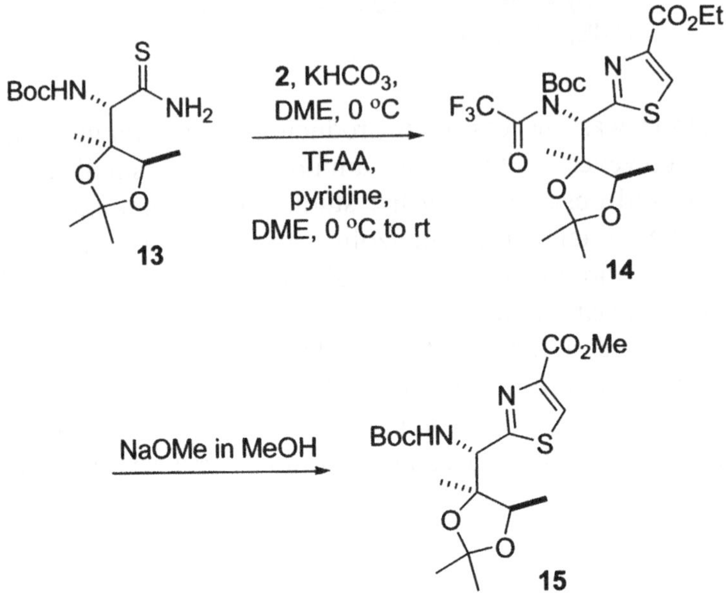

Bagley and co-workers have been the latest to reinvestigate the Hantzsch synthesis of the valine-derived thiazole during their synthetic efforts toward the natural product micrococcin P1.[44,45] This reinvestigation stemmed from either the low yield of the desired product on small scale or the compromised optical purity on larger scale reactions, findings that were also noted in Pattenden's earlier work.[46] The Meyers and Nicolaou conditions were the main focus of these subsequent studies. Meyers modification of the Hantzsch thiazole synthesis was performed at –18 °C but surprisingly delivered the trifluroroacetamide. However, subsequent treatment with sodium ethoxide in ethanol, according to Nicolaou's procedure, led to the isolation of **10** as a single enantiomer. Although the length of the sequence had been increased, this approach proved to be a reliable and high-yielding route to optically pure **10**. Efforts to circumvent the formation of the *N*-trifluoroacetyl thiazole and thus remove the need for a subsequent solvolysis step were tried using the elimination conditions of methanesulfonyl chloride and triethylamine but met with failure due to poor reaction yield and compromised enantiomeric excesses at higher reaction temperatures.

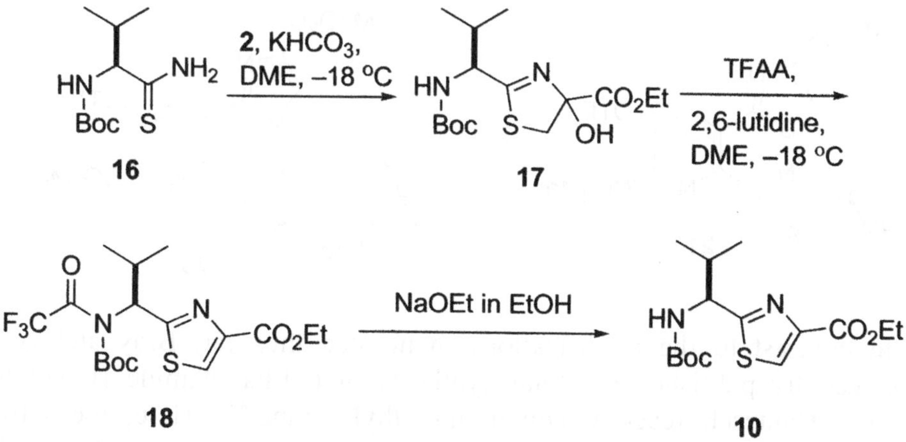

The Holzapfel–Meyers–Nicolaou modification is a significant improvement over the traditional Hantzsch conditions. However, racemization can still occur in some circumstances. For example, the two-step Hantzsch reaction of thioamide **20** with bromide **19** furnishes thiazole **21** with 60–85% enantiomeric excess due to partial epimerization at the α-stereogenic center. The epimerization issue is obviated when thioamide **20** is replaced with the ketal-protected thioamide **22**. Reaction of **19** with **22** delivers **23** with good optical purity (> 96% *ee*). This thiazole represents the core structure of the potent thiopeptide antibiotic nosiheptide.[47,48]

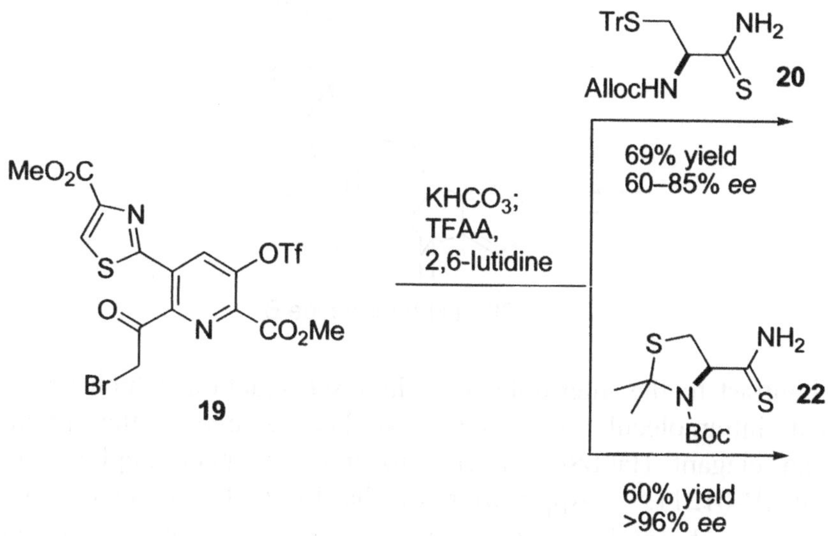

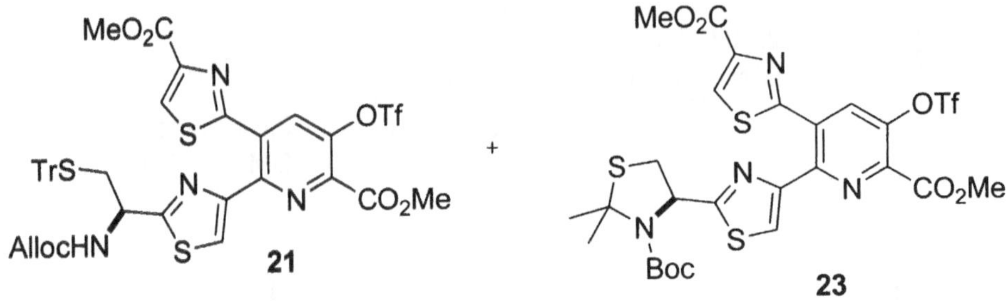

In contrast to the modifications mentioned thus far, Bray and co-workers recently published the total synthesis of (+)-bacillamide B which relied on a Hantzsch reaction containing ethyloxirane.[49] Here, the HBr generated *in situ* was trapped by the ethyloxirane, forming a bromohydrin, and prevented any epimerization. It is interesting to note the elevated temperatures in this synthesis without any loss of chiral integrity.

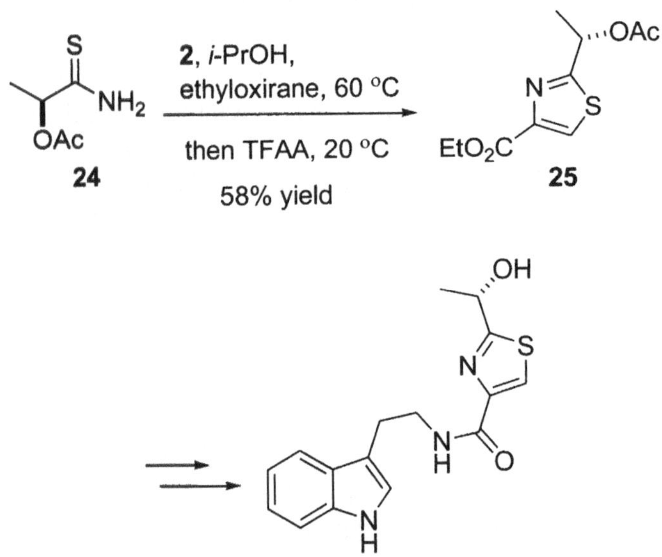

26 (+)-bacillamide B

In contrast to the intermolecular Hantzsch reaction, which has been widely used, intramolecular examples have been scarce in the literature. Recently, an elegant Hantzsch macrocylization has been applied to the synthesis of IB-01211.[50] Application of the Holzapfel–Meyers–Nicolaou modification to α-bromoketone-thioamide **27** brings about intramolecular thiazole formation with concomitant dehydration of the primary alcohol to give IB-01211 in moderate yield. This reaction is remarkable, especially considering the fact that the macrocyclization through amide formation at either bond a or b fails to produce any cyclized product.

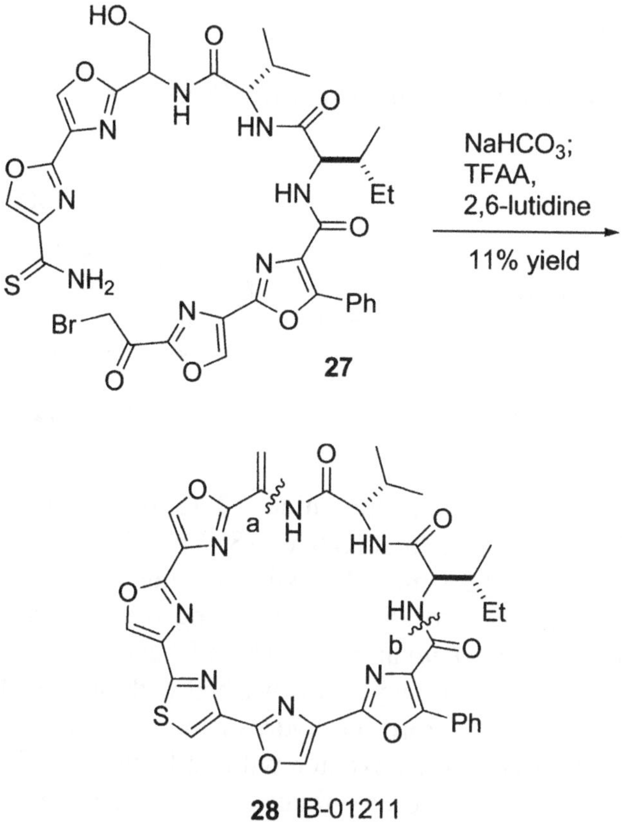

Other progress in the Hantzsch thiazole synthesis includes α,α-dibromoketones as a superior alternative to α-bromoketones.[51] Indeed, the increased reactivity of the α,α-dibromoketone **30** was able to obtain an 82% yield of **31** in 15 min at room temperature, compared to refluxing in ethanol for 5 h with the corresponding α-bromoketone to obtain the same yield of thiazole **31**.

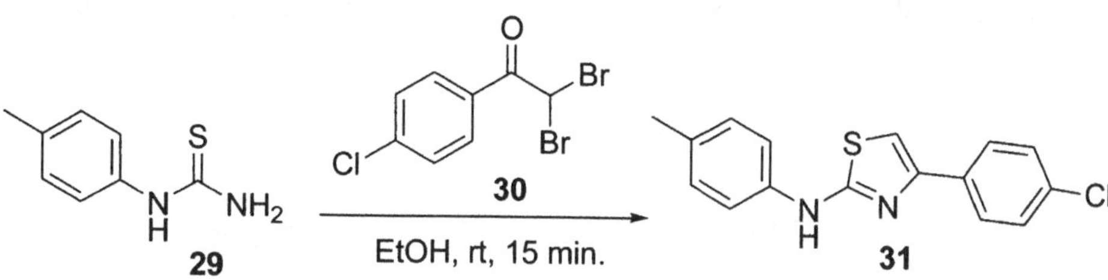

5.4.4 *Experimental*

Holzapfel–Meyers–Nicolaou Modified Hantzsch Procedure

Step1

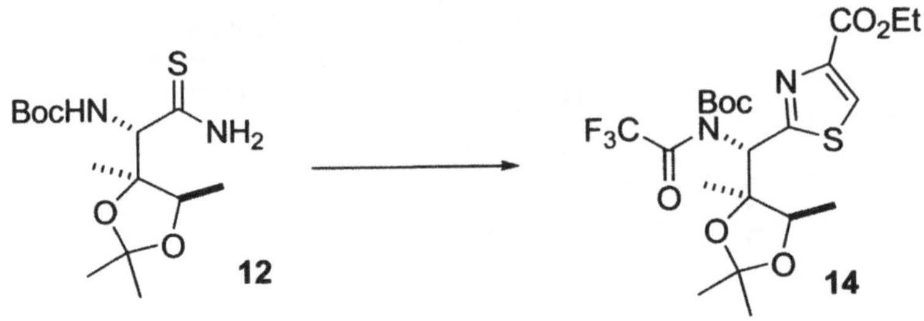

To a cooled (0 °C) solution of thioamide **12** (346 mg, 1.09 mmol) in anhydrous DME (5.4 mL) was added $KHCO_3$ (873 mg, 8.72 mmol) and ethyl bromopyruvate (0.41 mL, 3.26 mmol). After stirring for 4 h at 0 °C, the reaction was allowed to warm to 25 °C, and stirring was continued for 16 h. The solvent was removed *in vacuo* and ether (50 mL) and water (50 mL) were added to the residue. The organic layer was washed with brine (50 mL), dried ($MgSO_4$) and concentrated to afford the hydroxy thiazoline intermediate as a viscous liquid that was used in the next step without further purification. To a cooled (0 °C) solution of the hydroxy thiazoline intermediate in DME (5.4 mL) was added pyridine (0.71 mL, 8.71 mmol) and trifluoroacetic anhydride (0.62 mL, 4.31 mmol) dropwise. After stirring for 2 h at 0 °C and then for 1 h at 25 °C, Et_3N (4 mL) was added to the reaction mixture (pH ~ 8), and the solvent was removed *in vacuo*. Ether (50 mL) and water (50 mL) were added to the residue and the organic layer was washed with aqueous 5% HCl solution (50 mL), saturated aqueous $NaHCO_3$ solution (50 mL) and brine (50 mL) and then dried ($MgSO_4$) and concentrated. Purification of the residue by flash column chromatography (silica gel, hexanes:EtOAc, 9:1) afforded thiazole ethyl ester **14** (456 mg, 82% over two steps from **12**) as a colorless oil.

Step 2

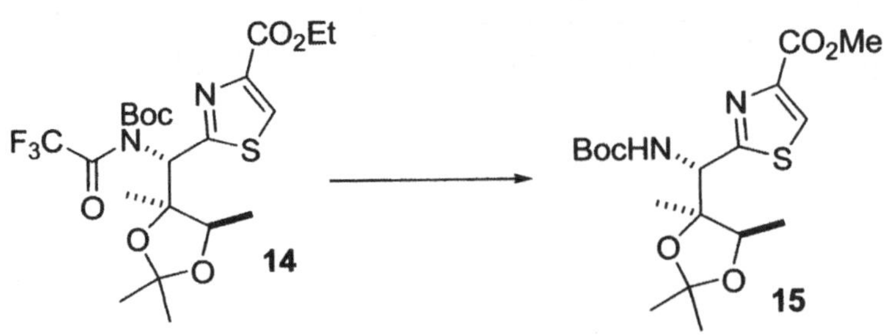

To a cooled (0 °C) solution of thiazole ethyl ester **14** (450 mg, 0.88 mmol) in dry MeOH (4.5 mL) was added NaOMe (153 mg, 2.82 mmol). After stirring for 5 h at 0 °C, the reaction mixture was quenched with saturated NH_4Cl solution (50 mL), and the resulting mixture was extracted with EtOAc (2 × 50 mL). The combined organic layers were washed with brine (50 mL), dried ($MgSO_4$) and concentrated. Purification of the residue by flash column chromatography (silica gel, hexanes/EtOAc, 6:4) afforded thiazole methyl ester **15** (320 mg, 91%) as a white foam.

5.4.5 References

1 Hantzsch, A.; Weber, J. H. *Ber. Dtsch. Chem. Ges.* **1887,** 3118–3119.
2 Traumann, V. *Liebigs Ann. Chem.* **1888,** *249*, 31–33.
3 [R] Vernin, G. *Thiazole and its Derivatives* Wiley, New York, **1979**; *Part 1*, Chapter 2.
4 De Kimpe, N.; Declercq, J. *Tetrahedron* **1993,** *49*, 3411–3415.
5 Sarkis, G. Y.; Al-Azawe, S. *J. Chem. Eng. Data* **1973,** *18*, 99–101.
6 Dash, B. C.; Nandi, B. B. *J. Indian Chem. Soc.* **1979,** *LVI*, 70–73.
7 Singh, H.; Ahuja, A. S. *J. Indian Chem. Soc.* **1979,** *18B*, 534–535.
8 Liebscher, J.; Pätzel, M.; Bechstein, U. *Synthesis* **1989,** 968–970.
9 Mohareb, R. M.; Shams, H. Z.; Aziz, S. I. *Sulfur Lett.* **1991,** *13*, 101–103.
10 Khazi, I. M.; Mahajanshetti, C. S. *Monatsh. Chem.* **1995,** *126*, 759–761.
11 Kidwai, M.; Kumar, R. *Gazz. Chim. Ital.* **1997,** *127*, 263–265.
12 Kasmi, S.; Hamelin, J.; Benhaoua, H. *Tetrahedron Lett.* **1998,** *39*, 8093–8095.
13 Binu, R.; Thomas, K. K.; Jenar-Danan, G. C.; Rajasekharan, K. N. *Org. Prep. Proced. Int.* **1998,** *30*, 93–98.
14 Demirayak, S.; Karaburun, A. C.; Mohsen, U. A.; Guven, K. *Acta Pharm. Turc.* **1999,** *41*, 78–81.
15 Rudolph, J. *Tetrahedron* **2000,** *56*, 3161–3165.
16 Suni, M. M.; Nair, V. A.; Joshua, C. P. *Synlett* **2001,** 409–411.
17 Behera, G. B.; Acharya, R. C.; Rout, M. K. *Indian J. Chem.* **1973,** *11*, 82–85.
18 Schäfer, V. H.; Gewald, K. *J. Prakt. Chem.* **1974,** *316*, 684–686.
19 Meslin, J. C.; Quiniou, H. *Tetrahedron* **1975,** *31*, 3055–3060.
20 Rajappa, S. *Heterocycles* **1977,** *7*, 507–510.
21 Lin, Y.; Seifert, C. M.; Kang, S. M.; Dusza, J. P.; Lang, S. A. *J. Heterocyclic Chem.* **1979,** *16*, 1377–1380.
22 Reliquet, A.; Meslin, J.; Reliquet, F. *Sulfur Lett.* **1978,** *7*, 49–51.
23 Harode, R.; Sharma, T. C. *J. Indian Chem. Soc.* **1989,** *66*, 282–284.
24 Rober, P.; Gully, D.; Courtemanche, G.; Gautier, C.; Geslin, M.; Wermuth, C. Eur. Pat. Appl. EP 659,747, **1995.**
25 Farag, A. M.; Dawood, K. M.; Kandeel, Z. E.; Algharib, M. S. *J. Chem. Res.* **1996,** 530–534.
26 Feng, Y.; Zhang, X.; Zhang, W. *Hecheng Huaxue.* **1997,** *5*, 269–275.
27 Amschler, H.; Martin, T.; Flockerzi, D.; Gutterer, B.; Thibaut, U.; Hatzelmann, A.; Boss, H.; Hafner, D.; Kley, H.; Beume, R.; Bar, T.; Ulrich, W. PCT Int. Appl. WO 9808,830, 1998.
28 Kalluraya, B.; Gunaga, P.; Ramana, M. V. *Indian J. Heterocyclic Chem.* **1999,** *8*, 241–245.
29 Dovlatyan, V. V.; Eliazyan, K. A.; Pivazyan, V. A.; Ghazaryan, E. A.; Engoyan, A. P.; Grigoryan, R. T.; Mirzoyan, R. G. *Chem. Heterocycl. Compd.* **2000,** *36*, 593–600.
30 Brown, M. D.; Gillon, D. W.; Meakins, G. D.; Whitman, G. A. *J. Chem. Soc., Chem. Commun.* **1982,** 444–448.
31 Fuchigami, T.; Yeh, M. Y.; Nonaka, T.; Tien, H. J. *Bull. Chem. Soc. Jpn.* **1983,** *56*, 3851–3854.
32 Fuchigami, T.; Nonaka, T. *J. Org. Chem.* **1983,** *48*, 3340–3345.
33 Schmidt, U.; Gleich, P.; Gleisser, H.; Utz, R. *Synthesis* **1986,** 992–995.
34 Kelly, R. C.; Gebhard, I.; Wicnienski, N. *J. Org. Chem.* **1986,** *51*, 4590–4593.
35 Bredankemp, M. W.; Holzapfel, C. W.; Vanzyl, W. J. *Synth. Commun.* **1990,** *20*, 2235–2236.

36 Aguilar, E.; Meyers, A. I. *Tetrahedron Lett.* **1994,** *35*, 2473–22476.
37 Aguilar, E.; Meyers, A. I. *Tetrahedron Lett.* **1994,** *35*, 2477–22480.
38 Moody, C. J.; Bagley, M. C. *J. Chem. Soc., Perkin Trans. 1* **1998,** 601–604.
39 Bagley, M. C.; Bashford, K. E.; Hesketh, C. L.; Moody, C. J. *J. Am. Chem. Soc.* **2000,** *122*, 3301–3306.
40 Bertram, A.; Blake, A. J.; Gonzàlez-López de Turiso, F.; Hannam, J. S.; Jolliffe, K. A.; Pattenden, G.; Skae, M *Tetrahedron* **2003,** *59*, 6979–6983.
41 Müller, H. M.; Delgado, O.; Bach, T. *Angew. Chem., Int. Ed.* **2007,** *46*, 4771–4776.
42 Nicolaou, K. C.; Safina, B. S.; Zak, M.; Lee, S. H.; Nevalainen, M.; Bella, M.; Estrada, A. A.; Funke, C.; Zécri, F. J.; Bulat, S. *J. Am. Chem. Soc.* **2005,** *127*, 11159–11175.
43 Nicolaou, K. C.; Zak, M.; Safina, B. S.; Estrada, A. A.; Lee, S. H.; Nevalainen, M. *J. Am. Chem. Soc.* **2005,** *127*, 11176–11180.
44 Merritt, E. A.; Bagley, M. C. *Synlett* **2007,** *6*, 954–957.
45 Merritt, E. A.; Bagley, M. C. *Synthesis* **2007,** *22*, 3531–3541.
46 Boden, C. D. J.; Pattenden, G.; Ye, T. *Synlett* **1995,** 417–420.
47 Lu, J.-Y.; Arndt, H.-D. *J. Org. Chem.* **2007,** *72*, 4205–4212.
48 Lu, J.-Y.; Riedrich, M.; Mikyna, M.; Arndt, H.-D. *Angew. Chem., Int. Ed.* **2009,** *48*, 8137–8140.
49 Bray, C. D.; Olasoji, J. *Synlett* **2010,** *4*, 599–601.
50 Hernández, D.; Vilar, G.; Riego, E.; Cañedo, L.; Cuevas, C.; Albericio, F.; Álvarez, M. *Org. Lett.* **2007,** *9*, 809–811.
51 Prakash, R.; Kumar, A.; Aggarwal, R.; Prakash, O.; Singh, S. P. *Synth. Commun.* **2007,** *37*, 2501–2505.

5.5 Huisgen Tetrazole Rearrangement

Jennifer Xiaoxin Qiao

5.5.1 Description

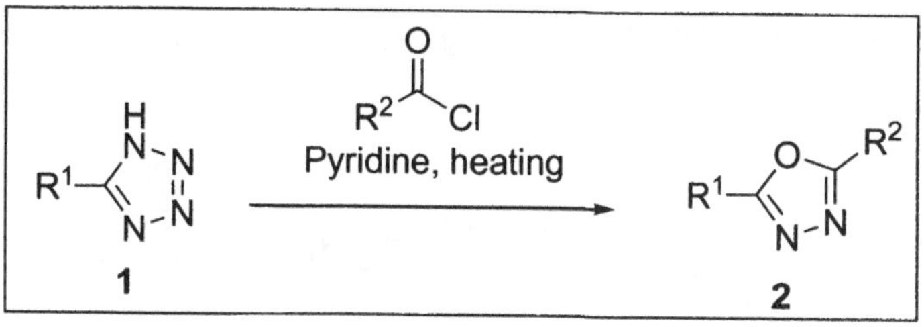

The Huisgen tetrazole rearrangement is the transformation of *C*-substituted tetrazole **1** to the corresponding 2,4-disubstituted 1,3,4-oxadiazole **2** by the reaction of carboxylic acid chloride in the presence of a base such as pyridine, at elevated temperature.[1–7]

5.5.2 Historical Perspective

Professor Rolf Huisgen discovered the titled reaction in 1958 at the University of Munich.[2–7] Huigen's great intuition drove him to ask his Ph.D. student Jurgen Sauer to repeat failed reactions of 5-phenyltetrazole with aroyl chlorides in pyridine that has been run by an an inept postdoctoral fellow, and hence led to the discovery of the named reaction. Since its discovery, the Huisgen tetrazole rearrangement reaction has been used for the synthesis of 1,3,4-oxadiazoles with aryl or alkyl substituents on the ring, and later on it was expanded for the preparation of 1,2,4-triazoles, 1,3,4-thiadiazoles or 3*H*-1,3,4-benzotriazepines.

5.5.3 Mechanism

Huisgen himself proposed that two electrocyclic steps of the pentadienyl type are involved in the Huisgen tetrazole rearrangement:[2] one is the electrocylic ring opening of acyltetrazole **3**, and the other is the electrocyclic ring closure of *N*-acyl nitrile imines **4**. The aryl- or alkyl-substituted tetrazole **5** is acylated with the acetyl chloride in the presence of pyridine as the base to produce *N*-acyltetrazole intermediate **3**, which is prone to ring opening because the aromaticity in **3** is weakened by a competing carboxamide resonance. The following steps after acylation are the rate-limiting steps, which involve the electrocyclic ring opening of **3** to the acylazo diazoalkane intermediate **6**, taking place in an equilibrium that is continuously disturbed by the loss of nitrogen. The loss of nitrogen obeys first-order kinetics and gives the corresponding *N*-acyl nitrile imine **4**, which closes to give the 1,3,4-oxadiazole ring **7**.

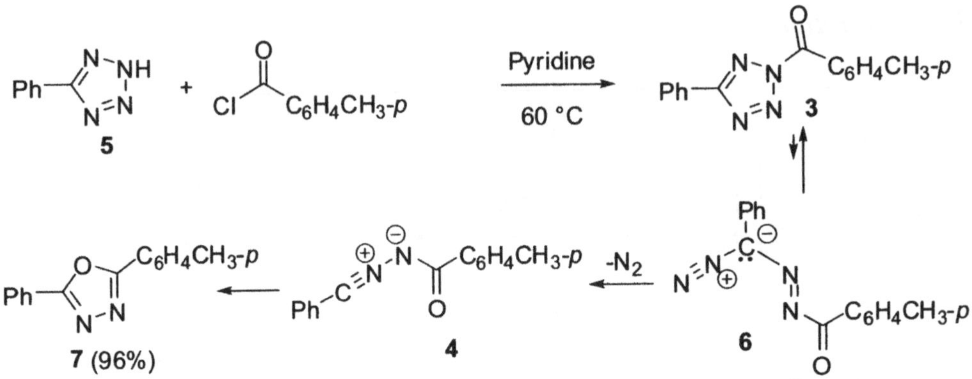

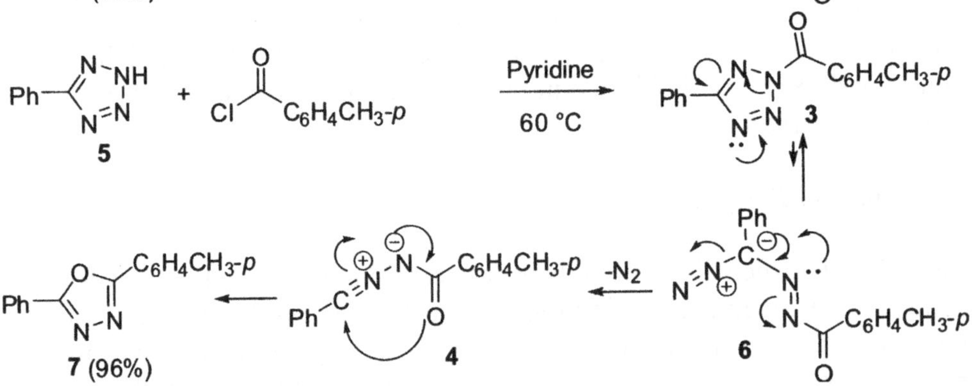

5.5.4 *Variations and Improvements*

It was found that addition of excess of benzoyl chloride in the absent of base provided 1,3,4-oxadiazoles **8–10** (R = CH_2Cl, CH_2COOEt, CH_2N_3, respectively) in good yields.[8] These substrates are not tolerated when pyridine was used as under the original Huisgen rearrangement conditions.

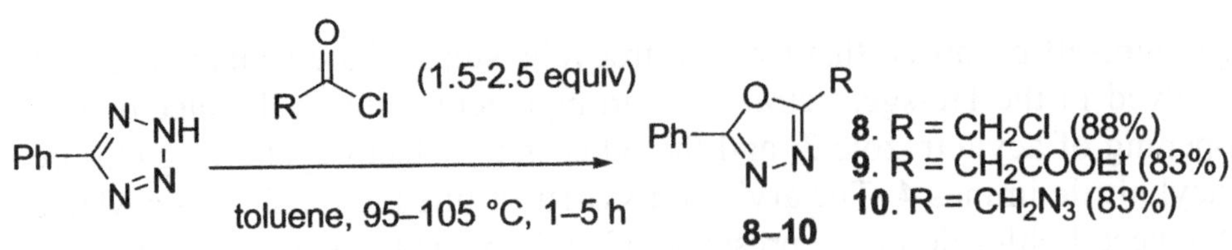

Besides acyl chloride, acetic anhydride and benzoyl anhydride were used as the acylating agents under microwave irradiation.[9] In addition, 2-anilino-5-aryl(hetaryl)-1,3,4-oxadiazoles **11** from 5-substituted tetrazoles **12**

and isocyanides **13** were synthesized by heating in nitrobenzene at 160 °C or under microwave irradiation.[10]

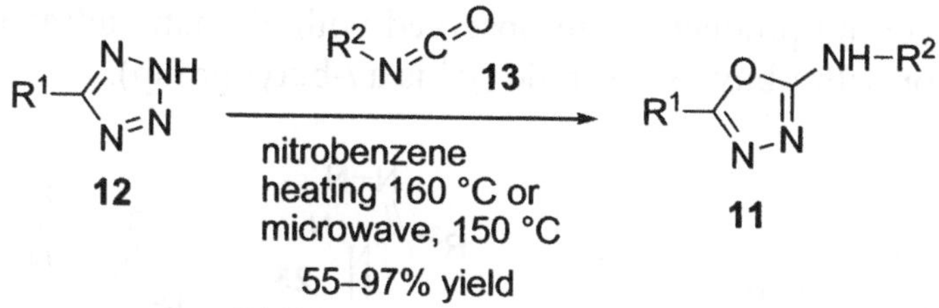

5-Phenyltetrazole and phenyl isothiocyanate **14** at elevated temperature gave the 1,3,4-thiadiazole derivative **15** in 64% yield via the *N*-thiocarbonyl nitrile imine intermediate **16**.[7]

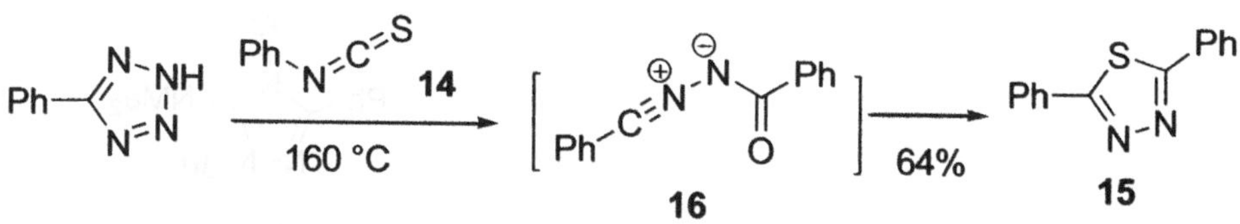

Iminoyl chloride **17** reacted with tetrazoles **18** analogously to afford the 1,2,4-triazoles **19** in excellent yield.[4] The electrocyclizaton of *N*-imino nitrile imines of the type **20** provided the key step for this general synthesis of 1,2,4-triazoles. Cyanuric chloride, as a triple aromatic imidoyl chloride, was treated with three equivalents of 5-phenyltetrazole to give **21** in 79% yield.[7]

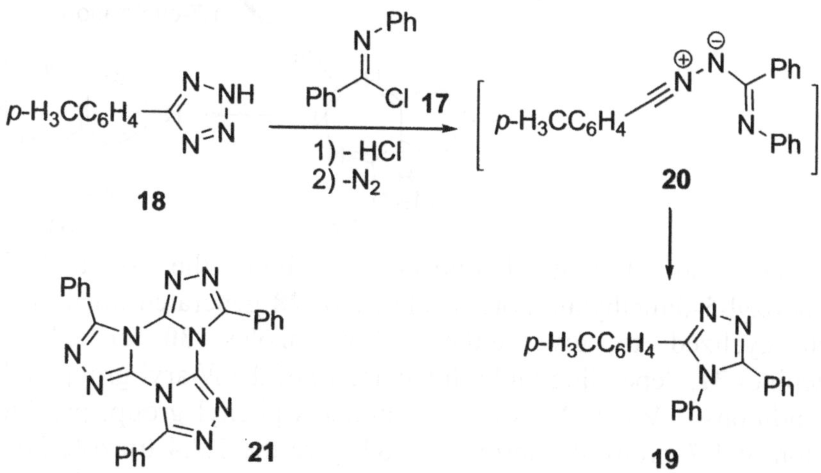

Recently, a three-component 1,2,4-triazole synthesis involving Nef–Huisgen cascade was developed.[11] Thus addition of acyl chloride **22** to isocyanide **23** (Nef reaction) generated the imidoyl chloride [nomenclature check]intermediate **24**, which was treated with aryltetrazole **25** in the presence of catalytic amount of $ZnCl_2$ (10 mol%). The resulting adduct is unstable and evolves 1,2,4-triazole **26** according to the Huisgen tetrazole rearrangement reaction. However, no products were observed with aliphatic tetrazoles (e.g., R^3 is a methyl) or with *t*-buyl isocyanide (R^1 is a *t*-butyl group).

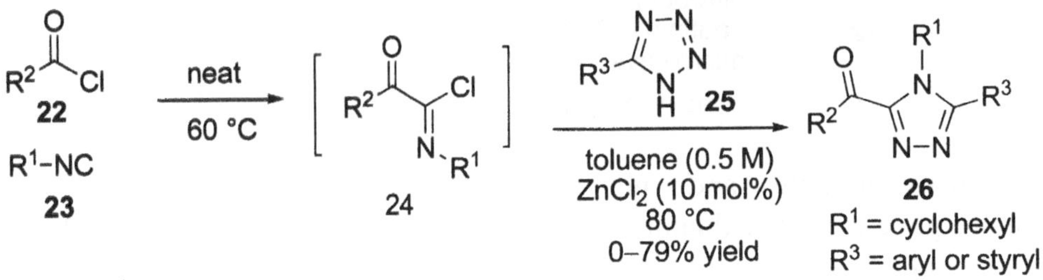

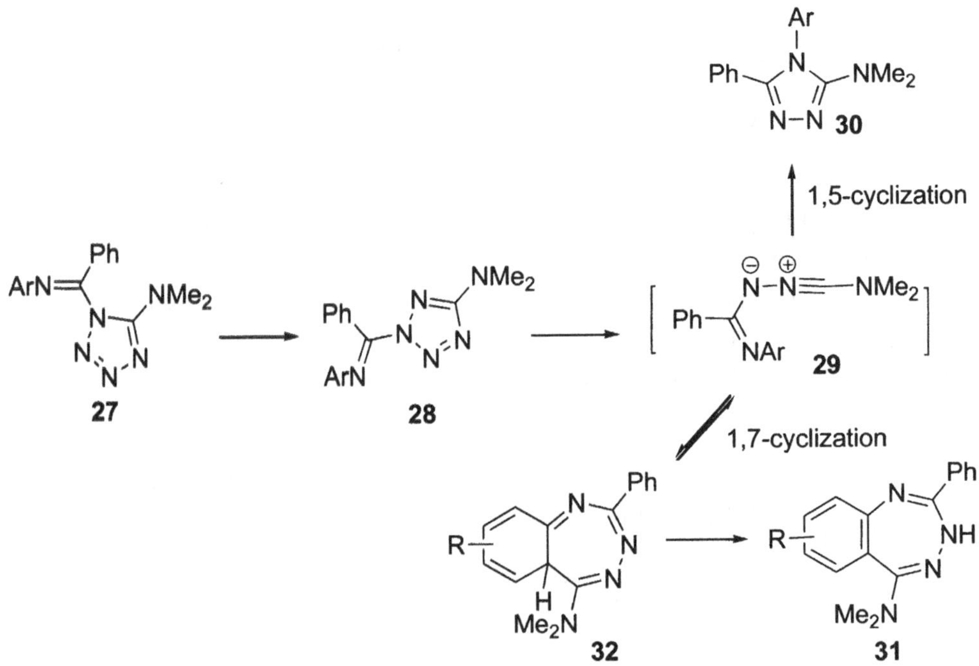

It was reported that thermolysis of both the 1- and 2-*N*-arylbenzimidozyl-5-dimethylaminotetrazole **27** or **28** generated nitrile imines **29**, which cyclized to give either 1,2,4-triazoles **30** or 3*H*-1,3,4-benzotriazepines **31**, depending on both the nature of the *N*-aryl group and the reaction conditions.[12] When Ar was a 2,6-dimethylphenyl group, precluding the formation of 1,7-cyclized intermediate **32** gave the 1,2,4-triazole **30**. For other aryl groups, thermolysis of both **27** and **28** gave the 1,7-cyclized products 3*H*-1,3,4-benzotriazepines **31**.

5.5.5 *Synthetic Utility*

The Huisgen tetrazole rearragentment was employed in the first-generation stage synthesis of dendrimers containing 1,3,4-oxadiazoles, in which the usual method starting from an aroyl hydrazide to construct the 1,3,4-oxadiazole ring was not successful.[13] On the other hand, several 9,9'-spirobifluorene-bridged bipolar systems containing 1,3,4-oxadiazole-conjugated oligoaryl and triarylamine moieties with remarkable solvent-polarity dependent fluorescence properties were synthesized using the Huisgen method.[14]

The carbon-chain polymers **33** with pendant 1,3,4-oxadiazole rings were prepared using the original Huisgen tetrazole rearrentment.[15] In the study of host-guest complexation by Cram and co-workers, the 1,3,4-thiadiazole containing host **34** was prepared in 89% yield using the Huisgen method.[16] Some of the lipophilic 1,3,4-oxadiazole derivatives showed interesting biological activities; therefore, several of such systems **35**–**38** were synthesized by using the Huisgen rearrangement starting from the corresponding tetrazole system.[17–20]

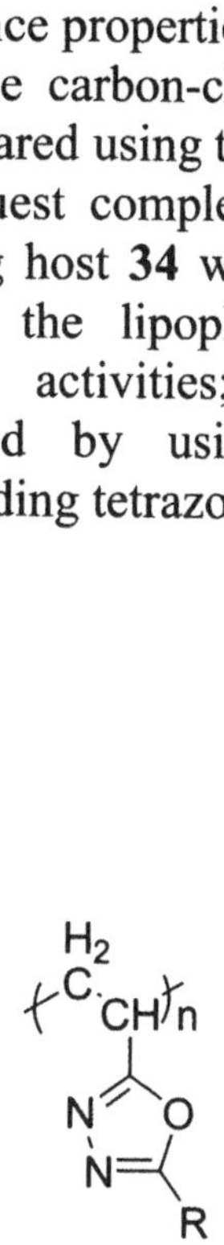

carbon-chain polymers (**33**)

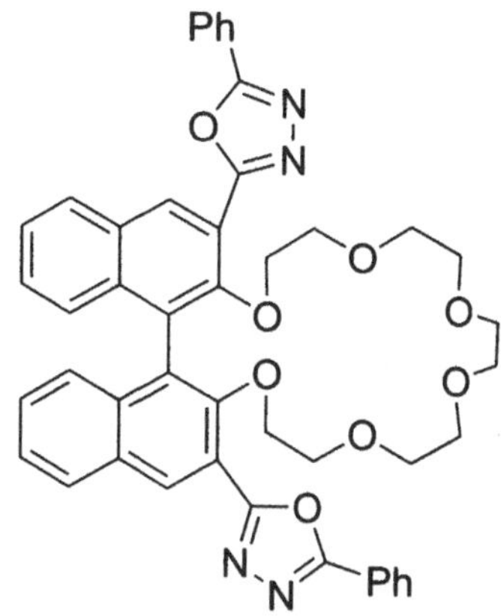

oxadiazole containing host molecule (**34**)

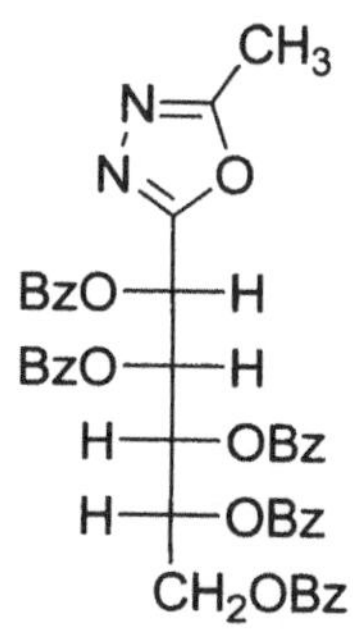

oxadiazole derivatives of D-Mannose (**35**)

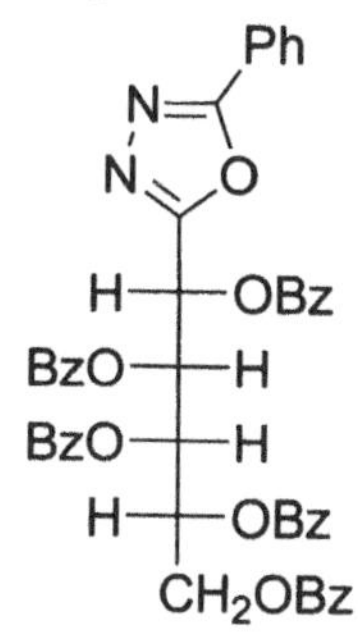

oxadiazole derivatives of D-Galactose (**36**)

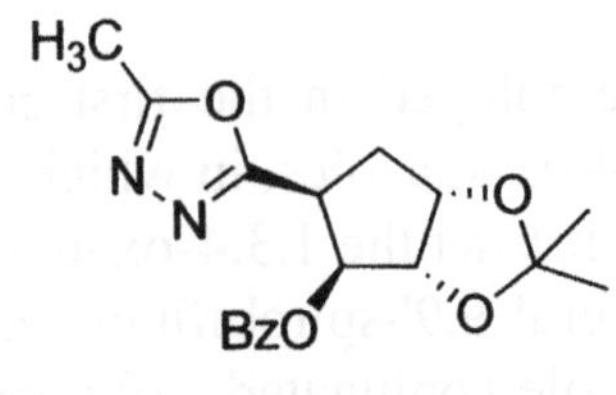

antiviral carbonucleoside analog (**37**)

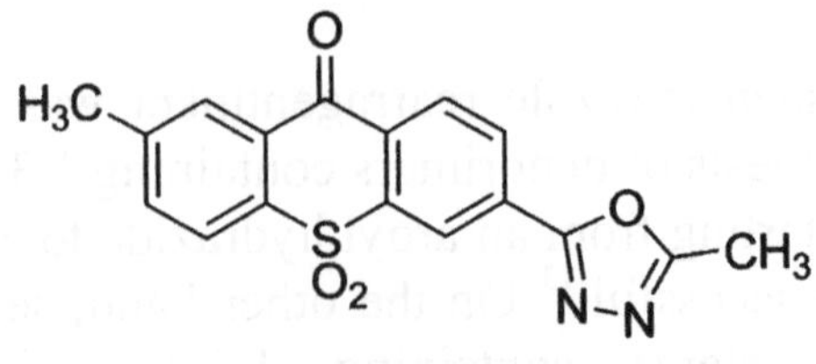

MAO inhibitor (**38**)

5.5.6 *Experimental*

Huisgen Reaction: **(*E*)-1,2-Bis[5-(4-methoxyphenyl)-1,3,4-oxadiazol-2-yl]ethene (40)**[21]

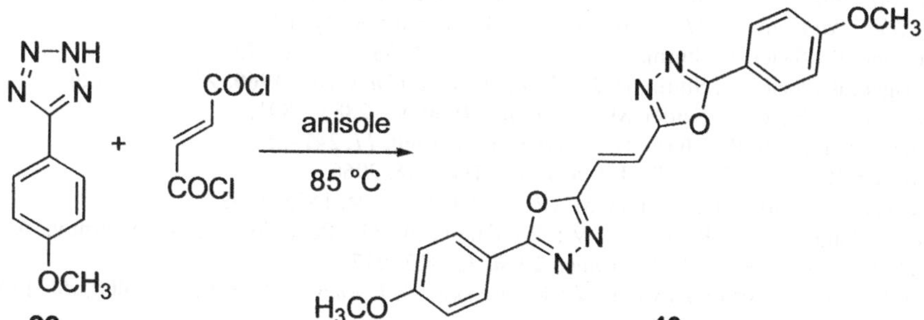

39 40

To a round-bottom flask equipped with a magnetic stirring bar, were added tetrazole **39** (352 mg, 2 mmol), anisole (20 mL) and fumaryl chloride (153 mg, 1 mmol). While stirring, collidine (242 mg, 2 mmol) diluted with anisole (2 mL) was added dropwise and the flask was connected to a gas burette. The flask was heated at 85 °C until gas evolution occurred and kept at this temperature. As the gas evolution ceased, the bath temperature was raised about 20 °C and kept for further 20 min. The mixture was cooled and poured on a chromatography column filled with silica gel and petroleum ether/toluene 1:1. After the solvent was eluted, toluene with gradually increased content of EtOAc was used as an eluent to give the titled compound (318 mg, 92%) as colorless crystals.

Three-Component Nef–Huisgen Access to 1,2,4-Triazoles: **(4-Cyclohexyl-5-phenyl-4*H*-1,2,4-triazol-3-yl)(4-fluorophenyl)methanone (41)**[11]

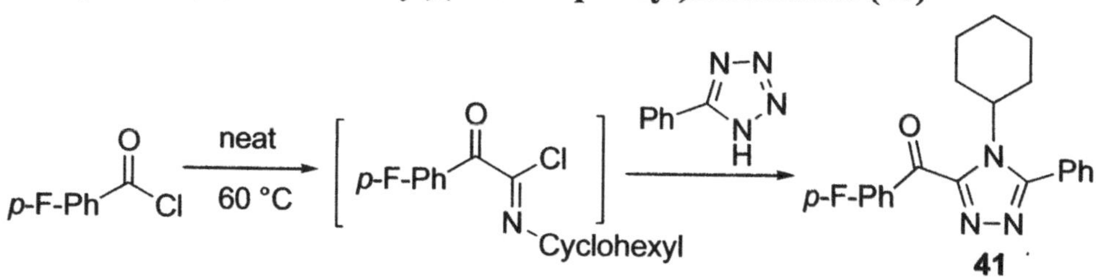

41

Cyclohexyl isocyanide (1 mmol) and *p*-fluorobenzoyl chloride (1 mmol) were heated neat for 1 h in a CEM microwave at 60 °C (50 W) and finally dissolved in toluene (2 mL). To this solution was added $ZnCl_2$ in THF (0.1 mL of a 1 M solution) followed by the addition of phenyl tetrazole (1 mmol, 1 equiv). The reaction mixture was then heated overnight at 80 °C. Hydrolysis followed by extraction and flash column chromatography afforded the tiled compound as a yellow oil (280 mg, 79%).

5.5.7 *References*

1. [R] (a) Huisgen, R. *Angew. Chem., Int. Ed. Engl.* **1980,** *19*, 947–1034. (b) Huisgen, R. In *Profiles, Pathways, and Dreams*, American Chemical Society, 1994.
2. Huisgen, R.; Sauer, J.; Sturm, H. J. *Angew. Chem.* **1958,** *70*, 272–273.
3. Huisgen, R.; Sauer, J.; Sfurm, H. J.; Markgraf, J. H. *Chem. Ber.* **1960,** *93*, 2106–2124.
4. Huisgen, R.; Sauer, J.; Seidel, M. *Chem. Ber.* **1960,** *93*, 2885–2891.
5. Sauer, J.; Huisgen, R.; Sfurm, H. J. *Tetrahedron* **1960,** *11*, 241–251.
6. Huisgen, R.; Axen, C.; Seidl, H. *Chem. Ber.* **1965,** *98*, 2966.
7. Huisgen, R.; Sturm, H. J.; Seidel, M. *Chem. Ber.* **1961,** *94*, 1555–1562.
8. Vereshchagin, L. I.; Petrov, A. V.; Proidakov, A. G.; Pokatilov, F. A.; Smirnov, A. I.; Kizhnyaev, V. N. *Russ. J. Org. Chem.* **2006,** *42*, 912–917.
9. Efimova, Y. A.; Artamonova, T. V.; Koldobskii, G. I. *Russ. J. Org. Chem.* **2008,** *44*, 1345–1347.
10. Efimova, Y. A.; Karabanovich, G. G.; Artamonova, T. V.; Koldobskii, G. I. *Russ. J. Org. Chem.* **2009,** *45*, 1241–1244.
11. El Kaim, L.; Grimaud, L.; Wagschal, S. *Synlett* **2009,** 1315–1317.
12. Boyd, G. V.; Cobb, J.; Lindley, P. F.; Mitchell, J. C.; Nicolaou, G. A. *J. Chem. Soc., Chem. Commun.* **1987,** 99–101.
13. Verheyde, B.; Dehaen, W. *J. Org. Chem.* **2001,** *66*, 4062–4064
14. Chien, Y.-Y.; Wong, K.-T.; Chou, P. T.; Cheng, Y.-M. *Chem. Commun.* **2002,** 2874–2875.
15. Kizhnyaev, V. N.; Pokatilov, A. F.; Vereshchagin, L. I.; Adamova, L. V.; Safronov, A. P.; Smirnov, A. I. *Russ. J. Appl. Chem.* **2006,** *79*, 1167–1173.
16. Lingenfelter, D. S.; Helgeson, R. C.; Cram, D. J. *J. Org. Chem.* **1981,** *46*, 393–406.
17. Fascio, M. L.; D'Accorso, N. B. *J. Heterocyclic Chem.* **1995,** *32*, 815–818.
18. (a) Cannizzaro, C. E.; Thiel, I. M. E.; D'Accorso, N. B. *J. Heterocyclic Chem.* **1998,** *35*, 481–484. (b) Faraco, A. A. G.; Prado, M. A. F.; D'Accorso, N. B.; Alves, R. J.; de Souza Filho, J. D.; Prado, R. F. *J. Heterocyclic Chem.* **1999,** *36*, 1129–1133.
19. Talarico, L. B.; García, C. C.; Damonte, E. B. *J. Heterocyclic Chem.* **2005,** *42*, 979–983.
20. Harfenist, M.; Heuser, D. J.; Joyner, C. T.; Batchelor, J. F.; White, H. L. *J. Med. Chem.* **1996,** *39*, 1857–1863.
21. Detert, H.; Schollmeier, D. *Synthesis* **1999,** *6*, 999–1004.

5.6 Knorr Pyrazole Synthesis

James Kempson

5.6.1 Description

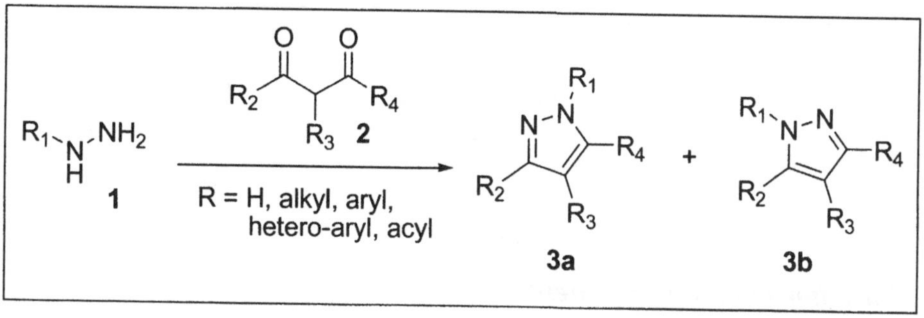

Similar to the Paal–Knorr pyrrole synthesis, the Knorr pyrazole synthesis[1,2] is the most common synthetic method for the preparation of pyrazoles. The Knorr pyrazole synthesis involves the cyclocondensation of an appropriate hydrazine, **1**, which acts as a bidentate nucleophile, with the three carbon unit of a 1,3-dicarbonyl moiety, **2**, featuring two electrophilic carbons. With unsymmetrical substrates having two electrophilic centers ($R_2 \neq R_4$), mixtures of regioisomers **3a** and **3b** are often obtained in reactions with substituted hydrazines ($R_1 \neq$ H). However, when R_1 = H, the prototropic tautomerism of pyrazoles renders **3a** equivalent to **3b**.

5.6.2 Mechanism

The reaction proceeds by way of nucleophilic attack by hydrazine nitrogen on the most electrophilc carbon atom of the 1,3-dicarbonyl unit. If the electrophilicity of these two carbon atoms is similar, then the selectivity in the initial hydrazone formation will be diminished. Each of the hydrazones, **4a** and **4b**, then cyclize to form a corresponding mixture of regioisomeric pyrazoles, **3a** and **3b**, respectively.

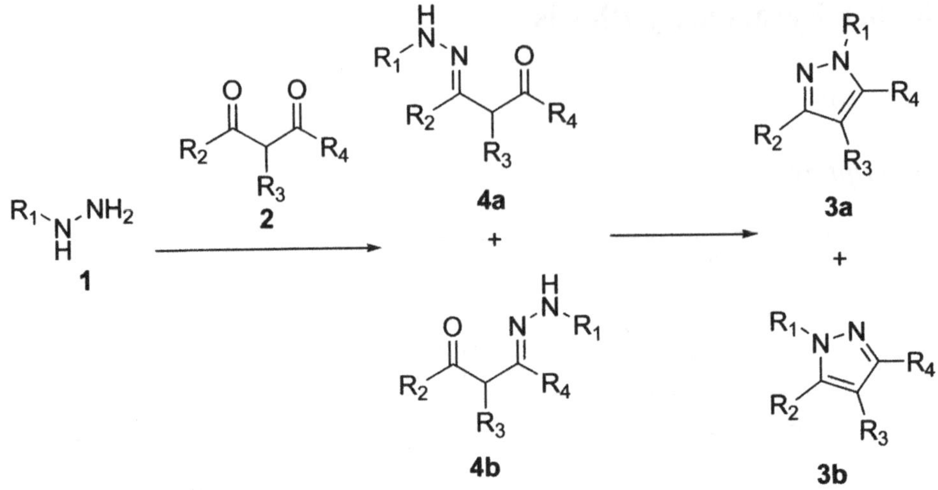

5.6.3 *Variations and Improvements*

Most of the 1,3-diketones used in the synthesis of pyrazoles must be previously prepared and purified, often being obtained as a mixture of condensation products. Recently, an efficient, rapid and general one-pot synthesis of 3,5-disubstituted pyrazoles **8** starting with enolates **5** and acid chlorides **6** has been reported. The 1,3-diketones thus formed were not isolated, but rather converted in situ into pyrazoles through the addition of hydrazines.[3] The process seems to tolerate a wide range of functional groups.

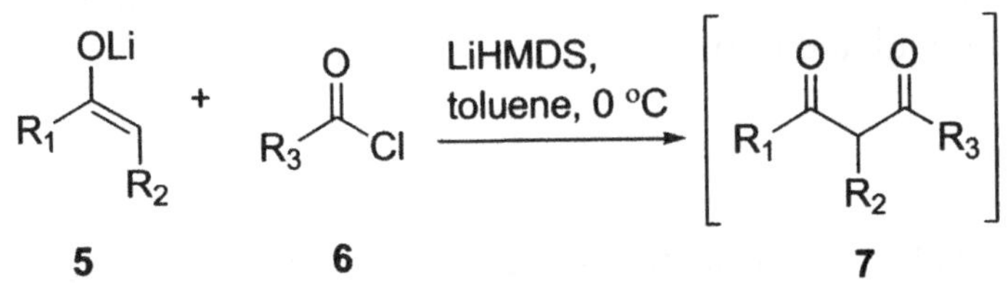

R_1 = aryl, heteroaryl; R_2 = H, *n*-propyl, Ph; R_1/R_2 = $CH_2CH_2OCH_2$, $CH_2CH_2N(Boc)CH_2$, $(CH_2)_4$; R_3 = aryl, *n*-C_5H_{11}; R_4 = H, Me, Ph.

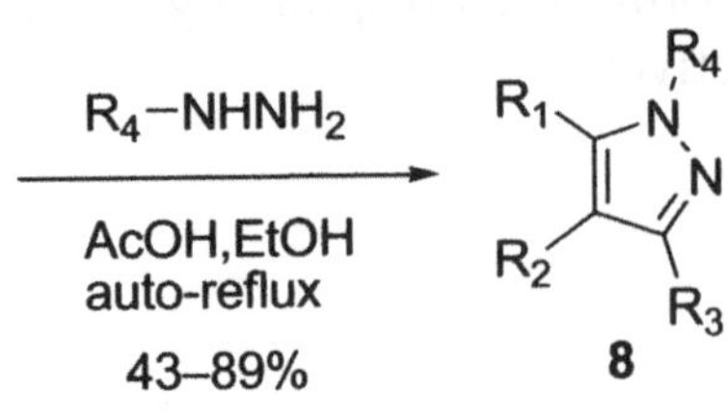

Compared to reactions in organic solvents, solventless reactions are often faster, occur in higher yields, and have both environmental and economic advantages. Thus, a solventless condensation of 1,3-diketone **9**

with hydrazine in the presence of a catalytic amount of sulfuric acid at room temperature afforded 3,5-disubstituted pyrazoles **10** in high yields.[4] The reactions were carried out in a mortar, in which the diketone and hydrazine hydrate were mixed with a drop of concentrated sulfuric acid.

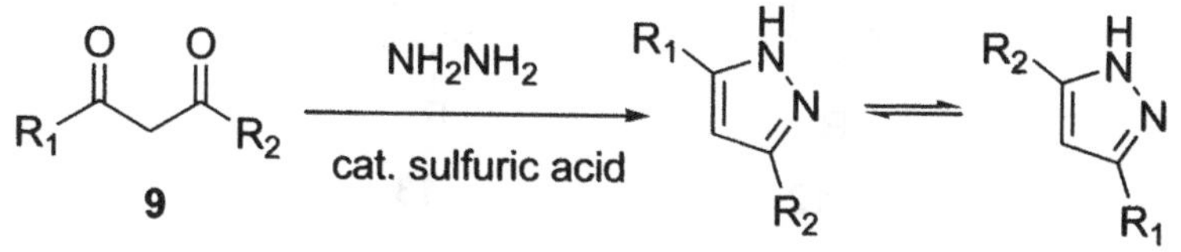

10a **10b**

$R_1 = R_2 =$ Me, 96%; $R_1 = R_2 =$ Bu^t, 92%; $R_1 =$ CO_2Et, $R_2 =$ Me, 90%; $R_1 =$ Me, $R_2 =$ Ph,96%

A systematic study correlating regiochemistry with the specific reaction conditions and the electronic/steric characteristics of 1,3-diketones and aromatic hydrazines in the synthesis of 1,5-diaryl-3-substituted pyrazoles was carried out by Singh *et al.*[5] Usually, cyclocondensations between arylhydrazines and 1,3-diketones are carried out on polar, protic solvents such as alcohols or acetic acid. However, excellent yields and regioselectivities were obtained in reactions of 1-arylbutane-1,3-diones **11** with arylhydrazine hydrochlorides in *N,N*-dimethylacetamide in the presence of 0.5 equiv of 10 N aqueous hydrochloric acid.[6] Nevertheless, when similar reaction conditions were performed in ethanol under reflux, the regioselectivities were lower (80:20–86:14).

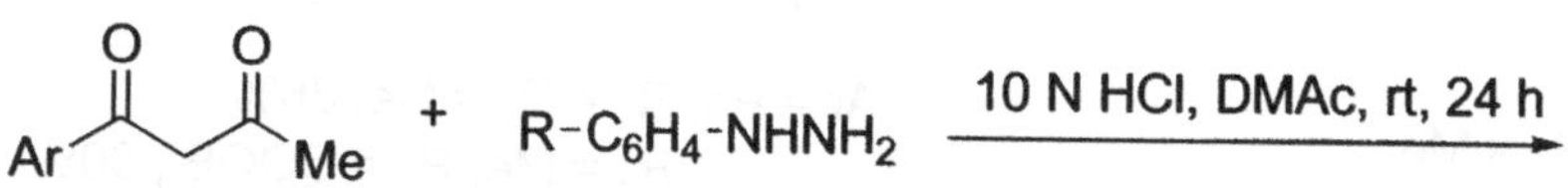

11 Ar = Ph, R= SO_2NH_2, 77%; Br, 98%; H, 83%
Ar = 4-$MeOC_6H_4$, R = SO_2NH_2, 91%; Br, 91%; H, 87%.

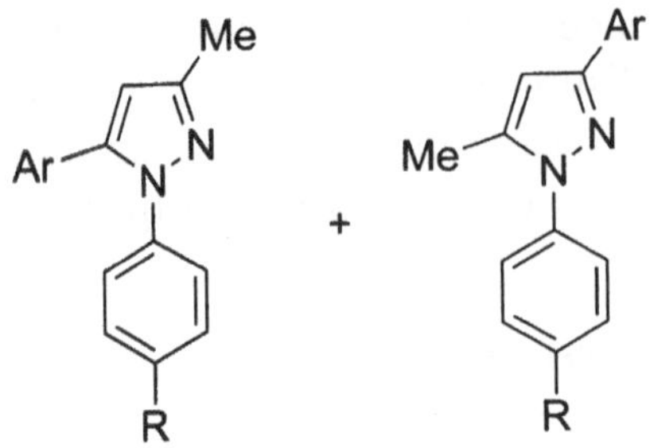

12 13

ratio **12/13** = 93:7–99.8:0.2

The condensation of unsymmetrical 1,3-diketones **14** (R_1 = Me, R_2 = CO_2Et, Ph) with phenyl- and *p*-tolylhydrazines in a solventless reaction catalyzed by sulfuric acid afforded mixtures of the two regioisomers **15** and **16**, generally in good to excellent yields. However, reactions of 1-phenylbutane-1,3-dione with acylhydrazines led to 4,5-dihydro-5-hydroxypyrazole derivatives **17** with complete regioselectivity. These compounds were then thermally dehydrated and deacylated (R = Ph) in the presence of a catalytic amount of sulfuric acid.

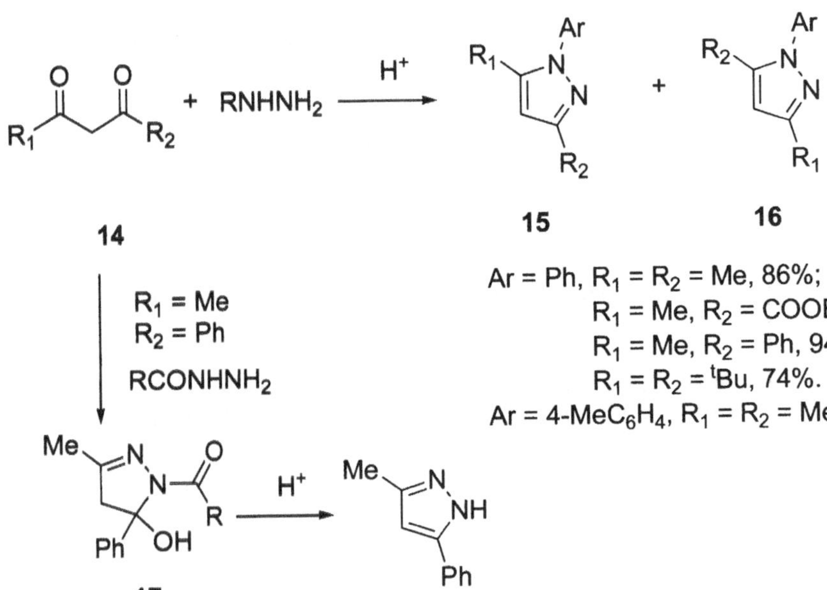

14 15 16 17

Ar = Ph, R_1 = R_2 = Me, 86%;
R_1 = Me, R_2 = COOEt, 90%;
R_1 = Me, R_2 = Ph, 94%;
R_1 = R_2 = ^{t}Bu, 74%.
Ar = 4-MeC_6H_4, R_1 = R_2 = Me, 86%.

R = Me, 99%
R = Ph, 91%

Strategies involving syntheses both in solution and in the solid phase have been developed to prepare libraries of 4-alkyl-1,3,5-triaryl and 5-alkyl-1,3,4-triarylpyrazoles, which have been widely investigated as ligands for the estrogen receptor (ER).[7–13] Initially, tetrasubstitued pyrazoles **20** were synthesized through the classic Knorr condensation route. To this end, two strategies were developed to introduce the C-4 alkyl group into the pyrazole derivatives: either through alkylation at C-2 of the 1,3-diketone precursor or through the use of an appropriate alkylphenone as a building block. The former method was described by Marzinzik and Felder[12] in the solid phase but is not general. In contrast, the latter method developed by Katzenellenbogen and co-workers[8,9] allows the introduction of a variety of alkyl groups in both solution and solid phase. In solution,[8] the synthetic sequence involves a crossed-Claisen condensation between an appropriate 4-methoxy alkylphenone **18** and methyl 4-nitrophenyl 4-methoxybenzoate, followed by reaction of the C2-alkylated 1,3-diketone **19** with an arylhydrazine. When this methodology was applied to unsymmetrical 1,3-diketones, a lack of regioselectivity was observed.

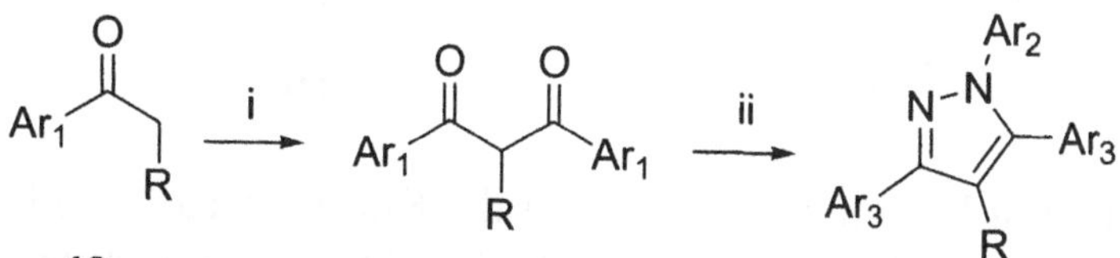

18 **19** **20**

i) Methyl 4-nitrophenyl 4-methoxybenzoate, LiHMDS/THF
ii) (a) $Ar_2NHNH_2 \cdot HCl$, THF/DMF (1:3), 53–87%; (b) BBr_3/CH_2Cl_2, 30–98%.
Ar_1 = 4-$MeOC_6H_4$, R = Me, Et, *n*-Pr, *i*-Bu, *n*-Bu;
Ar_2 = Ph, 4-OHC_6H_4; Ar_3 = 4-OHC_6H_4

Pyrazole-3(5)-carboxylic acid esters derivatives are also another important motif of considerable pharmacological relevance and also represent useful synthetic building blocks in both organic and medicinal chemistry. The approach most often used to prepare these involves [3 + 2] cycloadditions between hydrazines and 1,3-diketoesters. In general, Claisen condensation of the enolate of a ketone **21** with diethyl oxalate is the method most commonly employed in the preparation of the 1,3-diketoesters, **22**. These, in turn, constitute important building blocks for most of the pyrazole-3(5)-carboxylic acid derivatives **23** reported in the literature, some of which have been found to display important pharmaceutical activities.[14–20]

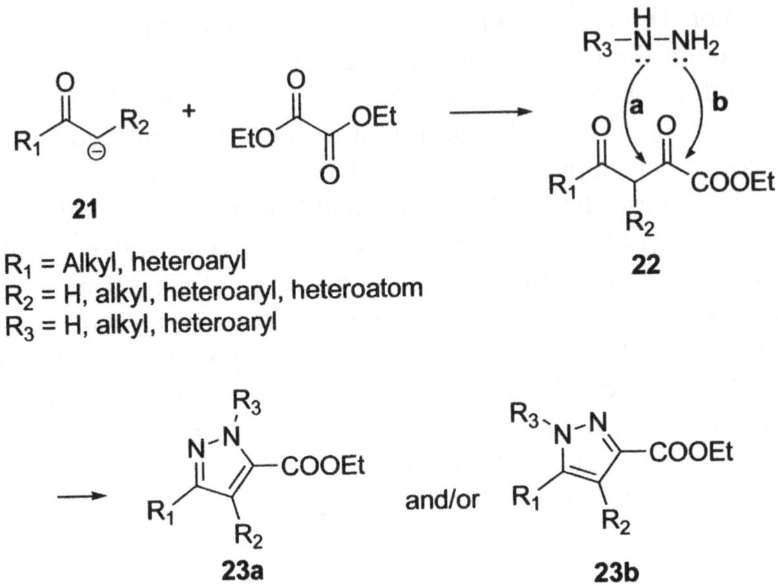

R_1 = Alkyl, heteroaryl
R_2 = H, alkyl, heteroaryl, heteroatom
R_3 = H, alkyl, heteroaryl

One of the best known applications of the Knorr pyrazole synthesis in drug discovery is the synthesis of celecoxib (Celebrex), a selective COX-2 inhibitor prescribed as an analgesic. As shown below, the substrate dione **26** was prepared by the Claisen condensation of 4-methylacetophenone **24** with ethyl trifluoroacetate **25** in the presence of NaOMe in methanol under reflux. Subsequent diarylpyrazole formation from the condensation of dione and 4-sulfonamidophenylhydrazine hydrochloride then delivered celecoxib **27**.[21]

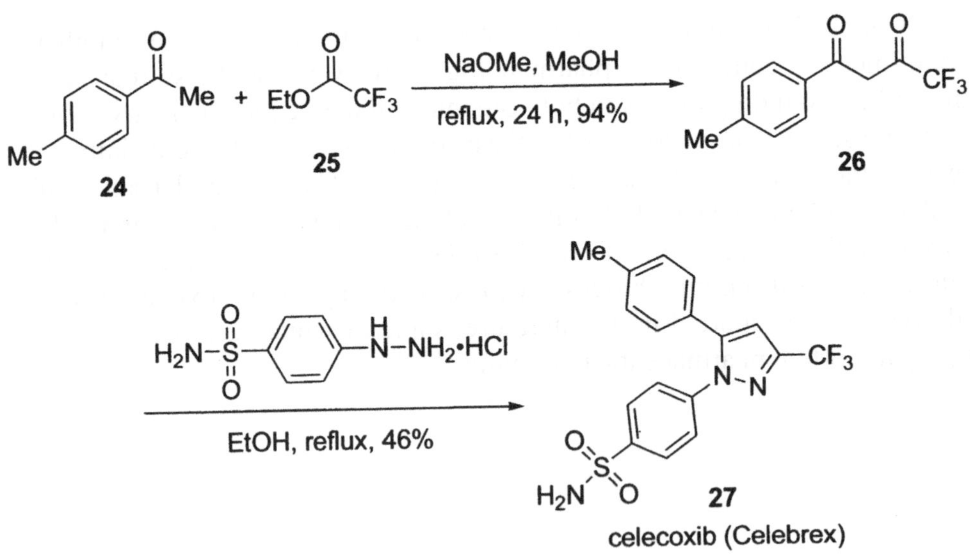

Another important application of the Knorr pryrazole synthesis is in the preparation of rimonabant (Acomplia), a potent and selective antagonist for cannabinoid type 1 (CB_1) receptor. Acomplia was marketed in 56 countries for the treatment of obesity by Sanofi–Aventis since 2006, but was withdrawn in 2009 due to an unfavorable toxicity profile. The synthesis of rimonabant in Sanofi's 1997 patent[22] commenced with the Claisen condensation of 4-chloropropiophenone **28** and ethyl oxalate in the presence of LiHMDS to give the enolate **29** of the corresponding bis-ketone-ester in 37% yield. However, simply switching the solvent from ether to methyl cyclohexane or THF, the yield for the condensation was improved to 70% and 59%, repectively. Pyrazole formation was promoted by hot acetic acid. The resulting pyrazole ester **31** was then converted to rimonabant **32** by hydrolysis and hydrazide formation.[structure 31 fix Et group]

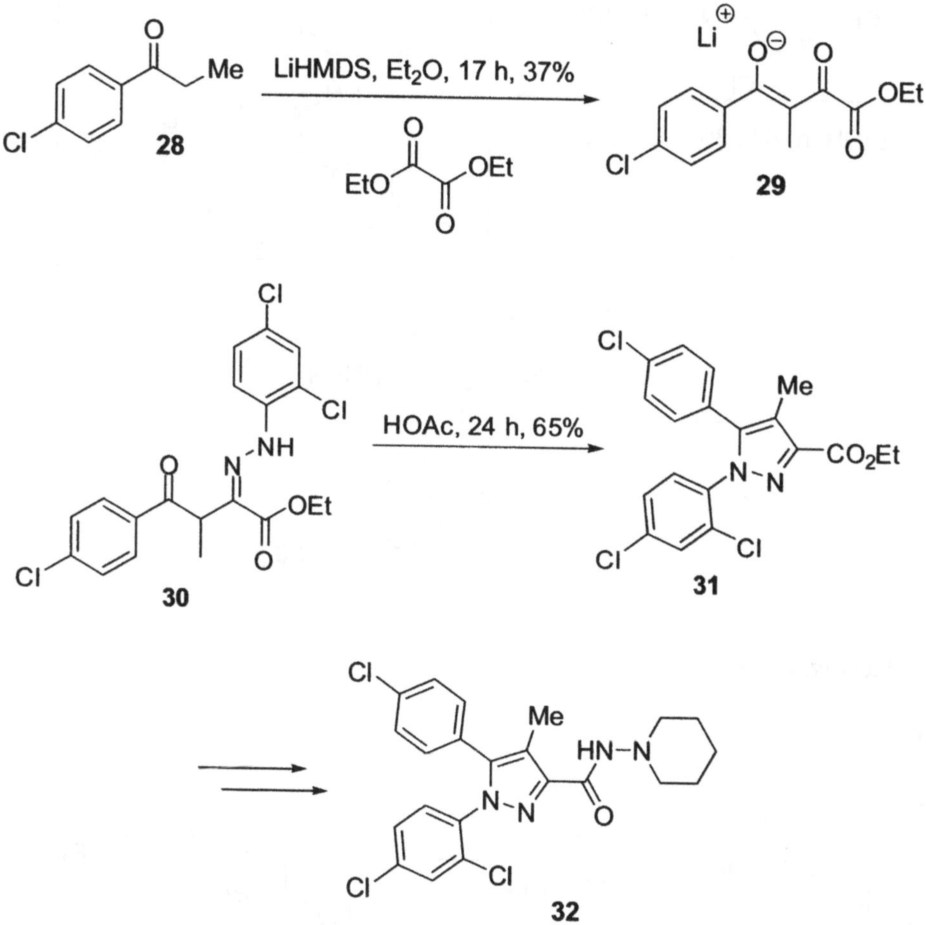

rimonabant (Acomplia)

It is interesting that fine tuning the chemical environment around the two carbonyl groups can completely alter the regiochemical outcome for the hydrazone formation.[23] Triethylamine mediated amide formation gave the corresponding hydrazide, **34**, which deactivated the neighboring carbonyl so that the hydrazone was formed at the distal carbonyl. Condensation with the hydrazine HCl salt then delivered the regioisomer **36** of rimonabant in 30% yield.

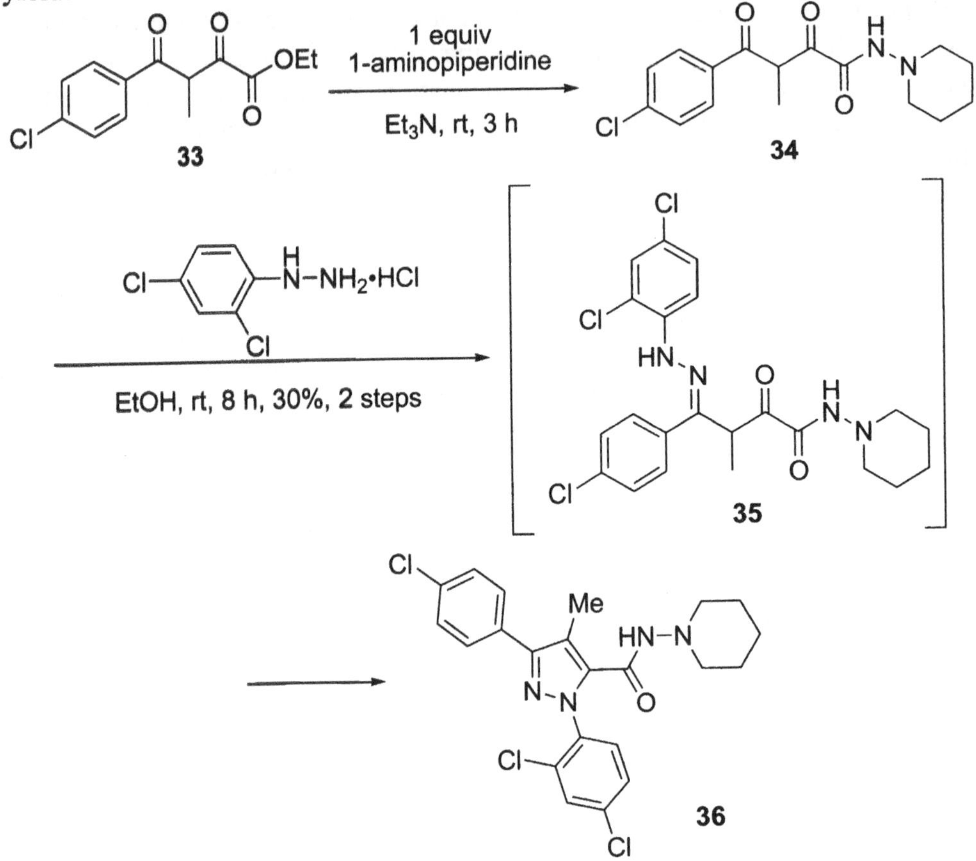

5.6.4 *Experimental*

Step 1

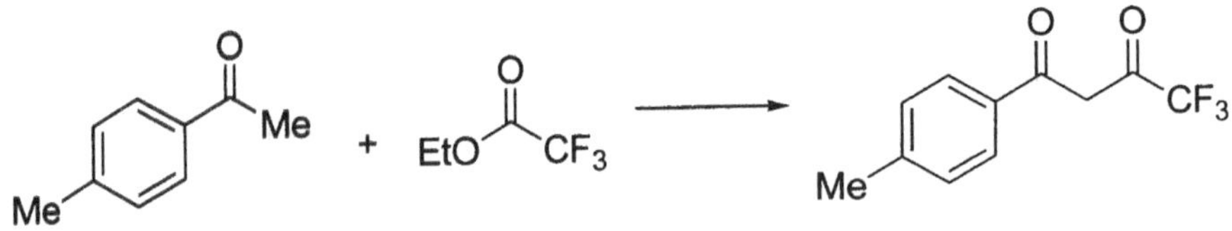

4-Methylacetophenone (5.26 g, 39.2 mmol) was dissolved in 25 mL of methanol under argon and 12 mL (52.5 mmol) sodium methoxide in methanol (25%) was added. The mixture was stirred for 5 min and 5.5 mL (46.2 mmol) ethyl trifluoroacetate was added. After refluxing for 24 h, the mixture was cooled to room temperature and concentrated. 100 mL 10% HCl was added, and the mixture extracted with 4 × 75 mL ethyl acetate. The extracts were dried over $MgSO_4$, filtered and concentrated to afford 8.47 g (94%) of a brown oil, which was carried on without further purification.

Step 2

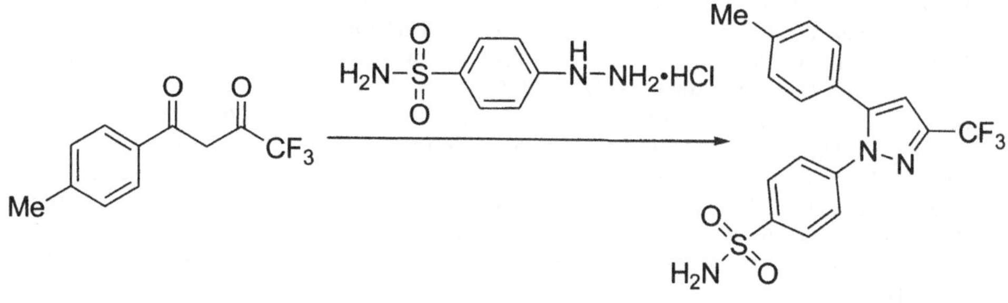

To the dione from Step 1 (4.14 g, 18.0 mmol) in 75 mL absolute ethanol was added 4.26 g (19.0 mmol) 4-sulphonamidophenylhydrazine hydrochloride. The reaction was refluxed under argon for 24 h. After cooling to room temperature and filtering, the reaction mixture was concentrated to afford 6.13 g of an orange solid. The solid was recrystallized from methylene chloride/hexane to give 3.11 g (8.2 mmol, 46%) of the product as a pale yellow solid: mp 157–159 °C.

5.6.5 References

1. Fustero, S.; Simón-Fuentes, A.; Sanz-Cervera, J. F. *Org. Prep. Proc. Int.* **2009,** 253–290.
2. [R] Janin, Y. L. *Mini-Rev. Org. Chem.* **2010,** *7,* 314–323.
3. Heller, S. T.; Natarajan, S. R. *Org. Lett.* **2006,** *8,* 2675–2679.
4. Wang, Z.-X; Qin, H.-L *Green Chem.* **2004,** *6,* 90–95.
5. Singh, S. K.; Reddy, M. S.; Shivaramakrishna, S.; Kavitha, D.; Vasudev, R.; Babu, J. M.; Sivalakshmidevi, A.; Rao, Y. K. *Tetrahedron Lett.* **2004,** *45,* 7679–7685.
6. Gosselin, F.; O'Shea, P. D.; Webster, R. A.; Reamer, R. A.; Tillyer, R. D.; Grabowski, E. J. J. *Synlett* **2006,** 3267–3269.
7. Fink, B. E.; Mortenson, D. S.; Stauffer, S. R.; Aron, Z. D.; Katzenellenbogen, J. A. *Chem. Biol.* **1999,** *6,* 205–210.
8. Stauffer, S. R.; Coletta, C. J.; Tedesco, R.; Nishiguchi, G.; Carlson, K.; Sun, J.; Katzenellenbogen, B. S.; Katzenellenbogen, J. A. *J. Med. Chem.* **2000,** *43,* 4934–4940.
9. Stauffer, S. R.; Katzenellenbogen, J. A. *J. Comb. Chem.* **2000,** *2,* 318–325.
10. Stauffer, S. R.; Huang, Y.; Coletta, C. J.; Tedesco, R.; Katzenellenbogen, J. A. *Bioorg. Med. Chem.* **2001,** *9,* 141–144.
11. Stauffer, S. R.; Huang, Y.; Aron, Z. D.; Coletta, C. J.; Sun, J.; Katzenellenbogen, B. S. Katzenellenbogen, J. A. *Bioorg. Med. Chem.* **2001,** *9,* 151–155.
12. Marzinzik, A. L.; Felder, E. R. *Tetrahedron Lett.* **1996,** *37,* 1003–1005.
13. Huang, Y.; Katzenellenbogen, J. A. *Org. Lett.* **2000,** *2,* 2833–2835.
14. Penning, T. D.; Talley, J. J.; Bertenshaw, S. R.; Carter, J. S.; Collins, P. W.; Docter, S.; Graneto, M. J.; Lee, L. F.; Malecha, J. W.; Miyashiro, J. M.; Rogers, R. S.; Rogier, D. J.; Yu,

S. S.; Anderson, G. D. ; Burton, E. G.; Cogburn, J. N.; Gregory, S. A.; Koboldt, C. M.; Perkins, W. E.; Seibert, K.; Veenhuizen, A. W.; Hang, Y. Y.; Isakson, P. C. *J. Med. Chem.* **1997,** *40*, 1347–1350.
15. Lan, R.; Liu, Q.; Fan, P.; Lin, S.; Fernando, S. R.; McCallion, D.; Pertwee, R.; Makriyannis, A. *J. Med. Chem.* **1999,** *42*, 769–775.
16. Katoch-Rouse, R.; Pavlova, O. A.; Caulder, T.; Hoffman, A. F. ; Mukhin, A. G.; Horti, A. G. *J. Med. Chem.* **2003,** *46*, 624–627.
17. Varano, F.; Catarzi, D.; Colotta, V.; Filacchioni, G.; Galli, A.; Costalgi, C.; Carlă, V. *J. Med. Chem.* **2002,** *45*, 1035–1038.
18. van Herk, T.; Brussee, J.; van den Nieuwendijk A.M. C. H.; van der Klein, P. A. M.; Ijzerman, A. P.; Stannek, C. ; Burmeister, A.; Lorenzen, A. J. *J. Med. Chem.* **2003,** *46*, 3945–3948.
19. Finn, J.; Mattia, K.; Morytko, M.; Ram, S.; Yang, Y.; Wu, X. ; Mak, E.; Gallant, P.; Kith, D. *Bioorg. Med. Chem. Lett.,* **2003,** *13*, 2231–2235.
20. Schmidt, A.; Habeck, T.; Kindermann, M. K.; Nieger, M. *J. Org. Chem.* **2003,** *68*, 5977–5979.
21. Penning, T. D.; Talley, J. J.; Bertenshaw, S. R.; Carter, J. S.; Collins, P. W.; Docter, S.; Graneto, M. J.; Lee, L. F.; Malecha, J. W.; Miyahiro, J. M.; Rogers, R. S.; Rogier, D. J.; Yu, S. S.; Anderson, G. D.; Koboldt, C. M.; Perkins, W. E.; Seibert, K.; Veenhuizen, A. W.; Zhang, Y. Y.; Isakson, P. C. *J. Med. Chem.* **1997,** *40*, 1347–1350.
22. Barth, F.; Casellas, P.; Congy, C.; Martinez, S.; Rinaldi, M.; Anne-Archard, G. US Pat. 5624941, 1997.
23. Kotagiri, V. K.; Reddy, J. M.; Suthrapu, S. K.; Rao, C. P.; Reddy, P. P.; Bhattacharya, A.; Bandichhor, R. *Monatsh. Chem.* **2008,** *139*, 1091–1093.

5.7 Pechmann Pyrazole Synthesis

Richard J. Mullins

5.7.1 Description

The 1,3-dipolar cycloaddition between diazoalkanes and alkynes resulting in pyrazole formation is known as the Pechmann pyrazole synthesis.

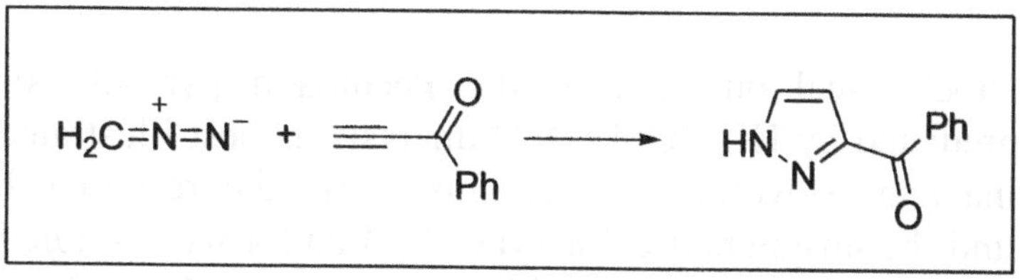

5.7.2 Historical Perspective

The Pechmann pyrazole synthesis was developed by the German chemist Hans von Pechmann. Following his doctoral studies at the University of Greifswald under Heinrich Limpricht, Pechmann became professor at the University of Munich. In 1894, he discovered diazomethane,[1,2] a compound that would lead to several notable discoveries in his career. Following a move to the University of Tübingen in 1895, Pechmann discovered the reaction of diazomethane and acetylene could produce pyrazoles, via a 1,3-dipolar cycloaddition that is the subject of this chapter.[3] He is also credited with the original discovery of the polymer ethylene, made while he was studying the thermal decomposition products of diazomethane. Pechmann remained at the University of Tübingen until his death in 1902.[4,5]

5.7.3 Mechanism

The Pechmann pyrazole synthesis features a 1,3-dipolar [3 + 2] cycloaddition between a diazoalkane and an alkyne followed by tautomerization, presumably via a 1,5-hydride shift, to give the pyrazole (**1**).

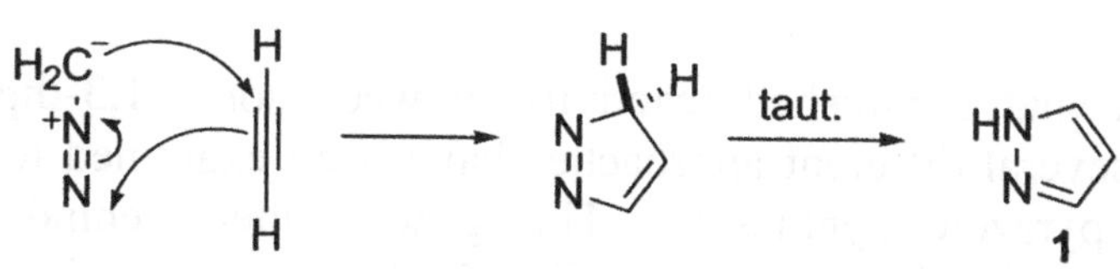

The concerted cycloaddition proceeds with a high level of regioselectivity, especially when using diazomethane along with an electron deficient alkyne. In this case, the reaction proceeds via HOMO$_{(diazomethane)}$–LUMO$_{(alkyne)}$ interaction. A detailed description of frontier molecular orbitals in 1,3-dipolar and other cycloadditions can be found here.[6]

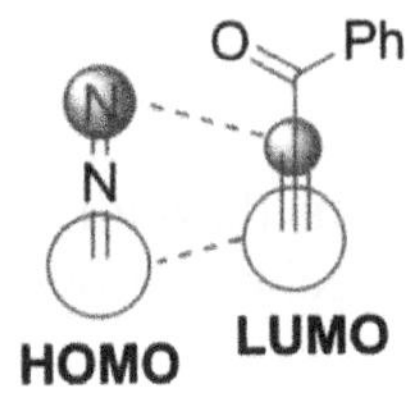

The regiochemical outcome of the Pechmann pyrazole synthesis, which can be predicted by HOMO–LUMO analysis, is best illustrated by the results of Fariña and co-workers.[7–9] In this work, the reactions between diazomethane and the unsymmetrical alkynes **2**, **3** and **4** were studied. With substrates containing a single electron withdrawing group (**2** and **3**) exclusive levels of regioselectivity could be achieved. However, the use of an alkyne containing two electron-withdrawing groups (**4**) resulted in a 60 : 40 mixture of regioisomers. As expected, the electron withdrawing aldehyde, while substantially enhancing the reactivity of the dipolarophile, resulted in the scrambling of the regiochemistry. Similar regioselectivity was obtained using ethyldiazoacetate with **2**, **3** and **4**.

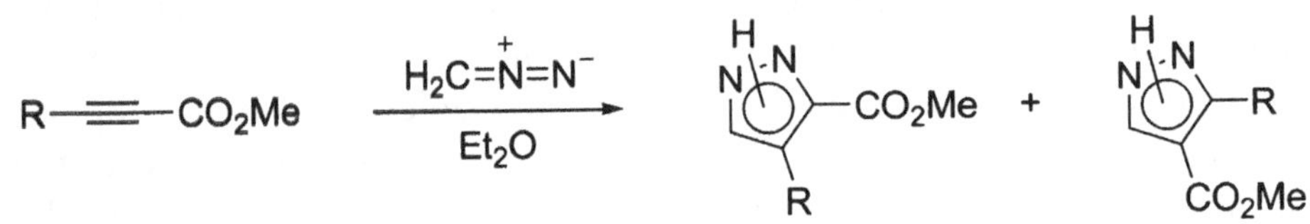

2 R = -CH(OMe)$_2$	4h, 0 °C, 92%	100	:	0
3 R = -CH$_2$OH	24h, 0 °C, 90%	100	:	0
4 R = -CHO	4h, –70 °C	60	:	40

5.7.4 Variations and Improvements

Given the somewhat diminished reactivity between some 1,3-dipoles and dipolarophiles, several different approaches have been examined to expedite the Pechmann pyrazole synthesis. The general idea behind the two approaches discussed below is very similar; both rely on manipulating the HOMO and LUMO of the reacting partners.[10,11] The first approach relies on the ability of Lewis acids to lower the energy of the dipolarophile's LUMO, thereby increasing the rate of the 1,3-dipolar cycloaddition. The resulting reaction then occurs via a $HOMO_{(1,3\text{-dipole})}$–$LUMO_{(\text{dipolarophile})}$ interaction. Activation of the reaction between **5** and **6** has been achieved in this manner, using $InCl_3$ in water to produce **7**.[12] Zeolite NaY has also been used to activate the reaction via similar principles.[13]

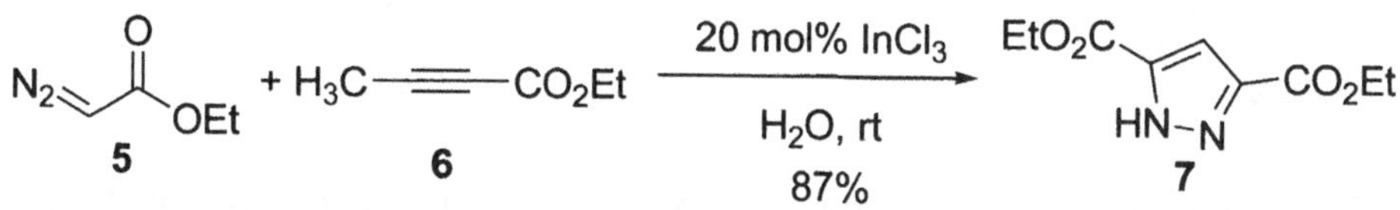

5 **6** **7**

Alternatively, increased electron density in the dipolarophile results in an increase in the energy of this compound's HOMO. As a result, pyrazole synthesis occurs through an inverse electron demand $HOMO_{(\text{dipolarophile})}$–$LUMO_{(1,3\text{-dipole})}$ interaction. This strategy has been demonstrated by Qi and Ready in the synthesis of **10**. Deprotonation of alkyne **8** and subsequent treatment with CuCN·6LiCl results in a copper acetylide species which undergoes facile cyclization with diazoacetate **9** to give **10**.[11] Numerous other examples of this method can be found in these studies, demonstrating it as a general and effective approach to the pyrazole moiety. A similar strategy uses $Zn(OTf)_2/Et_3N$ for generating the metal acetylide species.[14]

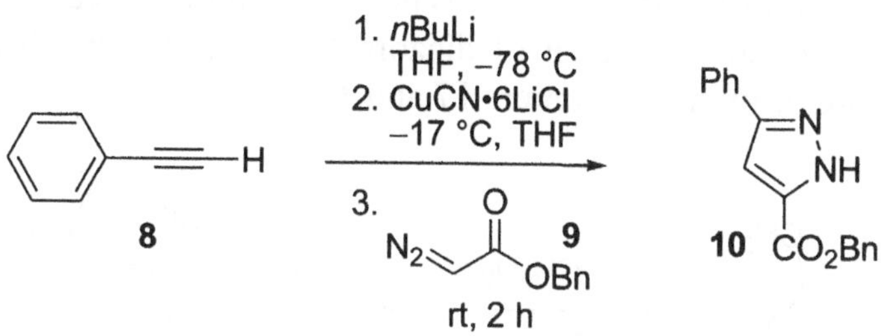

8 **9** **10**

5.7.5 *Synthetic Utility*

The original reaction discovered by Pechmann involved the cycloaddition of diazomethane and acetylene. Although a better understanding of the reaction has led to the common use of more electron-deficient alkynes, diazomethane continues to be synthetically useful.[15] A recent, elegant example of the use of diazomethane as the 1,3-dipole was demonstrated in the preparation of 2,3-benzodiazepine derivatives, potential noncompetitive AMPA antagonists.[16] Beginning with the alkyne **11**, the pyrazole moiety could be incorporated into the benzodiazepine structure, using the Pechmann pyrazole synthesis, to produce **12**.

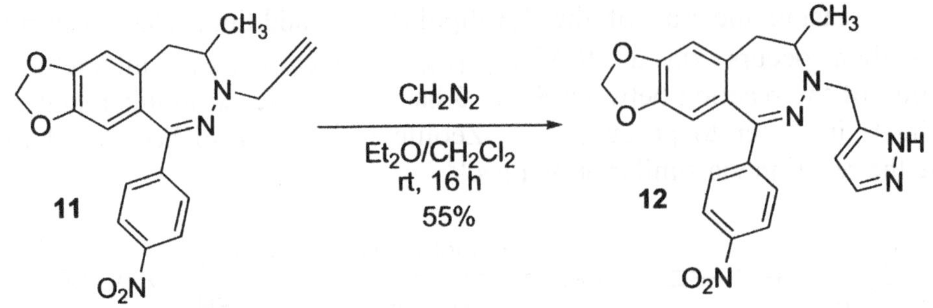

In a number of studies, the Pechmann pyrazole synthesis has found utility in the synthesis of fluorinated heterocycles.[17] As an example, the cyclization between **13** and diazomethane resulted in the efficient preparation of trifluoroalkyl pyrazole **14**.[18] As is the case with the majority of isolated pyrazoles, the product **14** is in equilibrium with the other N–H tautomer.

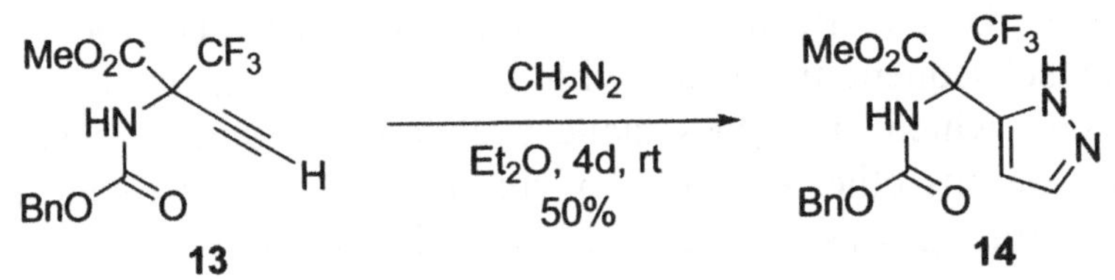

The trifluoromethyl substituted pyrazole **16** is synthesized in high yield following a similar protocol.[19] While **16** could be easily isolated, treatment with a second equivalent of diazomethane results in alkylation to give a 1 : 1 mixture of pyrazole isomers **17** and **18**.

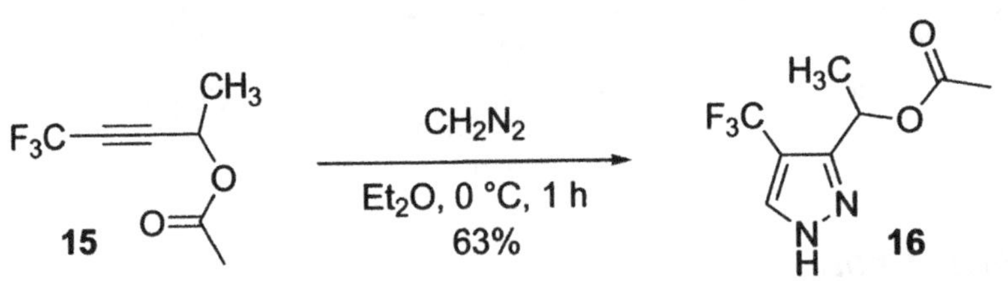

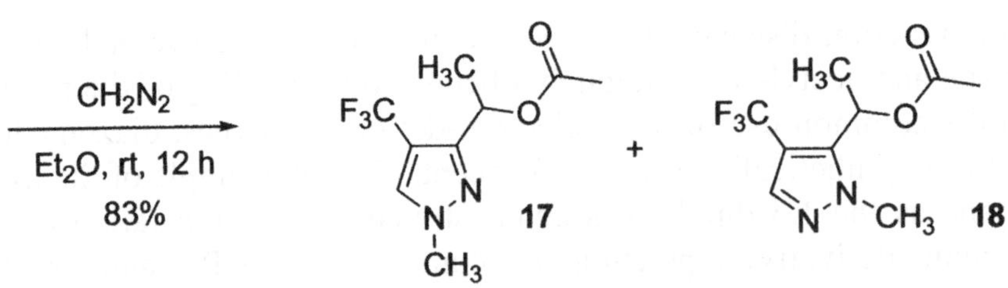

Complexes between **19** and Cu(hfac)$_2$ have been show to undergo specific magnetic–structural phase transitions. A known approach to the synthesis of mononitroxide **19** uses aldehyde **20**, a compound typically produced via formylation of an alkylpyrazole. This reaction, however, is plagued by low yields and the use of extremely harsh conditions, both of which might limit the generality of the synthesis. An alternate approach makes use of the Pechmann pyrazole synthesis between alkyne **21** and diazomethane.[20] When the reaction was performed in ether a mixture of regioisomers resulted. These regioisomers could then be carried on in the synthesis to yield **24** and **25** as a separable mixture.

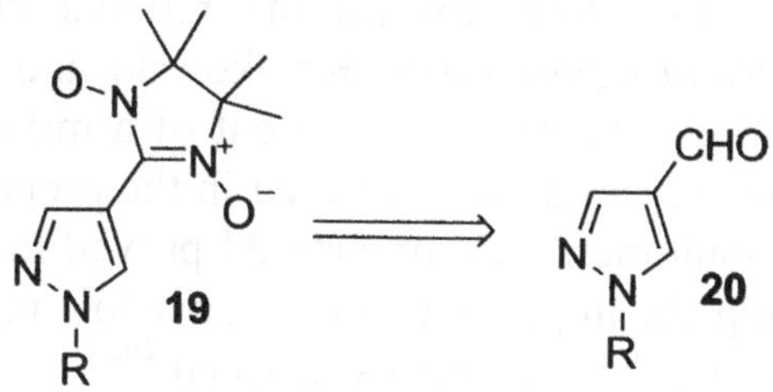

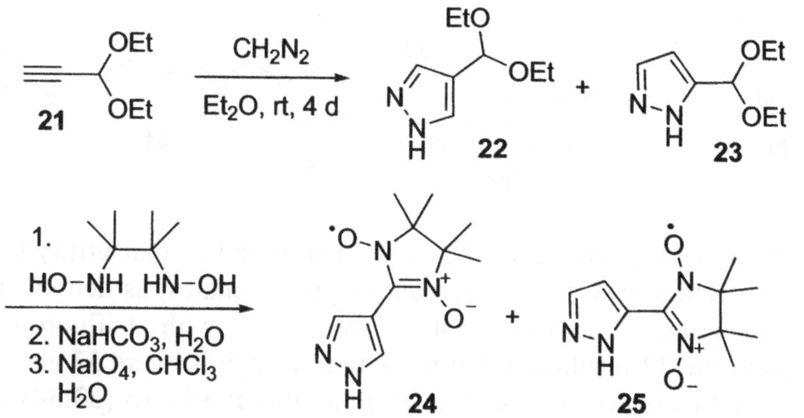

Because the 1,3-dipolar cylcoaddition reactions of diazomethane and alkynes occur between the HOMO of diazomethane and LUMO of the alkyne, electron rich alkynes with a high LUMO energy do not react with diazomethane in the Pechmann pyrazole synthesis. However, introduction of carbonyl functionality into the diazomethane moiety lowers the orbital energies such that concerted cyclization becomes feasible, even with highly electron-rich alkynes such as ynamines. As demonstrated by Huisgen and co-workers,[21] the reaction between methyldiazoacetate and 1-diethyl-aminopropyne occurs through $HOMO_{(dipolarophile)}$–$LUMO_{(diazoalkane)}$ interaction to give a high yield of the single regioisomer **28**. While the majority of electron-deficient diazo compounds still react via their HOMO, this example demonstrates that this preference can be reversed when the alkyne is highly electron-rich.

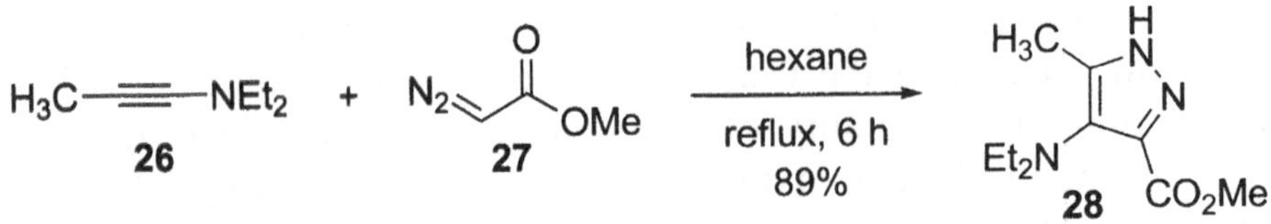

As a result of the observations discussed above, ethyldiazoacetate (EDAC) (**30**) has become a commonly used 1,3-dipole in the Pechmann pyrazole synthesis.[22–24] In efforts toward the rational design of growth inhibitors of *Mycobacterium tuberculosis*, Kozikowski and co-workers used this strategy for the synthesis of **31**.[25] Treatment of a mixture of alkyne **29** and EDAC under microwave conditions resulted in the preparation of **31** as a mixture of 1-*H* and 2-*H* tautomers. Compound **31** proved inactive in the anti-TB assay, thereby proving the importance of an isoxazole moiety for anti-TB activity, as previous work by these authors suggested.[26,27]

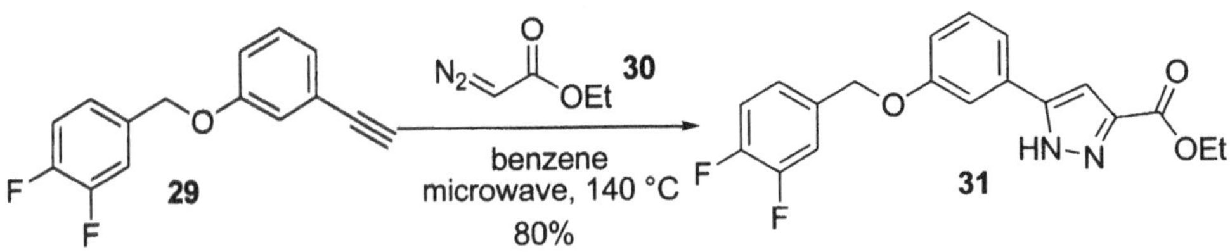

The Pechmann pyrazole synthesis has also found similar utility for the preparation of perfluoroalkylated heterocyclic phosphonates, as demonstrated by the Shen group.[28] Reaction of EDAC (**30**) with trifluoromethyl alkynylphosphonate **32** resulted in the formation of regioisomers **33** and **34** in high yield. The two regioisomers, isolated in a ratio of 85 : 15 (**33**:**34**) could be separated by column chromatography.

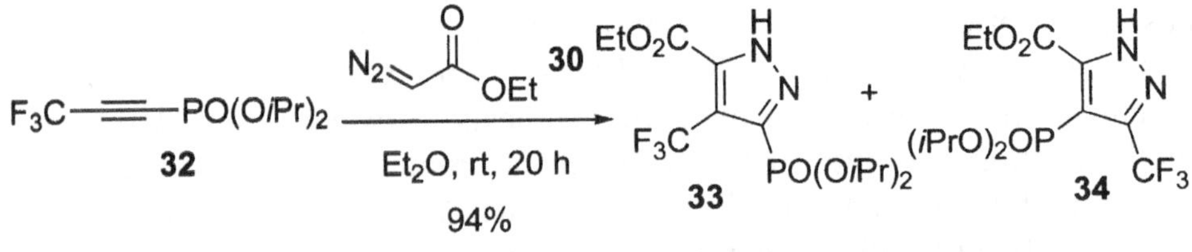

The use of ethyldiazoacetate in the Pechmann pyrazole synthesis has been adapted for use on solid support.[29] Treatment of solid-supported acetylenic sulfone **35** with an excess of EDAC, followed by cleavage from the solid support, provided **36** as a single regioisomer. Notably, when the reaction was done in solution phase with **37**, the isomers **38** and **39** were isolated in a 4 : 1 ratio.

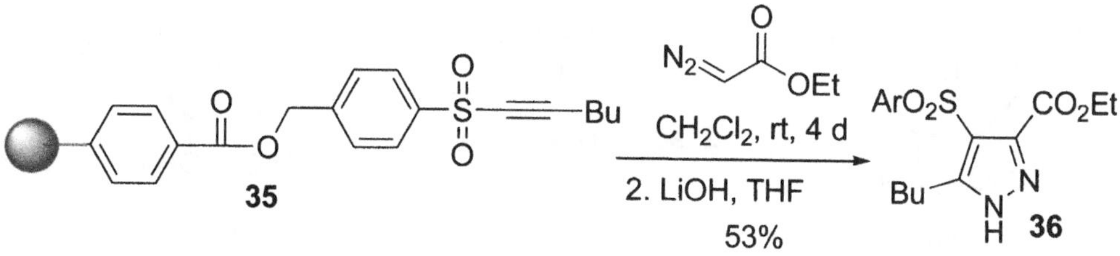

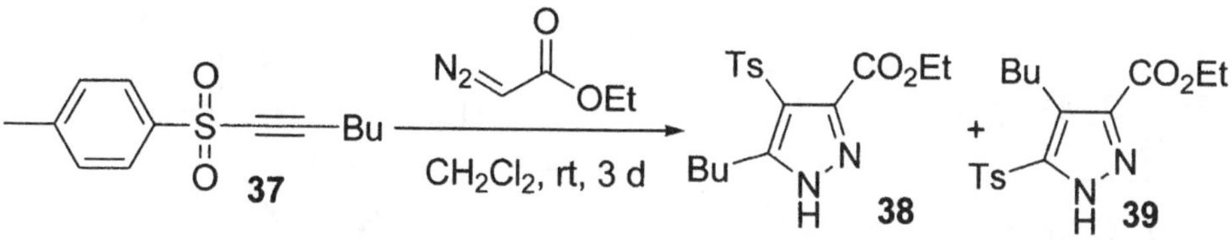

A common dipolarophile for the Pechmann pyrazole synthesis is dimethyl acetylene dicarboxylate (DMAD, **41**), a highly electron-deficient alkyne. Spring and co-workers used the 1,3-dipolar cycloaddition between DMAD and a fluorous-tagged diazoacetate (**40**) as part of studies aimed at production of structurally diverse scaffolds.[30] As expected, the reaction proceeded efficiently to produce **42** in high yield.

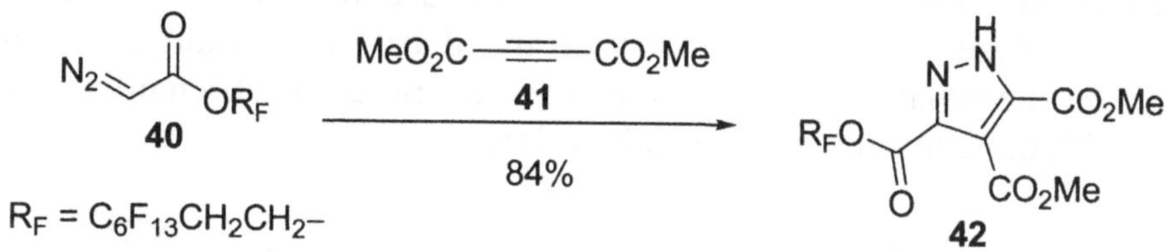

While the Pechmann pyrazole synthesis routinely features the use of simple diazo compounds such as diazomethane and ethyldiazoacetate, several complex diazo compounds have also found utility in the title reaction. One such reagent has been used to produce an analog of pyrazofurin (**48**)[31] in order to evaluate the importance of intramolecular hydrogen bonding in its intracellular conversion to the 5′-monophosphate, a process that contributes to its antitumor and antiviral properties. The diazo compound **44** was produced in a two-step process from **43**. Reaction of **43** with nitrogen dioxide and acetic acid to produce the *N*-nitrosamide was followed by treatment with aqueous potassium hydroxide to yield **44**. The reaction of **44** and methylpropiolate (**45**) resulted in efficient formation of **46**, in a 76% yield over the three steps. Completion of the synthesis eventually resulted in formation of the pyrazofurin analog (**47**) in high yield. Similar work in the same laboratory used the Pechmann pyrazole synthesis for preparation of an acyclic analogue of 4-deoxypyrazofurin.[32]

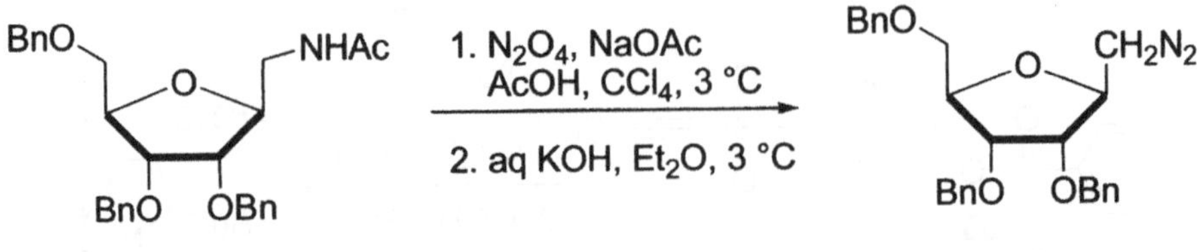

43 44

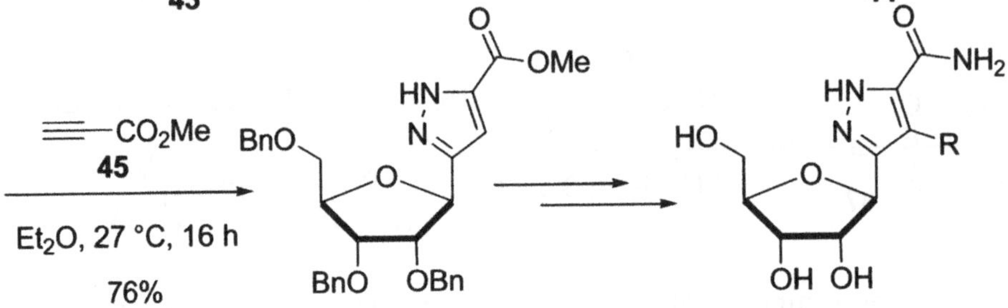

46 47 R = H 48 R = OH

Although unintended, it was found that the α-diazobenzamide **49** readily participates in the Pechmann pyrazole synthesis with dimethyl acetylene dicarboxylate (**41**).[33] While studying the reactivity of carbenes generated at the benzylic position of **49**, the use of $Cu(acac)_2$ did not effectively catalyze the decomposition of the diazo functionality, thereby allowing the Pechmann pyrazole synthesis to proceed efficiently upon treatment with dimethyl acetylene dicarboxylate.

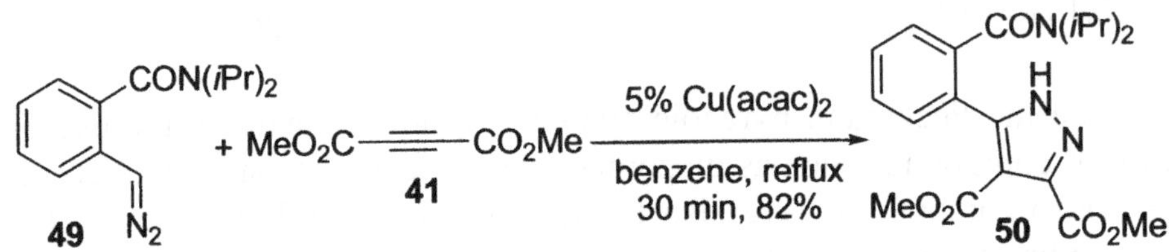

49 41 50

5.7.6 Experimental

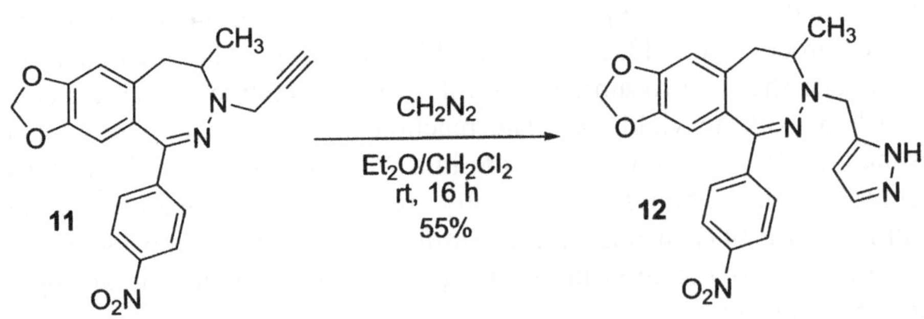

11 12

Synthesis of 12[16]

To a stirred and cooled (0 °C) solution of **11** (0.70 g, 2.0 mmol) in dichloromethane (14 mL) a solution of diazomethane (0.14 g, 4.0 mmol) in dry ether (17 mL) was added dropwise for 20 min. After the addition was complete the reaction mixture was stirred overnight at room temperature. The ether was evaporated to give an oily residue, which was chromatographed on a silica gel column. Elution was made with chloroform–methanol (98 : 2) and the main fraction gave after evaporation **12** as a solid residue (0.45 g, 55%).

5.7.7 References

1. Pechmann, H. v. *Ber. Dtsch. Chem. Ges.* **1894,** *27*, 1888–1891.
2. Pechmann, H. v. *Ber. Dtsch. Chem. Ges.* **1895,** *28*, 855–861.
3. Pechmann, H. v. *Ber. Dtsch. Chem. Ges.* **1898,** *31*, 2950–2951.
4. Koenigs, W. *Ber. Dtsch. Chem. Ges.* **1903,** *36*, 4417.
5. [R] Partington, J. R. *A History of Chemistry*; *vol. 4*, Macmillan: London, 1964; 838–839.
6. [R] Fleming, I. *Frontier Orbitals and Organic Chemical Reactions*, Wiley: New York, 1976.
7. Fariña, F.; Fernández, P.; Fraile, M. T.; Martin, M. V.; Martin, M. R. *Heterocycles* **1989,** *29*, 967–974.
8. Fariña, F.; Martin, M. V.; Sánchez, F. *An. Quim.* **1982,** *78C*, 332.
9. Fariña, F.; Fernández, P.; Martin, M. R.; Martin, M. V.; Sánchez, F. *An. Quim.* **1983,** *79C*, 333.
10. Krishna, P. R.; Sekhar, E. R.; Mongin, F. *Tetrahedron Lett.* **2008,** *49*, 6768–6772.
11. Qi, X.; Ready, J. M. *Angew. Chem., Int. Ed.* **2007,** *46*, 3242–3244.
12. Jiang, N.; Li, C.-J. *Chem. Commun.* **2004,** 394–395.
13. Kobayashi, K.; Igura, Y.; Imachi, S.; Masui, Y.; Onaka, M. *Chem. Lett.* **2007,** *36*, 60–61.
14. He, S.; Chen, L.; Niu, Y.-N.; Wu, L.-Y.; Liang, Y.-M. *Tetrahedron Lett.* **2009,** *50*, 2443–2445.
15. [R] Padwa, A.; Pearson, W. H., eds. *Synthetic Applications of 1,3-Dipolar Cycloaddition toward Heterocycles and Natural Products*, Wiley: New York, 2002.
16. Kertész, M.; Pallagi, I.; Sólyom, S. *J. Heterocyclic Chem.* **2007,** *44*, 431–435.
17. Froissard, J.; Greiner, J.; Pastor, R.; Cambon, A. *J. Fluorine Chem.* **1984,** *26*, 47–57.
18. Sewald, N.; Burger, K. *Liebigs Ann. Chem.* **1992,** 947–952.
19. Sibous, L.; Tipping, A. E. *J. Fluorine Chem.* **1993,** *62*, 39–49.
20. Tretyakov, E. V.; Tolstikov, S. E.; Romanenko, G. V.; Shvedenkov, Y. G.; Sagdeev, R. Z.; Ovcharenko, V. I. *Russ. Chem. Bull.* **2005,** *54*, 2169–2181.
21. Huisgen, R.; Reissig, H.-U.; Huber, H. *J. Am. Chem. Soc.* **1979,** *101*, 3647–3648.
22. Nenajdenko, V. G.; Statsuk, A. V.; Balenkova, E. S. *Chem. Heterocycl. Cmpds.* **2003,** *39*, 598–603.
23. Vuluga, D.; Legros, J.; Crousse, B.; Bonnet-Delpon, D. *Green Chem.* **2009,** *11*, 156–159.
24. Zrinski, I.; Juribasic, M.; Eckert-Maksic, M. *Heterocycles* **2006,** *68*, 1961–1967.
25. Lilienkampf, A.; Pieroni, M.; Wan, B.; Wang, Y.; Franzblau, S. G.; Kozikowski, A. P. *J. Med. Chem.* **2010,** *53*, 678–688.
26. Mao, J.; Wan, B.; Wang, Y.; Franzblau, S. G.; Kozikowski, A. P. *ChemMedChem* **2007,** *2*, 811–813.
27. Lilienkampf, A.; Mao, J.; Wan, B.; Wang, Y.; Franzblau, S. G.; Kozikowski, A. P. *J. Med. Chem.* **2009,** *52*, 2109–2118.
28. Shen, Y.; Zheng, J.; Xin, Y.; Lin, Y.; Qi, M. *J. Chem. Soc., Perkin Trans. 1* **1995,** 997–999.
29. Gao, D.; Zhai, H.; Parvez, M.; Back, T. G. *J. Org. Chem.* **2008,** *73*, 8057–8068.
30. Wyatt, E. E.; Fergus, S.; Galloway, W. R. J. D.; Bender, A.; Fox, D. J.; Plowright, A. T.; Jessiman, A. S.; Welch, M.; Spring, D. R. *Chem. Commun.* **2006,** 3296–3298.
31. Sauer, D. R.; Schneller, S. W. *Synthesis* **1991,** 747–750.

32. Sauer, D. R.; Schneller, S. W. *J. Org. Chem.* **1990,** *55*, 5535–5538.
33. Chen, C.-W.; Beak, P. *J. Org. Chem.* **1986,** *51*, 3325–3334.

6.1 Baeyer Pyridine Synthesis

David A. Conlon

6.1.1 *Description*

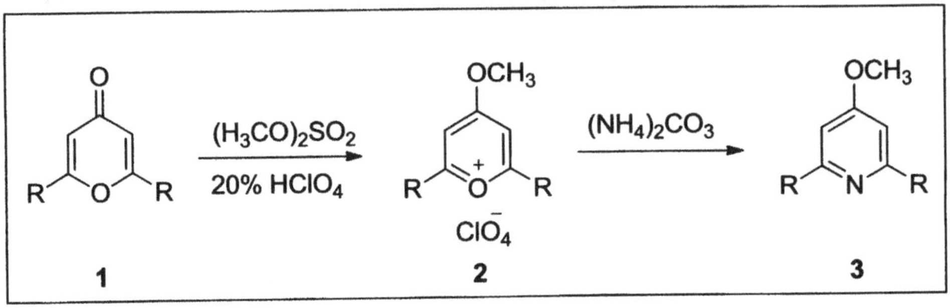

The reaction of pyrylium salts **2** with ammonia or ammonium salts to produce pyridine derivatives **3** is referred to as the Baeyer pyridine synthesis.[1] There are several methods available for the preparation of pyrylium salts, the alkylation of pyran-4-one derivatives **1** with dimethylsulfate to generate the prerequisite pyrylium salts was reported by Baeyer in his 1910 paper.

Addition of an ammonia source to pyrylium salts readily affords pyridine derivatives and provides a good method for the preparation of the pyridine moiety if the corresponding pyrylium salt is accessible. The carbon oxygen double bond present in the pyrylium salt is an oxonium ion however, owing to aromatic stabilization they are easily formed by a variety of methods. The reactivity of pyrylium salts toward nucleophiles makes them useful reagents for the preparation of structurally diverse heterocyclic compounds. Thus pyrylium salts afford pyridines by reaction with ammonia, pyridine-*N*-oxides by reaction with hydroxylamine and pyridinium salts by reaction with primary amines.

6.1.2 *Historical Perspective*

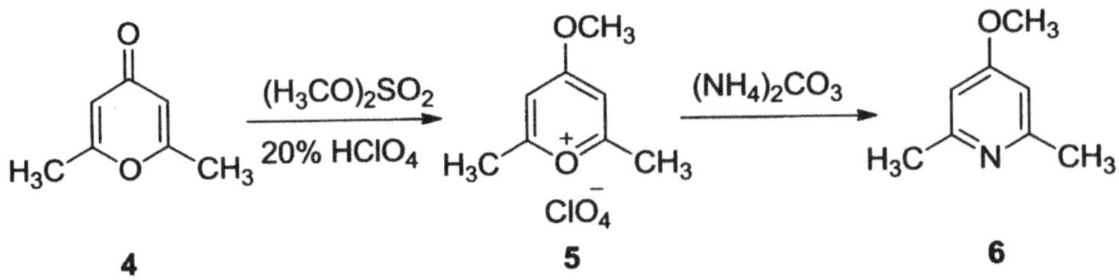

In 1910, Adolf von Baeyer at the University of Munich reported that the product from the reaction of dimethylsulfate and 2,6-dimethyl-4*H*-pyran-4-one **4** at 50 °C formed an insoluble pyrylium salt **5** when treated with perchloric acid. Treatment of the isolated solid with aqueous ammonium carbonate generated 4-methoxy-2,6-dimethylpyridine (**6**).

In 1912, Baeyer and Picard reported that 2,4,6-trimethylpyrylium perchlorate (**7**) could be prepared by reacting 2,6-dimethylpyrone (**4**) with methylmagnesium iodide followed by treatment of the reaction mixture with perchloric acid.[2] 2,4,6-Trimethypyridine (**8**) was readily formed upon treatment of **7** with ammonia.

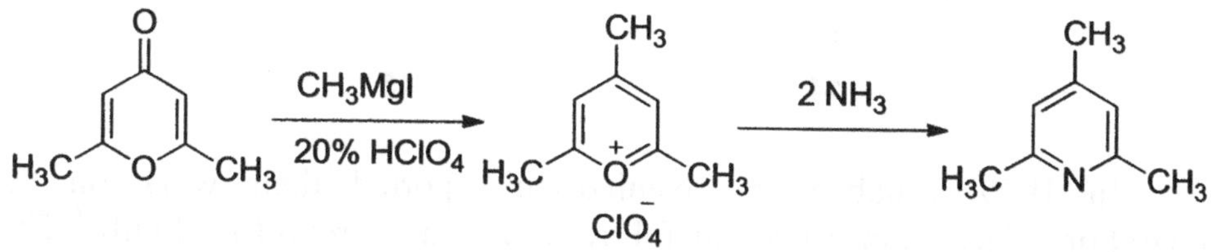

4 **7** **8**

2,6-Dimethyl-4-phenylpyridine was produced using an analogous sequence and this process has expanded the access to a variety of 2,4,6-substituted pyridines. This synthetic sequence had limited utility however, due to the reactivity of the intermediate pyrylium ions with residual organomagnesium reagents during the quench.

Two efficient procedures for synthesizing the requisite pyrylium salts were reported by Dilthey and Balaban. Symmetrical pyrylium salts with aromatic substituents such 2,4,6-triphenylpyrylium ferrochloride were prepared by Dilthey from 2 mol of acetophenone and 1 mol of benzaldehyde in the presence of acetic anhydride and ferric chloride.[3] By preforming the intermediate chalcone **11** it is also possible to prepare unsymmetrically substituted 2,4,6-triarylpyrylium salts **14**.[4] Once again, these pyrylium salts are readily converted to the corresponding triarylpyridines **15** following treatment with ammonium acetate under mild conditions.[5] In 2005, Wang reported the preparation and subsequent conversion to of a series of pyrylium salts to 2,4,6-triarylpyridines with diverse aryl ring substituents.[6] The conversion of the pyrylium salts to pyridines was performed under microwave irradiation in aqueous ammonia using PEG-400 as a phase transfer catalyst.

9 **10** **11**

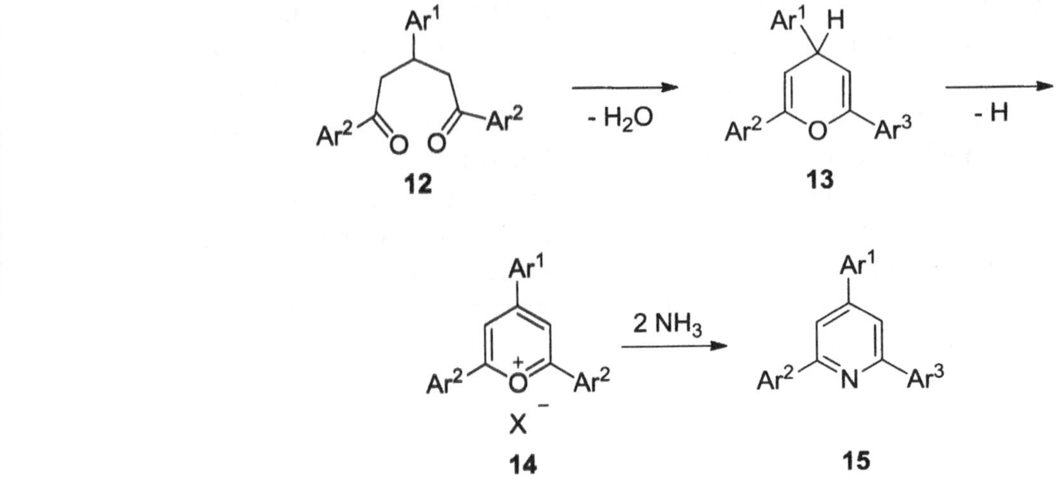

In 1959, Balaban and Nenitzescu reported their work on the preparation of pyrylium salts that followed an earlier report by Praill.[7] This process is often referred to as the Balaban–Nenitzescu–Praill synthesis and is an efficient method for the preparation of alkyl substituted pyrylium salts and therefore alkyl substituted pyridines. The process involves the diacylation of olefins and has become a common method for the synthesis of 2,4,6-trialkylsubstituted pyrylium salts with identical substituents in the 2- and 6-positions.[8] Stang reported an improved procedure to prepare 2,6-di-*tert*-butyl-4-methylpyridine (**17**) from the corresponding 2,6-di-*tert*-butyl-4-methyl-pyrylium triflate salt (**16**) in 1976.[9]

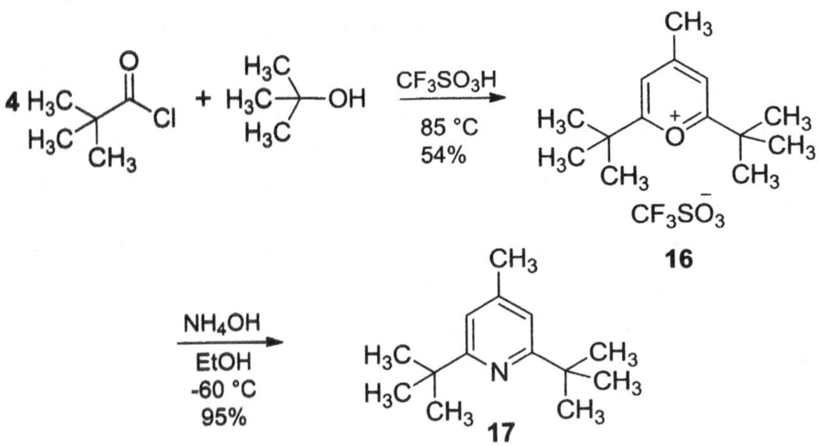

6.1.3 *Mechanism*

The initial nucleophilic attack of ammonia on a 2,4,6-trisubstituted pyrylium cation **18** occurs at the at α-oxonium carbon that has the smaller substituent. This is followed by deprotonation with a second equivalent of ammonia and an electrocyclic ring opening to produce a vinylogous amide **20** and an ammonium salt. The vinylogous amide is converted to the product pyridine **6** with the loss of a mole of water.

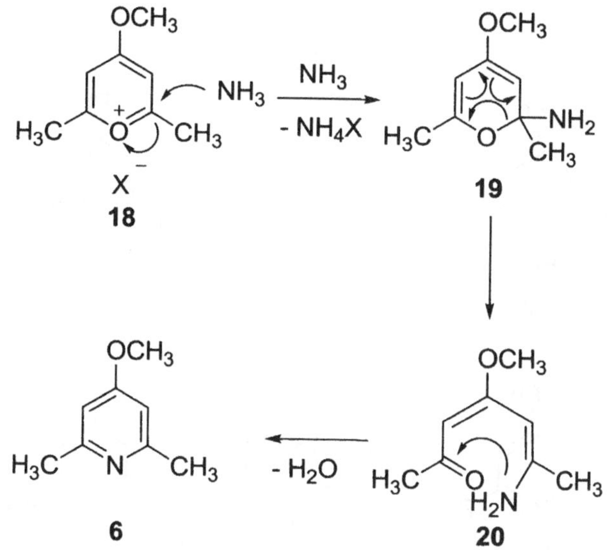

6.1.4 Variations

The reaction of pyrylium salts with hydroxylamines and primary amines produces pyridine *N*-oxides and pyridinium salts, respectively. These pyridine derivatives have interesting properties and are also useful intermediates for the preparation of a variety of new materials. For example, nucleophilic substitution on the pyridine ring is facilitated by the formation of an *N*-oxide or pyridinium salt.

It was reported in the late 1950s, that the reaction of pyrylium salts with hydroxylamine produces pyridine *N*-oxides.[10] This procedure has been used to prepare 2,6-disubstituted pyridine *N*-oxides in better yield than obtained from the oxidative procedure.[11]

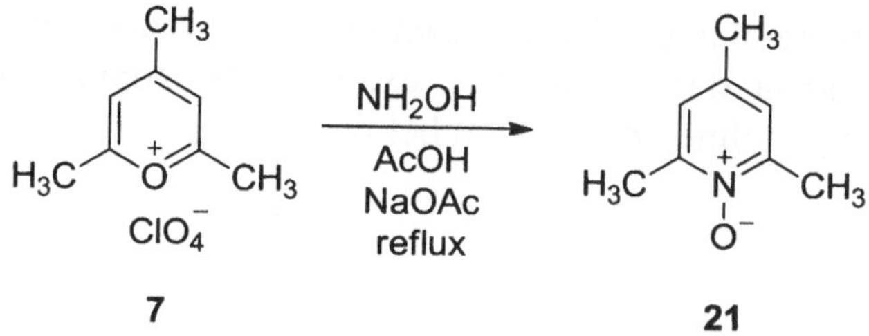

The reactions of primary amines with pyrylium salts produce pyridinium salts. Pyridinium salts like **23** are useful intermediates that have been used to prepare a diverse array of compounds by nucleophilic displacement of the pyridine.[12] Pyridinium salts such as **25** have found use as carbonic anhydrase activators[13] and derivatives like **26** exhibit anti-cholinesterase activity.[14]

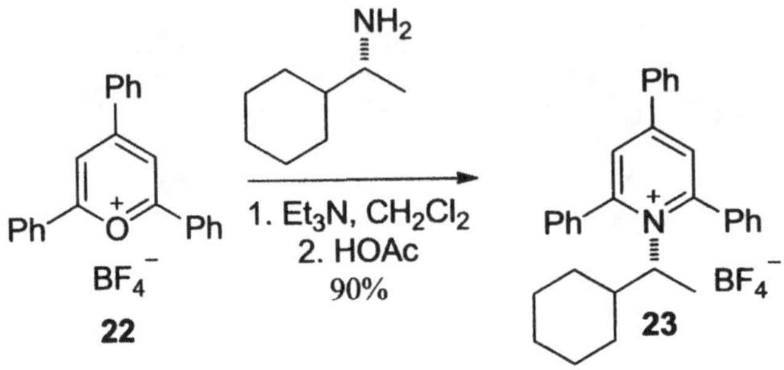

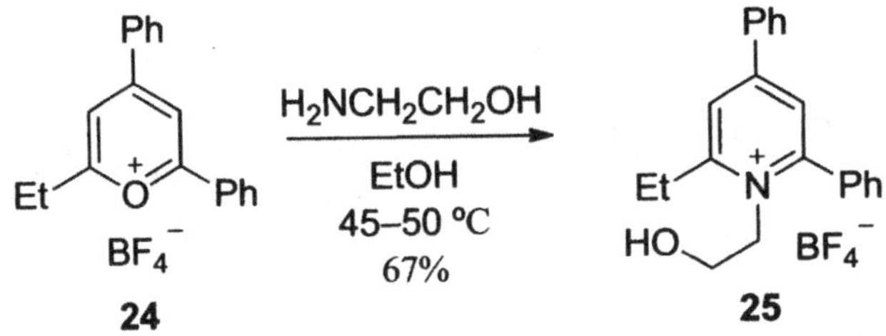

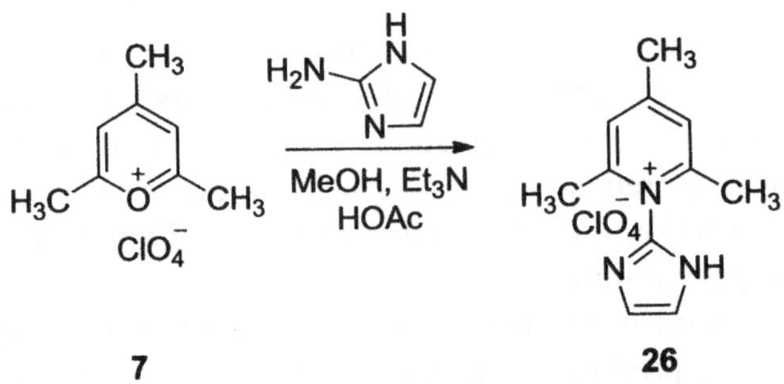

The Baeyer pyridine procedure has been extended to the preparation of other six-membered heterocycles, such as quinolines, isoquinolines, and acridines which are obtained by the reactions of the corresponding pyrylium salts with ammonia derivatives. For example, benzopyrylium salt **27** is converted to isoquinoline **28** as shown below.[15]

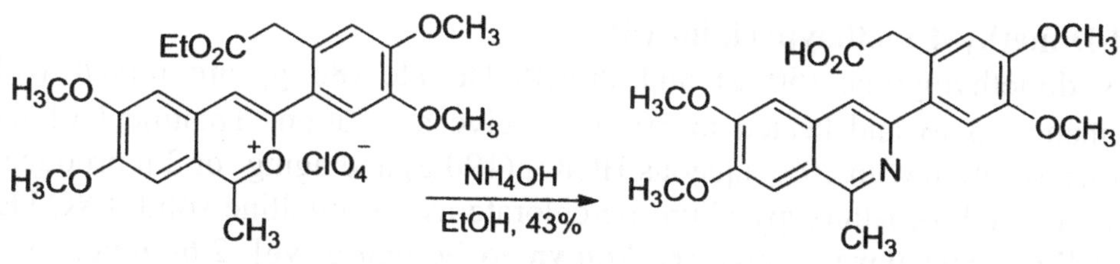

27 28

6.1.5 *Synthetic Utility*

The pyridine ring is ubiquitous in compounds of pharmaceutical and agrochemical interest. Six-membered heterocycles such as pyridines, quinolines, isoquinolines, and acridines can also be obtained by the reactions of the pyrylium salts with ammonia derivatives.

β-Carbolines are a class of indole alkaloids, which are structurally similar to the amino acid L-tryptophan and have a diverse biological activity. Several clinical investigations have indicated that β-carboline derivatives are potentially useful for a variety of neuroscience applications.[16] The Baeyer pyridine synthesis procedure was used during the preparation of the novel β-carboline derivative, ambocarb (**30**), as shown below.[17]

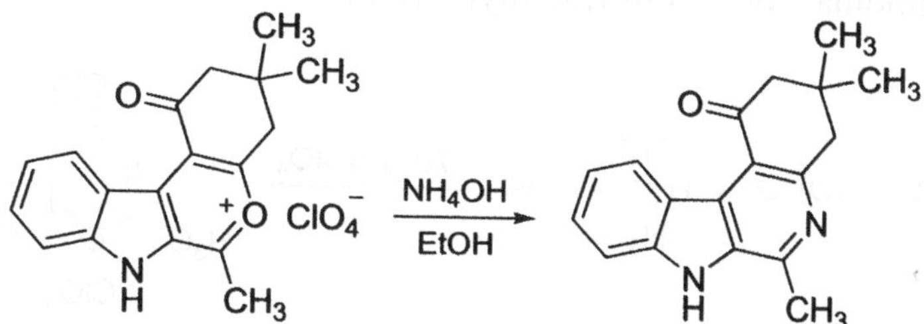

29 ambocarb **30**

Pyrylium salts have been shown to be versatile synthetic intermediates for the preparation of many novel and naturally occurring heterocyclic ring systems.[18]

6.1.6 *Experimental*

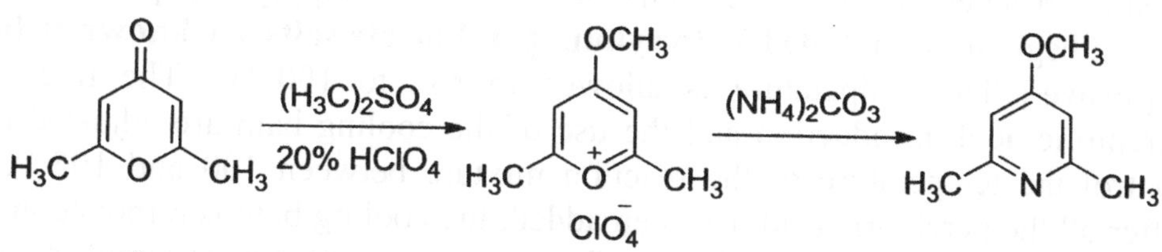

4 **5** **6**

2,4-Dimethy-4-methoypyridine (6)

Dry dimethylpyrone (50 g) and dimethylsulfate (80 g) are mixed with methanol (5 g) and heated to 50 °C to give a solution. Treatment of the cooled solution with 20% aqueous $HClO_4$ (190 g) and aging for 2 h produced 2,6-dimethyl-4-methoxypyrylium perchlorate as a crystalline solid. [**NOTE: Pyrylium perchlorate salts are known to be explosive**]. 2,6-Dimethyl-4-methoxypyrylium perchlorate is instantly converted into 2,6-dimethyl-4-methoxypyridine by the action of aqueous ammonium carbonate.

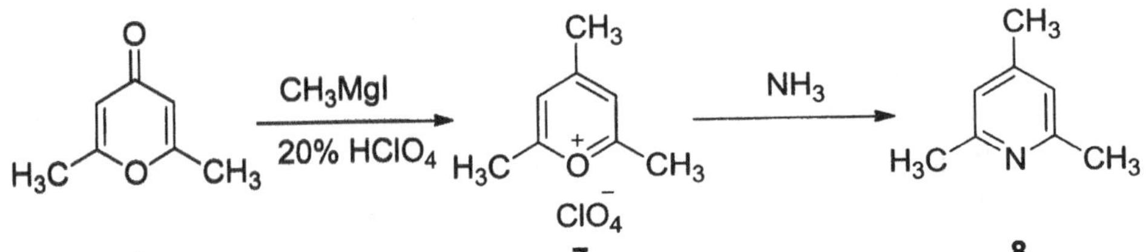

2,4,6-Trimethypyridine (8)

The addition of methylmagnesium iodide to dimethylpyrone in a mixture of ethyl ether and anisole followed by treatment with 20% aqueous $HClO_4$ generated 2,4,6-trimethylpyrylium perchlorate. Treatment of the pyrylium salt with ammonia gave 2,4,6-trimethylpyridine.

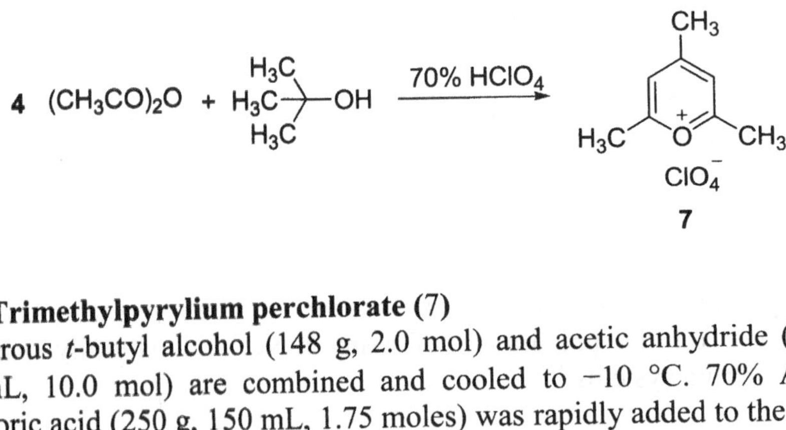

2,4,6-Trimethylpyrylium perchlorate (7)

Anhydrous *t*-butyl alcohol (148 g, 2.0 mol) and acetic anhydride (1020 g, 945 mL, 10.0 mol) are combined and cooled to −10 °C. 70% Aqueous perchloric acid (250 g, 150 mL, 1.75 moles) was rapidly added to the mixture of *t*-butyl alcohol and acetic anhydride. When the temperature of the reaction mixture reaches 40–50 °C, crystals of 2,4,6-trimethylpyrylium perchlorate should begin to form. [**NOTE: Pyrylium perchlorate salts are known to be explosive.**] The temperature is allowed to rise to 100 °C. The rate of perchloric acid introduction and the use of the cooling bath are adjusted to control the temperature of the reaction mixture between 100 and 105 °C. After all the perchloric acid has been added, the cooling bath is removed and stirring of the mixture is continued. The temperature remains at about 90° C for 10 or 15 min and then cools to about 75 °C after 30 min. The dark brown stirred mixture is cooled to 15 °C. The crystalline 2,4,6-trimethylpyrylium perchlorate, which has precipitated, is collected on a Büchner funnel and is washed with a 1:1 mixture of acetic acid and ethyl ether and then washed twice with ethyl ether. Suction is stopped before the crystals are dry. The product can be air dried to give 195–210 g (50–54%) of yellow crystals.

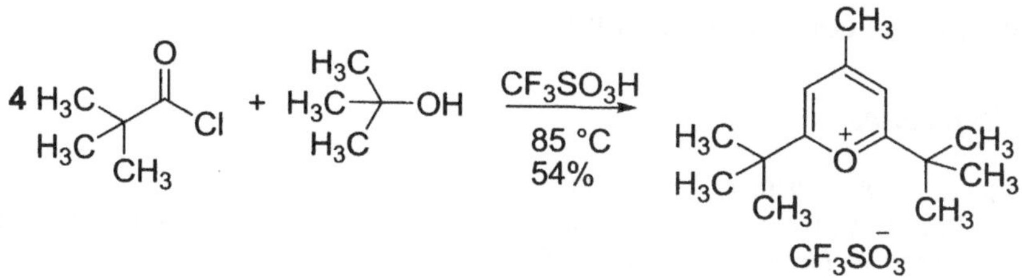

16

2,6-Di-*tert*-butyl-4-methylpyrylium Triflate (16)
Pivaloyl chloride (24.2 g, 0.2 mol) and *tert*-butyl alcohol (3.7 g, 0.05 mol) were combined and the mixture was heated to 85 °C. Triflic acid (15 g, 0.1 mol) was added over a period of 15 min. After the triflic acid addition was completed the mixture was stirred for an additional 10 min at 85 °C. The light brown reaction mixture was then cooled in an ice bath and poured into 100 mL of cold ethyl ether. The light tan precipitate was collected by filtration and air dried to give 9.6 g (54%) of pyrylium salt that was used without further purification in the next step. **[Note: The pyrylium trifluoromethanesulfonate salt does not have the explosive hazard of the corresponding perchlorate]**

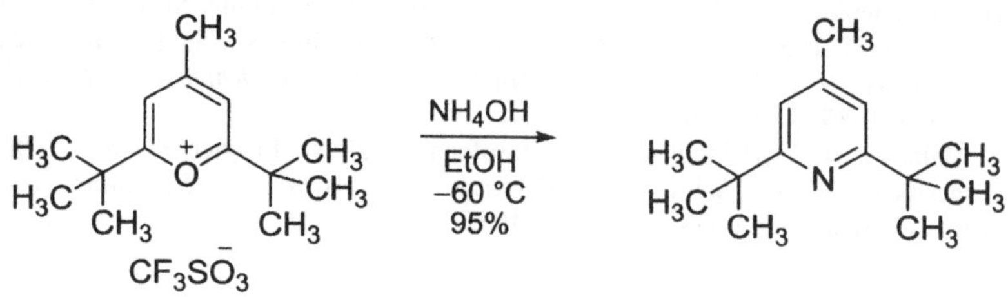

16 17

2,6-Di- tert-butyl-4-methylpyridine (17)
A slurry of crude pyrylium salt **16** (10 g, 0.028 mol) in 200 mL of 95% ethanol was cooled to –60 °C and added to concentrated ammonium hydroxide (100 mL) at –60 °C. The yellow reaction mixture was held at –60 °C for 30 min., then maintained at –40 °C for 2 h, during which time the slurry dissolved. The reaction mixture was then allowed to slowly warm up to room temperature. The reaction mixture was poured into 500 mL of a 2%

NaOH solution, and the resulting emulsion was extracted with four 100-mL portions of pentane; the combined extracts were washed with 25 mL of saturated NaCl, and the pentane was removed on a rotary evaporator. The residual light yellow oil was purified by column chromatography on a 50 × 0.5-cm activated alumina column using pentane as the eluent. The pentane was removed on a rotary evaporator to yield 5.46 g (95%) of a colorless oil that solidifies on cooling or standing.

6.1.7 References

1 Baeyer, A. *Ber.* **1910,** *43,* 2337–2343.
2 (a) Baeyer, A.; Piccard, J. *Liebigs Ann. Chem.* **1912,** *384,* 208–224. (b) Baeyer, A.; Piccard, J.; Gruber, W. *Liebigs Ann. Chem.* **1915,** *407,* 332–369.
3 Dilthey, W. *J. Prakt. Chem.* **1916,** *94,* 53–76.
4 Lin, S. S.; Li, C. Y.; Wang, X. *Chin. Chem. Lett.* **2002,** *13,* 605–606.
5 Simalty-Siemiatycki, M. *Bull. Soc. Chim. Fr.* **1965,** *7,* 1944–1950.
6 Huang, X. Q.; Li, H. X.; Wang, J. X.; Jia, X. F. *Chin. Chem. Lett.* **2005,** *16,* 607–608.
7 (a) Balaban, A. T.; Nenitzescu, C. D. *Tetrahedron Lett.* **1960,** *1,* 7–10. (b) Balaban, A. T.; Nenitzescu, C. D *Org. Synth., Coll. Vol 5,* **1973,** 1106–1110. (c) Balaban, A. T.; Nenitzescu, C. D. *Org. Synth., Coll. Vol 5,* **1973,** 1114–1116.
8 Anderson, A. G.; Stang, P. J. *J. Org. Chem.* **1976,** *41,* 3034–3036.
9 Pedersen, C.; Harrit, N.; Buchardt, O. *Acta. Chem. Scand* **1970,** *24,* 3435–3443.
10 Uncuţa, C.; Căproiu, M. T.; Câmpeanu, V.; Petride, A.; Dănilă, M. G.; Plăveţi, M.; Balaban, A. T. *Tetrahedron* **1998,** *54,* 9747–9764.
11 Katritzky, A. R. *Tetrahedron* **1980,** *36,* 679–699.
12 Said, S. A; Fiksdahl, A. *Tetrahedron: Asymmetry* **2001,** *12,* 1947–1951.
13 Ilies, M.; Banciu, M. D.; Ilies, M. A.; Scozzafava, A.; Caproiu, M. T.; Supuran, C. T. *J. Med. Chem.* **2002,** *45,* 504–510.
14 Golikov, A. G.; Reshetov, P. V.; Kriven'ko, A. P.; Safonova, A. A. *Pharm. Chem. J.* **2005,** *39,* 473–475.
15 Blaskó, G.; Cordell, G. A. *J. Heterocyclic Chem.* **1989,** *26,* 1601–1603.
16 Cao, R.; Peng, W.; Wang, Z.; Xu, A. *Curr. Med. Chem.* **2007,** *14,* 479–500.
17 (a) Luk'yanenko, V. I.; Komissarov, I. V.; Larina, T. F.; Brainina, M. E.; Komissarov, S. I.; Dulenko, V. I. U.S.S. R., 753093, 30 Nov. 1982. (b) Dulenko, V. I.; Luk'yanenko, V. I.; Kibal'nii, A. V.; Malienko, A. A.; Nikolyukin, Yu. A. *Khimiya Geterotsiklicheskikh Soedinenii* **1985,** *3,* 363–366.
18 [R] Balaban, A. T.; Dinculescu, A.; Dorofeenko, G. N.; Fischer, G. W.; Koblik, A. V.; Mezheritskii, V. V.; Schroth, W. *Pyrylium Salts: Syntheses, Reactions, and Physical Properties.* Academic Press: New York, **1982.**

6.2 Katritzky Pyridine Synthesis

David A. Conlon

6.2.1 Description

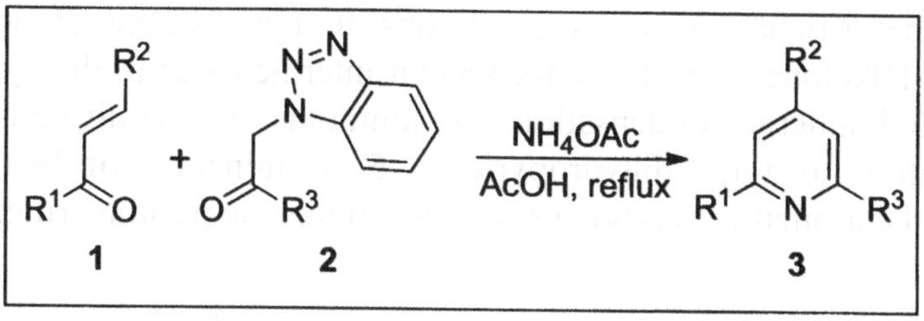

The Katritzky pyridine synthesis is similar to the Kröhnke pyridine synthesis, because both involve the Michael addition of α-substituted ketones **2** to α,β-unsaturated carbonyl compounds **1** in the presence of ammonium acetate, followed by the loss of the α-substituent to generate the pyridine **3**. The α-substituent on the ketone in the Kröhnke pyridine synthesis is typically a pyridinium salt. The α-substituent on the ketone in the Katritzky pyridine synthesis is the benzotriazolyl moiety.

6.2.2 Historical Perspective

In 1999, Katritzky reported a novel [3+2+1] synthesis of 2,4,6-trisubstituted pyridine derivatives that used the Michael addition of α-benzotriazolyl ketones to α,β-unsaturated carbonyl compounds.[1] This reaction resembles the Kröhnke pyridine synthesis and is an extension of Katritzky's earlier studies with benzotriazolyl derivatives that provided access to pyridones, 2-thiopyridones, 5-alkyl-2,4-diphenylpyridines and 2-aminopyridines.[2–5] This approach is attractive as both components are readily synthesized or commercially available. The availability of these starting materials allows for an efficient access to structurally diverse 2,4,6-triaryl pyridines when combined with ammonium acetate in acetic acid at reflux. In addition, it is possible to access fused 2,3,4,6-tetrasubstituted pyridines from the requisite fused bicyclic ketone starting material. The preparation of the pyridine ring via benzotriazole methodology has resulted in improved yields for many compounds and the opportunity to synthesize molecules with a substitution pattern that would be difficult to prepare by other methods.

6.2.3 *Mechanism*

The first step in the Katritzky pyridine synthesis is believed to be the Michael addition of a α-benzotriazolyl ketone **2** to the α,β-unsaturated carbonyl compound **1** to generate a 1,5-diketone derivative **4**. The 1,5-diketone is not typically isolated although its formation has been confirmed via preparation under typical Michael reaction conditions in the absence of ammonium acetate.1,5-Diketone derivatives are known intermediates in the synthesis of pyridines and undergo condensation with ammonia or its equivalent followed by cyclization to form dihydropyridine **5**. Elimination of benzotriazole completes the aromatization process and generates the pyridine ring.

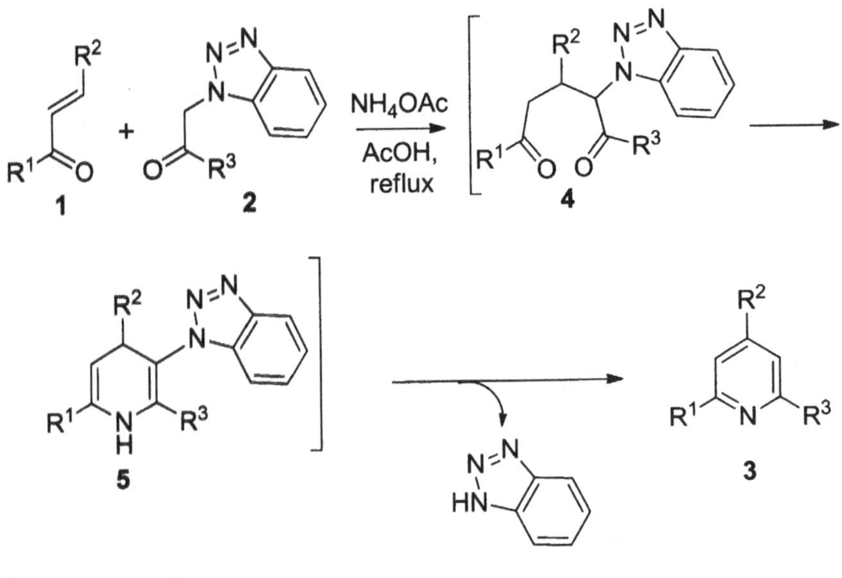

6.2.4 *Variations and Improvements*

Fused 2,3,4,6-tetrasubstituted pyridines **7**, **9** and **10** can be prepared by two complimentary procedures from fused bicyclic ketones.[1] In the first synthetic approach it is the α,β-unsaturated ketone **6** that is part of the fused ring system.

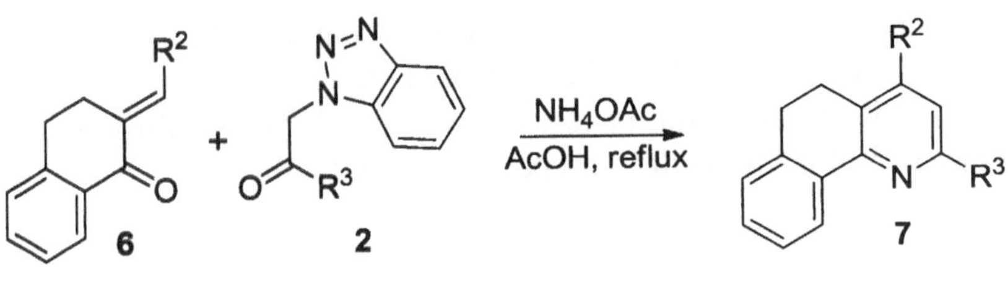

In the second approach, the α-benzotriazolyl ketone **8** coupling partner is part of the fused ring system. This approach was demonstrated for the preparation of the dihydrobenzoquinoline ring system (**9**, *n* = 2) and the indenopyridine ring (**10**, *n* = 1).

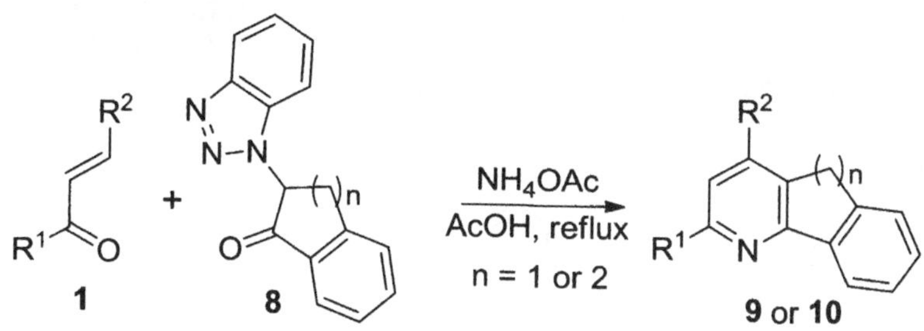

In 1997 Katritzky reported that the Michael addition of α-benzotriazole nitrile **12** to α,β-unsaturated ketones **11** in the presence of a secondary amine generated 2-aminopyridines, **13**.[4] This innovative [3+3] strategy provides an alternative to the typical method of preparing 2-aminopyridines by reacting 2-halopyridines with aliphatic amines. Following the initial Michael addition reaction, nucleophilic attack by the secondary amine on the nitrile moiety initiates the cyclization. This is followed by loss of benzotriazole and water to generate the 2-aminopyridine **13**.

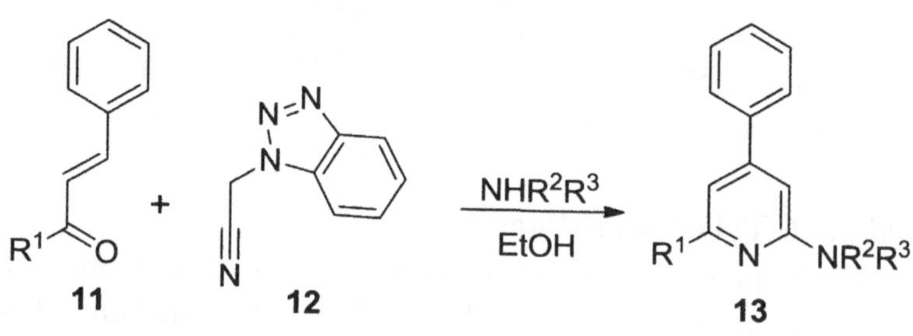

6.2.5 *Synthetic Utility*

The pyridine moiety is found in many biologically active compounds that have demonstrated activity in a wide range of pharmaceutical applications. Pyridines are also important intermediates for the construction of a diverse array of novel and natural molecules with pharmacologically important activity.

2-Amino-4,6-diarylpyridines have been reported to exhibit activity as estrogen receptors.[5] The synthesis of **16** uses a variation of the methodology developed by Katritzky. Addition of α-benzotriazole nitrile **12** to the enone **14** in the presence of secondary amine **15** resulted in the formation of **16**.

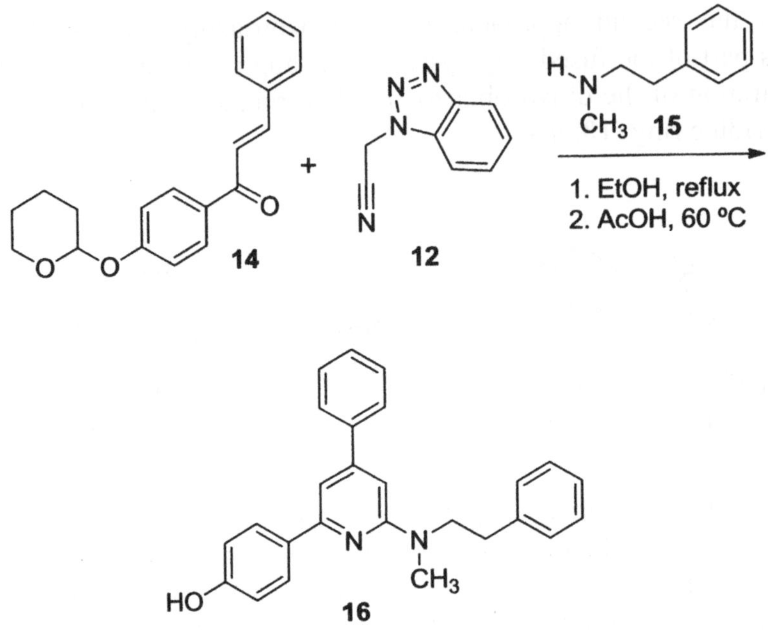

6.2.6 *Experimental*

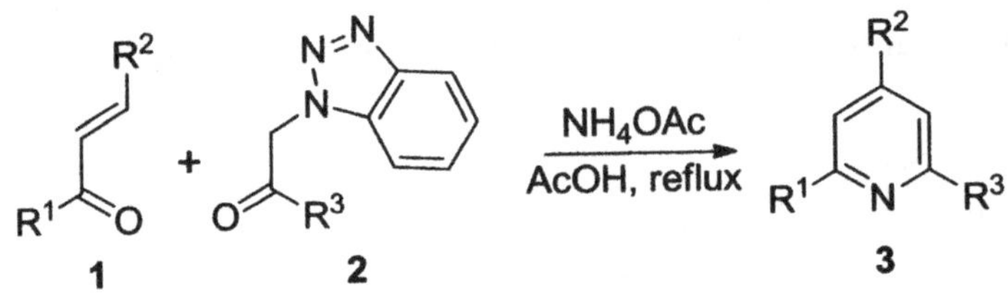

Synthesis of 2,4,6-Triarylpyridines 3

A solution the α-(benzotriazol-1-yl) ketone **2**, chalcone **1** and ammonium acetate in glacial acetic acid was refluxed for 30 h. Addition of ice–water resulted in the precipitation of pyridine **3** which was purified by column chromatography or by recrystallization from hexanes.

6.2.7 *References*

1. Katritzky, A. R.; Abdel-Fattah, A. A. A.; Tymoshenko, D. O.; Essawy, S. A. *Synthesis* **1999,** 2114–2118.
2. Katritzky, A. R.; Mazurkiewicz, R.; Stevens, C. V.; Goordeev, M. F. *J. Org. Chem.* **1994,** *59,* 2740–2742.
3. Katritzky, A. R.; Sheherbakove, I. V. *J. Heterocyclic Chem.* **1996,** *33,* 2031–2036.
4. Katritzky, A. R.; Belyakov, S. A.; Sorochinsky, A. E. Henderson, S. A. *J. Org. Chem.* **1997,** *62,* 6210–6214.
5. [R] Katritzky, A. R; Rachwal, S. *Chem. Rev.* **2010,** *110,* 1564–1610.
6. (a) Henke, B. R.; Drewry, D. H.; Jones, S. A.; Stewart, E. L; Weaver, S. L.; Wiethe, R. W. *Bioorg. Med. Chem. Lett.* **2001,** *11,* 1939–1942. (b) Drewry, D. H.; Henke, B. R. US Pat. 7, 276,523 Oct. 2, 2**007**.

Chapter 7 Quinolines and Isoquinolines 351

7.1 Betti Reaction

Matthew J. Fuchter

7.1.1 Description

The multicomponent reaction between 2-naphthol (**1**), aryl aldehydes and amines is known as the Betti reaction.[1,2]

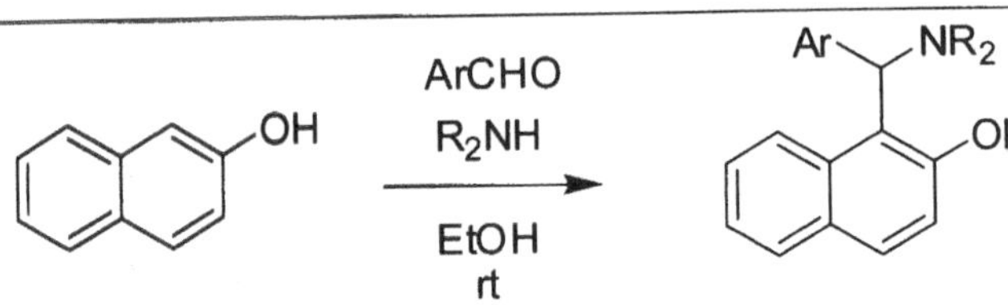

1 2

Either ammonia or a variety of amine substrates can be used to prepare the product **2**, widely called a Betti base.[2] Various substitution patterns are tolerated on both the naphthol and aryl aldehyde component. While the reaction was classically performed in ethanol, a variety of solvents, and using the substrates neat are also possible. Increased rates have been observed using acid catalysis.[3] The reaction results in a product **2** with a benzylic chiral center, which as such can be resolved into its enantiomers. Alternatively, chiral amines can be used to control the stereoselectivity of the process. Enantiomerically pure Betti bases have shown potential as chiral auxiliaries and as ligands in asymmetric reactions.[2]

7.1.2 Historical Perspective

In the early 1900s, the distinguished Italian chemist Mario Betti demonstrated that 2-naphthol (**1**) can be a good carbon nucleophile toward imine **3**, produced from benzaldehyde and aniline. Although the Betti reaction is mechanistically related to the Mannich reaction, Betti's work preceded the development of this more widely known reaction.[4–6]

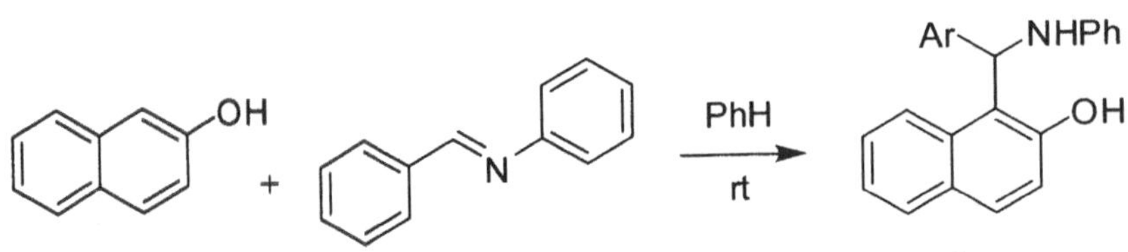

1 3 4

Betti later reported that ammonia was also a suitable reaction component when 2 equiv of benzaldehyde are used (followed by HCl treatment to hydrolyze the initial product—the imine/oxazine derivative of the Betti base, see **7.1.5**).[7] In addition to his work on the reaction that now bears his name, Betti also studied the relationship between molecular constitution and optical rotatory power and is thus considered a pioneer in the field of asymmetric synthesis. In 1939 he was appointed as the senator of the Kingdom of Italy.

7.1.3 Mechanism

Although detailed mechanistic studies on the Betti reaction have not been carried out, it is widely assumed the reaction follows a similar course to the Mannich reaction.[8] This principally involves the condensation of the aldehyde and amine constituents to form the active iminium electrophile 7. 2-Naphthol (**1**) functions as a nucleophile, giving the intermediate represented by the resonance structures **8** and **9**, which subsequently loses a proton to yield the product. It is easy to appreciate from this mechanism why acid catalysis may facilitate the reaction, and indeed increased rates have been observed using such catalysis.[3]

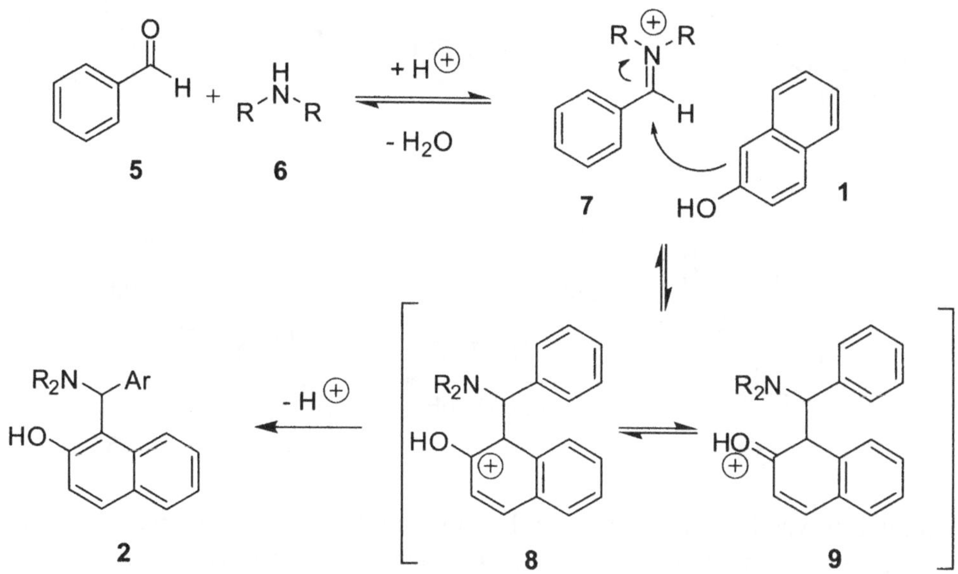

7.1.4 Synthetic Utility

Substrate Scope

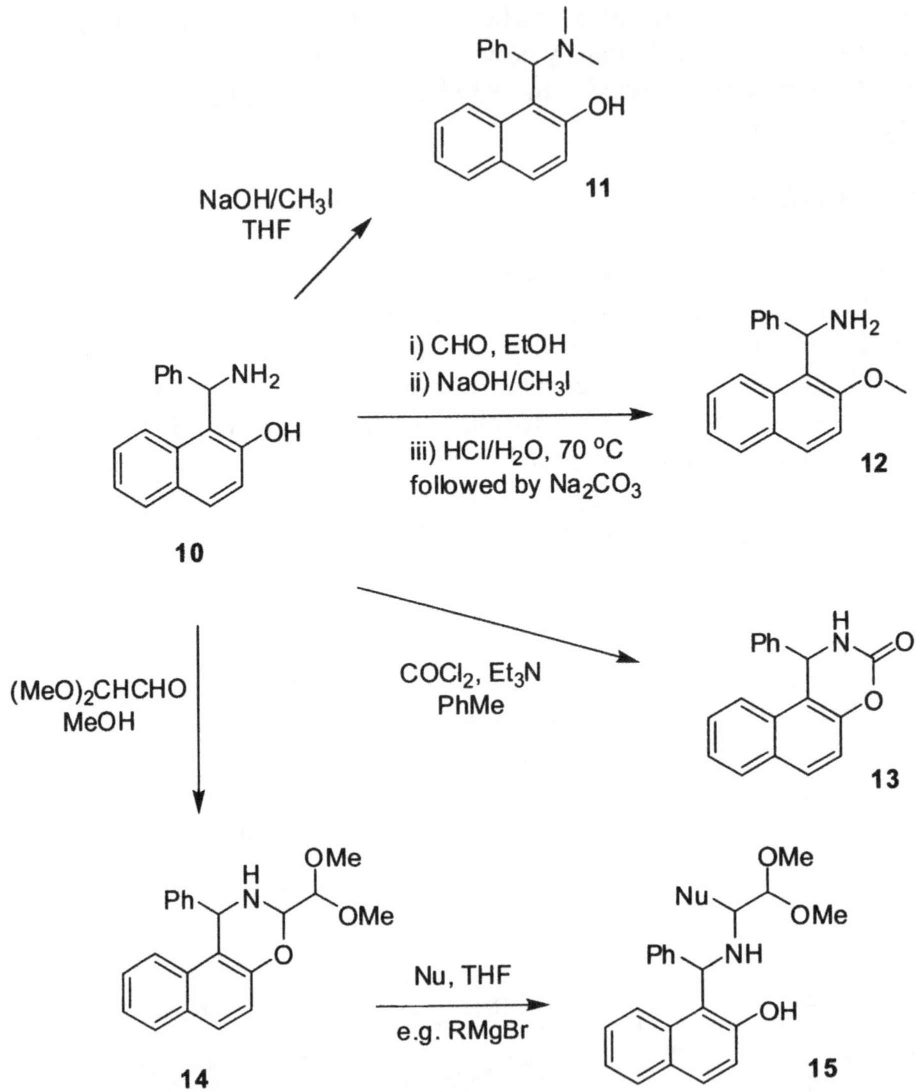

Although it was Betti himself who demonstrated that ammonia was a suitable reagent to form the parent Betti base **10**, it has more recently been shown that ammonium carbonate or ammonium hydrogen carbonate under microwave irradiation are useful alternatives.[9] Following the formation of product **10**, subsequent studies have demonstrated that it is feasible to convert it into a number of derivatives.[2] For example, it is possible to methylate the nitrogen functionality to form alkylated derivatives **11**,[10] selectively methylate the oxygen atom (to give **12**) by transiently protecting the nitrogen moiety as an imine,[10] and to form cyclic structures such as product **13**.[11] Liu *et al.* reported a useful procedure whereby cyclic aminal **14** can be formed upon treatment of **10** with an aldehyde substrate. They subsequently showed that this aminal can function as an electrophile upon treatment with nucleophiles such as Grignard reagents to give the ring-opened products **15**.[12]

Early work by Littman and Brode established that it was possible to vary the amino constituent of the Betti reaction.[13] It has subsequently been shown that a wide variety of amines are suitable, depending on the reaction conditions.[2] Indeed, primary and secondary aliphatic amines, aromatic amines and heteroaromatic amines have been used. For example, morpholine (**17**) was a suitable substrate for a Betti reaction when the reaction constituents were neat in the presence of a catalytic amount of *p*-toluenesulfonic acid under microwave irradiation.[3] This reaction protocol is especially notable for its fast reaction rates, with the products obtained after only one minute of irradiation. Other useful reaction procedures to form diverse Betti bases have also been reported including the use of water as a solvent,[14] and the addition of nonionic surfactants to the reaction mixture.[15] In addition to variation of the amine component, a number of alternative aromatic and heteroaromatic aldehydes are suitable substrates.[2] Finally, although most of the examples in the literature focus on 2-naphthol, a few alternatives have been reported, including the use of 8-quinolinol.[16]

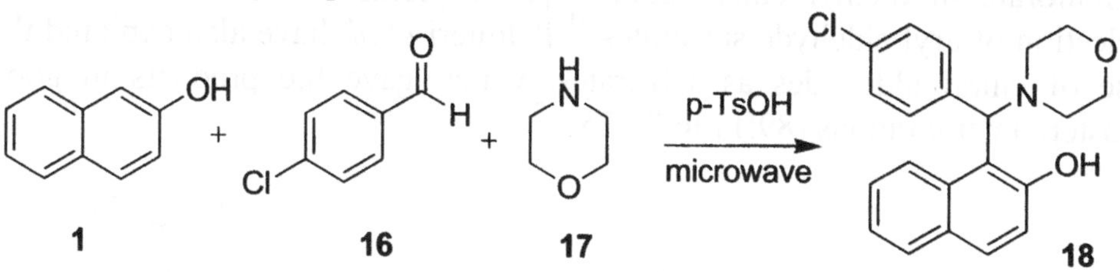

1 **16** **17** **18**

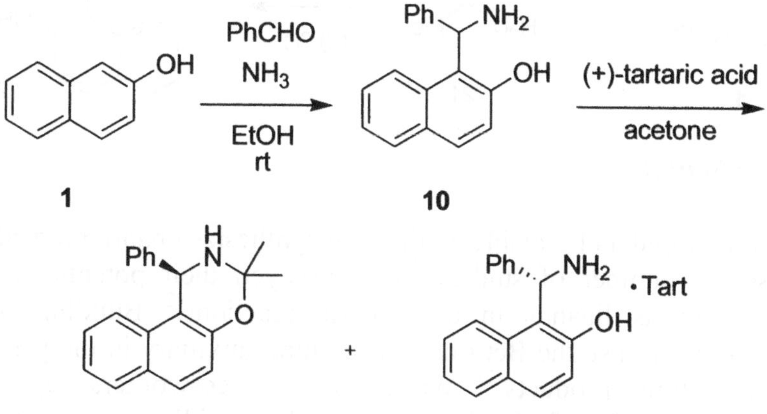

1 **10**

(*R*)-**19**
47% yield
>99% *ee*

(*S*)-**20**
48% yield
>99% *ee*

Since the Betti reaction results in a product with a benzylic chiral centre, several studies have been reported whereby the product enantiomers are resolved.[2] Betti reported in 1906 that his base **10** could be classically resolved following separation of the corresponding tartaric acid salts.[6] More recently Hu and co-workers have reported a high yielding and scaleable modification to this procedure by using tartaric acid in acetone.[17] In acetone, two different compounds were formed, the (*R*)-enantiomer giving the aminal **19** and the (*S*)-enantiomer giving the tartaric acid salt **20**.

The use of chiral, enantiopure amines can facilitate the stereoselective synthesis of Betti base derivatives.[2] Palmieri *et al.* demonstrated that the Betti reaction of naphthol **1**, enantiopure secondary amines **21** and aryl aldehydes can give products with a high level of stereocontrol.[18–20] For example, performing the reaction solvent free, and then adding a small amount of ethanol at the end favored the precipitation of the (*R,R*)-enantiomer, invoking a crystallization-induced asymmetric transformation. When benzaldehyde was used as a substrate, high yields (93%) and excellent diastereomeric ratios (99:1) were obtained, although lower yields and selectivities were observed for other aryl aldehydes. A number of other groups have reported similar procedures.[2] Notably, Saidi and Azizi demonstrated that the addition of trimethylsilyl chloride and lithium perchlorate in diethyl ether could improve yields and selectivities for a selection of aryl aldehyde substrates.[21] Palmieri *et al.* have also explored the use of chiral aldehydes as substrates, which gave the products in good diastereomeric rations (89:11 to 75:25).[22]

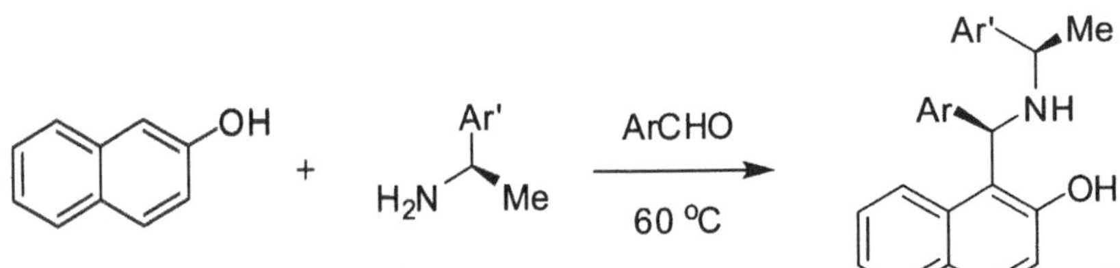

1 21 22

Uses of the Betti Bases

In light of the rapid and scalable methods to synthesize enantiomerically pure Betti bases, a number of studies have surveyed their potential as chiral auxiliaries and as ligands in asymmetric reactions.[2] Building on their previous studies to use the Betti base as a chiral auxiliary in the preparation of several natural products,[23] Wang, Hu and co-workers reported the asymmetric synthesis of chiral 2-substituted piperidines *en route* to the naturally occurring alkaloids (+)-pelletierine (**28**), (–)-lasubine II (**29**), and (+)-cermizine C (**30**).[24] Thus (*R*)-Betti base was condensed with 1,5-pentanedial in the presence of 1,2,3-benzotriazole (BtH) to give product **24** in excellent yield (93%). The benzotriazole moiety was stereoselectively displaced using allyl trimethyl silane as the nucleophile in the presence of boron trifluoride etherate. At low temperature, product **25** was isolated as a single stereoisomer. Reductive cleavage of the C–O bond using lithium aluminum hydride, followed by exposure to sodium hydroxide in methanol gave the product **27** in good yield (85%), following protection of the amine functionality. The formation of **26** was consistent with an *o*-quinone methide mechanism for the elimination reaction. Product **27** was then used as a versatile building block to synthesize all three alkaloids.

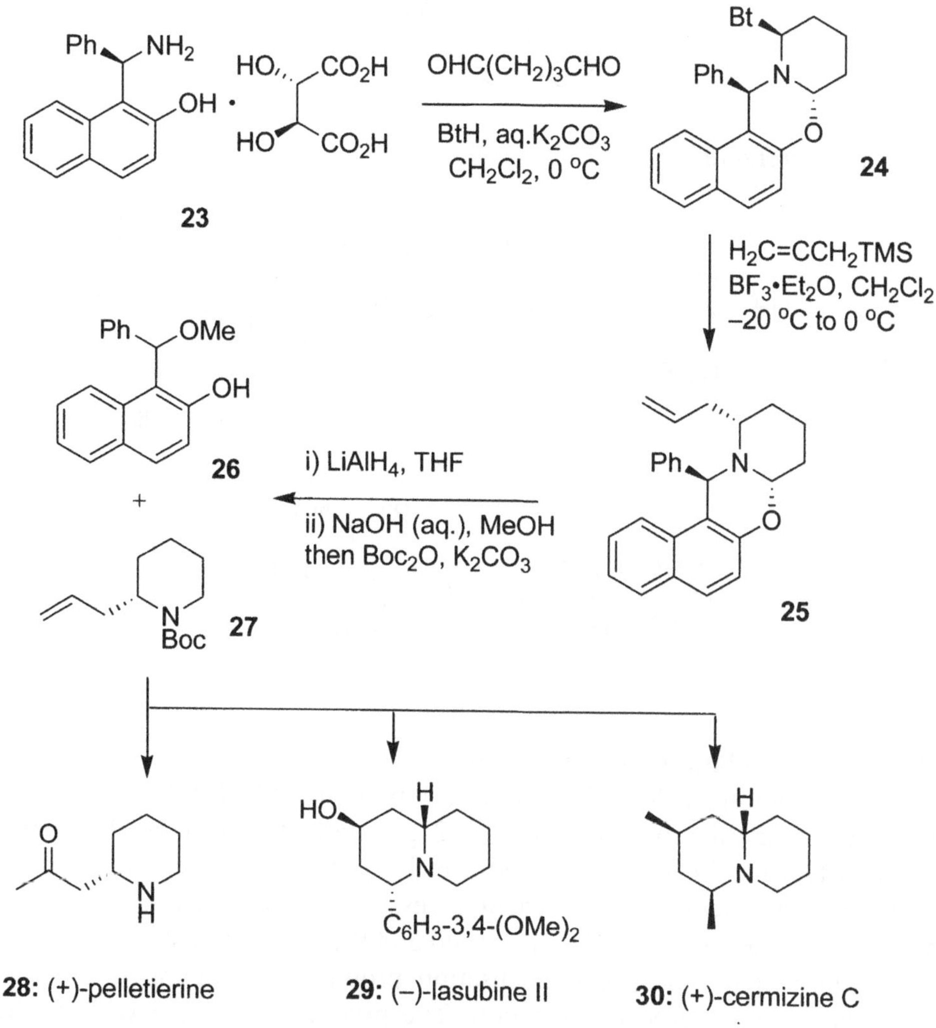

A number of studies have also been reported using enantiopure Betti base derivatives as chiral ligands in asymmetric reactions.[2] For example, Chan *et al.* reported that Betti base **33** was a useful ligand for the additional of diethyl zinc to aryl aldehydes, with excellent enantiomeric excesses obtained in certain cases.[25] Betti base derived chiral ligands have also shown promise as ligands for the alkenylation or arylation of aldehydes,[2] Tsuji–Trost allylations,[26] and hydrosilylation.[27]

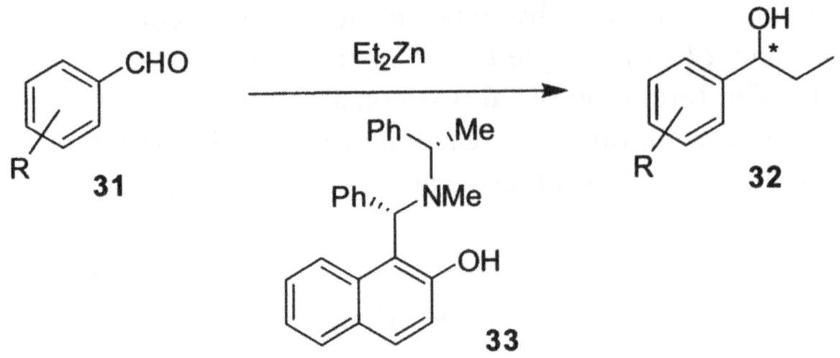

7.1.5 *Experimental*

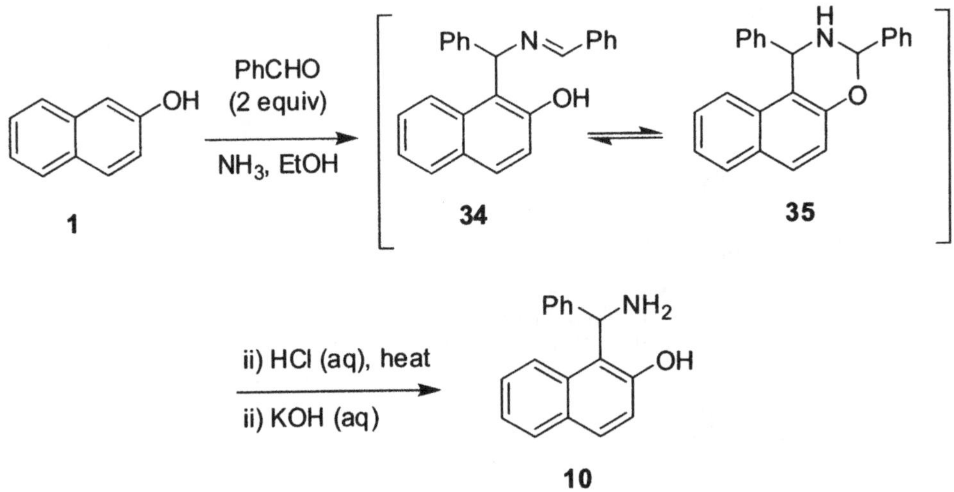

β-Naphtol phenylaminomethane (10)[28]
To a cold solution of 2-napthol (144 g, 1 mol) in 95% EtOH (200 mL) is added freshly distilled benzaldehyde (212 g, 2 mol), followed by 95% EtOH (200 mL) that has been saturated with NH_3. The solution becomes red and warms up spontaneously. The reaction vessel is stoppered and allowed to stand for 2 h. The stopper is then removed and the mixture is left for a further 12 h, after which time the condensation product has separated as white needles and is recovered by filtration. The product is washed with further EtOH (50 mL). Upon standing for three days the mother liquor yields another crop of the product. This gives product **34**/**35** (284 g, 84%) as a white solid. This intermediate is then treated with 4–5 times its volume of 20% aqueous HCl. The mixture is then steam distilled to remove all the benzaldehyde formed during the hydrolysis. During the distillation, the product precipitates

as white needles and is separately by filtration. This gives the product Betti base as a hydrochloric acid salt (240 g, 84%). To obtain the free base the hydrochloric salt (200 g) is stirred into a fine paste upon addition of water (300 mL). To this is added crushed ice (50 g) and the mixture is cooled in an ice bath; 5% Aqueous KOH (800 mL) is added slowly until a clear solution is obtained. This mixture is then extracted with Et_2O (8 × 300 mL). The combined organic extract is dried (Na_2SO_4), filtered and concentrated to 300 mL. This solution is then cooled in an ice bath, which results in crystallization of the product **10**. Several crops of the product are isolated by filtration to give **10** (127 g, 73%): mp 124–125 °C.

7.1.6 References

1. [R] Szatmari, I.; Fulop, F. *Curr. Org. Synth.* **2004,** *1*, 155–165.
2. [R] Cardellicchio, C; Capozzi, M. A. M.; Naso, F. *Tetrahedron: Asymmetry* **2010,** *21*, 507–517.
3. Jha, A.; Paul, N. K.; Trikha, S.; Cameron, T. S. *Can. J. Chem.* **2006,** *84*, 843–853.
4. Betti, M. *Gazz. Chim. Ital.* **1900,** *30II*, 301–309.
5. Betti, M. *Gazz. Chim. Ital.* **1900,** *30II*, 310–316.
6. Betti, M. *Gazz. Chim. Ital.* **1906,** *36II*, 392–394.
7. Smith, H. E.; Cooper, N. E. *J. Org. Chem.* **1970,** *35*, 2212–2215.
8. [R] Galatsis, P. In *Name Reactions for Homologations Part II*; Wiley: New York, **2009**; pp 653–670.
9. Szatmari, I.; Fulop, F. *Synthesis* **2009,** 775–778.
10. Cardellicchio, C.; Ciccarella, G.; Naso, F.; Schingaro, E.; Scordari, F. *Tetrahedron: Asymmetry* **1998,** *9*, 3667–3675.
11. Szatmari, I.; Hetenyi, A.; Lazar, L.; Fulop, F. *J. Heterocycl. Chem.* **2004,** *41*, 367–373.
12. Liu, B.; Su, D.; Cheng, G.; Liu, H.; Wang, X.; Hu, Y. *Synthesis* **2009,** 3227–3232.
13. Littman, J. B.; Brode, W. R. *J. Am. Chem. Soc.* **1930,** *52*, 1655–1659.
14. Gandhi, M.; Olyaei, A.; Raoufmoghaddam, S. *Synth. Commun.* **2008,** *38*, 4125–4138.
15. Kumar, A.; Gupta, M. K.; Kumar, M. *Tetrahedron Lett.* **2010,** *51*, 1582–1584.
16. [R] Philips, J. P. *Chem. Rev.* **1956,** *56*, 286–287.
17. Dong, Y.; Li, R.; Lu, J.; Xu, X.; Wang, X.; Hu, Y. *J. Org. Chem.* **2005,** *70*, 8617–8620.
18. Palmieri, G. *Tetrahedron: Asymmetry* **2000,** *11*, 3361–3373.
19. Cimarelli, C.; Mazzanti, A.; Palmieri, G.; Volpini, E. *J. Org. Chem.* **2001,** *66*, 4759–4765.
20. Cimarelli, C.; Palmieri, G.; Volpini, E. *Tetrahedron: Asymmetry* **2002,** *13*, 2417–2426.
21. Saidi, M. R.; Azizi, N. *Tetrahedron: Asymmetry* **2003,** *14*, 389–392.
22. Cappannini, L.; Cimarelli, C.; Giuli, S.; Palmieri, G.; Petrini, M. *Tetrahedron: Asymmetry* **2007,** *18*, 1022–1029.
23. Wang, X.; Dong, Y.; Sun, J.; Li, R.; Xu, X.; Hu, Y. *J. Org. Chem.* **2005,** *70*, 1897–1900.
24. Cheng, G.; Wang, X.; Su, D.; Liu, H.; Liu, F.; Hu, Y. *J. Org. Chem.* **2010,** *75*, 1911–1916.
25. Liu, D.-X.; Zhang, L.-C.; Wang, Q.; Da, C. S.; Xin, Z.-Q.; Wang, R.; Choi, M. C. K.; Chan, A. S. C. *Org. Lett.* **2001,** *3*, 2733–2735.
26. Wang, Y.; Li, X.; Ding, K. *Tetrahedron: Asymmetry* **2002,** *13*, 1291–1297.
27. Li, X.; Song, J.; Xu, D.; Kong, L. *Synthesis* **2008,** 925–931.
28. Betti, M. *Org. Synth. Coll. Vol. I* **1941,** *1*, 381–383.

7.2 Bernthsen Acridine Synthesis

Thomas Andrew Wynn

7.2.1 Description

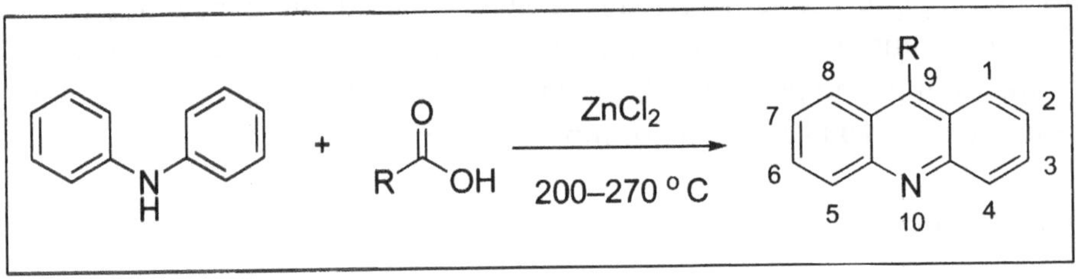

The Bernthsen reaction involves the cyclization of a diphenyl amine with a carboxylic acid at high temperatures for extended times in the presence of a stoichiometric amount of a Zinc Chloride catalyst.[1,2]

7.2.2 Historical Perspective

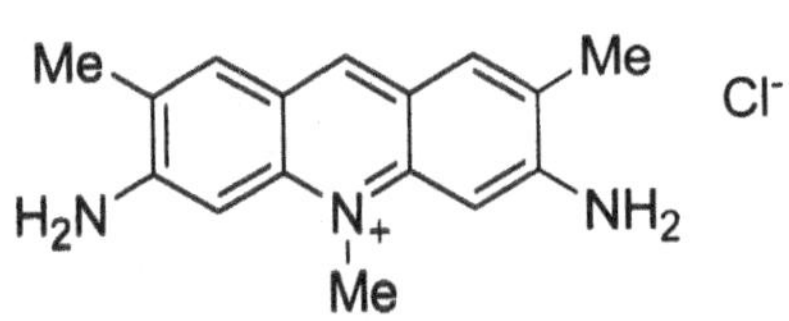

1

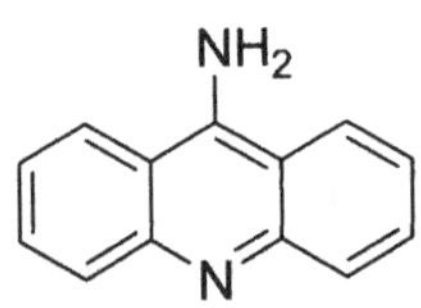

2

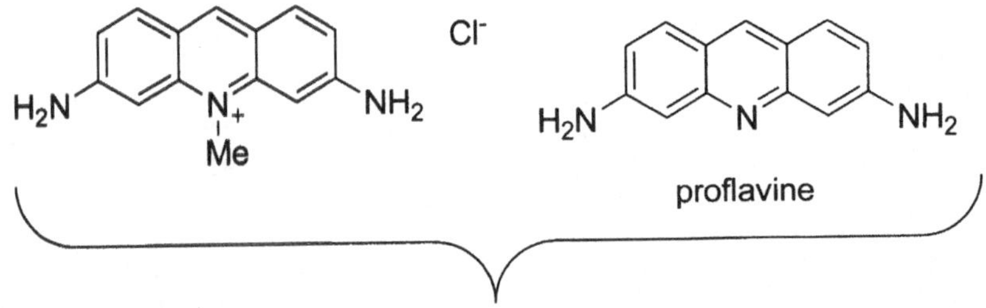

acrilavine–a mixture of acridines

Bernthsen's synthesis of 9-phenyl acridine in 1884 is the first reported method for making acridines and was used as proof of their structure.[3] At the turn of the 19th century, there were several syntheses of acridines reported following Bernthsen's procedure with a variety of substitutions, mainly in the 9 position through use of a variety of carboxylic acids.[4] Early in the 20th century, acridines were of interest because it was found that 3,6-diamino-2,7,10-trimethylacridinium chloride and 3,6-diamino-10-methylacridinum chloride had antibacterial activity against *Trypanosomona brucei.*[5] Furthermore, 3,6-diamino-2,7-dimethylacridine (acridine yellow) inhibited the growth of a variety of bacteria including *Escherichia coli, Staphylococcus aureus,* and *Bacillus anthracis* paracirtes, which caused many thousands of deaths in World War I.[6] Proflavine (3,6-diaminoacridine), Acriflavine (a mixture of 3,6-diamino-10-methylacridinum chloride and proflavine), and 9-aminoacridine are still used today as a topical antiseptics. Much of the early development and use of the Bernthsen reaction went into synthesizing these types of antibiotics. However, the generation of acridine analogs using the original Bernthsen conditions was generally low yielding, and reaction conditions varied from substrate to substrate.[2]

Acridines were also found to have potent activity against malaria.[7] This discovery was driven by the limited availability of Quinine during World War II and quinacrine became the main malarial treatment in both England and the United States. The development work done with quinacrine lead directly to the development of chloroquine, one of the most successful treatments for malaria.

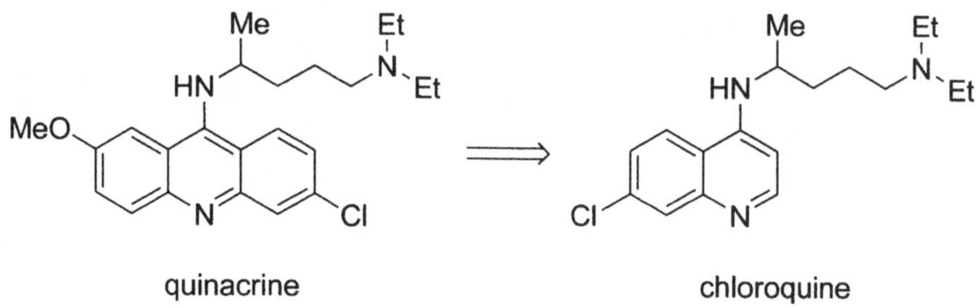

Concurrently with the identification of the biological activity of acridines, benzacridines were investigated in the 1930s and 1940s as carcinogens.[8] In those studies, simple acridines did not show significant carcinogenicity but their highly conjugated analogues, specifically angular benzacridines, showed significant activity.[9] A few additional trends were observed in the studies by Lacassagne and Buu-Hoi. For example the polymethylated analogues were especially potent carcinogens.

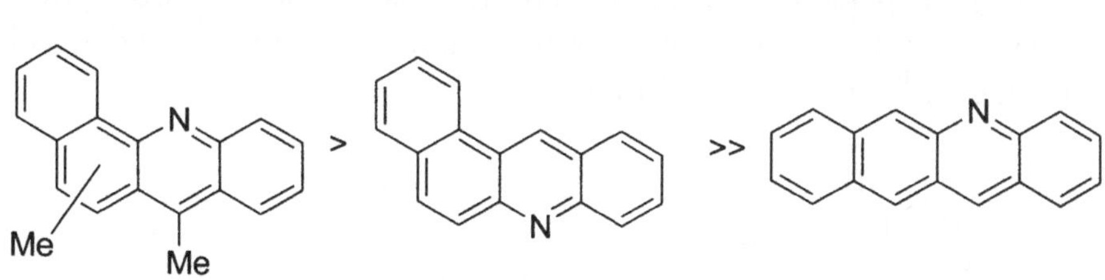

Relative carcinogenicity

It was later determined that acridine and proflavine are potent DNA intercalators.[10] The ability to intercalate DNA has been proposed as a basis for the carcinogenicity seen with several acridine analogues. In the early 20th century, acridines played a major role in the understanding of how small molecules interact in biological systems. The Bernthsen reaction was one of the work horse reactions necessary to generate those acridines.

7.2.3 *Mechanism*

Although investigations into the reaction mechanism are limited[11] it is believed to proceed through an acylation *ortho* to the aniline followed by acid catalyzed cyclization and dehydration of the resulting ketone. The observation of 4 and 4′ acyldiphenyl amine side products in many Bernthsen reactions support this mechanism.

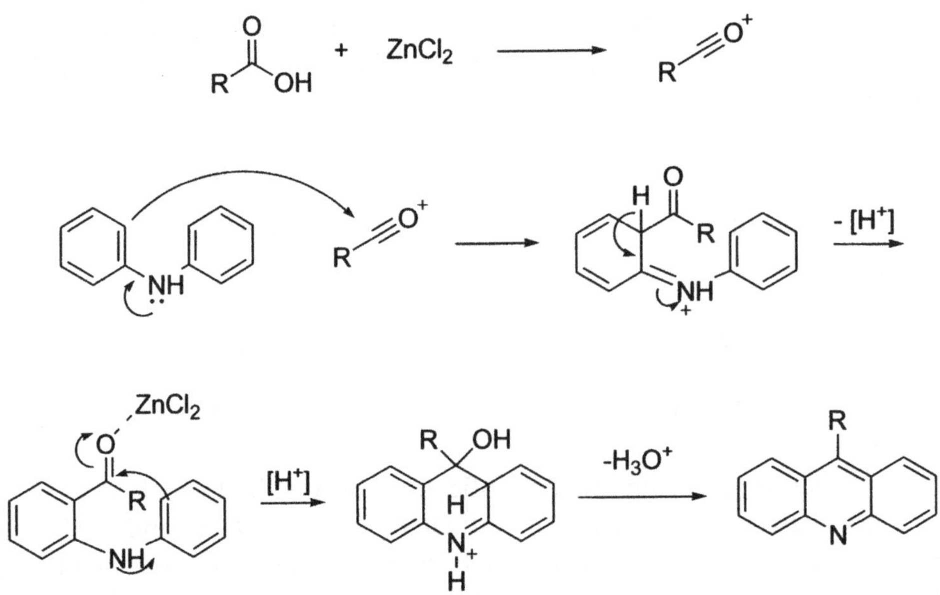

7.2.4 *Variations and Improvements*

In his original papers, Bernthsen reported using other carboxylic acid derivatives, including phenyl nitriles to generate acridines[1a] and trichloromethylbenzene,[1b] albeit in very low yields.

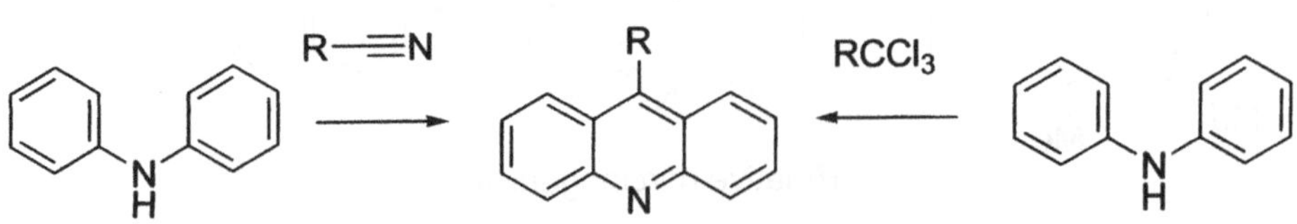

Acid anhydrides have also seen some success under $ZnCl_2$ catalyzed conditions especially with simple anhydrides.[12] Mixtures of more complex carboxylic acids and conjugate anhydrides showed poor (< 5%) yields under similar conditions.[13]

Popp published a major advance in the reaction in 1961 in which he used polyphosphoric acid as both catalyst and solvent.[14] Under these conditions the reaction times and temperatures were reduced to typically 15 min at 200 °C. Under these conditions several electron-rich analogs that were not compatible with the original conditions were generated.

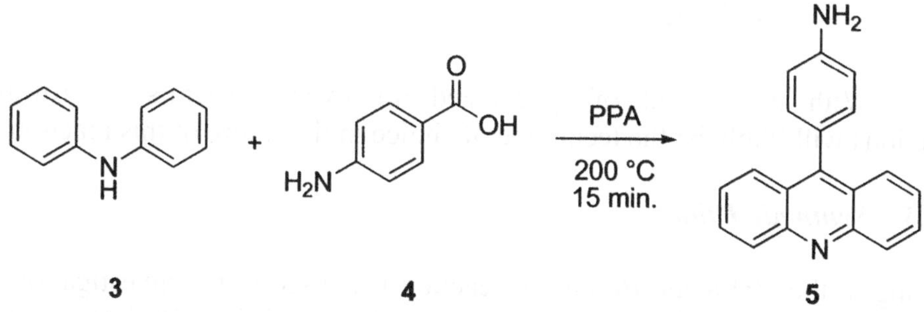

More recently, these conditions have been modified through use of microwave heating.[15] Sejias and co-workers reported the use of a commercial microwave oven to affect the cyclization albeit with lesser control over the reaction temperature.[15c]

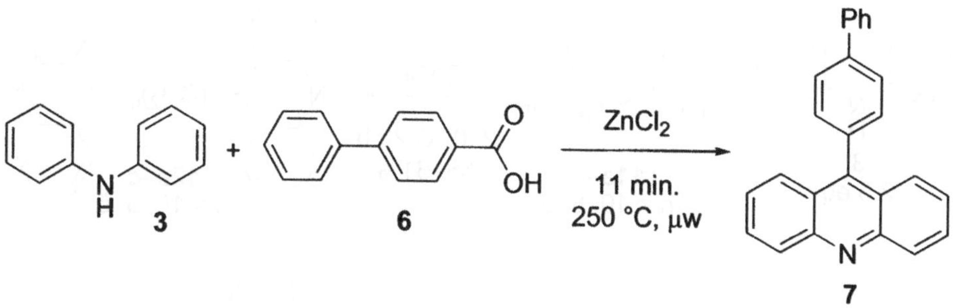

Tanaka and co-workers in their work on helical aromatic compounds were able to decrease reaction time and control racemization when heated in a commercial microwave oven.[16] When (*S*)-2-methylglutaric acid was reacted under classic heating conditions complete epimerization of the chiral methyl groups was observed. Using microwave heating a 50% enantiomeric excess was observed. The retention of stereochemistry was attributed by the authors to the markedly decreased reaction times.

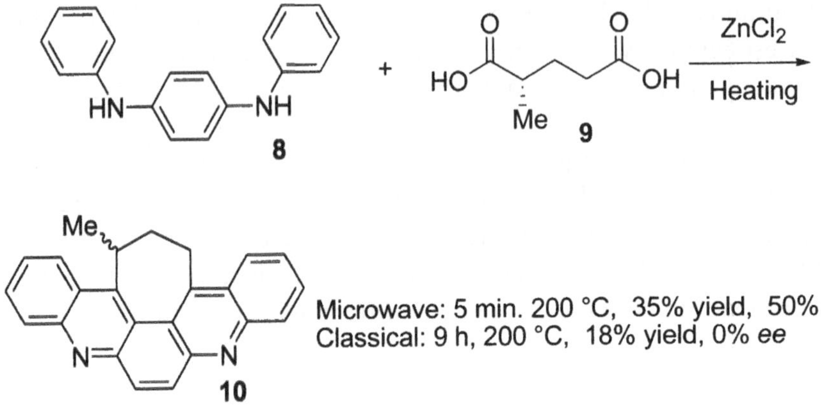

Microwave: 5 min. 200 °C, 35% yield, 50% *ee*
Classical: 9 h, 200 °C, 18% yield, 0% *ee*

With the advent of specialized microwave reactors, microwave reactions will likely be the technique of choice in the future of this reaction.

7.2.5 *Synthetic Utility*

Staring in the 1940s the Bernthsen reaction was used in the investigation of the origins of the carcinogenicity of benzacridines.[17] It has also been used to generate linked bis acridines with use of α,ω-diacids[18] It is interesting that in those studies when succinic acid was used under standard conditions only the mono acridine was observed. The authors postulated that sterics hindered the formation of the second acridine ring system.

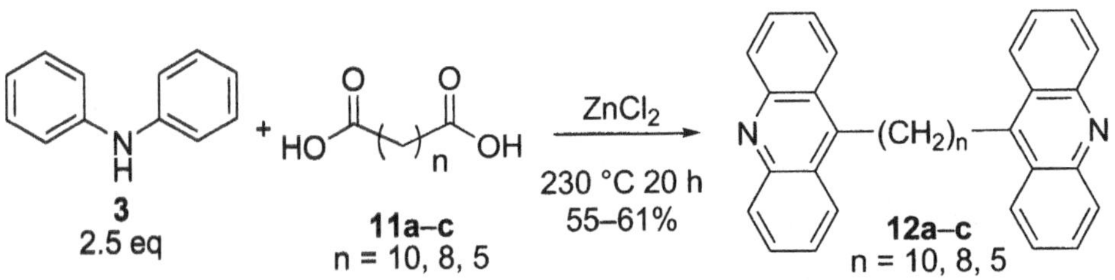

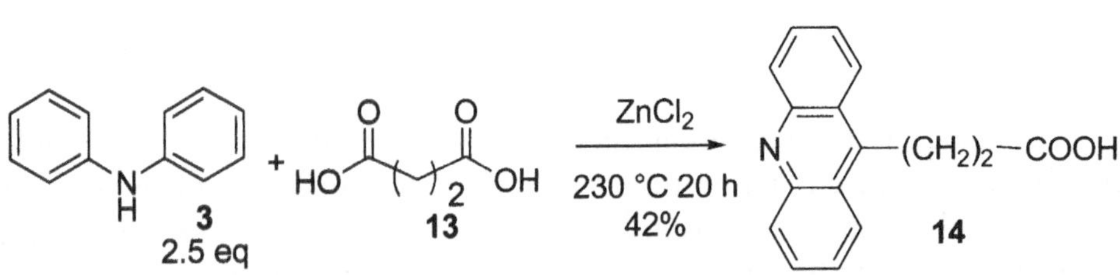

A more recent article used the Bernthsen reaction to generate a potent but not selective DYRK2 kinase inhibitor **19**.[19] It is interesting that the authors chose to use $InCl_3$ as the Lewis acid in this sequence. They also took advantage of the facile oxidation of a 9-methyl group to generate an aldehyde used in further modifications.

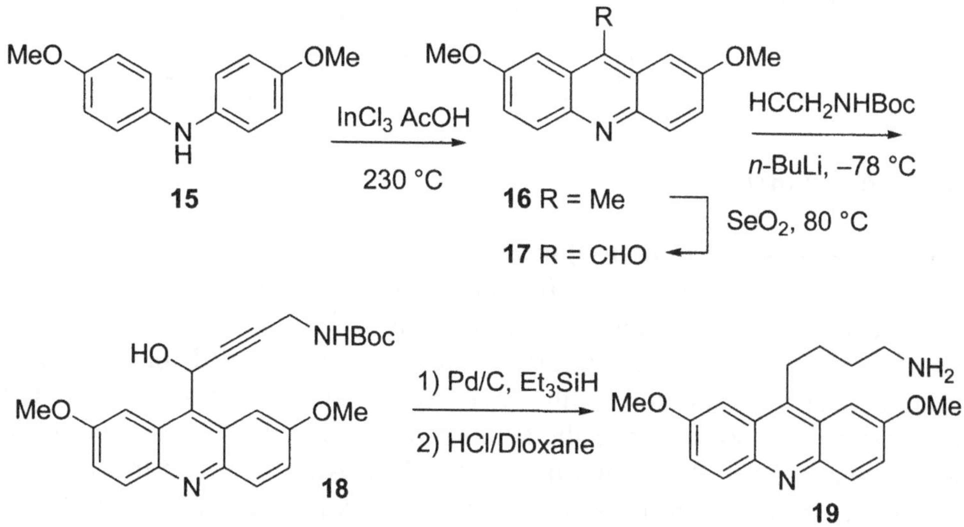

Using an interesting modification Prasad and co-workers used the Vilsmeier–Haack reagent at affect ring closure to generate indoloacridines **21**.[20] The authors propose a reaction mechanism similar to the $ZnCl_2$ catalyzed reaction.

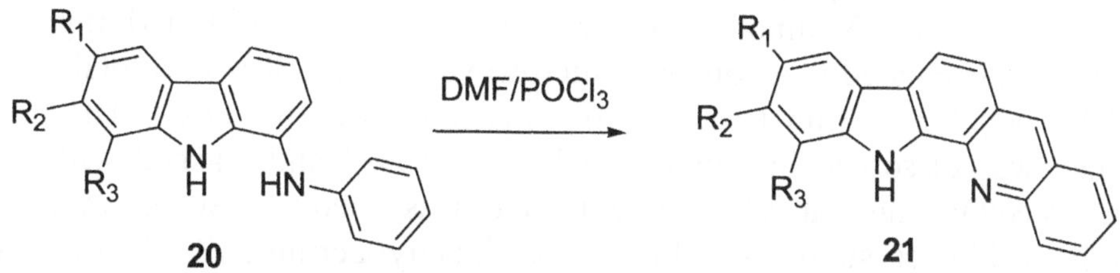

Several groups have used the Bernthsen reaction to access acridines tethered to other molecues to take advantage of photo-physical,[21] electron-transfer[22] or DNA intercalating propensity.[23]

Acridines have continued to be of interest but the Bernthsen reaction has generally been replaced by a stepwise cyclization using transition metal catalyzed amination of aryl ketones or acids followed by acid catalyzed cyclization to form the acridine ring system.[24]

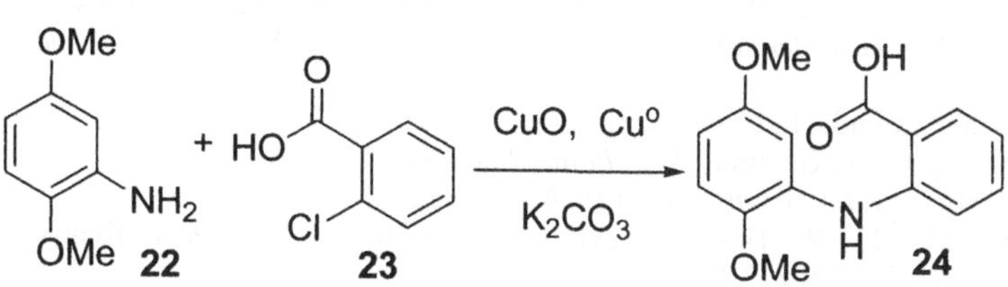

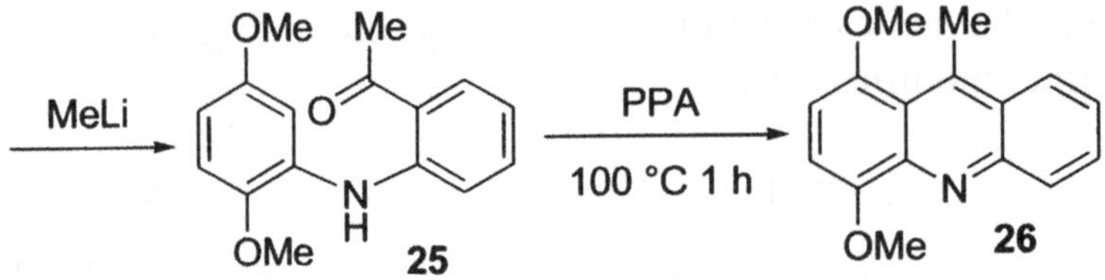

7.2.6 *Experimental*

Synthesis of 9-(4-aminophenyl)acridine (5)[14]

A mixture of *N,N*-diphenylamine (0.03 mol), 4-aminobenzoic acid (0.06 mol) was heated in 230 g of polyphosphoric acid at 200 °C for 15 min. The reaction mixture was then poured onto ice and filtered or decanted. Treatment of the solution with 25% NaOH (aq) solution caused the precipitation of a solid (while the solution was still acidic) presumed to be the phosphate salt of 9-(4-aminophenyl)acridine. After filtration the solution was made strongly basic with NaOH and extracted with chloroform. Concentration of the chloroform yielded only trace amounts of **5**. The phosphate salt was shaken with 25% NaOH (aq) and chloroform and the chloroform concentrated to give a 24% yield of **5**.

Synthesis of 9-(4-biphenyl)acridine (7)[17c]

A mixture of *N,N*-diphenylamine (100 mg, 0.592 mmol), 4-phenylbenzoic acid (234 mg, 1.183 mmol), and recrystallized zinc chloride (81 mg, 0.59 mmol) was irradiated in a domestic microwave oven (at 80% of a total output of 1000 W) for 11 min (the reaction was monitored by TLC). The crude product was dissolved in dichloromethane (20 mL) and washed with 10% aqueous NaOH and water. The organic layer was dried (Na_2SO_4), evaporated and the residue crystallized yielding 9-(4-biphenyl)acridine (**7**, 172 mg, 88% yield).

7.2.7 *References*

1 (a) Bernthsen, A. *Liebigs Ann. Chem.* **1878,** *192*, 1. (b) Bernthsen, A. *Ann.* **1884,** *224*, 1.

2 [R] For a thorough review of the history of acridines see Acheson, R. M. In *Chemistry of Heterocyclic Compounds Vol. 9*, Interscience: New York, 1973.

3 Acridine was first isolated from coal tar in 1871. Graebe, C.; Caro, H. *Ber. Dtsch. Chem. Ges.* **1880,** *13*, 99–103.

4 (a) Decker, H.; Hoch, T. *Chem. Ber.* **1904,** *37*, 1002. (b) Landauer, E. *Bull. Soc. Chem. Fr.*, **1904,** *31*, 1083, (c) Decker, H.; Hoch, T. *Chem. Ber.* **1904,** *37*, 1564. (d) Schmidt, A.; Decker, H. *Chem. Ber.* **1906,** *39*, 933. (e) Dunstan, A. E.; Stubbs, J. A. *Chem. Ber.* **1906,** *39*, 2402.

5 Ehrlich, P.; Benda, L. *Chem. Ber.*, **1913,** *46*, 1931.

6 Browning, C. G.; Gilmour, W. *J. Pathol.Bacteriol.*, **1913,** *18*, 144.

7 Mühlens, P. *Naturwissenschaften* **1926,** *14*, 1162.

8 Barr, C.; Cook, J.W.; Haslewood, G. A. D.; Hewett, C. L.; Hieger, I.; Kennaway, E. L. *Proc. R. Soc. Lond.* **1935,** *117* 318–351.

9 Lacassagne, A.; Buu-Hoi, N. P..; Daudel, R.; Zajdela, F. *Ad. Cancer Res.* **1956,** *4*, 315–369.

10 (a) Lerman, L. S. *J. Mol. Biol.***1961,** *3*, 18–30. (b) Eldho, N. V., Joseph, J., and Ramaiah, D. *Chem. Lett.* **2001,** 438–439.

11 For studies using polyphosphoric acid are the catalyst, see Birchall, J. M.; Thorpe, D. H. *J. Chem. Soc. C* **1967,** 2071–2076.

12 (a) Buu-Hoi, N. P. *J. Chem. Soc.* **1946,** *68*, 792–795. (b) Thu-Cuc, T. T.; Buu-Hoi, N. P.; Xuong, N. D. *J. Chem. Soc., Perkin 1* **1966,** *1*, 87–88. (c) Buu-Hoi, N. P.; Lecocq, J. *Compt. Rend.* **1944,** *218*, 792–793. (d) Buu-Hoi, N. P., Lecocq, J. *Rec. Trav. Chim.* **1945,** *64*, 250–254.

13 Klemm, L.H.; Chiang, E.; O'Bannon, G. W. *J. Heterocyclic Chem.* **1992,** *29*, 571–574.

14 Popp, F.D. *J. Org. Chem.* **1962,** *27*, 2658–2659.

15 (a) Veverkova, E.; N., Marika; Toma, S. *Synth. Commun.* **2002,** *32*, 729–733. (b) Koshima, H.; Kutsunai, K. *Heterocycles* **2002**, *57*, 1299–1302. (c) Seijas, J. A.; Vazquez-Tato, M. P.; Martinez, M. M.; Rodriguez-Parga, J. *Green Chem.* **2002,** *4*, 390–391.

16 Watanabe, M.; Suzuki, H.; Tanaka, Y.; Ishida, T.; Oshikawa, T.; Tori-i, A. *J. Org. Chem.* **2004,** *69*, 7794–7801.

17 Barry, G.; Cook, J. W.; Haslewood, G. A. D.; Hewett, C. L.; Hieger, I.; Kennaway, E. L. *Proc. R. Soc. Lond. B*. **1935,** *117*, 318–351.

18 Eldho, N. V.; Saminathan, M.; Ramaiah, D. *Synth. Commun.* **1999,** *29*, 4007–4014.

19 Cuny, G. D.; Robin, M.; Ulyanova, N. P.; Patnaik, D.; Pique, V.; Casano, G.; Liu, J.-F.; Lin, X.; Xian, J.; Glicksman, M. A.; Stein, R. L.; Higgins, J. M. G. *Bioorg. Med. Chem. Lett.* **2010,** *20*, 3491–3494.

20 Sridharan, M.; Prasad, K. J. R. *J. Chem. Res.* **2007,** 164–169.

21 Lambert, C.; Schelter, J.; Fiebig, T.; Mank, D.; Trifonov, A. *J. Am. Chem. Soc.* **2005,** *127*, 10600–10610.

22 Hariharan, M.; Neelakandan, P., P.; Ramaiah, D. *J. Phys. Chem. B* **2007,** *111*, 11940–11947.

23 David-Cordonnier, M.-H.; Hildebrand, M.-P.; Baldeyrou, B.; Lansiaux, A.; Keuser, C.; Benzschawel, K.; Lemster, T.; Pindur, U. *Eur J. Med. Chem.* **2007,** *42*, 752–771.

24 Delfourne, E.; Kiss, R.; Le Corre, L.; Merza, J.; Bastide, J.; Frydman, A.; Darro, F. *Biorg. Med. Chem.* **2003,** *11*, 4351–4356.

7.3 Lehmstedt–Tănăsescu Reaction

Alexandros L. Zografos

7.3.1 Description

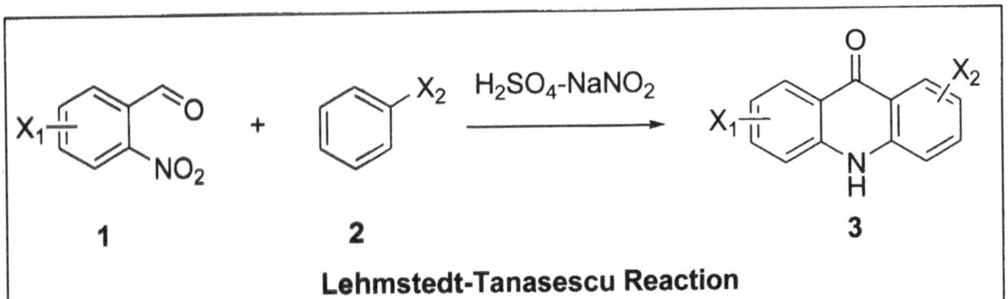

Lehmstedt–Tănăsescu reaction also known as Lehmstedt–Tănăsescu–acridone synthesis is a general reaction for the synthesis of acridone derivatives **3** from substituted *ortho*-nitrobenzaldehydes **1** and arene compounds **2** in the presence of a mixture of sulfuric acid and sodium nitrite.

7.3.2 Historical Perspective

The Lehmstedt–Tănăsescu reaction is named after the Romanian inventor Ioan Tănăsescu and the German chemist Kurt Lehmstedt, who devoted part of their career to expand and improve the described reaction. It was in 1928 when first Tănăsescu discovered that a mixture of sulfuric acid and sodium nitrite can promote the reaction between *ortho*-nitrobenzaldehyde and benzene for the synthesis of acridone.[1] Years later Kurt Lehmstedt realized the practicality of the described method improved and expanded its scope, transforming it to the known Lehmstedt–Tănăsescu reaction.[2,3]

7.3.3 Mechanism[4]

The reaction mechanism of Lehmstedt–Tănăsescu reaction is generally based on an aromatic electrophilic addition of an *ortho*-nitro-substituted benzaldehyde to electron-rich arenes, yielding substituted 3-phenyl 1,2-benzisoxazoles, which are then rearranged to the desired acridones. In the first step of the reaction mechanism the benzaldehyde is protonated by sulfuric acid to form intermediate **5**, followed by electrophilic addition to an electron rich arene (benzene in the following scheme), to give compound **6**. Intramolecular attack of the formed hydroxyl group to the *ortho*-positioned nitro group provides the five-member intermediate **7**, which in the aqueous reaction conditions forms 3-phenyl 1,2-benzisoxazole (anthranil) **8**.

Treatment of this compound with nitrous acid (formed *in situ* from sulfuric acid and sodium nitrite), leads to *N*-nitroso acridone **11**, through intermediates **9** and **10**. Finally, *N*-nitroso-acridone **11** liberates nitro group under acidic conditions to form acridone **12**.

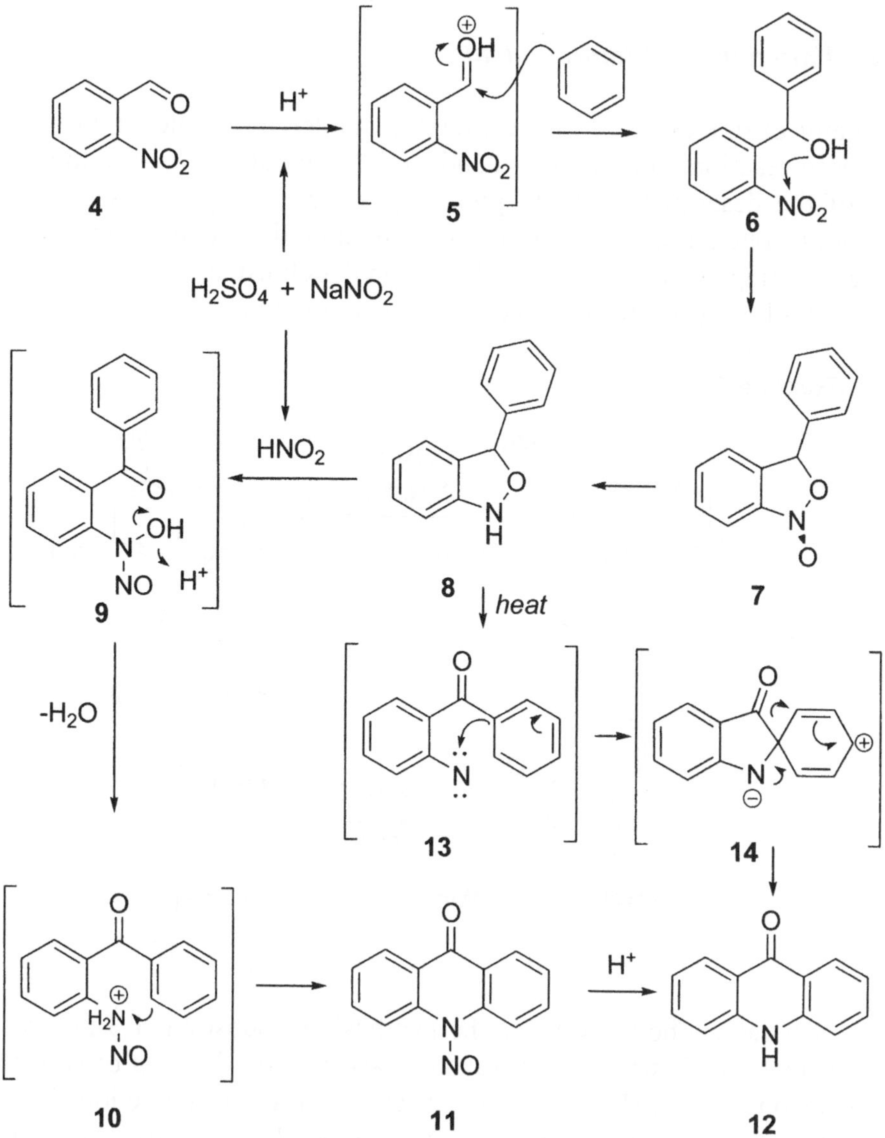

In late 1960s, the long known anthranil–acridone (**8** to **12**) transformation has come to be regarded in terms of a nitrene participation.[5]

In accordance to that, the described transformation can be further generalized as a nitrene-induced aromatic rearrangement, possessing the ability to prepare several cyclized heteroaromatics such as phenothiazines, phenoxazines, and dibenzooxazepines.[6]

The general mechanism for the subsequent conversion of **8** to **12** is believed to be realized through a five-membered spirodienyl intermediate **14**, something that is evidenced from the isolation of isomerized products in case of preexistent substituents on the arene reactant (*vide infra*).[4]

7.3.4 Variations and Improvements

Not many examples exist in the literature under the name of Lehmstedt–Tănăsescu reaction. On the other hand, several researchers are following the logic of the described reaction, preparing acridones by accessing first the appropriate phenyl benzisoxazole structure and then rearranging it either through heating or by the use of the original Lehmstedt–Tănăsescu nitrous acid protocol to the corresponding acridone compound.

Example 1

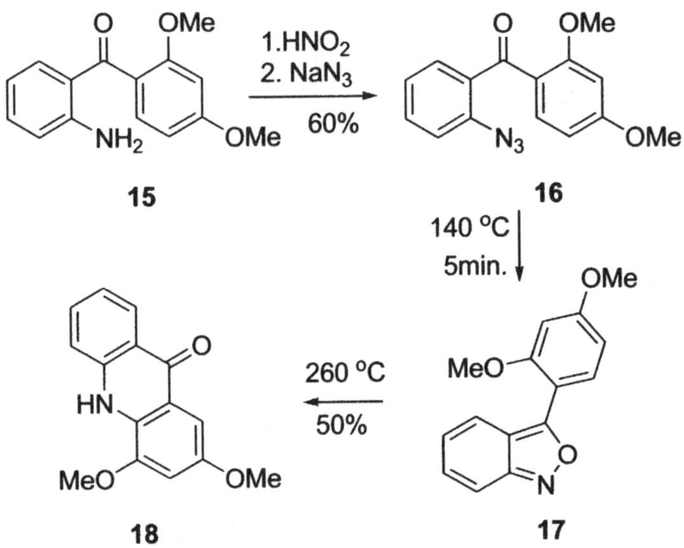

Several methods exist for the synthesis of substituted 3-phenyl 1,2-benzisoxazoles or hetero-benzisoxazoles varying from heating of substituted benzophenone azides (Example 1),[7,8] triethyl phosphite reduction of *ortho*-nitro benzophenone or heterophenone compounds (Example 2),[9] even to nucleophilic substitution of nitro compounds with aryl acetonitriles (Example 3).[10]

Example 2

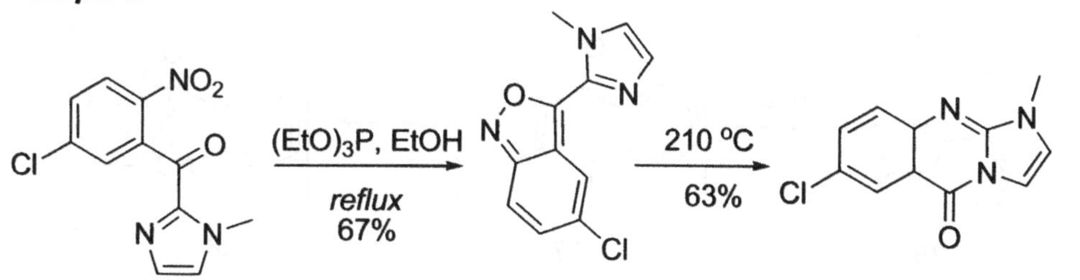

19 20 21

Example 3

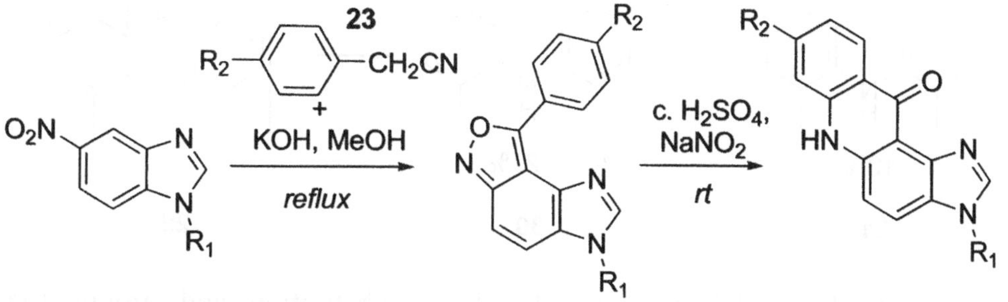

R_1= alkyl, R_2=H, Cl

22 24 25

7.3.5 *Synthetic Utility*

Lehmstedt–Tănăsescu reaction can be used with a variety of substituted arenes as coupling partners. Electron-donating groups in the aromatic ring are accelerating the reaction leading usually to good yields of anthranils. Further conversion of benzisoxazole structure to acridones usually lead to mixture of isomers, depending on the electron-donating efficiency of the substituents.[7] Alkyl group *para*-positioned to the benzisoxazole ring provide almost equimolar amounts of the two isomers,[7] while methoxy-substitutents leads usually exclusively to isomerized compounds (see Example 1). It should be outlined that thermolysis of anthranil compounds to acridone derivatives or its reaction with nitrous acid is remarkably sensitive to several conditions like temperature, solvents, metal catalysts and usually leads to mixtures of isomeric products when a substituted arene compound is used as starting reactant.[4] The products formed are always predicted through the intermediacy of a five-membered spirodienyl specie. An example is provided in the case of 2,6-dimethylphenylanthranil (**26**) producing a mixture of two major acridones, compound **30** and **31**, both generated by methyl cleavage of intermediate compound **29**.

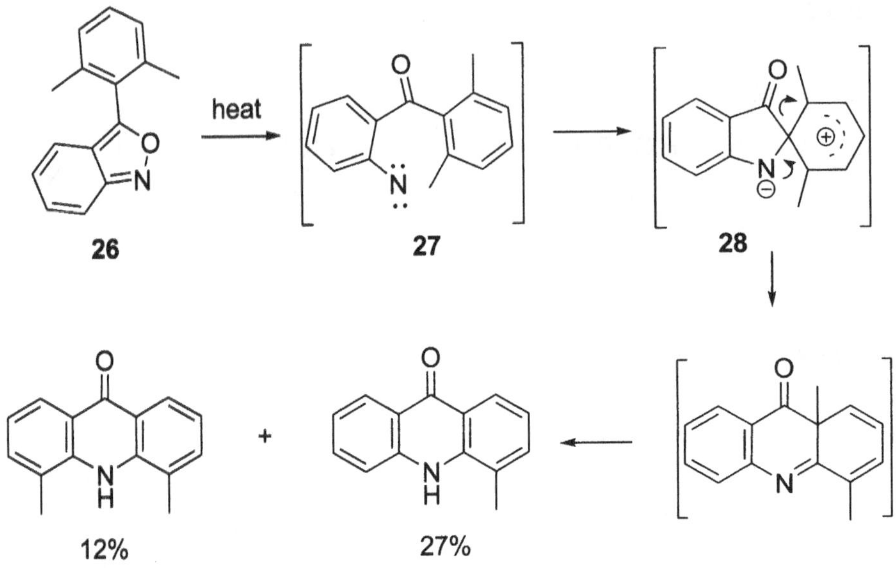

26 27 28 29 30 31

Nitro-benzaldehydes can also bear substitution and groups that stabilized the formed cation are leading to good yields of intermediate benzisoxazoles. Substituents on the benzaldehyde portion are not generally interfering in the rearrangement process.[7]

A wide range of heterocyclic compounds can also be formed based on the logic of Lehmstedt–Tănăsescu reaction. Thus formation of nitrene intermediates can in general rearrange to provide phenothiazines, dihydrophenazines, azepindoles, *etc.*[6]

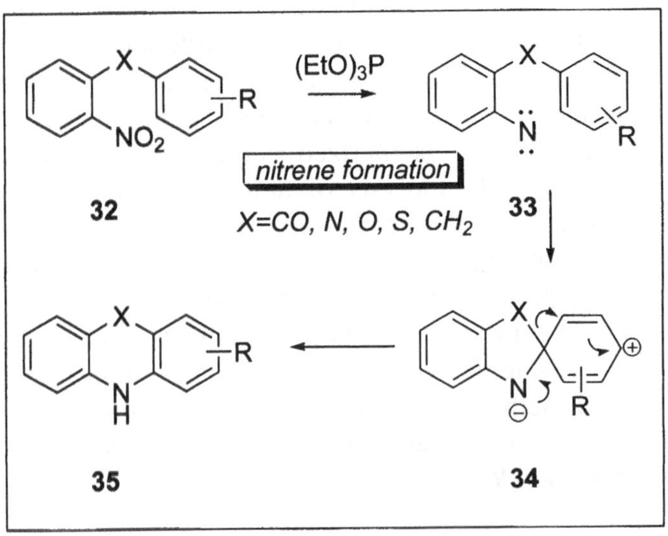

32 33 34 35

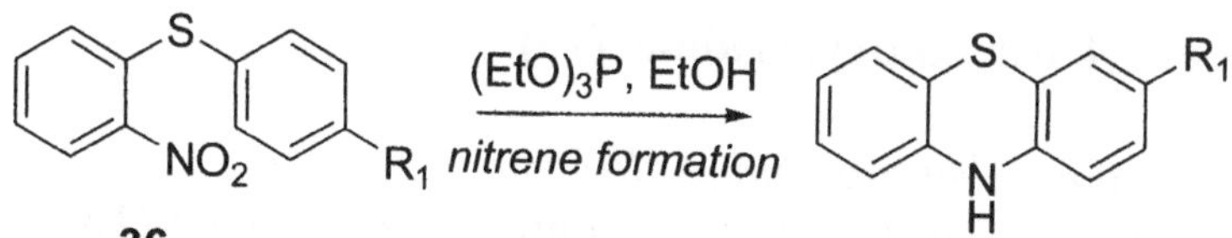

36 37

Synthesis of phenothiazines

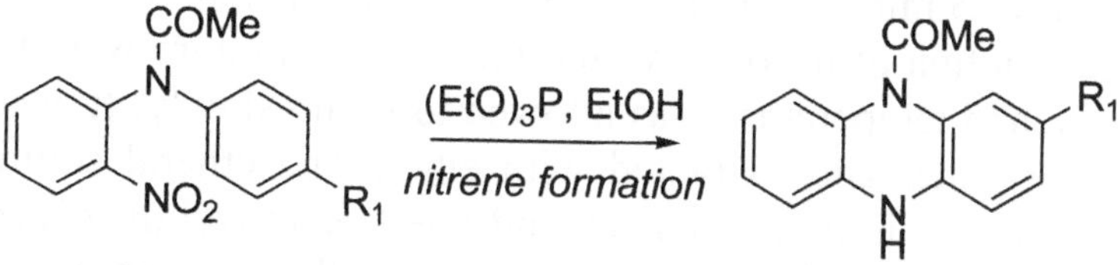

38 39

Synthesis of dihydrophenazines

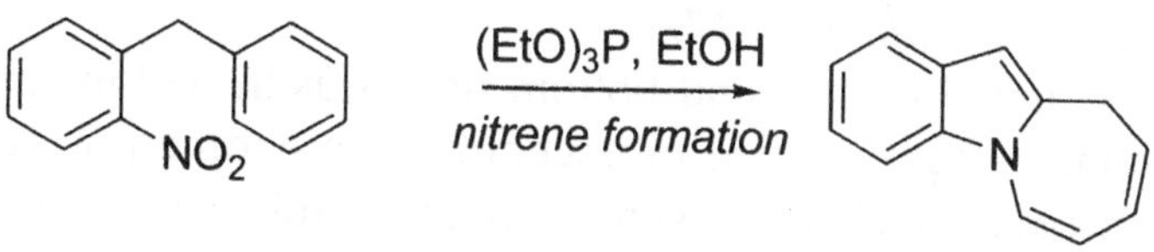

40 41

Synthesis of azepinoindoles

7.3.6 Experimental

Preparation of acridone (12)[11]

A total of 15.1 g of *o*-nitrobenzaldehyde (1 equiv) is dissolved in 62 mL of a thiophene-benzene mixture (7 equiv of benzene) under stirring. Then a freshly prepared solution of 37.5 mL of conc. sulfuric acid in which 0.35 g of sodium nitrite (5 equiv) was previously dissolved was added cautiously under stirring. During the addition, the brown reaction mixture was carefully controlled to prevent overheated from the exothermic reaction, and it was placed in an ice bath from time to time. After 5 days of stirring at room temperature, the aldehyde was consumed by TLC. At this point, 200 mL of water was added, and the remaining benzene and nitro-benzaldehyde was azeotropically distilled with steam. The mixture was cooled and further worked with an extra amount of water, filtered and washed to remove the remaining sulfuric acid. Acridone was purified by recrystallization from ethyl alcohol, yielding 8.2 g of light yellow acridone.

Preparation of 2,4-Dimethoxyacridone (18)[7]

*2-Azido-2',4'-dimethoxybenzophenone (**16**)*

A solution of 3.0 g of 2,4-dimethoxy-2'-aminobenzophenone (**15**) in 30 mL of acetic acid and 15 mL of 1 M sulfuric acid was cooled to –20 °C (dry ice, acetone) with stirring. To this solution was added 0.7 g of sodium nitrite in 6 mL of water. After 5 min, 1.3 g of sodium azide in 6 mL of water was added as fast as the evolution of nitrogen would allow. The mixture was stirred for another 30 min, and temperature was allowed to go up slowly. It was diluted with 100 mL of water and extracted with ether. The etheral solution was washed with water, sodium bicarbonate, and dried. Removal of solvent under reduced pressure gave an oily residue which crystallized when washed with ether. The product was recrystallized from ether to give 2.0 g of colorless crystals.

*3-(2,4-Dimethoxyphenyl)anthranil (**17**)*

The azide **16** (2.0 g) in a 50-mL round-bottom flask was heated in an oil bath at 140 °C for 5 min. (Large runs led to uncontrollable exothermic reaction.) The products from seven such runs were combined and recrystallized from hexane to give 13.0 g of greenish yellow needles.

*2,4-Dimethoxyacridone (**18**)*

A 100-mL, round-bottom flask containing 5.0 g of 3-(2,4-dimethoxyphenyl) anthranil (**17**) was heated at 255–260 °C for 10 min. The crude product was chromatographed on 700 g of silica gel and was eluted with benzene-ethyl acetate mixture (4:1). The product isolated was recrystallized from benzene-hexane, to give 2.5 g of yellow crystals.

Preparation of imidazo [4,5-α]acridones[10]

*General Procedure for the Synthesis of **24***

Compounds **22** (10 mmol) and **23** (12 mmol) were added with stirring to a solution of 20 g KOH (357 mmol) in 80 mL of methanol. The mixture was refluxed with stirring for 4 h and then poured into water. The precipitate was collected by filtration, washed with water, and air-dried to give **24**.

*General Procedure for the Synthesis of **25***

Sodium nitrite (5.0 g, 150 mmol) was added with stirring over half an hour period to a solution of **24** (7 mmol) in 100 mL concentrated sulfuric acid maintained at –10 °C. After the addition was completed, the mixture was allowed to warm to room temperature and to stand at room temperature for 17 h. After pouring this mixture into 500 mL crushed ice and water, the solid that precipitated was removed by filtration, was washed with water, and dried to give **25**.

7.3.7 References

1. Tănăsescu, I. *Bull. Soc. Chim. Fr.* **1927,** *41*, 528–532.
2. Lehmstedt, K. *Chem. Ber.* **1932,** *65*, 999–1005.
3. Lehmstedt, K. DE Pat. 581328.
4. Hawkins, D. G.; Meth-Cohn, O. *J. Chem. Soc., Perkin Trans. I* **1983,** 2077–2087.
5. Coe, P. L.; Jukes, A. T.; Tatlow, J. C. *J. Chem. Soc., (C)* **1966,** 2020.
6. [R] Cadogan, J. G. *Acc. Chem. Res.* **1972,** 303–310.
7. Kwok, R.; Pranc, P. *J. Org. Chem.* **1968,** *33*, 2880–2883.
8. Smith, P. A. S.; Braun, B. B.; Putney, R. K.; Reinisch, R. F. *J. Am. Chem. Soc.* **1953,** *75*, 6335–6337.
9. Ning, R. Y.; Blount, J. F.; Madan, P. B.; Fryer, R. I. *J. Org. Chem.* **1977,** *42*, 1791–1794.
10. Rahimizadeh, M.; Pordel, M.; Bakavoli, M.; Bakhtiarpoor, Z.; Orafaie, A. *Monatsh. Chem.* **2009,** *140*, 633–638.
11. Lehmstedt, K. *Chem. Ber.* **1932,** *65*, 834–839.

7.4 Niementowski Quinoline Synthesis

Richard A. Hartz

7.4.1 Description

The Niementowski quinoline synthesis is the reaction of an anthranilic acid (**1**) with an aldehyde or ketone containing an α methylene group that can undergo deprotonation (**2**) followed by cyclodehydration to produce a 4-hydroxyquinoline (**3**). The reaction occurs under thermal conditions at temperatures generally ranging from 120–250 °C, depending on the nature of the substrates.

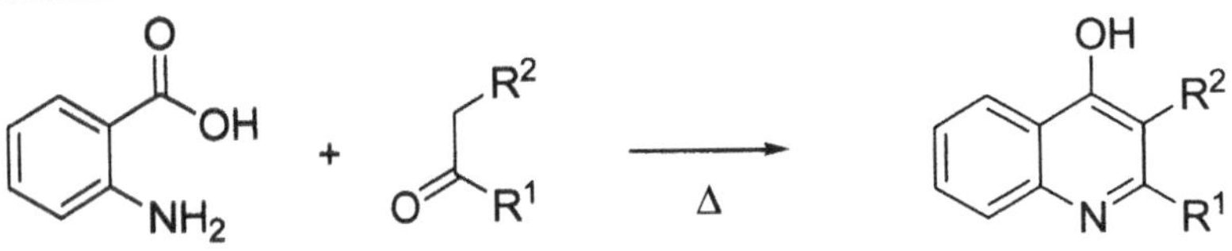

7.4.2 Historical Perspective

The last two decades of the 19th century proved to be an important era for the development of quinoline chemistry. It was during this time period that various methods for synthesizing quinolines, such as the Friedländer quinoline synthesis,[1,2] Pfitzinger quinoline synthesis,[3] Doebner–Miller quinoline synthesis,[4] Conrad–Limpach quinoline synthesis,[5] Combes quinoline synthesis,[6] and Camps quinoline synthesis[7] were first reported. In 1882, Friedländer showed that the condensation of 2-aminobenzaldehyde (**4**) with acetaldehyde (**5**) provided quinoline (**6**).[1]

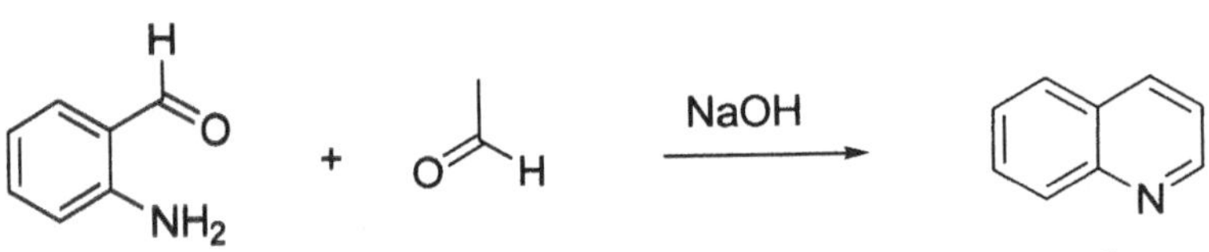

A variation of the Friedländer reaction was reported by Niementowski in 1894.[8] He found that when anthranilic acid (**1**) was heated in the presence of acetophenone (**7**) at 120–130 °C, 2-phenyl-4-hydroxyquinoline (**8**) was formed. In the following year, he reported that, in some cases, the Schiff's base of anthranilic acid (**1**) and heptaldehyde (**9**) yielded a small amount of 4-hydroxy-3-pentylquinoline (**10**) when heated at 200 °C.[9,10]

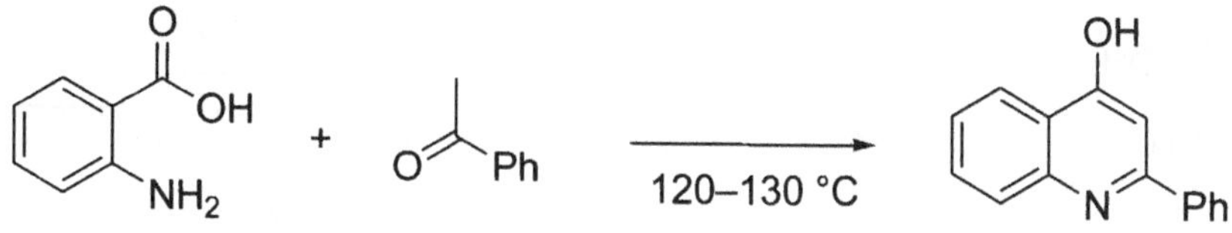

1 7 8

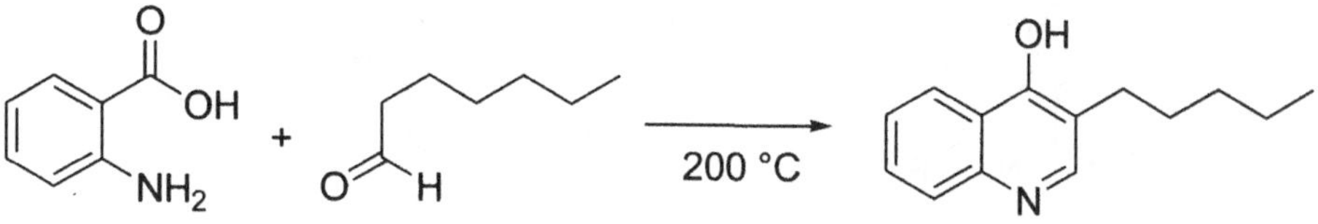

1 9 10

7.4.3 Mechanism

The mechanism for the Niementowski quinoline synthesis is presumably similar to that of the closely related Friedländer reaction. The mechanism for the Friedländer reaction has been studied extensively[11] and two possible mechanistic pathways exist, as illustrated below.[2b] There is support for both pathways. Most of the evidence, however, tends to favor initial formation of the Schiff's base intermediate (**13**) followed by cyclization to give quinoline **15**; however, the reaction conditions and structures of the reactants may influence the pathway by which the reaction proceeds.[2a]

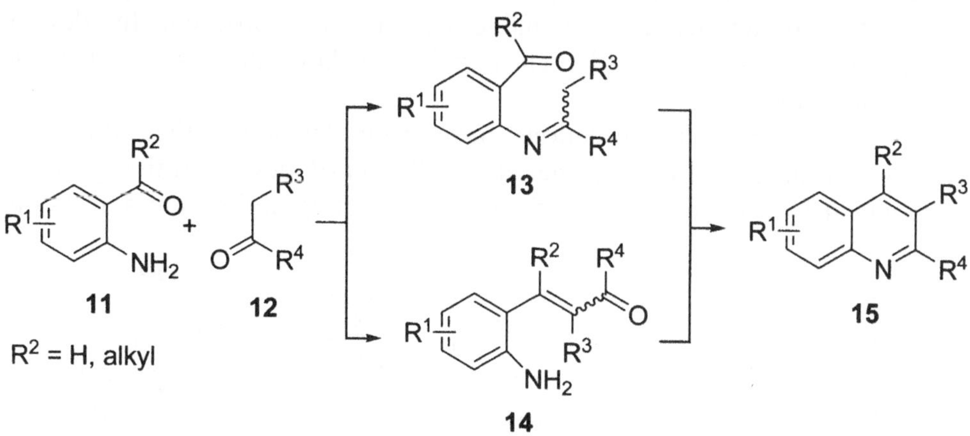

Since the Niementowski quinoline synthesis involves the condensation of an aldehyde or ketone (**2**) with an anthranilic acid (**1**), it is likely that the mechanistic pathway involving initial formation of a Schiff's base is favored, as shown below.

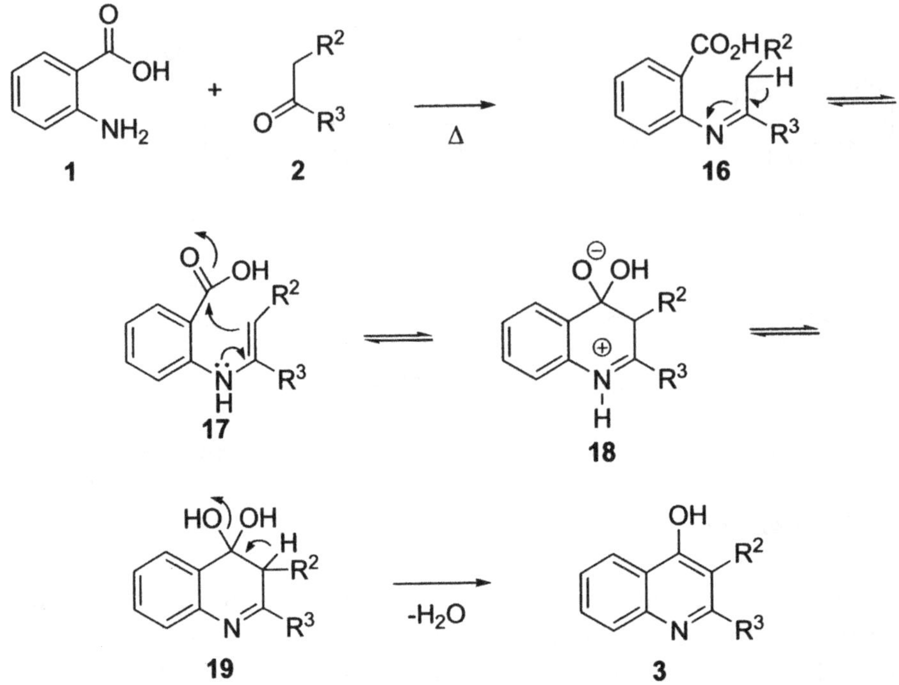

7.4.4 *Variations and Improvements*

The scope of the Niementowski reaction has been somewhat limited due to the reaction conditions required to effect this transformation. Traditionally, the reaction is carried out by simply heating the two reaction components at sufficiently high temperatures to enable the reaction to proceed. In addition to heating, sometimes an acid such as polyphosphoric acid is added to the reaction mixture.[12]

A number of modifications have been developed that allow the reaction to proceed at lower temperatures. Son *et al.* described the synthesis of various 4-hydroxyquinoline derivatives by heating anthranilic acid (**1**) in

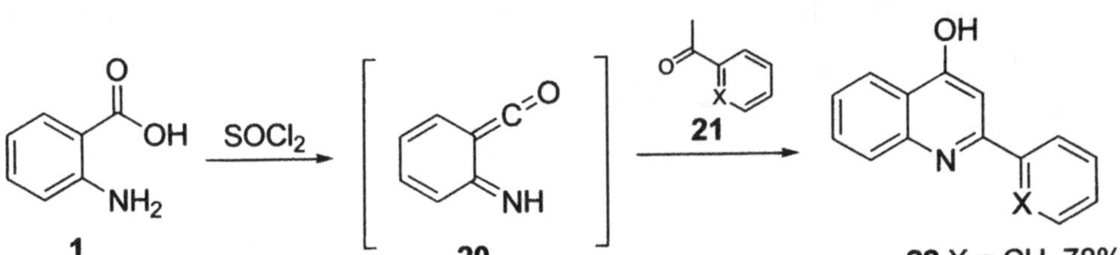

22 X = CH, 78%
23 X = N, 83%

the presence of a ketone and thionyl chloride.[13] It was proposed that the reaction proceeds via iminoketene **20** followed by a Diels–Alder reaction with, presumably, the enol tautomer of ketone **21** to furnish **22** and **23**.

A similar set of reaction conditions was reported wherein the Niementowski reaction was carried out in the presence of phosphorous oxychloride.[14] This phosphorous oxychloride-mediated condensation between **24** and **25** afforded a nearly 1:1 ratio of the two possible 4-hydroxyquinoline isomers **26** and **27**.

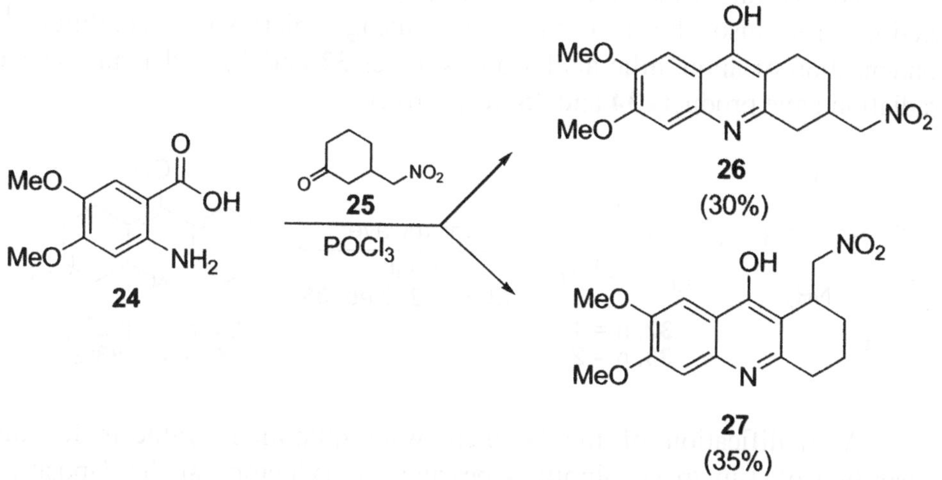

In another variant of the Niementowski reaction, it was found that this transformation can be carried out under relatively mild, base-catalyzed conditions.[15] Since a variety of substituted anthranilamides (**29**) can be prepared by a regiospecific ortho metalation-amination sequence,[16] this method appears to be a very versatile modification of the Niementowski quinoline synthesis. Lithiation of **28** with *s*-butyllithium was followed by treatment with tosyl azide. Reduction of the azide with sodium borohydride under phase transfer conditions furnished **29**. After conversion of **29** into the corresponding imine **30**, treatment of **30** with LDA afforded **31** in good yield.

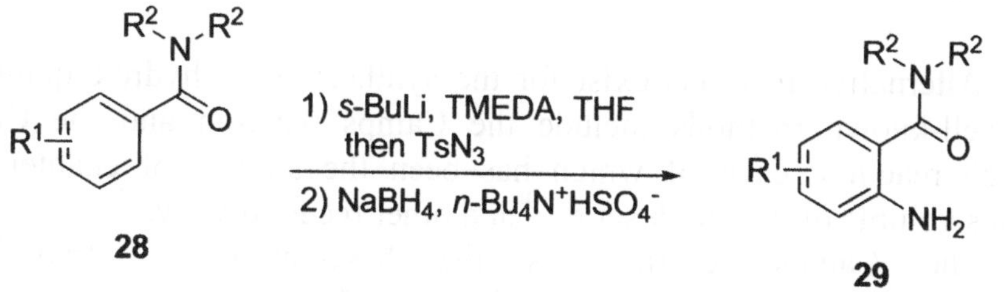

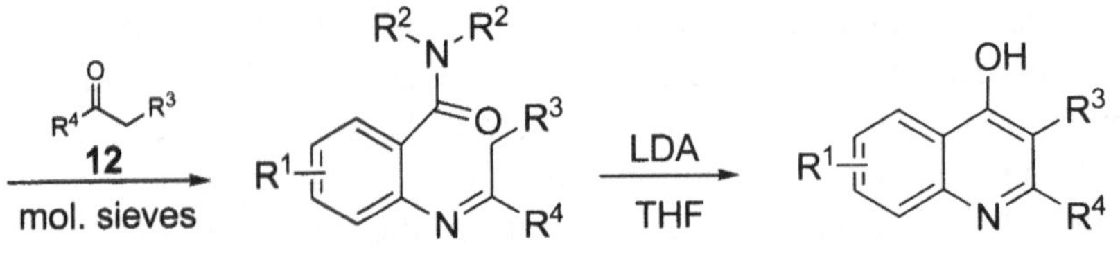

30

31
(45-95%)

The synthesis of quinolines and quinazolines via the Niementowski reaction has also been carried out using microwave conditions.[17] Condensation of anthranilic acid **1** and ketones **32** and **33** under microwave irradiation gave products **34** and **35**, respectively.

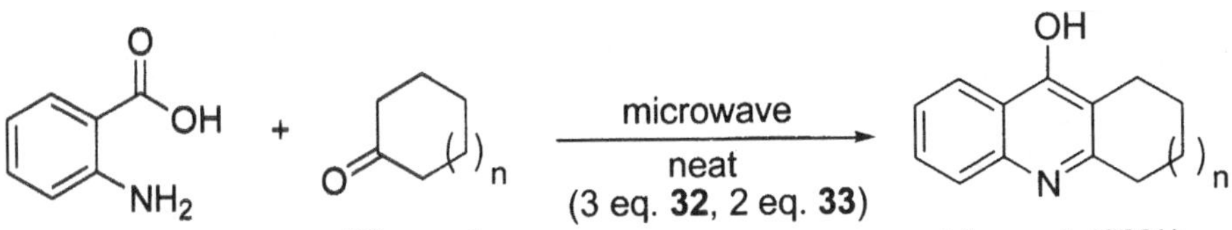

1

32, n = 1
33, n = 2

34, n = 1 (92%)
35, n = 2 (83%)

A modification of the Niementowski quinoline synthesis for the preparation of 4-hydroxyquinolines bearing an arylketone at the 3-position was reported by Wang *et al.*[18] Reaction of **36** with **37** afforded 4-hydroxyquinoline **38**. This 4-hydroxyquinoline with a 2-thiomethyl substituent can then serve as a useful intermediate for further chemical manipulations.

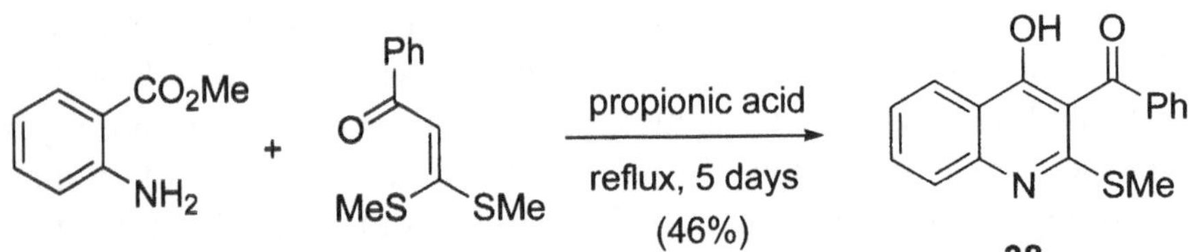

36 **37** **38**

Alternative methods exist for the synthesis of 4-hydroxyquinolines. Two well-known methods include the Camps reaction and the Conrad-Limpach reaction, each of which has been the subject of chapters in a previous volume of this book series[19] and other recent reviews.[20]

The Camps reaction[7] is the base-catalyzed intramolecular condensation of an *N*-acyl *o*-acylaniline **39** to form either a 4-hydroxyquinoline **40** or a 2-hydroxyquinoline **41** or mixtures of both depending on the nature of the substrate. Due to the possibility of the formation of two cyclization products, in general, the Camps reaction is synthetically most useful when only one of the carbonyl groups is enolizable or an electron withdrawing group, such as CN, $COCH_3$, COC_6H_5, CO_2Et, or C_6H_5, exists at either R^1 or R^2 to favor formation of one of the two isomers.[10]

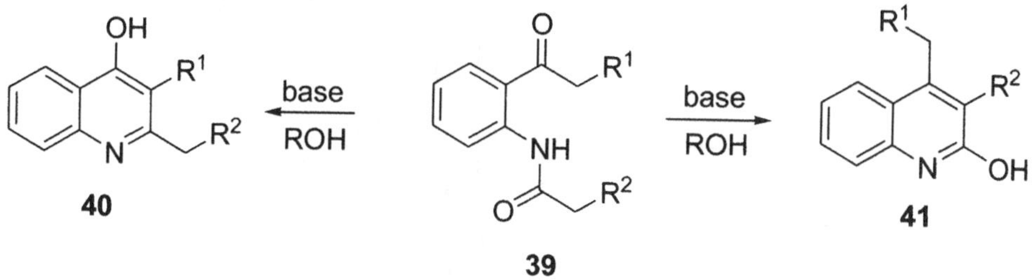

4-Hydroxyquinolines can also be formed by the Conrad–Limpach reaction.[5] This reaction typically involves the condensation of an aniline (**42**) with a β-ketoester (**43**). The resultant enamine **44** then undergoes cyclization followed by dehydration to afford a 4-hydroxyquinoline (**45**). This reaction is synthetically most useful for the synthesis of 4-hydroxyquinolines when the aniline is either symmetrical or contains one *ortho* substituent; otherwise, mixtures of products are often obtained.

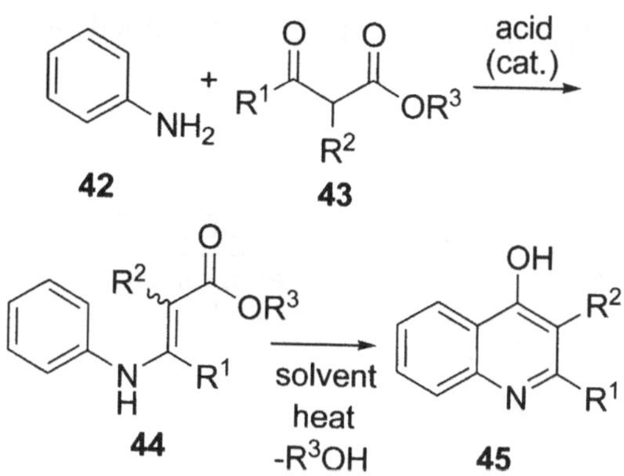

7.4.5 *Synthetic Utility*

The Niementowski reaction has been used to synthesize a series of 4-substituted-2,3-polymethylenequinolines that were then studied for their CNS effects—namely for local anesthetic activity, analaleptic activity, and antihistaminic activity.[21] The initial 4-hydroxyquinoline product **47** was converted to the chloride, which could then be displaced by various alkylamines to give **48**.

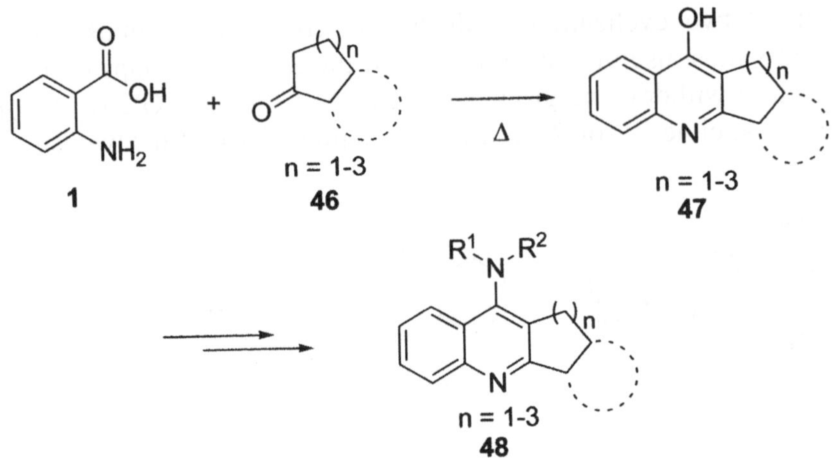

4-Aminoquinolines have been synthesized in a manner analogous to the Niementowski reaction, wherein the carboxylic acid in **1** was replaced with a nitrile (**49**). This transformation has been carried out under either acidic or basic conditions. Under the former conditions, a variety of Lewis acids have been employed to effect this transformation, including zinc chloride, aluminum trichloride, boron trifluoride diethyl etherate, and titanium tetrachloride.[2a,14] In a representative example, **49** was combined with **50** and heated at reflux in the presence of boron trifluoride diethyl etherate to furnish **51**, which contains an embedded 4-aminoquinoline ring system.[22]

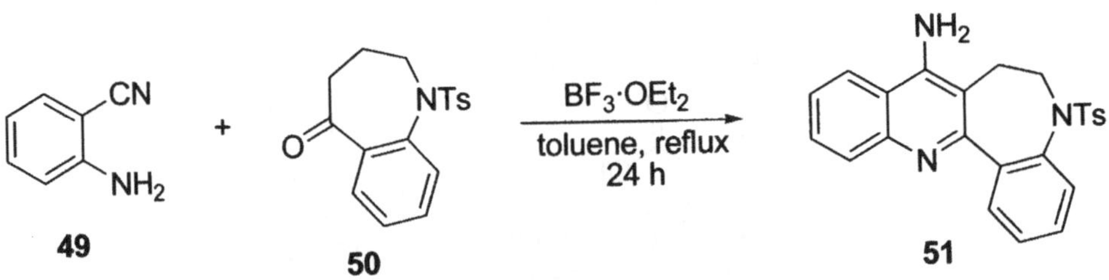

Alternatively, a Schiff's base (**53**), obtained from condensation of 2-aminobenzonitrile **49** and an aryl or heteroaryl methyl ketone **52**, was deprotonated with LDA, and the resulting carbanion underwent cyclization to form a 4-aminoquinoline product **54** in high yield.[23] It is interesting that when this reaction was carried out with the ethyl ketone analog of **52** it did not produce the analogous quinoline product. Instead, it was determined that rather than undergoing deprotonation, the more sterically hindered ethyl ketone derivative underwent an addition reaction with LDA to form **55**.[23]

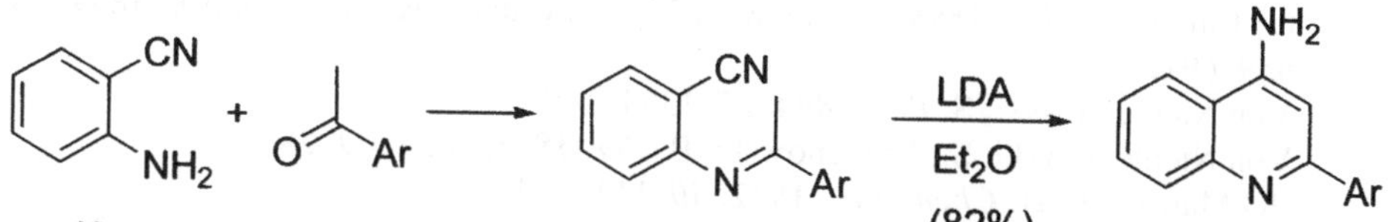

49 52 53 54

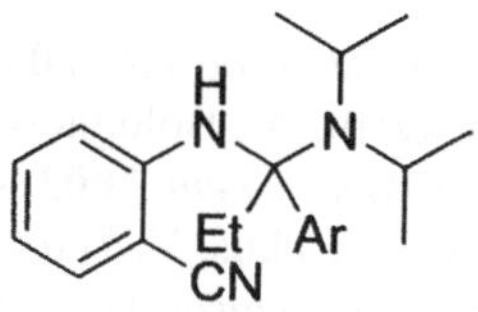

55

7.4.6 *Experimental*

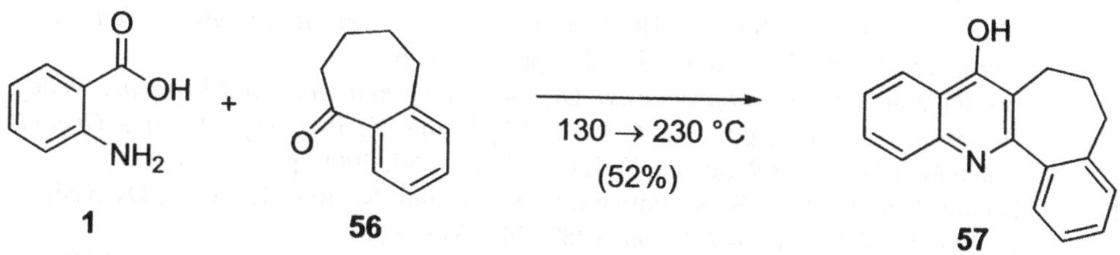

1 56 57

1′,2′-Benzo-4-hydroxy-2, 3-pentamethylenequinoline (57)[21]
Anthranilic acid (14.0 g, 102 mmol) was mixed with benzo-suberone (**56**) (16.3 g, 102 mmol), and the mixture was heated in an oil bath at 130 °C for 1 h. The temperature of the bath was increased to 180 °C over 0.5 h and then to 220–230 °C over an additional 0.5 h. The contents of the flask solidified during this time period. Heating was continued for an additional 0.5 h. The cooled product was washed with hot benzene to give **57** as a colorless powder (13.8 g, 52% yield). Crystallization from pyridine/benzene gave colorless plates (mp > 310 °C).

7.4.7 *References*

1. Friedländer, P. *Ber.* **1882,** *15*, 2572.
2. For recent reviews see: (a) [R] Marco-Contelles, J.; Pérez-Mayoral, E.; Samadi, A.; do Carmo Carreiras, M.; Soriano, E. *Chem. Rev.* **2009,** *109*, 2652–2671. (b) [R] Pflum, D. In *Name Reactions in Hetrocyclic Chemistry*, Wiley: Hoboken, NJ, **2005,** pp. 411–415. (c) [R] Wang, Z. In *Comprehensive Organic Name Reactions and Reagents*, Wiley: Hoboken, NJ, **2009,** pp. 1137–1142.
3. Pfitzinger, W. *J. Prakt. Chem.* **1886,** *33*, 100.
4. Doebner, O.; von Miller, W. *Ber.* **1881,** *14*, 2812.
5. (a) Conrad, M.; Limpach, L. *Ber.* **1887,** *20*, 944. (b) Conrad, M.; Limpach, L. *Ber.* **1887,** *20*, 948. (c) [R] Reitsema, R. H. *Chem. Rev.* **1948,** *43*, 43–68.
6. Combes, A. *Bull. Soc. Chim. Fr.* **1888,** *49*, 89.

7. (a) Camps, R. *Ber.* **1899,** *32*, 3228–3234. (b) Camps, R. *Arch. Pharm.* **1899,** *237*, 659–691.
8. Von Niementowski, S. *Ber.* **1894,** *27*, 1394–1403.
9. Von Niementowski, S.; Orzechowski, B. *Ber.* **1895,** *28*, 2809–2822.
10. [R] Manske, R. H. *Chem. Rev.* **1942,** *30*, 113–144.
11. For discussions on the mechanism of the Friedländer reaction see reference 2a.
12. Nandha Kumar, R.; Suresh, T.; Mythili, A.; Mohan, P. S. *Heterocyl. Commun.* **2001,** 7, 193–198.
13. Son, J. K.; Kim, S. I.; Jahng, Y. *Heterocycles* **2001,** *55*, 1981–1986.
14. Rosini, M.; Antonello, A.; Cavalli, A.; Bolognesi, M. L.; Minarini, A.; Marucci, G.; Poggesi, E.; Melchiorre, C. *J. Med. Chem.* **2003,** *46*, 4895–4903.
15. Chong, R. J.; Siddiqui, M. A.; Snieckus, V. *Tetrahedron Lett.* **1986,** *27*, 5323–5326.
16. (a) Reed, J. N.; Snieckus, V. *Tetrahedron Lett.* **1983**, *16*, 3795–3798. (b) [R] Beak, P.; Snieckus, V. *Acc. Chem. Res.* **1982,** *15*, 306–312.
17. Khajavi, M. S.; Afshani, P.; Rad Moghadam, K. *Iran J. Chem. Eng.* **1998,** *17*, 29–32.
18. Wang, M.-X.; Liu, Y.; Huang, Z.-T. *Tetrahedron Lett.* **2001,** *42*, 2553–2555.
19. (a) [R] Pflum, D. In *Name Reactions in Hetrocyclic Chemistry*, Wiley: Hoboken, NJ, 2005; pp. 386–389. (b) [R] Curran, T. T. In *Name Reactions in Hetrocyclic Chemistry*, Wiley: Hoboken, NJ, 2005; pp. 398–406.
20. (a) [R] Wang, Z. In *Comprehensive Organic Name Reactions and Reagents*, Wiley: Hoboken, NJ, 2009; pp. 598–601. (b) [R] Wang, Z. In *Comprehensive Organic Name Reactions and Reagents*; Wiley: Hoboken, NJ, 2009, pp. 692–696.
21. Bindra, J. S.; Rastogi, S. N.; Patnaik, G. K.; Anand, N.; Rao, K. G. G.; Dwivedi, P. C.; Rao, C. N. R. *Indian J. Chem.* **1987,** *26B*, 318–329.
22. McKenna, M.; Proctor, G. R.; Young, L. C.; Harvey, A. L. *J. Med. Chem.* **1997,** *40*, 3516–3523.
23. Strekowski, L; Kong, S.-B.; Cegla, M. T.; Harden, D. B. *Heterocycles* **1989,** *29*, 539–545.

7.5 Povarov reaction

Ji Zhang

7.5.1 Description

The Povarov reaction[1] is the inverse electron-demand aza-Diels–Alder reaction, a [4 + 2] cycloaddition between an *N*-arylimine (as the diene) and an electron-rich olefin (as the dienophile), which gives tetrahydroquinolines **3** or substituted quinolines **4** as the product. This reaction also called as imino-Diels–Alder reaction,[2] usually catalyzed by Lewis or Brønsted acids. Since the *N*-arylimine can be prepared *in situ* from aniline and aldehyde, thus the Povarov reaction can be performed in a one-pot fashion.[3]

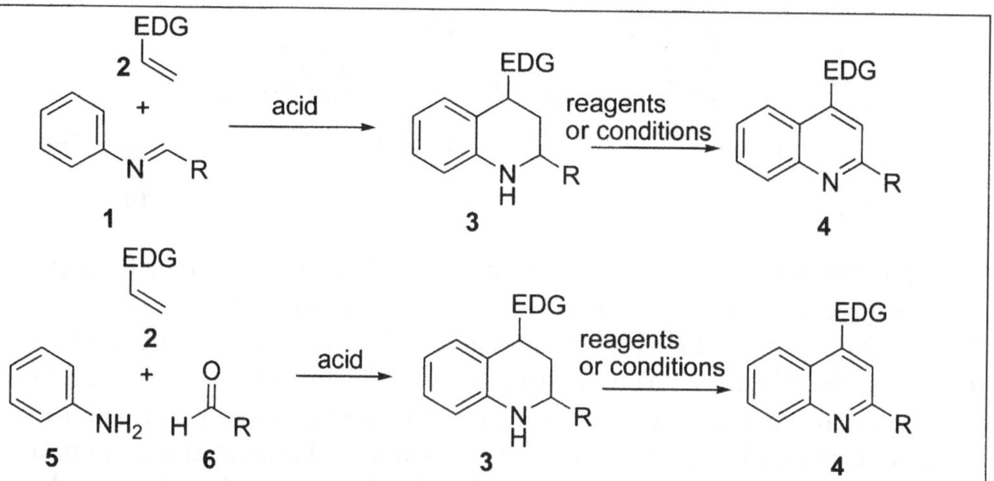

EDG = Electron Donating Group
acids: Lewis acids or Brønsted acids
reagents or conditions: 1: DDQ, 2:TsOH/distillation, 3: Pd/C, 4: air/heat, 5: $Mn(OAc)_3$

Normally, the *N*-arylimine is obtained by reaction of aldehyde and aniline in acidic condition. Either tetrahydroquinoline or its corresponding substituted quinoline can be generated in the Povarov reaction, depending on the reaction conditions. For instance, DDQ-promoted dehydrogenation,[4] vacuum distillation under acidic condition,[5] oxidation by air[6] or $Mn(OAc)_3$,[7] and Pd/C-catalyzed aromatization[8] of tetrahydroquinoline, provides the corresponding substituted quinolines in good to excellent yield. Since some tetrahydroquinolines are unstable under the reaction conditions, the corresponding substituted quinolines could be isolated as the sole products. Electron-rich olefin, such as vinyl enol ethers, vinyl sulfides, and silyl enol ethers, are widely used as dienophiles in the cycloaddition of *N*-aryl aldimines to obtain substituted tetrahydroquinolines. To access natural products and drug candidates, vinyl enamides, enamines and *N*-vinylcarbamate have been employed as dienophiles in the Povarov reaction. Simple alkene and terminal alkynes, such as styrene, phenylacetylene and ethoxyacetylene were found to have limited use in the Povarov reaction.[9]

7.5.2 *Historical Perspective*

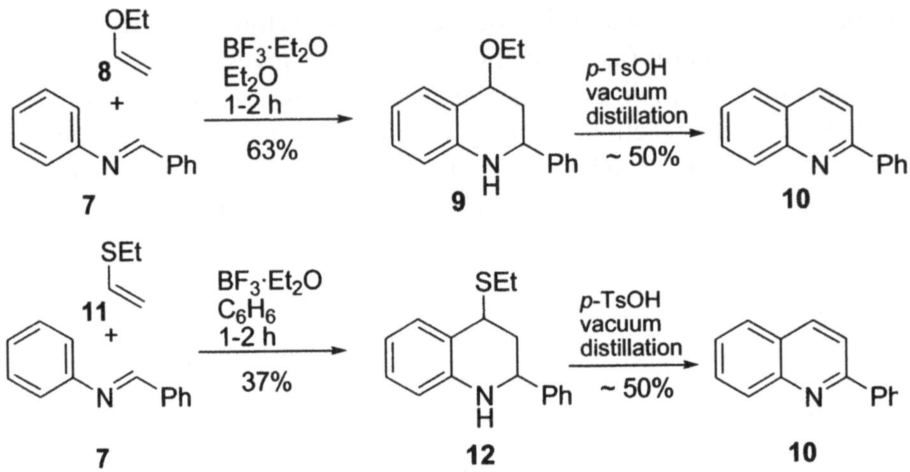

In 1963, Povarov first disclosed the reaction of ethyl vinyl ether **8** or ethyl vinyl sulfide **11** and *N*-aryl aldimine **7** in the presence of the Lewis acid $BF_3 \cdot Et_2O$ gave **9** and **12** which were converted into corresponding quinoline **10** in ~ 50% yield.[5,10] Tetrahydroquinolines **9** and **12** were isolated and characterized. Povarov viewed this reaction as a Diels–Alder cycloaddition. Later, Povarov outlined the scope of this reaction and indicated the important role of the Lewis acid catalyst.[11] By using dicobalt octacarbonyl as catalyst, Joh and Hagihara reported that 2,4-substituted tetrahydroquinolines **9** and quinoline **10** were prepared from the same starting materials.[12] In 1993, Narasaka reported that Schiff bases derived from various butanals can be employed and reacted with vinyl sulfide, giving adducts in 70–83% isolated yield.[4] In 1995, Kobayashi described that aliphatic aldehydes can be used in the three-component Povarov reaction, using lanthanide triflate as catalyst,[13] solving the problem associated with using aliphatic aldehydes with enolizable protons in the Povarov reaction. Currently several Lewis acids ($BF_3 \cdot Et_2O$, $SnCl_4$,[14] $InCl_3$,[15] $EtAlCl_2$,[16] $ZrCl_4$,[17] $LiClO_4$[18] and $BiCl_3$[19]), Brønsted acids (*p*-TsOH,[20] TFA,[21] CF_3SO_3H,[22] oxalic acid), and lanthanide triflates ($Yb(OTf)_3$[23] $Sc(OTf)_3$[24] and $Dy(OTf)_3$)[25] have been known to promote these reactions. Recently, iodine,[26] TMSCl[27] and cerium ammonium nitrate (CAN)[28] have been used to promote the Povarov reaction. Today, the Povarov reaction is becoming a very attractive approach to synthesize complex *N*-polyheterocycles, including some alkaloids and drug candidates due to its numerous advantages, including atom economy, readily available starting materials, operational simplicity, easily automatable and ecologically benign.

7.5.3 Mechanism[29]

The mechanism of tetrahydroquinoline formation by the Povarov reaction has been debated. A stepwise mechanism involves ionic intermediate **13**, followed by an intramolecular electrophilic substitution. A concerted hetero Diels–Alder reaction was proposed where a concerted asynchronous transition state **14** was suggested.

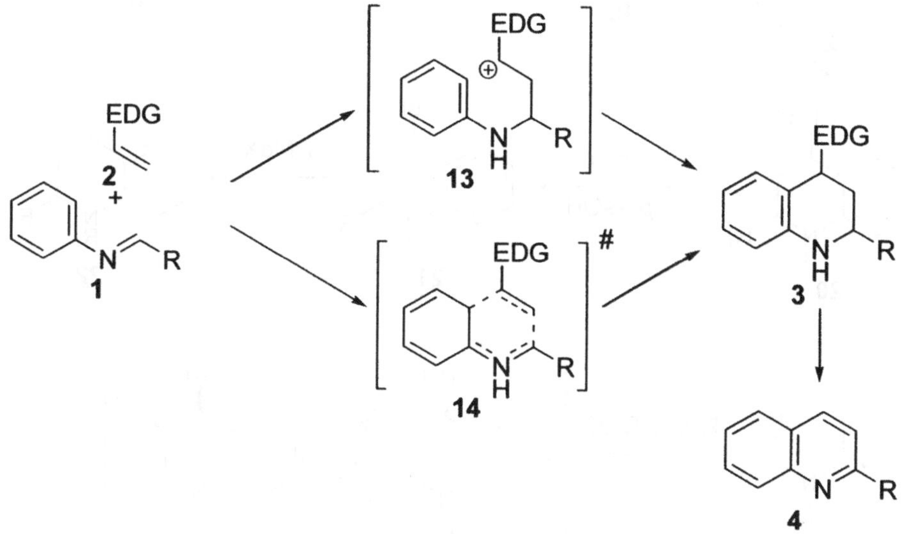

7.5.4 Variations and Improvements: Katritzky Methodology[30]

In 1993, Katritzky demonstrated that benzotrizole (BtH) **17** reacts with enamines, enamides, and vinyl ethers, giving iminium ion **19** and benzotriazolate anion, and providing substituted tetrahydroquinolines **20** *via* the aza-Diels–Alder reaction.[31] The Katritzky methodology has several advantages over the Povarov reaction since the moisture-sensitive Lewis acid can be avoided and benzotriazole adduct **18** is stable enough, and usually **18** can be isolated as a solid by crystallization. Alternatively, instead of using an alkene, aldehydes react with the Bt adduct **23**, generating intermediates **26** and **27**, gave final product **28** in good to excellent yield when treatment with RMgX.[32]

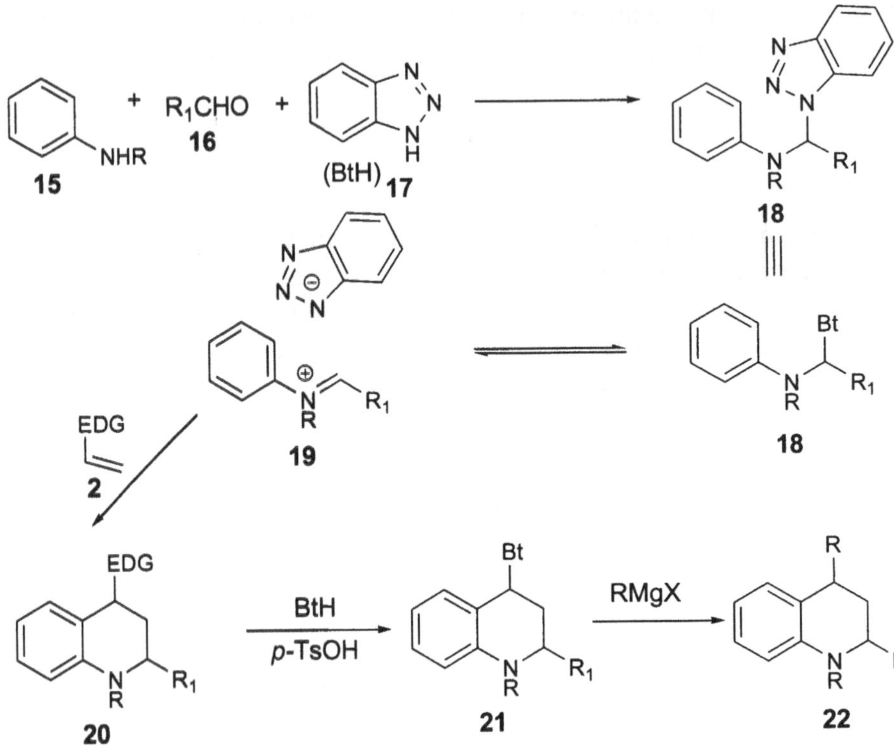

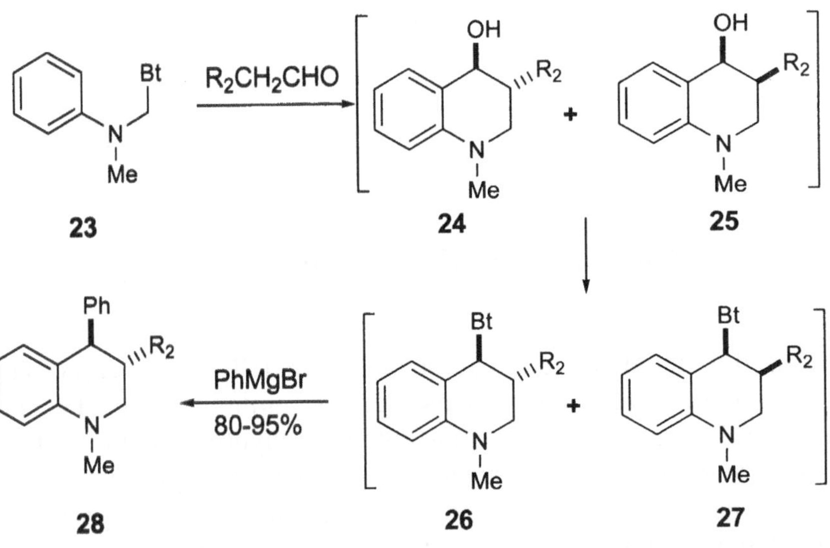

Dugger and co-workers of Pfizer applied the Katritzky methodology for the synthesis of a key intermediate for the CETP inhibitor **35**, torcetrapib (CP-529,414). Benzotriazole adduct **31** was prepared in 91% isolated yield in two crops, and racemic tetrahydroquinoline **34** was synthesized in 78%

isolated yield when benzotriazole adduct **31** was reacted with vinyl carbamate **32**.[33]

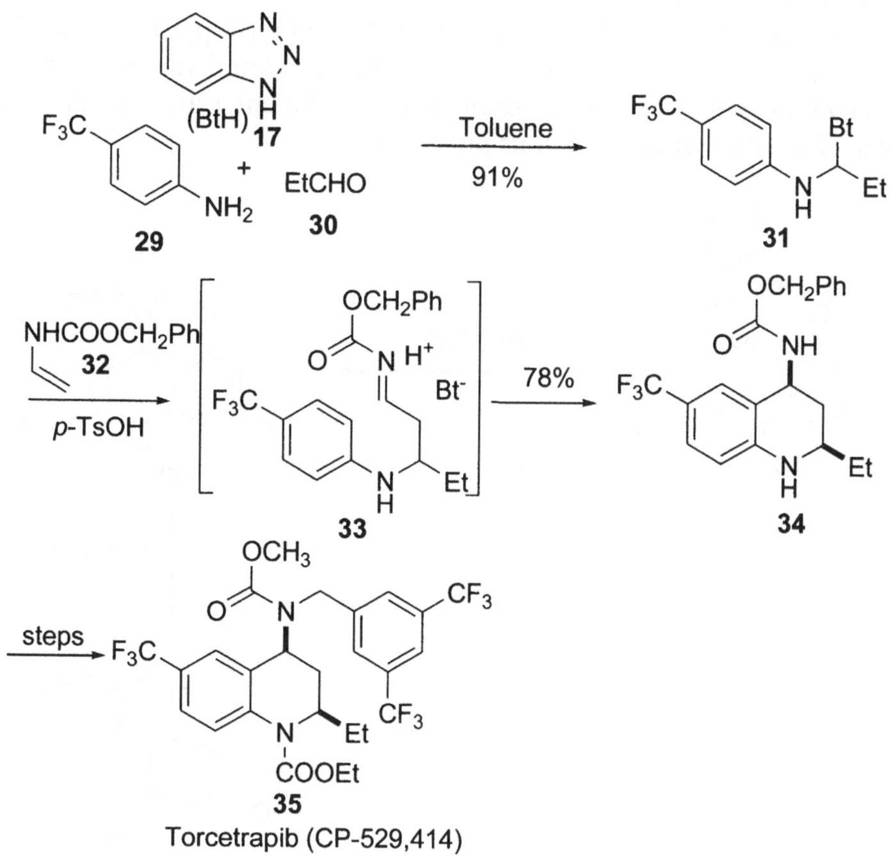

7.5.5 Synthetic Utility

Tetrahydroquinolines[34]

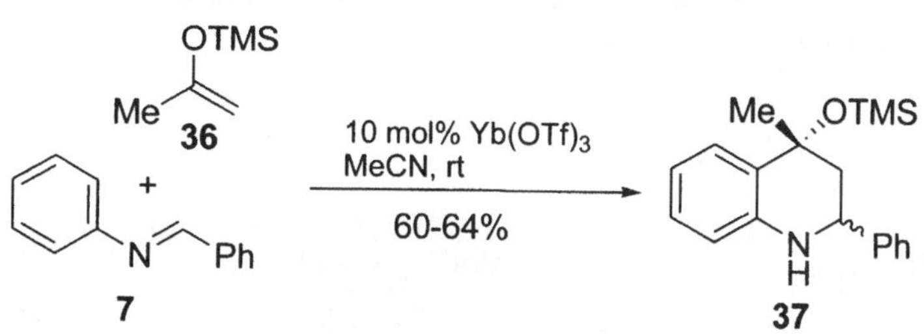

Using Kobayashi's protocol, 4-trimethylsilyloxy-substituted 2-phenyltetrahydroquiniline **37** was obtained in 60–64% yield as a mixture of stereoisomers when silyl ethers were used as dienophiles in the cycloaddition with *N*-aryl aldimine.[35]

N-Aryl aldimine **38** was treacted with *N*-vinyl-2-pyrrolidinone **39** to provide a 1:1 mixture of isomeric (*cis*/*trans*) tetrahydroquinoline **40**. Epimerization with methoxide resulted in enrichment (20:1) of the *trans*-isomer **41** which was isolated in its pure form.[36]

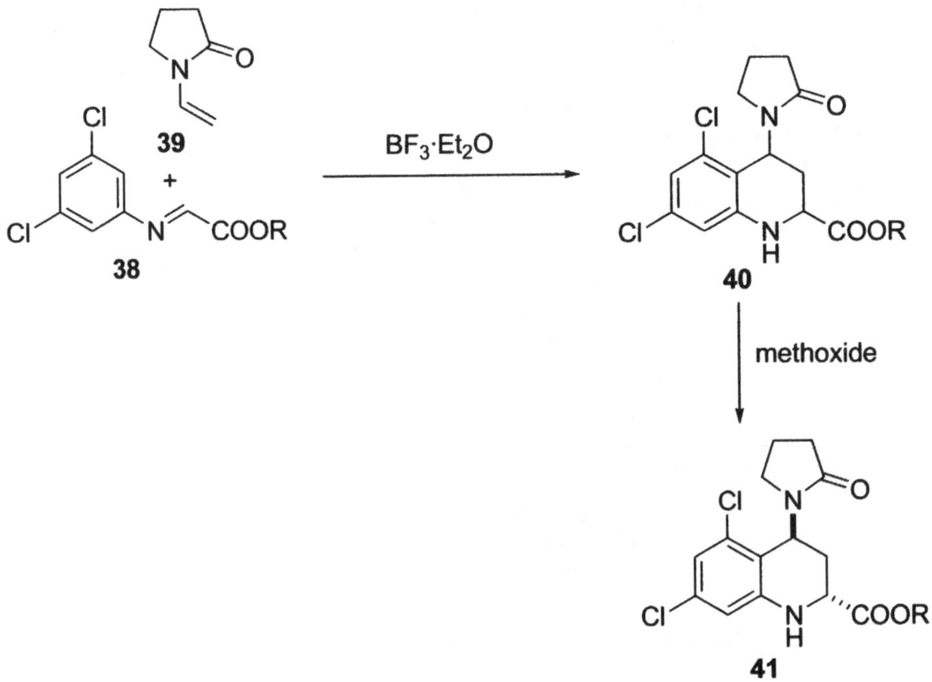

Using Ln(OTf)$_3$ catalyst, Batey demonstrated that the three-component Povarov reaction generated highly functionalized *C*-2 aliphatic-substituted tetrahydroquinolines **45**, which were inaccessible previously using the traditional Povarov reaction, due to the instability of the aliphatic *N*-arylaldimine. It is interesting that cyclopentadiene was used as the dienophile in this new process.[37]

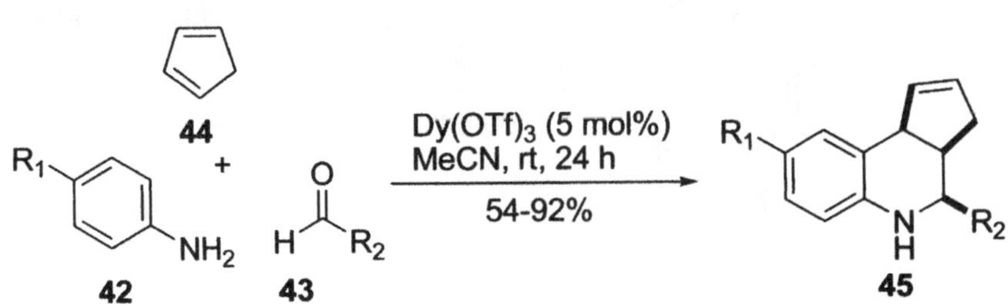

Recently, it was found that cerium(IV) ammonium nitrate (CAN) catalyzed the one-pot, three-component imino Diels–Alder reaction,[28] giving *cis*-4-alkoxy-2-aryl-1,2,3,4-tetrahydroquinoline derivatives **49** with almost complete diastereoselectivity (up to 97:3).[38] Noteworthy, cinnamaldehyde can be used for reaction, but diastereoselectivity (55:45) is dropped significantly.

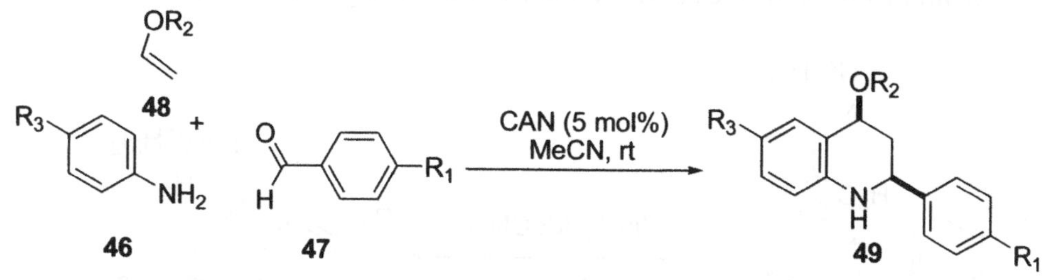

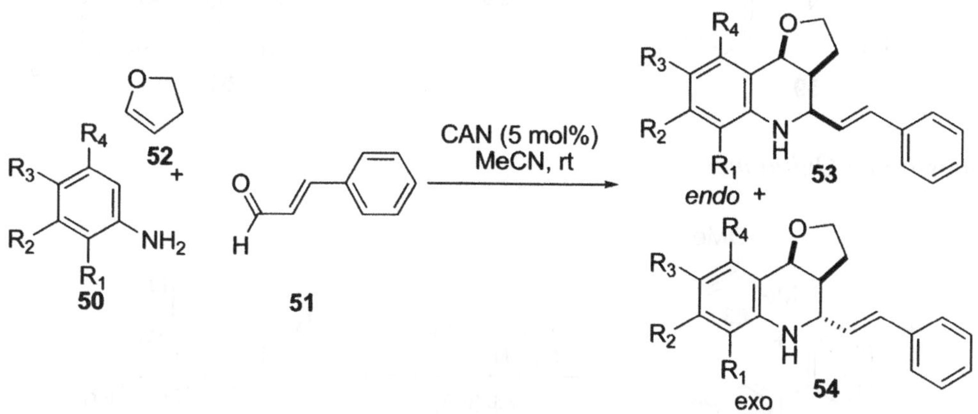

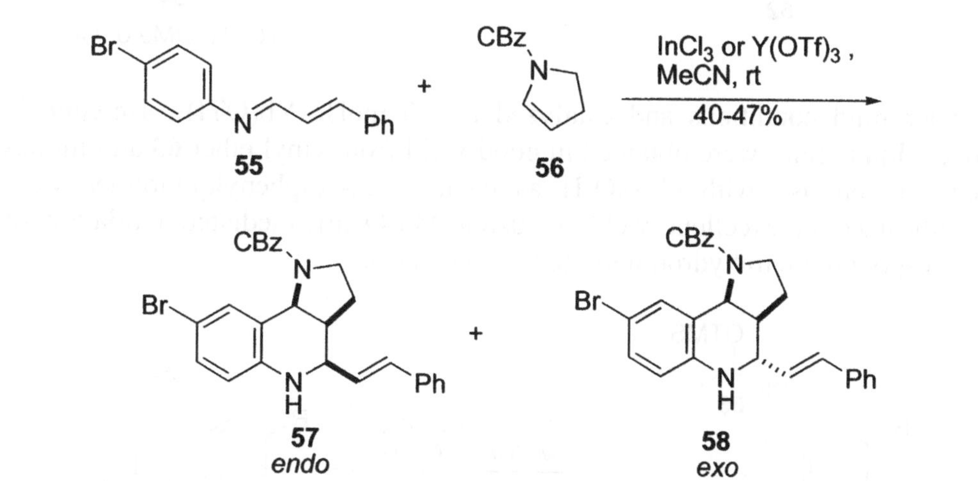

The heterocyclic core of martinelline and martinellic acid was prepared via the Povarov reaction between an aromatic imine derived from cinnamaldehyde and a cyclic enamide.[39] Although the *exo*/*endo* selectivity was poor for the cycloaddition, fortunately, the diasteroisomers were readily separated by flash chromatography.

α,β-Unsaturated hydrazone **60** can be used as dienophile, with 10% $InCl_3$, C-4 functionalized 1,2,3,4-tetrahydroquinolines **61** containing a quaternary stereocenter were synthesized in good to excellent yield (70–93%) from aromatic imines and methacroleindimethylhydrazone.[40]

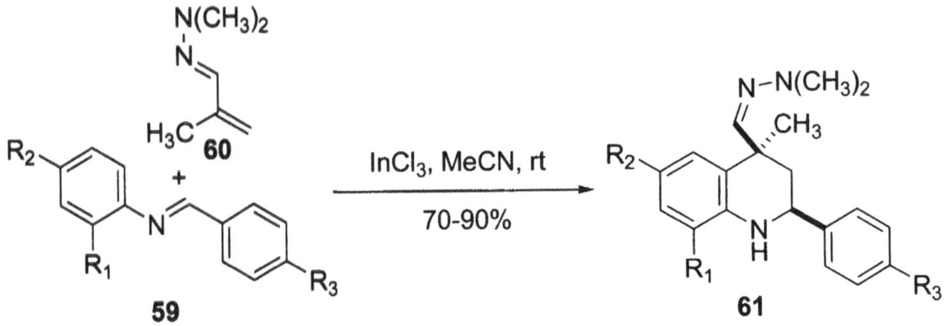

Substituted Quinolines

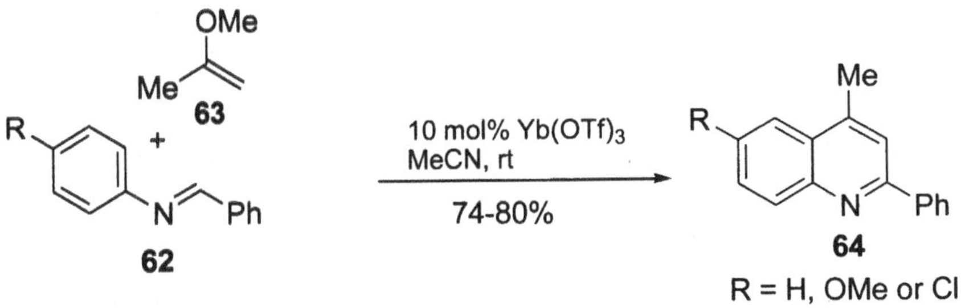

Under mild conditions and catalyzed by 10 mol% $Yb(OTf)_3$, 4-methyl-2-phenylquinolines were obtained in good yield from vinyl ether **63** and imines **62**. In contrast, with CF_3SO_3H as catalyst, 2,4-diphenylquinolines were synthesized in excellent yield by using $Mn(OAc)_3$-mediated oxidation of corresponding tetrahydroquinoline intermediates.[7]

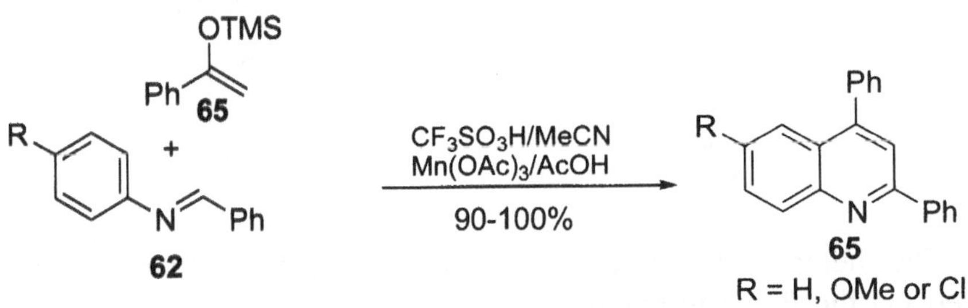

Mediated by CF_3SO_3H, ethynyl ketene-*S,S*-acetals **66** and **69**, the highly reactive dienophiles can react in a one-pot condition with various arylamines and aldehydes, giving the corresponding quinolines in good yield *via* regiospecific aza-Diels–Alder reaction.[41]

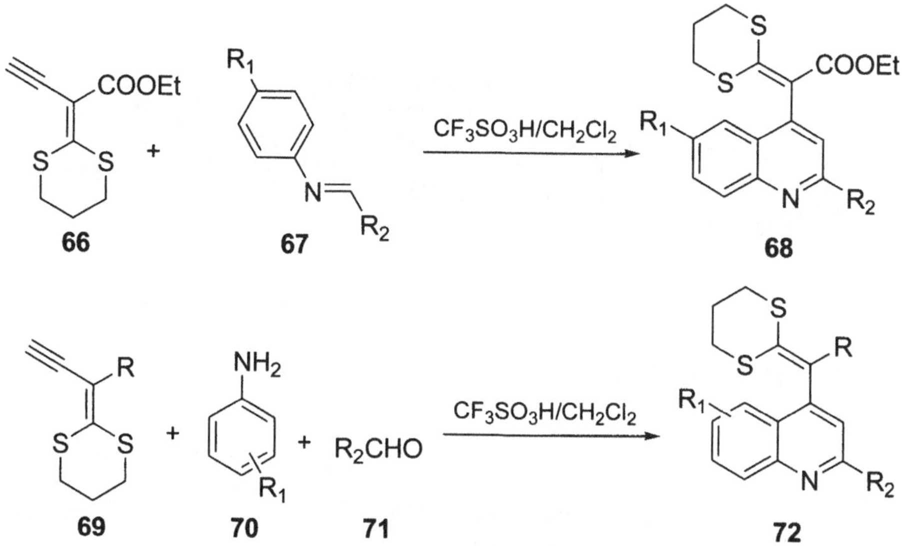

Applications in Alkaloids and N-Polyheterocycles Synthesis

Batey demonstrated that the Povarov reaction is the most valuable approach to the synthesis of alkaloids luotonin A, camptothecin and martinelline.[42]

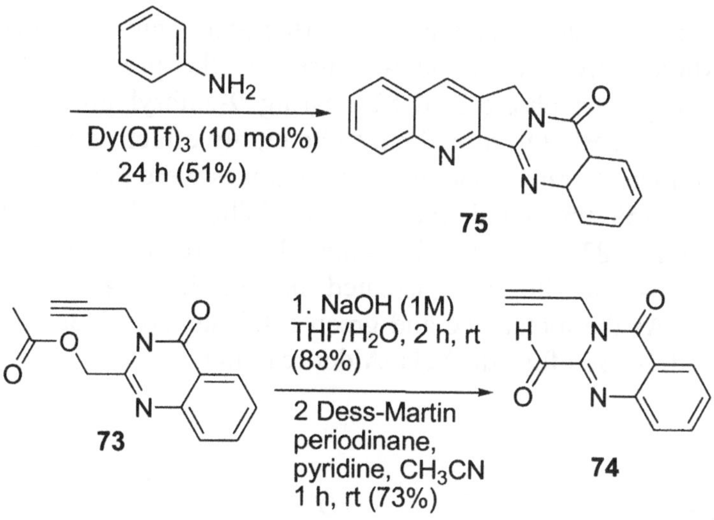

Structurally interesting and bioactive important indolopyrroloquinolines were successfully assembled *via* a one-pot cycloaddition of *N*-

prenylated-2-formyl-3-chloroindole **77** and aniline. Promoted by 10 mol% $La(OTf)_3$ in 1,4-dioxane, single diastereomers of indole-annulated pyrroloquinolines **78**, having the *cis*-configuration were isolated in good to excellent yield (up to 92%).[43]

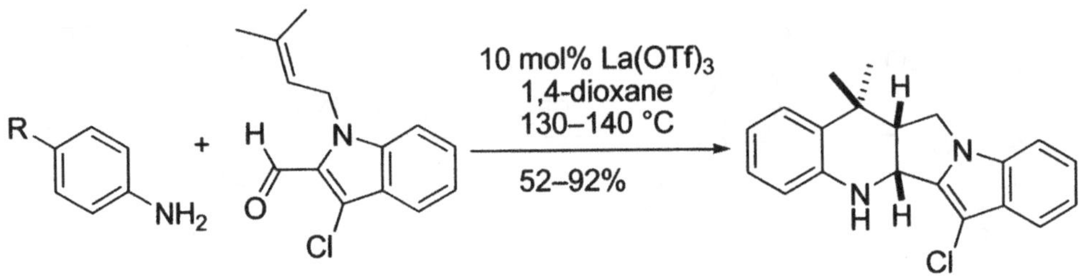

76 **77** **78**

Asymmetric Povarov Reaction

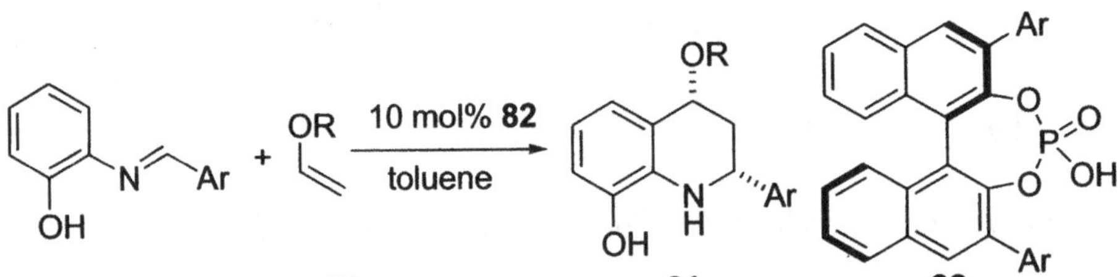

79 **80** **81** **82**

yield: 59–95%
ee: 88–97%

(*R*), Ar = 9-anthryl

In 2006, Akiyama developed a chiral Brønsted acid-promoted Povarov reaction, where *o*-hydroxyaniline was used as the precursor to make *N*-arylaldimines. A phosphoric acid **82** bearing 9-anthryl group on the 3,3′-position of (*R*)-(+)-BINOL turned out to be highly effective as the catalyst.[44] More than a dozen tetrahydroquinoline derivatives were generated in good to excellent yield (59–95%) and *ee* (87–97%). When cyclic vinyl ethers, such as dihydrofuran **83** and dihydropyran **85** were used as dienophiles, cycoladducts **84** and **86** were isolated in excellent *ee* (90% and 97%, respectively). As the metal-free and green catalysts, chiral Brønsted acid **82** should be widely used for aza Diels–Alder reaction.

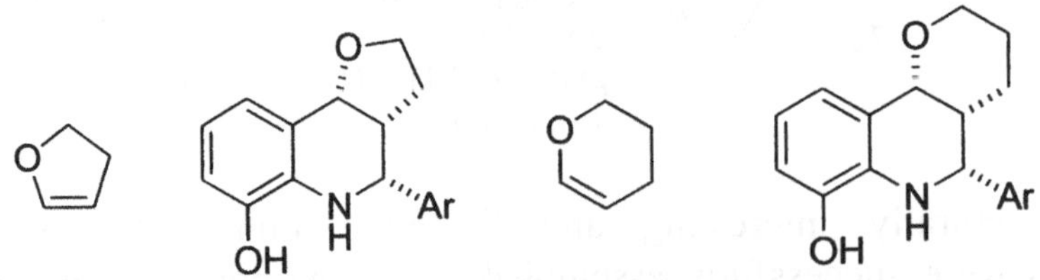

83 **84** **85** **86**

A practical approach using chiral Brønsted acids for the enantioselective three-component Povarov reaction has been successfully developed by Zhu and co-workers recently.[45] When 10 mol% chiral acid **91** was used as catalyst, a reaction of aniline, aldehyde and enecarbamate **88** in one-pot gave tetrahydroquinoline derivatives in good yield and excellent enantioselectivities (up to > 99% *ee*). To illustrate the power of this novel catalytic enantioselective three-component Povarov reaction, toprcetrapib was prepared in only 4 steps, so far the most effective approach to this target.

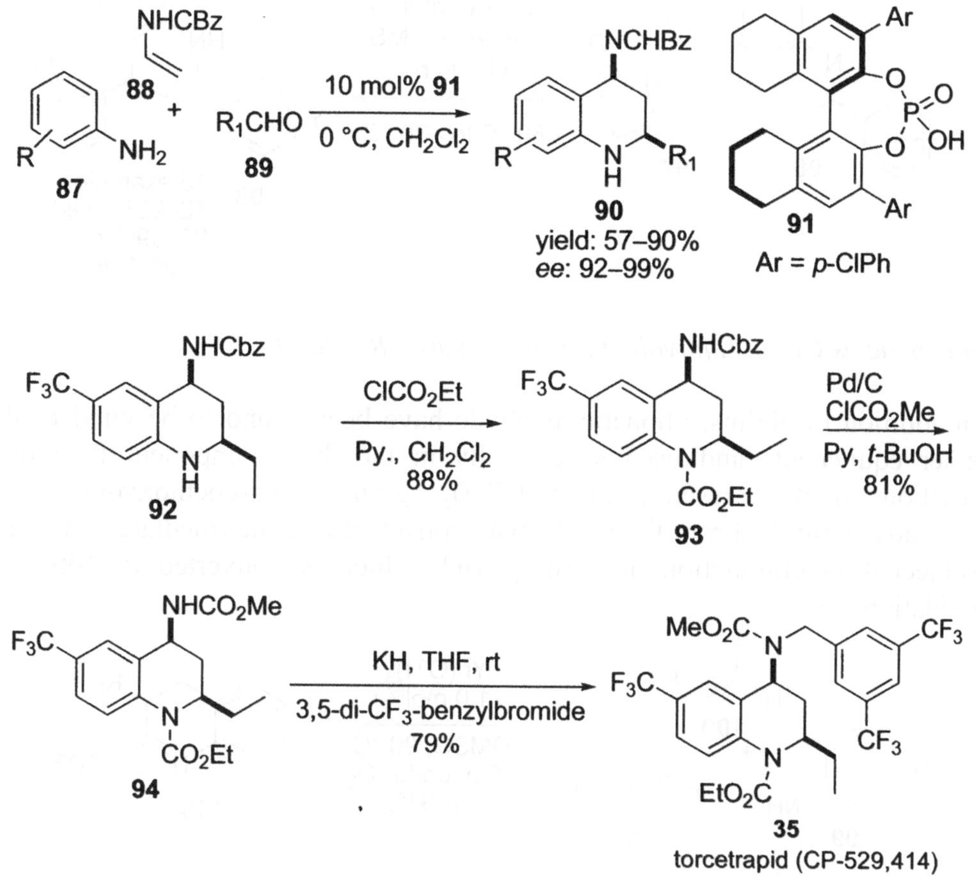

Recently, Jacobsen developed a new strategy for inducing enantioselectivity in reactions of protio-iminium ions, wherein a chiral urea **95** interacts with the highly reactive intermediate through a network of noncovalent interactions. This interaction leads to an attenuation of the reactivity of the iminium ion and allows high enantioselectivity in the Povarov reaction, giving lactam-substituted tetrahydroquinoline derivatives **98** in excellent enantioselectivities and diastereoselectivities.[46]

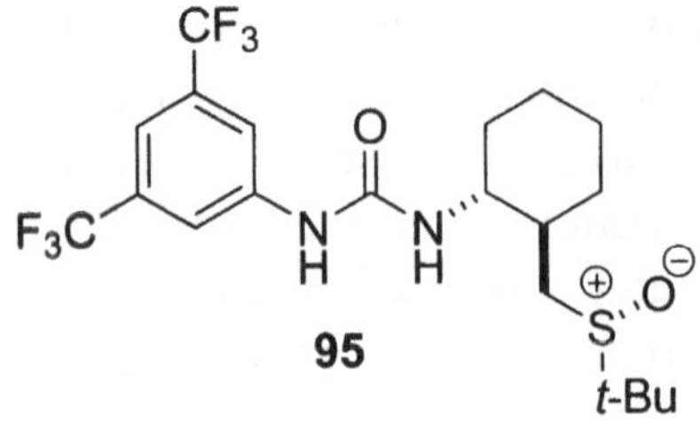

95

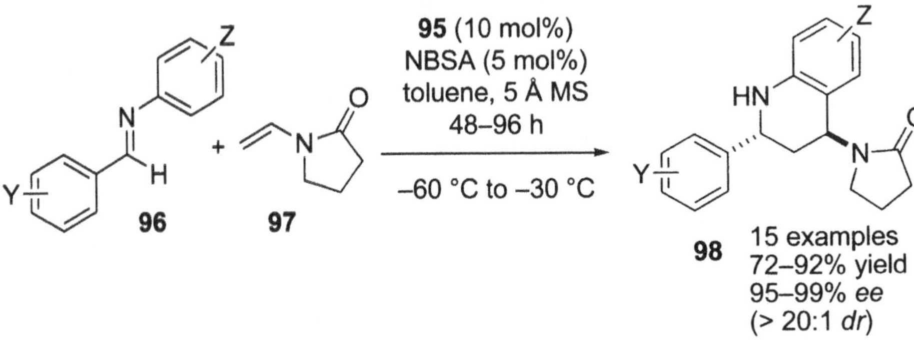

Aldehyde as Olefin Equivalent for the Povarov Reaction

In addition to olefins, aliphatic aldehyde have been found to be vinyl enol ether equivalents and used successfully in the three-component Povarov reaction in the presence of Yb(OTf)₃, giving 2-carboethoxyquinoline derivatives **105**.[47] From the mechanism consideration, intermediate **103** was subjected to elimination, generating **104**, which is converted to **105** by oxidation.

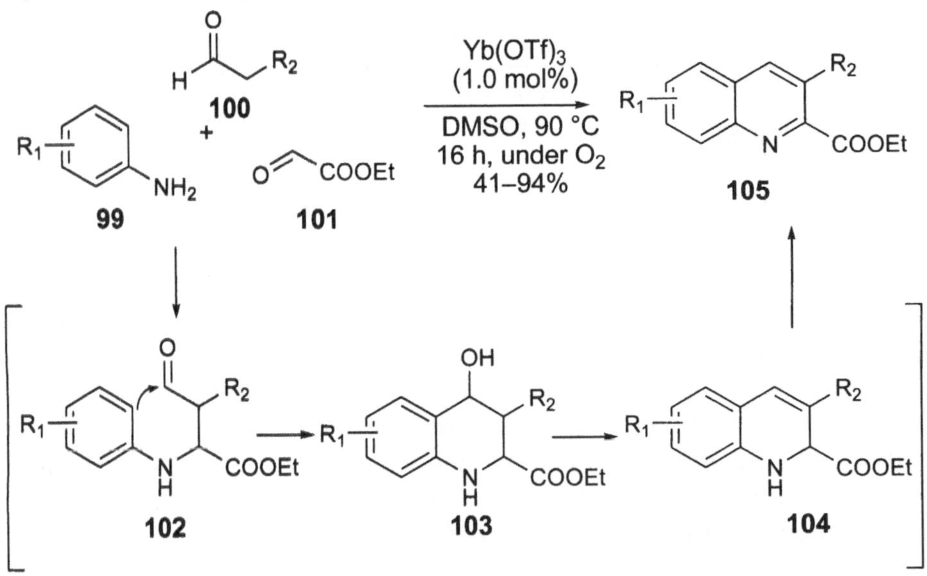

In DMSO, a catalytic amount of HCl activates aldehyde effectively under aerobic condition to give quinolines **107**, in which water is the only by-product.[48]

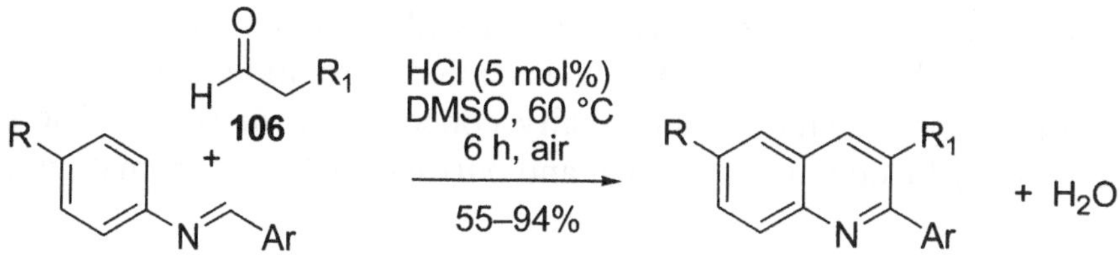

105 107

7.5.6 *Experimental*

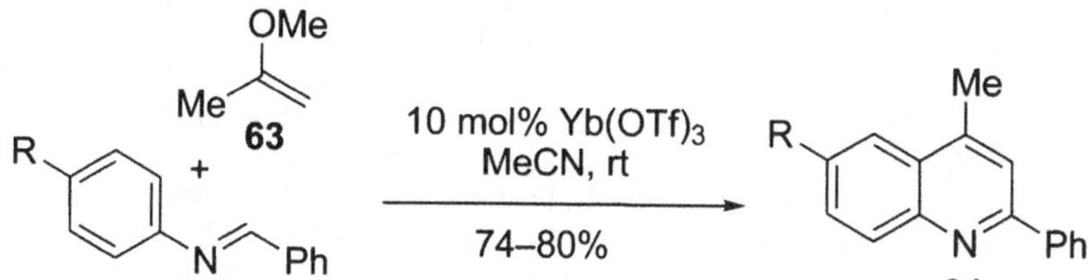

62 64

R = H, OMe or Cl

General Procedure[35]

A solution of *N*-arylaldimine 62 (0.4 mmol) in MeCN (1 mL) was added to a solution of $Yb(OTf)_3$ (25 mg, 0.04 mmol) in MeCN (1 mL) and the mixture was stirred at room temperature for 10 min. Then 2-methoxypropene (1.0 mmol) was added to the mixture and stirring was continued at room temperature for 0.5 to 4 h. The mixture was quenched with 2 M HCl (1 mL) and extracted with $CHCl_3$ (3×5 mL). The combined organic layer were washed with brine, dried ($MgSO_4$), and concentrated *in vacuo*. The residue was chromatographed on silica gel using hexane/EtOAc as eluent to give pure quinoline product.

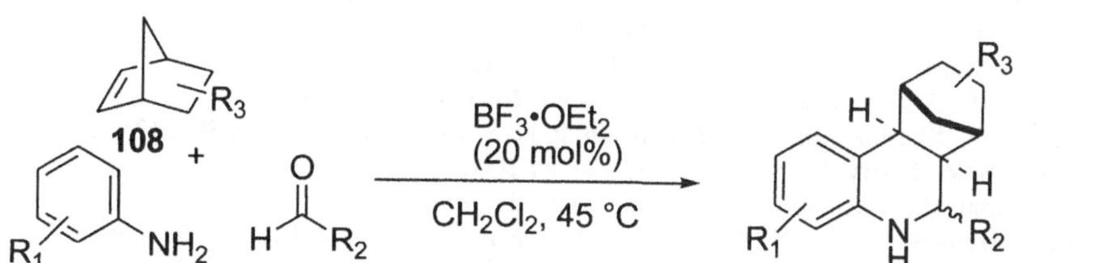

109 110 111

35–91% yield
up to 98:2 *dr*

General Procedure[49]

To a solution of aniline (0.26 mmol, 1 equiv), aldehyde (0.29 mmol, 1.1 equiv), and norbornene-based dienophil (0.39 or 0.52 mmol, 1.5 or 2.0 equiv), and freshly distilled CH_2Cl_2 (1.3 mL) was added $BF_3 \cdot OEt_2$ (0.056 mmol, 0.2 equiv). The mixture was purged with nitrogen and was heated to 45 °C for 22 h. The reaction mixture was cooled to room temperature, treated with a few drops of MeOH, and finally concentrated *in vacuo*. The crude residue was dry loaded onto silica gel and purified by flash chromatography.

7.5.7 *References*

1. (a) [R] Povarov, L. S. *Russ. Chem. Rev.* **1967,** *36*, 656–670. (b) [R] Kouznetsov, V. V. *Tetrahedron* **2009,** *65*, 2721–2750. (c) [R] Glushkov, V. A.; Tolstikov, A. G. *Russ. Chem. Rev.* **2008,** *77*, 137–159.
2. [R] Buonora, P.; Olsen, J.-C.; Oh, T. *Tetrahedron* **2001,** *57*, 6099–6138.
3. Narasaka, K.; Shibata, T. *Heterocycles* **1993,** *35*, 1039–1053.
4. Shindoh, N.; Tokuyama, H.; Takemoto, Y.; Takasu, K. *J. Org. Chem.* **2008,** *73*, 7451–7456.
5. Povarov, L. S.; Mikhailov, B. M. *Izv. Akad. Nauk SSR, Ser. Khim.* **1963,** 953–956.
6. Inada, T.; Nakayuki, T.; Shimizu, I. *Heterocycles* **2005,** *66*, 611.
7. Akiyama, T.; Nakashima, S.; Yokota, K.; Fuchibe, K. *Chem. Lett.* **2004,** *33*, 922–923.
8. Sridharan, V.; Avendańo, C.; Meněndez, J. C. *Tetrahedron* **2007,** *63*, 673–681.
9. Baudelle, R.; Melnyk, P.; Dėprez, B.; Tartar, A. *Tetrahedron* **1998,** *54*, 4125–4140.
10. Povarov, L. S.; Grigos, V. I.; Mikhailov, B. M. *Izv. Akad. Nauk SSR, Ser. Khim.* **1963,** 2039–2041.
11. (a) Povarov, L. S.; Mikhailov, B. M. *Izv. Akad. Nauk SSR, Ser. Khim.* **1964,** 2221–2222. (b) Grigos, V. I.; Povarov, L. S.; Mikhailov, B. M. *Izv. Akad. Nauk SSR, Ser. Khim.* **1965,** 2163–2171.
12. Joh, T.; Hagihara, H. *Tetrahedron Lett.* **1967,** *8*, 4199–4200.
13. Kobayashi, S.; Ishitani, H.; Nagayama, S. *Synthesis* **1995,** 1195–1202.
14. Akiyama, T.; Suzuki, M.; Kagoshima, H. *Heterocycles* **2000,** *52*, 529–532.
15. Babu, G.; Perumal, P. T. *Tetrahedron Lett.* **1998,** *39*, 3225–3228.
16. Laschat, S.; Lauterweine, J. *J. Org. Chem.* **1993,** *58*, 2856–2861.
17. Mahesh, M.; Venkateswar Reddy, C.; Srinivasa Reddy, K.; Raju, P. V. K.; Narayana, Reddy, V. V. *Synth. Commun.* **2004,** *34*, 4089–4104.
18. Yadav, J. S.; Subba, Reddy, B. V.; Srinivas, R.; Madhuri, C. H.; Ramalingam, T. *Synlett* **2001,** *2*, 240–242.
19. Toyata, M.; Komori, C.; Ihara, M. *Heterocycles* **2002,** *56*, 101–103.
20. Kouznetsov, V. V.; Cruz, U. M.; Zubkov, F. I.; Nikitina, E. V. *Synthesis* **2007,** *3*, 375–384.
21. Grieco, P. A.; Bahsas, A. *Tetrahedron Lett.* **1988,** *29*, 5855–5858.
22. Zhao, Y.-L.; Zhang, W.; Wang, S.; Liu, Q. *J. Org. Chem.* **2007,** *72*, 4985–4988.
23. Zhang, D.; Kiselyov, A. S. *Synlett* **2001,** 1173–1175.
24. Kobayashi, S.; Nagayama, S. *J. Am. Chem. Soc.* **1996,** *118*, 8977–8978.
25. Powell, D. A.; Batey, R. A. *Tetrahedron Lett.* **2003,** *44*, 7569–7573.
26. Rai, N. P.; Shashikanth, S.; Arunachalam, P. N. *Synth. Commun.* **2009,** *39*, 2125–2136.
27. More, S. V.; Sastry, M. N. V.; Yao, C. *Synlett* **2006,** 1399–1403.
28. Sridharan, V.; Avendaño, C.; Menēndez, J. C. *Synlett* **2007,** 1079–1082.
29. (a) Hermitage, S.; Jay, D. A.; Whiting, A. *Tetrahedron Lett.* **2002,** *43*, 9633–9636. (b) Alves, M. J.; Azoia, N. G.; Fortes, A. G. *Tetrahedron* **2007,** *63*, 727–734.
30. [R] Katrizky, A. R.; Belyakov, S. A. *Aldrichimica Acta* **1998,** *31*, 35–45.
31. Katrizky, A. R.; Rachwal, B.; Rachwal, S. *J. Org. Chem.* **1993,** *58*, 812–813.
32. Katrizky, A. R.; Rachwal, B.; Rachwal, S. *J. Org. Chem.* **1995,** *60*, 2588–2596.
33. Damon, D. B.; Dugger, R. W.; Magnus-Aryitey, G.; Ruggeri, R. B.; Wester, R. T.; Tu, M.; Abramov, Y. *Org. Process Res. Dev.* **2006,** *10*, 464–471.
34. [R] Katritzky, A. R.; Rachwal, S.; Rachwal, B. *Tetrahedron* **1996,** *48*, 15031–15070.

35. Makioka, Y.; Shindo, T.; Taniguchi, Y.; Takaki, K.; Fujiwara, Y. *Synthesis* **1995,** *7*, 801–804.
36. Leeson, P. D.; Carling, R. W.; Moore, K. W.; Moseley, A. M.; Smith, J. D.; Stevenson, G.; Chan, T.; Baker, R.; Foster, A. C.; Crimwood, S.; Kemp, J. A.; Marshall, G. R.; Hoogsteen, K. *J. Med. Chem.* **1992,** *35*, 1954–1968.
37. Powell, D. A.; Batey, R. A. *Tetrahedron Lett.* **2003,** *44*, 7569–7573.
38. Sridharan, V.; Avendaño, C.; Menēndez, J. C. *Synthesis* **2008,** *7*, 1039–1044.
39. Hadden, M.; Nieuwenhuyzen, M.; Osborne, D.; Stevenson, P. J.; Thompson, N.; Walker, A. D. *Tetrahedron* **2006,** *62*, 3977–3984.
40. Sridharan, V.; Perumal, P. T.; Avendaño, C.; Menēndez, J. C. *Org. Biomol. Chem.* **2007,** *5*, 1351–1353.
41. Zhao, Y.-L.; Zhang, W.; Wang, S.; Liu, Q. *J. Org. Chem.* **2007,** *72*, 4985–4988.
42. Twin, H.; Batey, R. A. *Org. Lett.* **2004,** *6*, 4913–4916.
43. Gaddam, V.; Nagarajan, R. *Tetrahedron Lett.* **2007,** *48*, 7335–7338.
44. Akiyama, T.; Morita, H.; Fuchibe, K. *J. Am. Chem. Soc.* **2006,** *128*, 13070–13071.
45. Liu, H.; Dagousset, G.; Masson, G.; Petailleau, P.; Zhu, J. *J. Am. Chem. Soc.* **2009,** *131*, 4598–4599.
46. Xu, H.; Zuend, S. J.; Woll, M. G.; Tao, Y.; Jacobsen, E. N. *Science* **2010,** *327*, 986–990.
47. Inada, T.; Nakayuki, T.; Shimizu, I. *Heterocycles* **2005,** *66*, 611–619.
48. Tanaka, S.; Yasuda, M.; Baba, A. *J. Org. Chem.* **2006,** *71*, 800–803.
49. Smith, C. D.; Gavrilyuk, J. I.; Lough, A. J.; Batey, R. A. *J. Org. Chem.* **2010,** *75*, 702–715.

Chapter 8 Six-Membered Heterocycles **401**

8.1 Balaban–Nenitzescu–Praill Reaction

Alexandru T. Balaban

8.1.1 Description

The diacylation of various alkenes that have three or more carbon atoms (**1**) in the presence of strong acids HY affording pyrylium salts (**2**) with identical substituents R in positions 2 and 6 (α-positions) was discovered independently in the late 1950s by Balaban and Nenitzescu in Bucharest and by Praill in London.[1–8] In both cases, it was a serendipitous finding.

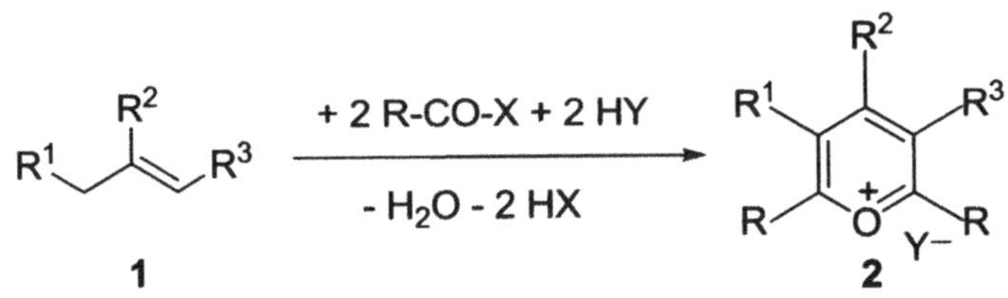

The acylation agent RCOY can be an acid chloride (Y = Cl) in the presence of a Lewis acid such as $FeCl_3$, $AlCl_3$, $SnCl_4$, $ZnCl_2$, and $SbCl_5$ or an anhydride (Y = OCOR) in the presence of a Brønsted acid HY such as $HClO_4$, HBF_4, HPF_6, and F_3C-CO-SO_3H. The alkene may be used as such; however, for compounds such as tertiary alcohols, esters or halides, which readily undergo elimination reactions under acid conditions, these substances are often more convenient starting materials than alkenes. The resulting pyrylium salt has identical substituents R in 2- and 6-positions (α-positions).

Although the index of named reactions in Eicher–Hauptmann's excellent book includes this reaction as "Balaban reaction,"[9] the more appropriate name should be that adopted in the present review and included also in the Hassner–Stumer book.[10]

8.1.2 Historical Perspective

Pyrones, benzopyrylium and pyrylium salts

Toward the end of the 19th century and in the early years of the 20th century, Collie and Tickle[11] had made the surprising observation that 2,6-dimethyl-4-pyrone afforded crystalline salts with acids or alkyl iodides; the discovery of a nitrogen-free base was unprecedented. It took several years till it was recognized by Baeyer and his co-workers that the exocyclic and not the endocyclic oxygen atom was involved,[12] and this fact contributed to the consolidation of the theory of the aromatic sextet, thanks to F. Arndt,[13] B. Eistert,[14] R. Robinson, Erich Hückel, and other chemists.[15]

The first pyrylium salt without hydroxy or alkoxy substituents made by the protonation or alkylation of 4-pyrones was obtained in 1911 by Baeyer and Piccard from 4-pyrones and Grignard reagents.[16] Isolation of these compounds was often difficult, and their characterization was confused by a lack of a clear understanding of their electronic structure. However, there was much interest in benzopyrones (coumarins, flavones, and chromones) and benzopyrylium (flavylium, chromylium) salts isolated from plants (anthocyanins, anthocyanidines) to which they conferred a large diversity of colors.

By 1916 Dilthey[17] had synthesized 2,4,6-triphenylpyrylium ferrichloride (tetrachloroferrate) from acetophenone and benzaldehyde in acetic anhydride in the presence of ferric chloride. He discovered other syntheses of pyrylium salts having aryl and alkyl substituents, as aryl groups enhanced the stability of pyrylium salts. Most often, the anions were perchlorate and tetrachloroferrate, but tetrafluoroborate began to be appreciated. A simple synthesis of alkyl-substituted pyrylium salts had to wait till the treatment of mesityl oxide with acetic anhydride was described by Schneider and Sack using sulfoacetic acid,[18] and by Diels and Alder using perchloric acid.[19] Then a lack of interest followed till the late 1950s, when Karl Dimroth[20,21] and Klaus Hafner[22] in Germany described new reactions of pyrylium salts, and new synthetic approaches were developed in Bucharest and in London.

After the 1930s there was a relative lull in work on pyrylium salts until the mid-1950s. Two independent research groups almost simultaneously observed the formation of simple pyrylium compounds under strong acid conditions.

Bucharest, Romania

In the 1930s, C. D. Nenitzescu was organic chemistry chairman at the largest Romanian university, the Bucharest Polytechnic. He had graduated in Zurich, Switzerland, had obtained his Ph.D. degree with Hans Fischer in Munich, Germany, and had returned to Romania where he had discovered two new indole syntheses. Moreover, by investigating $AlCl_3$-catalyzed reactions of aliphatics and advocating hydride transfer reactions of alkanes, he had achieved international recognition.

The Friedel–Crafts acylation of aromatics had been extended to alkenes by several chemists, starting with Kondakov (1892), Blanc (1898), Zelinsky (1898), Krapivin (1908), Harries (1904), and Darzens (1910), followed by Wieland (1922), Meerwein (1927), Colonge (1939), and Dilthey (1938). During the usual workup involving quenching the reaction mixture by adding ice water, any diacylation products that are water-soluble salts remained undetected and were thrown away, although in some cases they were probably the major reaction products.

In 1931, H. Hopff working for I. G. Farbenindustrie in Germany had published results concerning the reaction of carbon monoxide at high pressure (around 100 atm) with alkanes and cycloalkanes in the presence of anhydrous aluminum chloride.[23] By analogy with the Gattermann–Koch formation of benzaldehyde from CO and benzene, Hopff initially reported that cyclohexane afforded cyclohexanecarboxylic acid and cyclohexane-carboxaldehyde. On the basis of data provided by Nenitzescu, Hopff revised the structure of the latter product, which was 2-methylcyclohexanone (it was known that in the presence of $AlCl_3$, an equilibrium between cyclohexane and methylcyclopentane occurs, and the product results by insertion of CO into the 5-membered ring). The analogous reaction of isobutane with CO afforded pivalic acid, isopropyl methyl ketone, and *tert*-butyl isobutyl ketone. In 1936, in a joint paper, Hopff, Nenitzescu and co-workers, again by analogy with the Gattermann–Koch reaction, proposed a mechanism involving the reaction of a formyl cation with an alkene formed from the cycloalkane.[24] The presumed aldehyde formed initially was the supposed to become oxidized to the observed acid, or to rearrange to the observed ketone. Indeed, Nenitzescu *et al.* showed that in the presence of $AlCl_3$, pivalaldehyde rearranged to isopropyl methyl ketone, and α-methylcyclopentane-carboxaldehyde rearranged to 2-methylcyclohexanone, apparently confirming the formyl cation hypothesis.

However, in the years after World War II, theoretical organic chemistry had progressed using electronic theories, so that it was clear that the regioselectivity postulated for the formylation was wrong: with R = Alk, the correct regioselectivity would lead to an aldehyde that did not rearrange to the observed ketone, whereas the aldehyde that did rearrange could not have been an intermediate. A paradoxical situation had emerged.

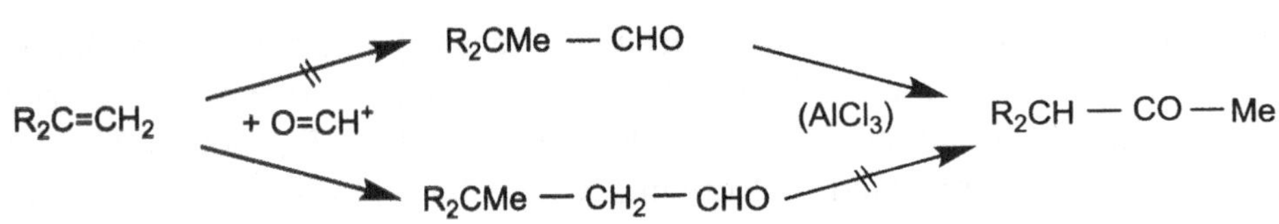

The solution to this puzzle was to assume a different mechanism: an electrophilic attack of an alkyl cation on carbon monoxide instead of an electrophilic attack of the formyl cation on an alkene. For proving this idea, Balaban's Ph.D. thesis with Nenitzescu as supervisor provided experimental evidence (around the same time that Koch and Haaf[25,26] were also investigating the reaction of CO with alkenes or alcohols in the presence of sulfuric acid) by using a different method for preparing carbocations, stating with *tert*-butyl chloride, $AlCl_3$, and CO under pressure.[27]

As expected, the reaction products after decompression and quenching with hydrochloric acid and ice water were pivalic acid, *tert*-butyl isobutenyl ketone (in the absence of a hydride donor, this unsaturated ketone replaced the saturated congener, and no isopropyl methyl ketone by rearrangement of the pivaloyl cation could result). An unexpected crystalline material also appeared in small yield when the aqueous solution was saturated with sodium chloride for facilitating the extraction of organic reaction products. This proved to be a pyrylium salt, as will be shown below.

As pointed out later by Balaban,[28] protonation of an alkene is energetically more favorable than protonation of carbon monoxide, according to the conjugated acid/base (A/B) character, as indicated in Scheme 1 (always only the weaker acid–base pairs are involved in reactions). Balaban's Ph.D. thesis also advanced new mechanisms for the Scholl reactions (dehydrogenating condensations of aromatics) and other reactions.[29–31]

Scheme 1

Basicity order ($H_2C{=}CHR$)	Aromatics (ArH) <	CO <	Alkenes
	Ar-B	**CO-B**	**Alk-B**
Acidity order	ArH_2^+ >	HCO^+ >	$MeCHR^+$
	Ar-A	**CO-A**	**Alk-A**

Scholl reaction	**Ar-A** + **Ar-B**
Koch–Haaf reaction	**Alk-A** + **CO-B** (not **CO-A** + **Alk-B**)
Gattermann–Koch reaction	**CO-A** + **Ar-B** (not **Ar-A** + **CO-B**)
Cationic dimerization	**Alk-A** + **Alk-B**
Friedel–Crafts alkylation	**Alk-A** + **Ar-B** (not **Ar-A** + **Alk-B**).

Although at that time, infrared and nuclear magnetic resonance spectra were not available in Bucharest, classical chemical methods pointed to the structure of a 2,6-di-*tert*-butyl-4-methylpyrylium salt (mixture of tetrachloro-aluminate and tetrachloroferrate), **3**. Anion exchange allowed the easy replacement of $AlCl_4^-$ by perchlorate or other anions. It became evident that five synthons (two CO and three $Me_2C{=}CH_2$) were involved in the formation of the cation. The result (**3**) of this one-pot stitching together of five synthons originating in Me_3CCl, $AlCl_3$, and CO deserves to be among "beautiful molecules."[31]

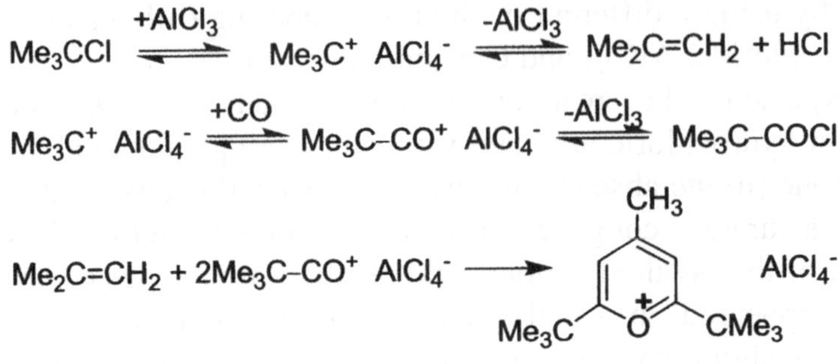

3

In 1970, Nenitzescu passed away, and Balaban continued working on pyrylium salts both at the Bucharest Polytechnic where he had become a professor, and at the Institute of Atomic Physics Bucharest, where he was head of the Laboratory of Isotopically Labeled Organic Compounds. By having a large collection of pyrylium salts, he reported on new reactions and physical properties of pyrylium salts. More recently, he prepared from pyrylium salts new pyridinium salts that were designed to be ionic liquids or nonviral gene transfer agents.

London, England

During early studies[35,36] on the acetylation properties of acetic anhydride-perchloric acid mixtures, Burton and Praill observed that if such mixtures were left for some time, solid material was deposited. Eventually, this solid was shown to contain mainly perchlorates of pyrano-pyrones **4** and **5**.[34,37]

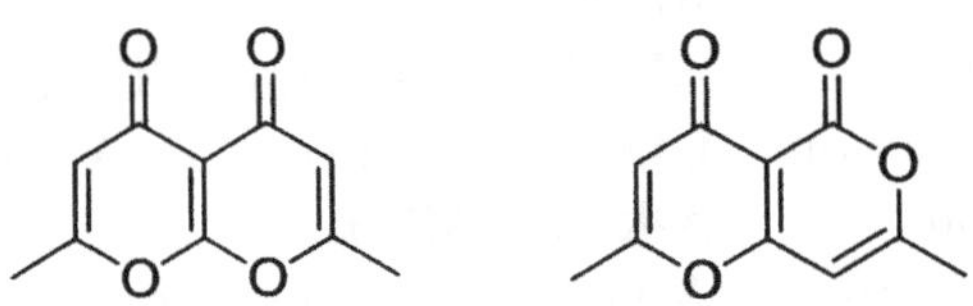

4 **5**

Praill, having discovered the efficiency of acylium perchlorates as acylating agents, decided to examine the esterification of *tert*-butanol as opposed to its dehydration to isobutene.[38] Using acetic anhydride and perchloric acid mixtures, both *tert*-butyl acetate and isobutene were rapidly produced and in accordance with the known alkyl oxygen fission of tertiary esters, the proportion of isobutene increased with time. In these reactions where acetic anhydride was in excess, crystalline material was deposited in the mixture. Later this was shown to be 2,4,6-trimethylpyrylium perchlorate, identified by its conversion to 2,4,6-trimethyl pyridine and its picrate.[39] When isobutene itself was acylated using acetic anhydride and perchloric acid, yields of 78% of the pyrylium salt were obtained. A number of other alkenes treated by the same method gave a range of pyrylium salts.[40] In those cases where the same alkenes were subjected to the Balaban and Nenitzescu procedure the products were in complete accord.

Reaction Mechanism

Both groups of workers found[33,41] that β,γ-unsaturated ketones played an important role on being attacked by an acyl cation and then underging the final cyclization process to form a pyrylium salt. By avoiding any contact with bases on starting from 4-methyl-4-penten-1-one (isomesityl oxide) instead of mesityl oxide **10** (the equilibrium mixture contains only 9% of isomesityl oxide), in the reaction with Ac_2O and $HClO_4$, Praill and Whitear raised the yield of 2,4,6-trimethyl-pyrylium perchlorate from 40% to 87%.[41] They argued that the reason diacylation of an alkene leads to higher yields and higher purity than the monoacylation of unsaturated ketones is that the first step in the alkene acylation leads to an enol (**8**) that affords the β,γ-enone (**9**), which equilibrates more slowly with the α,β-enone. The formation of the enol proceeds via a cyclic transition state (**7**), as postulated by Praill and Saville.[41]

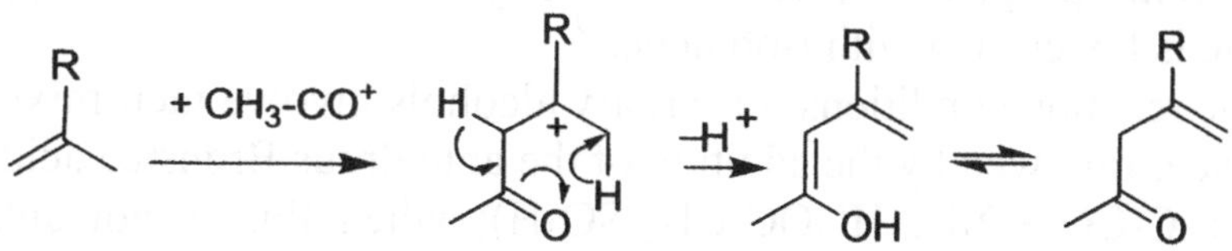

6 7 8 9

Comments on Serendipity

Pyrylium chloroaluminates are more soluble in water than chloferrates, but $AlCl_3$ contains sometimes small amounts of $FeCl_3$ and this unsuspected impurity was helpful in producing crystals of **3** in Bucharest. Around the same time (1958–1959) in London, when the mixture of excess Ac_2O and Me_3COH was treated with 50% $HClO_4$ and the exothermal reaction subsided, white crystals of 2,4,6-trimethylpyrylium perchlorate appeared.

By mutual agreement between the Bucharest and London research groups, it was decided to publish in 1961 back-to-back the results of their latest investigations in the *Journal of the Chemical Society*.

8.1.3 *Diacylation of Various Alkenes Affording Pyrylium Salts*

The diacetylation of isobutene (**6**, R = Me) introduced either as gas, or as *tert*-butanol with Brønsted acids such as perchloric, tetrafluoroboric, trifluoromethanesulfonic or sulfoacetic acids) or as *tert*-butyl chloride with Lewis acids such as $AlCl_3$ or $FeCl_3$ proceeds via the monoacetylation products, mesityl and isomesityl oxides (**10** and **9**, respectively), and yields the corresponding 2,4,6-trimethylpyrylium salts (**11**, R = Me).

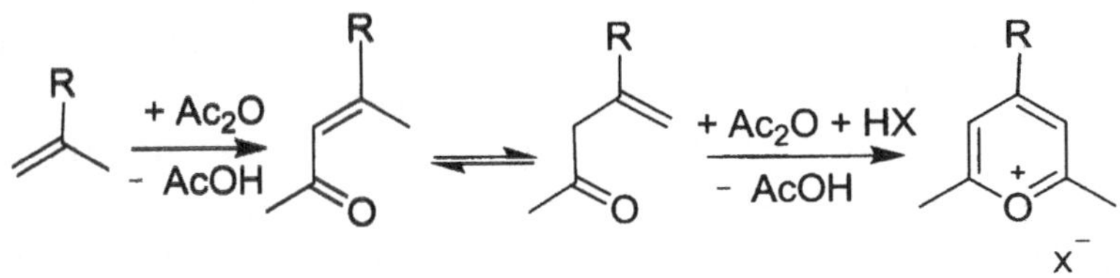

6 **10** **9** **11**

Detailed procedures were published for synthesizing 2,4,6-trimethyl-pyrylium perchlorate,[42] tetrafluoroborate,[43] triflate,[44] and sulfoacetate.[45] Being salts, these compounds are insoluble in ether and are therefore easily purified from side products such as mesityl oxide, so that yields are at least 50–60%. With propene (**6**, R = H) the yield in 2,6-dimethylpyrylium perchlorate is lower than with isobutene.[49]

The reaction conditions for tertiary alcohols involve their mixing with an anhydride, followed by the addition of the anhydrous Brønsted acid (HBF_4 in Et_2O, $F_3C\text{-}CO\text{-}SO_3H$, $HOOC\text{-}CH_2\text{-}SO_3H$); when this is not anhydrous (HPF_6, $HClO_4$), one has to take into account the amount of water that will react exothermally with the anhydride.

With acid chlorides and Lewis acids, one prepares first the acylation mixture of these two reagents according to the Perrier method, followed by gradual addition of the alkene or tertiary alkyl chloride, in the presence of a solvent (carbon disulfide, nitromethane, or dichloromethane) or without a solvent.

The pivaloyl cation that can be formed from Me_3CCl, $AlCl_3$, and CO at high pressure undergoes decarbonylation at normal pressure when it is prepared from pivaloyl chloride and most Lewis acids. Only in the presence of $SnCl_4$ can one diacylate isobutene (from Me_3CCl) with pivaloyl chloride to form **3** as chlorostannate, which can then be converted into other salts on treatment with $HClO_4$, HBF_4, etc.[46] It is interesting that it is also possible to use a Brønsted anhydride—namely, trifluoromethanesulfonic anhydride—for obtaining **3** as triflate.[47,48]

On adding Me_3CCl to a mixture of butyryl chloride and aluminum chloride, followed by hydrolyzing the mixture with ice and hydrochloric acid and adding perchloric acid, the result was a liquid dihydrate of the pyrylium perchlorate (mp 11 °C) which could be converted into the anhydrous salt with mp 85 °C.[50] Similar ionic liquids with mp < 100 °C can be prepared from higher alkanoic acids.[51] As will be described in the last section, pyrylium salts are easly converted by primary amines into pyridinium salts that are less reactive so that one can thus prepare conveniently pyridinium ionic liquids.[51]

Diisobutene affords on diacetylation with Ac_2O and $ZnCl_2$ a chlorozincate of **3** with R = CH_2CMe_3 (which can be converted into the perchlorate,[52] with a structure proved by ^{1}H-NMR,[53] and not the isomeric 3-*tert*-butyl-2,4,6-trimethylpyrylium) due to steric reasons. This reaction had been reported earlier as yielding a zinc complex of a unsaturated ketone (the monoacetylation product).[54]

Diacylations using aromatic acid derivatives (acid chlorides or anhydrides) usually result in lower yields, but for obtaining pyrylium salts with aromatic substituents in 2, 4, and/or 6 positions the Dilthey syntheses afford better results.

With an *N*-protected aminomethyl group, **12** affords different yields of pyrylium perchlorates **13**: 64% for X = CH_2, and 36% for X = S.[55]

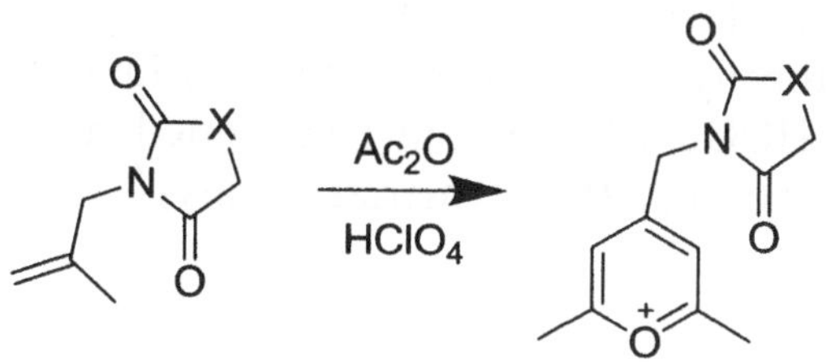

12 **13**

Diisobutyrylation of triethylcarbinol (**14**) in the presence of 60% hexafluorophosphoric acid affords 4-ethyl-2,6-diisopropyl-3,5-dimethyl-pyrylium hexafluorophosphate (**16**).[56] Owing to a double Janus effect due to buttressing by the β-methyl groups, the α-isopropyl groups adopt a conformation that mimics the effect of *tert*-butyl groups, so that the corresponding pyridine is nonnucleophilic, like 2,6-di-*tert*-butyl-4-methylpyridine, which can be prepared from **3** with ammonia.

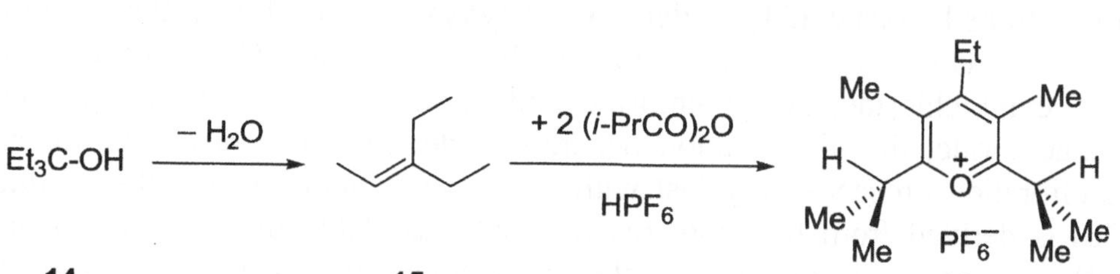

14 **15** **16**

Dulenko *et al.* prepared 4-chloromethyl-pyrylium salts (R = CH_2Cl) and investigated their reactions.[57]

With α-methylstyrene (**6**, R = Ph), the yield of the diacetylation with acetic anhydride and perchloric or sulfoacetic acid affording 2,6-dimethyl-4-phenylpyrylium salts is lowered because of polymerization side products.[58] Whereas the resulting phenyl group in position 4 (γ) is deactivated towards electrophilic acylation,[59] no such effect is exerted on a β-phenyl group, so that allylbenzene with acetic anhydride and perchloric acid affords a mixture of 2,6-dimethyl-3-phenylpyrylium and 2,6-dimethyl-3-(*para*-acetyl-phenyl)-pyrylium perchlorates.[60] However, starting from a tertiary alcohol (**17**), 2,4,6-trimethyl-3-phenylpyrylium perchlorate (**18**) can be obtained without *para*-acetylation.[61]

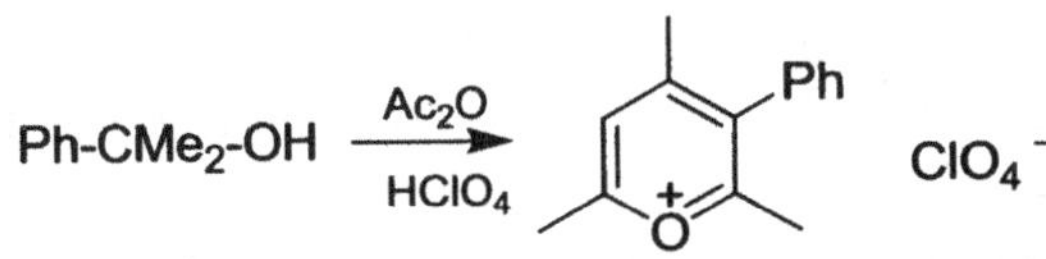

17 **18**

On adding sulfoacetic acid to a mixture of acetic anhydride with the commercial 1,3-diisopropenylbenzene (**19**), a bis-pyrylium sulfoacetate (**20**) was obtained with X = $HOOC\text{-}CH_2\text{-}SO_3$, then converted into the diperchlorate (yield 2% because **19** polymerizes easily).[62]

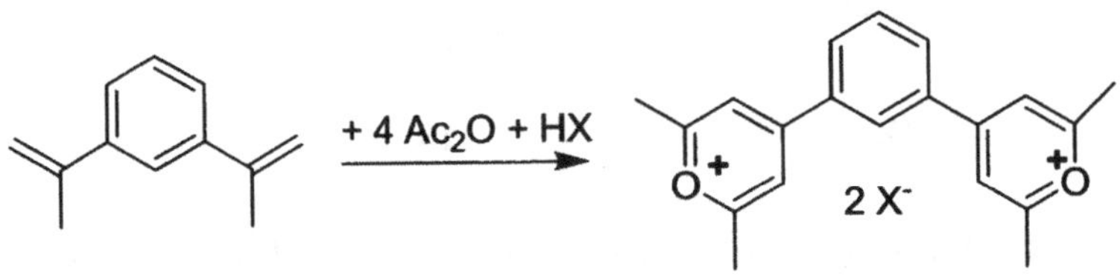

19 **20**

Regioselectivity of the Alkene Diacylation

When *tert*-pentyl alcohol was reacted with acetic anhydride and perchloric or tetrafluoroboric acids, the reaction proceeded via the more stable trisubstituted alkene (**21**) under thermodynamic control and the reaction product was mainly a 2,3,4,6-tetramethylpyrylium salt (**25**). However, with *tert*-pentyl chloride, acetyl chloride and aluminum chloride or antimony pentachloride, the diacetylation occurred under kinetic control: the alkene equilibration proceeds very fast with these Lewis acids so that the product (**26**) is derived from the more reactive but less stable disubstituted alkene (**22**).[63,64] In the formulas below, the intermediate monoacetylation products (**23, 24**) are written in the reactive β,γ-unsaturated isomeric form.

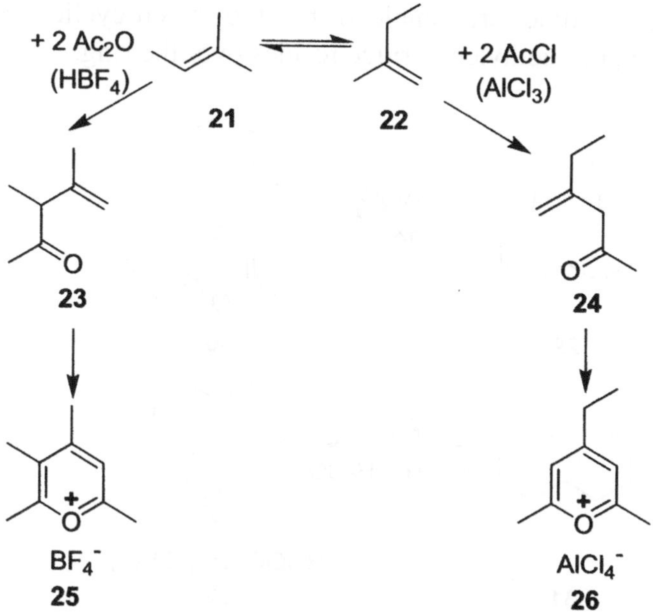

Using appropriate reaction conditions, one may obtain either one or the other of these two pyrylium salts. The highest regioselectivity for obtaining tetramethylpyrylium salts (**26**) among Lewis acids was obtained with acetyl chloride and beryllium chloride.[65,66]

Pyrylium salts with 2- or 4-chloro substituents (**28**) have reactive chlorine atoms, and their synthesis in the presence of antimony hexachloride must be carried out under cooling (below –5 °C).[67]

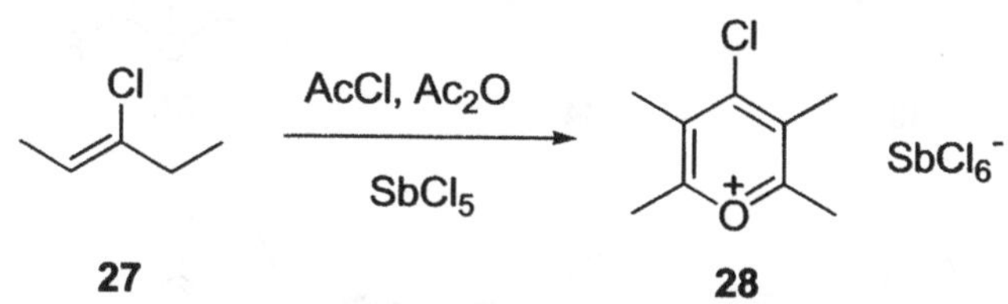

Diacylation of Cycloalkenes

Diacylation of cycloalkenes such as cyclododecene and cyclodecene affords the corresponding pyrylium salts.[68,69] Whereas the diacetylation of cyclododecene (**29**) with acetic anhydride in the presence of 70% perchloric acid proceeds smoothly in satisfactory yield,[69] the analogous reaction with cyclodecene (**31**) succeeds only with sulfoacetic acid (prepared by heating acetic anhydride with sulfuric acid under carefully controlled conditions), and the yield is lower than 10%.[68] The central CH_2 groups of the resulting 9,11-dimethyl-[7]-(2,6)-pyrylophanium perchlorate (**32**) and of the corresponding pyridine are shielded by the heterocyclic aromatic ring proving that the polymethylene bridge resides over the ring.

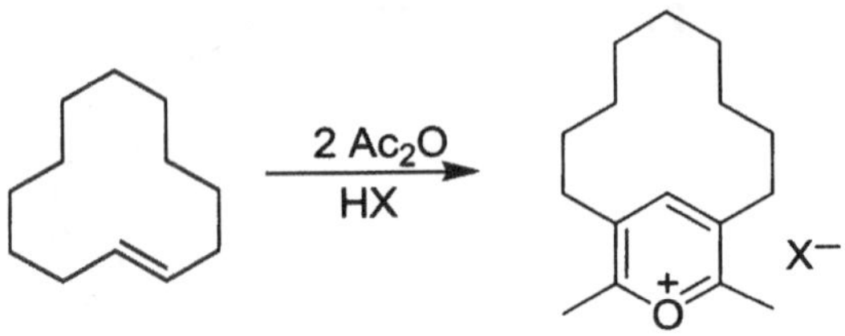

29 **30**

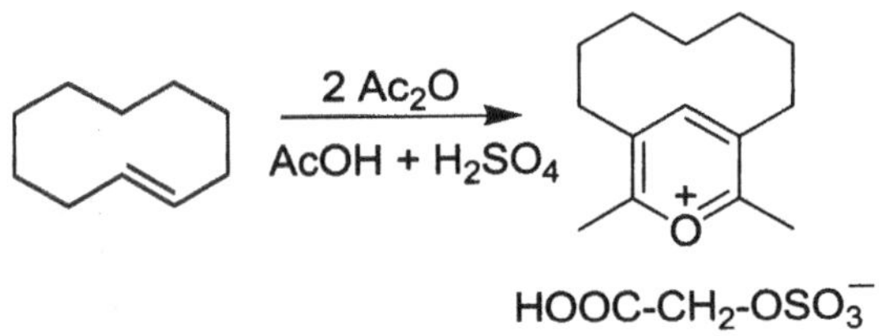

31 **32**

The reaction proceeds normally with 5- or 6-membered 1-methylcycloalkenes **33**, **36** (or for better yields with the isomeric methylenecycloalkanes **34**, **37**).[11,70] In the case of 6-membered rings, the products (**35**) may be converted by dehydrogenation into benzo[*b*]pyrylium salts and then isoquinolines.

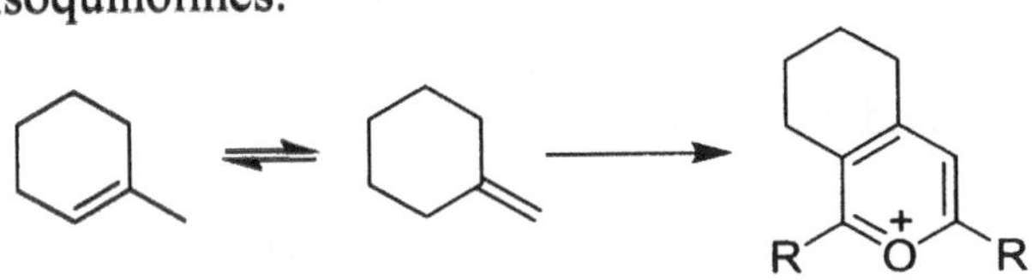

33 **34** **35**

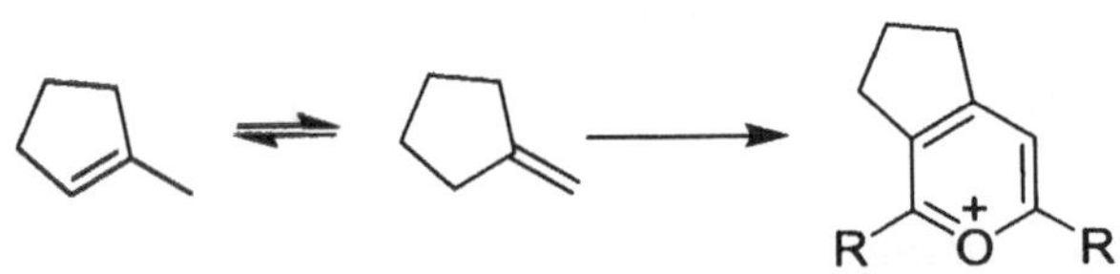

36 **37** **38**

A tricyclic system **40** was obtained analogously.[71]

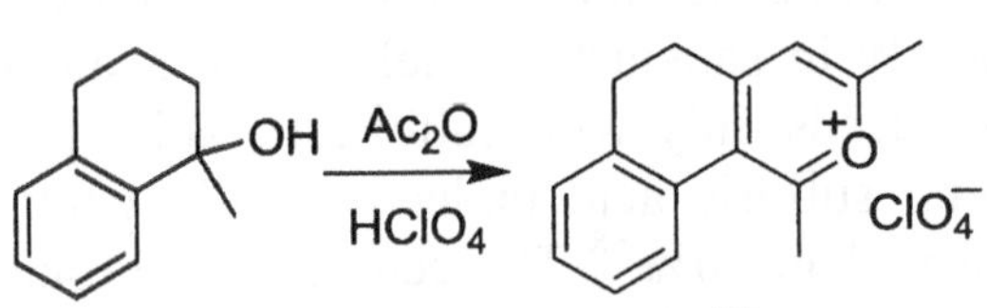

39 **40**

A methyl-tetrahydrophthalic anhydride (**41**) was diacylated to **42** with acetyl, butyryl, and valeryl anhydrides using perchloric acid.[72]

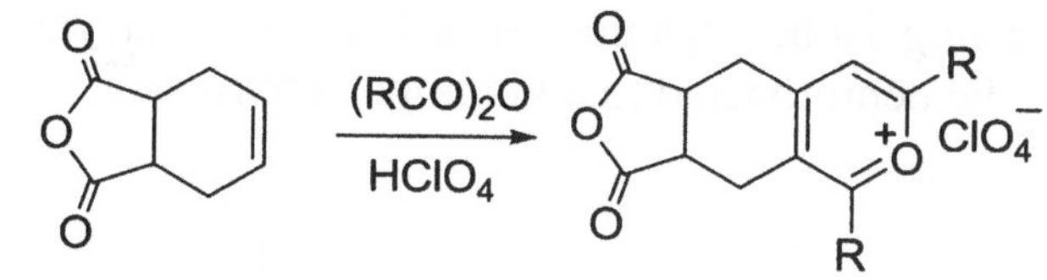

41 **42**

8.1.4 Triacylation of Alkenes

A crystallized side product in the $AlCl_3$-catalyzed diacetylation of isobutene with acetyl chloride proved to be a vinylogous 4-pyrone (**43**) formed by triacylation.[73] The ^{1}H-NMR spectrum reveals a strongly deshielded proton, proving the *s-cis* conformation.[74]

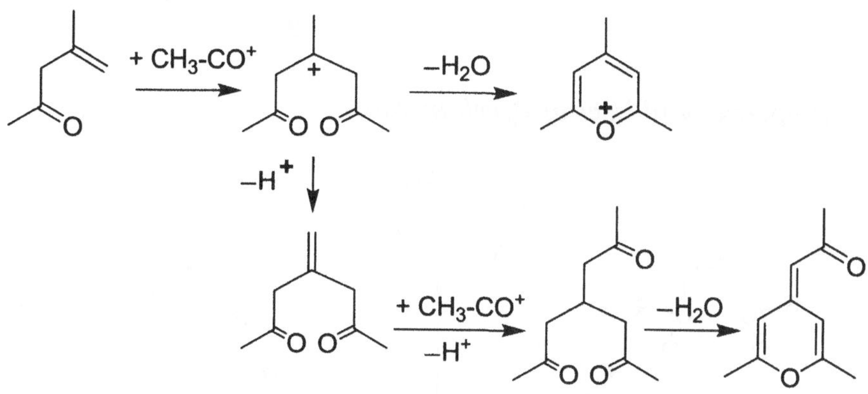

43

Like pyrones, **43** was treacted with acids and in this case a tautomeric mixture of two pyrylium salts (**44, 45**) is obtained, in ratios depending on the solvent.[74]

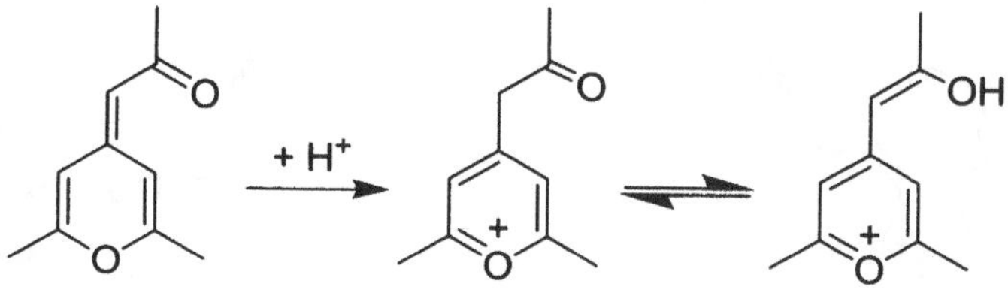

43 **44** **45**

2,3-Dimethyl-2-butene (tetramethylethylene) reacts with acetyl chloride and aluminum chloride in the isomeric form of 2,3-dimethyl-1-butene affording a 4-isopropyl-2,6-dimethylpyrylium salt (**48**).[75] However, the corresponding isobutyrylation affords also a triacylation product (**49**), which cannot be deprotonated to a vinylogous pyrone.[76]

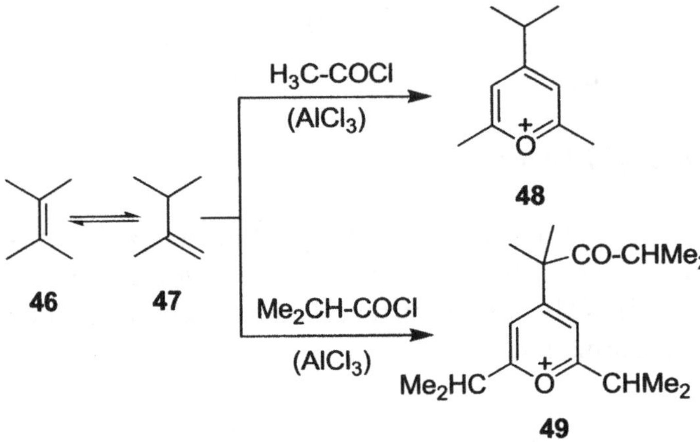

8.1.5 Properties and Uses of Pyrylium Salts

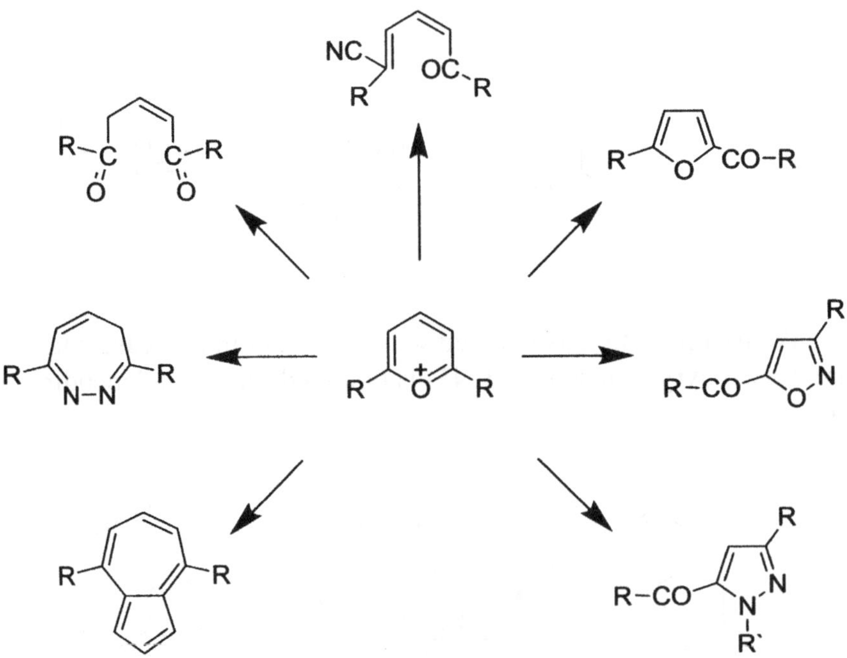

Owing to the high electronegativity of the oxygen heteroatom, pyrylium cations represent the strongest possible single perturbation of a benzene ring. As a consequence, the chemical behavior of such cations differs considerably from that of benzene or pyridine. The resonance energy is appreciably lower than that of benzene, but the aromatic character allows pyrylium salts to be stable in aqueous media at pH ≤ 7, unlike oxonium salts. With nucleophiles a variety of reactions occurs: in most cases, α-addition is followed by ring opening. With cyanide anions, the resulting cyanodienones are stable. 2,4,6-Triarylpyrylium cations afford stable 1,5-pentenediones (pseudobases). The alkyl-substituted pseudobases obtained from α-methyl- or α-ethylpyrylium salts undergo with hot alkali an intamolecular ring closure forming phenols. With cyclopentadiene, azulenes result. With hydrogen proxide, one obtains 2-acylfurans. With hydrazine, pyrylium salts form 4*H*-1,2-diazepines, but with phenylhydrazine or methylhydrazine one obtains pyrazole derivatives. With hydroxylamine under certain conditions, one converts pyrylium salts into isoxazoles.

When the nucleophile has hydrogen atoms, *e.g.*, with ammonia, primary amines, hydroxylamine, hydrazine derivatives, phosphine, hydrogen sulfide, nitromethane (EWG = NO_2), acetonitrile (EWG = CN), an ANRORC (attack by nucleophile, ring opening, ring closure) reaction results in the conversion of substituted pyrylium salts into a large variety of six-membered aromatic carbocyclic or heterocyclic compounds.

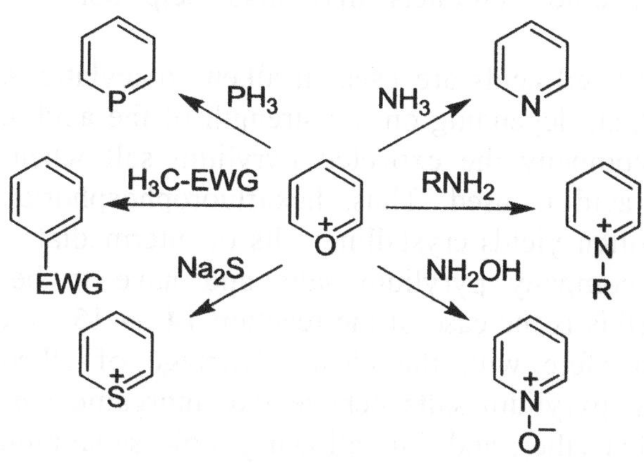

The aromaticity of pyrylium salts (manifested by ^{1}H-NMR chemical shifts) is decreased by the strong electronegativity of the oxygen heteroatom, but in all the above reactions it increases. Other properties of pyrylium cations (bond lengths, IR and UV-Vis spectra, ^{13}C-NMR and mass spectra) also indicate aromatic character. However, the higher energy inherent in pyrylium cations allows them to form easily from acyclic starting materials, and to ring-open readily on reacting with nucleophiles, resulting in the fact that pyrylium salts are excellent synthons for obtaining a large variety of aromatic and nonaromatic conjugated structures, as shown in reviews.[1–8]

8.1.6 Future Directions

Several developments and future research directions may be foreseen. So far, the alkene diacylation leading to the formation of pyrylium salts is the simplest method for obtaining rapidly a large variety of alkyl-substituted pyrylium salts with two identical substituents in α-positions. For aryl substituents, however, other methods afford better yields, mainly because in diacylations of alkenes with aryl groups these aryl substituents are also being acylated during the course of the reaction. It remains to be seen whether alkenes with aryl groups having electron-withdrawing substituents or with hetaryl substituents such as pyridine or azole groups have sufficient electronic density in the alkenic double bond to be diacylated without competing acylation of the aromatic substituents.

It will be interesting to explore the reactions using Lewis acids such as $ZnCl_2$, $SnCl_4$, and $TiCl_4$) by means of physical methods centered on transition metal nuclei of these Lewis acids (NMR or Mössbauer spectrometry) to discover whether true pyrylium salts are obtained directly or whether crystalline or liquid complexes of 1,5-enediones are precursors that become converted into salts only in contact with polar solvents. X-ray diffractometry of crude products may also help solve some of these problems.

When Brønsted acids are used in alkene diacylations, experimental evidence shows that, depending on the strength of the acid, crystalline side products can accompany the expected pyrylium salt when an extremely strong Brønsted acid is used. Thus, hexafluorophosphoric acid (but not perchloric acid) often yields crystalline salts of intermediate enones as side products that accompany pyrylium salts and have to be separated by recrystallization (this is the case of the reaction **14** → **15** → **16**). These side products may interfere with the clear advantage of alkene diacylation, namely that most pyrylium salts behave like inorganic compounds being insoluble in diethyl ether, and thus allowing rapid separation from organic side-products.

When using commercially available Brønsted acids (70% $HClO_4$, 50% HBF_4, 60% HPF_6) most of the anhydride $(RCO)_2O$ is lost by hydrolysis with the water accompanying the Brønsted acid in higher molar concentration than this acid. Therefore the use of commercial anhydrous HBF_4 in diethyl ether should be more extensively explored. With the numerous Brønsted acids used in manufacturing ionic liquids, other such anhydrous acids are at present available.

8.1.7 Aknowledgment

The help of Dr. Martin Šala with the newer version of Chemdraw is gratefully acknowledged.

8.1.8 References

1. [R] Balaban, A. T.; Dinculescu, A.; Dorofeenko, G. N.; Fischer, G. W.; Koblik, A. V.; Mezheritskii, V. V.; Schroth, W. In *Advances in Heterocyclic Chemistry*, suppl. vol. 2 Academic Press, New York, 1982.
2. [R] Balaban, A. T.; Schroth, W.; Fischer, G. In *Advances in Heterocyclic Chemistry*, Academic Press, New York, 1969, vol. 10, pp. 241–326.
3. [R] Balaban, A. T. In *New Trends in Heterocyclic Chemistry*, Elsevier, Amsterdam, 1979, pp. 79–111.
4. [R] Balaban, A. T. In *Organic Synthesis: Modern Trends* Blackwell, Oxford, 1987, pp. 263–274.
5. [R] Kuznetsov, E. V.; Shcherbakova, I. V.; Balaban, A. T. In *Advances in Heterocyclic Chemistry*, Academic Press, New York, 1990, **50**, pp. 157–254.
6. Schroth, W.; Balaban, A. T. In *Methoden der Organischen Chemie (Houben-Weyl), vol. E7b, Hetarene II (Teil 2)*, Thieme, Stuttgart, 1992: Introduction (pp. 755–766); Synthesis (pp. 767–963) (in German).
7. [R] Balaban, A. T. "Trimethylpyrylium" in vol. 7, pp. 5224–5227; "Triphenylpyrylium" in vol. 8, pp. 5407–5411, monographs in *Encyclopedia of Reagents for Organic Synthesis*, (Paquette, L. A. Ed.), Wiley, New York, 1996; electronic version (EROS), including also 2,4,6-tri-(*p*-methoxyphenyl)pyrylium tetrafluoroborate, Wiley, New York, 2004.
8. [R] Balaban, T. S.; Balaban, A. T. In *Science of Synthesis; Houben-Weyl Methods of Molecular Transformations* Thieme, Stuttgart, 2003, pp. 11–200.
9. [R] Eicher, T.; Hauptmann, S. *The Chemistry of Heterocycles*, 2nd ed., Wiley-VCH, Weinheim, 2003, p. 554.
10. [R] Hassner, A.; Stumer, C. *Organic Syntheses Based on Name Reactions*, 2nd ed., Pergamon – Elsevier Science, Amsterdam, 2002, p. 17.
11. Collie, J. N.; Tickle, T. *J. Chem. Soc.* **1899,** *75*, 710–717.
12. Baeyer, A.; Villiger, V. *Ber. dtsch. chem. Ges.* **1901,** *34*, 2679–2698.
13. Arndt, F.; Lorenz, L. *Ber. dtsch. chem. Ges.* **1930,** *63*, 3121–3132.
14. [R] Eistert, B. *Tautomerie und Mesomerie, Gleichgewicht und Resonanz*, Enke, Stuttgart, 1938, p. 64.
15. [R] Balaban, A. T.; Schleyer, P. v. R.; Rzepa, H. S. *Chem. Rev.* **2005,** *105*, 3436–3447.
16. Baeyer, A.; Piccard, J. *Liebigs Ann. Chem.* **1915,** *407*, 332–369.
17. Dilthey, W. *J. prakt. Chem.* **1917,** *95*, 107–120.
18. Schneider, W.; Sack, A. *Ber. Dtsch. Chem. Ges.* **1923,** *56*, 1786–1787.
19. Diels, O.; Alder, K. *Ber. Dtsch. Chem. Ges.* **1927,** *60*, 716–723.
20. Dimroth, K. *Angew. Chem.* **1960,** *72*, 331–342.
21. [R] Dimroth, K.; Wolf, K. H. In *Newer Methods of Preparative Organic Chemistry* Academic Press, New York, 1964, *vol. 3*, p. 357.
22. Hafner, K. *Org. Synth. Coll. Vol. V*, 1973, 1088–1090.
23. Hopff, H. *Ber. Dtsch. Chem. Ges.* **1931,** *64*, 2739–2748.
24. Hopff, H.; Nenitzescu, C. D.; Isacescu, D.; Cantuniari, I. P. *Ber. Dtsch. Chem. Ges.* **1936,** *69*, 2244–2251.
25. Koch, H.; Haaf, W. *Liebigs Ann. Chem.* **1958,** *618*, 251–266.
26. Koch, H.; Haaf, W. *Angew. Chem.* **1958,** *70*, 311.
27. Balaban, A. T.; Nenitzescu, C. D. *Tetrahedron* **1960,** *10*, 55–64.
28. Balaban, A. T. *Rev. Chim. Acad. Romania* **1962,** *7*, 675–682.
29. Nenitzescu, C. D.; Balaban, A. T. *Chem. Ber.* **1958,** *91*, 2109–2116.
30. [R] Balaban, A. T.; Nenitzescu, C. D. In *Friedel–Crafts and Related Reactions*, Wiley-Interscience, New York, 1964, vol. 2. pp. 979–1047.

31. [R] Nenitzescu, C. D.; Balaban, A. T., "Aliphatic Acylation", in *"Friedel-Crafts and Related Reactions"*, Olah, G. A. (ed.), Wiley-Interscience, New-York, 1964, vol. 3, pp.1033–1152.
32. Balaban, A. T.; Nenitzescu, C. D. *Liebigs Ann. Chem.* **1959,** *625*, 66–73.
33. Balaban, A. T.; Nenitzescu, C. D. *Liebigs Ann. Chem.* **1959**, *625*, 74–88.
34. [R] Balaban, A. T. *Chem. Intelligencer* **1998,** 55–56.
35. Burton, H.; Praill, P. F. G. *J. Chem. Soc.* **1950**, 1203–1206.
36. Burton, H.; Praill, P. F. G. *J. Chem. Soc.* **1950,** 2034–2038.
37. Praill, P. F. G.; Whitear, A. L. *Proc. Chem. Soc.* **1961,** 112.
38. Praill, P. F. G. *Chem. Ind.* **1959,** 1123–1124.
39. Praill, P. F. G.; Whitear, A. L. *Proc. Chem. Soc.* **1959,** 312.
40. Praill, P. F. G.; Whitear, A. L. *J. Chem. Soc.* **1961,** 3573–3579.
41. Praill, P. F. G.; Saville, B. *Chem. Ind.* **1960,** 495–496.
42. Balaban, A. T.; Nenitzescu, C. D. *Org. Synth.* **1964,** *44*, 98–103; Reprinted in *Org. Synth. Coll. Vol.* 5, 1973, 1106–1108.
43. Balaban A. T.; Boulton, A. J. *Org. Synth. Coll. Vol. 5*, 1973, 1112–1113.
44. Balaban A. T.; Boulton, A. J. *Org. Synth. Coll. Vol. 5*, 1973, 1114–1116.
45. Dinculescu, A.; Balaban, A. T. *Org. Prep. Proc. Int.* **1982,** *14*, 39–44.
46. Balaban, A. T. *Org. Prep. Proced. Int.* **1977,** *9*, 125–130.
47. Anderson, A. G.; Stang, P. J. *J. Org. Chem.* **1976,** *41*, 3034–3036.
48. Anderson, A. G.; Stang, P. J. *Org. Synth.* **1981,** *60*, 34–40.
49. Balaban, A. T.; Farcasiu, D.; Nenitzescu, C. D. *Tetrahedron* **1962,** *18*, 1075–1078.
50. Balaban, A. T.; Romas, E.; Rentea, C. C. *Tetrahedron* **1966,** *22*, 1–5.
51. Balaban, A. T.; Lesko, M. J.; Ghiviriga, I.; Seitz, W. W. *Rev. Chim. (Bucuresti)* **2006,** *57*, 25–28.
52. Balaban, A. T.; Ghenea, A.; Nenitzescu, C. D. *Bull. Acad. Sci. USSR, Div. Chem. Sci. (Engl. Transl.)*, **1961,** 1018–1023.
53. Balaban, A. T.; Bedford, G. R.; Katritzky, A. R. *J. Chem. Soc.* **1964,** 1646–1649.
54. Byrns, A. C.; Doumani, T. F. *Ind. Eng. Chem.* **1943,** *35*, 349–353.
55. Voshchula, V. N.; Tolkunov, S. V.; Zubritskii, M. Yu.; Dulenko, V. I. *Chem. Heterocycl. Compd. (Engl. Transl.)* **1988,** *24*, 1175–1180.
47. Anderson, A. G.; Stang, P. J. *J. Org. Chem.* **1976,** *41*, 3034–3036.
48. Anderson, A. G.; Stang, P. J. *Org. Synth.* **1981,** *60*, 34–40.
56. Balaban, A. T., Ghiviriga, I.; Czerwinski, E. W.; De, P.; Faust, R. *J. Org. Chem.* **2004,** *69*, 536–542.
57. Dulenko, V. I.; Golyak, V. M.; Alekseev, N. N.; Gubar, V. I. *Chem. Heterocycl. Compd. (Engl. Transl.)* **1983,** *19*, 723–724.
58. Balaban, T. S.; Dinculescu, A.; Balaban, A. T. *Org. Prep. Proc. Internat.* **1988,** *20*, 289–292.
59. Balaban, A. T.; Gavat, M.; Frangopol, P. T.; Mocanu, M.; Nenitzescu, C. D. *Rev. Roum. Chim.* **1964,** *9*, 79–92.
60. Balaban, A. T.; Nenitzescu, C. D.; Gavat, M.; Mateescu, G. D. *J. Chem. Soc.* **1961,** 3564–3566.
61. Balaban, A. T.; Uncuta, C.; Chiraleu, F. *J. Labelled Comp. Radiopharm.* **1982,** *19*, 783–794.
62. Balaban, A. T.; Balaban, T. S. *Rev. Roum. Chim.* **1989,** *34*, 41–50.
63. Balaban, A. T.; Nenitzescu, C. D. *Tetrahedron Lett.* **1960,** *2*, 7–10.
64. Balaban, A. T.; Nenitzescu, C. D. *J. Chem. Soc.* **1961,** 3553–3560.
65. Balaban, A. T.; Barabas, E.; Farcasiu, M. *Chem. Ind.* **1962,** 781–782.
66. Balaban, A. T.; Bota, A. *Org. Prep. Proc. Int.* **1982,** *14*, 31–38.
67. Zubkova, O. V.; Borodaev, S. V.; Lukyanov, S. M. *J. Org. Chem. USSR (Engl. Transl.)* **1991,** *27*, 2109–2112.
68. Balaban, A. T.; Badilescu, I. I. *Rev. Roum. Chim.* **1976,** *21*, 1339–1343.
69. Balaban, A. T. *Tetrahedron Lett.* **1968,** 4643–4644.
70. Balaban, A. T.; Nenitzescu, C. D. *J. Chem. Soc.* **1961,** 3561–3564.
71. Dorofeenko, G. N.; Zhdanov, Yu. A.Yarova, V. N.; Palchkov, V. A. *J. Org. Chem. USSR (Engl. Transl.)* **1967,** *3*, 917.
72. Dulenko, V. I. Golyak, V. M.; Alekseev, N. N.; Gorenko, N. F. *Chem. Heterocycl. Compd. (Engl. Transl.)* **1984,** *20*, 1078–1080.

73. Balaban, A. T.; Frangopol, P. T.; Katritzky, A. R.; Nenitzescu, C. D. *J. Chem. Soc.* **1962,** 3889–3895.
74. Balaban, A. T.; Gheorghiu, M. *Rev. Roum. Chim.* **1978,** *23*, 1065–1077.
75. Balaban, A. T.; Bota A.; Chiraleu, F.; Sliam, E.; Hanes, A.; Draghici, C. *Rev. Roum. Chim.* **1977,** *22*, 1003–1015.
76. Balaban, A. T.; Bota, A.; Stanoiu, I. I. *Rev. Roum. Chim.* **1976,** *21*, 1183–1187.

8.2 Borsche Cinnoline Synthesis

Stephen W. Wright

8.2.1 Description

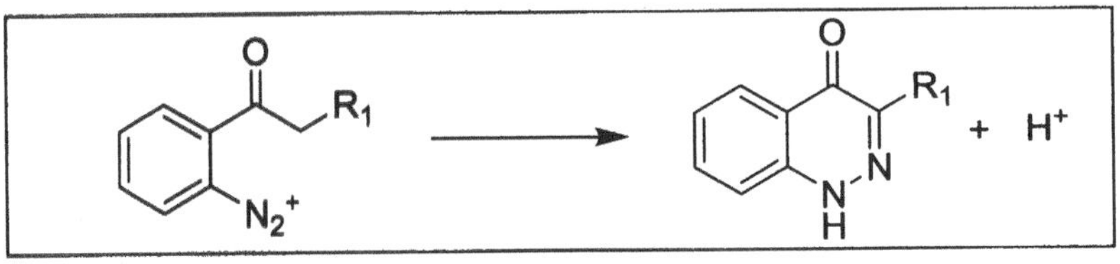

The Borsche (or Borsche–Herbert) cinnoline synthesis constitutes the cyclization of an *ortho*-diazonium aryl ketone to form a 4-cinnolone, or 4-hydroxycinnoline. It is mechanistically related to the Widman–Stoermer and Richter cinnoline syntheses, which involve the cyclization of *ortho*-diazonium aryl olefins and *ortho*-diazonium aryl alkynes, respectively.

8.2.2 Historical Perspective

Walther Borsche and Alfred Herbert first reported the spontaneous cyclization of an *ortho*-diazonium acetophenone in 1941, as part of a larger study on the synthetic transformations of 2-bromo-5-nitroacetophenone carried out at the University of Frankfurt am Main.[1] The authors did not draw especial attention to this observation, and it was not until four years later that Schofield and Simpson suggested that this reaction might offer a general route to 4-cinnolones.[2] After 70 years, this general reaction yielding 4-cinnolones remains one of the most useful methods for the synthesis of cinnolines.

8.2.3 Mechanism

The mechanism of the Borsche cyclization reaction has been studied and is well understood. Cyclization occurs *via* enolization of the ketone, followed by addition of the enol to the diazonium group. Cyclization is facilitated by the conjugation of electron withdrawing substituents to the diazonium group, which render the diazonium group more electrophilic. Electron-releasing substituents in conjugation to the diazonium group facilitate the solvolysis of the diazonium salt to afford the corresponding *ortho*-acyl phenol.

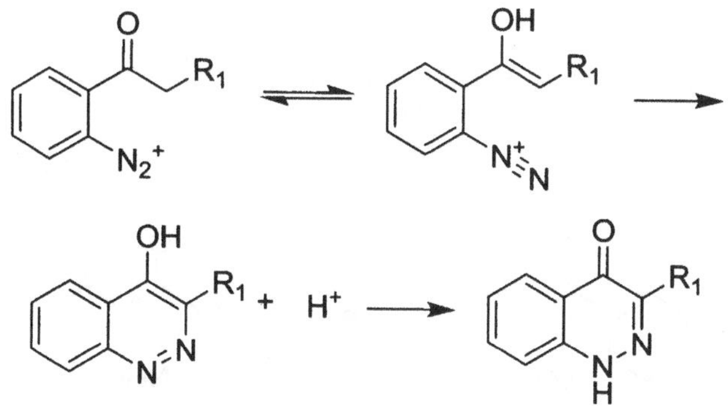

In as much as enolization of the ketone is the rate-limiting step, it is usually advantageous to conduct the reaction in very strongly acidic solution to facilitate the enolization of the ketone.[3] Examples of the use of heteroatom substituents (R_1) on the acyl group are relatively scarce, even though they might be expected to facilitate enolization. For example, halogen substitutents on the acyl group, such as chloroacetyl or bromoacetyl, are only rarely employed.[4] Instead, 3-halo-4-cinnolones are more often prepared by cyclization of the *ortho*-acetyl aniline, followed by halogenation of the 4-cinnolone.[5] Alkyl[6] and aryl substituents on the acyl group are tolerated—for example, (2-phenyl)acetyl anilines give rise to 3-phenyl-4-cinnolones upon diazotization.[7]

The tautomerism of the 4-cinnolone products of the Borsche synthesis has also been studied in detail. Studies have been carried out using infrared spectroscopy,[8] ultraviolet spectroscopy,[9] and more recently by NMR spectroscopy.[10] All of these studies have shown that 4-cinnolones exist predominantly or exclusively as the 4-cinnolone tautomers in the solid state and in solution.

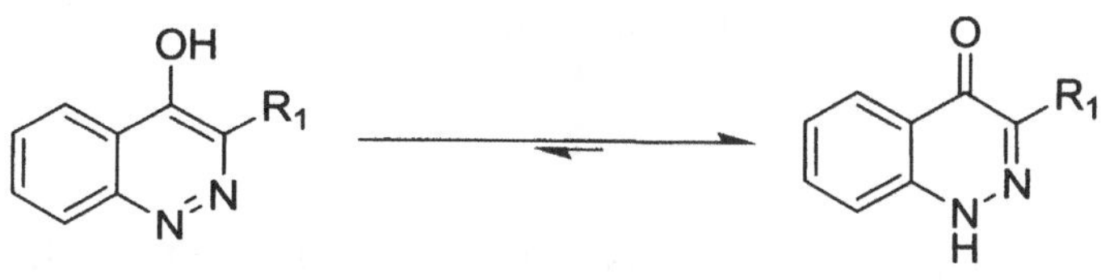

8.2.4 Variations and Improvements

A deficiency of the Borsche cinnoline synthesis lies in the fact that the *ortho*-acyl aniline starting materials are sometimes not available by convenient means. In the older literature, the starting materials were prepared by the nitration of acetophenones, followed by a difficult separation of the *ortho*- and *meta*-nitro isomers, followed last by reduction of the nitro group to afford the aniline.[11]

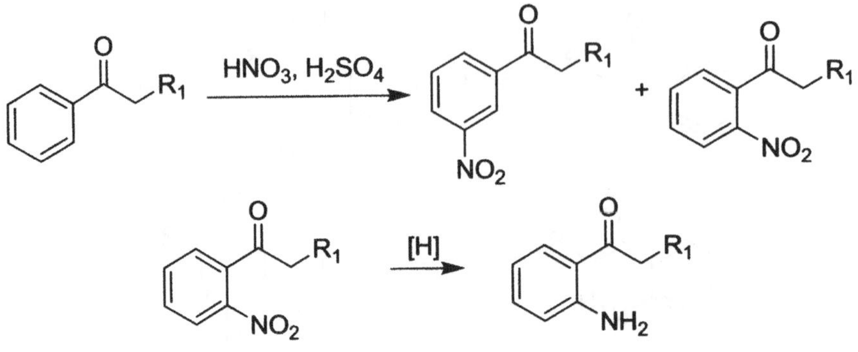

Another commonly employed route was Friedel–Crafts acylation of an appropriately substituted acyl aniline; however, yields may range from poor to good depending on the reactants employed.[12]

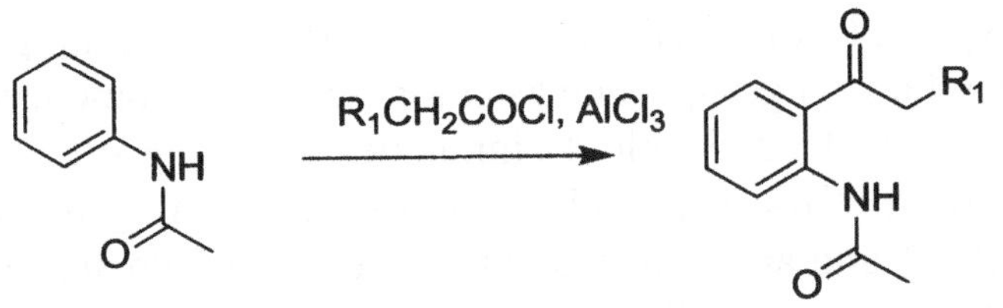

A modification of the Borsche synthesis was disclosed in 1986, in which triphenylphosphoranylidene ketones **1** are diazotized to provide triphenylphosphorylidene cinnolones **2**. Cleavage of the triphenylphosphorane with aqueous alkali affords the 4-cinnolone product **3**.[13]

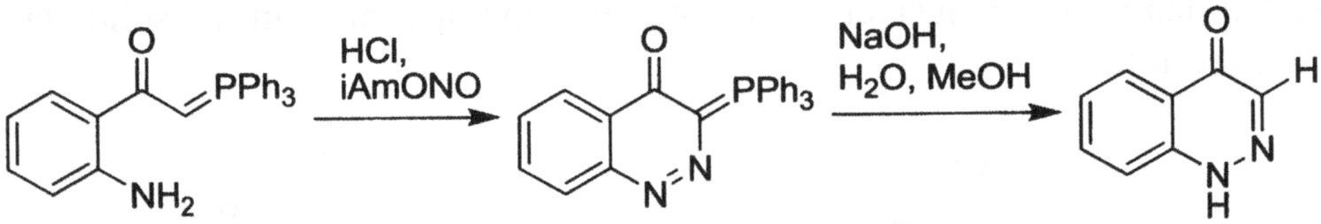

1 2 3

An advantage of this method is that the starting triphenylphosphoranylidene ketones **1** are prepared from the more commonly available anthranilate esters **4**.

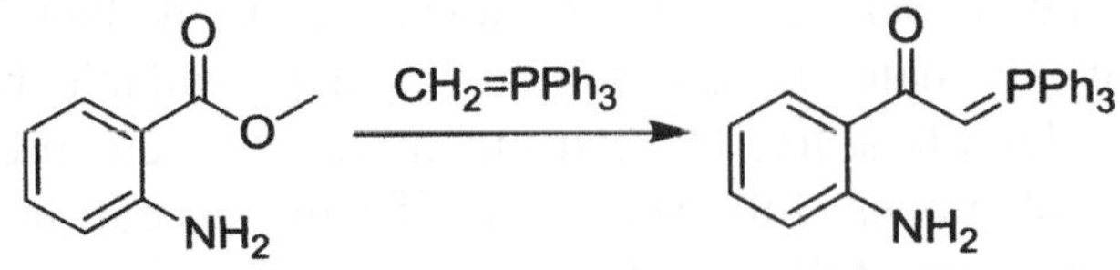

4 1

It should be noted that the somewhat related Baumgarten method has not been successfully employed to produce 3-nitro-4-cinnolones.[14] In this method, an *ortho*-amino acetophenone (or benzaldehyde, R_1 = H) is diazotized, and the resulting diazonium salt is intercepted with the sodium salt of nitromethane to afford a nitroformaldehyde phenylhydrazone (**5**), which may be isolated. Cyclization of **5** may be viewed as occurring by a Knoevenagel condensation between the methylene group and the ketone carbonyl. However, attempts to extend this chemistry to anthranilate esters were not successful.

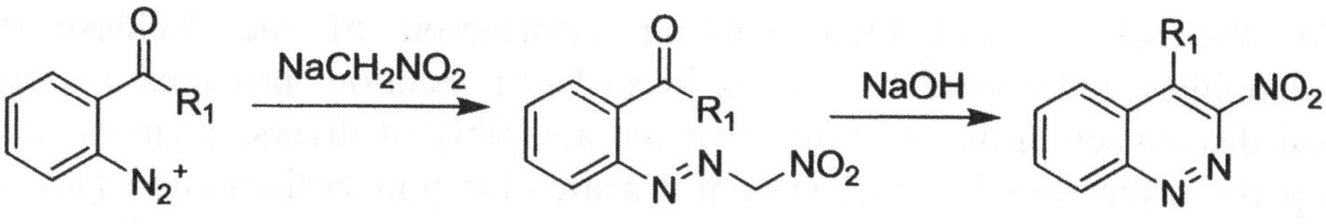

5

Somewhat surprisingly, these modifications have generally seen relatively little use.[15] This may be due in part to the extensive development of palladium catalyzed cross-coupling chemistry and, in particular, the refinement of the Sonagashira reaction, which entails the coupling of a terminal alkyne with an aryl halide under catalysis by copper and palladium. The exceptionally mild conditions, high yields, high functional group tolerance, and availability of appropriately substituted *ortho*-halo anilines for the Sonogashira reaction can make the Richter reaction a viable alternative.[16]

8.2.5 *Synthetic Utility*

General Utility

In general, the Borsche cinnoline synthesis allows the preparation of a wide variety of cinnoline derivatives containing substituents at all possible positions of the benzene ring and the 3-position of the pyridazine ring. Yields are generally above 70%, making the reaction a useful route for preparative work. *ortho*-Acyl anilines having alkyl and aryl substituents on the acyl group have been used most frequently. *ortho*-Acyl anilines having halogen substituents on the acyl group polymerize easily and have been little used. Numerous substituents have been employed in the aniline ring, such as ester, nitro, alkyl, alkoxy, trifluoromethyl, aryl, heteroaryl, fused heteroaryl, fused cycloalkyl, and of course halogen, singly or in combination.[17]

Diazotization is most often carried out using sodium nitrite in a mineral acid such as hydrochloric acid or dilute sulfuric acid. Yields appear to be the highest when concentrated hydrochloric acid is used. Occasionally, sodium nitrite may be used in a weaker acid such as acetic acid solution, and the use of a buffered biphasic reaction medium has also been investigated.[18] Alkyl nitrite esters such as butyl nitrite or isoamyl nitrite may also be used to effect diazotization of the aniline.[19]

Applications in Synthesis

It is not surprising that the 4-cinnolone products of the Borsche synthesis have been most frequently used as intermediates in the synthesis of biologically active compounds, both pharmaceuticals and agrochemicals. The facile conversion of 1*H*-4-cinnolones to 4-chlorocinnolines, coupled with the reliable and high yielding substitution of the 4-chloro by nucleophiles under mild conditions in an S_NAr reaction, provides a useful bond disconnection that permits the rapid assembly of diverse analogs. This property of the cinnoline ring system is shared by quinazolines and stands in contrast to the more difficult construction of similar quinoline and particularly naphthalene derivatives.

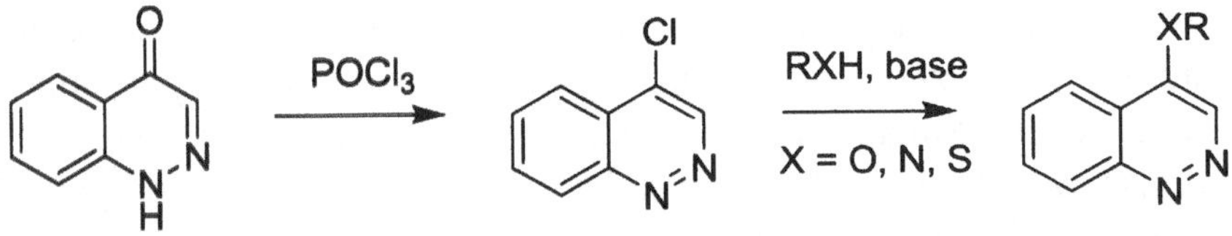

6 7

In particular, 4-aminocinnolines prepared in this manner have often synthesized to compete for nucleotide co-factors in kinases and phosphodiesterases.[20] Thus 4-cinnolones have been used as key intermediates in the synthesis of inhibitors of these families of enzymes.

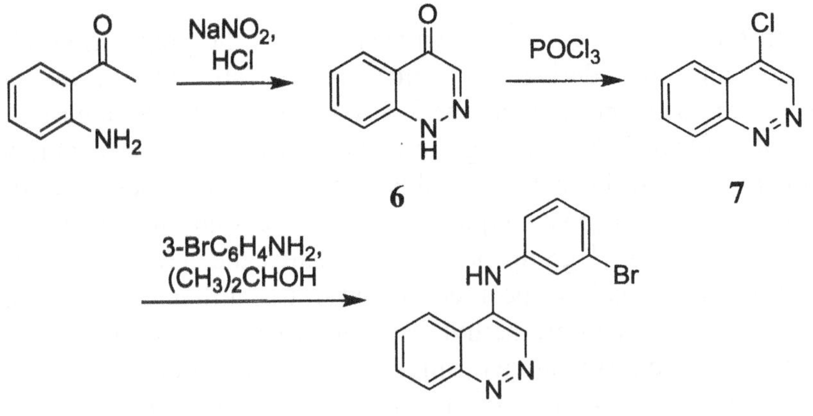

6 7

8

For example, the cinnoline **8** was synthesized for evaluation as an inhibitor of the tyrosine kinase domain of the epidermal growth factor receptor (EGFR).[21] This was prepared from 1*H*-4-cinnolone **6** by chlorination and subsequent displacement of chlorine by 3-bromoaniline, as shown in the previous page. The 1*H*-4-cinnolone **6** was prepared in turn from (2-amino)acetophenone by the Borsche synthesis.

Similarly, cinnoline inhibitors of the tyrosine kinase of the vascular endothelial growth factor receptor (VEGF), such as **10**, have also been described. Again, the Borsche synthesis was used to construct the appropriately functionalized 1*H*-4-cinnolone **9**, which was then further elaborated to afford **10.**[22]

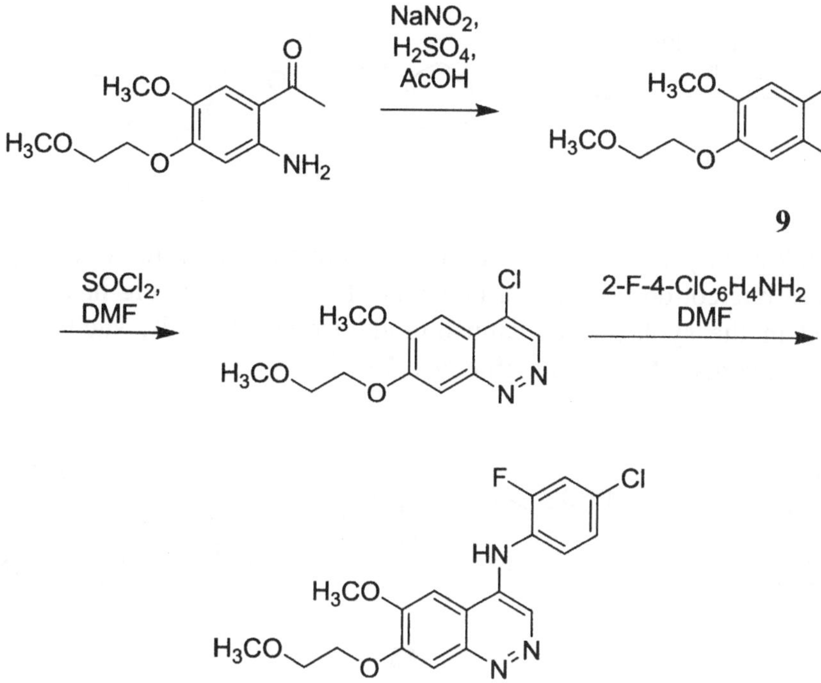

9

10

More recently, a cinnoline containing inhibitor of protein kinase B (**13**) has been prepared through the use of the Borsche cinnoline synthesis.[23] The origin of this product in the Borsche synthesis may not be immediately apparent, because there is no functional group at the 4-position of the cinnoline. In this case, the 1*H*-4-cinnolone product **11** of the Borsche synthesis was reduced to afford the desired intermediate 6-bromocinnoline **12**, which was subjected to a Stille coupling to afford **13**.

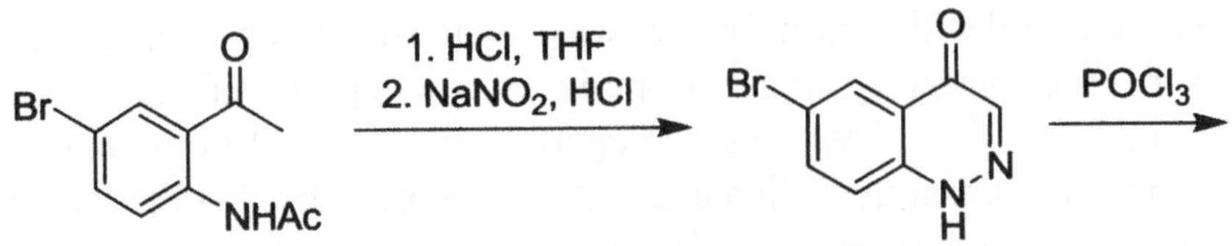

11

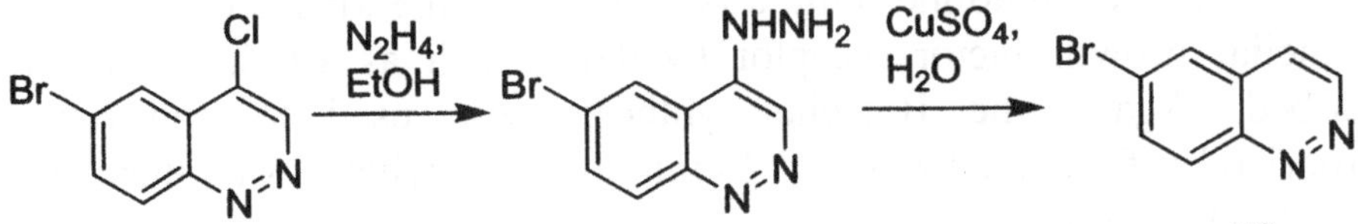

12

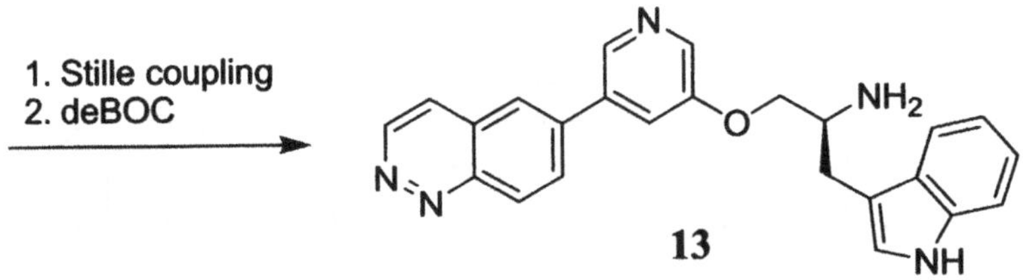

13

Biologically active cinnolines are not limited to inhibitors of enzymes that require nucleotide co-factors, however. The cinnoline ring has recently been used in the synthesis of cannabinoid receptor ligands, such as **14**,[24] antimalarial agents (**15**),[25] and herbicides (**16**).[26]

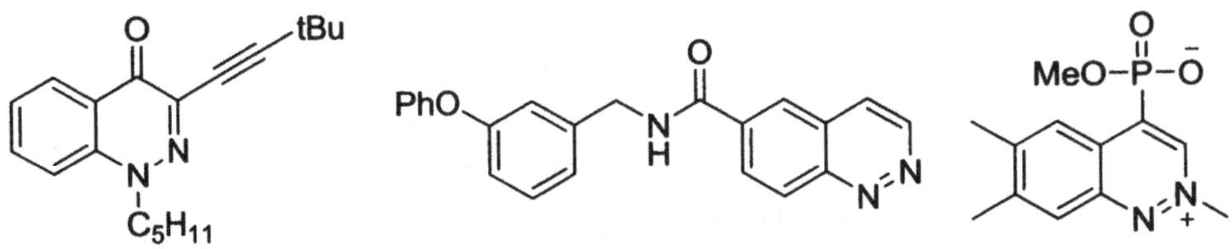

14 **15** **16**

In each of these compounds, the final targets shown were derived from 1*H*-4-cinnolones prepared by the Borsch synthesis. For example, **14** was prepared from 1*H*-4-cinnolone by iodination, *N*-alkylation, and Sonogashira coupling as shown below.

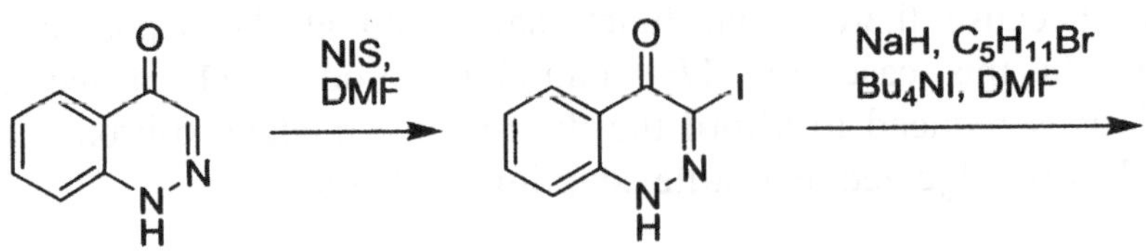

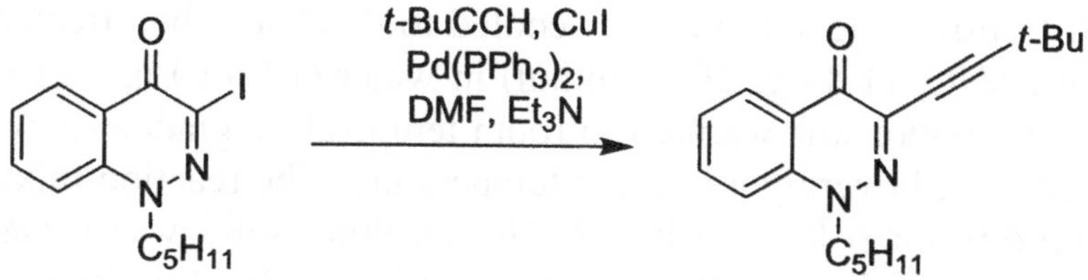

14

A review of cinnolines with biological activity, not limited to those prepared by the Borsche synthesis, has been published recently.[27]

8.2.6 *Experimental*

Borsche and Herbert's Original Preparation of 6-Nitro-1*H*-4-cinnolone[1]
2-Amino-5-nitroacetophenone (1.8 g) was dissolved in 25 mL of glacial acetic acid, then treated with 5 mL of sufuric acid and 1 mL of water, and then treated with 0.7 g finely powdered sodium nitrite added in portions with stirring and cooling to 5 °C in a cooling mixture. The whole was kept in a refrigerator until the next morning and thereupon warmed for a short time (until completion of the diazotization) in a water bath and left to stand at room temperature. During the course of the next day, the crude nitro-oxycinnoline precipitated in the form of compact, brown rhombs (1.5 g). It crystallized from nitrobenzene in brownish needles with mp 343–344 °C, which dissolved in dilute sodium hydroxide to give an orange–yellow solution from which it was precipitated in the form of yellowish needles by hydrochloric acid.

5-Methoxy-1*H*-4-cinnolone[28]
The above amine (2-amino-6-methoxyacetophenone, 9.9 g) in concentrated hydrochloric acid (260 mL) was diazotised at 0 °C with sodium nitrite (6 g) in water (10 mL). Concentrated hydrochloric acid (250 mL) was added, and after 4 days at room temperature, the solution was rapidly (15 min) evaporated to dryness. The residue in boiling water was treated with excess of hot aqueous sodium acetate, and the precipitate was recrystallized (charcoal) from boiling water. 4-Hydroxy-5-methoxycinnoline (5.80 g) gave pale cream needles from water, mp 275 °C (decomp.).

6-Bromo-1*H*-cinnolin-4-one[23]
A solution of *N*-(2-acetyl-4-bromo-phenyl)-acetamide (6.28 g, 24.4 mmol) in THF (75 mL) was treated with concentrated HCl (aq, 15 mL) and water (15 mL). The reaction was heated at reflux for 1 h then concentrated to remove the THF. The aqueous solution was treated with additional water (5 mL) and

conc. HCl (5 mL). The solution was cooled to 0 °C and then treated with a solution of $NaNO_2$ (1.85 g, 26.84 mmol) in water (10 mL) in five portions. The reaction mixture was warmed to room temperature gradually over a 2-h period, then stirred overnight at room temperature. The reaction mixture was heated to reflux for 6 h and filtered. The resulting solid was washed with water (50 mL) and diethyl ether (50 mL) and then dried under vacuum to provide the desired product (3.0 g, 54%).

8.2.7 References

1. Borsche, W.; Herbert, A. *Ann. Chem.* **1941,** *546*, 293–303.
2. Schofield, K.; Simpson, J. C. E. *J. Chem. Soc.* **1945,** 520–524.
3. (a) Keneford, J. R.; Simpson, J. C. E. *J. Chem. Soc.* **1947,** 917–920. (b) Keneford, J. R.; Simpson, J. C. E. *J. Chem. Soc.* **1948,** 354–358. (c) Schofield, K.; Theobald, R. S. *J. Chem. Soc.* **1949,** 2404–2408.
4. Schofield, K.; Simpson, J. C. E. *J. Chem. Soc.* **1948,** 1170–1174.
5. (a) Chlorination with SO_2Cl_2, see Chambers, R. D.; MacBride, J. A. H.; Musgrave, W. K. R. *J. Chem. Soc., Chem. Commun.* **1970,** 739–740. (b) Bromination with Br_2, see Kaushal, A. N.; Narang, K. S. *Ind. J. Chem.* **1968,** *6*, 350–352.
6. Maeba, I.; Castle, R. N. *J. Heterocyclic Chem.* **1980,** *17*, 407–408.
7. (a) Hu, B.; Unwalla, R.; Collini, M.; Wrobel, J.; Quinet, E.; Nambi, P.; Feingold, I.; Goos-Nilsson, A.; Wihelmsson, A. *Bioorg. Med. Chem.* **2009,** *17*, 3519–35233. (b) Egner, U.; Gerbling, K. P.; Hoyer, G.-A.; Krueger, G.; Wegner, P. *Pest. Sci.* **1996,** *47*, 145–158.
8. Mason, S. F. *J. Chem. Soc.* **1957,** 4874–4880.
9. Ames, D. E.; Chapman, R. F.; Kucharska, H. Z.; Waite, D. *J. Chem. Soc.* **1965,** 5391–5401.
10. (a) Bruce, J. M.; Knowles, P.; Besford, L. S. *J. Chem. Soc.* **1964,** 4044–4046. (b) Holzer, W.; Eller, G. A.; Schoenberger, S. *Heterocycles* **2008,** *75*, 77–86. Similarly, cinnoline-4-thiol exists in the thione form, whereas cinnolin-4-amine exists in the amino form. Despite the predominance of the 4-cinnolone tautomer, alkylation of 4-cinnolones often affords mixtures. For example, methylation of 4-hydroxy-3-methylcinnoline gives alkylation at N-1 and N-2; see Ames, D. E.; Chapman, R. F.; Waite, D. *J. Chem. Soc. (C)* **1966,** 470–476.
11. (a) Keneford, J. R.; Simpson, J. C. E. *J. Chem. Soc.* **1947,** 227–232. (b) Leonard, N. J.; Boyd, S. N. Jr. *J. Org. Chem.* **1946,** *11*, 405–418; (c) Waters, W. A. *J. Chem. Soc.* **1945,** 629; (d) Simpson, J. C. E.; Atkinson, C. M.; Schofield, K.; Stephenson, O. *J. Chem. Soc.* **1945,** 646–657.
12. (a) Schofield, K.; Swain, T.; Theobald, R. S. *J. Chem. Soc.* **1949,** 2399–2404. (b) Buu-Hoi, N. P.; Eckert, B.; Royer, R. *J. Org. Chem.* **1952,** *17*, 1000–1004.
13. (a) Buttero, P. D.; Licandro, E.; Maiorana, S.; Papagni, A. *Synthesis* **1986,** 1059–1061. (b) Baldoli, C.; Licandro, E.; Maiorana, S.; Menta, E.; Papagni, A. *Synthesis* **1987,** 288–290.
14. Baumgarten, H. E.; DeBrunner, M. R. *J. Am. Chem. Soc.* **1954,** *76*, 3489–3493.
15. See, for example, Jones, P.; Kinzel, O.; Koch, U.; Ontoria Ontoria, J. M.; Pescatore, G.; Scarpelli, R.; Torrisi, C. WO 200927730, 2009 (*Chem. Abstr.* 150:306510).
16. See, for example: (a) Hodgetts, K. J. WO 2005099710, 2005; *Chem. Abstr.* 143:416271. (b) Chapoulaud, V. G.; Ple, N.; Turck, A.; Queguiner, G. *Tetrahedron* **2000,** *56*, 5499–5508. (b) Fur, N. L.; Mojovic, L.; Turck, A.; Ple, N.; Queguiner, G.; Reboul, V.; Perrio, S.; Metzner, P. *Tetrahedron* **2004,** *60*, 7983–7994.
17. (a) Ester: Koelsch, C. F. *J. Org. Chem.* **1943,** *8*, 295–299. (b) Nitro: Alford, E. J.; Schofield, K. *J. Chem. Soc.* **1953,** 609–612. (c) Alkyl: Keneford, J. R.; Morley, J. S.; Simpson, J. C. E. *J. Chem. Soc.* **1948,** 1702–1707. (d) Alkoxy: Maeba, I.; Castle, R. N. *J. Heterocyclic Chem.* **1980,** *17*, 407–408. (e) Trifluoromethyl: Hu, B.; Unwalla, R.; Collini, M.; Wrobel, J.; Quinet, E.; Nambi, P.; Feingold, I.; Goos-Nilsson, A.; Wihelmsson, A. *Bioorg. Med. Chem.* **2009,** *17*, 3519–3527. (f) Aryl: Atkinson, C. M.; Sharpe, C. J. *J. Chem. Soc.* **1959,** 2858–2864. (g) Heteroaryl: Singh, B. *J. Heterocyclic Chem.* **1991,** *28*, 881–883. (h) Fused heteroaryl: Case, F. H.; Brennan, J. A. *J. Am. Chem. Soc.* **1959,** *81*, 6297–6301. (i) Fused cycloalkyl:

Schofield, K.; Swain, T. *J. Chem. Soc.* **1949,** 2393–2399. (j) Halo: Schofield, K.; Simpson, J. C. E. *J. Chem. Soc.* **1945,** 512–520. (k) Schofield, K.; Theobald, R. S. *J. Chem. Soc.* **1950,** 395–402.

18. Denes, L. R. US 4620000, 1986 (*Chem. Abstr.* 106:84624).
19. Vinogradova, O. V.; Balova, I. A. *Chem. Heterocycl. Compd.* **2008,** *44*, 501–522.
20. (a) Cee, V. J.; Deak, H. L.; Geuns-Meyer, S. D.; Du, B.; Hodous, B. L.; Martin, M. W.; Nguyen, H. N.; Olivieri, P. R.; Panter, K.; Romero, K.; Schenkel, L.; White, R. WO 2008/124083 A2, 2008 (*Chem. Abstr.* 149:471489). (b) Hitchcock, S. A.; Liu, R.; Arrington, M. P.; Hopper, A. T.; Conticello, R. D.; Nguyen, T. M.; Danca, M. D. WO 2007/98214 A1, 2007 (*Chem. Abstr.* 147:322996). (c) Hitchcock, S. A.; Arrington, M. P. WO 2007/100880 A1, 2007 (*Chem. Abstr.* 147:322998). (d) Arrington, M. P.; Liu, R.; Conticello, R. D.; Gauss, C. M.; Hopper, A.; Nguyen, T. M.; Tehim, A. WO 2006/28957 A1, 2006 (*Chem. Abstr.* 144:292767).
21. (a) Rewcastle, G. W.; Denny, W. A.; Bridges, A. J.; Zhou, H.; Cody, D. R.; McMichael, A.; Fry, D. W. *J. Med. Chem.***1995,** *38*, 3482–3487. (b) Leonard, N. J.; Boyd, S. N. Jr. *J. Org. Chem.* **1946,** *11*, 419–428.
22. Hennequin, L. F.; Thomas, A. P.; Johnstone, C.; Stokes, E. S. E.; Plé, P. A.; Lohmann, J.-J. M.; Ogilvie, D. J.; Dukes, M.; Wedge, S. R.; Curwen, J. O.; Kendrew, J.; Lambert-van der Brempt, C. *J. Med. Chem.* **1999,** *42*, 5369–5389.
23. Woods, K. W.; Fischer, J. P.; Claiborne, A.; Li, T.; Thomas, S. A.; Zhu, G.-D.; Diebold, R. B.; Liu, X.; Shi, Y.; Klinghofer, V.; Han, E. K.; Guan, R.; Magnone, S. R.; Johnson, E. F.; Bouska, J. J.; Olson, A. M.; de Jong, R.; Oltersdorf, T.; Luo, Y.; Rosenberg, S. H.; Giranda, V. L.; Li, Q. *Bioorg. Med. Chem.* **2006,** *14*, 6832–6846.
24. Muthupplaniappan, M.; Margal, S.; Sukeerthi, K.; Khairatkar-Joshi, N.; Karnik, P. WO 2009/53799 A1, 2009 (*Chem. Abstr.* 150:472732).
25. (a) Nakamoto, K.; Matsukura, M.; Tanaka, K.; Inoue, S.; Tsukada, I.; Haneda, T.; Ueda, N.; Abe, S.; Sagane, K. WO 2006/016548, 2006 (*Chem. Abstr.* 144:232928). (b) Nakamoto, K.; Tsukada, I.; Tanaka, K.; Matsukura, M.; Haneda, T.; Inoue, S.; Ueda, N.; Abe, S.; Hata, K.; Watanabe, N. WO 2005/033079 A1, 2005 (*Chem. Abstr.* 142:392428).
26. (a) Egner, U.; Gerbling, K. P.; Hoyer, G.-A.; Krueger, G.; Wegner, P. *Pesticide Sci.* **1996,** *47*, 145–158. (b) Ponte, J. R.; Steffens, J. J.; Estreicher, H. US Pat. 4666499, 1987 (*Chem. Abstr.* 107:78082).
27. Lewgowd, W.; Stanczak, A. *Arch. Pharm. Chem. Life Sci.* **2007,** *340*, 65–80.

8.3 Gutknecht Pyrazine Synthesis

Stephen W. Wright

8.3.1 Description

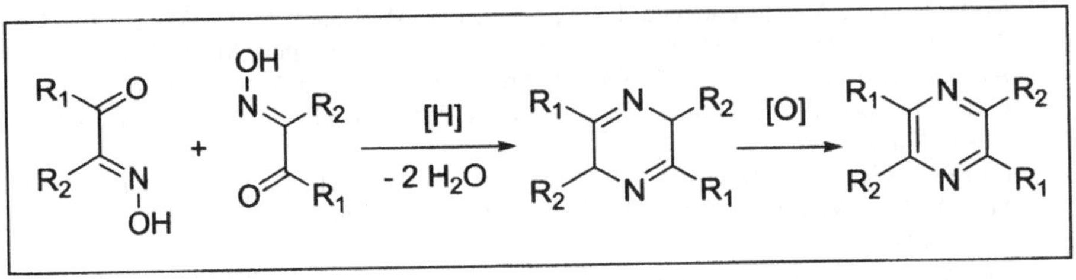

The Gutknecht pyrazine synthesis constitutes the cyclization of an α-oximino carbonyl compound to ultimately form a pyrazine. The α-oximino carbonyl compound is reduced to afford an α-amino carbonyl compound, which undergoes cyclocondensation to a dihydropyrazine. Oxidation of the dihydropyrazine to the pyrazine may occur spontaneously or may be carried out in a second step. It is mechanistically related to the Stadel and Rugheimer and Erdmann syntheses, which involve the cyclization of α-haloketones with ammonia and α-hydroxyketones with ammonia, respectively.

8.3.2 Historical Perspective

H. Gutknecht first reported the reduction of an α-oximino ketone in 1879, as part of a study on the chemistry of α-isonitroso ethyl methyl ketone carried out in the laboratories of Viktor Meyer at what is now the ETH Zurich.[1] At the time it was not known that these substances exist primarily as the oxime tautomer and not as the nitroso tautomer.

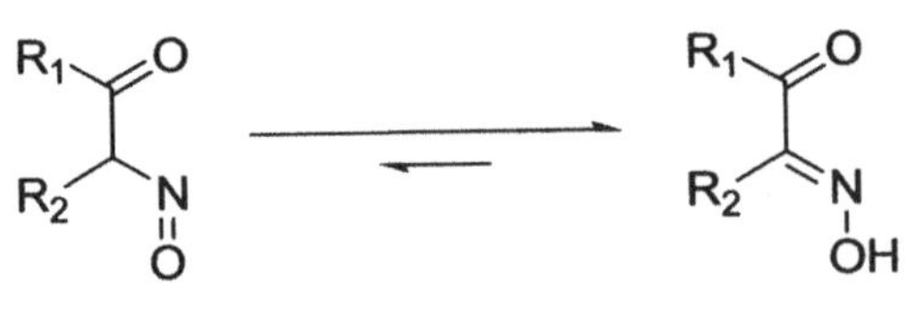

Reduction of "α-isonitroso ethyl methyl ketone" (**1**, 3-(hydroxyimino)butan-2-one or α-oximinobutanone) with tin and hydrochloric acid afforded a product that, after addition of sodium carbonate and extraction with ether, melted at 80 °C and failed to react with phthalic anhydride as anticipated. Elemental analysis and molecular weight determination indicated that the product contained no oxygen and was derived from two molecules of the α-isonitroso ketone. The expected α-amino ketone was obtained as the hexachloroplatinate salt when the ethereal extract was promptly extracted with hydrochloric acid and treated with platinic chloride. This salt was transformed into the hexachloroplatinate salt of the unexpected product, with loss of water, upon heating or standing under vacuum over concentrated sulfuric acid. Gutknecht proposed that the substance was an "inner anhydride" (**2**) of the expected amino ketone, by analogy to the hypothesis proposed by Stadel and Rugheimer for the structure of a product they obtained from the reaction of ammonia with 2-chloroacetophenone.[2]

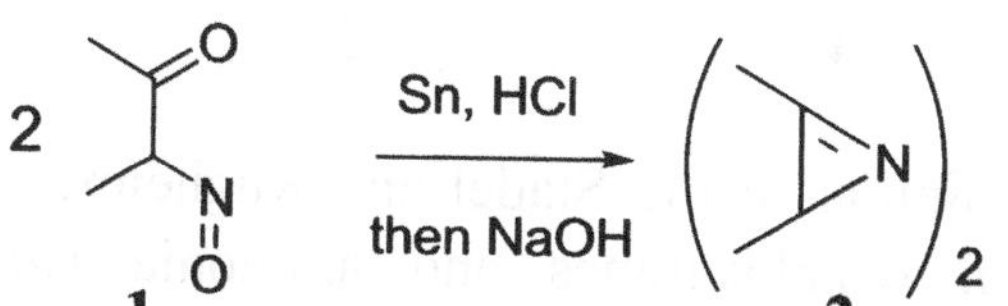

1 **2**

The following year, Gutknecht reported that treatment of α-isonitrosopropionic acid with tin and hydrochloric acid afforded alanine as expected, and dimeric material was not obtained.[3] In 1882, S. Wleugel first suggested that these products contained a six-membered ring and were analogous to the pyridines.[4] In 1886, L. Oeconomides correctly proposed the mechanism by which these substances were formed.[5]

8.3.3 Mechanism

The mechanism of Gutknecht pyrazine synthesis has been studied and is well understood. Reduction of the α-oximino ketone affords an α-amino ketone. If the reduction is carried out under acidic conditions, the α-amino ketone may be isolated as an acid salt.[6] These acid addition salts are entirely stable. In these salts the ketone carbonyl may be hydrated, and this is particularly true for α-amino aldehydes.[7] However, as soon as the free base of the amine is generated, either from the salt or during reduction of the oxime if this is carried out under neutral or basic conditions, rapid bimolecular imine formation occurs, which is then followed by rapid intramolecular formation of a second imine to afford a dihydropyrazine. Oxidation to the pyrazine may occur spontaneously upon exposure to air, particularly in the presence of transition metals, and it is this facile aerobic oxidation that doubtless accounts for the isolation of pyrazines by early workers in the field.

It is also possible that under certain conditions, reduction of the oxime to an imine may occur, and subsequent reduction of the imine to an amine may be slow enough to permit the imine to accumulate. Reaction of the resulting α-imino ketone with α-amino ketone also present may afford a pyrazine. Under these conditions a pyrazine would be obtained directly, without need for the subsequent oxidation of a dihydropyrazine intermediate. While this could happen in any reduction of an α-oximino ketone under neutral or basic conditions, it could be experimentally difficult to distinguish between this pathway and the adventitious oxidation of a dihydropyrazine.

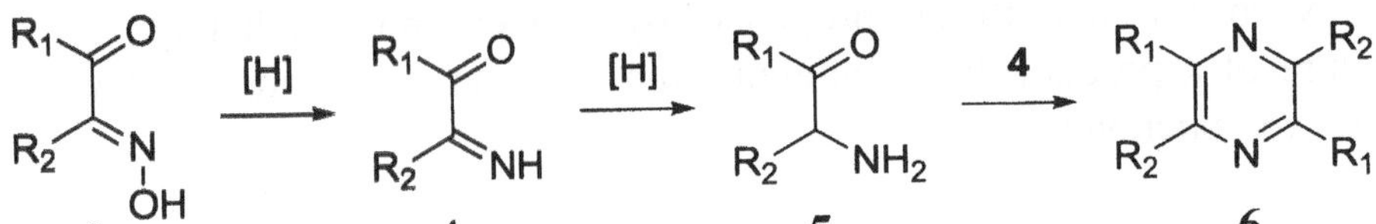

3 4 5 6

While very similar to the Stadel and Rugheimer pyrazine synthesis that proceeds from α-haloketones and ammonia *via* α-amino ketone intermediates, a key difference is that the Gutknecht synthesis starts with the key C–N bond already installed. The Stadel and Rugheimer synthesis has long been known to be capable of affording mixtures of pyrazine isomers, with the unexpected regioisomers often resulting from *di*alkylation of ammonia followed by reaction of ammonia with the resulting 1,5-diketone **9**.[8]

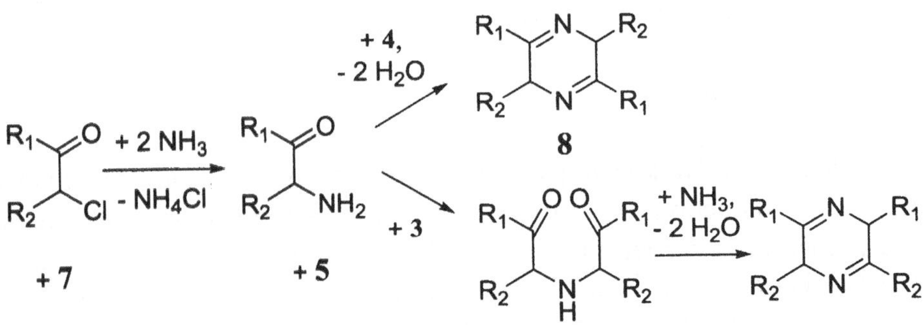

9 10

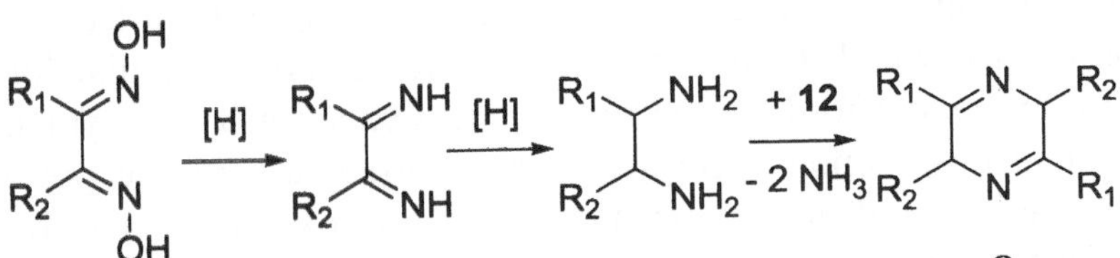

11 12 13 8

Less frequently, dioximes have been used as starting materials in the Gutknecht synthesis.[9] Formation of a dihydropyrazine from a dioxime could occur by hydrolysis of an oxime or imine intermediate to a diketone under the reaction conditions or by a pathway involving imine exchange between an amine and a partially reduced oxime.

8.3.4 Variations and Improvements

The starting materials for the Gutknecht synthesis are readily available from the corresponding ketones. The nitrosation step is usually simple and high yielding. Many oximino ketones have been prepared using nitrous acid generated *in situ* from sodium nitrite and acetic acid. This procedure tends to afford the best yields with ketones that are readily enolized, particularly β-diketones and β-ketoesters. A more general method that works well in nearly all cases uses a nitrite ester, typically isoamyl nitrite, butyl nitrite or *t*-butyl nitrite, with either a strong acid catalyst or an alkoxide catalyst.[10] Occassionally, the oximes are prepared using other sources of the nitrosyl cation (NO^+), such as nitrosyl chloride (NOCl),[11] nitrosyl sulfuric acid ($NOSO_3OH$),[12], or nitrogen dioxide (NO_2).[13]

The reduction of the oximino ketone may be carried out with a variety of reagents. Most commonly, the reduction is effected using either tin metal, stannous chloride, or zinc in acid solution. Other reducing agents that have been used include zinc in alkaline solution,[14] aluminum amalgam,[15] sodium dithionite,[16] titanous chloride,[17] and electrolysis.[18] Oximes are readily hydrogenated over platinum group metals, and the Gutknecht synthesis has been conducted using hydrogen with palladium[19] and Raney nickel[20] catalysts. Care must be taken when catalytic hydrogenation is used to reduce the oxime because ketones may be reduced to alcohols over these metals, and the resulting α-aminoalcohol will be unable to undergo cyclization to a dihydropyrazine. This has been most frequently observed with Raney nickel.[21] Hydrogenation of aryl α-oximino ketones in acid solution over palladium affords good yields of α-amino ketones if the reduction is stopped after 2 equiv of hydrogen have been consumed, despite the otherwise facile hydrogenolysis of aryl ketones over this metal.[22] However, prolonged hydrogenation of aryl α-oximino ketones can result in further reduction to the amino alcohol.[23]

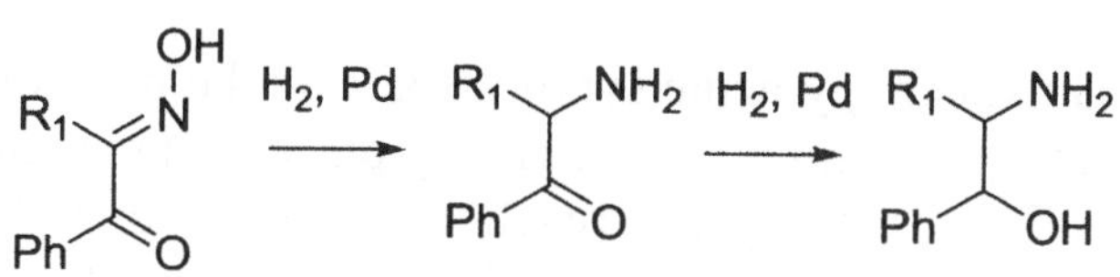

14 **15** **16**

Oxidation of the resulting dihydropyrazine has been effected with numerous oxidants. In the older literature, mercuric ion and cupric ion were commonly employed.[24] Bromine or sodium hypobromite,[25] nitric acid,[26] and of course air or oxygen[27] have been employed.

A limitation of the Gutknecht synthesis lies in the fact that the pyrazine products are necessarily symmetrically substituted. Thus the 2 and 5 substituents must be the same, as must the 3 and 6 substituents. Thus the reaction is not applicable to those situations in which an unsymmetrical pyrazine is the desired product. In addition, the Gutknecht synthesis does not provide a useful preparation of pyrazine itself.[28]

It should be noted that there are several other means by which the key α-aminocarbonyl intermediates of the Gutknecht synthesis may be prepared besides by the reduction of α-oximino ketones. The oxidation of α-aminoalcohols, reduction of α-amino acids and their derivatives, hydrolysis of α-acetamido ketones formed from α-amino acids and acetic anhydride by the Dakin–West reaction,[29] and the reduction of α-azido, α-diazo, and α-nitro ketones all lead to dihydropyrazines by way of α-amino ketones. The Gastaldi synthesis provides an alternate use of α-oximino ketones, to afford dicyano pyrazines.[30]

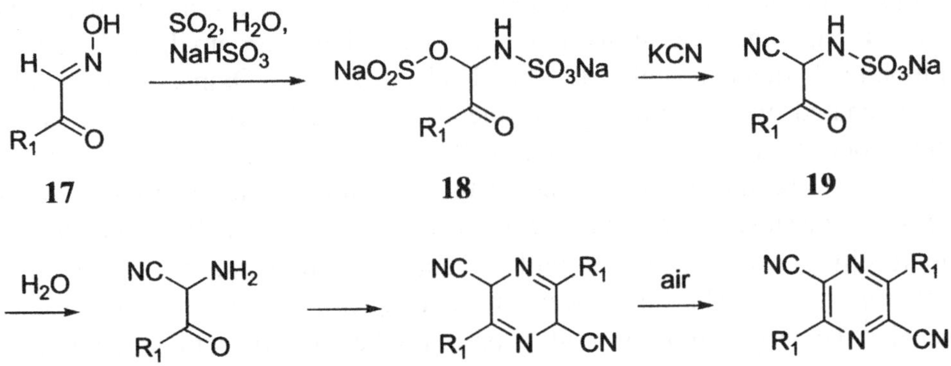

17 18 19

20 21 22

8.3.5 *Synthetic Utility*

General Utility

In general, the Gutknecht pyrazine synthesis allows the preparation of a wide variety of pyrazines containing symmetrically placed substituents. Yields are generally above 75%, making the reaction a useful route for preparative work. Oximino ketones derived from aralkyl ketones and β-keto esters have perhaps been used most frequently due the ease with which the ketones may be prepared and nitrosated.

Applications in Synthesis

It is not surprising that the pyrazine products of the Gutknecht synthesis have been frequently used as intermediates in the synthesis of biologically active compounds, both pharmaceuticals and agrochemicals. For example, a series of pyrazine derivatives (**23**, R_1 = methyl, ethyl, propyl, *iso*-propyl, and *n*-butyl) have recently been examined as heterocyclic nonsteroidal estrogens with selectivity for estrogen receptor subtype ERα over ERβ.[31]

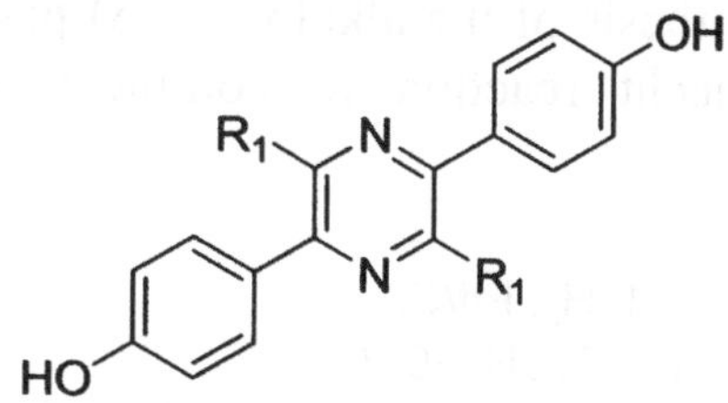

23

The Gutknecht synthesis has also lent itself to the inexpensive, large-scale preparation of pyrazines for specialty engineered materials. Recently a series of 6,12-dihydro-diindeno[1,2-b;1',2'-*e*]pyrazine derivatives with electroluminescent properties were prepared and studied as deep blue light-emitting substances for use in organic light-emitting diodes.[32] More recently the pyridazine **24** was used as a starting material for the preparation of polyalkylated indenopyrazines with tunable luminescent properties.[33]

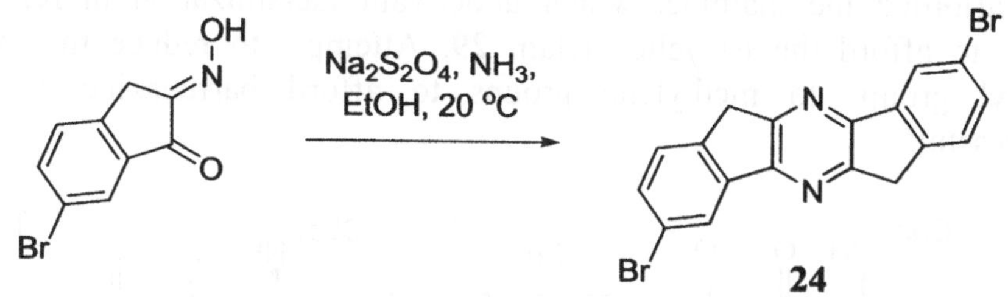

24

The symmetrically substituted pyrazine 2,5-bis(4-hydroxyphenyl)-pyrazine (**25**) and related high molecular weight phenolic compounds have been investigated as monomers for the preparation of high-performance polymers.[34] Nitrosation of 4-hydroxyacetophenone was accomplished using nitrosyl chloride generated from sodium nitrite and hydrochloric acid.

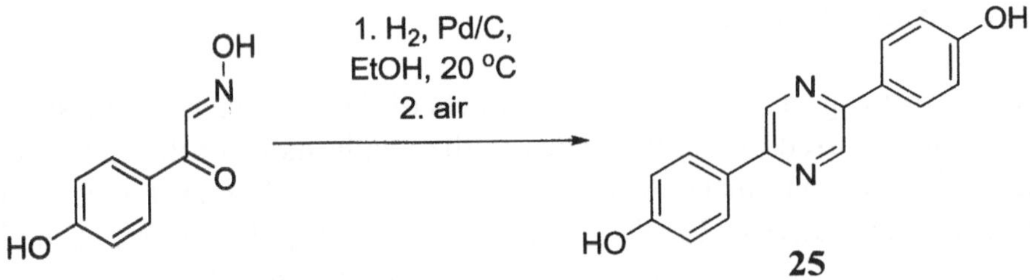

25

The Gutknecht synthesis has been used in the synthesis of natural products as well. A synthesis of the alkaloid (*S,S*)-palythazine (**26**) from D-glucose used a Gutknecht reaction to complete the synthesis of this substance.[35]

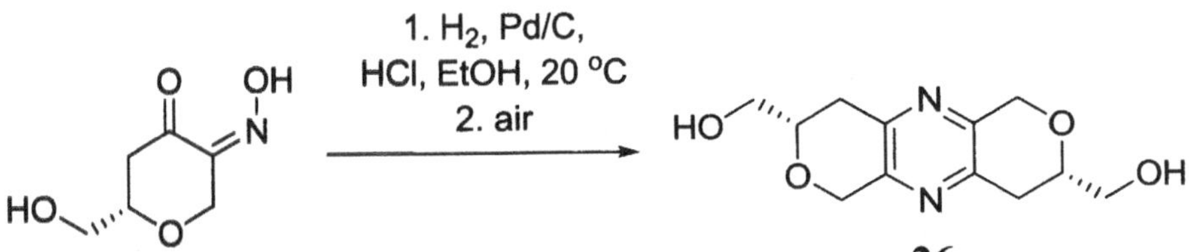

26

Likewise, a biomimetic synthesis of an analog of the cytotoxic marine alkaloid barrenazine A was completed by the use of a Gutknecht synthesis. Again, the C_2 symmetry of the natural product pointed to the Gutknecht synthesis as the most efficient means to construct the central core pyrazine in the key intermediate **28**.[36] Hydrogenolytic deprotection of the amine residues in **28** afforded the diamine, which underwant lactamization in refluxing toluene to afford the tricyclic lactam **29**. Attempts to reduce the lactam carbonyl groups to methylene groups to afford barrenazine A were unsuccessful.

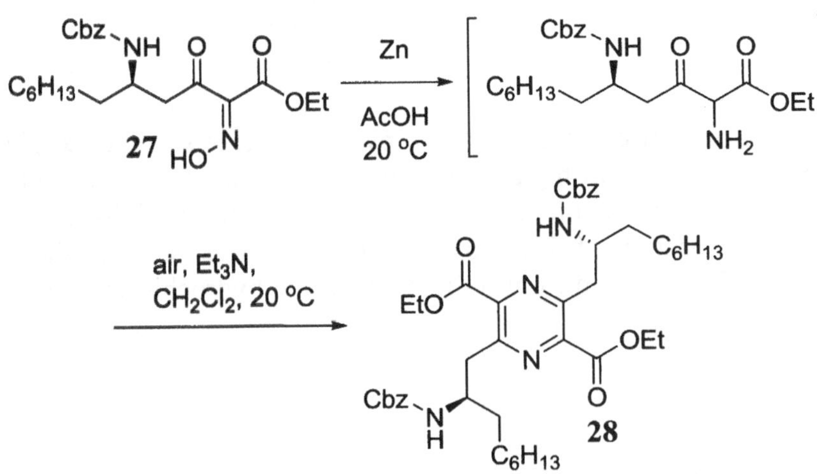

27

28

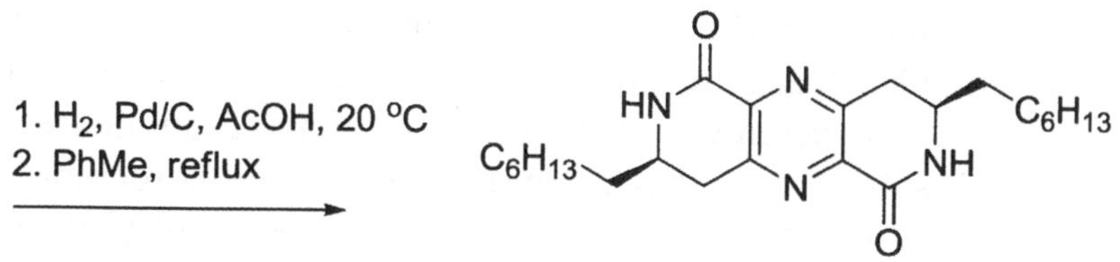

29

8.3.6 *Experimental*

2,5-Dimethyl-3,6-diphenylpyrazine[14]

Zinc dust (80 g, 1.2 mol) was added in portions to a stirred solution of 1-hydroxyimino-1-phenylacetone (60 g, 0.37 mol) in 600 mL of 5 N sodium hydroxide at 25–30 °C. The mixture was stirred for 2 h, then diluted with 600 mL of water and filtered. The product was extracted with 500 mL of hot chloroform and air was bubbled through the extract for 15 min. After being dried ($MgSO_4$), the solution was evaporated and the residue was distilled at 1 torr. The gummy distillate was rubbed with a little ether and the resulting solid was crystallised from dilute acetic acid, giving 2,5-dimethyl-3,6-diphenylpyrazine (19.1 g, 40%), mp 126 °C.

Diethyl 3(R),6(R)-3,6-Di(benzyloxycarbonylamino)-pyrazine-2,5-dicarboxylate (28)[36]

To a solution of ethyl 5(R)-2-(hydroxyimino)-5-(benzyloxy carbonylamino)-3-oxododecanoate (**27**, 600 mg, 1.47 mmol) in 10 mL acetic acid was added zinc dust (965 mg, 14.7 mmol) in small portions. The reaction mixture was stirred for 12 h at room temperature, then filtered through a pad of Celite, and the solid washed with CH_2Cl_2. The solvents were evaporated, and the residue dissolved in 20 mL dichloromethane. To this solution was added triethylamine (0.62 mL, 4.44 mmol) and the mixture was stirred for 1 h at room temperature. Additional water (10 mL) was added, and the solution extracted with dichloromethane (3 × 50 mL). The organic layer was washed with brine (30 mL), dried over anhydrous $MgSO_4$ and the solvent evaporated. The crude product was purified by column chromatography (petroleum ether–AcOEt, 7:3), giving diethyl 3(*R*),6(*R*)-3,6-di(benzyloxycarbonyl-amino)-pyrazine-2,5-dicarboxylate (400 mg, 70%), as a white solid, mp 89–91 °C.

Diethyl 3,6-dimethylpyrazine-2,5-dicarboxylate[37]

Ethyl 2-oximinoacetoacetate (16 g, 0.10 mol) in absolute ethanol (160 mL) containing dry hydrogen chloride (10 g, 0.27 mol) was hydrogenated over a 5% palladium on carbon catalyst for 2 h at room temperature. After the theoretical amount of hydrogen had been absorbed, the reaction was filtered to remove the catalyst and concentrated under reduced pressure to yield crude ethyl 2-aminoacetoacetate hydrochloride. Recrystallization of the crude product from acetone afforded 15 g (83%) of ethyl 2-aminoacetoacetate hydrochloride. A solution of ethyl 2-aminoacetoacetate hydrochloride (18.2 g, 0.1 mol) and sodium acetate (10 g, 0.12 mol) in ethanol (250 mL) was subjected to oxidation by bubbling air at 75–80 °C over 4 h. The solution was concentrated under reduced pressure. Water was added to the residue. The solution was shaken vigorously, and allowed to stand to crystallize crude diethyl 3,6-dimethylpyrazine-2,5-dicarboxylate. Recrystallization of the crude product from ethanol afforded 10.7 g (85%) of white crystalline diethyl 3,6-dimethylpyrazine-2,5-dicarboxylate.

8.3.7 References

1. Gutknecht, H. *Chem. Ber.* **1879,** *12*, 2290–2292.
2. Stadel, W.; Rugheimer, L. *Chem. Ber.* **1876,** *9*, 563–564.
3. Gutknecht, H. *Chem. Ber.* **1880,** *13*, 1116–1119.
4. Wleugel, S. *Chem. Ber.* **1882,** *15*, 1050–1056.
5. Oeconomides, L. *Chem. Ber.* **1886,** *19*, 2524–2527. See also [R] *Chem. Rev.* **1947,** *40*, 279–358.
6. Hoy, K. L.; Hartung, W. H. *J. Org. Chem.* **1958,** *23*, 967–971.
7. Myers, A. G; Kung, D. W.; Zhong, B. *J. Am. Chem. Soc.* **2000,** *122*, 3236–3237.
8. Tutin, F. *J. Chem. Soc.* **1910,** *97*, 2495–2524.
9. (a) Winans, C. F.; Adkins, H. *J. Am. Chem. Soc.* **1933,** *55*, 2051–2058; (b) Winans, C. F.; Adkins, H. *J. Am. Chem. Soc.* **1933,** *55*, 4167–4176.
10. [R] Krems, I. J.; Spoerri, P. E. *Chem. Rev.* **1947,** *40*, 279–358.
11. Kataoka, M.; Ohno, M. *Bull. Chem. Soc. Jpn.* **1973,** *46*, 3474–3477.
12. Ponzio, G. *Gazz. Chim. Ital.* **1921,** *51*, 213–225.
13. Russell, G. A.; Norris, R. K.; Metcalfe, A. R. *J. Am. Chem. Soc.* **1972,** *94*, 4959–4963.
14. Beech, W. F. *J. Chem. Soc.* **1955,** 3094–3098.
15. Albertson, N. F.; Tullar, B. F.; King, J. A.; Fishburn, B. B.; Archer, S. *J. Am. Chem. Soc.* **1948,** *70*, 1150–1153.
16. Kang, J. S.; Kim, S.-K.; Jung, J.-Y.; Park, J.-W.; Lee, J.-H.; Jaung, J.-Y. *Mol. Cryst. Liq. Cryst.* **2009,** *514*, 171–179.
17. Zercher, C. K.; Miller, M. J. *Heterocycles* **1988,** *27*, 1123–1126.
18. Marcot, B.; Rabaron, A,; Viel, C,; Bellec, C.; Deswarte, S.; Maitte, P. *Can. J. Chem.* **1981,** *59*, 1224–1234.
19. See, for example (a) Zupet, R.; Tisler, M.; Golic, L. *J. Heterocyclic Chem.* **1991,** *28*, 1731–1740. (b) Diels, O.; Poetsch, W. *Chem. Ber.* **1921,** *54*, 1585–1591. (c) Buron, F.; Turck, A.; Ple, N.; Bischoff, L.; Marsais, F. *Tetrahedron Lett.* **2007,** *48*, 4327–4330.
20. (a) Winans, C. F.; Adkins, H. *J. Am. Chem. Soc.* **1933,** *55*, 2051–2058. (b) Baumgartner, H.; O'Sullivan, A. C. *Tetrahedron* **1997,** *53*, 2775–2784.
21. Adkins, H.; Reeve, E. W. *J. Am. Chem. Soc.* **1938,** *60*, 1328–1331.
22. (a) Ginsburg, D.; Pappo, R. *J. Chem. Soc.* **1953,** 1524–1536; (b) Heinzelmann, R. V.; Kolloff, H. G.; Hunter, J. H. *J. Am. Chem. Soc.* **1948,** *70*, 1386–1390.
23. Chang, Y.-t.; Hartung, W. H. *J. Am. Chem. Soc.* **1953,** *75*, 89–91.
24. (a) Künne, H. *Chem. Ber.* **1895,** *28*, 2036–2044; (b) Gabriel, S.; Posner, T. *Chem. Ber.* **1894,** *27*, 1037–1045.
25. See reference 14.
26. Braun, E.; Meyer, V. *Chem. Ber.* **1888,** *21*, 1269–1282.
27. See references 14 and 19c.
28. The Stadel and Rugheimer synthesis is also poorly suited for the preparation of pyrazine, see Tschitschibabin, A. E.; Schtschukina, M. N. *Chem. Ber.* **1929,** *62*, 1075–1080.

29. (a) Dakin, H. D.; West, R. *J. Biol. Chem.* **1928,** *78*, 91–104. (b) Matsushita, H.; Lee, S.-H.; Yoshida, K.; Clapham, B.; Koch, G.; Zimmermann, J.; Janda, K. D. *Org. Lett.* **2004,** *6*, 4627–4630. (c) Levene, P. A.; Steiger, R. E. *J. Biol. Chem.* **1928,** *79*, 95–103. (d) Moldvai, I.; Temesvari-Major, E.; Incze, M.; Platthy, T.; Gacs-Baitz, E.; Szantay, C. *Heterocycles* **2003,** *60*, 309–320. [R] (e) For a review of the Dakin–West reaction, see Buchanan, G. L. *Chem. Soc. Rev.* **1988,** *17*, 91–109.

30. (a) Gastaldi, G. *Gazz. Chim. Ital.* **1921,** *51*, 233–255; (b) Sharp, W.; Spring, F. S. *J. Chem. Soc.* **1948**, 1862–1864. (c) Golombok, E.; Spring, F. S. *J. Chem. Soc.* **1949,** 1364–1367.

31. Ghosh, U.; Ganessunker, D.; Sattigeri, V. J.; Carlson, K. E.; Mortensen, D. J.; Katzenellenbogen, B. S.; Katzenellenbogen, J. A. *Bioorg. Med. Chem.* **2003,** *11*, 629–657.

32. Park, Y.-I.; Son, J.-H.; Kang, J.-S.; Kim, S.-K.; Lee, J.-H.; Park, J.-W. *Chem. Commun.* **2008,** 2143–2145.

33. See reference 16.

34. Kvakovszky, G.; Vicari, R.; Tafesh, A. M.; Juneau, K. N.; Fruchey, O. S.; Mcdonough, J. A.; Kuila, D. US Pat. 5459266, 1995 (*Chem. Abstr.* 124:147109).

35. (a) Jarglis, P.; Lichtenthaler, F. W. *Angew. Chem., Int. Ed.* **1982,** *94*, 140–141. (b) Brehm, M.; Goeckel, V. H.; Jarglis, P.; Lichtenthaler, F. W. *Tetrahedron: Asymmetry* **2008,** *19*, 358–373.

36. Buron, F.; Turck, A.; Ple, N.; Bischoff, L.; Marsais, F. *Tetrahedron Lett.* **2007,** *48*, 4327–4330.

37. Iida, H.; Hayashida, K.; Yamada, M.; Takahashi, K.; Yamada, K. *Synth. Commun.* **1973,** *3*, 225–230.

8.4 Niementowski Quinazoline Synthesis

Richard A. Hartz

8.4.1 Description

The Niementowski quinazoline synthesis is the condensation of an anthranilic acid (**1**) with an amide (**2**) to produce a 4-keto-3,4-dihydroquinazoline product (**3**). The reaction occurs under thermal conditions, and reaction temperatures > 100 °C are generally required. This reaction is also often referred to as the Niementowski reaction, and occasionally as the Niementowski 4-quinazolone synthesis or von Niementowski synthesis.[1]

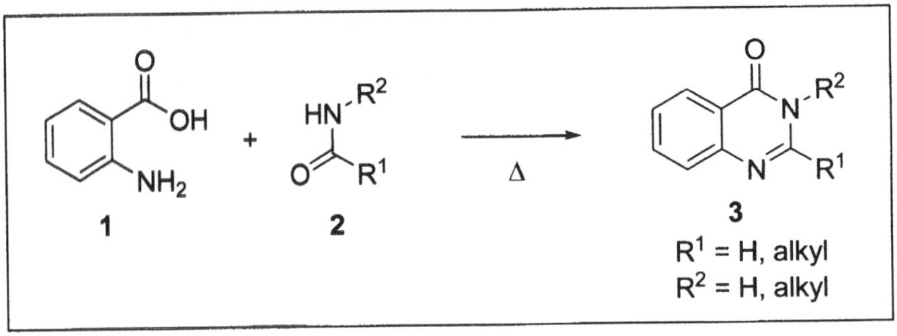

Spectroscopic data indicate that 4-hydroxyquinazoline exists as an equilibrium mixture of **4**, **5**, and **6**, with the keto form (**5**) being the most favored and the lactam (**6**) being the least favored.[2] 4-Hydroxyquinazoline, which is tautomeric with 4-keto-3,4-dihydroquinazoline, is commonly referred to as 4-quinazolone. The name 4-quinazolone will be used throughout this chapter.

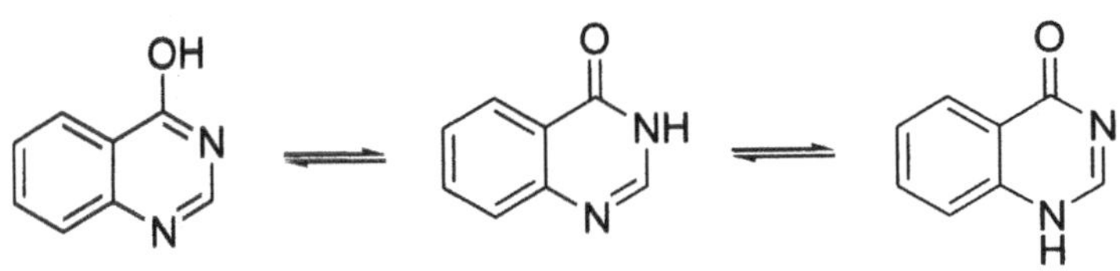

4 **5** **6**

8.4.2 Historical Perspective

In 1895, Niementowski described the reaction of anthranilic acid and its homologs with amides of fatty acids.[3] For example, he found that heating anthranilic acid (**1**) with formamide (**7**) for 2–3 h at 120–130 °C produced 4-quinazolone (**5**) in excellent yield. In a similar fashion, he also synthesized

2-methyl-, 2-ethyl-, and 2-isopropyl-4-quinazolones from acetamide, propionamide, and isobutyramide, respectively. An excess of the amide component was used in this reaction. While the yields of the reactions of anthranilic acid (**1**) with formamide and acetamide were excellent, as the molecular weight of the amide increased, the yields progressively decreased. Higher reaction temperatures were required likely resulting in more decomposition and the formation of more by-products.[4] This reaction has since undergone numerous adaptations and has been used extensively for the synthesis of 4-quinazolones.

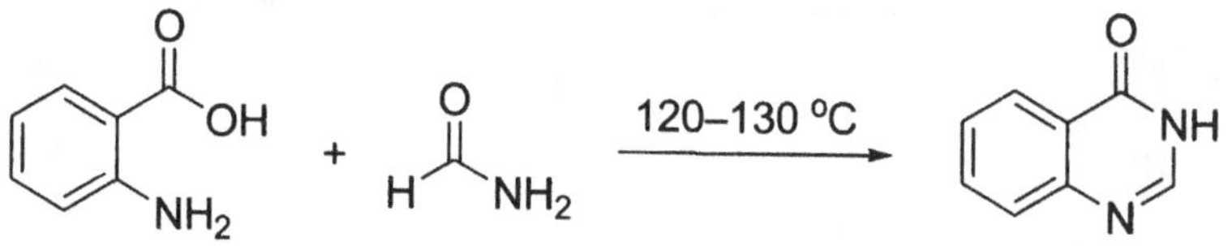

1 **7** **5**

8.4.3 Mechanism

At first glance, one might assume that the reaction between anthranilic acid and acetamide to give 2-methyl-4-keto-3,4-dihydroquinazoline proceeds by way of a simultaneous condensation between the amino group of anthranilic acid and the carbonyl group of the amide and concurrently between the carboxyl group of anthranilic acid and the amido group of the amide to give the observed product. However, in 1900 Bogart and Gotthelf proposed the reaction mechanism summarized below.[5] Heating anthranilic acid (**1**) in the presence of excess acetamide (**8**) leads to the formation of *N*-acetylanthranilic acid **10** along with evolution of ammonia. Subsequent conversion of **10** to amide **12**, followed by cyclodehydration furnishes the 4-quinazolone product **14**. Experimental results consistent with this reaction sequence were later provided by Meyer and Wagner in 1943.[6] Experiments conducted by heating the methyl ester of **1** in the presence of acetamide resulted in the formation of the methyl ester corresponding to **10**, during this time evolution of ammonia from the reaction mixture was observed, and upon prolonged heating the appearance of a 4-quinazolone product occurred. In addition, conversion of ammonium-*N*-acetylanthranilate to 2-methyl-3,4-dihydro-4-quinazolone was reported earlier by Bischler and Burkhart.[7]

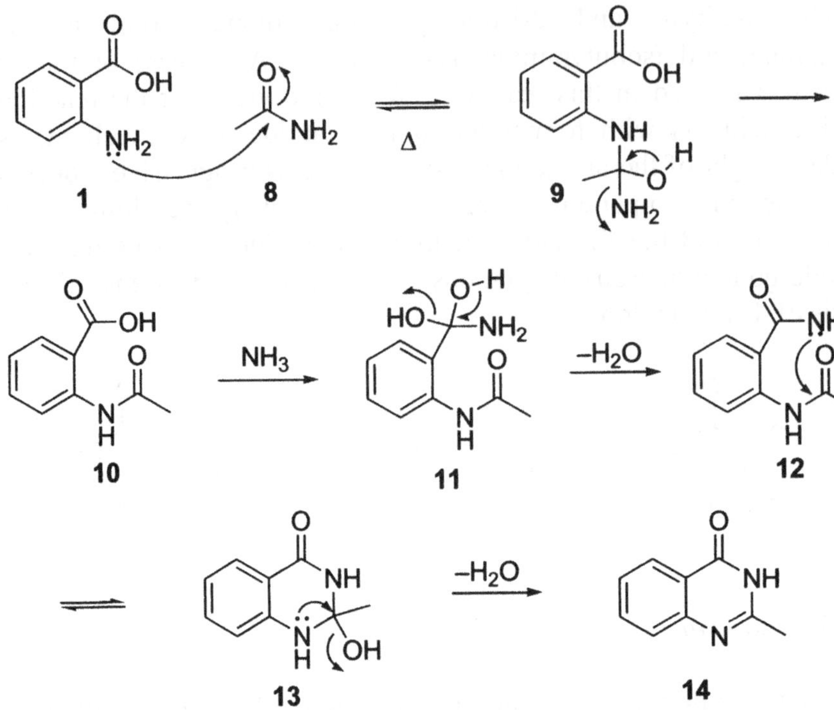

More recent studies of the mechanism of this reaction by Patel and Patel[8] in 1965 suggested an alternate reaction pathway for the conversion of **10** to **12**. They found that heating *N*-acetylanthranilic acid (**10**) in the presence of a slight excess of formamide or acetamide resulted in the formation of 4-quinazolone **14**. Since ammonia is not generated under these reaction conditions, they proposed that the conversion of **10** to amide **12** proceeded by way of a transamidoylation reaction. Experimental support for this reaction pathway was found in the observation that formic acid was detected when heating **10** with formamide at 160 °C (shown below), and acetic acid was detected when heating **10** with acetamide at 180 °C. In both cases, the same 4-quinazolone product (**14**) was formed.

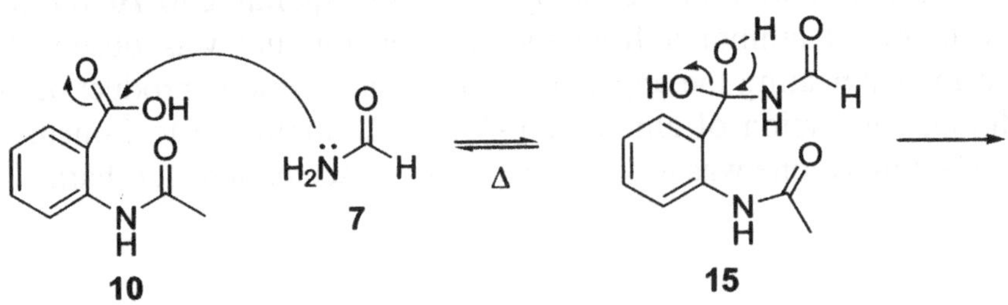

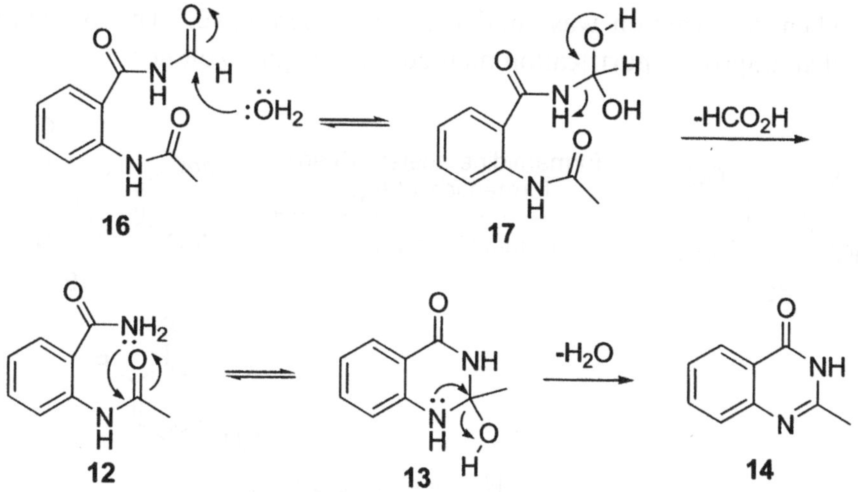

8.4.4 *Variations and Improvements*

When the Neimentowski reaction is carried out under conventional reaction conditions, the yield and purity of the product are significantly affected by the reaction temperature, reaction time, and reagent ratios. As a result, numerous variations and modifications of this reaction have appeared in the literature since its initial discovery over 100 years ago, resulting in improved yields. Variations of the anthranilic acid component include the use of *o*-acylaminobenzamides,[9] ammonium *o*-acylaminobenzoates,[7] *o*-acetamino-benzonitriles,[10] methyl anthranilates,[6] acetanthranils,[11] and isatoic anhydrides[6,12] in place of anthranilic acid. Conditions wherein the amide component has been replaced with either a nitrile,[4,5] imidate,[13,14] thioamide,[15,16] or amidine[6] have been reported. Reaction conditions including acetic anhydride[4] and alkaline peroxide[10] were also developed. More recently, reaction conditions under microwave irradiation have been reported.[17,18] Details regarding these modifications are summarized below.

Reaction of Anthranilic Acids and Derivatives with Amides

Slight modifications to the conventional reaction conditions have resulted in improved yields for this transformation. For example, **19**, a key intermediate in the synthesis of gefitinib (Iressa)[19] was synthesized in only 20% yield by the conventional fusion of formamide with 4,5-dimethoxyanthranilic acid (**18**).[20] Örfi *et al.* reported that heating a variety of anthranilic acids in the presence of formamidine acetate and formamide, either under thermal conditions or microwave conditions, produced excellent yields of 4-quinazolone products.[17] A quantitative yield was reported for the synthesis of **19** when this reaction was conducted in a microwave. The authors also reported an improved purification method for 3*H*-quinazolones.

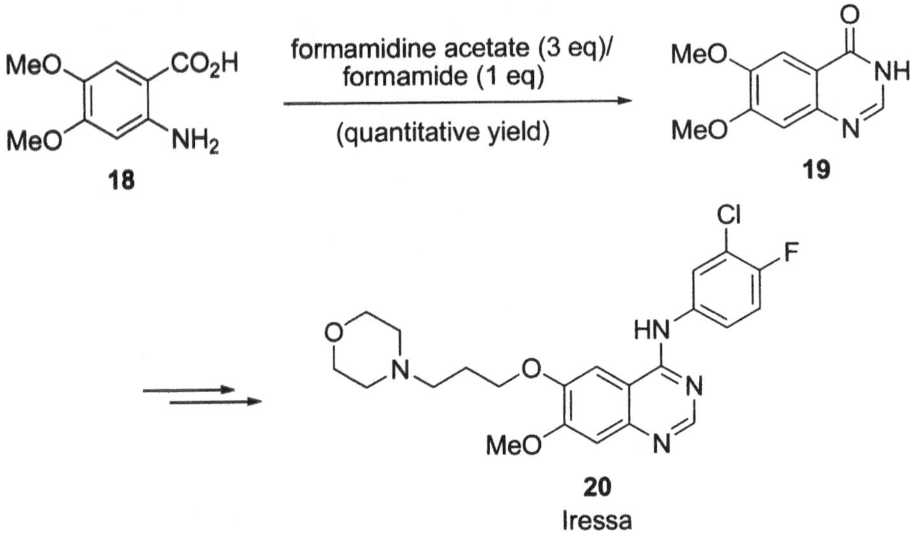

Excellent yields have been reported for a transformation consisting of the reaction of 2-nitrocarboxylic acids, and derivatives thereof, with formamide or *N*-methylformamide in the presence of indium(III) or bismuth (III) salts.[21]

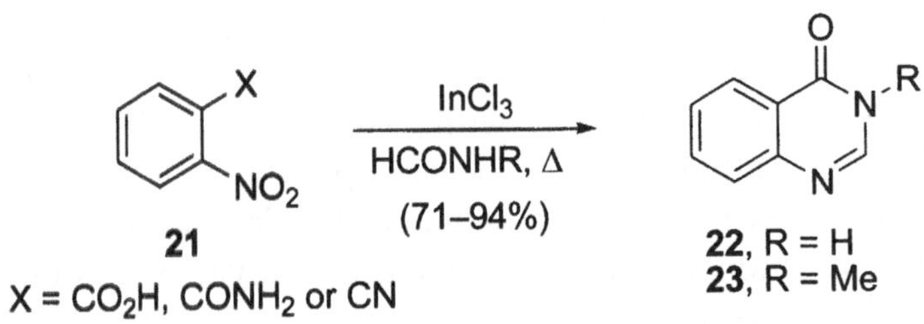

The Niementowski reaction has also been carried out using DMF as a catalyst. For example, heating **24** with formamide (6 equiv) in the presence of a catalytic amount of DMF afforded **25**, which served as an intermediate in the synthesis of Src kinase inhibitors.[22]

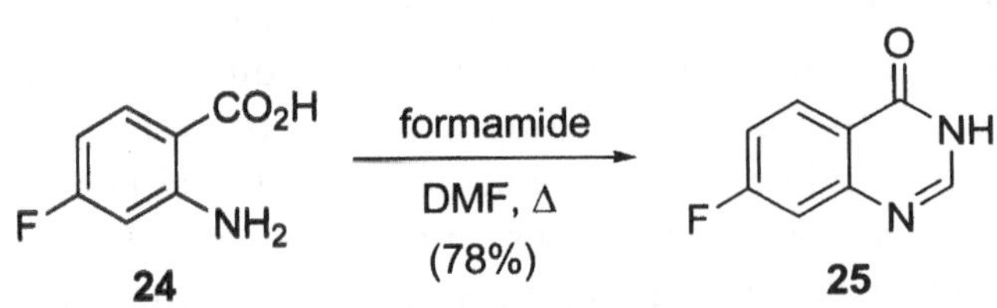

The use of methyl anthranilate (**26**) in the Niementowski reaction was reported by Meyer and Wagner in 1943.[6,23] One of the factors limiting the usefulness of the Niementowski synthesis under conventional conditions is the propensity of anthranilic acid to undergo thermal decarboxylation at sustained heating of 150 °C or above. Methyl anthranilate, however, is stable at temperatures approaching its boiling point of 260 °C. Moreover, it was found to react with formamide at about 200 °C to form the expected 4-quinazolone product in 49% yield. Extension of this procedure to amides larger than acetamide, however, did not result in improved yields compared to the use of anthranilic acid.

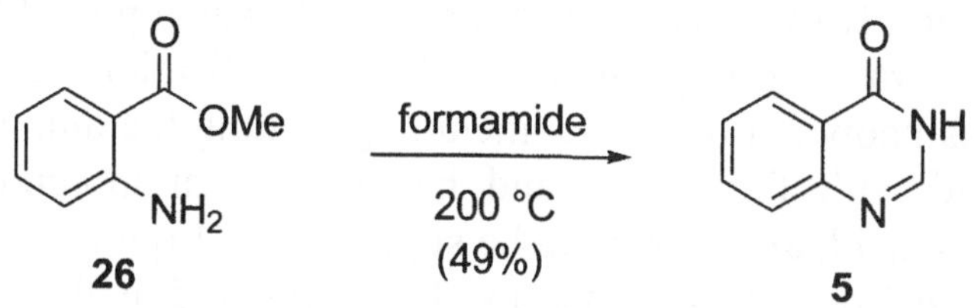

Reaction of N-*Acylanthranilic Acids with Amines*

A modification of the Niementowski reaction involving the use of *N*-acylanthranilic acids facilitated the synthesis of 4-quinazolones **28**, which have a variety of groups at the 2-position, including Me, Et, Pr, Ph, and substituted phenyl.[8] *N*-Acylanthranilic acids **27** were heated in the presence 1.5 equiv of formamide at temperatures ranging from 150–180 °C, depending on the substrate. Yields ranged from 65–83%.

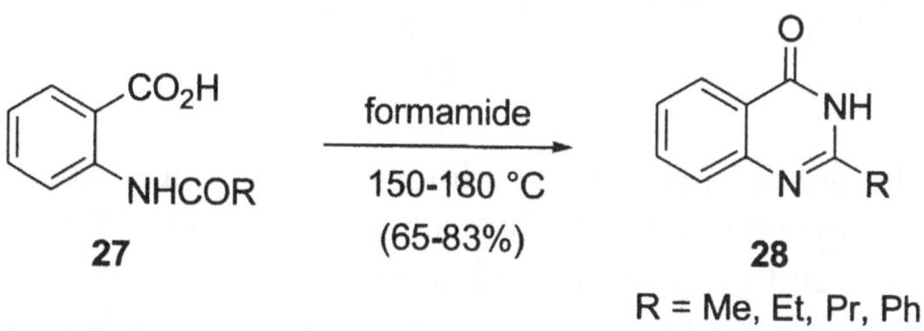

In addition, 2,3-disubstituted-4-quinazolones (**29**) can be synthesized from *N*-acylanthranilic acids by treating **27** with various amines. Generally, addition of the amine is carried out in the presence of an agent to activate the carboxylic acid, such as phosphorous trichloride (PCl_3)[24] or 1,1'-carbonyldiimidazole (CDI).[25]

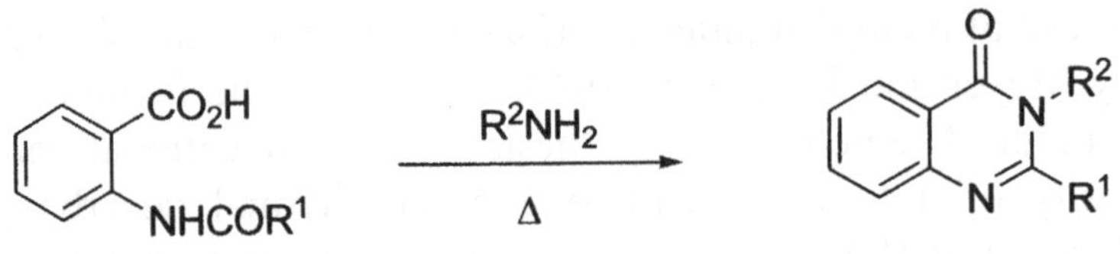

27 29

Reaction of Acylanthranils with Amines

In 1905, Bogert and Chambers described the synthesis of **32** via the acetanthranil intermediate **31**, which was formed by heating 2-amino-6-nitrobenzoic acid (**30**) with acetic anhydride.[11] It was reported that the yield for this reaction was nearly quantitative. This modification was compared to other variations known at that time, including heating the ammonium salt of anthranilic acid with formamide and heating the ammonium salt of *N*-acetylanthranilic acid, and proved to be superior. A 4-quinazolone without a substituent at the 3-position (**35**) has also been prepared by this method.[26]

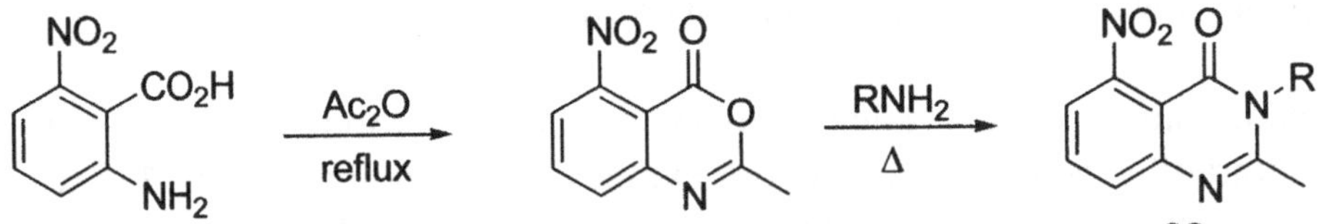

30 31 32

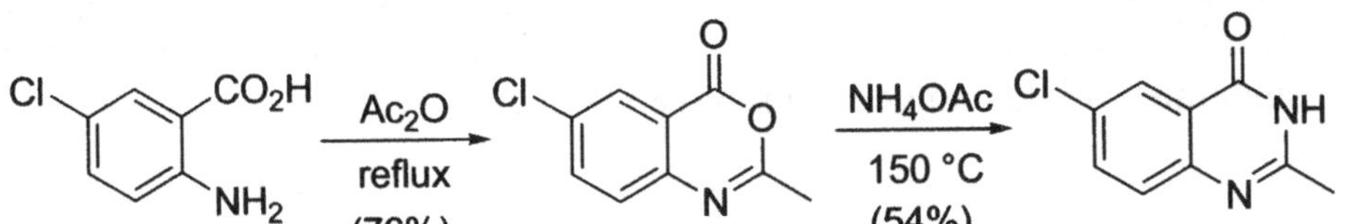

33 34 35

Alternative reaction conditions have since been developed to synthesize acylanthranils **36** from anthranilic acid. Treatment of **1** with benzoic acid and triphenyl phosphite in pyridine produced acylanthranil **36** (R = Ph) in 84% yield.[27] It was not necessary to isolate acylanthranil **36**. Instead the reaction mixture could be treated directly with aniline to give **37** (R = Ph) in 90% yield. Alternatively, it was found that treatment of *N*-acylanthranilic acid **27** (R = Me, Ph) with triphenyl phosphite and aniline in pyridine furnished **37** in 84% yield (R = Me) and 90% yield (R = Ph). Experimental evidence suggests that formation of **37** from acylanthranil **36** occurs via formation of an amidine as a result of attack of the aniline nitrogen on the C2 carbon of **36** followed by cyclization.[28]

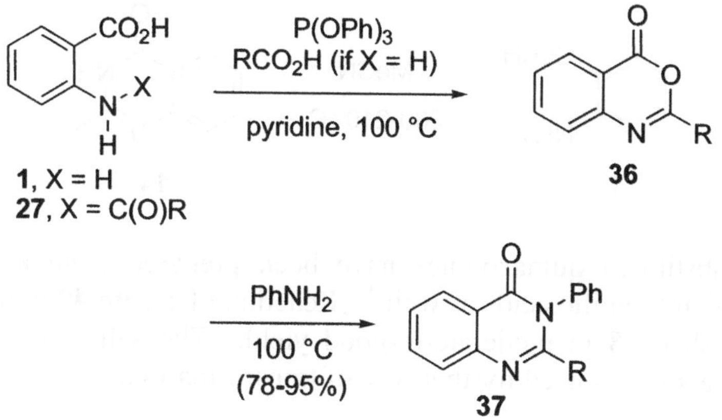

Reaction of Isatoic Anhydrides with Amines

In an extension of the Niementowski synthesis, it was found that 4-quinazolones could be also be synthesized via an isatoic anhydride intermediate.[6,12] Isatoic anhydride has been prepared from anthranilic acid using either phosgene or ethyl chlorocarbonate.[12] Clark and Wagner reported that when isotoic anhydride **38** was heated with a primary amine and triethyl orthoformate, **39** was formed. The reaction likely proceeds by way of initial attack of the amine on the carbonyl group at the 4-position of **38** to form the corresponding anthranilamide followed by ring closure with triethyl orthoformate. If triethyl orthoformate is not added to the reaction mixture, the anthranilamide that is formed can be isolated and used as an intermediate for the synthesis of quinazolones by various methods.[1b]

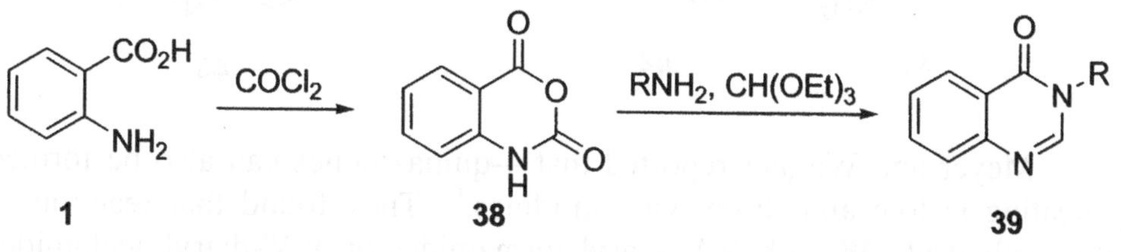

Reaction of Anthranilic Acids or Isatoic Anhydrides with Nitriles, Imidates, Thioamides, or Amidines

In 1900 Bogert and Gotthelf synthesized 2-methyl-4-quinazolone by heating anthranilic acid (**1**) and acetonitrile in a sealed tube at 200–210 °C for 6 h.[5] They were able to confirm the identity of the product by comparison of the analytical data with the product produced by the original method published by Niementowski.

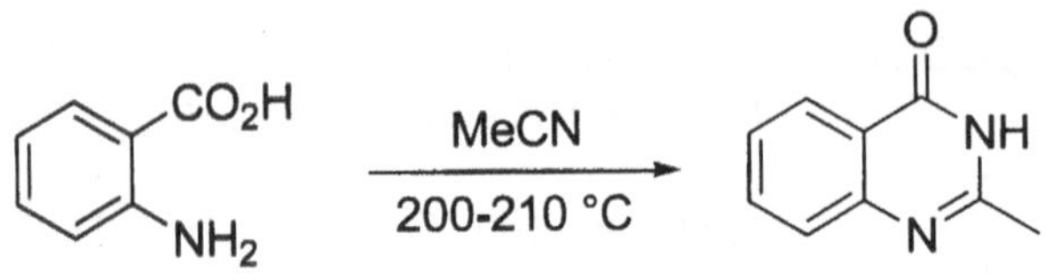

1 14

2-Substituted quinazolones have been prepared from a variety of imidates and anthranilic acids as well.[13] Reaction of **1** with **40** in methanol at 80 °C furnished **28** in moderate-to-good yield. The salt form of a cyclic imidate has also been used for this type of transformation.[14]

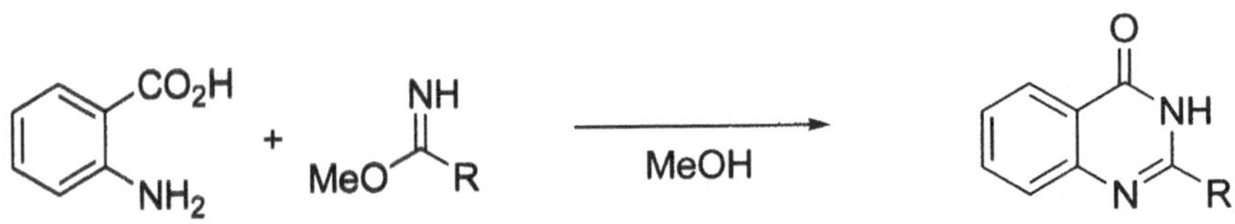

1 40 28

Thioamides have also been used in place of amides for the synthesis of 4-quinazolones. For example, heating 2-amino-5-methylbenzoic acid (**41**) with thioacetamide (**42**) at 135–150 °C for 2 h yielded **43** as the product in 73% yield.[29]

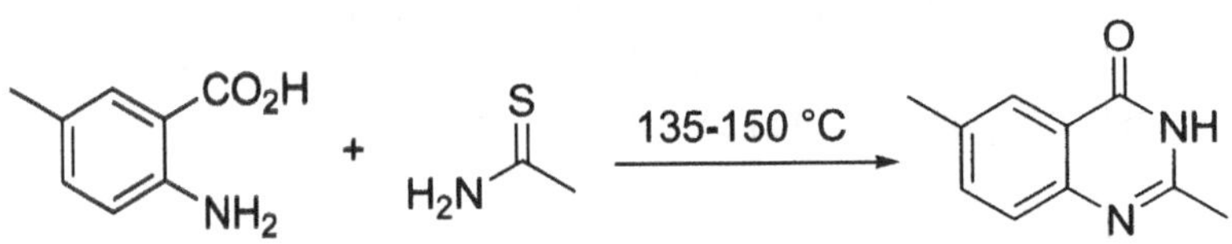

41 42 43

Meyer and Wagner reported that 4-quinazolones can also be formed by heating isatoic anhydride with amidines.[6] They found that reaction of isatoic anhydride **38** with *N,N′*-diaryl formamides or *N,N′*-diaryl acetamides **44** proceeded smoothly at moderate temperatures (120–140 °C) to give **45** along with the evolution of carbon dioxide.

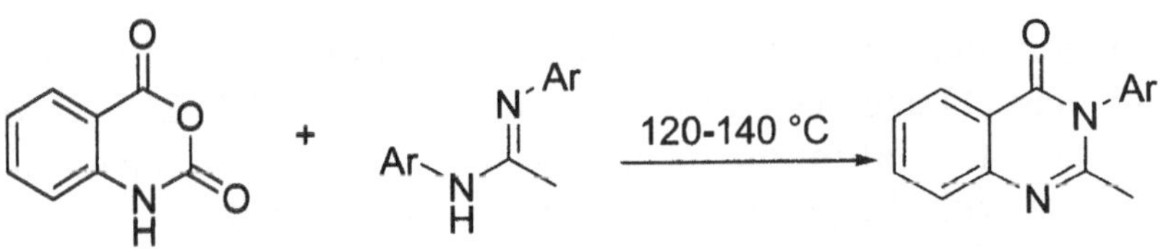

38 44 45

Reaction of Anthranilic Acids with Lactams

An extension of the Niementowski reaction that has been used for the condensation of anthranilic acids with lactams has been reported by Sastry et al.[30] Addition of phosphorous oxychloride at room temperature to a solution of lactam **46** was followed by addition of anthranilic acid (**1**). Subsequent stirring at room temperature followed by heating at reflux furnished products **48** in 60–90% yield. The reaction presumably proceeds via formation of **47** followed by condensation with anthranilic acid (**1**) and cyclodehydration to give **48**. Additional examples of phosphorous oxychloride mediated 4-quinazolone syntheses with either anthranilic acid[31] or the corresponding methyl ester[32] have been reported. A thionyl chloride mediated one-pot reaction of anthranilic acids with lactams to give 4-quinazolone derivatives has also been reported, although a different mechanism for the reaction has been proposed.[33]

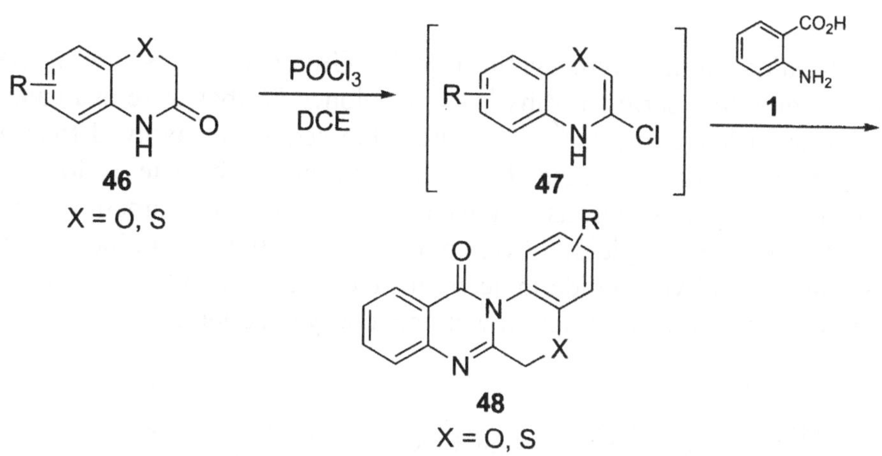

8.4.5 *Synthetic Utility*

The quinazolone moiety is a common subunit of numerous biologically active agents. The biological activity of both naturally occurring and synthetic analogs span a variety of therapeutic areas, including antimalarial, anticonvulsant, antibacterial, antidiabetic, and anticancer applications. In addition, the quinazolone moiety is a building block for approximately 150 naturally occurring alkaloids.

The Niementowski reaction and variations thereof have found widespread use in the pharmaceutical and agrochemical fields for the synthesis of compounds containing 4-quinazolones. This reaction was used by Xu *et al.* in the first step of the synthesis of a series of fluoroquinoline derivatives (**51**) that were evaluated for their antifungal activities.[34] The

Niementowski reaction was also used in the first step of the synthesis of a series of 4-aminoquinazoline derivatives (**54**), which were evaluated for their anti-proliferative properties against several types of tumor cells.[35]

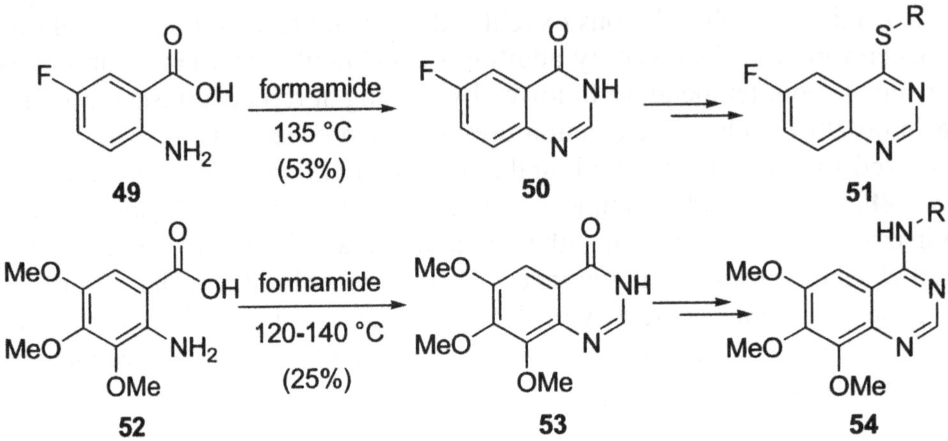

4-Quinazolones were synthesized by Grosso et al. as intermediates along the route to a series of dihydroquinazolones **56** that were evaluated as potential dopamine agonists.[36] The three-step sequence involved treatment of substituted anthranilic acid **18** with ethyl chloroformate followed by treatment with phosphorous tribromide to give 4,5-dimethoxyisatoic anhydride **55** in 83% yield. Compound **55** was then combined with *S*-methylisothiourea hydroiodide in acetonitrile or dioxane and heated at reflux to give **56**. Yields for this reaction were generally excellent.

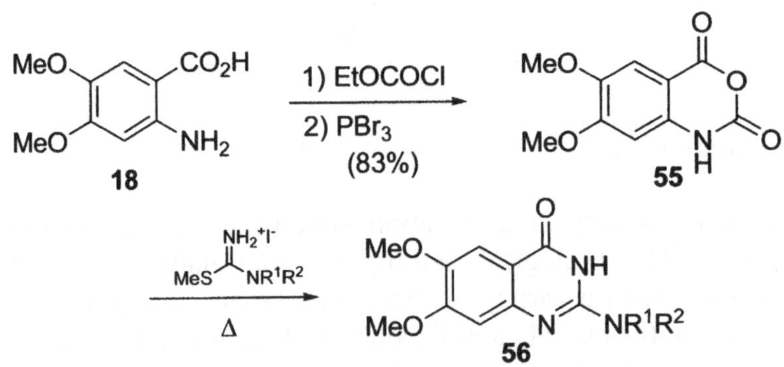

A similar three-step procedure was used by Ban *et al.* in their synthesis of **58** from **57** on the route to 4-anilinoquinazolones, which were evaluated as EGFR tyrosine kinase inhibitors.[37] The overall yield was reported to be 70% for this three-step transformation.

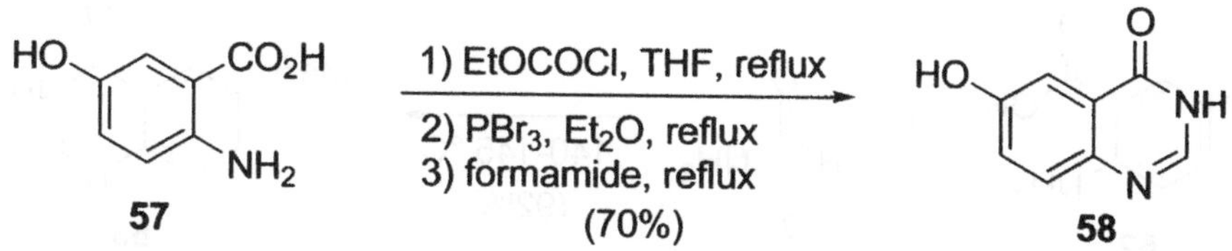

A one-pot, two-step microwave assisted method to prepare 2,3-disubstituted 4-quinazolones was developed by Lui *et al.*[38] and was subsequently used for the synthesis of several natural products.[39] By this method, heating **59** and either a carboxylic acid or an acid chloride in the presence of triphenyl phosphite in pyridine followed by addition of an amine produced **61** via acylanthranil **60** in generally good yield. The reaction produced a lower yield of **61** when the reaction mixture was heated at 120 °C under conventional heating conditions, providing < 50% conversion of cyclized product **61** along with multiple side products.

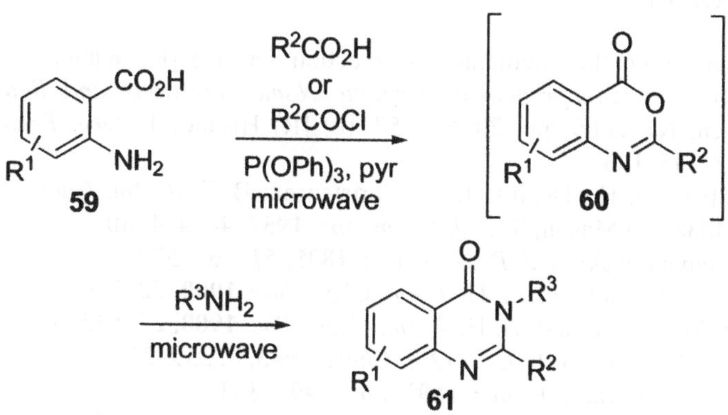

8.4.6 *Experimental*

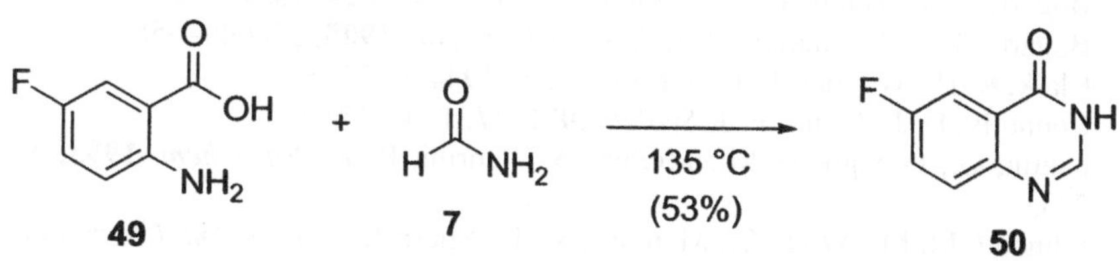

6-Fluoroquinazolin-4(3*H*)-one (50)[34]

A mixture of formamide (**7**, 5.4 g, 120 mmol) and 2-amino-5-fluorobenzoic acid (**49**, 2.4 g, 15 mmol) was heated with stirring at 135 °C for 7 h. The mixture was then poured into ice water (10 mL). The crude product was collected by filtration and recrystallized from ethanol to afford **50** (2.46 g, 53% yield): mp 258–261 °C.

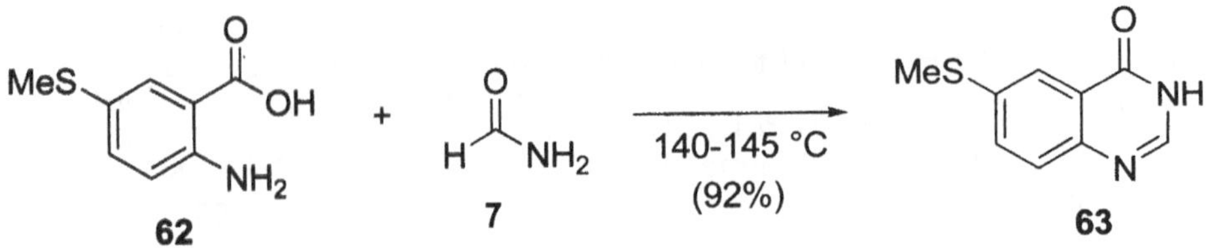

6-(Methylthio)quinazolin-4(3*H*)-one (63)[40]

A mixture of 2-amino-5-(methylthio)benzoic acid (**62**, 55.0 g, 0.30 mol) and formamide (**7**, 48 mL, 1.2 mol) was stirred mechanically and heated at 140–145 °C under an argon atmosphere. During this time the reaction mixture became a solid mass that was difficult to stir. After heating for 4 h, the reaction mixture was cooled, and water (1 L) was added. The solid was collected by filtration and was recrystallized from methanol to give **63** (53.2 g, 92% yield): mp 202–203 °C.

8.4.7 References

1. For reviews on the Niementowski reaction see: (a) [R] Wang, Z. *Niementowski Reaction.* In *Comprehensive Organic Name Reactions and Reagents*, Wiley: Hoboken, NJ, 2009, pp. 2054–2057. (b) [R] Hisano, T. *Org. Prep. Proced. Int.* **1973,** *5*, 145–193.
2. (a) Culbertson, H.; Decius, J. C.; Christensen, B. E. *J. Am. Chem. Soc.* **1952,** *74*, 4834–4838. (b) Mason, S. F. *J. Chem. Soc.* **1957,** 4874–4880.
3. Von Niementowski, S. *J. Prakt. Chem.* **1895,** *51*, 564–572.
4. Bogart, M. T.; Gotthelf, A. H. *J. Am. Chem. Soc.* **1900,** *22*, 129–132.
5. Bogart, M. T.; Gotthelf, A. H. *J. Am. Chem. Soc.* **1900,** *22*, 522–535.
6. Meyer, J. F.; Wagner, E. C. *J. Org. Chem.* **1943,** *8*, 239–252.
7. Bischler, A.; Burkart, E. *Ber.* **1893,** *26*, 1349–1353.
8. Patel, V. S.; Patel, S. R. *J. Indian Chem. Soc.* **1965,** *42*, 531–535.
9. (a) Weddige, J. *J. Prakt. Chem.* **1885,** *31*, 124. (b) Knape, E. J. *J. Prakt. Chem.* **1891,** *43*, 209.
10. Bogart, M. T.; Hand, W. F. *J. Am. Chem. Soc.* **1902,** *24*, 1031–1050.
11. Bogart, M. T.; Chambers, V. J. *J. Am. Chem. Soc.* **1905,** *27*, 649–658.
12. Clark, R. H.; Wagner, E. C. *J. Org. Chem.* **1944,** *9*, 55–67.
13. Connolly, D. J.; Guiry, P. J. *Synlett* **2001,** *11*, 1707–1710.
14. Fantin, G.; Fogagnolo, M.; Medici, A.; Penrini, P. *J. Org. Chem.* **1993,** *58*, 741–743.
15. Edincott, M. M.; Wick, E.; Mercury, M. L.; Sherrill, M. L. *J. Am. Chem. Soc.* **1946,** *68*, 1299–1301.
16. For example see: Cao, S.-L.; Feng, Y.-P.; Jiang, Y.-Y.; Liu, S.-Y.; Ding, G.-Y.; Li, R.-T. *Bioorg. Med. Chem. Lett.* **2005,** *15*, 1915–1917.
17. Őrfi, L.; Wáczek, F.; Pató, J.; Varga, I.; Hegymegi-Barakonyi, B.; Houghten, R. A.; Kéri, G. *Curr. Med. Chem.* **2004,** *11*, 2549–2553.
18. (a) Alexandre, F.-R.; Berecibar, A.; Wrigglesworth, R.; Besson, T. *Tetrahedron* **2003,** *59*, 1413–1419. (b) Yang, Y.-L.; Chang, F.-R.; Wu, Y.-C. *Tetrahedron Lett.* **2003,** *44*, 319–322. (c) Alexandre, F.-R.; Berecibar, A.; Besson, T. *Tetrahedron Lett.* **2002,** *43*, 3911–3913. (d) Domon, L.; Le Coeur, C.; Grelard, A.; Thiéry, V.; Besson, T. *Tetrahedron Lett.* **2001,** *42*, 6671–6674. (e) Desai, A. R.; Desai, K. R.

ARKIVOC **2005,** (*xiii*), 98–108. (f) Li, F.; Feng, Y.; Meng, Q.; Li, W.; Li, Z.; Wang, Q.; Tao, F. *ARKIVOC* **2007,** (*i*) 40–50. (g) Laddha, S. S.; Bhatnagar, S. P.; *ARKIVOC* **2008,** (*xvii*), 212–220. (h) Kostakis, I. K.; Elomri, A.; Seguin, E.; Iannelli, M.; Besson, T. *Tetrahedron Lett.* **2007,** *48*, 6609–6613.

19. Barker, A. J.; Gibson, K. H.; Grundy, W.; Godfrey, A. A.; Barlow, J. J.; Healy, M. P.; Woodburn, J. R.; Ashton, S. E.; Curry, B. J.; Scarlett, L.; Henthorn, L.; Richards, L. *Bioorg. Med. Chem. Lett.* **2001,** *11*, 1911–1914.
20. Gazit, A.; Chen, J.; App, H.; McMahon, G.; Hirth, P.; Chen, I.; Levitzki, A. *Bioorg. Med. Chem.* **1996,** *4,* 1203–1207.
21. Kundu, A. K.; Mahindaratne, M. P. D.; Quintero, M. V.; Bao, A.; Negrete, G. R. *ARKIVOC* **2008,** (*ii*), 33–42.
22. Lee, K.; Kim, J.; Jeong, K.-W.; Lee, K. W.; Lee, Y.; Song, J. Y.; Kim, M. S.; Lee, G. S.; Kim, Y. *Bioorg. Med. Chem.* **2009,** *17*, 3152–3161.
23. For an additional example, see Lippa, B.; Kauffman, G. S.; Arcari, J.; Kwan, T.; Chen, J.; Hungerford, W.; Bhattacharya, S.; Zhao, X.; Williams, C.; Xiao, J.; Pustilnik, L.; Su, C.; Moyer, J. D.; Ma, L.; Campbell, M.; Steyn, S. *Bioorg. Med. Chem. Lett.* **2007,** *17*, 3081–3086.
24. Giri, R. S.; Thaker, H. M.; Giordano, T.; Williams, J.; Rogers, D.; Sudersanam, V.; Vasu, K. K. *Eur. J. Med. Chem.* **2009,** *44*, 2184–2189.
25. Brunton, S. A.; Stibbard, J. H. A.; Rubin, L. L.; Kruse, L. I.; Guicherit, O. M.; Boyd, E. A.; Price, S. *J. Med. Chem.* **2008,** *51*, 1108–1110.
26. Jiang, J. B.; Hesson, D. P.; Dusak, B. A.; Dexter, D. L.; Kang, G. L.; Hamel, E. *J. Med. Chem.* **1990,** *33*, 1721–1728.
27. [R] Rabilloud, G.; Sillion, B. *J. Heterocyclic Chem.* **1980,** *17*, 1065–1068.
28. Errede, L. A. *J. Org. Chem.* **1976,** *41*, 1763–1765.
29. Cao, S.-L.; Feng, Y.-P.; Jiang, Y.-Y.; Liu, S.-Y.; Ding, G.-Y.; Li, R.-T. *Bioorg. Med. Chem. Lett.* **2005,** *15*, 1915–1917.
30. Sastry, C. V. R.; Rao, K. S.; Krishnan, V. S. H.; Rastogi, K.; Jain, M. L. *Synthesis* **1988,** 336–337.
31. Tan, J.-H.; Ou, T.-M.; Hou, J.-Q.; Lu, Y.-J.; Huang, S.-L.; Luo, H.-B.; Wu, J.-Y.; Huang, Z.-S.; Wong, K.-Y.; Gu, L.-Q. *J. Med. Chem.* **2009,** *52*, 2825–2835.
32. Pan, L.; Tan, J.-H.; Hou, J.-Q.; Huang, S.-L.; Gu, L.-Q.; Huang, Z.-S. *Bioorg. Med. Chem. Lett.* **2008,** *18*, 3790–3793.
33. (a) Jahng, K. C.; Kim, S. I.; Kim, D. H.; Seo, C. S.; Son, J.-K.; Lee, S. H.; Lee, E. S.; Jahng, Y. *Chem. Pharm. Bull.* **2008,** *56*, 607–609. (b) Chen, Z.; Hu, G.; Li, D.; Chen, J.; Li, Y.; Zhou, H.; Xie, Y. *Bioorg. Med. Chem.* **2009,** *17*, 2351–2359.
34. Xu, G.-F.; Song; B.-A.; Bhadury, P. S.; Yang, S.; Zhang, P.-Q.; Jin, L.-H.; Xue, W.; Hu, D.-Y.; Lu, P. *Bioorg. Med. Chem.* **2007,** *15*, 3768–3774.
35. Liu, G.; Hu, D.-Y.; Jin, L.-H.; Song, B.-A.; Yang, S.; Liu, P.-S.; Bhadury, P. S.; Ma, Y.; Luo, H.; Zhou, X. *Bioorg. Med. Chem.* **2007,** *15*, 6608–6617.
36. Grosso, J. A.; Nichols, D. E.; Kohli, J. D.; Glock, D. *J. Med. Chem.* **1982,** *25*, 703–708.
37. Ban, H. S.; Usui, T.; Nabeyama, W.; Morita, H.; Fukuzawa, K.; Nakamura, H. *Org. Biomol. Chem.* **2009,** *7*, 4415–4427.
38. Liu, J.-F.; Lee, J.; Dalton, A. M.; Bi, G.; Yu, L.; Baldino, C. M.; McElroy, E.; Brown, M. *Terahedron Lett.* **2005,** *46*, 1241–1244.
39. Liu, J.-F.; Ye, P.; Zhang, B.; Bi, G.; Sargent, K.; Yu, L.; Yohannes, D.; Baldino, C. M. *J. Org. Chem.* **2005,** *70*, 6339–6345.
40. LeMahieu, R. A.; Carson, M.; Nason, W. C.; Parrish, D. R.; Welton, A. F.; Baruth, H. W.; Yaremko, B. *J. Med. Chem.* **1983,** *26*, 420–425.

8.5 Pechmann Coumarin Synthesis

Brian A. Lanman

8.5.1 Description

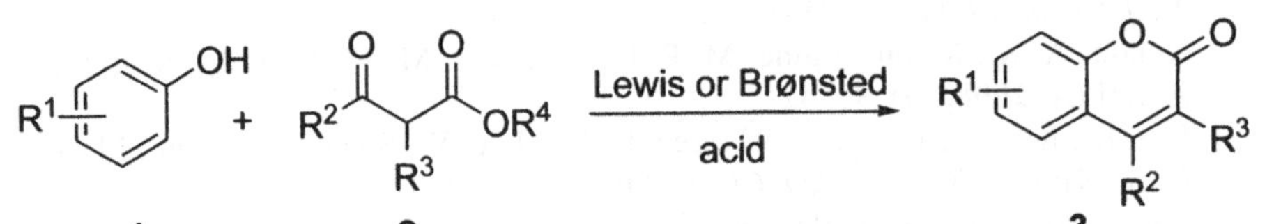

1 2 3

The Pechmann coumarin synthesis, also known as the Pechmann condensation or Pechmann reaction, is the Lewis or Brønsted acid-promoted condensation of phenols (**1**) and β-keto (or other 3-oxo aliphatic) esters (**2**) to produce coumarins (2*H*-chromen-2-ones, **3**).[1–6]

8.5.2 Historical Perspective

The synthesis of 4-methylcoumarin (**3**; R^2 = Me) from the condensation of phenol and ethyl acetoacetate in the presence of concentrated sulfuric acid was first reported in 1883 by Hans von Pechmann (1850–1902) and his student Friedrich Carl Duisberg (1861–1935) from the laboratories of Adolf von Baeyer at the University of Munich.[7] Pechmann later extended this methodology to encompass the preparation of coumarins lacking pyrone ring substituents by employing malic acid in place of the β-keto ester coupling partner (vide infra).[8]

8.5.3 Mechanism

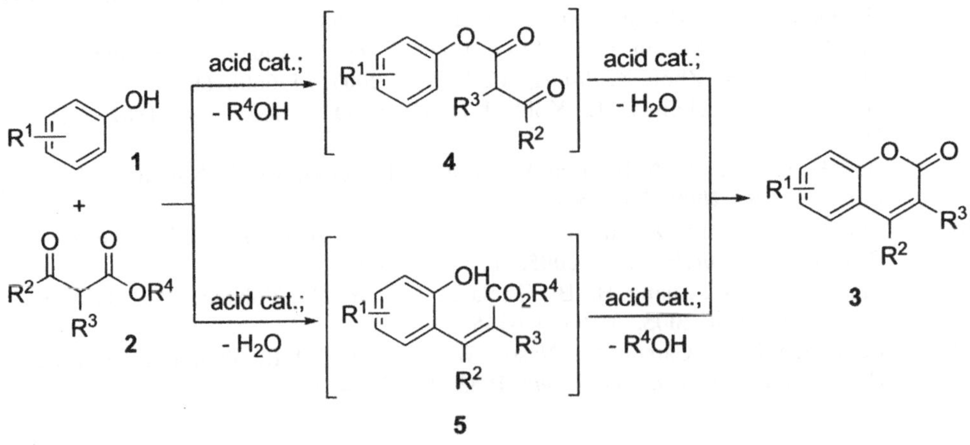

Two mechanisms can be envisioned for the Pechmann condensation: (1) acid-catalyzed transesterification of the β-keto ester followed by acid-catalyzed cyclodehydration of the resulting aryl acetoacetate (**4**), or (2) acid-catalyzed Friedel–Crafts-like phenol alkylation (*ortho* to the phenolic hydroxyl group) followed by intramolecular transesterification/cyclization of the resulting *o*-hydroxycinnamic acid ester (**5**).

In 1954, Lacey demonstrated that aryl acetoacetates **4** prepared from the condensation of a range of phenols with diketene could be cyclized in warm sulfuric acid to the corresponding coumarins (**3**) in yields comparable to those obtained from the corresponding phenols under conventional Pechmann condensation conditions, thereby establishing **4** as a viable intermediate in the Pechmann condensation.[9]

The intermediacy of *o*-hydroxycinnamic acid esters (**5**) in the Pechmann condensation was first proposed by Pechmann in 1884.[8] In 1932, Robertson and co-workers provided support for this proposal through their examination of the reactions of *m*-methylanisole and dimethylresorcinol with ethyl acetoacetate under the conditions of the Pechmann condensation.[10] Allowing these anisoles to stand with ethyl acetoacetate in sulfuric acid resulted in the formation of 4,7-dimethylcoumarin and 7-methoxy-4-methylcoumarin, respectively. 3-(2-Methoxy-4-methylphenyl)-2-butenoic acid could likewise be converted to 4,7-dimethylcoumarin under similar conditions. These observations revealed cinnamate intermediates such as **5** to be viable intermediates in the Pechmann condensation and demonstrated that the formation of aryl acetoacetate intermediates such as **4** is not required for coumarin formation.

In 1913, Simonis reported that the condensation of phenols and ethyl acetoacetate in the presence of phosphorus pentoxide (P_2O_5) produced chromones (4*H*-chromen-4-ones) rather than the corresponding coumarins.[11] This transformation has subsequently become known as the *Simonis reaction*. This result initially suggested that the nature of the acid catalyst could profoundly impact the outcome of the Pechmann condensation. Subsequent research, however, has revealed that regioselectivity in the condensation of phenols and β-keto esters is chiefly controlled by the nature of phenol and β-keto esters coupling partners (vide infra). Dann provided a mechanistic rationale for this observation through his studies of the acid-promoted cyclization of β-phenoxyacrylate esters—putative intermediates in the Simonis reaction.[12] Dann observed that the exposure of several β-phenoxyacrylate esters to sulfuric acid or P_2O_5 resulted in the formation of coumarins in yields similar to those obtained from the condensation of the corresponding phenolic fragments with ethyl acetoacetate in sulfuric acid. Thus the formation of β-phenoxyacrylate esters during the course of the Pechmann condensation does not impair coumarin formation; the regiochemical outcome of the acid-promoted condensation of phenols and β-keto esters is therefore principally controlled by the nature of the phenolic and β-keto ester coupling partners rather than by the nature of the acid catalyst.

8.5.4 Variations and Improvements

Due to the forcing conditions of the classical Pechmann reaction—superstoichiometric sulfuric acid, extended reaction times, and, on many occasions, elevated temperatures—a range of alternative acidic promoters have been investigated in efforts to expand the scope of this reaction, simplify product isolation, reduce acidic waste streams, and reduce reaction times. Table 1 provides an overview of many of the commercially available acidic promoters that have been employed in the Pechmann condensation as of mid-2010. Soluble acidic promoters have been grouped to indicate whether the acid catalyst was employed in superstoichiometric quantities ($\geq$ 1 equiv relative to β-keto ester) or substoichiometric quantities ($<$ 1 equiv relative to β-keto ester). As seen in the Table, a range of solid-supported acidic catalysts has also been successfully employed in the Pechmann reaction. Within each column, promoters are listed in chronological order by the date of their first reported use.

The majority of the promoters listed in Table 1 demonstrate little fundamental difference in their catalysis of the Pechmann condensation, the chief distinction being that more strongly acidic agents generally provide coumarin products in shorter reaction times and promote the condensation of less reactive phenols and β-keto esters. Agents exhibiting unique characteristics are discussed below.

As previously noted, the production of chromones rather than coumarins in the P_2O_5-promoted condensation of phenols and β-keto esters (Simonis reaction) was initially believed to indicate that the nature of the acid catalyst strongly influenced the regiochemical outcome of the Pechmann condensation. Subsequent research revealed that regioselectivity in the condensation of phenols and β-keto esters is chiefly controlled by the nature of phenol and β-keto esters coupling partners. Phenols that afford coumarins when treated with β-keto esters in sulfuric acid similarly afford coumarins when treated with β-keto esters and P_2O_5. In the case of phenol, which are poor substrates for the sulfuric acid-catalyzed Pechmann reaction, employing P_2O_5 as the promoter results in chromone formation.[26] Likewise, α-substituted β-keto esters, which retard traditional Pechmann condensations with less-reactive phenols, afford chromones when condensed with such phenols in the presence of P_2O_5.[10,54]

Table 1: Alternative Acidic Promoters

Superstoichiometric	Substoichiometric	Heterogeneous Catalyst
$ZnCl_2$[13]	TsOH (5%[a])[14]	Amberlite® IR120[15,b]
HCl[16,17]	$InCl_3$ (10%[a])[18]	Zeokarb 225[15,b]
$POCl_3$[19,20]	$Yb(OTf)_3$ (5%[a])[21]	Nafion®-H[22,c]
P_2O_5[23–26]	$ZrCl_4$ (2–10%[a])[27,28]	Amberlyst®-15[29,30,b]
H_3PO_4[31]	H_2NSO_3H (50%[a])[32]	Zeolite H-beta[29,30]
B_2O_3[31]	$TiCl_4$ (50%[a])[33]	Montmorillonite K-10 or KSF[34]
$AlCl_3$[35,36]	$BiCl_3$ (5%[a])[37,38]	Sulfated zirconia[39,40]
$FeCl_3$[41]	AgOTf (10%[a])[42]	
$TiCl_4$[41]	I_2 (5%[a])[42]	
$SnCl_4$[41]	$ZrOCl_2{\cdot}8H_2O$ (1%[a])[43]	
CH_3SO_3H[44]	LiBr (10%[a])[45]	
HF[46]	$Sc(OTf)_3$ (10%[a])[47]	
CF_3CO_2H[48]	$SnCl_2{\cdot}2H_2O$ (10%[a])[49]	
CF_3SO_3H[50]	$CuCl_2$ or $CuBr_2$ (10%[a])[51]	
$BF_3{\cdot}OEt_2$[52]		
$HClO_4$[53]		

[a] Mol% catalyst employed. [b] Sulfonated polystyrene resin. [c] Sulfonated tetrafluoroethylene-based co-polymer.

Phosphorus oxychloride ($POCl_3$) has proven useful as a promoter in a range of Pechmann condensations employing sterically encumbered β-keto ester coupling partners and electronically deactivated phenols, reactions which fail using sulfuric acid as a promoter. Thus $POCl_3$ has been found to successfully promote the condensation of α-benzylacetoacetate and 1-naphthol[19] as well as the reaction of ethyl acetoacetate with 4-acyl-resorcinols.[55,56]

Aluminum trichloride exhibits several unique characteristics as a promoter of the Pechmann condensation. Not only does $AlCl_3$ lead to significantly enhanced yields in condensations that proceed either poorly or not at all in sulfuric acid (*e.g.*, the synthesis of 4-methylcoumarin from phenol and ethyl acetoacetate[35]) but $AlCl_3$ also exerts an unusual influence on the regiochemical outcome of the condensation of resorcinol derivatives and β-keto esters. Whereas other condensing agents favor the formation of

7-hydroxycoumarin products (*cf.* **8**), $AlCl_3$ leads predominantly to the regioisomeric 5-hydroxycoumarin derivatives (*cf.* **9**).[36]

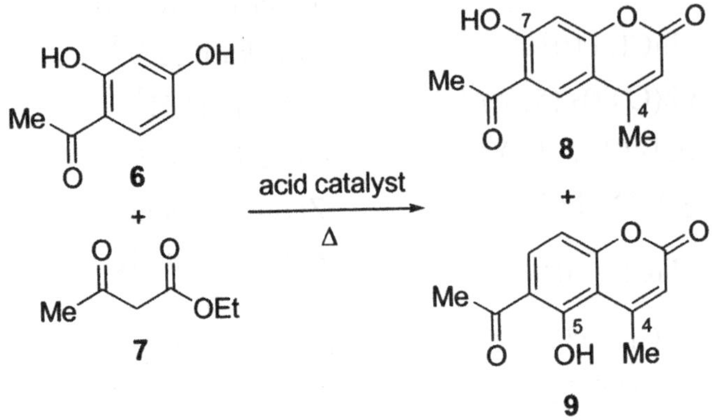

Acid	Product
H_2SO_4[36]	no reaction
$POCl_3$[54]	**8**
$AlCl_3$[36]	only **9**

Base-promoted catalysis of the Pechmann condensation has also briefly been examined. Sodium acetate and sodium ethoxide both promote the Pechmann condensation of resorcinol and ethyl acetoacetate in good yield when employed in superstoichiometric quantities.[31]

In 1960, Israelstam reported that reduced quantities of sulfuric and phosphoric acid could be successfully employed in the Pechmann condensation of electron-rich phenols.[57] A wide range of substoichiometric catalysts for the Pechmann condensation have subsequently been investigated (see Table 1). As with superstoichiometric acid promoters, differences among these catalysts can largely be traced to differences in their Lewis or Brønsted acidities, with stronger acids resulting in reduced reaction times and lower catalyst loadings. Thus $TiCl_4$ (50 mol%) brings about the high-yielding Pechmann condensation of reactive phenols in minutes at ambient temperature,[33] and $ZrCl_4$-catalyzed reactions are complete in less than a hour at 80 °C, even at 2 mol% catalyst loadings.[27] A wide range of weak Lewis acid catalysts have also been successfully employed in the Pechmann condensation of electron-rich phenols, including lithium bromide,[45] silver trifluoromethanesulfonate,[42] ytterbium triflate,[21] and bismuth(III) chloride.[37] The reactivity of these mild Lewis acid catalysts appears to be comparable, with 5–10 mol% catalyst generally bringing about complete starting material conversion after a few hours at 60–85 °C.

Nonconventional heating methods, including microwave heating[58,59] and ultrasonication,[38] have also been successfully employed in accelerating the Pechmann condensation.

In an alternative approach toward reducing the acidic waste stream generated from the Pechmann condensation, a range of heterogeneous catalysts have been investigated (see Table 1). Resin sulfonic acids,[15,22,29,30] acidic zeolites,[29,30] acidic clays,[34] and acid-treated metal oxides[39,40] have all been successfully employed, affording coumarinic products in comparable yields to those obtained with traditional soluble acid catalysts, albeit frequently only at higher reaction temperatures and longer reaction times. The use of heterogeneous acid catalysts additionally simplifies product isolation, as the catalyst can easily be removed from the coumarin product by postreaction filtration. Microwave heating methods have also been employed in conjunction with heterogeneous acid catalysts in efforts to further expedite the Pechmann condensation.[60,61]

Lewis acid-containing ionic liquids such as 1-butyl-3-methylimidazolium (bmim) chloroaluminate,[62] butylpyridinium chloro-aluminate,[63] and $FeCl_3$-doped 1-butyl-3-methylimidazolium triflimide[64] have likewise proven effective promoters for the Pechmann condensation. [Bmim]Cl•$2AlCl_3$ has proven to be one of the most reactive ionic liquid promoters, effecting the Pechmann condensation of ethyl acetoacetate and a range of electron-rich phenols in 10–45 min at 30 °C.[62]

The highly acidic conditions traditionally employed in the Pechmann condensation are incompatible with a wide range of phenolic heterocycles and limit the scope of this methodology. In 1969, Kappe reported that the condensation of resorcinol and ethyl 3-aminocrotonate (**11**, R^4 = Et) could be accomplished by heating these reactants at 180 °C for 30 min in the absence of an acid catalyst, providing 7-hydroxy-4-methylcoumarin in 85% yield.[65] Kappe subsequently employed this methodology in an extension of the Pechmann reaction to the preparation of a range of pyrone-containing heteroaromatics that had proven difficult to access using ethyl acetoacetate and sulfuric acid.[66,67]

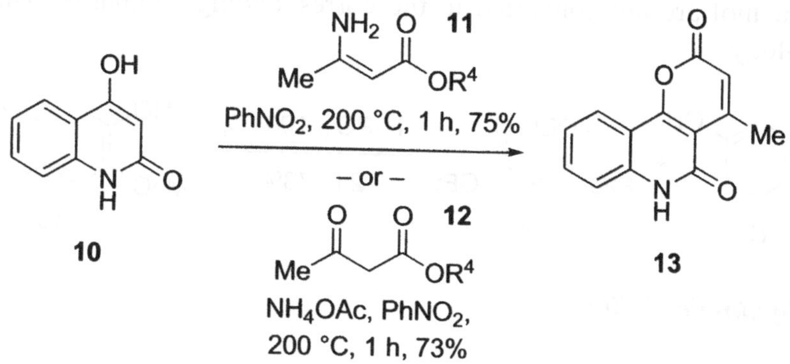

10 **13**

Kappe later determined that the 3-aminoacrylate species (**11**) need not be independently prepared before the reaction, but could be formed under the reaction conditions by the condensation of a β-keto-ester with ammonium acetate.[68] In cases where yields from the two procedures could be directly compared, yields in reactions where the 3-aminoacylate species was formed *in situ* were generally only slightly reduced (~ 20%) relative to those obtained using preformed **11**.

An additional limitation of the traditional Pechmann condensation is its restriction to the preparation of 4-substituted coumarins. Several approaches have subsequently been devised to access the unstable 3-oxopropionate substrates required for the preparation of coumarins lacking 4-substituents. In 1884, Pechmann reported a modification of his coumarin synthesis employing malic acid (**15**) in place of the previously used β-keto-esters. In the presence of concentrated sulfuric acid, **15** extrudes water and carbon monoxide, generating 3-oxopropionic acid, which reacts analogously to previously employed β-keto-esters, affording coumarinic products lacking substituents at the 3- or 4-positions (*cf.*, **16**).[8] Microwave heating has been reported to accelerate this reaction, allowing it to be conducted at reduced sulfuric acid concentrations.[69,70]

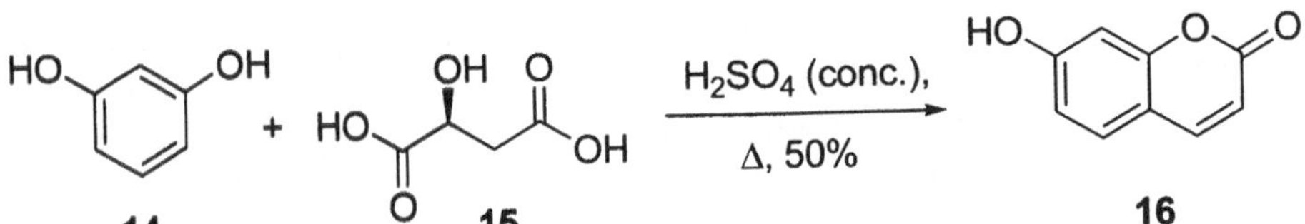

14 **15** **16**

Crosby[71] subsequently reported an alternative approach toward the synthesis of 3,4-unsubstituted pyrones employing ethyl 3,3-diethoxy-propionate (**18**) in place of malic acid. This procedure allowed ready access to the phytochemical scopoletin (**19**). This method is highly sensitive to the nature of the phenolic coupling partner, however; resorcinol and phloroglucinol are not converted to the corresponding coumarins using this methodology.

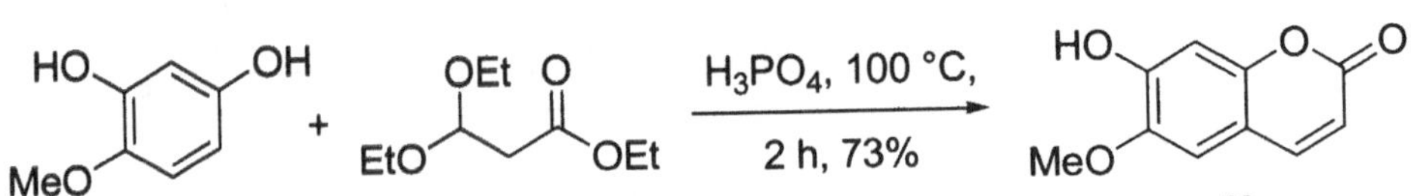

17 **18** **19**

8.5.5 *Synthetic Utility*

Scope and Limitations

The Pechmann condensation provides a unified synthetic approach toward the preparation of many poly-substituted coumarins. Electron-rich phenols (*e.g.*, resorcinol, pyrogallol, phloroglucinol) are readily converted to the corresponding coumarins. Less electron-rich phenols (including phenol itself) provide reduced yields of coumarin products, and monohydric phenols bearing electron-withdrawing substituents (*e.g.*, cyano, nitro, acyl, carboxyl, *etc.*) do not undergo the Pechmann condensation.[72]

The substitution pattern of the phenolic coupling partner significantly influences its reactivity in the Pechmann condensation. For simple phenols, electron-donating groups at the 3-position greatly facilitate reaction.[73] 2-Substitution (except in the case of strongly electron-donating groups) impedes condensation, as illustrated in the syntheses of the isomeric dimethylcoumarins **21**.[74]

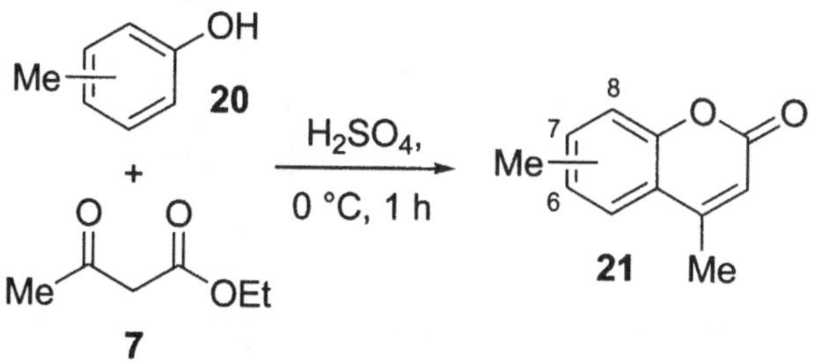

Phenol	Product (yield)
2-Me	8-Me (20%)
3-Me	7-Me (65%)
4-Me	6-Me (60%)

For phenols bearing multiple electron-donating substituents, the impact of electron-withdrawing groups is more complicated. Both 2- and 4-nitroresorcinols condense with ethyl acetoacetate in sulfuric acid, however 4-nitroresorcinol does not afford coumarinic products when condensed with α-alkyl β-keto esters.[75] α-Alkyl substitution of the β-keto ester coupling partner likewise impedes its condensation with 2-nitroresorcinol, however α-ethyl substitution is required to completely suppress condensation.[76] These observations illustrate a broader trend that electron-withdrawing groups exert a greater deactivating influence in the 4-position than in the 2-position of the resorcinol scaffold.[3]

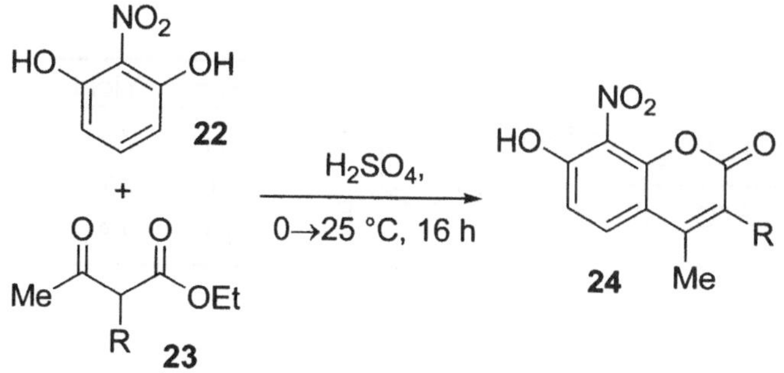

R =	Yield
H	60%
Me	15%
Et	0%

Steric effects likewise affect the regiochemical outcome of Pechmann condensations with poly-substituted phenols. Whereas resorcinol typically condenses with β-keto-esters to provide 7-hydroxycoumarin products (*cf.*

26), 5-alkylresorcinols condense with β-keto esters to afford the corresponding 5-hydroxycoumarin isomers (*cf.* **27**).[3,18,47]

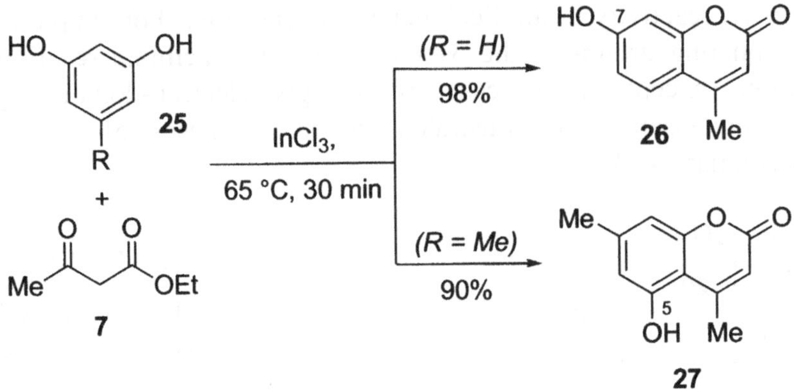

As illustrated in the foregoing transformations, polyhydric phenols do not typically undergo multiple Pechmann condensations in a single reaction; however, resubmission of isolated hydroxycoumarin products to the Pechmann reaction can lead to bis(pyrone) derivatives.

A diverse collection of α-substituted β-keto-esters—including esters wherein the α-and γ-positions are connected via a ring—has been successfully employed in the Pechmann condensation. Whereas reactive phenols may be condensed with a wide variety of α-substituted β-keto-esters, as the reactivity of phenol decreases, sterically demanding α-substituents retard condensation, leading to decreased yields and ultimately inhibiting the reaction altogether (*cf.* synthesis of **29**, R = H and R = Ph).[3,73,77] Limited results also suggest that β-keto ester γ-substituents impede the Pechmann reaction to a greater degree than corresponding α-substituents.[78]

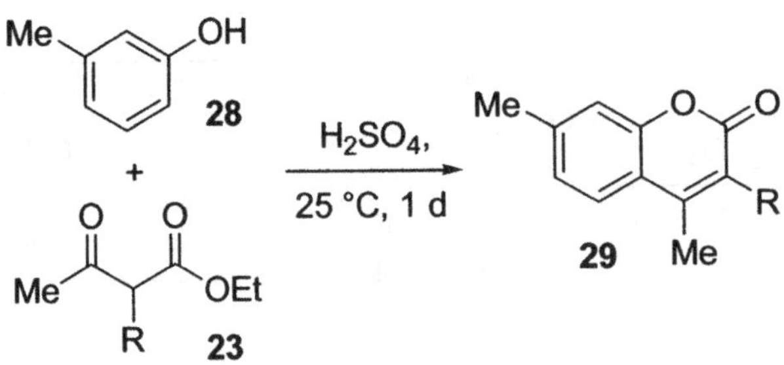

R =	Yield
H	71%
Me	40%
Ph	0%

The reactivity of the β-keto-ester coupling partner toward less electron-rich phenols is enhanced by substitution of the α-carbon with electron-withdrawing groups (*e.g.*, the benzoyl group of benzoylacetoacetate **30**). As first reported by Pechmann and Hanke,[79] α-acyl or α-carboxy substituents may be cleaved under the conditions of the Pechmann reaction, however, resulting in the formation of 3-unsubstituted products. Thus the condensation of resorcinol (**14**) and ethyl α-benzoylacetoacetate (**30**) in sulfuric acid provides 7-hydroxy-4-phenylcoumarin (**31**) rather than the corresponding 3-acetyl-4-phenylcoumarin product.[80] The formation of **31** rather than the corresponding 4-methylcoumarin product suggests that this reaction likely proceeds via the ethyl cinnamate tautomer of **30**.[80]

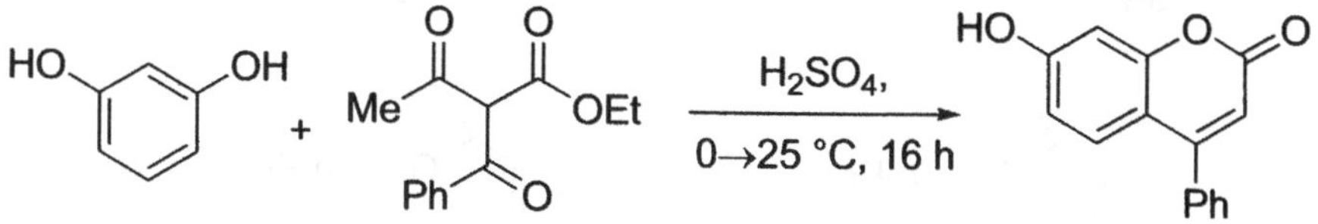

14 **30** **31**

Electron-donating α-substituents are also tolerated in the Pechmann reaction, as demonstrated by Corrie's synthesis of *N*-(7-(dimethylamino)-4-methyl-coumarinyl)acetamide (**34**);[81] such β-keto ester coupling partners have not been the subject of extensive investigation, however.

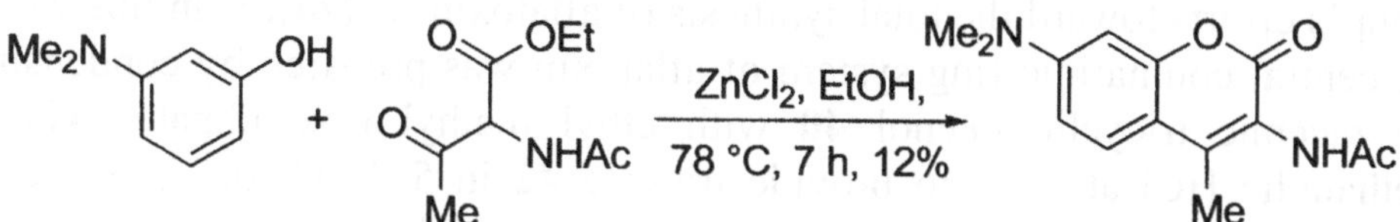

32 **33** **34**

The regiochemical course of Pechmann condensations of poly-functionalized phenolic substrates generally appears to be controlled by the nature of the coupling partners. However, the choice of acidic promoter has been demonstrated to have some influence on product regiochemistry, particularly in the case of aluminum chloride-promoted condensations of resorcinols, as discussed above.[35]

Applications in the Total Synthesis of Natural Products

The Pechmann condensation served as a key ring-forming reaction in Xu and co-workers's synthesis of the HIV-1-selective nonnucleoside reverse transcriptase inhibitor calanolide A (**39**).[82,83] In this work, phloroglucinol (**35**) and ethyl butyrylacetate (**36**) were condensed in sulfuric acid at 90 °C to provide 5,7-dihydroxy-4-propylcoumarin (**37**) in nearly quantitative yield on 150 g scale. Subsequent Friedel–Crafts acylation of **37** followed by condensation of the acylcoumarin product with 4,4-dimethoxy-2-methylbutan-2-ol provided pyranylcoumarin **38**, which was converted to calanolide A (**39**) by condensation with acetaldehyde and subsequent Luche reduction.

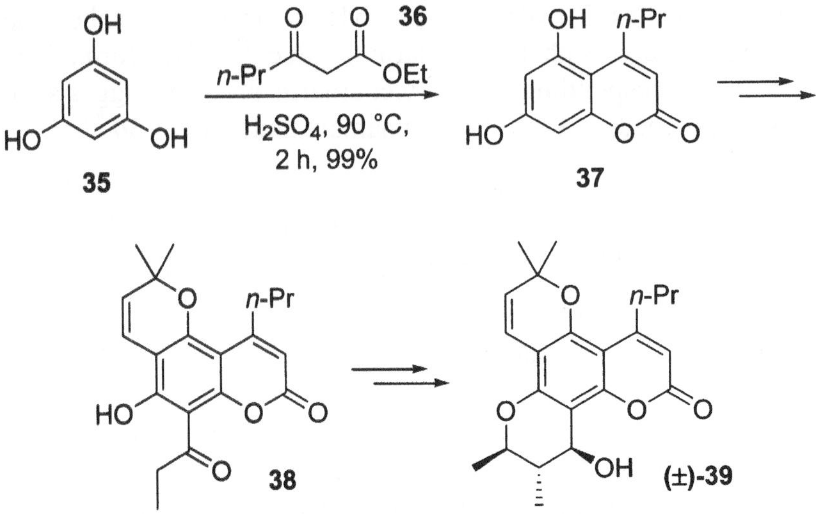

The Pechmann condensation also played a key role in the Büchi group's efforts toward the total synthesis of aflatoxin B_1 (**44**).[84] In this work, the central coumarinic ring system of aflatoxin was prepared by condensing the racemic tricyclic phenol **40** with ethyl methyl-3-oxoadipate (**41**) in methanolic HCl at 5 °C to provide tricycle **42** in 57% yield. The use of methanol as a solvent was crucial to the success of this reaction, as attempts to perform this reaction in sulfuric acid led to the rapid isomerization of **40** to the corresponding benzofuran-3-acetic acid, which could be converted to the analogous coumarin product only in very poor yield. Subsequent cyclization of the pendant ester moieties of **42** afforded pentacyclic lactone **43**. Treatment of **43** with disiamylborane resulted in selective reduction of the dihydrofuranone to the corresponding hemiaminal, which was then acetylated and pyrolyzed (240 °C) to provide (±)-aflatoxin B_1 (**44**).

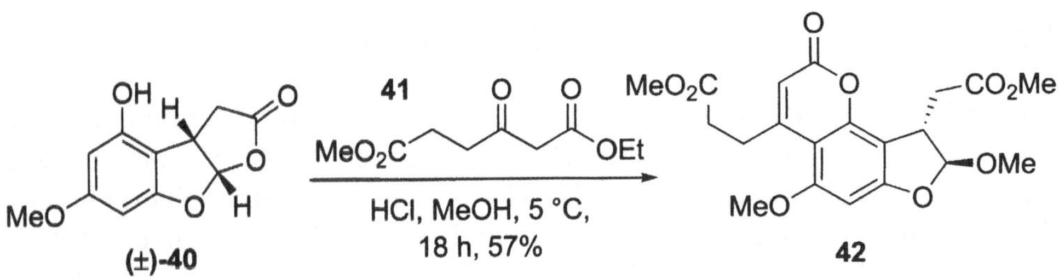

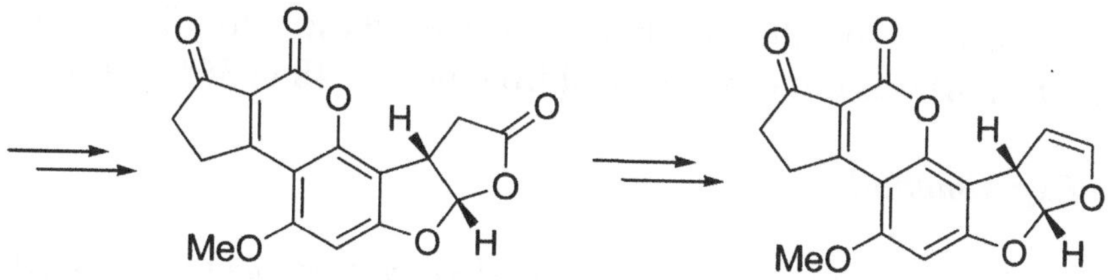

43 (±)-44

In subsequent efforts toward an improved synthesis of the aflatoxins, Büchi and Weinreb reported a late-stage Pechmann condensation between phenol **45** and cyclic β-keto ester **46** using phosphorous pentoxide as a catalyst.[85]

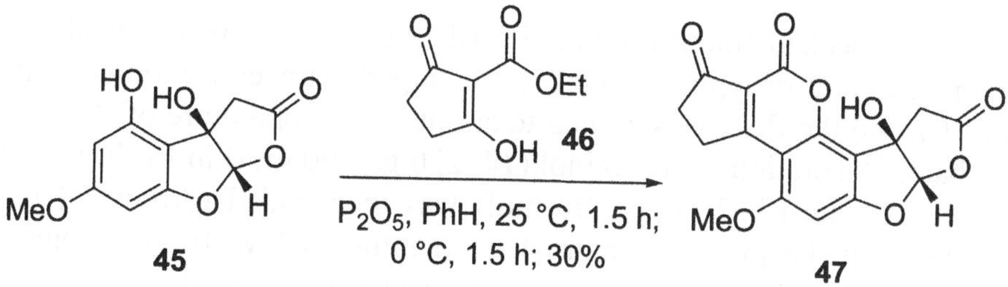

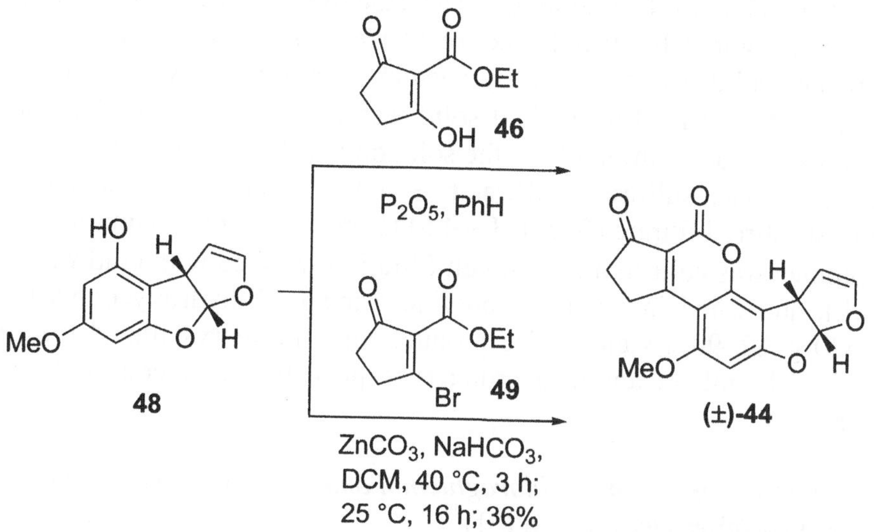

Attempts to apply this methodology to a second-generation synthesis of aflatoxin B_1 (**44**) through the condensation of **46** with the related phenol **48** in the presence of P_2O_5 were unsuccessful, however, due to the marked acid-sensitivity of **48**. This difficulty was surmounted by employing the more highly reactive β-bromoacrylate **49** in a zinc carbonate-promoted annulation (sodium bicarbonate was added as a scavenger for HBr generated during the course of the reaction). Under these modified conditions, aflatoxin B_1 (**44**) was directly obtained from phenol **48** in 36% yield.

8.5.6 Experimental

Pechmann Condensation of Resorcinol and Ethyl Acetoacetate: **4-Methyl-7-hydroxycoumarin** (26)[86]

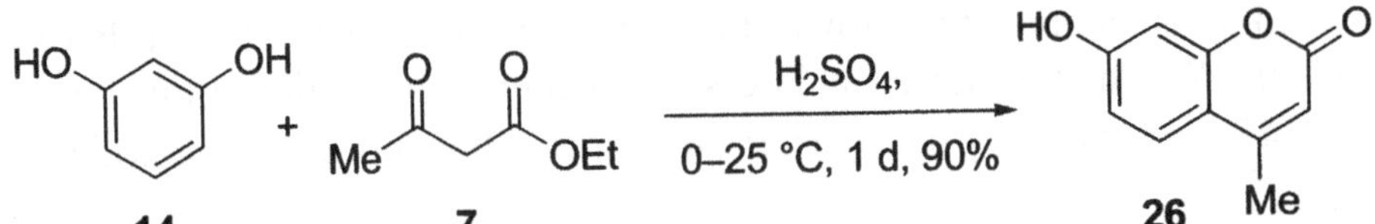

14 **7** **26**

A 5-L, three-necked round-bottomed flask fitted with a mechanical stirrer, thermometer, and dropping funnel was charged with concentrated sulfuric acid (2 L), and the flask was cooled to an internal temperature of < 10 °C on an ice bath. A solution of resorcinol (**14**; 220 g, 2.00 mol) in freshly distilled ethyl acetoacetate (**7**; 260 g, 2.00 mol) was then added (dropwise) to the stirred reaction, keeping the internal temperature below 10 °C. Once the addition was complete (~ 2 h), the reaction mixture was allow to stand for 12–24 h without further cooling. The reaction mixture was subsequently added to a vigorously stirred mixture of ice (4 kg) and water (6 L). The precipitated solid was collected by vacuum filtration and washed with cold water (3 × 50 mL). The washed solid was then dissolved in 5% aqueous sodium hydroxide solution (3 L), the solution was filtered, and dilute sulfuric acid (1:10 conc. sulfuric acid/water; ~ 1.1 L) was slowly added to the vigorously stirred filtrate (final pH ~ 4.5) to precipitate the coumarin product. The product was collected by vacuum filtration, washed with cold water (4 × 50 mL), and dried in vacuo to provide 4-methyl-7-hydroxycoumarin (**26**; 290–320 g, 82–90% yield). The product could be recrystallized from 95% ethanol (~ 15 mL ethanol/5 g product) to provide nearly colorless needles (mp 185 °C).

Pechmann Condensation of Phloroglucinol and Ethyl Benzoylacetate: **5,7-Dihydroxy-4-phenylcoumarin** (51)[47]

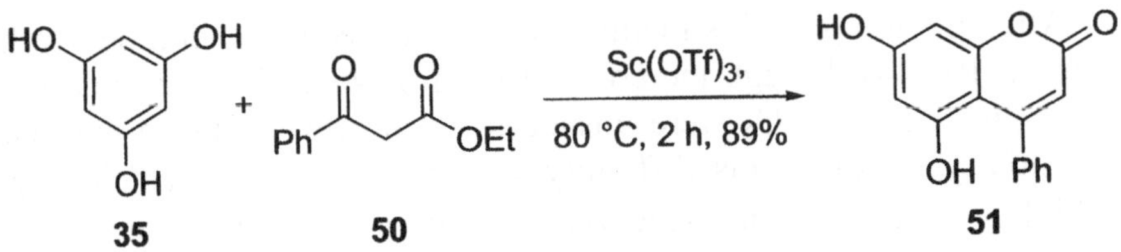

35 **50** **51**

A mixture of phloroglucinol (**35**; 252 mg, 2.00 mmol), ethyl benzoylacetate (**50**; 0.42 mL, 2.4 mmol), and scandium(III) trifluoromethanesulfonate (98 mg, 0.2 mmol) was heated at 80 °C for 2 h. The reaction mixture was then added to water (20 mL), and the precipitated product was collected by vacuum filtration. Chromatographic purification of the collected solid (silica gel, 50% ethyl acetate/hexanes) furnished 5,7-dihydroxy-4-phenylcoumarin (**51**; 452 mg, 89% yield).

8.5.7 References

1. [R] Sethna, S. M.; Shah, N. M. *Chem. Rev.* **1945,** *36*, 1–62.
2. [R] Wawzonek, S. In *Heterocyclic Compounds, Vol. 2: Polycyclic Five- and Six-Membered Compounds Containing One O or S Atom;* Wiley: New York, 1951, pp. 173–216.
3. [R] Sethna, S.; Phadke, R. *Org. React.* **1953,** *7*, 1–58.
4. [R] Livingstone, R. In *Rodd's Chemistry of Carbon Compounds, Vol. IV, Part E*; Elsevier, Amsterdam, 1977, pp. 1–346.
5. [R] Kürti, L.; Czakó, B. *Strategic Applications of Named Reactions in Organic Synthesis*; Elsevier: Boston, 2005, pp. 472–473.
6. [R] Wang, Z. *Comprehensive Organic Name Reactions and Reagents, Volume 2*; Wiley, 2009, pp. 2151–2156.
7. Pechmann, H. v.; Duisberg, C. *Ber. Dtsch. Chem. Ges.* **1883,** *16*, 2119–2128.
8. Pechmann, H. v. *Ber. Dtsch. Chem. Ges.* **1884,** *17*, 929–936.
9. Lacey, R. N. *J. Chem. Soc.* **1954,** 854–860.
10. Robertson, A.; Waters, R. B.; Jones, E. T. *J. Chem. Soc.* **1932,** 1681–1688.
11. Petschek, E.; Simonis, H. *Ber. Dtsch. Chem. Ges.* **1913,** *46*, 2014–2020.
12. Dann, O.; Illing, G. *Liebigs Ann. Chem.* **1957,** *605*, 158–167.
13. Pechmann, H. v. *Ber. Dtsch. Chem. Ges.* **1899,** *32*, 3681–3690.
14. Sugino, T.; Tanaka, K. *Chem. Lett.* **2001,** 110–111.
15. John, E. V. O.; Israelstam, S. S. *J. Org. Chem.* **1961,** *26*, 240–242.
16. Bülow, C.; Siebert, E. *Ber. Dtsch. Chem. Ges.* **1905,** *38*, 474–486.
17. Appel, H. *J. Chem. Soc.* **1935,** 1031–1032.
18. Bose, D. S.; Rudradas, A. P.; Babu, M. H. *Tetrahedron Lett.* **2002,** *43*, 9195–9197.
19. Naik, R. G.; Desai, R. D.; Trivedi, R. K. *J. Indian Chem. Soc.* **1929,** *6*, 801–802.
20. Goodall, I.; Robertson, A. *J. Chem. Soc.* **1936,** 426–428.
21. Wang, L.; Xia, J.; Tian, H.; Qian, C.; Ma. Y. *Indian J. Chem., Sect. B* **2003,** *42*, 2097–2099.
22. Chaudhari, D. D. *Chem. Ind.* **1983,** 568–569.
23. Robertson, A.; Sandrock, W. F.; Hendry, C. B. *J. Chem. Soc.* **1931,** 2426–2432.
24. Chakravarti, D. *J. Indian Chem. Soc.* **1931,** *8*, 129–136.
25. Chakravarti, D. *J. Indian Chem. Soc.* **1931,** *8*, 407–411.
26. Chakravarti, D. *J. Indian Chem. Soc.* **1932,** *9*, 31–35.
27. Smitha, G.; Reddy, C. S. *Synth. Commun.* **2004,** *34*, 3997–4003.
28. Sharma, G. V. M.; Reddy, J. J.; Lakshmi, P. S.; Krishna, P. R. *Tetrahedron Lett.* **2005,** *46*, 6119–6121.
29. Hoefnagel, A. J.; Gunnewegh, E. A.; Downing, R. S.; Bekkum, H. v. *J. Chem. Soc., Chem. Commun.* **1995,** 225–226.
30. Gunnewegh, E. A.; Hoefnagel, A. J.; Downing, R. S.; Bekkum, H. v. *Recl. Trav. Chim. Pays-Bas* **1996,** *115*, 226–230.
31. Chakravarti, D. *J. Indian Chem. Soc.* **1935,** *12*, 536–539.
32. Singh, P. R.; Singh, D. U.; Samant, S. D. *Synlett* **2004,** 1909–1912.
33. Valizadeh, H.; Shockravi, A. *Tetrahedron Lett.* **2005,** *46*, 3501–3503.
34. Li, T.-S.; Zhang, Z.-H.; Yang, F.; Fu, C.-G. *J. Chem. Res., (S)* **1998,** 38–39.
35. Sethna, S. M.; Shah, N. M.; Shah, R. C. *Current Sci.* **1937,** *6*, 93–94.
36. Sethna, S. M.; Shah, N. M.; Shah, R. C. *J. Chem. Soc.* **1938,** 228–232.
37. De, S. K.; Gibbs, R. A. *Synthesis* **2005,** 1231–1233.

38. Patil, S. B.; Bhat, R. P.; Raje, V. P.; Samant, S. D. *Synth. Commun.* **2006,** *36*, 525–531.
39. Rodríguez-Domínguez, J. C.; Kirsch, G. *Tetrahedron Lett.* **2006,** *47*, 3279–3281.
40. Tyagi, B.; Mishra, M. K.; Jasra, R. V. *J. Mol. Catal. A* **2007,** *276*, 47–56.
41. Horii, Z. *Yakugaku Zasshi* **1939,** *59*, 201–203.
42. Wu. J.; Diao, T.; Sun, W.; Li, Y. *Synth. Commun.* **2006,** *36*, 2949–2956.
43. Rodríguez-Domínguez, J. C.; Kirsch, G. *Synthesis* **2006,** 1895–1897.
44. Boekelheide, V.; Pennington, F. C. *J. Am. Chem. Soc.* **1952,** *74*, 1558–1562.
45. Kumar, S.; Saini, A.; Sandhu, J. S. *ARKIVOC* **2007,** (*15*), 18–23.
46. Dann, O.; Mylius, G. *Liebigs Ann. Chem.* **1954,** *587*, 1–15.
47. Jung, K.; Park, Y.-J.; Ryu, J.-S. *Synth. Commun.* **2008,** *38*, 4395–4406.
48. Woods, L. L.; Sapp, J. *J. Org. Chem.* **1962,** *27*, 3703–3705.
49. Upadhyay, K. K.; Mishra, R. K.; Kumar, A. *Catal. Lett.* **2008,** *121*, 118–120.
50. Chenera, B.; West, M. L.; Finkelstein, J. A.; Dreyer, G. B. *J. Org. Chem.* **1993,** *58*, 5605–5606.
51. Wang, Y.; Xu, F.; Tian, Y.-P.; Li, H.-L.; Wang, J.-J. *J. Chem. Res.* **2009,** 339–341.
52. Starkov, S. P.; Goncharenko, G. A.; Panasenko, A. I. *Zh. Obshch. Khim.* **1993,** *63*, 1111–1115.
53. Kolancilar, H.; Oyman, U. *J. Indian Chem. Soc.* **2003,** *80*, 853–857.
54. Robertson, A.; Sandrock, W. F. *J. Chem. Soc.* **1932,** 1180–1184.
55. Desai, R. D.; Hamid, S. A. *Proc. Indian Acad. Sci., Sect. A.* **1937,** *6*, 185–190.
56. Desai, R. D.; Ekhlas, M. *Proc. Indian Acad. Sci., Sect. A* **1938,** *8*, 567–577.
57. Barris, B. E.; Israelstam, S. S. *J. South African Chem. Inst.* **1960,** *13*, 125–128.
58. Singh, V.; Singh, J.; Kaur, K. P.; Kad, G. L. *J. Chem. Res., Synop.* **1997,** 58–59.
59. Manhas, M. S.; Ganguly, S. N.; Mukherjee, S.; Jain, A. K.; Bose, A. K. *Tetrahedron Lett.* **2006,** *47*, 2423–2425.
60. Hoz, A. d. l.; Moreno, A.; Vázquez, E. *Synlett* **1999,** 608–610.
61. Frère, S.; Thiéry, V.; Besson, T. *Tetrahedron Lett.* **2001,** *42*, 2791–2794.
62. Potdar, M. K.; Mohile, S. S.; Salunkhe, M. M. *Tetrahedron Lett.* **2001,** *42*, 9285–9287.
63. Khandekar, A. C.; Khadilkar, B. M. *Synlett* **2002,** 152–154.
64. Kumar, V.; Tomar, S.; Patel, R.; Yousaf, A.; Parmar, V. S.; Malhotra, S. V. *Synth. Commun.* **2008,** *38*, 2646–2654.
65. Kappe, T.; Ziegler, E. *Org. Prep. Proced. Int.* **1969,** *1*, 61–62.
66. Kappe, T.; Baxevanidis, G.; Ziegler, E. *Monatsh. Chem.* **1971,** *102*, 1392–1399.
67. Kappe, T.; Linnau, Y. *Liebigs Ann. Chem.* **1972,** *761*, 25–33.
68. Kappe, T.; Mayer, C. *Synthesis* **1981,** 524–526.
69. Helavi, V. B.; Solabannavar, S. B.; Salunkhe, R. S.; Mane, R. B. *J. Chem. Res., (S)* **2003,** 279–280.
70. Symeonidis, T.; Chamilos, M.; Hadjipavolu-Litina, D. J.; Kallitsakis, M.; Litinas, K. E. *Bioorg. Med. Chem. Lett.* **2009,** *19*, 1139–1142.
71. Crosby, D. G.; Berthold, R. V. *J. Org. Chem.* **1962,** *27*, 3083–3085.
72. Clayton, A. *J. Chem. Soc.* **1908,** *93*, 2016–2022.
73. Fries, K.; Klostermann, W. *Ber. Dtsch. Chem. Ges.* **1906,** *39*, 871–875.
74. Riveiro, M. A.; Moglioni, A.; Vazquez, R.; Gomez, N.; Facorro, G.; Piehl, L.; Celis, E. R. d.; Shayo, C.; Davio, C. *Bioorg. Med. Chem.* **2008,** *16*, 2665–2675.
75. Chakravarti, D.; Banerji, B. C. *J. Indian Chem. Soc.***1937,** *14*, 37–38.
76. Chakravarti, D.; Ghosh, B. *J. Indian Chem. Soc.***1935,** *12*, 622–626.
77. Fries, K.; Klostermann, W. *Liebigs Ann. Chem.* **1908,** *362*, 1–29.
78. Kotwani, N. G.; Sethna, S. M.; Advani, G. D. *Proc. Indian Acad. Sci., Sect. A* **1942,** *15*, 441–444.
79. Pechmann, H. v.; Hanke, E. *Ber. Dtsch. Chem. Ges.* **1901,** *34*, 354–362.
80. Jacobson, S.; Ghosh, B. *J. Chem. Soc.* **1915,** *107*, 1051–1058.
81. Corrie, J. E. T. *J. Chem. Soc., Perkin Trans. 1* **1990,** 2151–2152.
82. Kucherenko, A.; Flavin, M. T.; Boulanger, W. A.; Khilevich, A.; Shone, R. L.; Rizzo, J. D.; Sheinkman, A. K.; Xu, Z.-Q. *Tetrahedron Lett.* **1995,** *36*, 5475–5478.
83. Flavin, M. T.; Rizzo, J. D.; Khilevich, A.; Kucherenko, A.; Sheinkman, A. K.; Vilaychack, V.; Lin, L.; Chen, W.; Greenwood, E. M.; Pengsuparp, T.; Pezzuto, J. M.; Hughes, S. H.;

Flavin, T. M.; Cibulski, M.; Boulanger, W. A.; Shone, R. L.; Xu, Z.-Q. *J. Med. Chem.* **1996,** *39*, 1303–1313.

84. Büchi, G.; Foulkes, D. M.; Kurono, M.; Mitchell, G. F.; Schneider, R. S. *J. Am. Chem. Soc.* **1967,** *89*, 6745–6753.
85. Büchi, G.; Weinreb, S. M. *J. Am. Chem. Soc.* **1971,** *93*, 746–752.
86. Russell, A.; Frye, J. R. *Org. Synth.* **1941,** *21*, 22–27.

8.6 Robinson–Schöpf Condensation

Kyle J. Eastman

8.6.1 Description

The Robinson–Schöpf condensation is a tandem process most notably recognized as the reaction between one molecule of succinaldehyde (**1**), methyl amine (**2**), and acetone dicarboxylic acid **3** to give the bicyclic product tropinone (**4**) in a one-pot procedure.

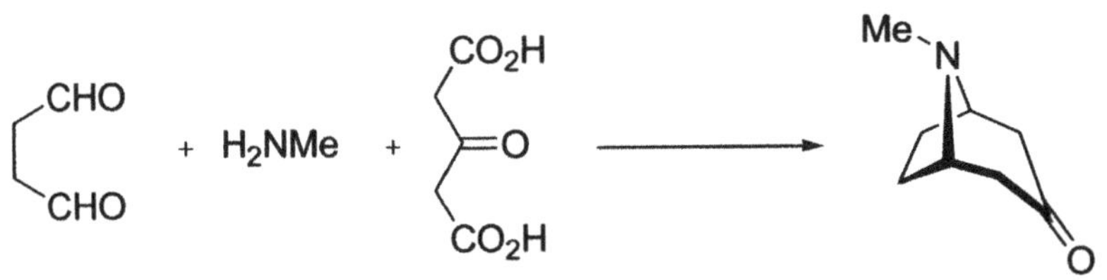

1 2 3 4

8.6.2 Historical Perspective

Sir Robert Robinson developed with striking success the remarkable concept that a fundamental reaction of biogenesis involves the condensation between an amino (or imino) group, a carbonyl group and a nucleophilic atom.[1,2] This concept was first put into practice in the beautiful synthesis of (±)-tropinone (**4**),[3] and was later improved upon by Clemens Schöpf.[4] The Robinson–Schöpf condensation might be considered the quintessential double Mannich condensation involving aldehydes other than formaldehyde.[5] The clever transformation has gained longstanding attention and proven useful in a number of contexts as demonstrated by the references herein.

8.6.3 Mechanism

In general, Robinson–Schöpf condensation is accepted as a double Mannich reaction. The reaction begins with nucleophilic addition of methyl amine (**2**) to succinaldehyde (**1**), which is followed by loss of water to generate imine **5**. Depending on steric constraints (*vida infra*), intramolecular attack of the imine on the proximal aldehyde function **6** effects the first ring closure and forms the reactive iminium **7**. Next, an intermolecular Mannich reaction between iminium **7** and the enolate of acetone dicarboxylate **8** ensues to give amino alcohol **9**, which dehydrates rapidly to give iminium **10** (shown in enol form). The second ring closure is accomplished via an intramolecular

Mannich reaction between the enol and iminium **10** to give bicycle **11**. Finally, action of HCl on bis-carboxylic acid **11** results in double decarboxylation to generate tropinone (**4**). As shown by Paquette,[5] slight variations in this mechanism may be operative depending on steric compression during ring-closing events.

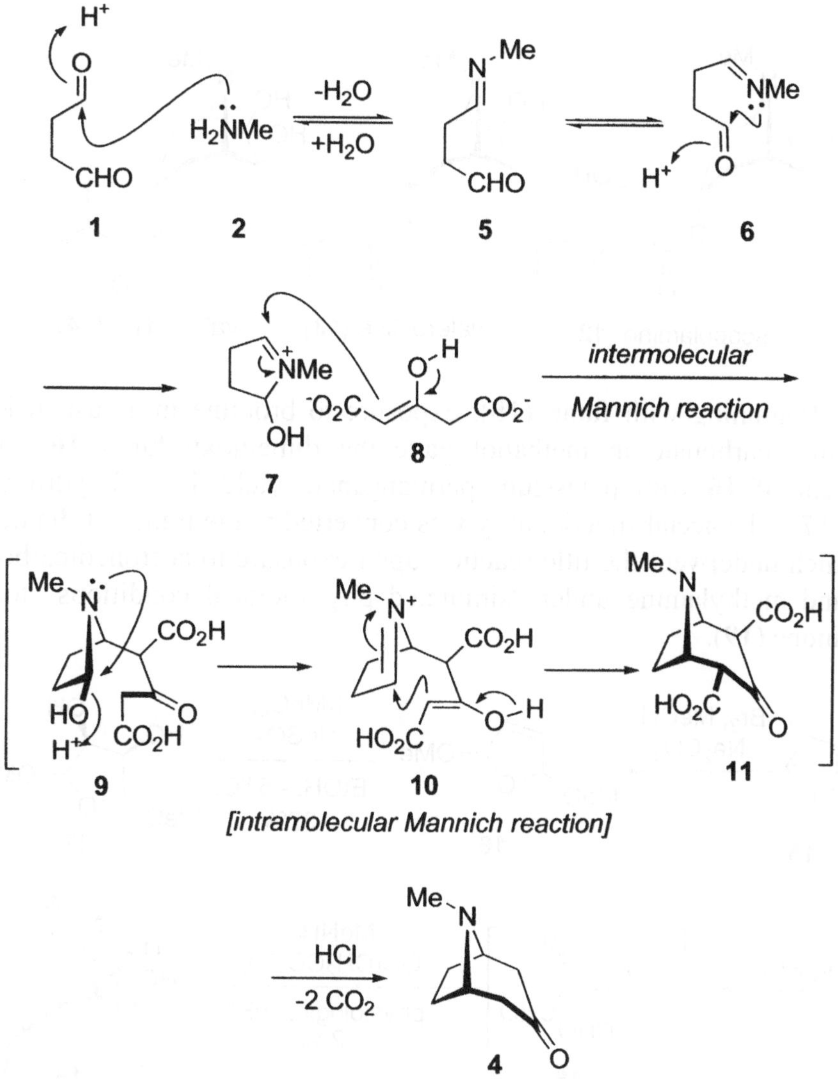

8.6.4 *Synthetic Utility*

General Utility

Due to its remarkable simplicity, many early approaches to tropane alkaloids utilized the Robinson–Schöpf condensation. As pointed out by Sheehan,[6] the major obstacle in accomplishing the synthesis of more complex tropane alkaloids, by way of the Robinson–Schöpf condensation, lies the preparation of the appropriate aldehyde. Specifically capturing the attention of Sheehan and Bloom was the structural similarity and difficulty in preparing requisite aldehyde *en route* to alkaloids such as scopolamine (**12**), valeroidine (**13**), and meteoloidine (**14**), embodying hydroxyl substituents at C6 and C7.[6]

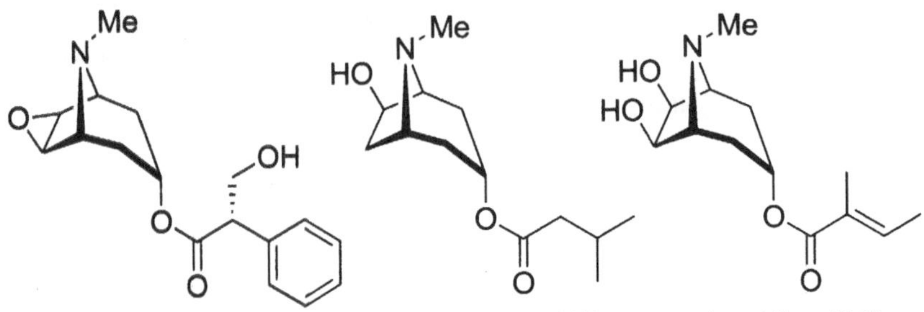

scopolamine (**12**) valeroidine (**13**) valeroidine (**14**)

Beginning with furan (**15**), exposure to bromine in a suspension of potassium carbonate in methanol gave the dimethoxy furan **16**. Next, treatment of **16** with potassium permanganate yielded *cis*-dihydroxylated acetal **17**. The acetal functionality was converted to the transient dialdehyde **18**, which underwent the title reaction upon exposure to acetonedicarboxylic acid and methylamine under "simulated physiological conditions" to give teloidinone (**19**).

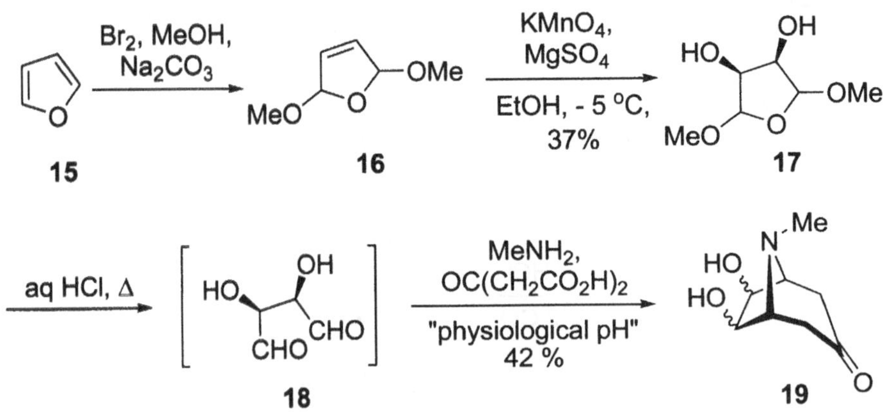

Stern and Wasserman[7] demonstrated for the first time the preservation of optical activity of an enolizable starting material under the conditions of the Robinson–Schöpf condensation. Mono-acetone-D-manitol (**20**), the source of optical activity, was first oxidized with pure lead tetraacetate to give the intermediate acetone-L-tartardialdehyde (**21**). Immediate hydrolysis of dialdehyde **21** with 0.1 N H_2SO_4 liberated the isopropylidene-protecting group and exposed the diol function of **22**, which after neutralization was subjected to slightly modified conditions of the Robinson–Schöpf condensation to afford levorotatory teloidinone **23**.

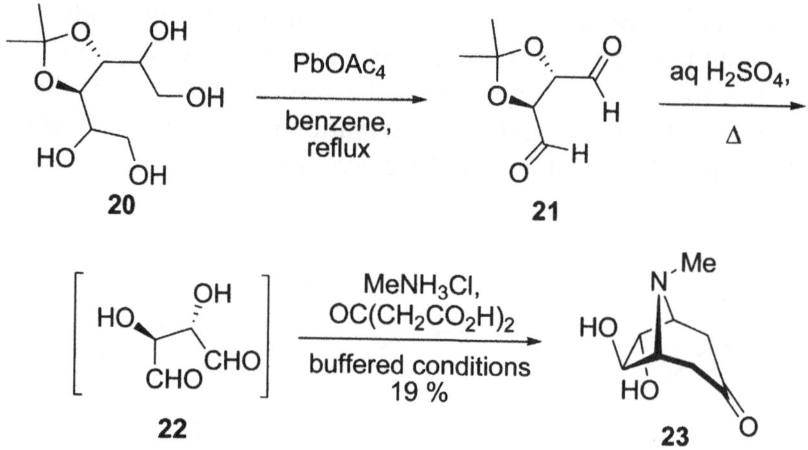

20 21 22 23

The conditions of the Robinson–Schöpf condensation allow for a wide breadth of nucleophilic amines. In addition to alkyl amines, hydroxyalkyl, benzylic, aromatic, heterocyclic amines, as well as α-amino acids and esters are sufficiently reactive to participate in this reaction.[8,9]

The dialdehyde component of the title reaction is not limited to alkyl chain aldehydes. Guthrie and McCarty[10] have shown oxidized glycosides **24** are competent partners for this tandem process to give oxo bridged pseudopelletierine analogues **26**.

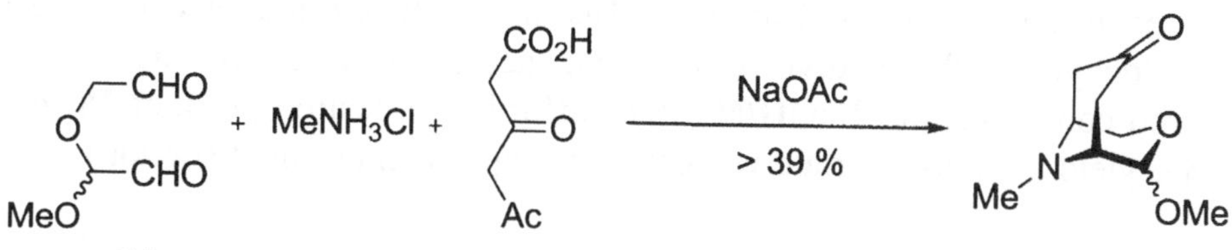

24 25 3 26

The Robinson–Schöpf condensation has been employed to construct structures other than the familiar tricyclic core. In addition, there are cases where once obtained, the triyclic core is fragmented into natural or synthetic scaffolds. Contexts for use of the Robinson–Schöpf in these ways include hydroazulenes,[11] betalains,[12] (±)-coniine,[13] pyrrolidine synthons,[14] and substituted piperidones.[15]

Applications in the Total Synthesis of Natural Products

As shown before (*vida supra*), the seminal application of the Robinson–Schöpf condensation to natural product synthesis was the concise preparation of tropinone (**4**).[3] In pondering the biogenesis of tropane alkaloids, Robinson put forth a bold hypothesis: "By imaginary hydrolysis at the points indicated by the dotted lines, the substance may be resolved into succindialdehyde, methyl amine, and acetone, and this observation suggested a line of attack of the problem which has resulted in a direct synthesis."[3]

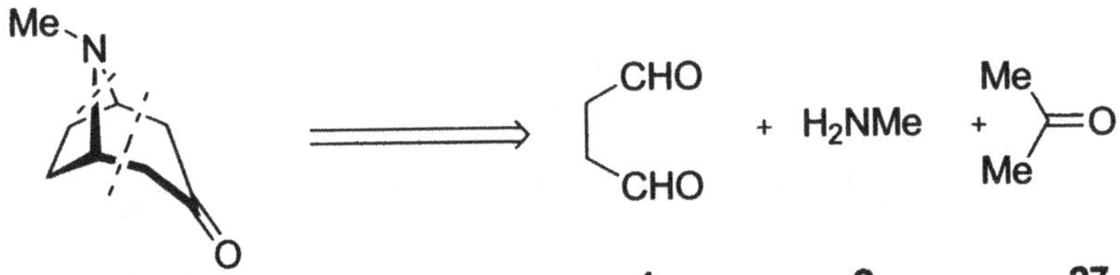

4 **1** **2** **27**

Indeed, Robinson's initial experiments demonstrated tropinone could be prepared in a single step, albeit in low yield, via condensation of succindialdehyde, with acetone and methyl amine in aqueous solution. Further work by Robinson demonstrated the replacement of acetone by a salt of acetonedicarboxylic acid or ethyl acetonediarboxylate gave some improvement in yield after decarboxylation. As previously mentioned, a critical contribution to this process was Schöpf's demonstration of greatly improved yields under buffered or physiological conditions.[4]

Tropinone's one carbon homolog, pseudopelletierine (**28**), can be prepared in an analogous fashion.[16] Heating a stirred aqueous mixture of glutaraldehyde (**29**) methyl amine and acetonedicarboxylic acid under buffered conditions (aq $NaOH/Na_2PO_4$) affords the desired product **28** in up to 68% yield after appropriate work up, see experimental (next section).

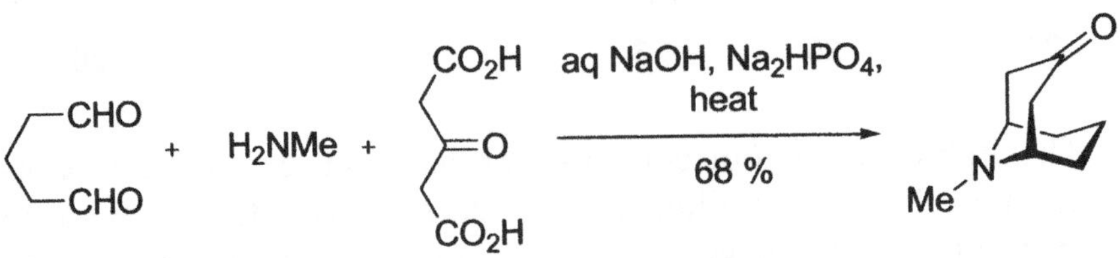

29 **2** **3** **28**

8.6.5 *Experimental*

Preparation of Pseudopelletierine 28

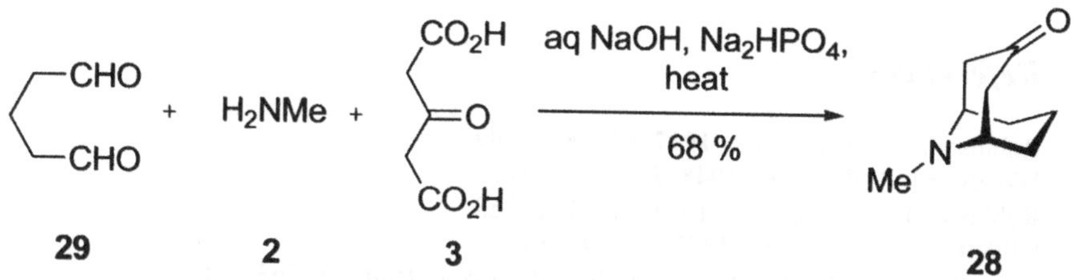

In a 3-L round-bottomed flask equipped with a mechanical stirrer and flushed with a slow stream of nitrogen are placed 22 mL (0.26 mol) of concentrated hydrochloric acid, 165 mL of deoxygenated water, and 64 g (0.5 mol) of 2-ethoxy-3,4-dihydro-2*H*-pyran. The mixture is stirred vigorously for 20 min and then allowed to stand for 1 h. To the resulting colorless solution of glutaraldehyde are added, in order: 350 mL of water, 50 g (0.74 mol) of commercial methylamine hydrochloride dissolved in 500 mL of water, 83 g (0.57 mol) of acetonedicarboxylic acid dissolved in 830 mL of water, and a solution of 88 g (0.25 mol) of disodium hydrogen phosphate dodecahydrate and 7.3 g (0.18 mol) of sodium hydroxide dissolved in 200 mL of water by heating. Carbon dioxide is evolved, and the pH of the solution, initially 2.5, increases to 4.5 after the mixture has been stirred under nitrogen for 24 h. Concentrated hydrochloric acid (33 mL) is added, and the solution is heated on the steam bath for 1 h to complete the decarboxylation. After the solution has been cooled to room temperature, 75 g of sodium hydroxide in 100 mL of water is added, and the basic mixture is extracted with eight 250 mL portions of methylene chloride. The combined methylene chloride extracts are dried over sodium sulfate, concentrated to about 500 mL and filtered through a layer of 400 g of alumina packed in a 50 mm column. The column is eluted with methylene chloride until about 1.5 L of eluate has been collected. The eluate is concentrated under reduced pressure to yield crystalline but yellow pseudopelletierine. The solid is sublimed at 40 °C and 0.3 mm to yield 47–55 g (61–73%) of crude, nearly colorless pseudopelletierine. The product is dissolved in 100 mL of boiling pentane, 3 mL of water is added, and the mixture is boiled until the aqueous layer disappears. After thorough chilling in a refrigerator, the crystals that separate are collected on a filter and washed well with ice-cold pentane. Evaporation of the combined filtrate and washings to 20 mL, followed by filtration and washing, yields a second crop of almost equally pure material. The combined pseudopelletierine hemihydrate weighs 47–55 g and melts at 47–48.5 °C. Sublimation of the hemihydrate as described above removes the water of hydration and yields 44–52 g (58–68 %) of pure, colorless pseudopelletierine, m.p. 63–64 °C (sealed tube). Anhydrous material which has been prepared in this manner does not decompose on storage under dry conditions.

8.6.6 References

1. Robinson, R. *J. Chem. Soc.* **1917,** *111*, 876–899.
2. Woodward, R. B. *Nature* **1948,** *162*, 155–156.
3. Robinson, R. *J. Chem. Soc.* **1917,** *111*, 762–768.
4. Schöpf, C. *Angew. Chem.* **1937,** *50*, 797–805.
5. Paquette, L. A.; Heimaster, J. W. *J. Am. Chem. Soc.* **1966,** *88*, 763–768.
6. Shehan, J. C.; Bloom, B. M. *J. Am. Chem. Soc.* **1952,** *74*, 3825–3828.
7. Stern, R.; Wasserman, H. H. *J. Org. Chem.* **1959,** *24*, 1689–1691.
8. Shimizu, M.; Uchimaru, F. *Chem. Pharm. Bull.* **1961,** *9*, 300–303; and references therein.
9. Shimizu, M.; Uchimaru, F. *Chem. Pharm. Bull.* **1961,** *9*, 313–315; and references therein.
10. Guthrie, R. D.; McCarthy, J. F. *J. Chem. Soc. C* **1967,** *1*, 62–66.
11. Chapman, O. L.; Koch, T. H. *J. Org. Chem.* **1966,** *31*, 1042–1045.
12. Büchi, G.; Fliri, H.; Shapiro, R. *J. Org. Chem.* **1978,** *43*, 4765–4769.
13. Guerrier, L.; Royer, J.; Grierson, D. S.; Husson, H.-P. *J. Am. Chem. Soc.* **1983,** *105*, 7754–7755.
14. Royer, J; Husson, H.-P. *Tetrahedron Lett.* **1987,** *28*, 6175–6178.
15. Langlois, M.; Yang, D.; Soulier, J.-L.; Florac, C. *Synth. Commun.* **1992,** *21*, 3115–3127.
16. Cope, A. C.; Dryden, Jr., H. L.; Howell, C. F. *Org. Synth.* **1957,** *37*, 73–74.

8.7 Simonis Chromone Synthesis

Timothy T. Curran

8.7.1 Description

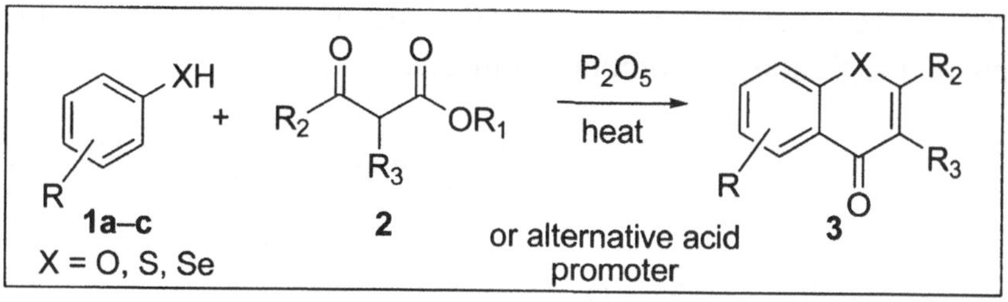

The Simonis chromone synthesis is the reaction of a phenol **1a–c** with a β-keto ester **2** using an appropriate acid promoter to generate a chromone or benzo-γ-pyrone **3** (also called a benzo-1,4-pyrone). While compound **3** is actually a chromenone, for this article, whether the double bond is present or not, the system will be characterized as a chromone. The condensation is related to the Pechmann–Duisberg reaction, which yields coumarins from the condensation of a phenol with a β-keto ester; and like its relative, the reaction conditions require the loss of water from the ketone moiety and alcohol from the ester moiety.

8.7.2 Historical Perspective

In 1913, Simonis and co-workers first reported the isolation of the 2,3-dimethyl chromone **5** by reaction of phenol **1a** with ethyl 2-methyl-acetylacetate.[1] While the product was reported in poor yield, the reaction was a change from the typically reported formation of the coumarin. Simonis did report using a different promoter (P_2O_5) instead of the traditionally used H_2SO_4 for the Pechmann reaction. For some time, the differences in the two reactions were thought to be that simple; sulfuric acid gave the coumarin as reported by Pechmann and phosphorous pentoxide gave the chromone as reported by Simonis.

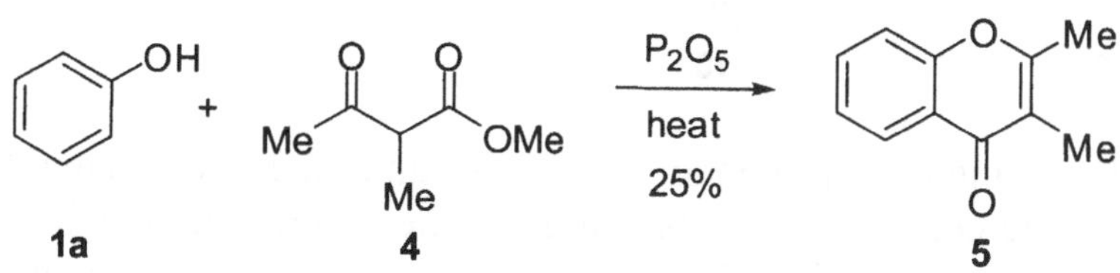

Simonis continued to prove the correct structural assignment by degradation and began exploring the scope and limitation of the reaction.[2] As more co-workers joined in the exploration, many were convinced that Simonis did not make the chromone, but actually made the coumarin and that the structure was merely misassigned. While descriptions of early work solely on the Simonis chromone synthesis are lacking, reviews on the synthesis of coumarins (the Pechmann reaction) typically include an overview on the Simonis chromone synthesis.[3,4]

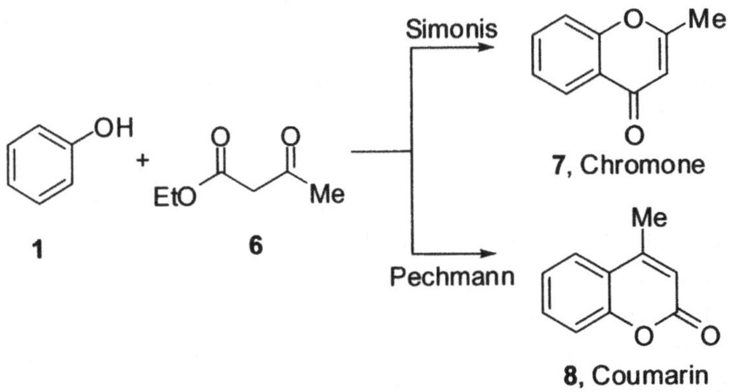

Some trends have been observed for the two reactions. The Simonis chromone synthesis is normally reported to occur when phenols are used that do not provide the coumarin (phenols containing electron-donating groups tend to provide coumarins). These are typically phenols containing electron withdrawing groups. Thus phenols, halogenated phenols. or nitrated phenols typically give the chromone product; β-naphthol also reportedly yields only the chromone. In addition, substituents on the α-position of the β-keto ester also reportedly favor chromone formation.[5] For example, *p*-cresol **9** reacts with ethyl acetoacetate **6** to give the coumarin even when promoted by P_2O_5. However, with the α-alkyl substituent such as ethyl (Et), the chromone was obtained as shown for the conversion of **9** and **10** into **11**.[6]

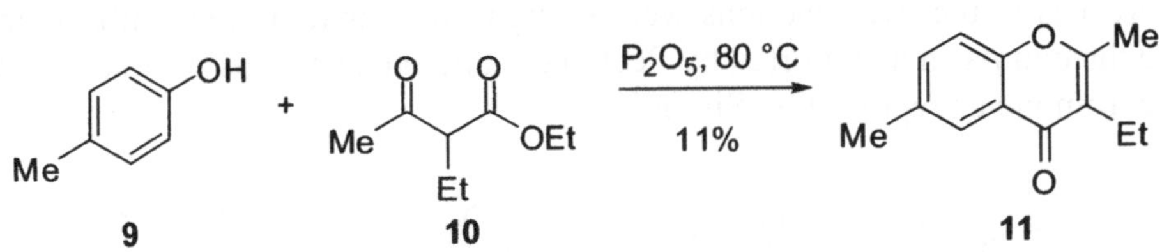

The Simonis chromone synthesis and the Pechmann–Duisberg reaction can be considered as *O*-analogs of the Conrad–Limpach and Knorr reactions for the preparation of quinolines from the corresponding aniline.[7]

8.7.3 Mechanism

Just as the reaction itself has been argued, the mechanism has also been disputed. Early, there were two main pathways proposed: (1) transesterification of the phenol with the ester followed by Fries rearrangement and cyclization,[8] or (2) direct attack of the phenol on the ketone moiety of the β-keto-ester, followed by subsequent thermal cyclization.[9] There was modification to proposal 2 by Delemagne and Martinet,[10] which suggested that both Simonis and Pechmann reactions went through the common intermediate **15**, which could eliminate (promoted by substitutions at the 2 position of the β-keto-ester and more powerful dehydrating agents) and cyclize to give the chromone (Simonis product) or rearrange to give ultimately the coumarin (Pechmann product).

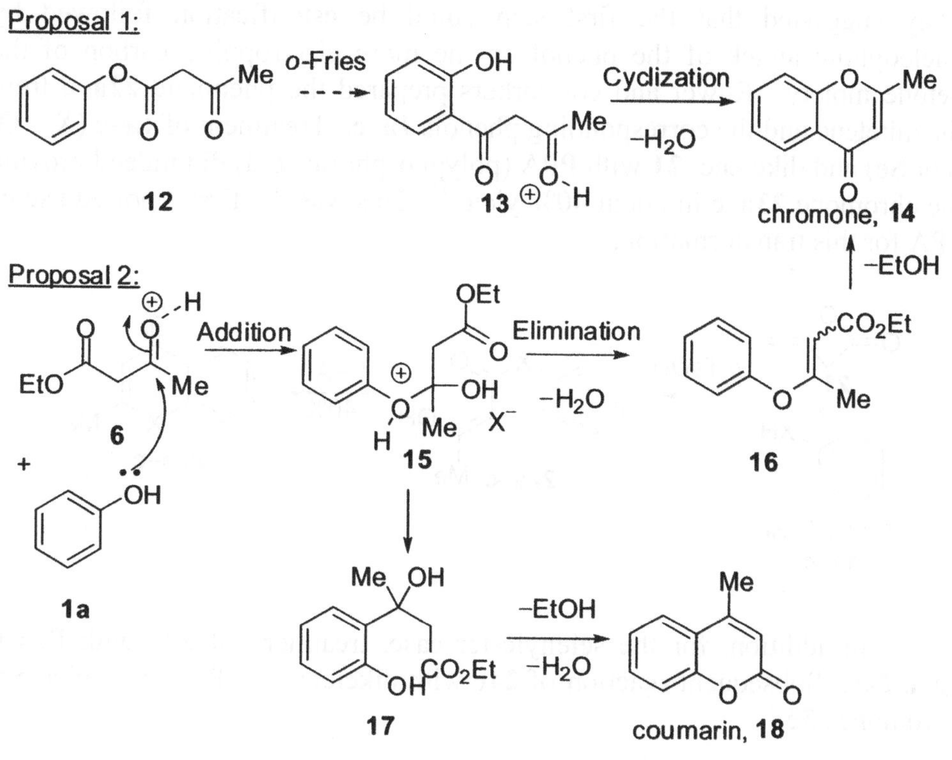

A key experiment reported by Lacey suggested[11] that the reaction could not go through the ester followed by Fries rearrangement as under such conditions, Lacey failed to isolate the chromone. Instead, when phenolate ester **19** was reacted with P_2O_5, he reported the isolation of the self-condensation product **20**. In addition, to have a selective *ortho*-Fries rearrangement seemed unlikely.

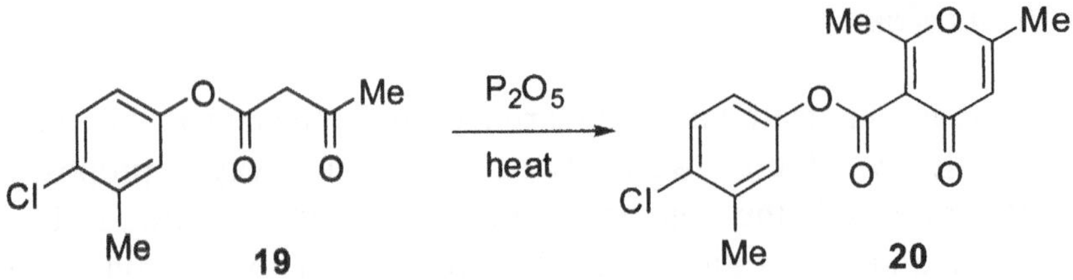

19 20

To prepare the phenolic esters of the β -keto ester **19**, Lacey used a new method of directly providing the β-ketoester from the phenol using diketene.

Subsequent work by Ruwet and co-workers proposed something similar to the proposal 1 mechanism, albeit they did not think that the phenolate ester rearranged via a Fries reaction. Ruwet and co-workers studied the heteroatom displacement of the phenol oxygen with S and Se. They suggested that the first step could be esterification followed by nucleophilic attack of the phenol on the more electrophilic carbon of the ketone moiety. Ruwet and co-workers prepared the phenolate **22a–c** from the diketene and the corresponding phenols **1a–c**. Treatment of **1a–c** (X = O, S or Se) and diketene **21** with PPA (polyphosphoric acid) did indeed provide the chromone **23a–c** in about 40% yield.[12] This was the first reported use of PPA for this transformation.

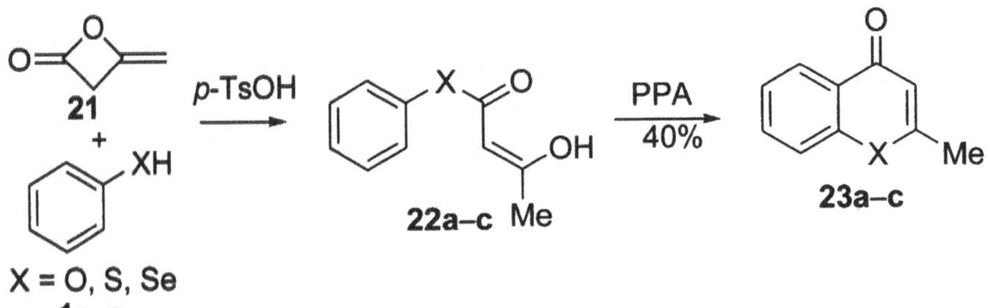

21 1a–c 22a–c 23a–c

In addition, for the selenylester case, treatment of **22c** with $POCl_3$ gave **24c**. Subsequent reaction of **24c** with diketene and PPA, provided Se-chromone **23c**.

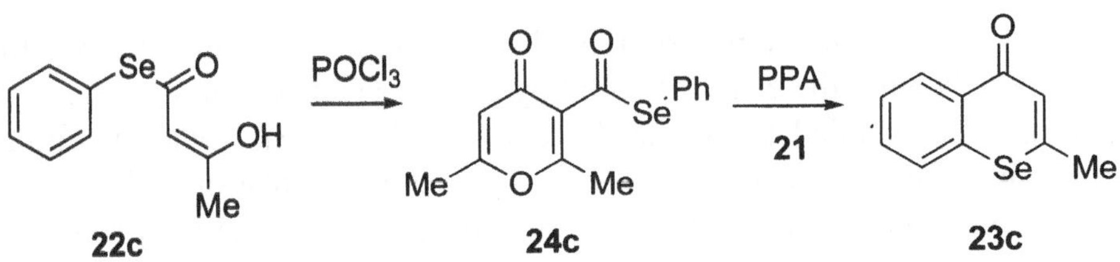

22c 24c 23c

In terms of the data from Ruwet and co-workers, the possibility of the ester leading to chromone needs to be reconsidered. In such consideration, Ruwet's team proposed the potential for intramolecular migration as shown for the conversion of **22a–c** into **23a–c**. Participation of the phenolate into the neighboring activated keto-group would be followed by electrophilic aromatic substitution (EAS) to provide intermediate.

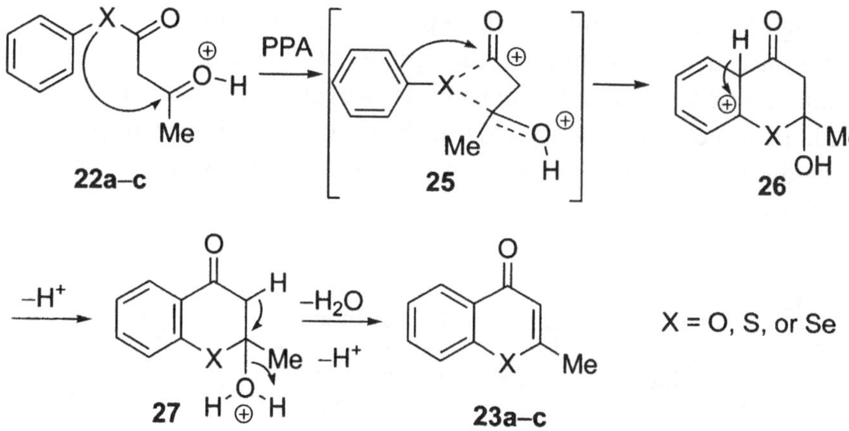

An alternative broadly accepted mechanism based on all of these data has also been proposed.[13] This mechanism suggests "protection" of the ester as its enol **28**, then attack of phenol **1a** on the P_2O_5-activated ketone, providing **29**. Dehydration provides **16** followed by electrophilic aromatic substitution of the activated ester to provide intermediate **30**. Re-aromatization and loss of EtOH completes the formation of chromone **23a**.

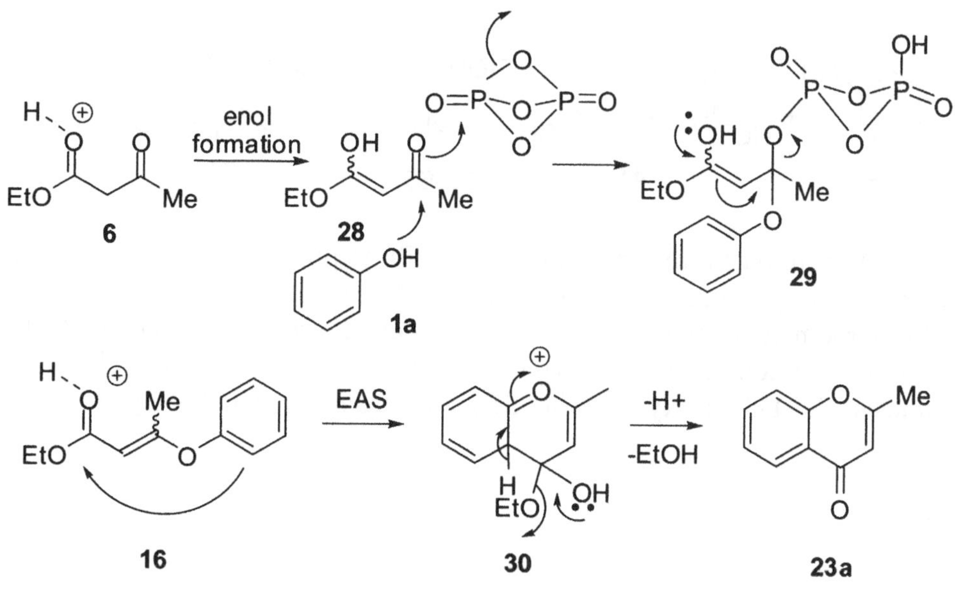

8.7.4 *Variations and Improvements*

Variations on Reagents to Promote the Reaction

In addition to P_2O_5, PPA (as shown in the previous section), H_2SO_4, PPE and thermal conditions have been shown to successfully promote the Simonis synthesis of chromones. One would suspect several additional acidic, dehydrating conditions should work to promote the reaction.

Sulfuric acid was initially thought to be the reagent to promote the Pechmann reaction to yield coumarins. However, it was subsequently found to provide in some cases good yields of the chromone. For example, reaction of chloro-dimethylphenol **31** with ethyl acetoacetate **6** with conc. H_2SO_4 gave a 35% yield of **32**.[14]

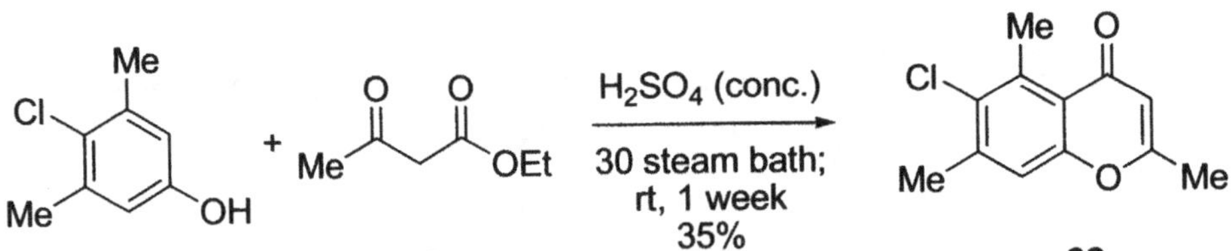

31 **6** **32**

A patent application by a group at Eli Lilly & Co. reported the use of dilute H_2SO_4 in EtOH promoted the Simonis reaction and esterified the acid to provide chromone **34** in 55–65% yield.[15] Compound **34** was used as an intermediate toward the preparation of agents for platelet aggregation inhibition.

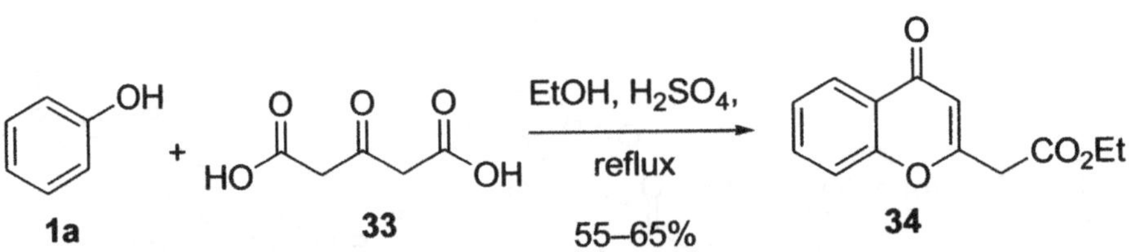

1a **33** **34**

A more recent report on the use of PPE to promote cyclization of a β-naphthol has appeared. The reagent allowed product formation in excellent yield at room temperature. Such results were a great improvement to the reaction conditions and yield typically obtained in the Simonis chromone synthesis. 2,6-Dihydroxy-naphthol **35** gave chromone **36** in 98% yield.[16]

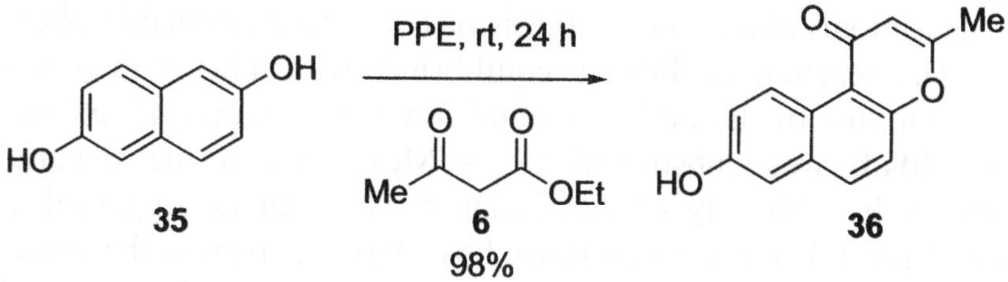

Thermal conditions with no acid or dehydrating agents were first reported by Mentzer and co-workers in 1952.[11] An example of applying these conditions will be displayed in Section 8.7.5.

Variation on β-Keto-Ester and Phenol

As natural products do not always have the double bond in the 1,4-pyran ring, there exists the need to have alternative building blocks at different oxidation states. For example, resorcinol **37** was reacted with acrylic acid **38** promoted by BF_3 to provide a high yield of chromone **39**, which was used to prepare a cannabinoid plant extract.[17]

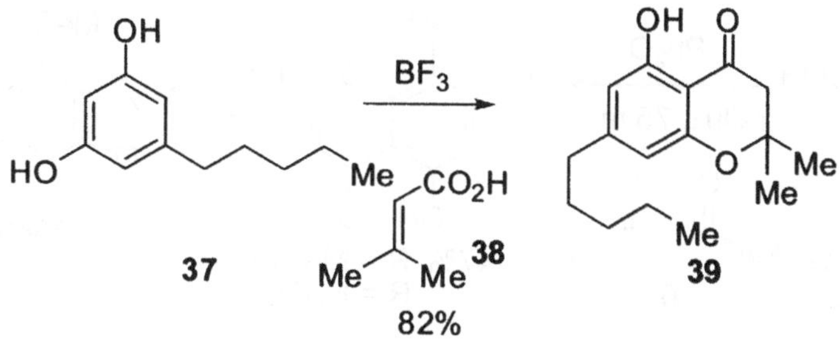

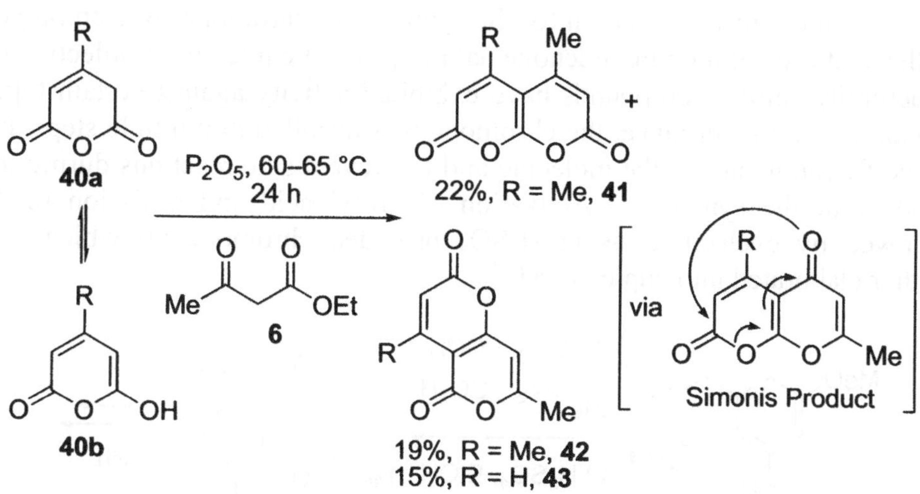

The Simonis chromone synthesis has also been applied to phenol-like systems. The reaction of **40b** (in equilibrium with **40a**) with **6** and P_2O_5 provided a mixture of the Pechmann and Simonis products **41** and **42** when anhydride **40a,b** was substituted (R = Me). When the material was unsubstituted (R = H) only **43** was isolated. The authors suggested that the formation of the 1,4-pyran was followed by rearrangement to the more stable product.[18] While the yields were low, entry was established into these highly oxidized bis-pyranyl systems.

8.7.5 *Synthetic Utility*

The Simonis chromone synthesis was reported as a key reaction in the preparation of the natural product visnagin **48**, a component found in fruits. Badawi and Fayez[19] reported the conversion of **44** and **46** using heat in diphenyl ether in satisfactory yield. The aldehydic product **47** was derived from phloroglucin aldehyde **45** via similar thermal reaction conditions and was subsequently converted into visnagin **48**. Note that the cyclization occurred under thermal conditions with no acid catalysis.

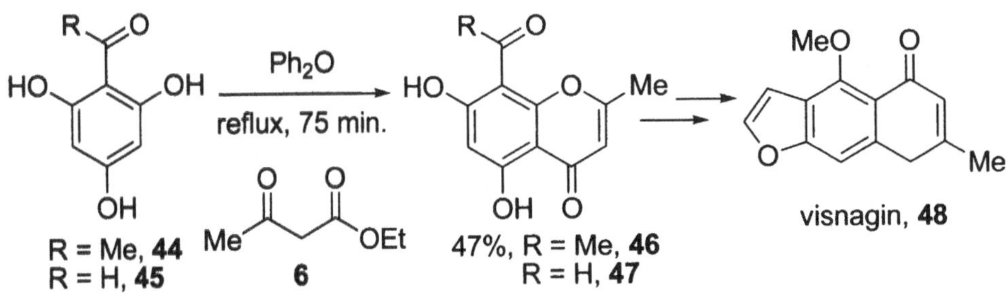

A more recent variation for the synthesis of chromones was employed in the synthesis of the rare indenone natural product euplectin. Euplectin and structurally similar compounds have exhibited activity against certain types of cancer. In this instance, the chromone was installed in multiple steps due to the fragile nature of the molecule and to prevent side reactions during the conduct of the synthesis. Activation of crotyl acid and acylation of **49** followed by cyclization using H_2SO_4 provided chromone **50** which was further elaborated into euplectin **51**.[20]

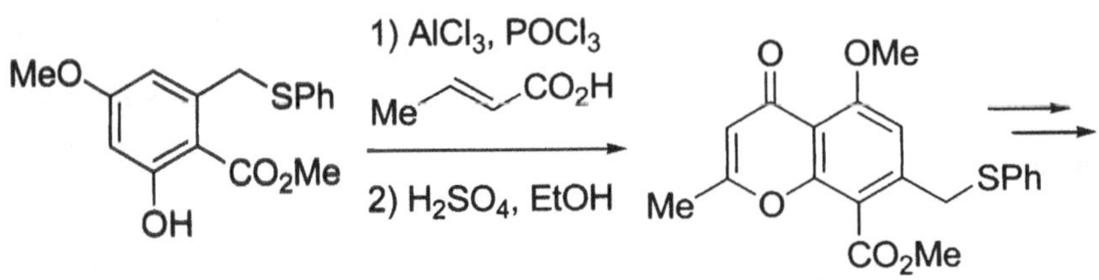

49 **50**

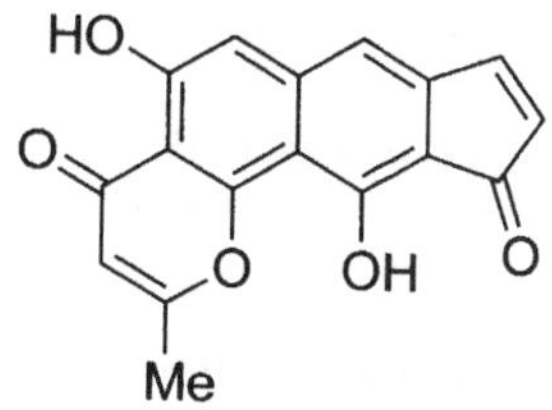

euplectin, **51**

While the chromone moiety is found in several natural products, the Simonis chromone synthesis has not been exclusively used to prepare chromones. Other methods exist that are oftentimes more easily utilized, run under more mild conditions, or provide higher yields. For example, the Kostanecki–Robinson reaction has found application in this arena.[21] However, alternative reaction types are not without their own issues; therefore, the Simonis chromone synthesis remains a tool for the synthetic chemist.

8.7.6 *Experimental*

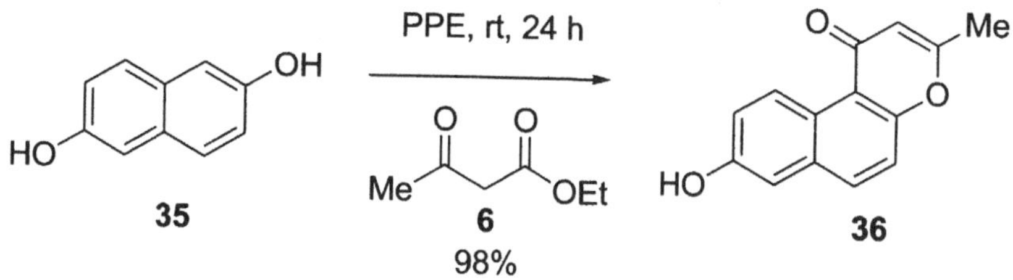

PPE-Promoted Condensation of Naphthalene 2,6-diol 35 with Ethyl Acetoacetate 6[16]

PPE was prepared by warming P_2O_5 (15 g) with absolute Et_2O (15 mL) and $CHCl_3$ (15 mL) at 60 °C for 15 h. This mixture was measured out and added directly to the reaction mixture.

A mixture of **35** (1.6 g, 0.01 mol) and **6** (2.68 g, 0.02 mol) was treated with PPE solution described above (25 mL) and stirred at room temperature for 24 h. The resulting reaction mixture was poured into crushed ice, and the resulting solid collected was then crystallized from EtOH to give **36** as violet needles, mp = 279 °C, 98%.

8.7.7 *References*

1 Petschek, E.; Simonis, H. *Ber. Dtsch. Chem. Ges.* **1913,** *46*, 2014–2020.

2 (a) Simonis, H.; Lehmann, C. B. A. *Ber. Dtsch. Chem. Ges.* **1914,** *47*, 692–699. (b) Simonis, H.; Remmert, P. *Ber. Dtsch. Chem. Ges.* **1914,** *47*, 2229–2233.

3 [R] Sethna, S.; Phadke, R. *Org. React. VII* **1953**, 1–58, Wiley: New York, NY.

4 [R] Wawzonek; S. In *Heterocyclic Compounds Vol 2*, Wiley: New York, NY; **1951**, pp. 229–276.
5 [R] Sethna, S. M.; Shah, N. M. *Chem. Rev.* **1945,** *36*, 1–62
6 Robertson, A.; Sandrock, W. F. *J. Chem. Soc.* **1932,** 1180–1184.
7 Curran, T. T. in *Name Reactions in Heterocyclic Chemistry* Wiley: New York, NY, **2005**; pp. 1398–406.
8 Ahmad, S. Z.; Desai, R. D. *Proceed. Indian Acad. Sci. Section A* **1937,** 6–11.
9 Robertson, A.; Waters, R. B.; Jones, E. T. *J. Chem. Soc.* **1932,** 1681–1688.
10 Dallemagne, M. J.; Martinet, J. *Bull. Soc. Chim. Fr.* **1950,** 1132–1133.
11 Lacey, R. N. *J. Chem. Soc.* **1954,** 854–860.
12 Ruwet, A.; Janne, D.; Renson, M. *Bull. Soc. Chim. Belges*, **1970,** *79*, 81–88.
13 [R] Li, J. J. In *Name Reactions* Springer-Verlag, Heidelberg, Germany, **2003,** pp. 376–377.
14 Adams, R.; Mecorney, J. W. *J. Am. Chem. Soc.* **1944,** *66*, 802–805.
15 Briggs, B.; Hansen, M.; Kanter, J.; Mullins, J. J.G.; Ruhter, G.; Udodong, U.; Verral, D.; Zmijewski, WO Pat. 2001094335, **2001.**
16 Oyman, U.; Gunaydin, K. *Bull. Soc. Chim. Belg.* **1994,** *103*, 763–764.
17 *Anand*, R. C.; Ranjan, H. *Monatsh. Chem.* **1983,** *114,* 123–125.
18 Tan, S. F. *Aust. J. Chem.* **1972,** *25*, 1367–1370.
19 Badawi, M. M.; Fayez, M. B. E. *Tetrahedron* **1965,** *21*, 2925–2929.
20 Mal, D.; De, S. R. *Org. Lett.* **2009,** *11*, 4398–4401.
21 Limberakis, C. In *Name Reactions in Heterocyclic Chemistry,* Wiley: New York, NY, **2005**; pp. 521–535.

8.8 Wessely–Moser Rearrangement

Ke Chen

8.8.1 Description

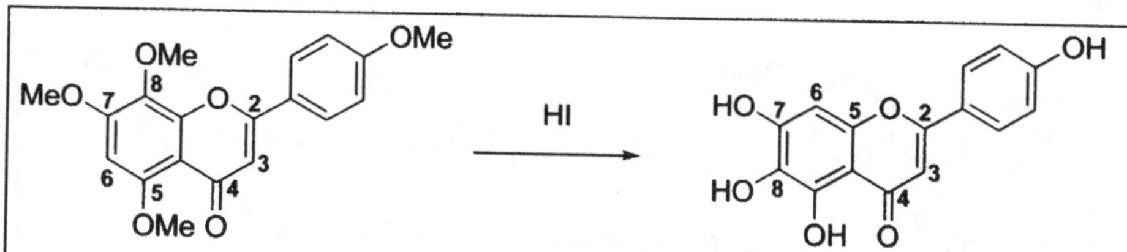

1

scutellarein (**2**)

Applicable substrates: flavanoids, isoflavanoids, bioflavonoids, chromones, xanthones

Wessely–Moser rearrangement[1,2] refers to the rearrangement of benzopyran-4-ones possessing a 5-hydroxyl group, such as the demethylation product of 5,7,8-methoxyflavone **1**, to yield scutellarein **2**, its isomeric product. Acidic conditions are most commonly used for such transformation; however, several examples of base-catalyzed reactions have also been reported. The reversibility of the rearrangement largely depends on the substituent pattern of the benzopyranone.

8.8.2 Historical Perspective

Fritz Wessely and Georg H. Moser published their work on the synthesis and structural elucidation of scutellarein **2** in 1930; one of the key steps was the titled reaction featuring the unprecedented rearrangement from **1** to **2**.[1] This unique isomerization method soon found application in the structural elucidation of flavanoids and further by extension to chromones, xanthones, and their derivatives.[3] In addition, Wessely–Moser rearrangement has been used as a surrogate of direct synthesis toward 5,6,7-substituted flavones.[4]

8.8.3 Mechanism

The generally accepted pathway for Wessely–Moser rearrangement involves the scission of the pyrone ring followed by isomeric ring formation.[2] For example, demethylation of **1** took place readily in hydroiodic acid yielding polyphenol **3**. Under such conditions, the pyrone ring in **3** could open to afford 1,3-dicarbonyl compound **4**. Ring closure by nucleophilic attack of the hydroxyl group at C-5 onto the carbonyl group at C-2, followed by dehydration, gave rise to scutellarein **2**, a Wessely–Moser rearrangement product. The research done by Barberán and co-workers suggested that under acidic conditions, demethylation of flavones (**1**→**3**) was much faster than the opening of the pyrone ring (**3**→**4**).[5] However, with alternative reaction conditions and substrates, the extent of demethylation could vary; as a consequence, the order of demethylation and ring-scission could be switched.

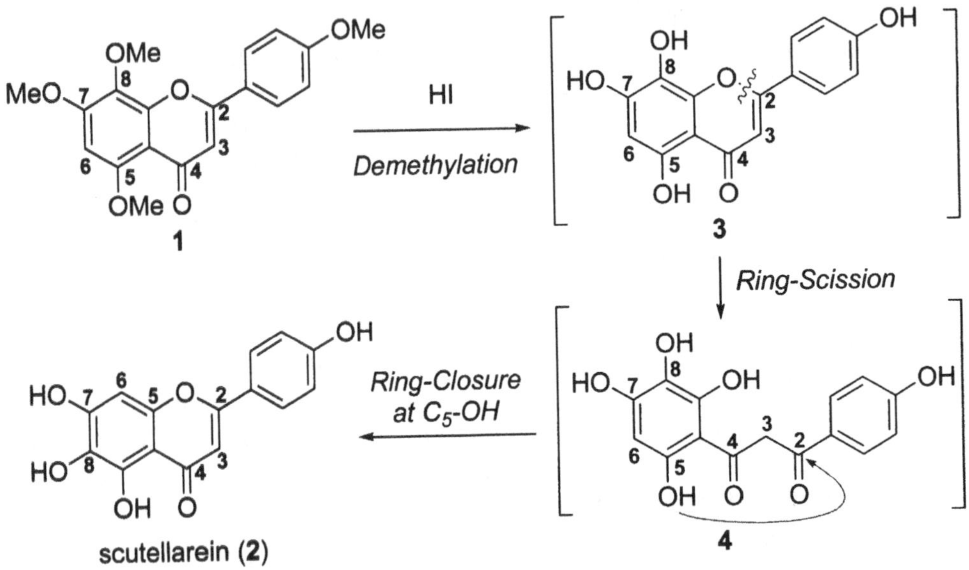

8.8.4 *Synthetic Utility*

General Isomerisation Pattern

Wessely–Moser rearrangement is generally regarded as a reversible isomerization process, favoring the most thermodynamically stable product.[1] For example, the preferential formation of the 5,6,7-arranged flavone **6** over its 5,7,8-arranged isomer **5** was attributed to the greater stability of **6**, a phenomenon supported by similar observation from the Allan–Robinson synthesis of flavones.[6] However, the substitution pattern on the benzopyranone could affect the extent and, in some cases, even reverse the direction of the rearrangement. As a case in point, Chopin *et al.* reported an interesting observation that the isomerization between **7** and **8** favored the 5,8-dihydroxy isomer **7**.[7] The unusual preference was attributed to intramolecular hydrogen bonding between the hydroxyl group at C-8 and the pyrone oxygen, stabilizing isomeric form **7**. Alkaline medium was used to avoid demethylation, since empirical data revealed that the methoxy group at C-7 was essential for this stabilization to be effective: compound **9** containing a hydroxyl group at C-7 underwent facile rearrangement, yielding isomer **10** as expected.

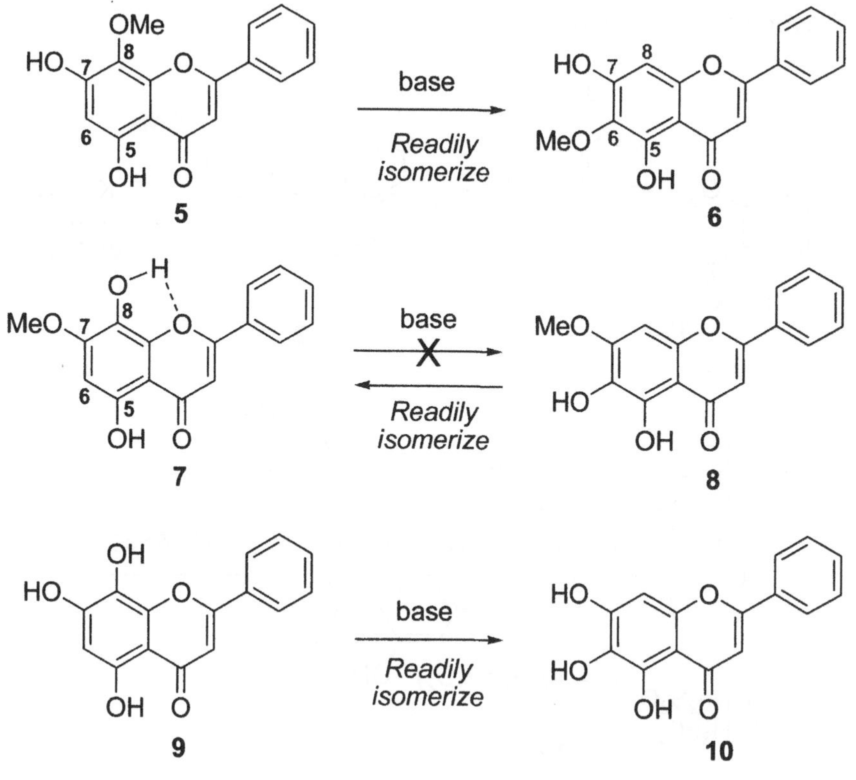

Application in Structural Elucidation

Since its discovery, Wessely–Moser rearrangement has been widely used in the structural elucidation of flavanones, chromones, xanthones, and their derivatives.[3] Its popularity should be largely attributed to the fact that differentiation between a 5,7,8-arranged benzopyranone and its 5,6,7-isomer is often difficult, and sometimes impossible, by the classical techniques such as NMR studies (even with shift reagents).[5]

By way of illustration, hinokiflavone **15** containing a unique structure of a bisapigenyl ether was synthesized by Nakazawa in 1967, one of the key steps involving an acid-catalyzed Wessely–Moser rearrangement.[8] The synthesis commenced with the condensation of intermediate **11** and **12**, two readily available building blocks, to give rise to bisflavone **13**. Reduction of the nitro group of **13** then afforded key intermediate **14**, which was originally thought to be the pentamethyl ether of natural hinokiflavone. However, chemical derivatization ruled out such hypothesis since melting point of **14** was different from the synthesized pentamethyl ether of natural hinokiflavone. Thus a Wessely–Moser rearrangement was affected by exposing intermediate **14** to hydroiodic acid, via demethylation, yielding natural product hinokiflavone **15**. These studies established the structure of hinokiflavone **15** un-ambiguously.

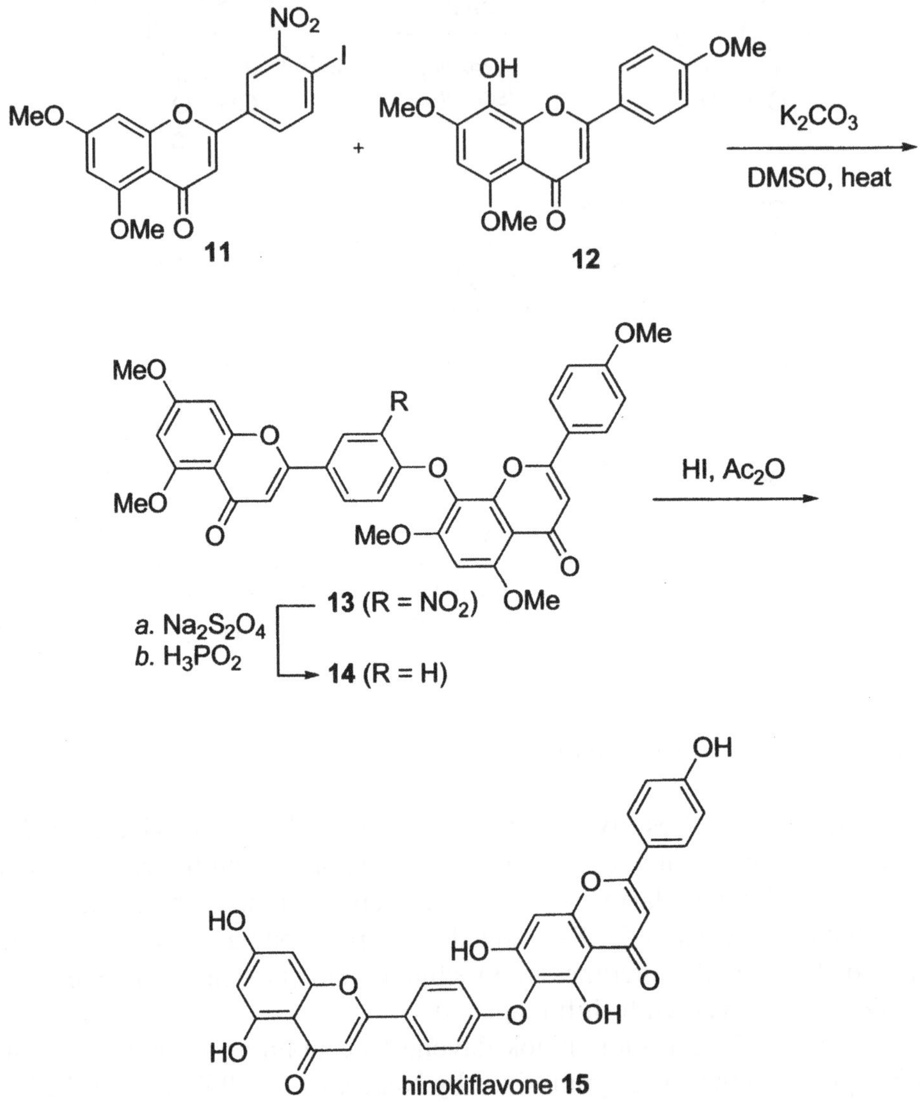

Application in Natural Product Synthesis

Studies in the laboratory of Tanaka toward the synthesis of bikaverin **20**, a fungal metabolite with a benzo-xanthen skeleton, led to the discovery of an elegant method to form ring C.[9] Upon treatment with aqueous oxalic acid, benzopyranone **16** was directly transformed into xanthone **17**, which could be potentially elaborated intro bikaverin **20**. The unexpected formation of intermediate **17** could be explained by the following sequence: hydrolysis gave arise to benzopyranone **18**, which underwent condensation followed by tautomerization to afford xanthone **19**; a Wessely–Moser type rearrangement then took place, leading to the formation of **17** as the observed major product.

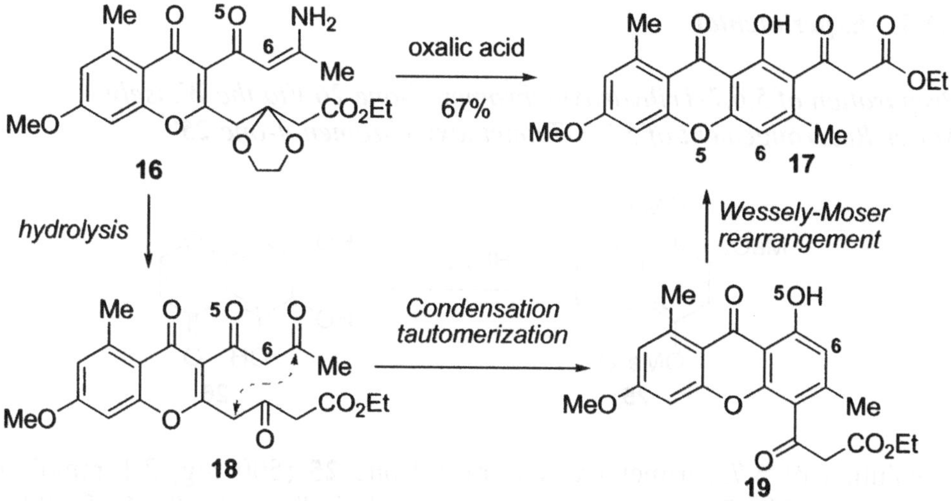

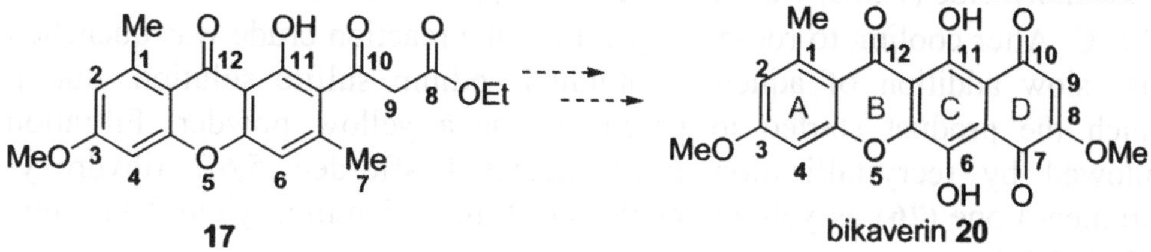

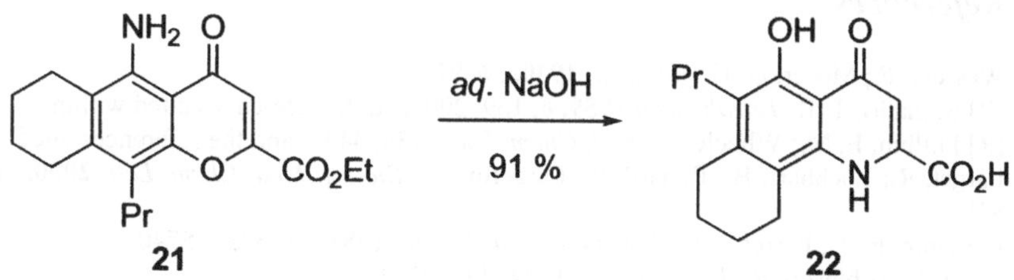

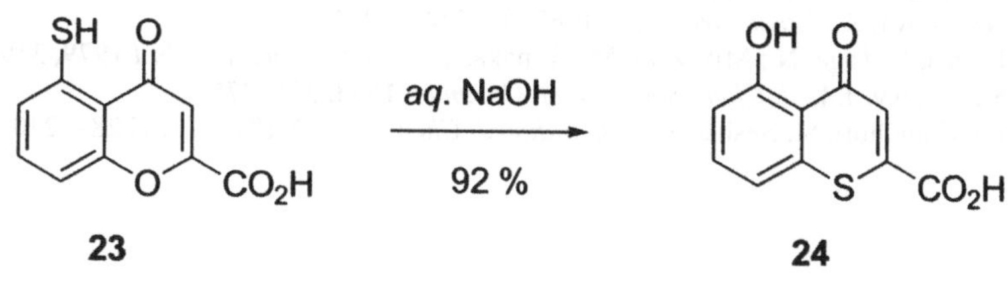

Research by Suschitzky *et al.* demonstrated that thia- and aza-version of Wessely–Moser rearrangements could be achieved under basic conditions

(**21**→**22**, **23**→**24**).[10] Furthermore, the yields of such transformations were generally improved due to the greater stability of the resulting rearrangement products. These studies represented a novel approach toward the synthesis of thio- and amino-chromones.

8.8.5 *Experimental*

Preparation of 5,6,7-Trihydroxy-chromen-4-one **26** *Via the Wessely–Moser Rearrangement of 5,7,8-Trimethoxy-chromen-4-one* **25**[11]

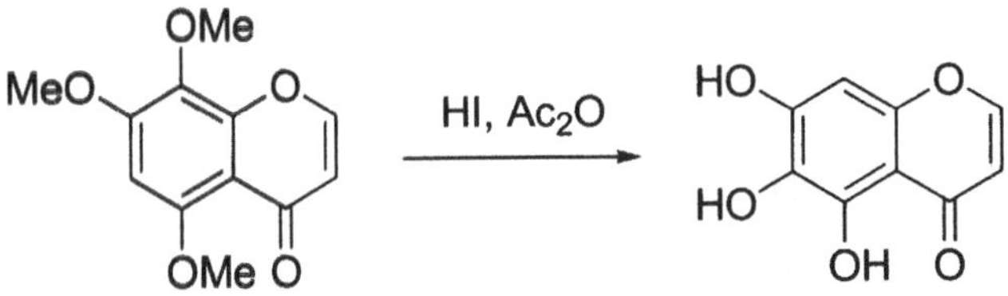

25 **26**

A solution of 5,7,8-trimethoxy-chromen-4-one **25** (500 mg, 2.1 mmol) in acetic anhydride (7 mL) was refluxed with hydroiodic acid (10 mL) for 2 h at 120 °C. After cooling to room temperature, the reaction crude was quenched with slow addition of aqueous saturated sodium sulfite solution, during which the product started to precipitate as a yellow powder. Filtration followed by recrystallization from methanol afforded 5,6,7-trihydroxy-chromen-4-one (**26**) as yellow needles (100 mg, 0.5 mmol, yield 25%). mp, 172–173 °C.

8.8.6 *References*

1. Wessely, F.; Moser, G. H. *Monatsh.* **1930,** *56*, 97.
2. [R] Seshadri, T. R. *Tetrahedron* **1959,** *6*, 169–200, and the references cited within.
3. [R] Philbin, E. M.; Wheeler, T. S. *J. Chem. Soc.* **1956,** 4455; and the references cited within.
4. Larget, R.; Lockhart, B.; Renard, P.; Largeron, M. *Bioorg. Med. Chem. Lett.* **2000,** *10*, 835–838.
5. Barberán, F. A.; Ferreres, F.; Tomás, F. *Tetrahedron* **1985,** *41*, 5733–5740.
6. Allan, J.; Robinson, R. *J. Chem. Soc.* **1924,** *125*, 2192.
7. Chopin, J.; Molho, D.; Pacheco, H.; Mentzer, C. *Acad. Sci. Paris* **1956,** *243*, 712.
8. Nakazawa, K. *Tetrahedron Lett.* **1967,** *51*, 5223–5225.
9. Lijima, I.; Taga, N.; Miyazaki, M.; Tanaka, T. *J. Chem. Soc., Perkin I* **1979,** 3190–3195.
10. Suschitzky, J. L. *J. Chem Soc., Chem. Commun.* **1984,** 275–276.
11. Raychaudhuri, S.; Seshadri, T. R. *Indian J. Chem., Sect B* **1973,** *11*, 1228–1230.

8.9 Widman–Stoermer Cinnoline Synthesis

Kyle J Eastman

8.9.1 Description

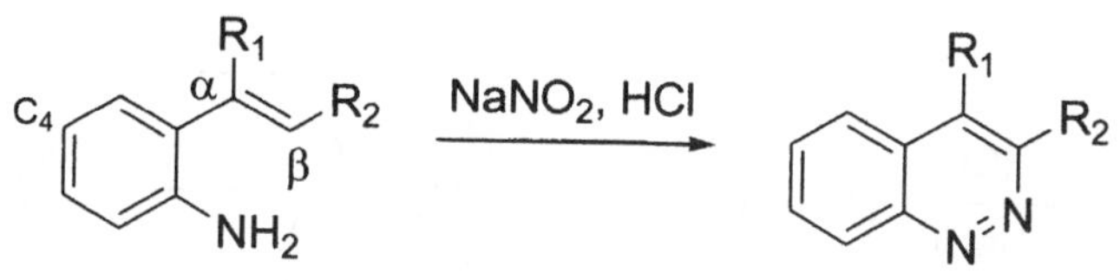

The Widman–Stoermer cinnoline synthesis refers to the conversion of an *o*-vinyl aniline **1** to the representative cinnoline **2** via an *in situ* generated diazonium intermiediate.[1,2] Cinnolines have shown utility as anticancer, fungicidal, bactericidal, antithrombotic, and antituberculosis agents as well as having been shown to embody anesthetizing and sedative activity.[3] Widman and Stoermer's approach to cinnolines represents a classical approach to this ubiquitous core structure.

8.9.2 Historical Perspective

The cinnoline ring system was first prepared by Richter via diazotization and subsequent cyclization of *o*-aminophenylpropionic acid. Subsequent efforts by Widman, and later Stoermer, demonstrated access to the cinnoline ring system via subjection of *o*-vinyl anilines to diazotization conditions. Other than the observation by Stoermer that *o*-aminocinnamic acid does not undergo cyclization when exposed to diazotization conditions, initially, little attention was paid to the influence of R_1 and R_2 on the cyclization of **1** to **2**. Sixty years after Widman's first disclosure, Simpson and Schofield[4–7] began to address the role of R_1 and R_2 (*vide infra*). There is a remarkable dearth of literature featuring the title reaction beyond the efforts of Simpson and Schofield, owing, in part, to many other useful methods to prepare cinnolines.

8.9.3 Mechanism

The first of many steps for the Widman–Stoermer cinnoline synthesis is diazotization. Of critical importance is the formation of dinitrogen trioxide (**6**) or a nitrosonium equivalent. The formation of **6** is followed by the addition of the amine function in aniline **1** to the electrophilic nitrogen of **6** with subsequent loss of anionic NO_2. The resultant addition product **7** then loses a proton to give **8**. Proton transfer gives tautomer **9**, from which the subsequent protonation/dehydration to the requisite diazonium salt **10** becomes clear. Following formation of diazonium salt **10**, 6-*endo-dig* cyclization of the olefin onto the terminal diazonium nitrogen gives the cyclized product **11**, which aromatizes to the observed cinnoline **2**.

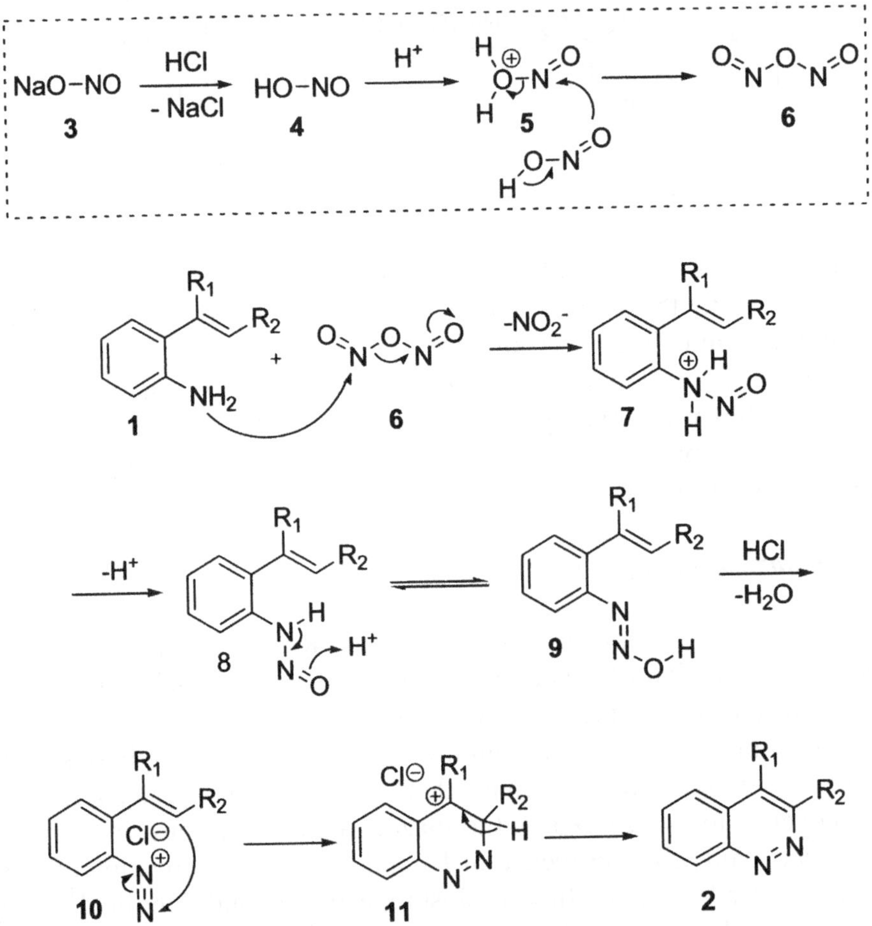

8.9.4 *Synthetic Utility*

Cumulative efforts of Stoermer, Simpson and Schofield[4–7] have led to a fairly comprehensive understanding of the circumstances in which the Widman–Stoermer reaction might be expected to succeed. Initial efforts of Simpson and Schofield,[4–5] demonstrated that cinnoline formation is not expected to succeed for *o*-vinyl anilines (**1**) where R_1 = H or CO_2H and R_2 = aryl, CO_2H, CO_2R, CN, 2-pyridyl, or 2-quinolyl. On the other hand, for compounds (**1**) with R_1 = aryl, cyclization is favored even when R_2 is also an aryl group. It seems that in the later case, cinnoline formation is independent of the stereochemical configuration of the olefin, although in the case of *cis*-compounds, it should be considered in competition with the Pschorr reaction (*vide infra*). Further observations by Schofield[6] showed that ethylenes (**1**), where R_1 = *p*-methoxy phenyl and R_2 = 2-pyrrolo or 2-pyridyl, give excellent conversion to cinnolines upon exposure to diazotization conditions. Conversely, for compounds (**1**) where R = Ph and R_2 = 2-pyrrolo or H the conversion to cinnolines was very poor, and olefins (**1**) embodying R_1 = Ph and R_2 = 2-pyridyl gave only tars. In 1953, Schofield[7] evaluated a number of 4-aryl ethylenes of type **1** with R_2 = pyridyl and quinolyl substitution. These findings showed the advantage of a C-4 *p*-methoxyphenyl substituent, which leads to rapid cinnoline formation independent of pH. In contrast, cinnoline formation from starting materials containing a C-4 phenyl substituent is greatly dependent on pH, with rapidly decreasing yields accompanying increasing acidity. Generally speaking, an essential feature for successful cinnoline formation is an electron donating group on the α-carbon atom of the olefinic side chain. In addition, the presence of a strongly electron withdrawing substituent at α- or the β-position has an inhibitory effect on cinnoline formation.[8]

8.9.5 Experimental

The Widman–Stoermer Cinnoline Synthesis[4]

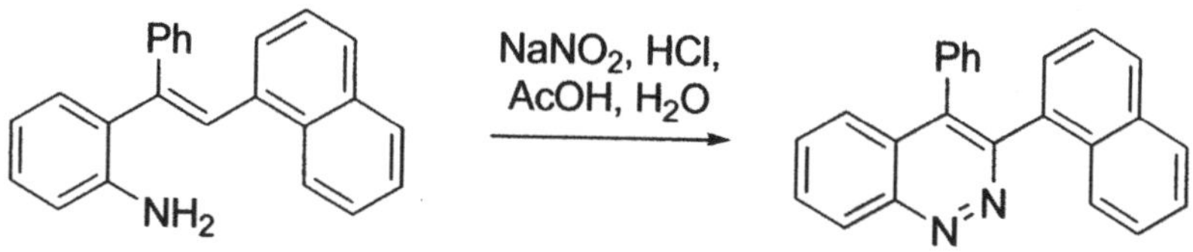

12 **13**

(a) A solution of the amino ethylene **12** (1.0 g) in acetic acid (10 mL) and concentrated hydrochloric acid (7 mL) was treated at 0 °C with 2.5% aqueous sodium nitrite (16 mL). (A brownish red solution of the diazonium salt was formed, which gave red insoluble product when a few drops of the solution were added to alkaline β-naphthol solution.) The clear diazonium solution was divided into two portions.

(b) (Intentional diversion to competing Pschorr chemistry.) One portion of the diazonium salt solution from above was poured slowly (5 min) into an excess of a slightly warm solution of sodium acetate in which was suspended 0.2 g of copper powder. After a few minutes the precipitate was collected, washed, dried, and extracted with hot ethyl acetate-alcohol (charcoal). The filtered extract, on concentration to a small volume, deposited 0.19 g of 2-phenylchrysene (Pschorr reaction product), mp 180–185 °C, which after two recrystallizations from ethyl acetate formed discolored leaflets, mp 192–193 °C. No other compound was isolated from the mother liquors; though an exhaustive search was not made.

(c) (Conversion to cinnoline.) One of the two diazonium salt solution portions was diluted with water (1.25 vol) and left for a few hours at room temperature. The precipitate that had gradually formed was collected, washed, and crystallized from aqueous alcohol after removal of a few milligrams of insoluble material, presumably 2-phenylchrysene; one further crystallization yielded the cinnoline **13** (270 mg) in almost pure condition, mp 174–176 °C. The pure cinnoline separated from aqueous alcohol in brittle yellow polyhedra.

8.9.6 References

1. Widman, O. *Ber.* **1884,** *17*, 722–727.
2. Stoermer, R.; Fincke, H. *Ber.* **1909,** *42*, 3115–3122.
3. Vinogradova, O. V.; Balova, I. A. *Chem. Heterocycl. Compounds* **2008,** *44*, 501–522.
4. Simpson, J. C. E. *J. Chem. Soc.* **1943,** 447–452.
5. Schofield, K.; Simpson, J. C. E. *J. Chem. Soc.* **1945,** 520–524.
6. Schofield, K. *J. Chem. Soc.* **1949,** 2408–2412.
7. Nunn, A. J.; Schofield, K. *J. Chem. Soc.* **1953,** 3700–3706.
8. Atkinson, C. M.; Biddle, B. N. *J. Chem. Soc.* **1966,** 2053–2060.

8.10 Wichterle Reaction

Michael T. Corbett

8.10.1 Description

The Wichterle reaction sequence is the annulation of a ketone to a cyclic enone using 1,3-dichloro-2-butene (**1**), the Wichterle reagent, via a three-step process consisting of an alkylation, hydrolysis, and intramolecular cyclization.

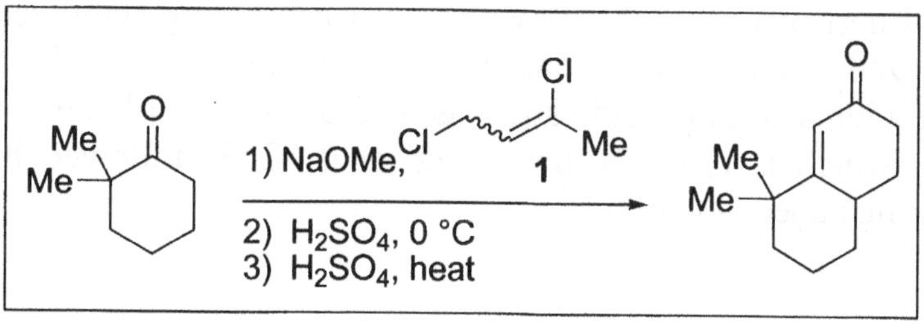

8.10.2 Historical Perspective

During the late 1930s, Otto Wichterle was working at the Czech Technical University in Prague when he first encountered 1,3-dicholo-2-butene, a byproduct in the production of the synthetic rubber monomer chloroprene.[1] The compound, which contains a reactive allylic chloride and an inert vinylic chloride, was first used in the bis-alkylation of diethyl malonate. When this compound was dissolved in sulfuric acid, evolution of HCl gas began immediately. The product obtained from this reaction was identified as being diethyl 3-acetyl-4-methyl-3-cyclohexene-1,1-dicarboxylate, which is the net result of the hydrolysis of the vinylic chloride moiety followed by a subsequent intramolecular aldol condensation.

Wichterle, however, was not able to fully realize the synthetic utility of this reaction due to the arrival of the Nazis in Czechoslovakia in 1938. Wichterle was forced to leave his university post in Prague and took exile in Zlín where he worked for Bat'a Shoes where he worked on the development of synthetic fibers. Despite imprisonment by the Gestapo in 1942, Wichterle began to publish his work in 1943.[2] After the war, Wichterle returned to Prague and began working in the area of hydrogels, which led to his most famous work, the invention of modern contact lenses.

Perhaps overshadowed by the earlier work of Sir Robert Robinson where methyl vinyl ketone was shown to perform the same annulation, the potential of the Wichterle reaction remained relatively untapped.[3] Despite some seminal works by Prelog and Julia, the full synthetic utility was not fully investigated until the 1960s and 1970s when ring-forming reactions became increasingly invaluable tools in the syntheses of polycyclic natural products, such as steroids and diterpenes.

8.10.3 Mechanism

There is no widely accepted mechanism for the Wichterle reaction and in-depth mechanistic studies have not been performed to fully probe the step-wise sequence. The first step in the Wichterle reaction is the base-mediated alkylation of carbonyl **2** with the allylic chloride **1** to afford the γ-chlorocrotyl derivative **3**. The alkylation proceeds using stabilized enolates or unstabilized enolates and enamines. The vinylic chloride **3** is then hydrolized with a strong acid, such as H_2SO_4, to afford the diketone **4**. Hydrolysis, which is typically performed at 0 °C, is followed by a rapid dilution during aqueous workup.[4,5]

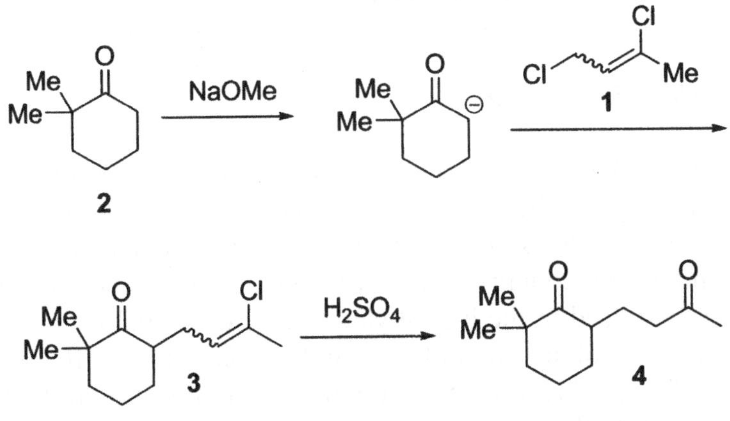

The pathway by which the hydrolysis proceeds is not fully understood. In Wichterle's original paper, the mechanism proposed for the formal hydrolysis consisted of the Markovnikov addition of H_2SO_4 across the olefin in γ-chlorocrotyl derivative **3** to reach the fleeting intermediate **5**.[2,6] This was followed by the concerted expulsion of SO_3 and HCl in a pericyclic pathway leading to the formation of the diketone **4**. This mechanistic pathway is consistent with reaction conditions where a molar equivalent of H_2O is not present.

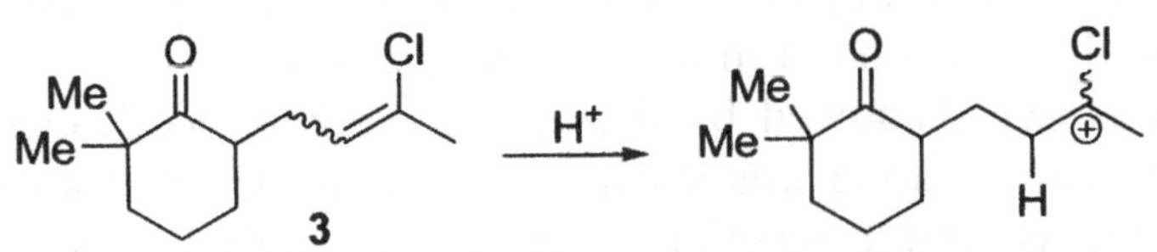

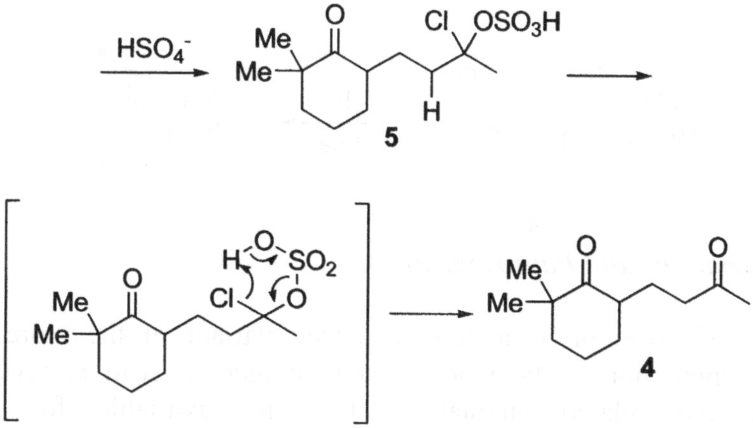

Studies by Matsumoto and Lansbury suggest that the hydrolysis proceeds through a different pathway.[7,8] Although initial protonation of the vinylic chloride **3** with H_2SO_4 produces the more stable carbocation **6**, the carbocation is not quenched with the bisulfate anion, but with a molecule of water present in solution to afford the transient alcohol **7**. The alcohol **7** then undergoes rapid expulsion of HCl to yield the diketone **4**. This mechanistic pathway relies on the ample presence of H_2O in solution and is inconsistent with observations that HCl evolution occurs instantaneously upon addition of acid.

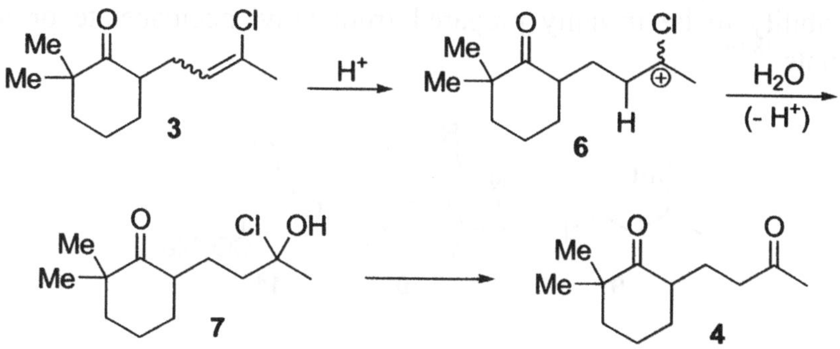

Following hydrolysis, the 1,5-diketone **4** undergoes an intramolecular aldol condensation to afford the cyclic enone **8** in a fashion analogous to the Robinson annulation.

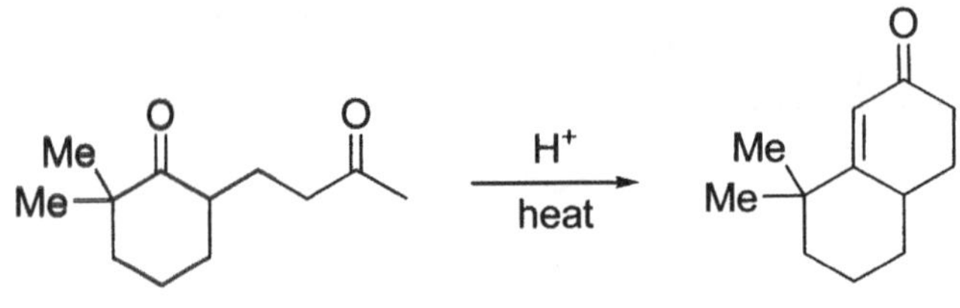

4 8

8.10.4 Variations and Improvements

The Wichterle reaction is a closely related variant of the more popular Robinson annulation. The reader is encouraged to read reviews of the Robinson and related annulations that are available for a more comprehensive overview of the variations that can be employed in synthesis.[9–11]

Due to the vast synthetic utility of annulation reactions, there are numerous analogous sequences that can be performed to reach the same class of products as the Wichterle reaction. The most notable of which is the Robinson annulation, which uses methyl vinyl ketone in place of 1,3-dichloro-2-butene.[3] Later research has elaborated on the Wichterle reagent by constructing an extensive library of methyl vinyl ketone surrogates **9–14** that can be employed in a manner analogous to 1,3-dichloro-2-butene.[12–17] Despite the development of annulation reagents like **9–14**, 1,3-dichloro-2-butene retains ample synthetic utility due to both its commercial availability and its ability to be trivially prepared from ethyl acetoacetate or methyl acetoacetate.[18,19]

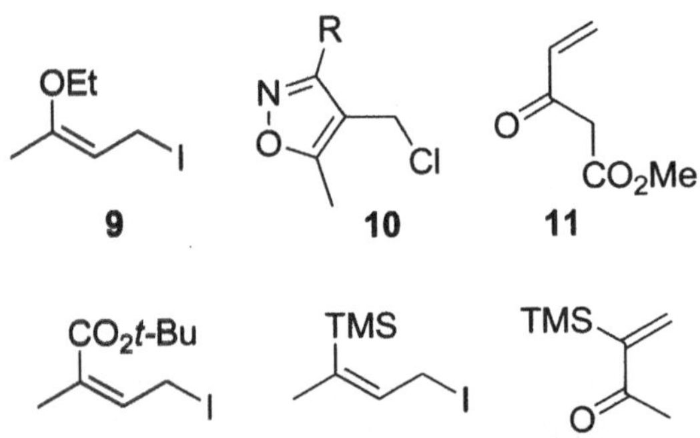

9 **10** **11** **12** **13** **14**

Despite the inherent increase in the rate of alkylation employing allyl iodides, such as **9** or **13**, or allyl chlorides in the presence of potassium iodide (*in situ* Finkelstein), the regioselectivity of the alkylation can be strongly influenced through the use of sterically hindered enolates.[20] To overcome this sluggish reactivity of the allyl chloride Wichterle reagent, Negishi and coworkers developed a Pd-mediated allylation of enoxyborates that is not only fast, but is extremely stereospecific.[21] Potassium cyclohexenoxytriethylborate (**15**) is treated with **1** in the presence of a catalytic amount of Pd(0) to afford the δ-chloro-γ,δ-unsaturated ketone **16** in 86% yield.

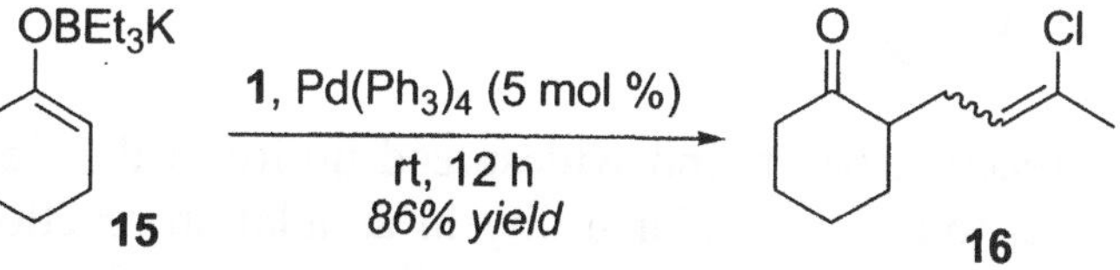

Various improvements to the Wichterle reaction have been made to enhance the utility of the reaction. One prevalent problem with the Wichterle reaction is the sometimes unpredictable nature of the hydrolysis and cyclization steps. During hydrolysis of the vinylic chloride, premature cyclization can occur in molecules that contain unsaturated ketones. Julia and Marshall both observed this phenomenon during the hydrolysis of 2,2′-dialkylcyclohexanone **17** with conc. H_2SO_4 to isolate the bicyclo[3.3.1]-nonanone **18** instead of the desired fused ring system.[4,22] The use of an unsaturated ketone, such as **19**, avoids this problem by affording an intermediate **20** that cannot undergo undesired cyclization to the bridged enone.

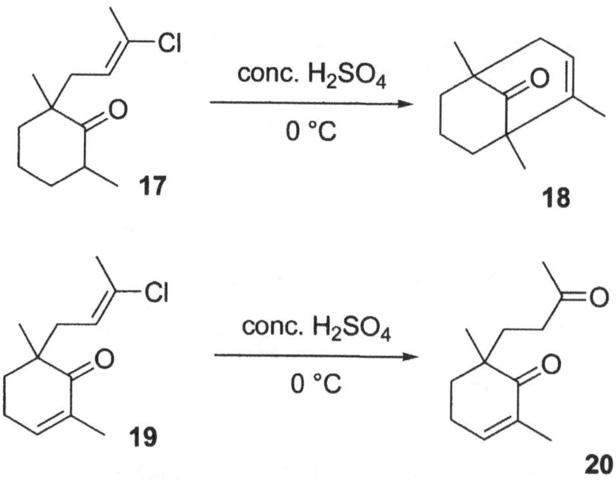

Another way to control the unpredictability of the hydrolysis is to use milder acidic conditions. The so-called modified Wichterle reaction was developed by Kobayashi and Matsumoto in the 1970s that changed the hydrolysis conditions from sulfuric acid to a 1:1 mix of formic acid and perchloric acid.[23] This system is compatible with saturated ketones and efficiently combines the hydrolysis and cyclization steps. The use of mercury(II) trifluoroacetate or titanium(IV) chloride is also a commonly used practice since it will only hydrolyze the vinylic chloride without inducing cyclization.[21,24] This allows for the intramolecular aldol condensation to occur in a separate step allowing for greater control.

8.10.5 Synthetic Utility

The Wichterle reaction has found widespread utility in the field of organic chemistry. It has found particular utility in annulation reactions where the use of methyl vinyl ketone is incompatible. Methyl vinyl ketone is susceptible to polymerization and is a relatively reactive Michael acceptor that can participate in undesired side reactions.

Lednicer and co-workers used the Wichterle sequence in the syntheses of mammalian antifertility agents derived from 5,6-diarylhydronaphthalenones, such as **24**.[25] The conversion of enamine **21** to bicyclic enone **24** proceeded in low yields when methyl vinyl ketone was used to perform the annulation; however, when the Wichterle sequence was employed to access the bicyclic enone **24** from the same starting material, the yield improved 10-fold. Double-bond isomerization to form an extended conjugated system was observed to occur during the hydrolysis of the vinylic chloride **22** to form the diketone **23**.

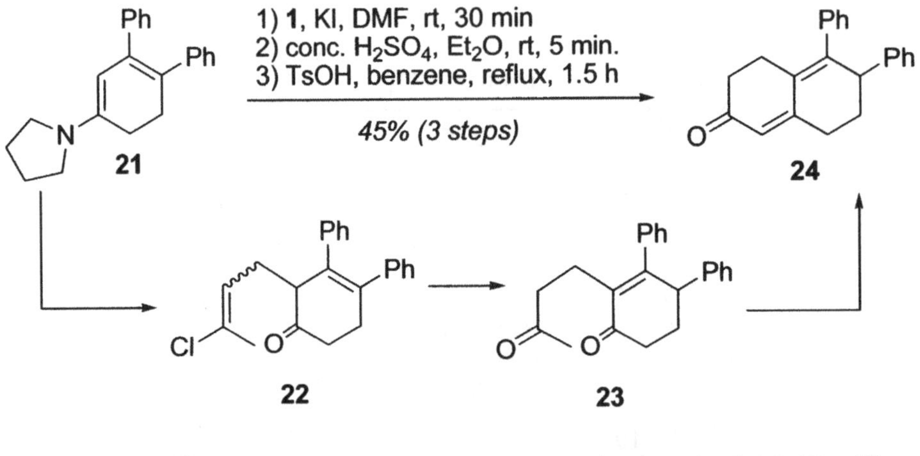

During their enantiocontrolled total synthesis of 18,18,18-trifluorosteroid **26**, Fukumoto and co-workers used a five-step Wichterle sequence to install the A-ring beginning from a previously prepared enantiopure precursor.[26] Alkylation of the enamine of ketone **25** was followed by subsequent reprotection and hydrolysis of the vinylic chloride to afford the diketone intermediate **26**, which underwent cyclization and acetylative isomerization to afford 18,18,18-trifluorosteroid **27** in 24% yield over the five-step sequence.

Fukumoto and co-workers used the Wichterle reaction in their formal synthesis of (+)-estrone.[27] The enone **28** was treated with pyrrolidine to form an enamine that was utilized in the alkylation with Wichterle's reagent, which then underwent subsequent hydrolysis and cyclization to afford (+)-dienedione (**29**). This dione intercepts Danishefsky's earlier total synthesis of (±)-estrone.

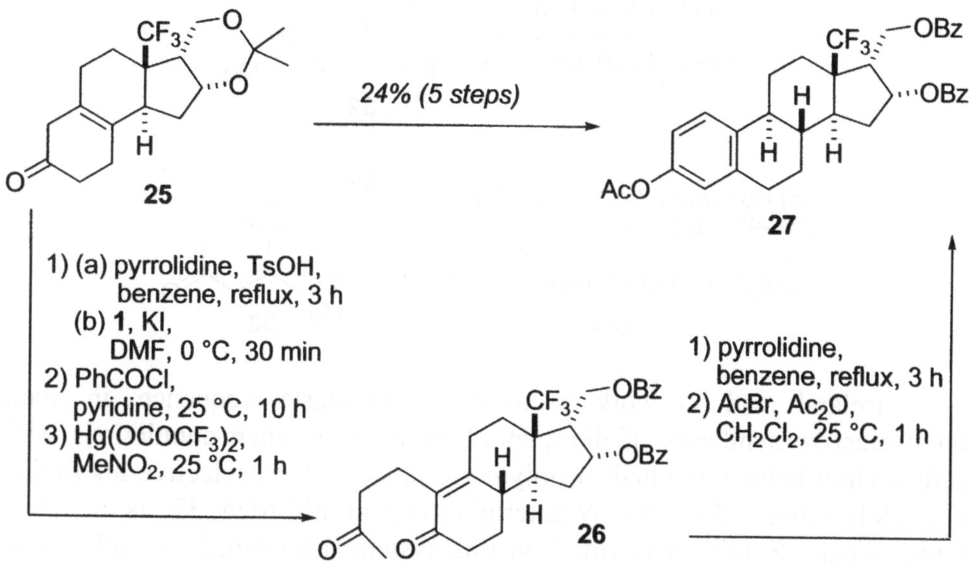

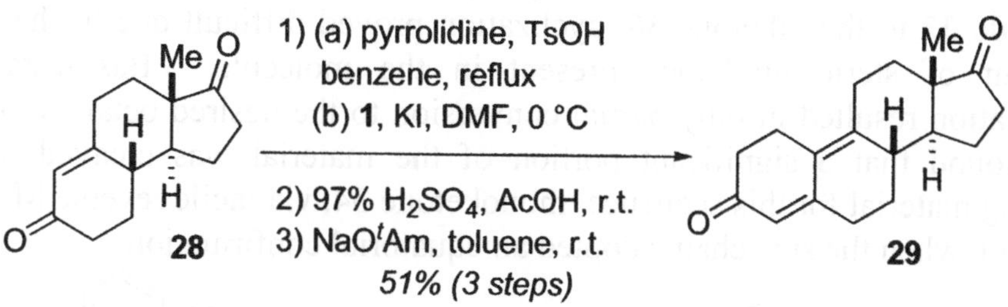

During the construction of the diterpene natural product (±)-4aβ,10β-doladiol acetate, Paquette and co-workers used a Wichterle sequence to construct the anthracenone **33**.[28,29] The β-keto ester **30** was alkylated with the Wichterle reagent to obtain the thermodynamic mixture of **31** (α:β = 3:2). Although the mixture of **31** can be separated via chromatography, both stereoisomers undergo a stereospecific methylation to obtain **32**, which undergoes subsequent hydrolysis and cyclization to provide the enone **33** in good overall yield (65%). When the aforementioned sequence was performed at the multigram scale, the yield was unfortunately subjected to a large decrease in efficiency (37%).

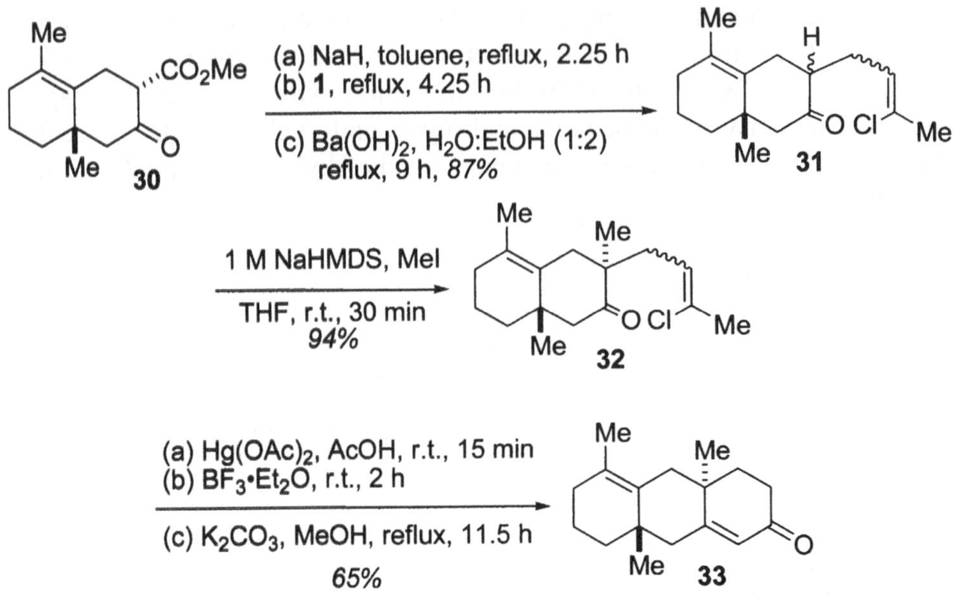

Ireland and co-workers used a Wichterle sequence in their stereoselective syntheses of diterpenoid resin acids when annulations with methyl vinyl ketone resulted in polymeric tars.[30] Stereoselective alkylation of cyclohexanone **34** with Wichterle's reagent afforded **35** as a single stereoisomer. Studies performed on this system determined that alkylation was favored *cis* to the C_2 methyl group. After hydrolysis of the vinylic chloride **35** to the diketone **36**, cyclization proved difficult due to the large amount of steric hindrance present in the molecule. Base-catalyzed cyclization resulted in only partial conversion to the desired octalone **37**. It was found that a significant portion of the material was cleaved to the starting material for this sequence, monoketone **34**, via facile reverse Michael addition when the side chain adopted an equatorial confirmation.

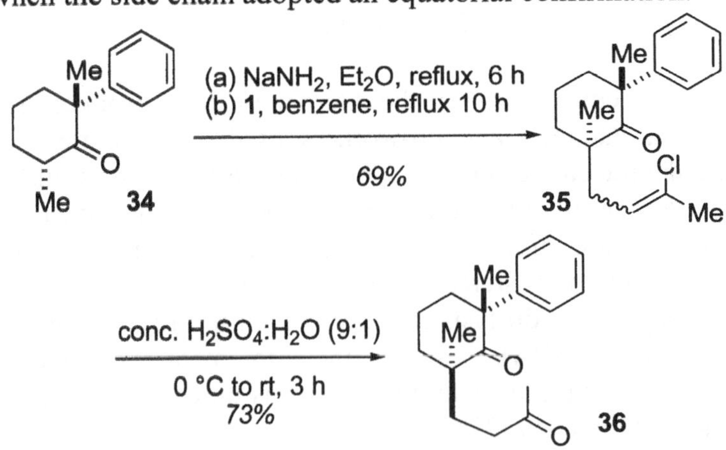

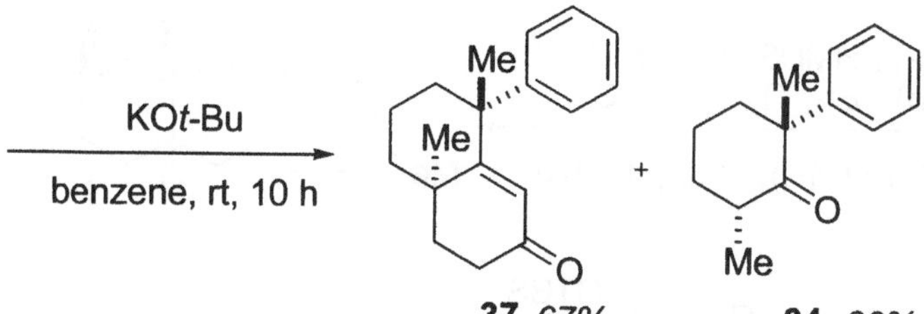

37 *67%* **34** *30%*

Covey and co-workers employed a stereo- and regio-specific alkylation of Wichterle's reagent in their enantiospecific total syntheses of the cyclopenta[β]anthracene and cyclopenta[β]phenanthrene ring systems.[32,33] The ketone **38**, which is derived from the Hajos–Parrish ketone, was alkylated with 1,3-dichloro-2-butene (**1**) to afford the respective chlorobutenyl isomers in excellent yield with complete regio- and stereocontrol following Negishi's protocol.[34–36] Subsequent hydrolysis and intramolecular aldol condensation provide the two benz[*f*]indene isomers, enone **39** and tetrasubstituted olefin **40**, in modest yields over five steps. Elaboration of **39** to the ketone **41** proceeds in two steps. During a second Wichterle sequence, however, alkylation of ketone **41** with **1** afforded a complex mixture of the regioisomers and stereoisomers **42** and **43**. This complex mixture co-eluted so the crude mixture was taken forward in hopes of separating the mixture in the future. Hydrolysis and aldol condensation of **42** and **43** proceeded to afford cyclopenta[*b*]anthracenone **44** and cyclopenta[*b*]phenanthrenone **45**, which were carefully separated via column chromatography and subsequent recrystallization.

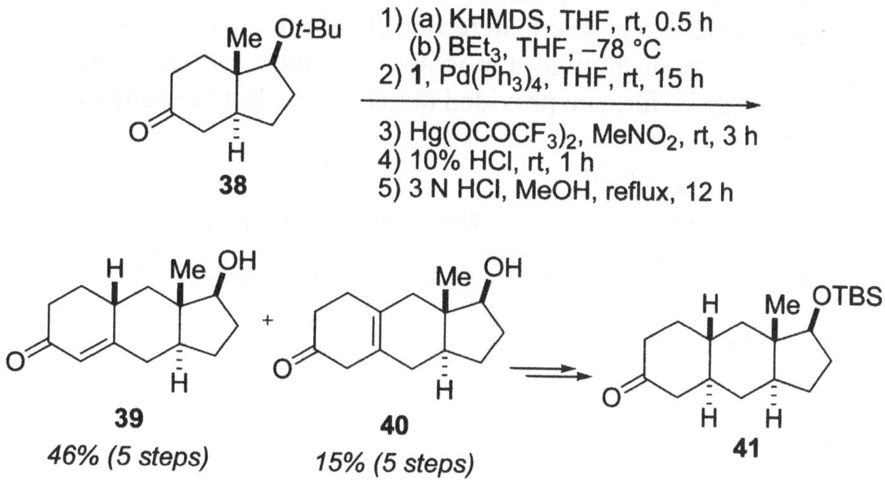

38

39 *46% (5 steps)*

40 *15% (5 steps)*

41

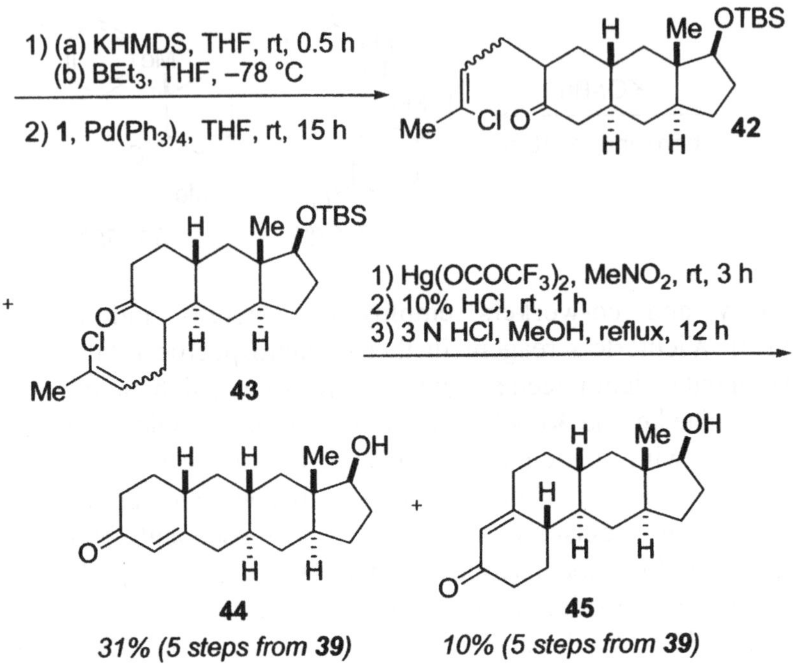

Reductive Alkylation

A method for the reductive alkylation of enediones was developed by Stork for incorporation in the synthesis of corticosteroids such as (±)-D-homoadrenosterone (**48**) and (±)-adrenosterone (**50**).[37,38] Attempted reductive alkylation of **46** by aprotic Michael addition using silyl enones was found to be ineffective thus equatorial alkylation with the modified Wichterle reagent was found to proceed stereoselectively to afford the dione **47**. Subsequent elaboration through a Wichterle sequence and hydrolysis/Jones oxidation of the *C*17a formate provided (±)-D-homoadrenosterone (**48**).

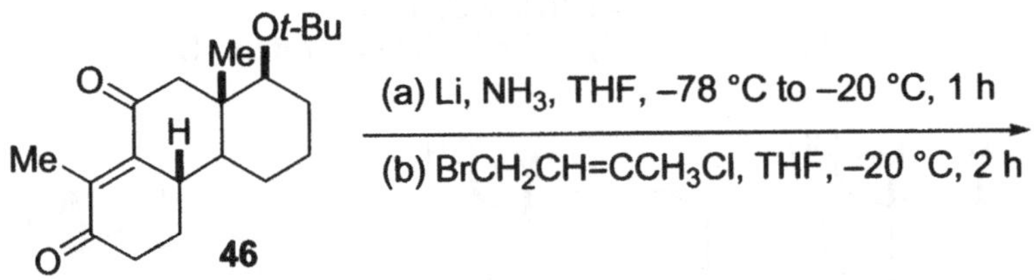

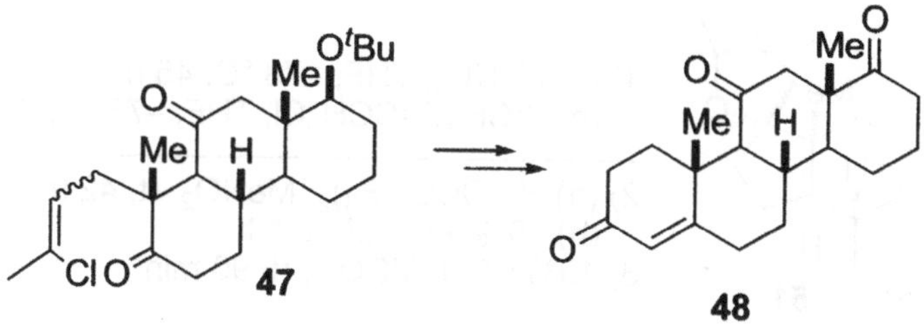

Stork applied the same strategy to the synthesis of the A-ring of (±)-adrenosterone (**50**) beginning from the enedione **49**.[39] Stereoselectivity of the alkylation is strongly governed by interactions between the *C*-10 methyl group and *C*-11 methoxy group that forces the transition state into a half-boat conformation to relieve this interaction and substantially increases the accessibility of the α-face. Alkylation from the β-face would force the methyl group to move past the methoxy group in a fully-eclipsed position, which is strongly disfavored. Subsequent syntheses of (±)-adrenosterone (**50**) have used analogous constructions of the A-ring using the Wichterle sequence by intercepting analogues of Stork's intermediate **49**.[27,40,41]

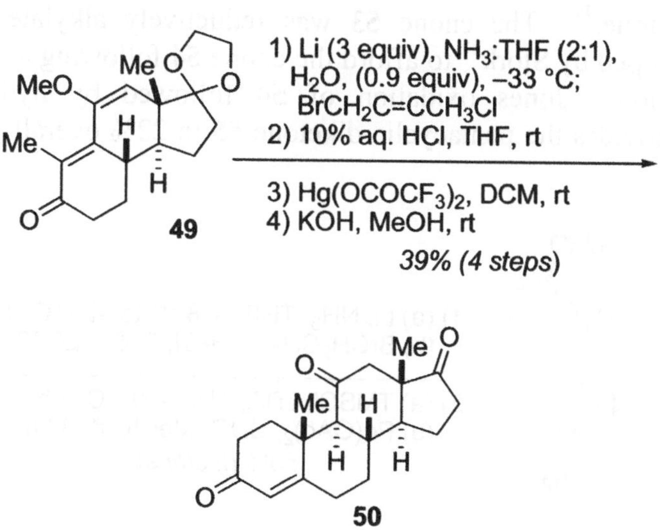

Stork's methodology was employed by Fukumoto and co-workers in the enantioselective construction of the steroidal skeletons of (+)-cortisone and (+)-11-deoxy-19-norcorticosterone, which contain nearly identical A, B, and C rings to (±)-adrenosterone (**50**).[42,43] The dione **51** underwent reductive alkylation followed by hydrolysis and base-catalyzed cyclization to afford the tetracyclic compound **52**, which contains the skeleton of (+)-coritsone. A similar protocol was followed in the construction of the core of (+)-11-deoxy-19-norcorticosterone.

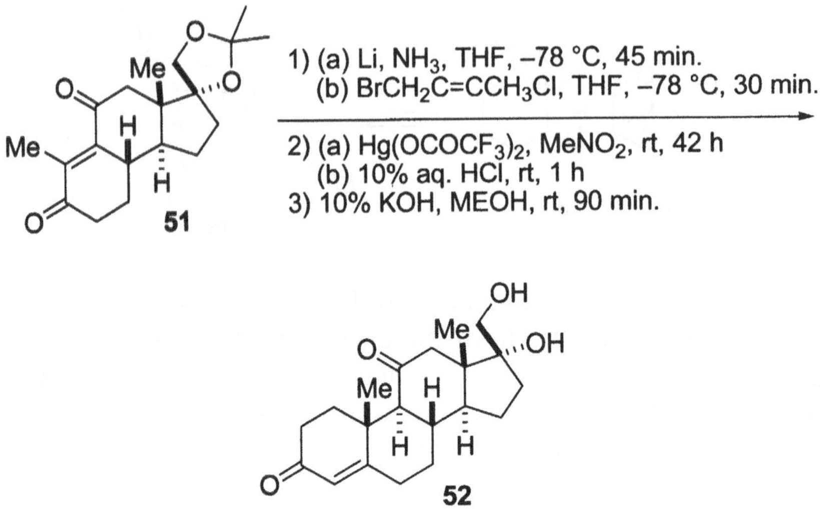

Fukumoto and co-workers also used the Wichterle sequence to construct the A-ring in their total synthesis of 19-norcanrenone (**55**), constituting a formal total synthesis of the spiro steroid 19-norspironolactone.[44] The enone **53** was reductively alkylated using the method developed by Stork[37] to afford the enone **54** following a Pd-mediated dehydrogenation. Jones oxidation of **54** followed by hydrolysis and cyclization provides the pentacyclic diketone **55** in 22% overall yield over 5 steps.

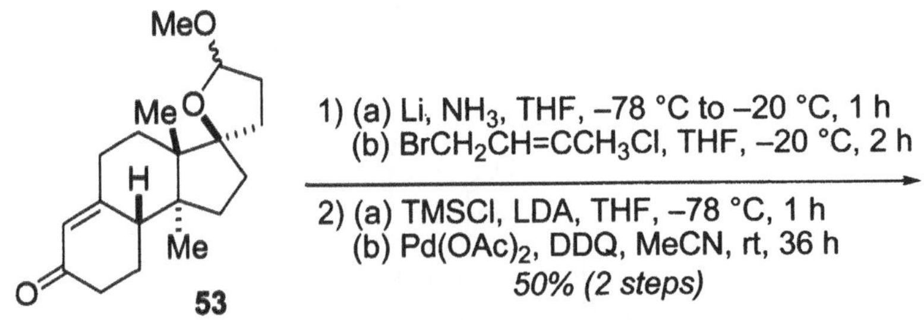

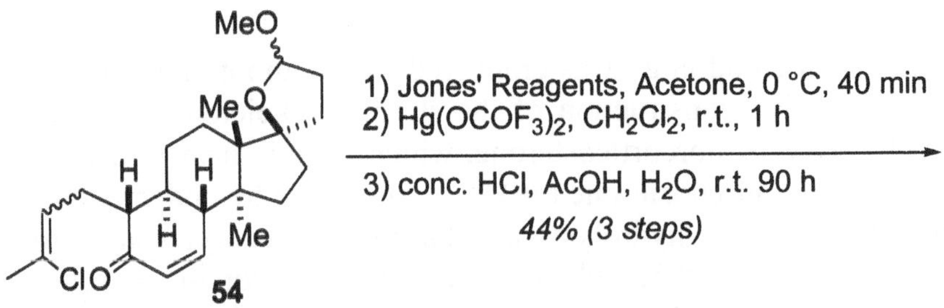

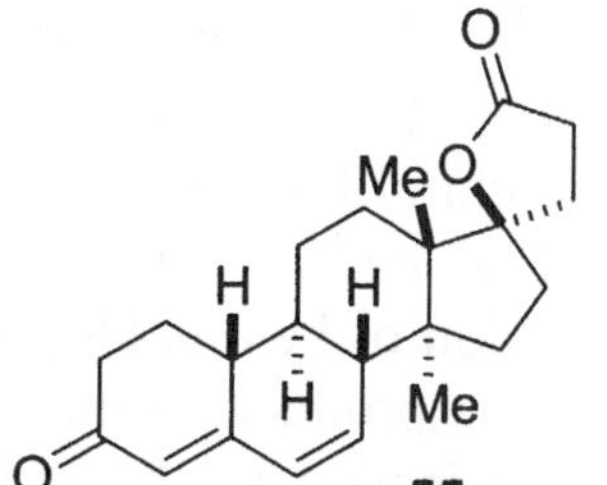

55

In their total synthesis of 6-methyl-5-androstene-3,11,17-trione 3-ethylene acetal, which is a building block for 6α-methylprednisolone, Daniewski and co-workers used a modified Wichterle reaction to install the A-ring of the natural product.[45] A racemic mixture of **56** was used in a reductive alkylation with Wichterle's reagent to afford the product **57** as a mixture that could be separated by chromatography. Upon treatment of each diastereomer of **57** with formic acid and perchloric acid to induce cyclization, it was observed that both diastereomers resulted in the same enantiopure product **58**. This preferred conformation of the steroidal skeleton was confirmed by comparison to an authentic sample of the product that can be obtained directly from natural cortisone.

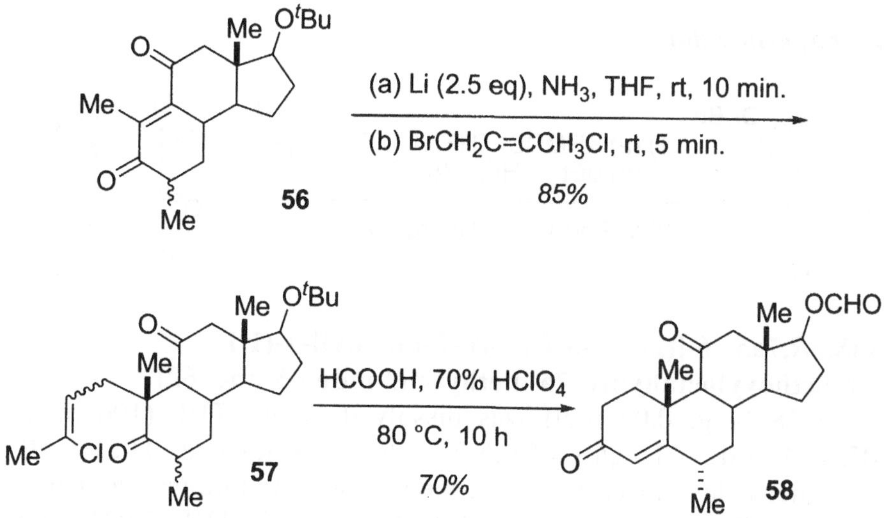

Enantioselective Catalysis

The Wichterle reagent has also been employed in enantioselective Robinson annulations involving the use of phase-transfer catalysis (PTC). Bhattacharya and co-workers had previously investigated the asymmetric alkylation of inadanones, such as **59**, using substituted *N*-benzylcinchoninium salts realizing their ability to produce enantio-enriched products in excellent yield.[46] This same strategy was applied to the alkylation of indanone **59** with **1** using *N*-(*p*-trifluoromethylbenzyl)-cinchoninium bromide as the chiral PTC, which afforded the desired ketone **60** in 99% yield and 92% *ee*.[47] The subsequent hydrolysis and intramolecular aldol condensation of ketone **60** provided the cyclic enone **61** in 97% yield from the alkylation product. Similar indanones have also been employed in the same enantioselective Wichterle sequence to obtain various analogues.[48,49]

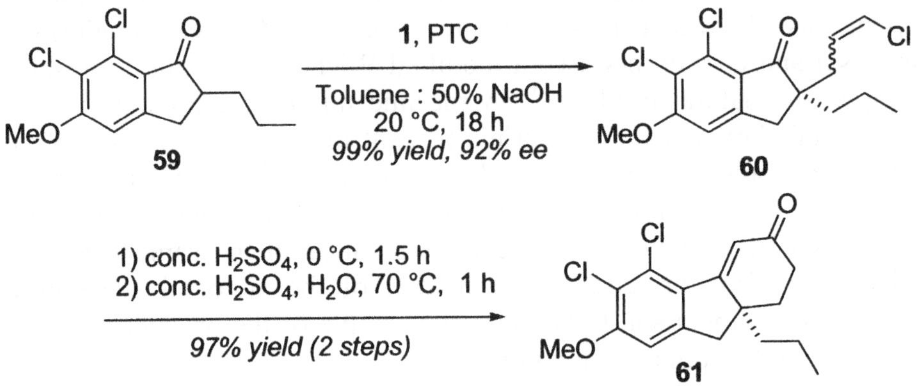

8.10.6 Experimental

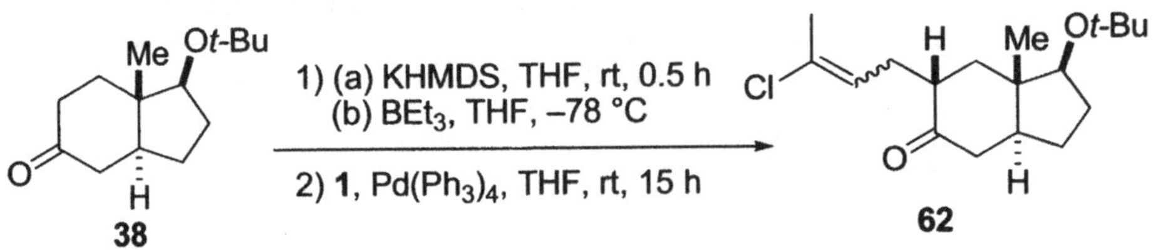

(1*S*,3a*S*,6*R*,7a*S*)-6-[(2*E*/*Z*)-3-Chloro-2-butenyl]-1-(1,1-dimethylethoxy)octahydro-7a-methyl-5*H*-inden-5-one (62)[33]

Indenone **38** (4 g, 0.02 mol) was dissolved in dry THF (100 mL), and KHMDS (42 mL, 1.1 equiv of 0.5 M solution in toluene, 0.021 mol) was added. After being stirred for 0.5 h, the reaction mixture was cooled to –78 °C, and BEt_3 (21 mL, 1.1 equiv of 1.0 M solution in THF, 0.021 mol) was added. This was followed by addition of a mixture of 1,3-dichloro-2-butene (**1**) (a mixture of cis and trans isomers, 2.3 mL, 1.1 equiv, 0.021 mol) and $Pd(PPh_3)_4$ (1.1 g, 5 mol %) in dry THF (30 mL). The reaction mixture was allowed to come to room temperature and was stirred for 15 h. At this time, 3 N HCl (15 mL) was added. The reaction was transferred to a separatory funnel, and the aqueous layer was extracted with Et_2O. The combined organic layers were washed with saturated $NaHCO_3$, dried, and evaporated.

Chromatography (silica gel, 5% EtOAc in hexanes) gave a mixture of cis and *trans* vinyl chlorides **62** as a yellow oil (5.45 g, 88%).

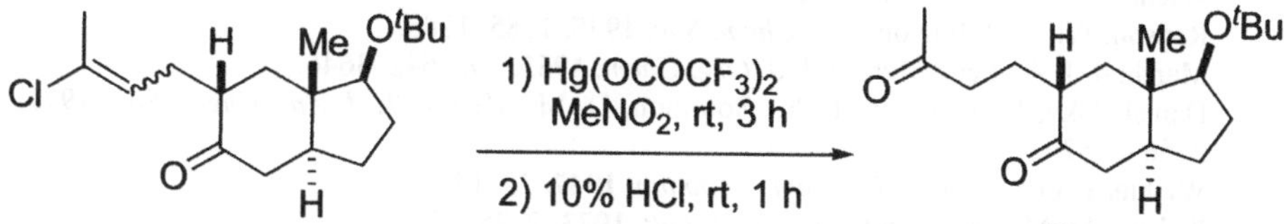

62 **63**

(1*S*,3a*S*,6*R*,7a*S*)-1-(1,1-Dimethylethoxy)octahydro-7a-methyl-6-(3-oxobutyl)-5*H*-inden-5-one (63)[33]

To a solution of mercury(II) trifluoroacetate (30 g, 0.070 mol, 4 equiv) in nitromethane (950 mL) was added the vinyl chloride mixture **62** (5.5 g, 0.017 mol) in dry THF (30 mL). The reaction mixture was stirred at room temperature for 3 h, at which time 10% HCl (450 mL) was added and the mixture was stirred for an additional 1 h. The mixture was transferred to a separatory funnel and the aqueous layer extracted with CH_2Cl_2. The combined organic fractions were washed with brine and dried, and the solvent was evaporated to give a yellow oil. Column chromatography (silica gel, 15% EtOAc in hexanes) yielded 3.7 g (72%) of diketone **63** as a colorless oil.

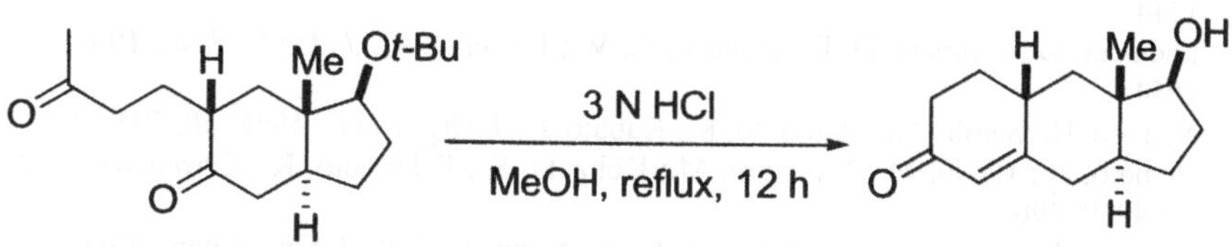

63 **39**

(1*S*,3a*S*,8a*R*,9a*S*)-1,2,3,3a,4,7,8,8a,9,9a-Decahydro-1-hydroxy-9a-methyl-6 *H*-benz[*f*]inden-6-one (39)[33]

Diketone **63** (3.4 g, 0.012 mol) was dissolved in MeOH (125 mL), and 3 N HCl (30 mL) was added. The reaction mixture was refluxed overnight and then cooled to room temperature, and the MeOH evaporated in vacuo. EtOAc and H_2O were added to the residue, and the aqueous layer was extracted with EtOAc. The combined organic fractions were then washed with saturated $NaHCO_3$ and dried, and the solvent was removed. Chromatography (silica gel, 35% EtOAc in hexanes) gave the previously reported enone **39** (1.84 g, 73%).

8.10.7 References

1. [R] Hudlický, M. *Coll. Czech. Chem. Commun.* **1993,** *58*, 2229–2244.
2. Wichterle, O. *Chem. Listy* **1943,** *37,* 180.
3. Rapson, W. S.; Robinson R. *J. Chem. Soc.* **1935,** 1285–1288.
4. Marshall, J. A.; Schaeffer, D. J. *J. Org. Chem.* **1965,** *30*, 3642–3646.
5. Danishefsky, S.; Crawley, L. S.; Solomon, D. M.; Heggs, P. *J. Am. Chem. Soc.* **1971,** *93*, 2356–2357.
6. Wichterle, O. *Coll. Czech. Chem. Commun.* **1947,** *12,* 193.
7. Kobayashi, M.; Matsumoto, T. *Chem. Lett.* **1973,** *2*, 957–960.
8. Lansbury, P. T.; Nienhouse, E. J.; Scharf, D. J.; Hilfiker, F. R. *J. Am. Chem. Soc.* **1970,** *92*, 5649–5657.
9. [R] Bergmann, E. D.; Ginsberg, D.; Pappo, R. *Org. React.* **1959,** *10*, 179–555.
10. [R] Gawley, R. E. *Synthesis* **1976,** 777–794.
11. [R] Hajos, Z. G.; Parrish, D. R. *J. Org. Chem.* **1974,** *39*, 1615–1621.
12. Stork, G.; Lowenthal, H. J. E.; Mukharji, P. C. *J. Am. Chem. Soc.* **1956,** *78*, 501–502.
13. Stork, G.; Danishefsky, S.; Ohashi, M. *J. Am. Chem. Soc.* **1967,** *89*, 5459–5460.
14. Weinkert, E.; Afonso, A.; B-son Bredenberg, J.; Kaneko, C.; Tahara, A. *J. Am. Chem. Soc.* **1964,** *86*, 2038–2043.
15. Stotter, P. L.; Hill, K. A. *J. Am. Chem. Soc.* **1974,** *96*, 6524–6526.
16. Stork, G.; Ganem, B. *J. Am. Chem. Soc.* **1973,** *95*, 6152–6153.
17. Stork, G.; Jung, M. E. *J. Am. Chem. Soc.* **1974,** *96*, 3682–3684.
18. Hatch, L. F.; Perry, Jr., R. H. *J. Am. Chem. Soc.* **1955,** *77*, 1136–1138.
19. Conrow, R. E.; Marshall, J. A. *Synth. Commun.* **1981,** *11*, 419–422.
20. Näf, F.; Decorzant, R. *Helv. Chim. Acta* **1974,** *57*, 1317–1327.
21. Negishi, E.; Luo, F.-T.; Pecora, A. J.; Silveira, A. *J. Org. Chem.* **1983,** *48*, 2427–2430.
22. Julia, S. *Bull. Soc. Chim. Fr.* **1954,** 780–789.
23. Kobayashi, M.; Matsumoto, T. *Bull. Chem. Soc. Jpn.* **1979,** *52*, 2978–2990.
24. Nemoto, H.; Satoh, A.; Ando, M.; Fukumoto, K. *J. Chem. Soc., Perkin Trans. 1* **1991,** 1309–1314.
25. Lednicer, D.; Emmert, D. E.; Duncan, G. W.; Lyster, S. C. *J. Med. Chem.* **1967,** *10*, 1051–1054.
26. Nemoto, H.; Satoh, A.; Fukumoto, K.; Kabuto, C. *J. Org. Chem.* **1995,** *60*, 594–600.
27. Nemoto, H.; Hagai, M.; Moizumi, M.; Kohzuki, K.; Fukumoto, K. *Tetrahedron Lett.* **1988,** *29*, 4959–4962.
28. Paquette, L. A.; Lin, H.-S.; Belmont, D. T.; Springer, J. P. *J. Org. Chem.* **1986,** *51*, 4807–4813.
29. Paquette, L. A.; Belmont, D. T.; Hsu, Y.-L. *J. Org. Chem.* **1985,** *50*, 4667–4672.
30. Ireland, R. E.; Kierstead, R. C. *J. Org. Chem.* **1966,** *31*, 2543–2559.
32. Scaglione, J. B.; Rath, N. P.; Covey, D. F. *J. Org. Chem.* **2005,** *70*, 1089–1092.
33. Jastrzebska, I.; Scaglione, J. B.; DeKoster, G. T.; Rath, N. P.; Covey, D. F. *J. Org. Chem.* **2007,** *72*, 4837–4843.
34. Negishi, E.; Idacavage, M. J. *Tetrahedron Lett.* **1979,** *10*, 845–848.
35. Negishi, E.; Matsushita, H.; Chatterjee, S.; John, R. *J. Org. Chem.* **1982,** *47*, 3188–3190.
36. Negishi, E.; Luo, F. *J. Org. Chem.* **1983,** *48*, 2427–2430.
37. Stork, G.; Logusch, E. W. *J. Am. Chem. Soc.* **1980,** *102*, 1218–1219.
38. Stork, G.; Logusch, E. W. *J. Am. Chem. Soc.* **1980,** *102*, 1219–1220.
39. Stork, G.; Winkler, J. D.; Shiner, C. S. *J. Am. Chem. Soc.* **1982,** *104*, 3767–3768.
40. Van Royen, L. A.; Mijngheer, R.; De Clercq, P. J. *Tetrahedron Lett.* **1983**, *24*, 3145–3148.
41. Van Royen, L. A.; Mijngheer, R.; De Clercq, P. J. *Tetrahedron* **1985,** *41*, 4667–4680.
42. Nemoto, H.; Matsuhashi, N.; Imaizumi, M.; Nagai, M.; Fukumoto, K. *J. Org. Chem.* **1990,** *55*, 5625–5631.
43. Nemoto, H.; Satoh, A.; Ando, M.; Fukumoto, K. *J. Chem. Soc., Perkin Trans. 1* **1991,** 1309–1314.
44. Nemoto, H.; Fujita, S.; Nagai, M.; Fukumoto, K.; Kametani, T. *J. Am. Chem. Soc.* **1988,** *110*, 2931–2938.
45. Daniewski, A. R.; Piotrowska, E. *Liebigs Ann. Chem.* **1989,** 571–576.

46. Dolling, U.-H.; Davis, P.; Grabowski, E. J. J. *J. Am. Chem. Soc.* **1984,** *106*, 446–447.
47. Bhattacharya, A.; Dolling, U.-H.; Grabowski, E. J. J.; Karady, S.; Ryan, K. M.; Weinstock, L. M. *Angew. Chem., Int. Ed. Engl.* **1986,** *25*, 476–477.
48. Bhattacharya, A.; Vasques, T.; Ramirez, T.; Plata, R. E.; Wu, J. *Tetrahedron Lett.* **2006,** *47*, 5581–5583.
49. Huffman, M. A.; Rosen, J. D.; Farr, R. N.; Lynch, J. E. *Tetrahedron* **2007,** *63*, 4459–4463.

Chapter 9 Miscellaneous Name Reactions 515

9.1 ANRORC Mechanism

Ian S. Young

9.1.1 Description

ANRORC (Addition of Nucleophile, Ring Opening, Ring Closing) reactions[1,2] represent a class of transformation that can occur with certain heterocyclic systems. This rearrangement cannot accurately be represented with a single scheme, as the reaction course/outcome is dependent on the heterocycle and nucleophile employed. Generally, nucleophilic addition to an activated heterocycle induces ring opening to an acyclic structure, and in many cases, re-closure incorporates the nucleophile into the newly formed heterocycle. The process finds the greatest utility in the rearrangement of easily prepared heterocycles to products that are difficult to access via conventional chemistry.

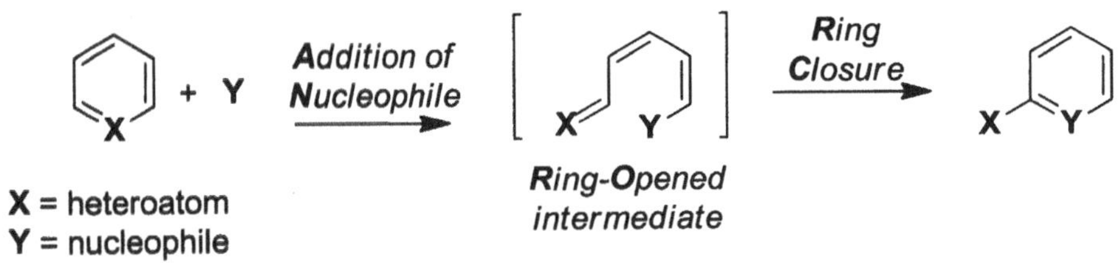

9.1.2 Historical Perspective

The ANRORC mechanism is operative in a number of named reactions (*e.g.*, Zincke–König[3] Dimroth[4] and in some instances Chichibabin),[5] that have been known for over 100 years. Seminal work in the 1960s by Henk van der Plas used isotopic labeling to deduce the mechanism of these transformations for a number of systems. Based on van der Plas' studies, the acronym ANRORC was coined.[1]

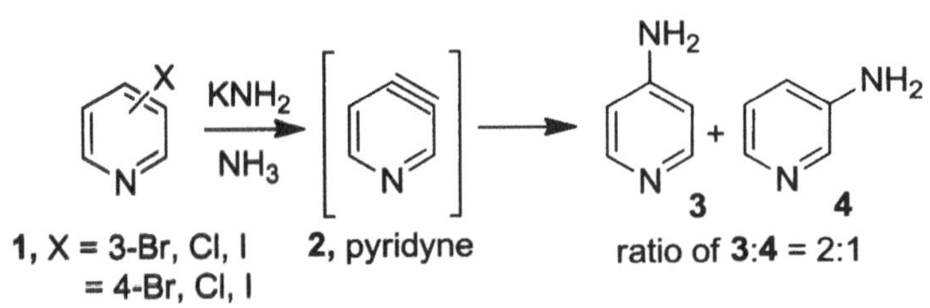

By the early 1960s there was strong evidence for the existence of pyridyne (**2**).[6,7] This was based on the observation that both 3- and 4-halosubstituted pyridines (**1**) yielded the same ratio of the 3- and 4-amination products (**3** and **4**) when reacted with potassium amide. The ratio was independent of the identity of the starting halogen.

Although not the discoverer of pyridyne, van der Plas began studying the addition of strong bases to halopyrimidines. He found that 6-bromo-4-arylpyrimidine **5** underwent exchange with potassium amide to yield **6**,[1] a transformation that could be envisioned to proceed through pyrimidyne **7**. Further study using the secondary amide base, lithium piperidide, provided unexpected ring-opened intermediate **8**. This led van der Plas to speculate that potassium amide reacts with similar regioselectivity, and the formation of **6** proceeds through a ring-opened, and not a pyrimidyne (**7**) intermediate.

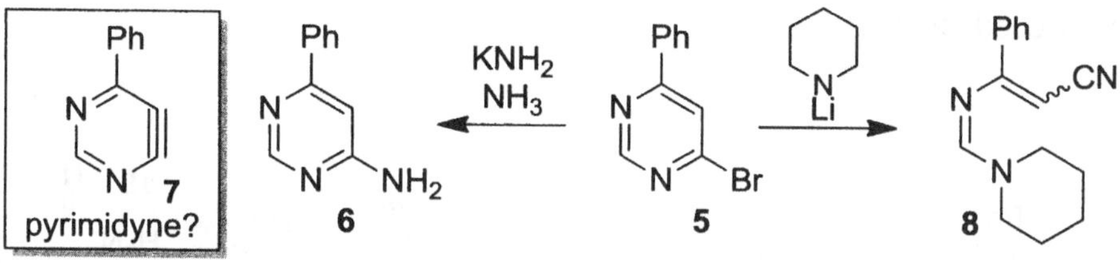

To support the occurrence of ring-opening with potassium amide, 6-bromo-4-phenyl pyrimidine (**9**) was prepared containing a 6% enrichment of ^{15}N distributed over both ring nitrogen atoms. Ring opening, followed by re-closure would transpose an internal nitrogen to the exterior, with incorporation of the nucleophile-derived nitrogen into the ring of **10**. After formation of, and conversion of **10** back to starting material (**12**), it was found that **12** derived from this sequence contained a 3.5% excess of ^{15}N. This indicated that extrusion of an original ring nitrogen had occurred, and supported an ANRORC mechanism for this pyrimidine system (in contrast to a pyrimidyne intermediate). Using similar labeling studies, van der Plas investigated the scope of the ANRORC mechanism with respect to heterocycle, nucleophile and leaving/activating group electronics.[1]

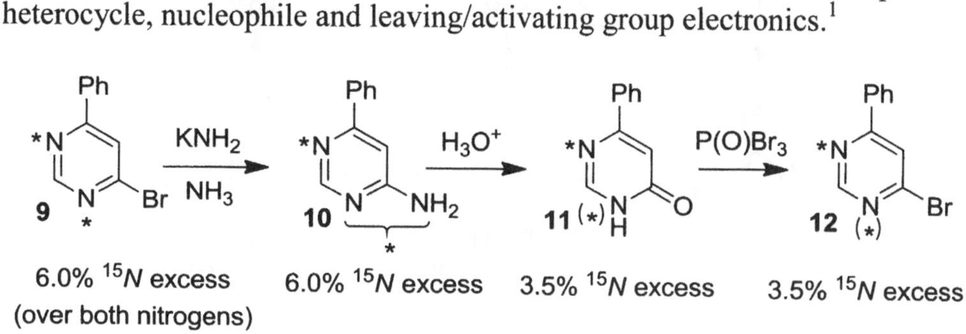

9.1.3 *Mechanism*

Proposed Mechanism Based on Isotopic Labeling

The exact mechanistic sequence of the ANRORC transformation is dependent on the system studied. Throughout this chapter, proposals are presented to provide an overview of the varied mechanistic progression.

Based on van der Plas' ^{15}N-labeling study with pyrimidines, the following mechanism was deduced for the conversion of **9** to **10**. Amide anion attack at the 2-position of pyrimidine **9** produces resonance stabilized anion **13**. Fragmentation yields iminoyl bromide **14**, which can directly form product (**10**), or alternatively can eliminate HBr to form nitrile **15**, which then proceeds to **10**.

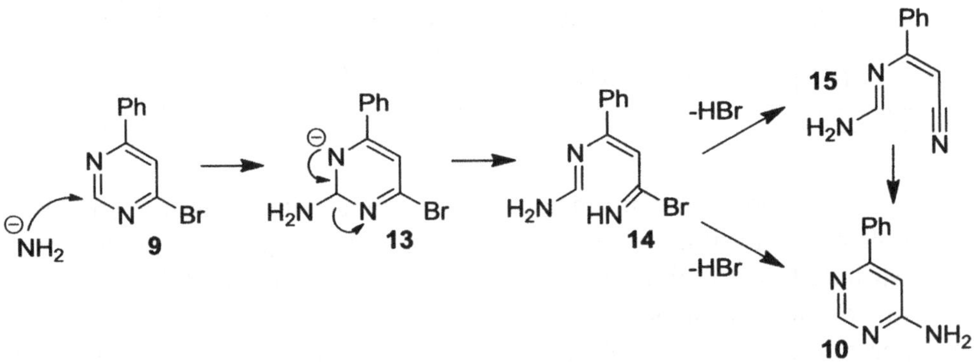

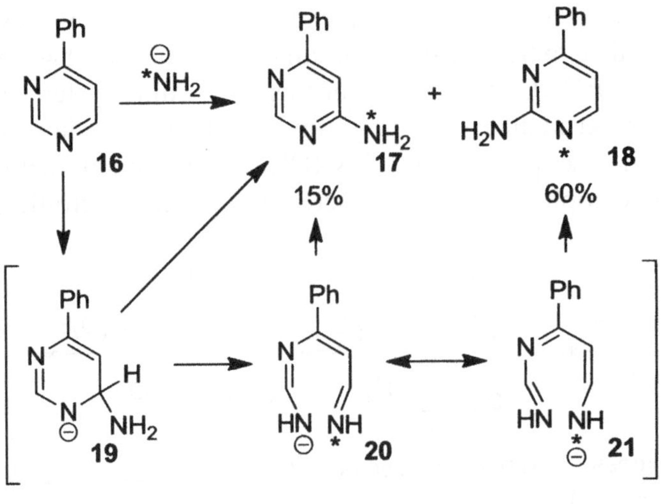

The Chichibabin reaction has been shown, at least in part, to proceed through an ANRORC pathway with pyrimidines.[8] After attack at the 6-position of **16** by ^{15}N-labeled amide anion, intermediate **19** could either collapse to the minor product **17** via the standard Chichibabin mechanism or undergo ring opening to resonance structures **20** and **21**. Ring closure produces the isomeric products (**17** or **18**), the major (**18**) having ^{15}N-incorporated into the heterocycle.

Isolation and Identification of Reaction Intermediates

While exploring the mechanism of the reaction of substituted pyridines with hydroxide, Haynes found that 2-chloro-5-nitropyridine (**22**) yielded expected product **24**, but the corresponding 2-chloro-3-nitropyridine **25** does not undergo reclosure after opening.[9] Using ^{1}H-NMR experiments, it was found that initially formed *cis*-**26** isomerizes to *trans*-**27**, which cannot undergo pyridine formation upon addition of hydroxide to the nitrile. Intermediate **23** undergoes this isomerisation to a lesser degree. Compound **27** was isolated as the sodium salt, and the structure was confirmed by *X*-ray diffraction.

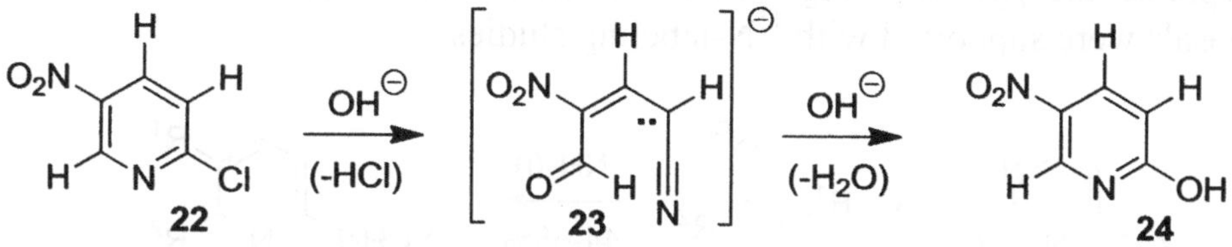

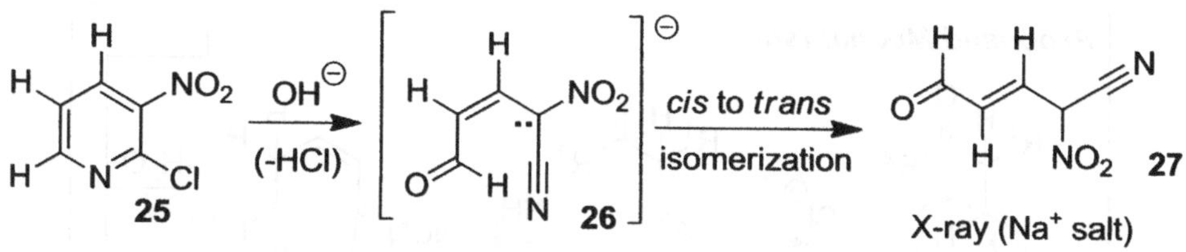

9.1.4 Variations and Improvements

ANRORC Reactions with Carbon Nucleophiles

ANRORC reactions are not limited to heteroatom nucleophiles, with the following examples being initiated by addition of carbanions.

During the course of studying the photostimulated addition of nitrile-stabilized carbanions to halopyridines and haloquinolines,[10] Wolfe found that alternative products were produced if the reaction was performed in the dark. In the absence of light, reaction of **28** with the potassium salt of acetonitrile produced **29** via the proposed ANRORC pathway.

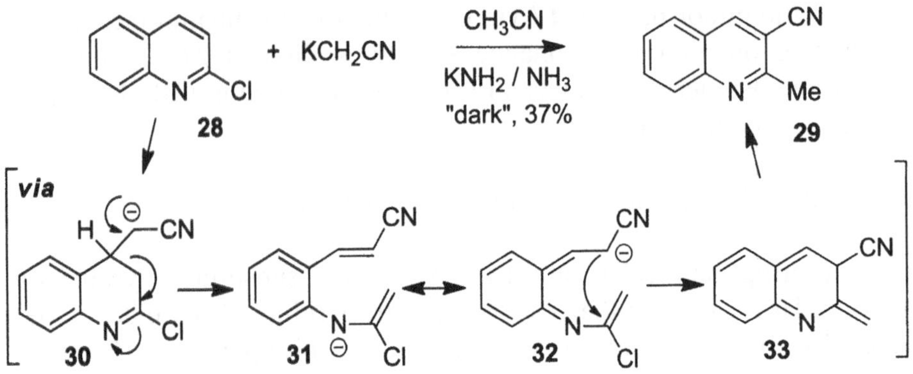

Wolfe also added ketone enolates (**35**) to 2,4-dichloropyrimidine (**34**) to prepare functionalized pyridines (**36**).[11] The proposed mechanism begins with the addition of the ketone enolate to the 6-position of the pyrimidine inducing fragmentation. Two potential pathways could be envisioned that incorporate the pyridine ring nitrogen from the ammonia solvent. These proposals were supported with ^{15}N-labeling studies.

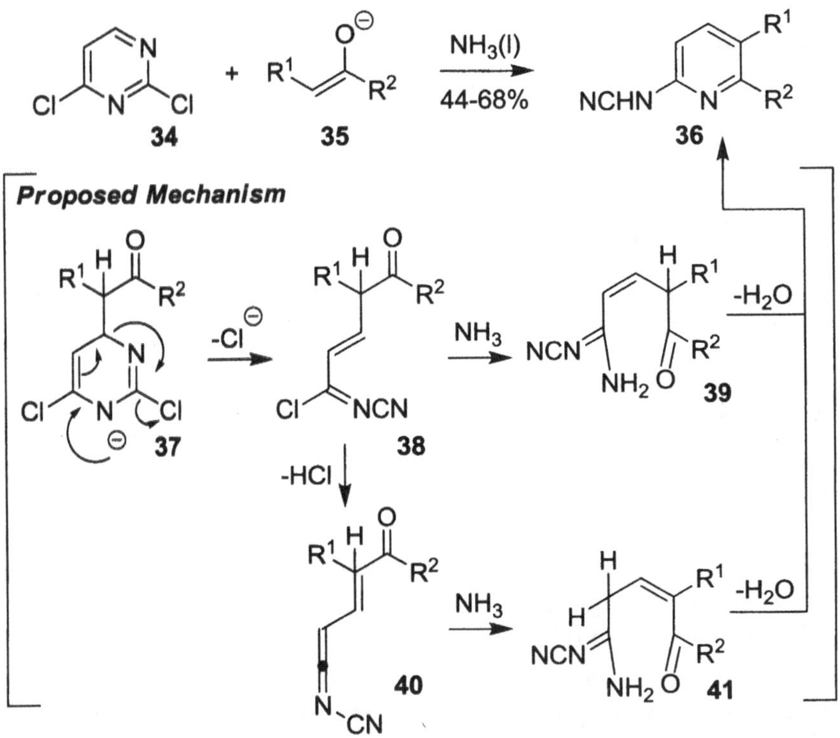

Reactions of Amines with Pyrylium Salts to form Pyridinium Salts

Katritzky converted pyrylium salts **44** (formed from the reaction of enone **42** and ethyl pyruvate **43**) to the corresponding pyridiniums (**45**) via ANRORC reaction with a primary amines.[12] This strategy serves as an alternative to the Zincke–König method for the formation of substituted pyridinium salts.

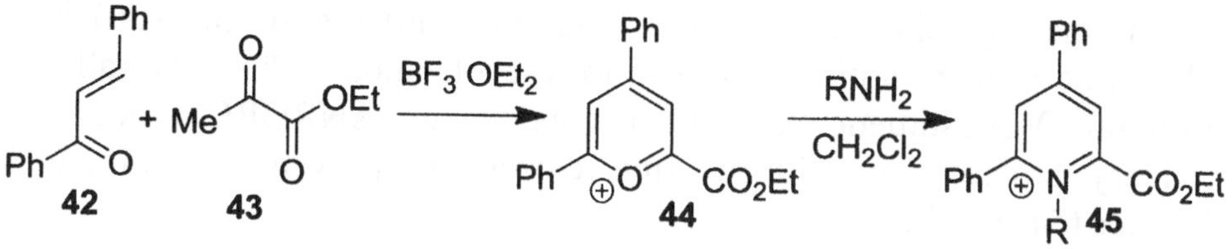

Rearrangement of Oxadiazoles with Hydroxylamines

The examples previously presented have featured ANRORC reactions on 6-membered rings. More recently, ANRORC-type rearrangements have been observed in their five-membered counterparts, although most examples feature oxadiazoles with bidentate nucleophiles. Spinelli has shown that oxadiazoles with perfluoro-substituents (**46**) undergo reaction with hydroxylamine to form oxadiazoles (**47**) with a net reversal in substitution pattern.[13] A perfluoro-group is required at the 5-position for reactivity, as the product 3-perfluoro-oxadiazole (**47**) is inert to the reaction conditions (preventing reversibility). Hydrazine also efficiently reacts with oxadiazoles to form 1,2,4-triazines.

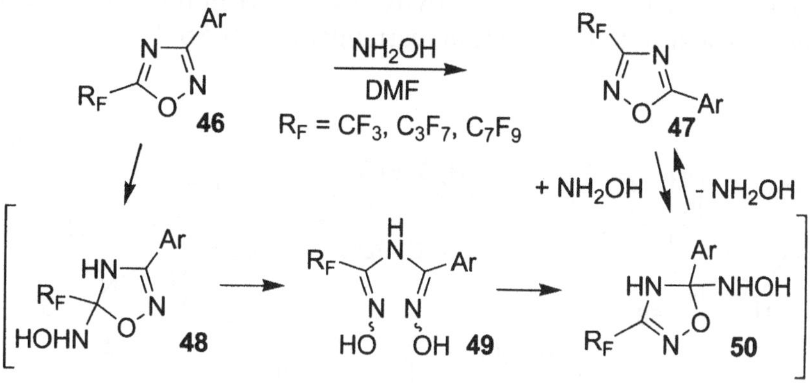

9.1.5 *Synthetic Utility*

Zincke–König Reaction

At the turn of the 20th century, Zincke[14] and König[15] independently discovered that pyridinium salts could be opened by amines. This process is facilitated by an electron-withdrawing group (commonly 2,4-dinitrophenyl, **53**) on the pyridinium nitrogen, and the utilization of sufficiently nucleophilic primary amines can lead to re-cyclization to **55**. Overall a net pyridinium *N*-substituent exchange occurs via an ANRORC process. If secondary amines are utilized, ring closure is prevented, and Zincke aldehydes (**54**) result.

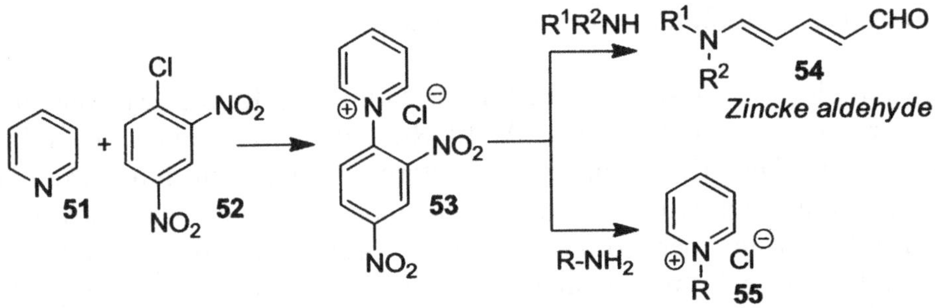

Dimroth Rearrangement to Prepare Pharmaceutically Active Pyrrolo[2,3-d]pyrimidines.

Novartis prepared over 275 g of pyrrolo[2,3*d*]pyrimidine **57** in excellent yield via the water initiated Dimroth rearrangement of **56**.[16]

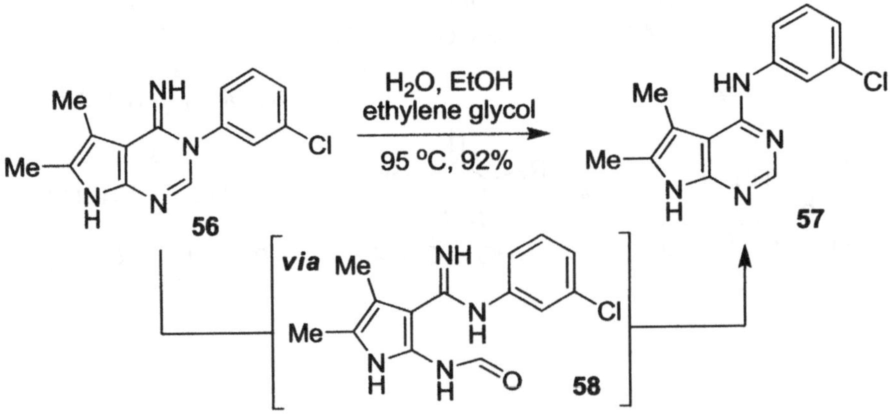

Capture with Nucleophilic Side Chain: Synthesis of Porothramycin

A variation of the ANRORC reaction uses an intramolecular nucleophile to open the activated heterocycle. Vanderwal accomplished a short, formal synthesis of porothramycin A and B (not depicted) via opening of the pyridinium salt with the amide nitrogen found within **59**.[17] Product **60** contains the key functionalized pyrroline of the natural product, and the expelled Zincke aldehyde was further functionalized.

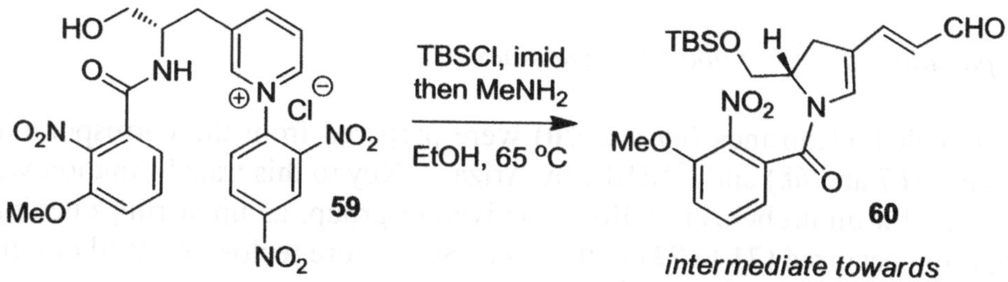

intermediate towards porothramycin A and B

Synthesis of Ergothioneine

The first synthesis of the natural amino acid L-ergothioneine (not depicted) was completed by Yadan and featured the direct conversion of an imidazole to an imidazole-2-thione.[18] Treatment of **61** with phenyl chlorothiono-formate followed by base led to a ring-opening/ring-closing cascade that proceeded through intermediate **62**. Compound **63** was elaborated to the target amino acid.

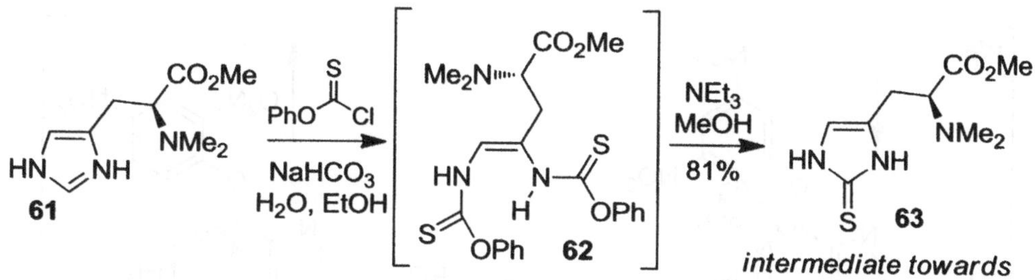

intermediate towards L-ergothioneine

Chiral Uracils

Walczak introduced chiral substituents onto the nitrogen of uracils (**64**)[19] via reaction with amino acid derivatives (**65**). This mild method did not lead to epimerization of the chiral center and offered an alternative to alkylation of the uracil nitrogen. An electron-withdrawing group (nitro or cyano) on the uracil ring was necessary for product (**66**) formation.

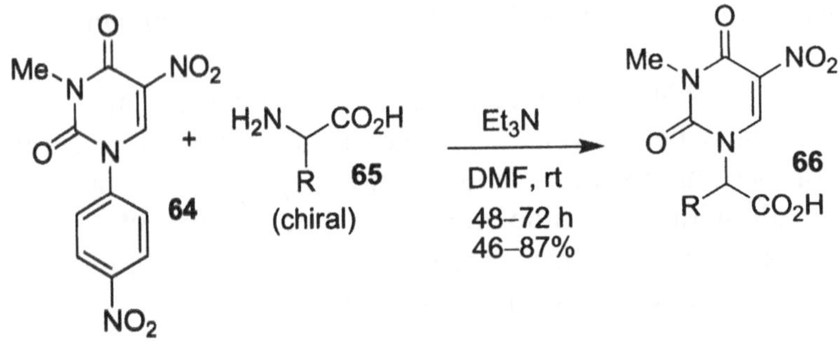

Preparation of 1-^{15}N-Labeled Adenosines

^{15}N-Labeled adenosines (**69** and **70**) were prepared from the corresponding inosines (**67** and **68**) and $^{15}NH_4Cl$ by Ariza.[20] Key to this transformation was the use of a dinitrobenzenesulfonyl activating group, as upon ring opening, migration occurred (**71** to **72**) with loss of SO_2 before reclosure. Without this migration, the corresponding ^{15}*N*-inosine would result.

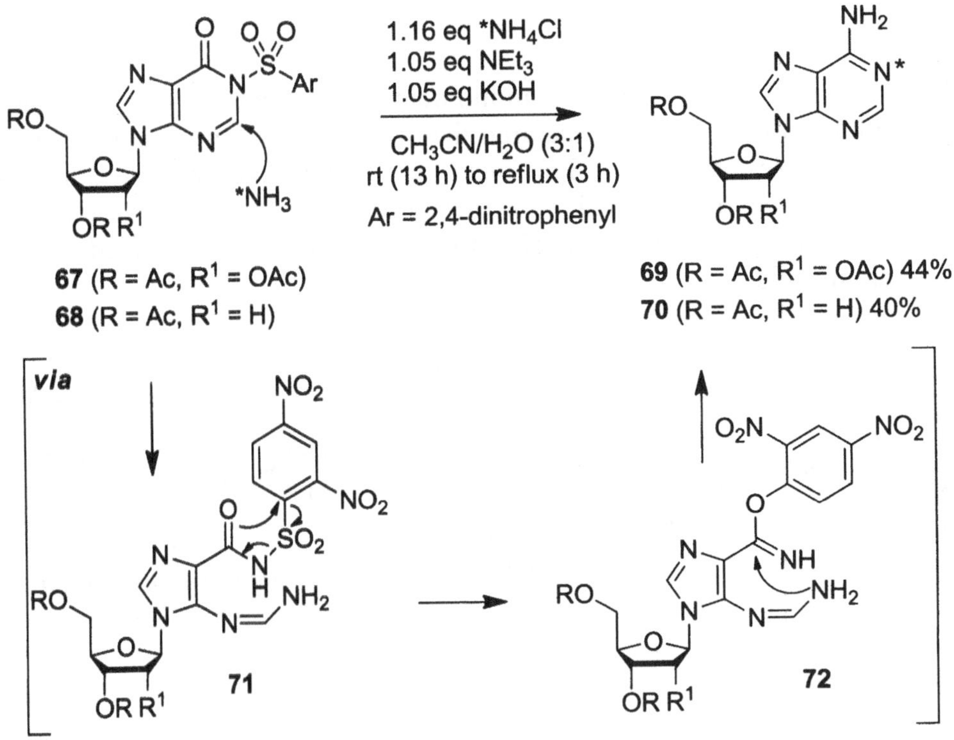

9.1.6 *Experimental*

Amination of Halonitropyridines[21]

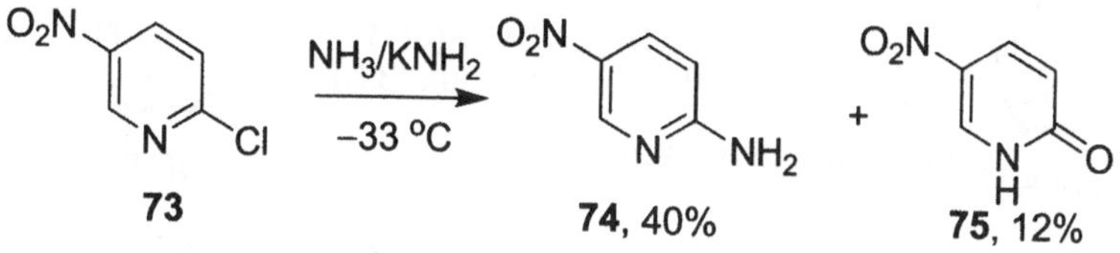

Compound **73** (318 mg, 2.0 mmol) was added to a solution of potassium amide (4.0 mmol) in liquid ammonia (50 mL). The reaction was stirred for 1 h at –33 °C, after which it was terminated by the addition of ammonium chloride. The ammonia was evaporated, and the residue was extracted with chloroform (3 × 25 mL). The chloroform extracts were dried with $MgSO_4$ and the solvent removed under reduced pressure. The residue was purified by column chromatography on silica gel using dichloromethane as eluent to yield **74** (40%) and **75** (12%).

General Procedure for the Reaction of Hydrazine with 3-Phenyl-5-perfluoro-alkyl-1,2,4-oxadiazoles[22]

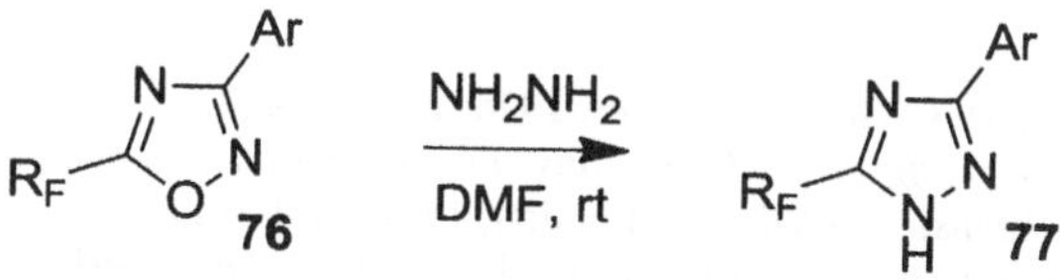

To a sample of oxadiazole **76** (1.5 mmol) in dry DMF (2 mL) was added an excess of hydrazine monohydrate (7.5 mmol), and the mixture was stirred at room temperature for 1–10 h. After dilution with water, the mixture was extracted with ethyl acetate, which was then dried with Na_2SO_4. The solvent was removed under reduced pressure, and the resulting crude material was purified by chromatography. Yields of the 1,2,4-triazine **77** ranged from 70–94%. The reaction can be performed in methanol, although longer reaction times are required (48 h) and the yields are reduced (44–88%).

General Procedure for the Preparation of 1-^{15}N-Labeled Adenosines[20]

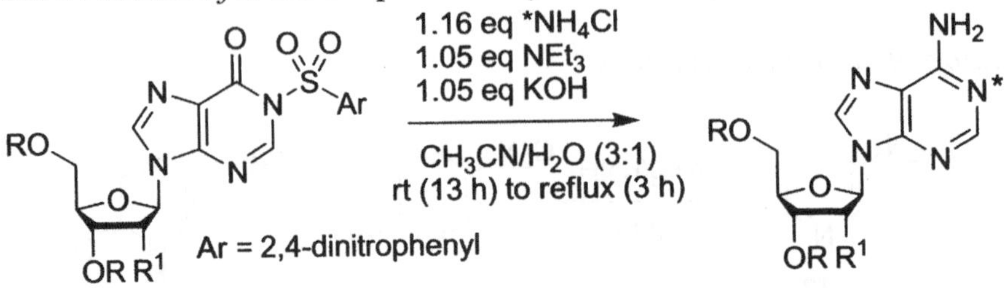

67 (R = Ac, R^1 = OAc)
68 (R = Ac, R^1 = H)

69 (R = Ac, R^1 = OAc) 44%
70 (R = Ac, R^1 = H) 40%

In a round-bottom flask sealed with a septum were placed $^{15}NH_4Cl$ (1.16 mmol) and KOH (1.05 mmol). Then water (5 mL), CH_3CN (14 mL), NEt_3 (1.05 mmol), and a solution of inosine **67** or deoxyinosine **68** (1.00 mmol) in CH_3CN (2 mL) were added sequentially via syringe. After vigorous stirring for 13 h, the reaction mixture was heated at reflux for 3 h. The resulting yellow solution was cooled to room temperature and then concentrated under reduced pressure. The product **69** or **70** was isolated by flash column chromatography (gradient of CH_2Cl_2/MeOH, 98:2 to 95:5) on silica gel.

9.1.7 References

1. [R] van der Plas, H. C. *Acc. Chem. Res.* **1978,** *11*, 462–468.
2. [R] *Degenerate Ring Transformations of Heterocyclic Compounds*. In *Advances in Heterocyclic Chemistry*, Academic: London, **1999**, *74*, 9–149.
3. [R] Kost, A. N.; Gromov, S. P.; Sagitullin, R. S. *Tetrahedron* **1981,** *37*, 3423–3454.
4. [R] El Ashry, E. H.; Nadeem, S.; Shah, M. R.; El Kilany, Y. *Adv. Heterocyclic Chem.* **2010,** *101*, 161–228.
5. [R] McGill, C. K.; Rappa, A. *Adv. Heterocyclic Chem.* **1988,** *44*, 1–79.
6. Levine, R.; Leake, W. W. *Science* **1955,** *121*, 780
7. Pieterse, M. J.; den Hertog, H. J. *Recl. Trav. Chim. Pays-Bas* **1961,** *80*, 1376–1386.
8. Brueker, K.; van der Plas, H. C. *J. Org. Chem.* **1979,** *44*, 4677–4680.
9. Haynes, L. W.; Pett, V. B. *J. Org. Chem.* **2007,** *72*, 633–635.
10. Moon, M. P.; Komin, A. P.; Wolfe, J. F.; Morris, G. F. *J. Org. Chem.* **1983,** *48*, 2392–2399.
11. Bell, H. M.; Carver, D. R.; Hubbard, J. S.; Sachdeva, Y. P.; Wolfe, J. F.; Greenwood, T. D. *J. Org. Chem.* **1985,** *50*, 3442–3444.
12. Katritzky, A. R.; Awartani, R.; Patel, R. C. *J. Org. Chem.* **1982,** *47*, 498–502.
13. Buscemi, S.; Pace, A.; Pibiri, I.; Vivona, N.; Zaira Lanza, C.; Spinelli, D. *Eur. J. Org. Chem.* **2004,** 974–980.
14. Zincke, T. *Ann. Chim.* **1903,** *330*, 361–374.
15. König, W. *J. Prakt. Chem.* **1904,** *69*, 105–137.
16. Fischer, R. W.; Misum, M. *Org. Process Res. Dev.* **2001,** *5*, 581–586.
17. Michels, T. D.; Kier, M. J.; Kearney, A. M.; Vanderwal, C. D. *Org. Lett.* **2010,** *12*, 3093–3095.
18. Xu, J.; Yadan, J. C. *J. Org. Chem.* **1995,** *60*, 6296–6301.
19. Gonela, A.; Walczak, K. *Tetrahedron: Asymmetry* **2005,** *16*, 2107–2112.
20. Terrazas, M.; Ariza, X.; Farràs, J.; Vilarrasa, J. *Chem. Commun.* **2005,** 3968–3970.
21. de Bie, D. A.; Geurtsen, B.; van der Plas, H. C. *J. Org. Chem.* **1985,** *50*, 484–487.
22. Buscemi, S.; Pace, A.; Pibiri, I.; Vivona, N. *J. Org. Chem.* **2003,** *69*, 605–608.

9.2 Boulton–Katritzky Rearrangement

Michael T. Corbett and Richard J. Mullins

9.2.1 Description

The Boulton–Katritzky rearrangement is the thermally induced rearrangement of one five-membered heterocycle into another, more stable five-membered heterocycle.

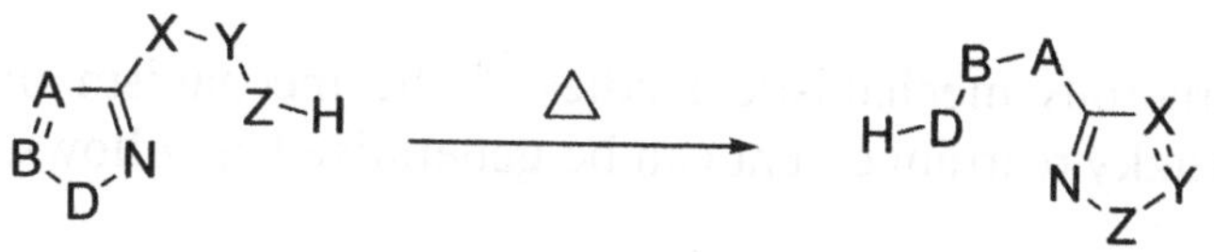

9.2.2 Historical Perspective

In 1966, Professor Alan Katritzky and A. J. Boulton, working at the University of East Anglia in Norwich, delineated the novel, generalized rearrangement of **1** to **2**.[1,2] This reaction, known as the Boulton–Katritzky rearrangement, has been exploited for the synthesis of a large number of heterocycles, including benzotriazoles, benzofurazans, and indazoles, among others. Because these rearrangements involve an aromatic ring that links the side chain, with the heterocycle undergoing rearrangement, they are currently referred to as bicyclic rearrangements of heterocycles (BRH).

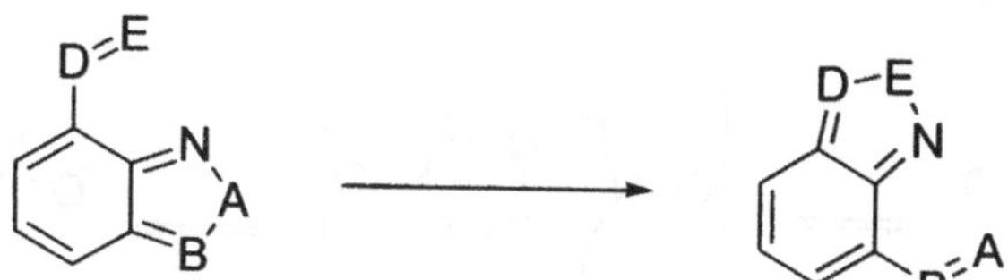

1 **2**

Shortly after this initial discovery, these same authors examined the analogous monocyclic rearrangement, generalized as **3** to **4**.[3] Although they were not the first to observe rearrangements of this type, they were the first to collectively identify them as being analogous to the Boulton–Katritzky rearrangement described above. Thus these reactions, because of their mechanistic similarity, have been referred to as monocyclic Boulton–Katritzky rearrangements, or monocyclic rearrangements of heterocycles (MRH). This reaction has been the subject of a number of reviews.[4–6]

3 4

In both BRH and MRH, the rearrangement is classified by both the starting hetereocycle and its three-atom side-chain, X–Y–Z in the case of **3**. It is also important to note that all BRH and MRH occur about a pivotal nitrogen that undergoes nucleophilic attack to initiate the rearrangement.

9.2.3 Mechanism

Based on numerous mechanistic studies,[7,8] the mechanism of the bicyclic Boulton–Katritzky rearrangement can be generalized as follows:

1 2

Although the mechanism above is generally accepted, several research groups have worked to gain a better understanding of the subtleties involved in the transformation. As an example, Eckert and Rauhut[9] have examined the well-studied[9–16] rearrangement of 4-nitrobenzofuroxan (**5**). Their computational study examined the hypothesis that the reaction proceeded in a single step via a symmetric transition state **TS**.

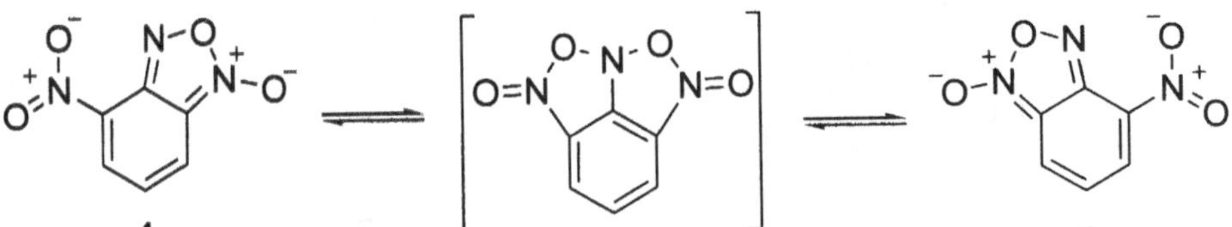

4 TS 4

In contrast to the BRH, the monocyclic Boulton–Katritzky does not require the reorganization of π-bonds, proceeding only with the formation and destruction of σ-bonds.[9] However, the exact mechanism has been a subject of great debate and depends on the manner in which the nucleophilic side chain is generated. Although a representative example is shown below,[10] the reader is directed to the numerous mechanistic studies available,[17–32] which have previously been done on the Boulton–Katritzky rearrangement.

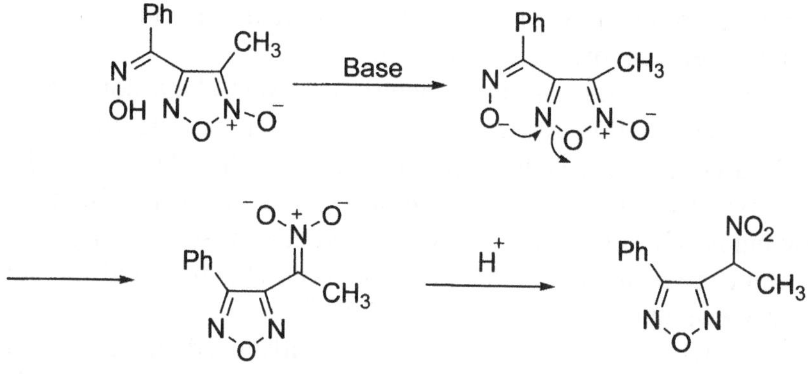

9.2.4 *Variations and Improvements*

The Boulton–Katritzky rearrangement typically occurs spontaneously; however, base-catalysis is often necessary to initiate less favorable rearrangements. To avoid base-catalyzed side reactions, Vivona and co-workers have shown that Cu(II) salts can also trigger the rearrangement of 3-benzoylazoles through Lewis acid activation without participating in any significant competitive redox processes.[33] When phenylhydrazone (*Z*)-**5** was treated with catalytic Cu(II), the desired 1,2,3-triazole **6** was obtained quantitatively. Similarly, rearrangement of phenylhydrazone (*Z*)-**7** to **8** occurred in excellent yield; however, a small amount of phenacyltriazole **9** was observed. Further studies concluded that this byproduct results from a parallel pathway and is not the result of oxidation of the desired product **8**. A more in-depth study by D'anna and Noto has investigated the role of the anion in the Cu(II)-mediated rearrangement of 1,2,4-oxadiazoles in ionic liquids.[34]

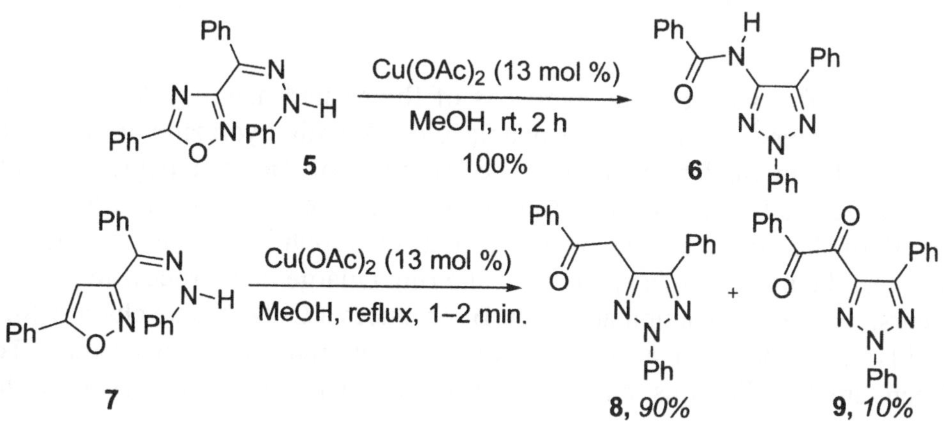

9.2.5 Synthetic Utility

The Boulton–Katritzky rearrangement has been widely used for the synthesis of interesting compounds for medicinal and materials purposes. Several reviews have been devoted to the subject.[4–6]

Although a wide variety of heterocycle/side-chain combinations can be used in the Boulton–Katritzky rearrangement, among these, the most commonly used heterocycle has been the 1,2,4-oxadiazole.[33,35–47] This is due to the relative ease with which it is synthesized as well as the fact that it has a very low index of aromaticity. Its heightened chemical reactivity has made it an important target for medicinal and materials chemists and the subject of multiple reviews in the last several years.[42,48,49]

Piccionello and co-workers have recently used the 1,2,4-oxadiazole nucleus in the Boulton–Katritzky rearrangement involving a nucleophilic side-chain carbon.[50] The precursor to rearrangement was prepared using an acid-catalyzed imine formation with amine **10** and 4-bromobenzaldehyde (**11**). Treatment of the resulting imine **12** with strong base induced the rearrangement to give **13** in high yield. The method proved to be somewhat general, as a number of aromatic aldehydes successfully participated in the sequence.

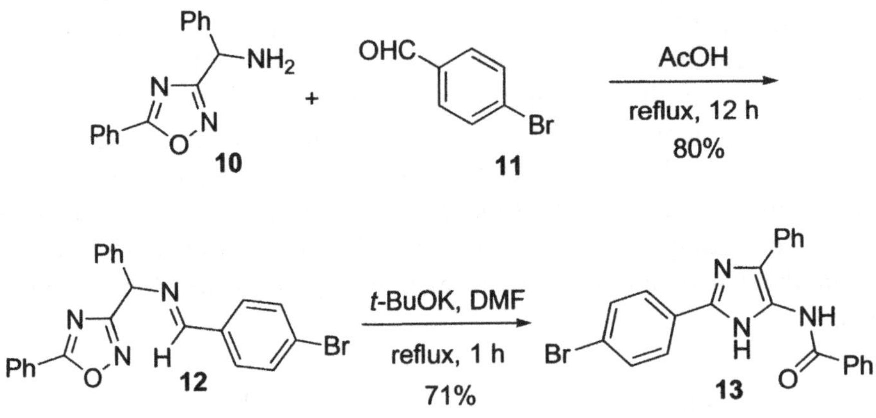

In efforts toward the synthesis of fluorinated heterocycles, Buscemi and co-workers[51] treated 1,2,4-oxadiazole **14** with hydrazine in DMF to produce **15** in moderate yield. The isolation of **16** as a byproduct was rationalized as forming via the Boulton–Katritzky rearrangement. Carbons A and B of the starting oxadiazole are both electrophilic; when nucleophilic attack of the hydrazine happens at A, the ring enlargement process to give **15** occurs. However, reaction at carbon B, produces a C–N–N side chain, which readily participates in a Boulton–Katritzky rearrangement to produce **16**. When smaller fluorous chains were used, attack at carbon A was the preferred pathway, suggesting steric factors differentiate between the two pathways.

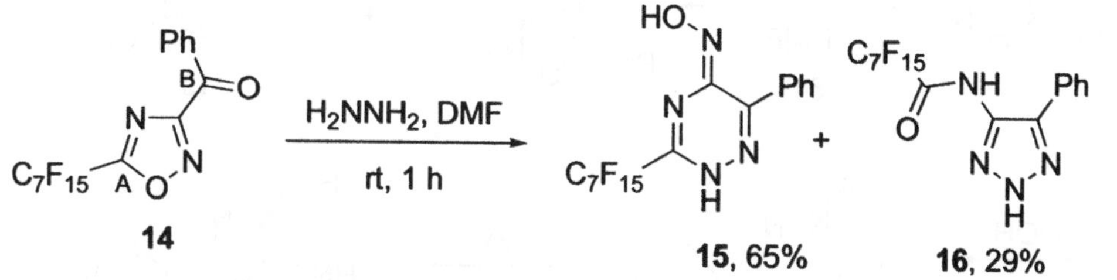

In related work, treatment of **17** with hydroxylamine resulted in nearly exclusive formation of the fluorinated heterocycle **15**.[52,53] On the other hand, when *N*-methylhydroxylamine was used, the Boulton–Katritzky rearrangement through intermediate **18** was the preferred pathway, giving **16**. A small amount of deacylated amine was also isolated.

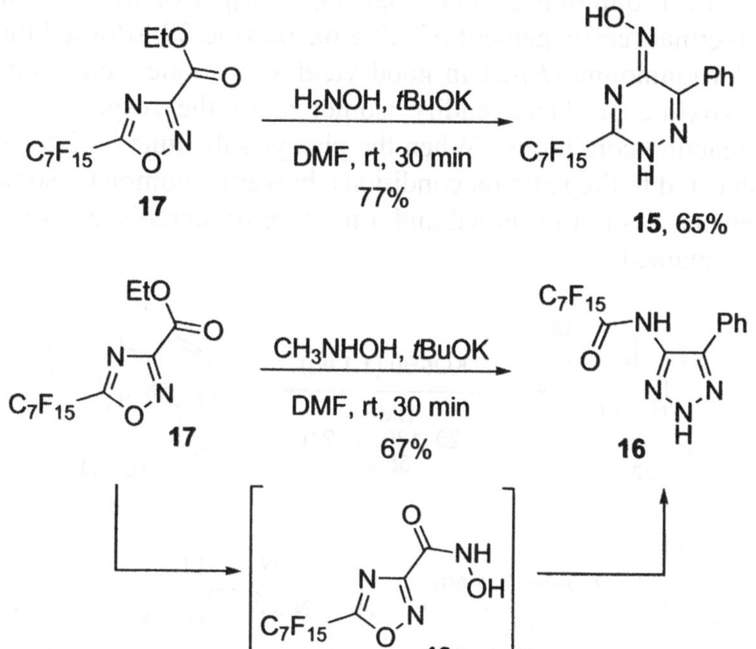

Piccionello and co-workers have also developed a versatile and efficient Boulton–Katritzky rearrangement involving a N–N–C side-chain.[54] Installation of the side chain was effected by S_NAr substitution with methylhydrazine and subsequent imine formation with 4-trifluoromethylbenzaldehyde (**19**), to provide **20**. Solvent free thermolysis gave **21** in nearly quantitative yield.

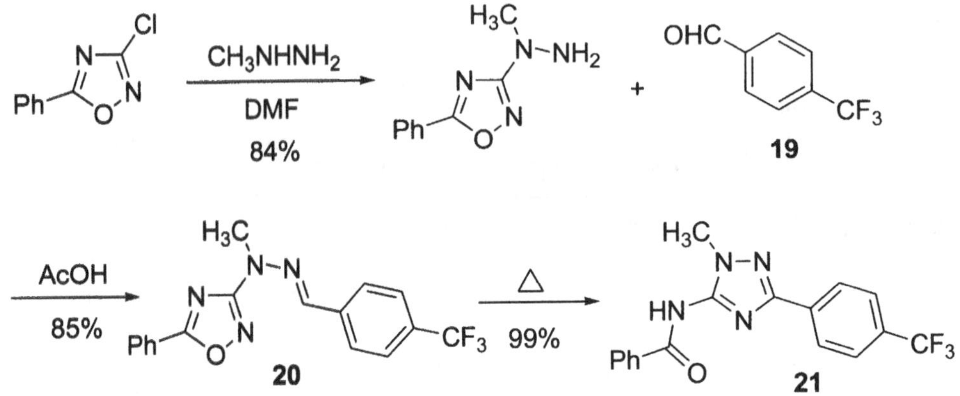

The Boulton–Katritzkiy rearrangement of 1,2,5-oxadiazoles has also been investigated.[11,55,56] The rearrangement of 1,2,5-oxadiazoles bearing a N–C–N side chain was used by Cusmano and co-workers in the preparation of triazolo[1,5-α]quinolines and triazolo[1,5-α]pyridines.[55] The base-catalyzed thermal rearrangement of 1,2,5-oxadiazole **22** afforded the desired triazolo[1,5-α]quinoline (*E*)-**23** in good yield as a single stereoisomer. The expected *Z*-oxime products readily isomerize to the respective *E*-oximes under the reaction conditions. When the phenyl-substituted 1,2,5-oxadiazole **24** was subjected to the reaction conditions, however, complete isomerization of the products was not observed and a mixture of stereoisomers (*E*)-**25** and (*Z*)-**25** was obtained.

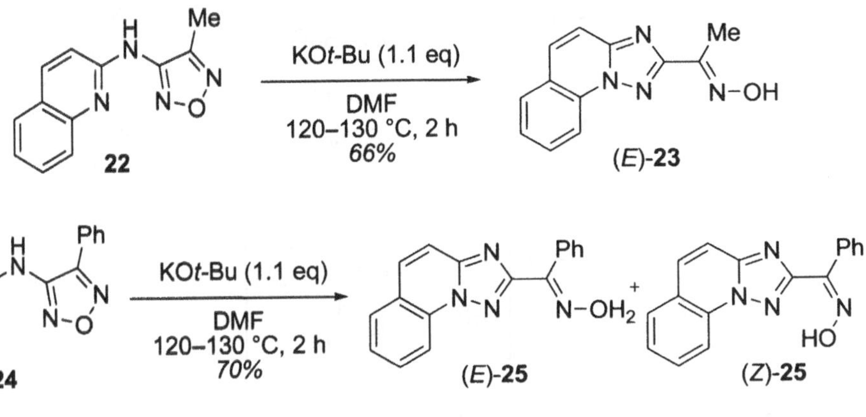

Based on the extensive utility of the Boulton–Katritzky rearrangement when employing the aforementioned 1,2,4- and 1,2,5-oxadiazole moieties, the application of analogous moieties such as 1,2,4-thiadiazoles has been investigated. Vivona and co-workers developed a general method for the synthesis of 1,2,4-thiadiazoles such as **28**.[57,58] The starting thioamide **27** can be trivially prepared from **26** upon treatment with Lawesson's reagent. The thioamide **27** can then undergo spontaneous rearrangement to afford the desired 5-aryl-1,2,4-thiadiazole **28** in moderate yield.

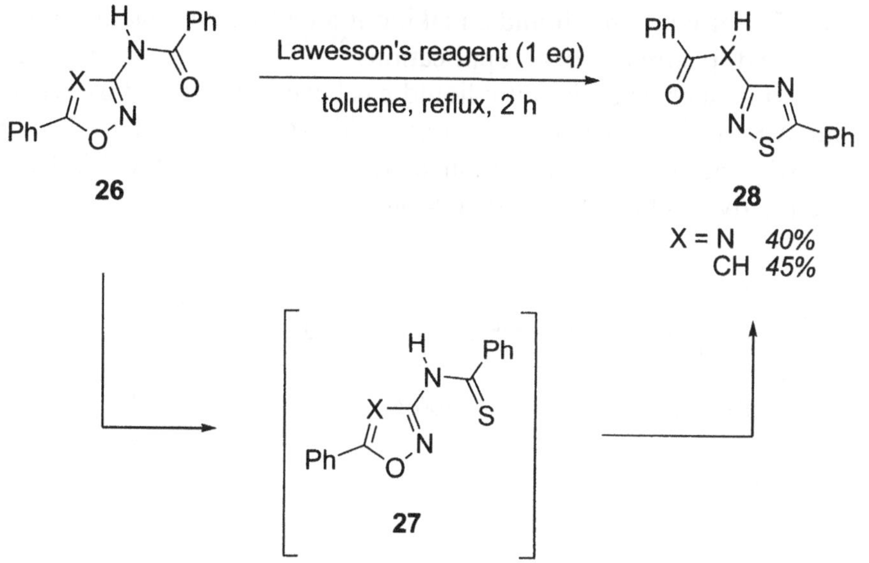

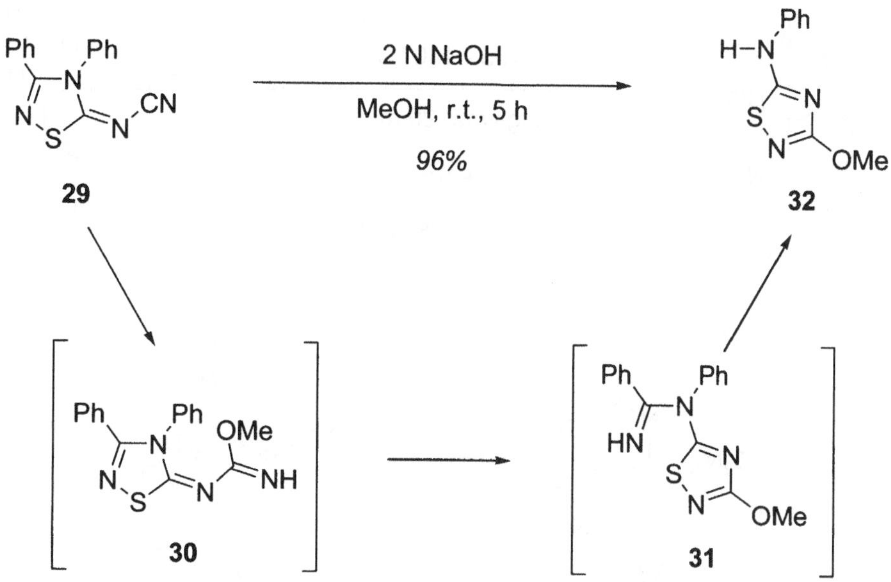

In addition to the aforementioned synthesis of 1,2,4-thiadiazoles, Sonnenschein and co-workers have demonstrated a Boulton–Katritzky rearrangement that is initiated through nucleophilic attack.[59] Nucleophilic addition of methanol (or dimethylamine) to 5-(cyanoimino)thiadiazoline **29** forms the intermediate **30** that can then undergo rearrangement to **31**. Elimination of a nitrile then affords the desired 3-methoxythiadiazole **32** in excellent yield. These products have been employed as redox switchable ionophores when the nucleophile is a crown ether, such as 1-aza-18-crown-6.[60]

The Boulton–Katritzky rearrangement has also been shown to proceed efficiently in ionic liquids, making it an attractive reaction for green chemistry and pharmaceutical applications.[25,34,61] The rearrangement of isocarboxazids, such as **33** in ionic liquid has been investigated by Niemczyk and Van Amum.[61] When the isocarboxazid **33** is warmed to 100 °C in [bmim]HSO_4^- the sole product obtained from extraction is the 1,2,3-triazole **34** in quantitative yield and analytically pure.

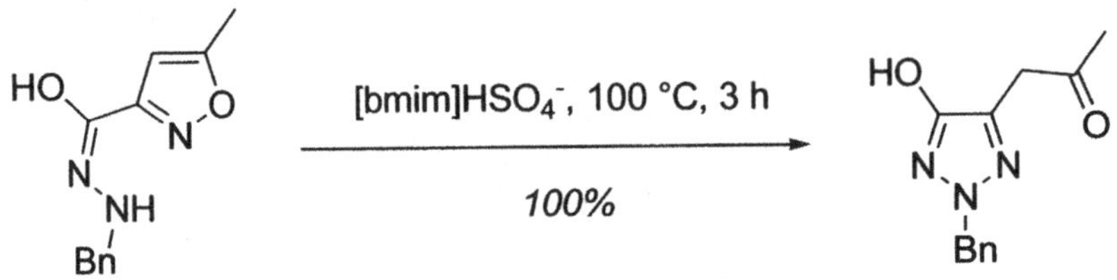

33 **34**

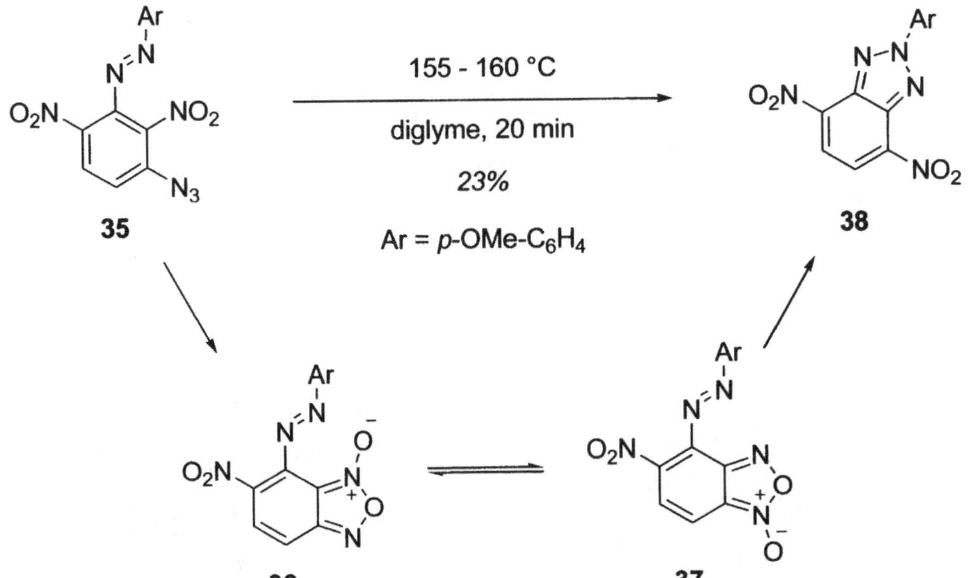

35 **38** **36** **37**

4-Nitrobenzofuroxanes and 4-nitrobenzofurazans have also served as significant substrates for studies into the mechanism and synthetic utility of the Boulton–Katritzky rearrangement.[62–64] In their early work, Boulton and Katritzky showed that when azide **35** underwent pyrolysis at 155 °C, nitrogen was expelled from the system and a colorless product **38** was obtained.[1] They propose that the reaction proceeds through two sequential rearrangements, the first being the formation of the intermediate **36**, which lies in equilibrium with **37**. This is followed by the bicyclic rearrangement of **37** to the desired 1,2,3-triazole **38**, in low yield.

In a similar reaction, Boulton and Katritzky had shown that anthranils, such as **41**, can be obtained through the rearrangement of 4-acetylbenzofuroxans.[2] When acetophenone **39** was heated to reflux, a thermal rearrangement occurred to give the anthranil **41** in good yield. As in the above example, the reaction proceeds through two sequential rearrangements via the intermediate **40**.

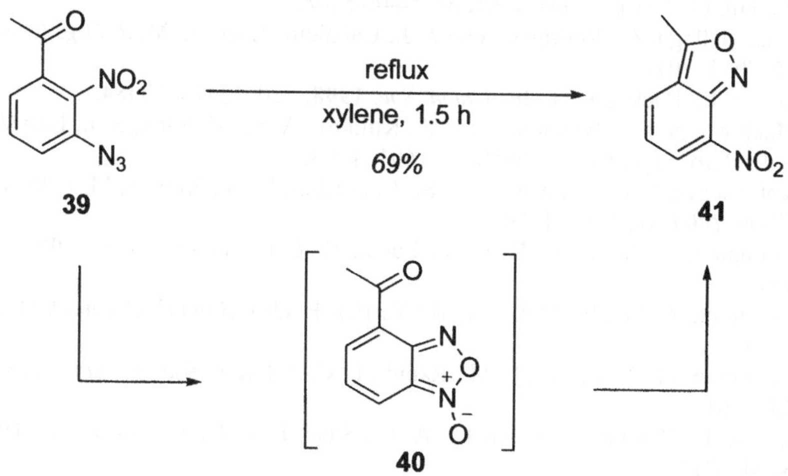

9.2.6 *Experimental*

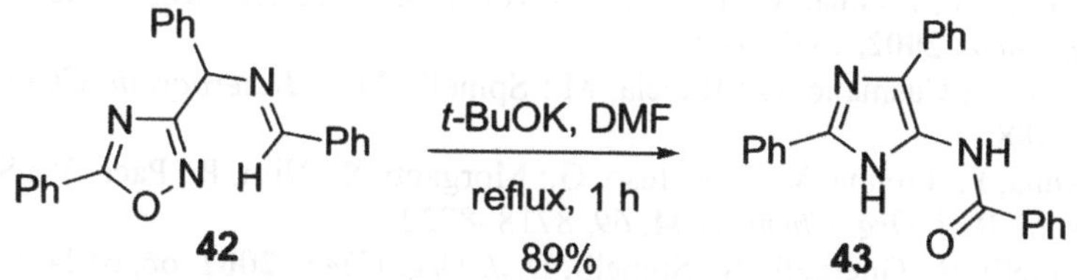

2,4(5)-Diphenyl-5(4)-*N*-benzoylamino-imidazole (43)[50]
To a solution of imine **42** (1 mmol) in DMF (5 ml), *t*-BuOK (123 mg, 1.1 mmol) was added and the solution refluxed for 1 h. After cooling, the mixture was reduced to dryness under vacuum and the residue treated with water. The resulting mixture was neutralized with HCl 0.1 M, extracted with EtOAc, which was dried and evaporated, and the residue was chromatographed giving 2,4(5)-diphenyl-5(4)-*N*-benzoylamino-imidazole (**43**) (302 mg, 89%).

9.2.7 *References*

1. Boulton, A. J.; Ghosh, P. B.; Katritzky, A. R. *J. Chem. Soc. B* **1966,** 1004–1011.
2. Boulton, A. J.; Ghosh, P. B.; Katritzky, A. R. *J. Chem. Soc. B* **1966,** 1011–1015.
3. Boulton, A. J.; Katritzky, A. R.; Hamid, A. M. *J. Chem. Soc. C* **1967,** 2005–2007.
4. [R] L'abbé, G. *J. Heterocyclic Chem.* **1984,** *21,* 627–638.
5. [R] Ruccia, M.; Vivona, N.; Spinelli, D. *Adv. Heterocyclic Chem.* **1981,** *29,* 141–169.
6. [R] Vivona, N.; Buscemi, S.; Frenna, V.; Cusmano, G. *Adv. Heterocyclic Chem.* **1993,** *56,* 49–154.
7. Rauhut, G. *J. Org. Chem.* **2001,** *66,* 5444–5448.
8. Peña-Gallego, A.; Rodríguez-Otero, J.; Cabaleiro-Lago, E. M. *J. Org. Chem.* **2004,** *69,* 7013–7017.
9. Eckert, F.; Rauhut, G. *J. Am. Chem. Soc.* **1998,** *120,* 13478–13484.
10. Makhova, N. N.; Ovchinnikov, I. V.; Kulikov, A. S.; Molotov, S. I.; Baryshnikova, E. L. *Pure Appl. Chem.* **2004,** *76,* 1691–1703.
11. Kotovskaya, S. K.; Romanova, S. S.; Charushin, V. N.; Kodess, M. I. *Russ. J. Org. Chem.* **2004,** *40,* 1167–1174.
12. Crampton, M. R.; Lunn, R. E. A.; Lucas, D. *Org. Biomol. Chem.* **2003,** *1,* 3438–3443.
13. Goumont, R.; Jan, E.; Makosza, M.; Terrier, F. *Org. Biomol. Chem.* **2003,** *1,* 2192–2199.
14. Crampton, M. R.; Pearce, L. M.; Rabbit, L. C. *J. Chem. Soc., Perkin Trans. 2* **2002,** 257–261.
15. Eckert, F.; Rauhut, G.; Katritzky, A. R.; Steel, P. J. *J. Am. Chem. Soc.* **1999,** *121,* 6700–6711.
16. Ghosh, P. B. *J. Chem. Soc. B* **1968,** 334–338.
17. Bottoni, A.; Frenna, V.; Lanza, C. Z.; Macaluso, G.; Spinelli, D. *J. Phys. Chem. A* **2004,** *108,* 1731–1740.
18. Buscemi, S.; Frenna, V.; Pace, A.; Vivona, N.; Cosimelli, B.; Spinelli, D. *Eur. J. Org. Chem.* **2002,** 1417–1423.
19. Vivona,N.; Cusmano, G.; Ruccia, M.; Spinelli, D. *J. Heterocyclic Chem.* **1975,** *12,* 985–988.
20. D'Anna, F.; Frenna, V.; Macaluso, G.; Morganti, S.; Nitti, P.; Pace, V.; Spinelli, D.; Spisani, R. *J. Org. Chem.* **2004,** *69,* 8718–8722.
21. Cosimelli, B.; Guernelli, S.; Spinelli, D. *J. Org. Chem.* **2001,** *66,* 6124–6129.
22. D'Anna, F.; Frenna, V.; Macaluso, G.; Marullo, S.; Morganti, S.; Pace, V.; Spinelli, D.; Spisani, R.; Tavani, C. *J. Org. Chem.* **2006,** *71,* 5616–5624.
23. D'Anna, F.; Ferroni, F.; Frenna, V.; Guernelli, S.; Lanza, C. Z.; Macaluso, G.; Pace, V.; Petrillo, G.; Spinelli, D.; Spisani, R. *Tetrahedron* **2005,** *61,* 167–178.
24. Cosimelli, B.; Frenna, V.; Guernelli, S.; Lanza, C. Z.; Macaluso, G.; Petrillo, G.; Spinelli, D. *J. Org. Chem.* **2002,** *67,* 8010–8018.
25. D'Anna, F.; Frenna, V.; Noto, R.; Pace, V.; Spinelli, D. *J. Org. Chem.* **2005,** *70,* 2828–2831.
26. Mezzina, E.; Spinelli, D.; Lamartina, L.; D'Anna, F.; Frenna, V.; Macaluso, G. *Eur. J. Org. Chem.* **2005,** 3980–3986.
27. Guernelli, S.; Laganà, M. F.; Spinelli, D. *J. Org. Chem.* **2002,** *67,* 2948–2953.
28. Horváth, K.; Korbonits, D.; Náray-Szábo, G.; Simon, K. *J. Mol. Struct. THEOCHEM* **1986,** *136,* 215–227.
29. Andrianov, V. G.; Makushenkov, S. V.; Eremeev, A. V. *Mendeleev Commun.* **1992,** 129–130.

30. La Manna, G.; Buscemi, S.; Vivona, N. *J. Mol. Struct. THEOCHEM* **1998,** *452,* 67–74.
31. Pace, A.; Pibiri, I.; Piccionello, A. P.; Buscemi, S.; Vivona, N.; Barone, G. *J. Org. Chem.* **2007,** *72,* 7656–7666.
32. Pace, A.; Pierro, P.; Buscemi, S.; Vivona, N.; Barone, G. *J. Org. Chem.* **2009,** *74,* 351–358.
33. Buscemi, S.; Frenna, V.; Vivona, N.; Spinelli, D. *J. Chem. Soc., Perkin Trans 1* **1993,** 2491–2493.
34. D'Anna, F.; Frenna, V.; Marullo, S.; Noto, R.; Spinelli, D. *Tetrahedron* **2008,** *64,* 11209–11217.
35. Piccionello, A. P.; Pace, A.; Buscemi, S.; Vivona, N.; Pani, M. *Tetrahedron* **2008,** *64,* 4004–4010.
36. Bata, I.; Héja, G.; Kiss, P.; Korbonits, D. *J. Chem. Soc, Perkin Trans. 1* **1992,** 3069–3074.
37. Vivona, N.; Buscemi, S.; Frenna, V.; Ruccia, M. *J. Chem. Soc., Perkin Trans. I* **1986,** 17–19.
38. D'Anna, F.; Frenna, V.; Guernelli, S.; Lanza, C. Z.; Macaluso, G.; Marullo, S.; Spinelli, D. *ARKIVOC* **2009,** 125–144.
39. Guernelli, S.; Noto, R.; Sbriziolo, C.; Spinelli, D.; Liveri, M. L. T. *J. Colloid Interf. Sci.* **2001,** *239,* 217–221.
40. Katayama, H.; Takatsu, N.; Sakurada, M.; Kawada, Y. *Heterocycles* **1993,** *35,* 453–459.
41. Korbonits, D.; Tóbiás-Héja, E.; Simon, K.; Krámer, G.; Kolonits, P. *J. Chem. Soc., Perkin Trans. I* **1992,** 3069–3074.
42. [R] Pace, A.; Pierro, P. *Org. Biomol. Chem.* **2009,** *7,* 4337–4348.
43. Korbonitis, D.; M.-Bakó, E.; Horváth, K. *J. Chem. Res., Synop.* **1979,** 64–65.
44. Macaluso, G.; Cusmano, G.; Buscemi, S.; Frenna, V.; Vivona, N.; Ruccía, M. *Heterocycles* **1986,** *24,* 3433–3439.
45. [R] Pace, A.; Pibiri, I.; Buscemi, S.; Vivona, N. *Heterocycles* **2004,** *63,* 2627–2648.
46. Buscemi, S.; Pace, A.; Frenna, V.; Vivona, N. *Heterocycles* **2002,** *57,* 811–823.
47. [R] Pace, A.; Buscemi, S.; Vivona, N. *Org. Prep. Proc. Int.* **2007,** *39,* 1–70.
48. [R] Kayukova, L. A. *Pharm. Chem. J.* **2005,** *39,* 539–547.
49. [R] Hemming, K. in *Comprehensive Heterocyclic Chemistry III, 3rd ed.,* vol. 5, Elsevier: London, 2008; 243–314.
50. Piccionello, A. P.; Buscemi, S.; Vivona, N.; Pace, A. *Org. Lett.* **2010,** *12,* 3491–3493.
51. Buscemi, S.; Pace, A.; Piccionello, A. P.; Macaluso, G.; Vivona, N.; Spinelli, D.; Giorgi, G. *J. Org. Chem.* **2005,** *70,* 3288–3291.
52. Piccionello, A. P.; Pace, A.; Buscemi, S.; Vivona, N.; Giorgi, G. *Tetrahedron Lett.* **2009,** *50,* 1472–1474.
53. Buscemi, S.; Pace, A.; Piccionello, A. P.; Pibiri, I.; Vivona, N. *J. Org. Chem.* **2006,** *71,* 8106–8113.
54. Piccionello, A. P.; Pace, A.; Buscemi, S.; Vivona, N. *Org. Lett.* **2009,** *11,* 4018–4020.
55. Cusmano, G.; Macaluso, G.; Gruttadauria, M. *Heterocycles* **1993,** *36,* 1577–1588.
56. Sheremetev, A. B.; Mantseva, E. V. *Mendeleev Commun.* **1996,** 246–247.
57. Buscemi, S.; Vivona, N. *Heterocycles* **1994,** *38,* 2423–2432.
58. Macaluso, G.; Cusmano, G.; Buscemi, S.; Frenna, V.; Vivona, N.; Ruccía, M. *Heterocycles* **1986,** *24,* 3433–3439.
59. Sonnenschein, H.; Schmitz, E.; Gründemann, E.; Schöder, E. *Liebigs Ann. Chem.* **1994,** 1177–1180.

60. Hennrich, G.; Sonnenschein, H.; Resch-Genger, U. *Eur. J. Org. Chem.* **2000,** 539–542.
61. Van Arnum, S. D.; Niemczyk, H. J. *J. Heterocyclic Chem.* **2009,** *46,* 909–913.
62. [R] Katritzky, A. R.; Gordeev, M. F. *Heterocycles* **1993,** *35,* 483–518.
63. Takakis, I. M.; Hadjimihalakis, P. M. *J. Heterocyclic Chem.* **1992,** *29,* 121–122.
64. Takakis, I. M.; Hadjimihalakis, P. M.; Tsantali, G. G. *Tetrahedron* **1991,** *47,* 7157–7170.

9.3 Chichibabin Amination Reaction

Noha S. Maklad

9.3.1 Description

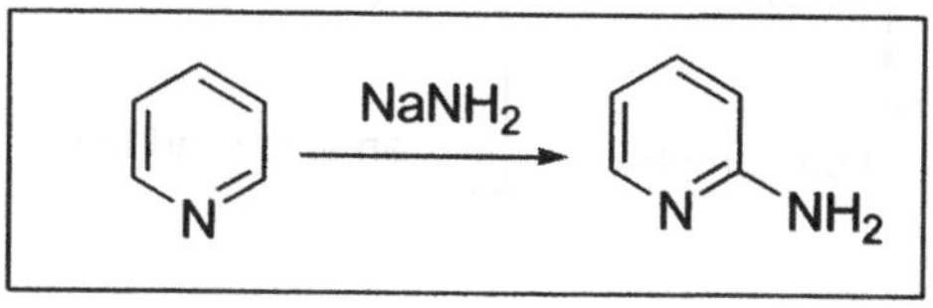

The Chichibabin amination reaction, first reported by Chichibabin and Zeide in 1914, is simply a reaction for the introduction of an amino group to azines. This is one of the first examples of nucleophilic hydrogen substitution (S_NH) reactions and is considered one of the most influential reactions in heterocyclic chemistry transformations of pyridines and azines in the 20th century.[1–3]

9.3.2 Historical Perspective

Alexej Euguenievitsch Chichibabin (or Tchitchibabine) was born at Kusemino, Russia on March 17, 1871.[4] He earned his science degree from the University of Moscow in 1892, then his "Magister Chimia" in 1902. After a long career in Moscow's most prestigious universities, Chichibabin moved to France in 1931, and he was given a laboratory in Collége de France where he worked until his death in Paris in 1945.

Chichibabin's pioneering work on pyridine chemistry has helped focus attention on the importance of this chemistry, and such attention has shown positive effects on the pharmaceutical industry. In 1914 Chichibabin and Zeide demonstrated their ability to synthesize 2-aminopyridine and 2-aminopicoline using sodium amide ($NaNH_2$) powder in toluene.[1] For decades this transformation has been used in the amination of different heterocyclic classes of compounds, e.g. quinolines, pyrazines, and pyrimidines. The reaction holds a great deal of potential for the synthesis of compounds of pharmaceutical interest, as it eliminates the need to introduce a good leaving group and the need for scavenging acidic by-products such as H_2SO_4 or HNO_2, since H_2O and H_2 are formed instead.

9.3.3 Mechanism

The Chichibabin amination reaction can progress through two distinct mechanisms. Through extensive experimental and spectroscopic data, it was

$S_N(AE)$ mechanism

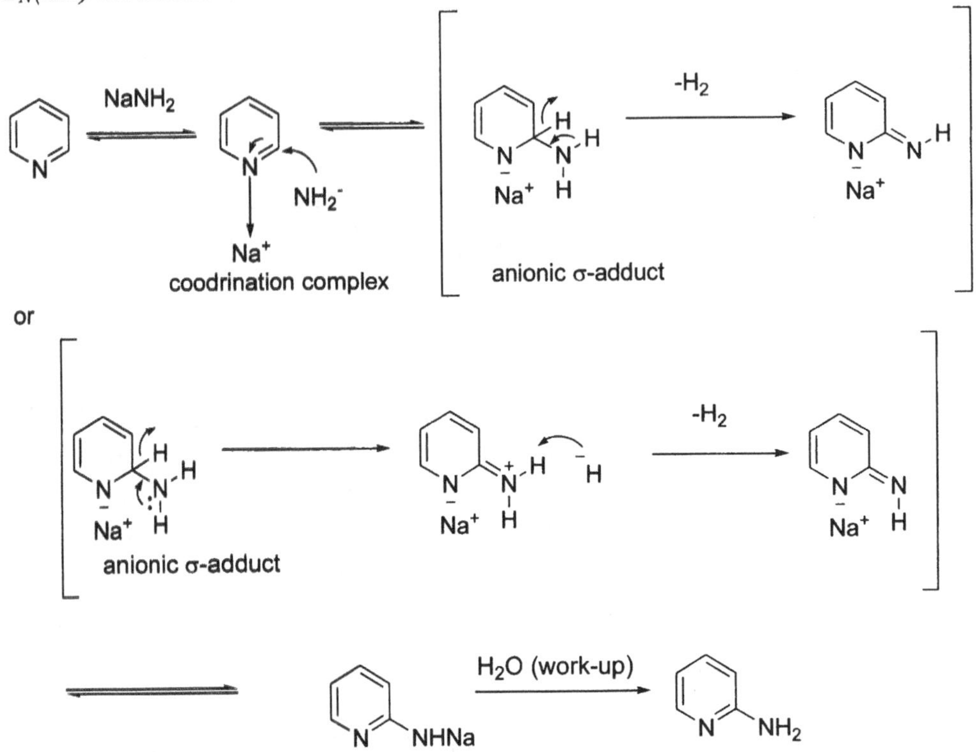

observed that the reaction course progresses through one of the two standard mechanisms, depending on reaction conditions, substitution and substitution pattern.[3] The first mechanism can simply be defined as a nucleophilic aromatic substitution of a hydride ion (S_NH substitutions) by an amino group, i.e., amino-dehydrogenation. The mechanism is explained by a nucleophilic addition–elimination process $S_N(AE)$, which involves σ^H-adduct formation followed by aromatization.[3,5–7] The classical Chichibabin is run under heterogeneous conditions—for example, heating the azine in powder sodium amide ($NaNH_2$), or potassium amide (KNH_2) in aprotic solvents, such as toluene, xylene, decalin, tetralin or *N,N*-dimethylaniline (DMA).[1] Sorption of the heterocyclic ring's nitrogen to the metal cation of the sodium amide is the first step (pyridine is usually used as a model). Sorption is then followed by the formation of a coordination complex between the sodium cation and the azine's nitrogen. To that effect, a partial positive charge forms on the α-carbon, which facilitates the amide ion attack. An anionic σ^H-adduct is then formed that has low solubility under such conditions and thus possesses a high activation energy (and thus the reaction runs at elevated temperature >100 °C), so that the addition step is the rate-determining step. Hydride ion is a poor leaving group because its elimination from an sp^3 carbon is a very difficult step. At the elimination step, the reaction proceeds by evolution of hydrogen gas, which is used to monitor the reaction progress. The hydrogen source could be due either to hydride ion elimination together with a proton from the amine group (from the σ^H-adduct) or to the loss of the hydride ion as metal hydride, which in turn reacts with the amino group to give hydrogen gas.[3]

SN(ANRORC) Mechanism

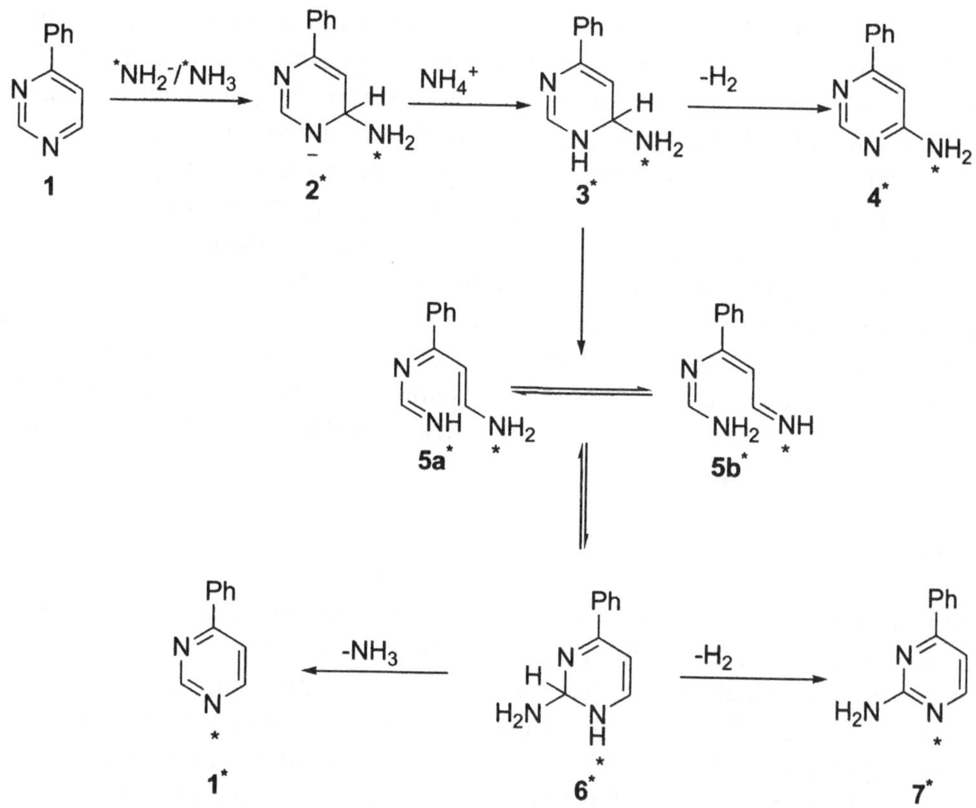

Under homogenous conditions (*i.e.*, amide salts in protic solvents such as $NaNH_2$ or KNH_2 in liquid ammonia), the σ^H-adduct formation is rapid due to solubility of the starting material. Oxidants such as KNO_3 and $KMnO_4$ are used in some cases (*i.e.*, oxidative amino-dehydrogenation process) to facilitate the elimination step and sometimes aid in lowering the reaction temperature; the presence of oxidant can allow the reaction to proceed under milder conditions and can improve its yield.

The second mechanism is an S_N(ANRORC)—addition of nucleophile to the heterocycle, ring opening and ring closure. It was proposed by and heavily investigated and reviewed by H. C. Van der Plas and co-workers as a possible mechanism for some Chichibabin amination cases under homogenous conditions.[8,9] The hypothesis comes from his extensive experimental data on the Chichibabin amination reaction for different azines, *e.g.*, quinolines, pyrimidines and phenanthridines.[10] Under this mechanism, site specificity for the amino group could be different from the S_N(AE) mechanism and hence provide a different regio-chemical outcome.

One of the early examples by Van der Plas is the amination of 4-phenylpyrimidine (**1**) using an ^{15}N-labeled potassium amide–ammonia (KNH_2–NH_3). The amination gives the ring degeneration S_N(ANRORC) adduct **7*** in 60% yield and the 6-amino-4-phenylpyridimine (**4***) adduct in 15% yield.[8b] The Chichibabin reaction is usually quenched with ammonium chloride to destroy the amide salts. It has been suggested by Van der Plas that quenching with ammonium salts favors the S_N(ANRORC) mechanism, since ammonium chloride behaves as a strong acid in liquid ammonia and so helps in formation of the intermediate 3*, which is more liable to ring opening.[8c]

Regio-chemical control and outcome are important aspects in terms of using the reaction. Aromatic azine's amination is subjected to variables such as the compound's class, substituents, substitution pattern, reaction condition and consequently the reaction mechanism. All variables are used for predicting and explaining the regio-chemical outcome for the reaction. Although it is outside the scope of this summary to provide examples from every heteroaromatic azine used synthetically, below is a sample of some representative examples.

Pyridines

Amination of the parent pyridine is the simplest example, and it is often used as a model molecule for this reaction. Substitution occurs at the α-position unless both are blocked.[3] Its reaction with sodium amide in toluene at reflux gives 2-aminopyridine in a 70% yield, while its heating at 170 °C gives 2,6-diaminopyridine (**8**) in a 55% yield; the deactivating influence of the electron-donating effect of the amino group is evident (higher temperature, lower yield). When the α-position is blocked, amination occurs at the 4-position with poor yield; an example is the amination of 2,6-dibenzyloxypyridine, which gives the γ-amino adduct in a 38% yield.[11]

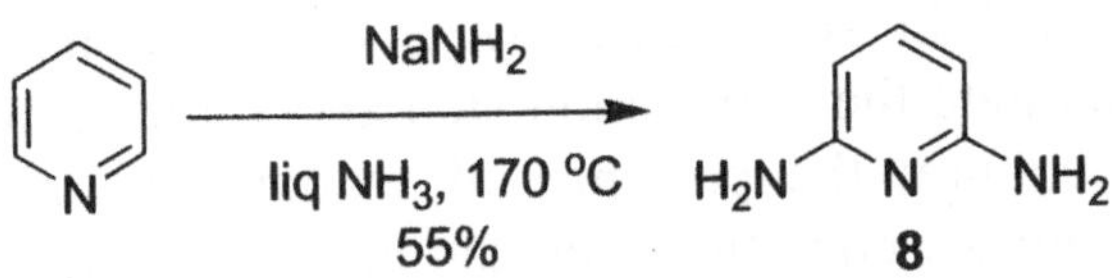

Amination of 3-substituted pyridines' occurs preferentially at the α-position, *i.e.*, 2(6) position. One example is the amination of 3-phenylpyridine (**9**) which gives the 2-amino isomer **10** in a 52% yield under the classical Chichibabin amination conditions.[12]

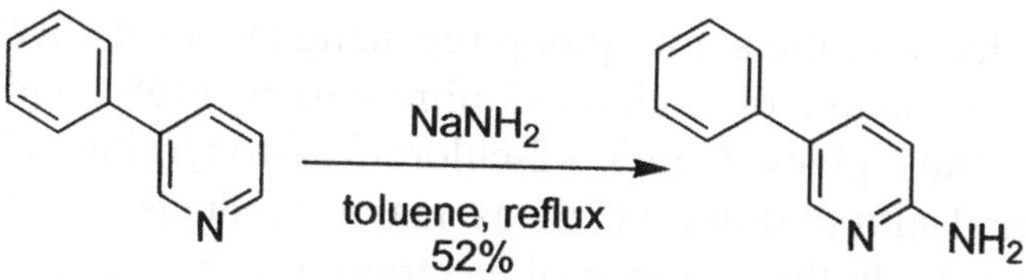

9 10

Sterics has an influence on the regio-chemical outcome for the amination of 3-subsituted pyridine. A bulky substituent, as in pyridine **11**, gives the *para* amino adduct as a major product, and so the amination of **11** gives **12** and **13** in a 20:1 ratio.[13]

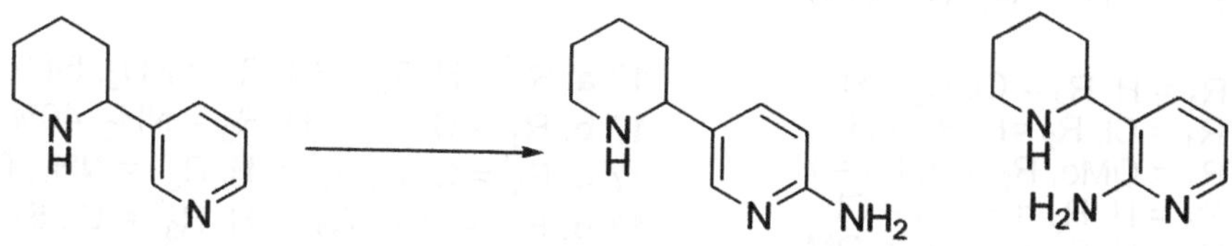

11 12 20 : 1 13

In the 3-substituted pyridines, the 2-amino regio-isomer can form selectively. In 3-dimethylaminopyridine (**14**) case, amination gives **15** exclusively in a 62% yield; the product is favored due to the strong ion–dipole interaction of the sodium amide with the dimethyl amino-substituent.[14]

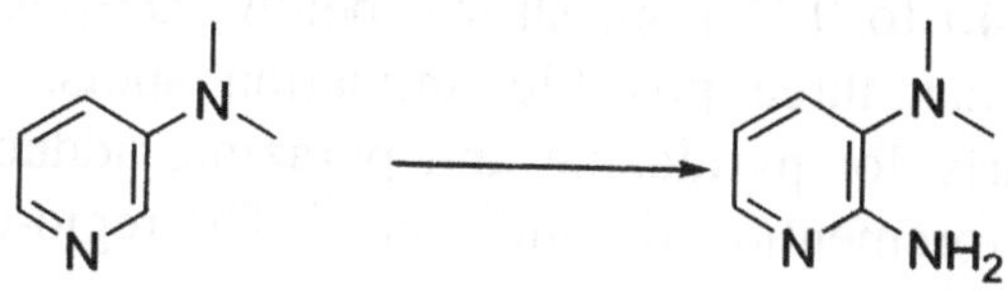

14 15

Under classical Chichibabin conditions, the amination of electron-deficient pyridines (for example, nitropyridines) is very difficult. The σ-adduct aromatization is challenging and thus requires elevated temperatures, which causes decomposition. Use of oxidants such as $KMnO_4$ oxidize the σ-adduct to the amino product easily and at lower temperatures. The amination of 3-nitropyridines still occurs on the 2(6) position (as mentioned above). In KNH_2–NH_3/$KMnO_4$, at low temperature (under kinetic control), charge distribution controls the subsequent regio-chemical outcome. 4-chloro-3-nitropyridine (**16a**) at –33 °C gives 2-amino-4-chloro-5-nitropyridine (**17a**) as the major product in 54% yields.[15] With compounds of R_1 or R_3 = Cl, or OMe such as **16b–e** (2-R_1-3-nitropyridines and 6-R_3-3-nitropyridines), amination occurs at the 6- and 2-positon, respectively. It is interesting that there is no evidence of the amino group replacing any of the leaving groups. For example, amination of 2-chloro-3-nitropyrdine (**16b**) and 2-chloro-5-nitropyridine (**16d**) gives 6-amino-2-chloro-3-nitropyridine (**17b**) and 2-amino-6-chloro-3-nitropyridine (**17d**) as major products in 40% and 57% yields, respectively. In the presence of electron-donating groups, as in 2-, 4-, 6-hydroxy and 4- and 6-amino-3-nitropyridines, Chichibabin amination is nonexistent, even in the presence of an oxidant.[15]

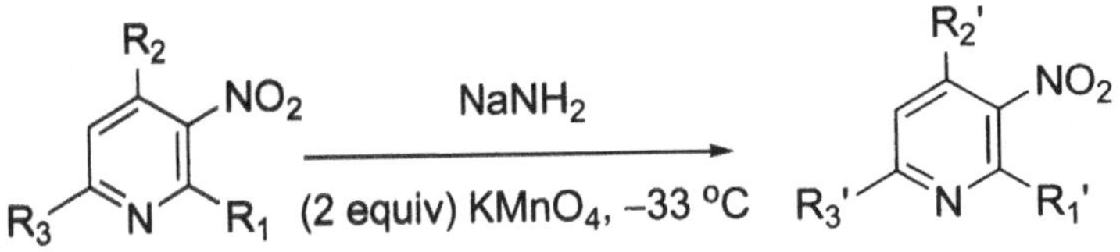

16 a: R_1 = H, R_2 = Cl, R_3 = H
16 b: R_1 = Cl, R_2 =H, R_3 = H
16 c: R_1 = OMe, R_2 = H, R_3 = H
16 d: R_1 = H, R_2 = H, R_3 = Cl
16 e: R_1 = H, R_2 = Cl, R_3 = OMe

17 a: R_1' = H, R_2' = Cl, R_3' = NH_2, 54%
17 b: R_1' = Cl, R_2' = H, R_3' = NH_2, 40%
17 c: R_1' = OMe, R_2' = H, R_3' = NH_2, 62%
17 d: R_1' = NH_2, R_2' = H, R_3' = Cl, 57%
17 e: R_1' = NH_2, R_2' = Cl, R_3' = OMe, 75%

Diazines: Pyrimidines, Pyridazines and Pyrazines

In 1972, Zoltewicz and Helmick reported the first evidence of anionic σ-adducts in the Chichibabin amination reaction. The parent diazines' σ-adduct complex (pyrimidine, pyridazine and pyrazine) in KNH_2–NH_3/$KMnO_4$ (at low temperature: –40 to 0 °C) are shown below. Despite the fact that the parent pyrimidine has three possible amination spots, **18** is the adduct formed; and similarly for pyridazine and pyrazine, adducts **19** and **20** are formed under the aforementioned conditions.[16] The regio-chemical outcome changes when these rings are substituted.

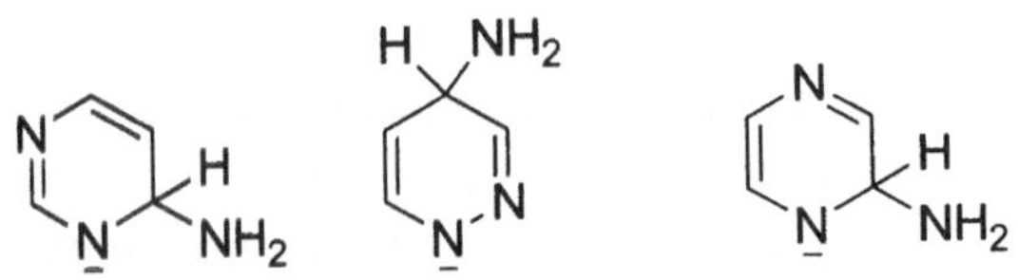

18 **19** **20**

It has been established by experimental data that the Chichibabin amination of pyrimidine proceeds either by an S_N(AE) or S_N(ANRORC) mechanism. Each mechanism is favored according to substituents, substitution pattern and reaction conditions. In turn, the region-chemical outcome for the reaction depends on the mechanism course the reaction follows. 2-bromo-4-phenylpyrimidine (**21**) reaction with KNH_2–NH_3

pyrimidine undergoes amination with an SN(ANRORC) mechanism. The reaction starts with amide addition at the C-6 position, which is followed by ring opening and bromide ion elimination to give **24**, a cyano enamine intermediate, which undergoes ring closure to give the 2-amino product **7**.[17]

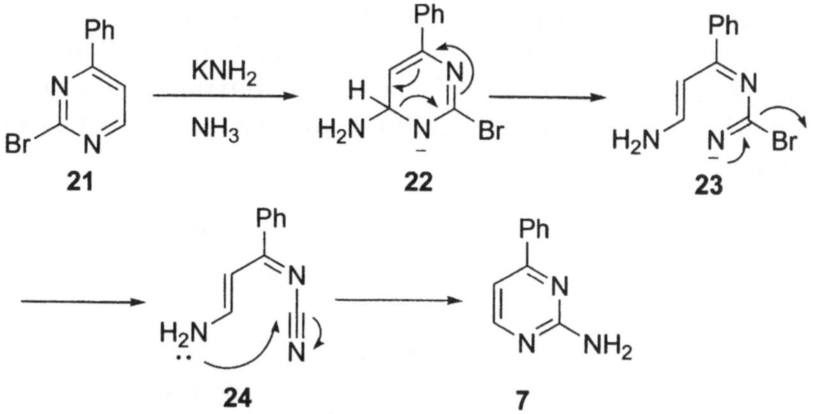

Van der Plas *et al.* showed an interesting example in the amination of pyrimidines, where alternation between different mechanisms is very clear. The different results depend on the variations of the reaction's conditions (temperature and presence of oxidizing agent). In the absence of an oxidant, the amination of 4-methoxy-5-nitropyrimidine (**25**) in liquid NH_3 at –20 °C gives **26** in a 65% yield through an S_N(ANRORC) mechanism (amino σ-adduct forms at the C-2 position first).[18a] In the presence of $KMnO_4$ at –60 °C, oxidative amination gives **27** in a 50% yield through an S_N(AE) mechanism.

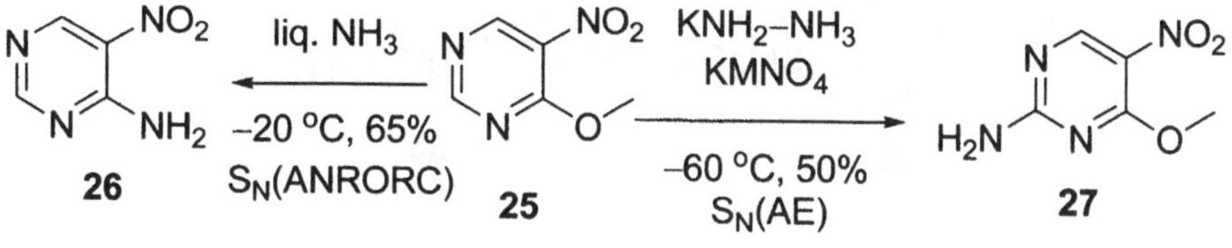

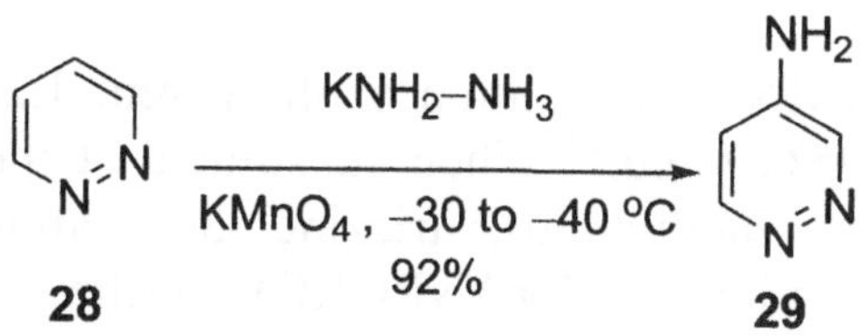

Pyridazine's oxidative amination with KNH_2–NH_3/$KMnO_4$ gives the 4-amino adduct **29** in 92% yield.[18b,19] When substituted, the amination is not guaranteed to be regioselective. Under the aforementioned conditions, 3-

phenylpyridazine gives a mixture of 4-, 5- and 6-amino derivatives in 49%, 18% and 5% yields, respectively, but 3-methoxy pyridazine forms the 4-amino adduct selectively.[18b]

If pyridazine is substituted at the 4-position, amination occurs at C-5; for example, **30a** gives **31a** in a 38% yield (variant 1). In the presence of an oxidant, electron-withdrawing groups facilitate oxidative amination of pyridazines and thus push the reaction even in the absence of amide salt. In liquid NH_3, 4-nitropyridazine gives 5-amino-4-nitropyridazine in an excellent 98% yield, while the nitropyridazine **30b** gives the amino adduct **31b** in 93% yield (variant 2).

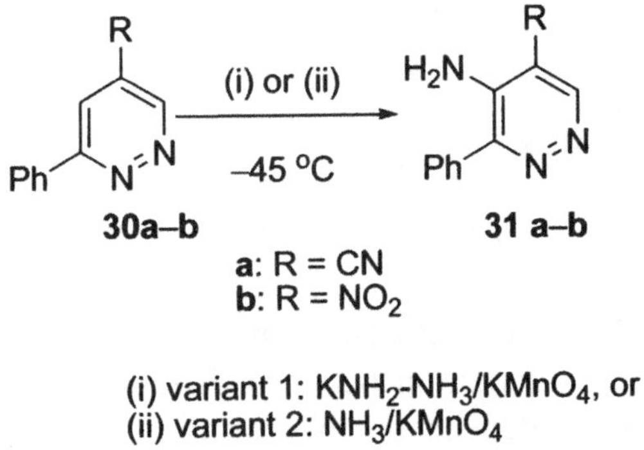

30a–b **31 a–b**

There is but one example of nonfused pyrazine Chichibabin amination in the literature, reported by Van der Plas in 1982. In $KNH_2–NH_3/KMnO_4$ at room temperature, the reaction gives 2-aminopyrazine (**33**) in a 65% yield.[19]

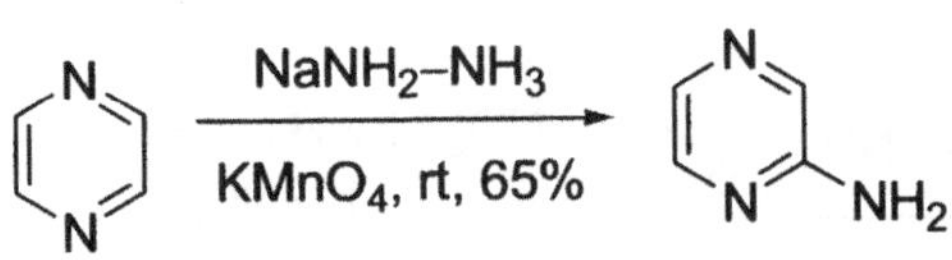

32 **33**

1,2,4-Triazine

In liquid ammonia/$KMNO_4$, amination of the parent 1,2,4-triazine gives 6-amino adduct in > 95% yield.[20] When substituted at the 3-position with a leaving group (*e.g.*, methylthio), triazine **36a** amination in $KNH_2–NH_3/KMnO_4$ occurs with an S_N(ANRORC) mechanism to give the 3-aminoadduct **37a** in a 93% yield. For 3-R_1-6-R_2-1,2,4-triazines where R_2 is a leaving group (*e.g.*, halo group: F, Cl, Br), regio selectivity of the reaction can be neither controlled nor predicted. It gives a mixture of amino adducts that renders it synthetically useless.

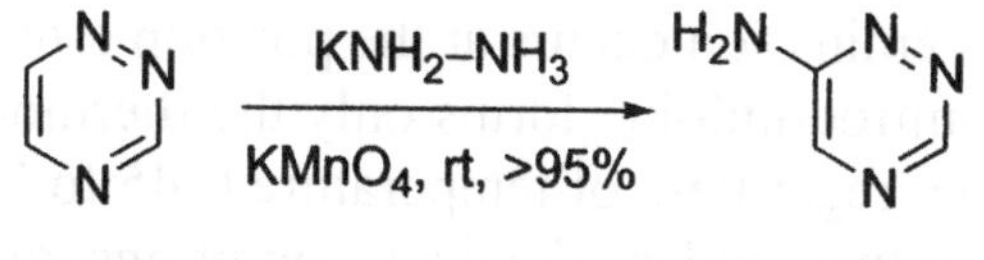

34 **35**

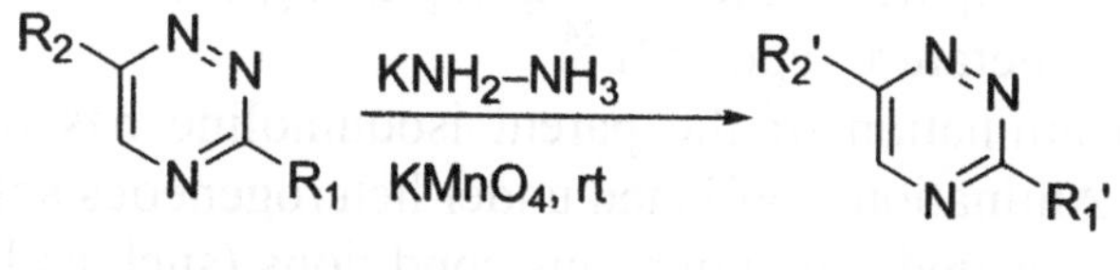

36 **37**

36a : R_1 = SMe, R_2 = H **37a** : R_1' = NH_2, R_2' = H, 93%
36b : R_1 = F, R_2 = Ph **37b** : R_1' = NH_2, R_2' = Ph, 26%

Fused Bicyclics: Quinolines, Isoquinolines

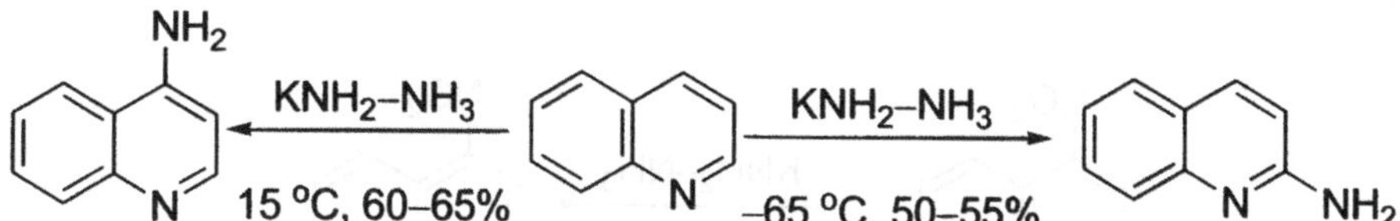

40 **38** **39**

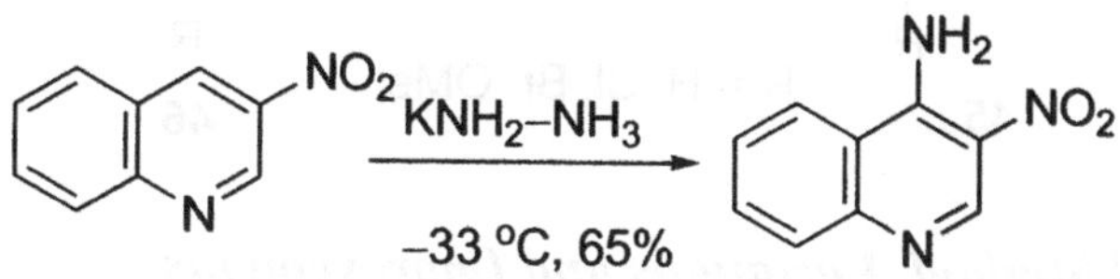

41 **42**

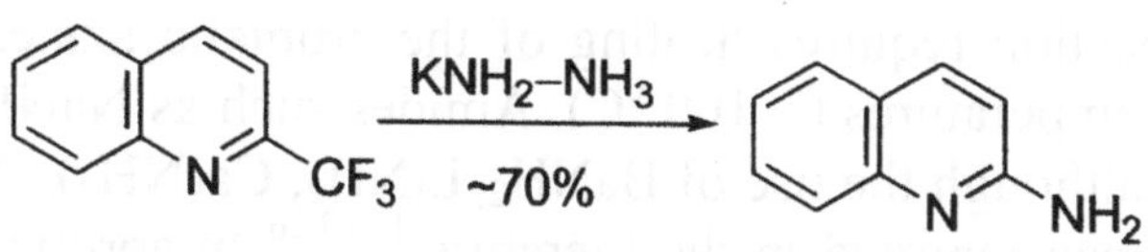

43 **44**

The standard Chichibabin amination of quinoline with sodium amide in DMA gives 2-aminoquinoline in a low 7% yield and 2-amino-3,4-dihydroquinoline in a 24% yield. Not until the introduction of the Chichibabin oxidative amination variant (*e.g.*, in the presence of $KMnO_4$) were these yields improved. Regio-chemical outcome depends on reaction temperature (kinetic or thermodynamic control). 2-aminoquinoline (**39**) forms at –65 °C, *i.e.*, the kinetic adduct, and is isolated in a 50–55% yield with no evidence of 4-amino adduct formation, whereas 4-aminoquinoline (**40**) forms in 60–65% yield at 15 °C, *i.e.*, the thermodynamic adduct (oxidant added at –45 °C), with 6–7% of **39**.[21,22] Substituted quinolines give various results. Nitroquinoline's amination occurs at the position α to the nitro group, not the ring nitrogen. 3-nitroquinloine forms only the thermodynamic adduct 4-amino-3-nitroquinoline, regardless of temperature (–45 to 20 °C), in a 65% yield, while 4-nitroquinoline under identical conditions gives 3-amino-4-nitroquinoline in an 86% yield.[22,23] 2-Trifluoromethylquinoline amination interestingly occurs by replacing the $-CF_3$ group and gives 2-aminoquinoline in ~70% yield (no temperature reported).[24]

Chichibabin amination of the parent isoquinoline has two possible sites for amination. Amination is affected under heterogeneous solvents (such as DMA, or toluene) or under homogenous conditions (such as liquid NH_3), to give 1-amino-isoquinoline. The product is formed in a 38% yield both in DMA or toluene and in an 86% yield in liquid NH_3. Nitroisoquinoline amination occurs *ortho* to the nitro group rather than the ring nitrogen, and so 1-R-5-nitroisoquinoline (R= Cl, Br, OMe) produces 6-amino-5-nitro-1-R-isoquinoline with no replacement of the leaving groups by an amino group.[7,25]

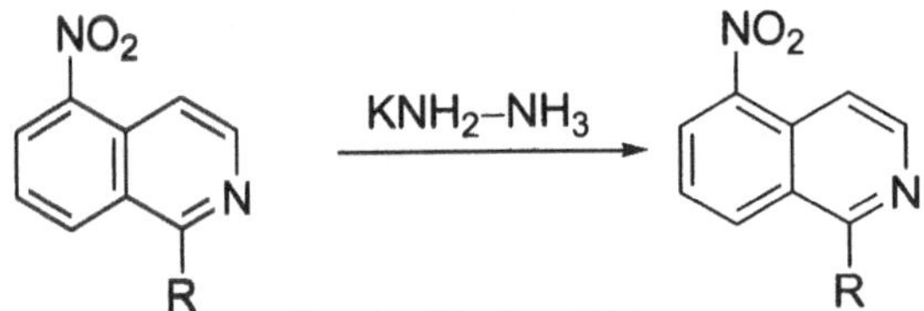

45 46

9.3.4 *Standard Method, Variations and Improvements*

Generally, the reaction requires heating of the aromatic azines with amide salts at elevated temperatures (> 100 °C). Amides such as $NaNH_2$ and KNH_2 are mainly used, although the use of $BaNH_2$, $LiNH_2$, $Ca(NH_2)_2$, $Sr(NH_2)_2$ and $Mg(NH_2)_2$ have been reported in the literature.[1–3,7,26] In aprotic solvents, the reaction requires elevated temperature due to the difficulty of the σ^H-adduct formation (as discussed earlier). Polar solvents such as liquid NH_3 render the reaction amenable to running under lower temperatures.

Introduction of an oxidative variant by Bergstrom in 1934 led to a great transformation of this reaction; an example of oxidants are $Ba(NO_3)_2$ and $NaNO_3$. It was not until Van der Plas's introduction of $KMnO_4$—which is as an excellent oxidant for the σ^H-adduct—was this variation influential (*vide supra*).[19,27–29]

In addition to intermolecular Chichibabin amination, there are some intramolecular cyclization examples that have been reported in the literature. The variation is useful in the synthesis for some interesting pharmaceutical cores. For example, in a 4-step synthesis, formation of 2-[3-aminopropyl]-

5,6,7,8-tetrahydronaphthyridine (**50**) from 2,5-dibromopyiridne is accomplished mainly by a double Suzuki reaction followed by Chichibabin cyclization in a 76% overall yield. Cyclization in $NaNH_2$ or $LiNH_2$ gives the best yields of ~84%, and no intermolecular amination adduct is detected.[30]

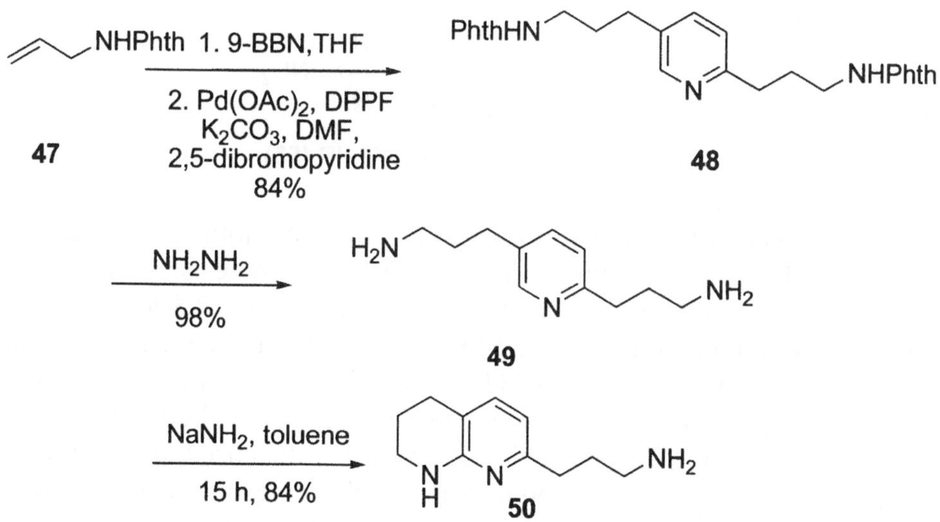

9.3.5 *Synthetic Utility*

In general, the reaction utility is for the amination of heteroaromatic azines. Providing an amine handle, it can further aid in the synthesis of a number of fused bicycles; consequently, the reaction has great utility in the pharmaceutical industry.

The fused bicycle phenylimidazo[4,5-*b*]pyridine and its different isomers, such as 2-amino-1-methyl-6-phenylimidazo[4,5-*b*]pyridine (1-Me-6-PhIP) and 1-Me-5-PhIP (**55**), are suspected procarcinogens. Chichibabin amination is used in the synthesis of these PhIP isomers.[12,31] Synthesis of 1-Me-5-PhIP is affected by the amination of 5-phenylpyridine (**51**) using $NaNH_2$ in DMA at 170 °C. This gives the 2-aminopyridine adduct **52** in an 81% yield. The 2-amino adduct is then protected and followed by various transformations to give **55**. 1-Me-6-PhIP isomer is also formed under the aforementioned conditions from 2-amino-5-phenylpyridine **10** (*vide supra*).

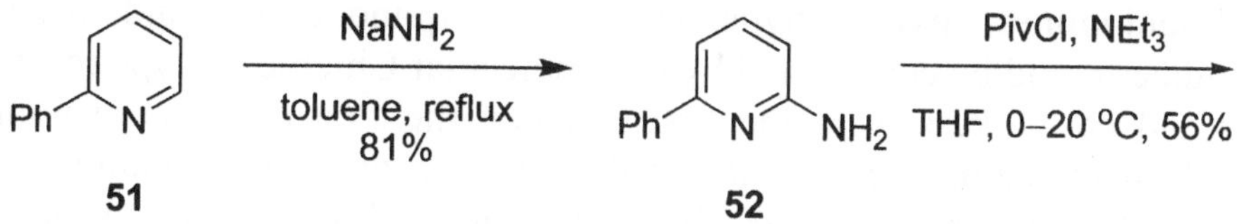

f

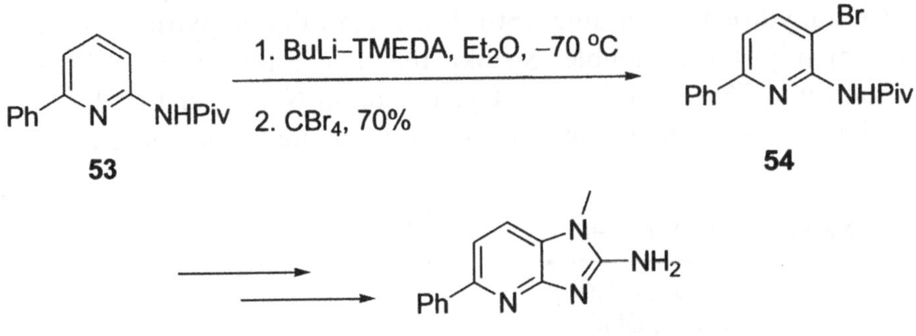

53 **54**

1-Me-5-Ph-IP (**55**)

Synthesis of tetrahydroimidazo[2,1-*a*]isoquinolines is another example of the reaction's utility in a fused core synthesis. Formation of compounds of structure **57** starts with 7-phenyl-5,6,7,8-tetrahydroisoquinoline (**56**) amination in either DMA or tetralin in autoclave at 220 °C for 18 h, which gives **53** in 41–48% yield, consecutively, although a higher yield is achieved when heating the starting material in DMA in the microwave for 12 h, which gives **58** in a 59% yield. The product is then used to construct the tetrahydroimidazo[2,1-*a*]isoquinoline framework using chloroacetone or 3-bromobutanone, which affords **58a,b** in 46% and 75% yields, respectively.[32]

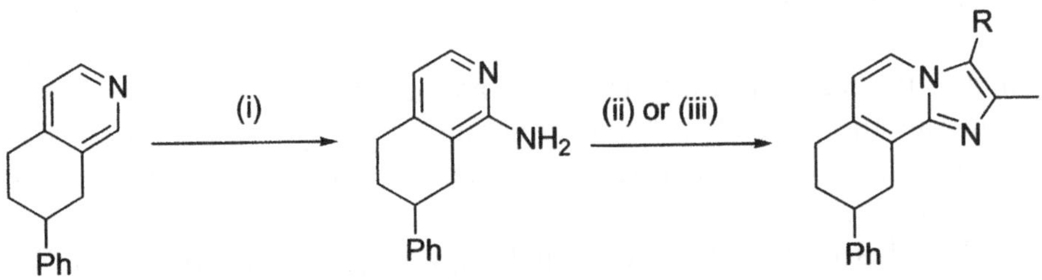

56 **57** **58a** R = H
58b R = Me

(i) $NaNH_2$/tetralin, 220 °C, 18 h, 48% or
$NaNH_2$/DMA, 220 °C, 18 h, 41%
or $NaNH_2$/DMA, 180 °C, 12 h, 59% (microwave)
(ii) chloroacetone, THF, 2.5 d, 46%
(iii) 3-bromobutanone, THF, 70 °C, 24h , 75%

Interesting constructs of molecules such as pyrrolo-pyridines, pyrrolo-quinolines, pyrrolo-pyrazines, pyrrolo-quinoxalines and pyrrolo-pyrimidines could be constructed by intramolecular Chichibain-type reaction. Davis and co-workers were the first to show the first of these examples, where β-methylazines react with nitriles to give the corresponding fused bicycle in the presence of excess base (2.5 equiv). For example, β-picoline reacts with benzonitrile in the presence of LDA to give the pyrrolo-pyridine adduct **61** in a 90% yield and traces of **62a**.[33] Under similar conditions and if the 2-position is blocked, reaction of **60b** with benzonitrile is specific to the 4-position and gives **62b** in a 61% yield.

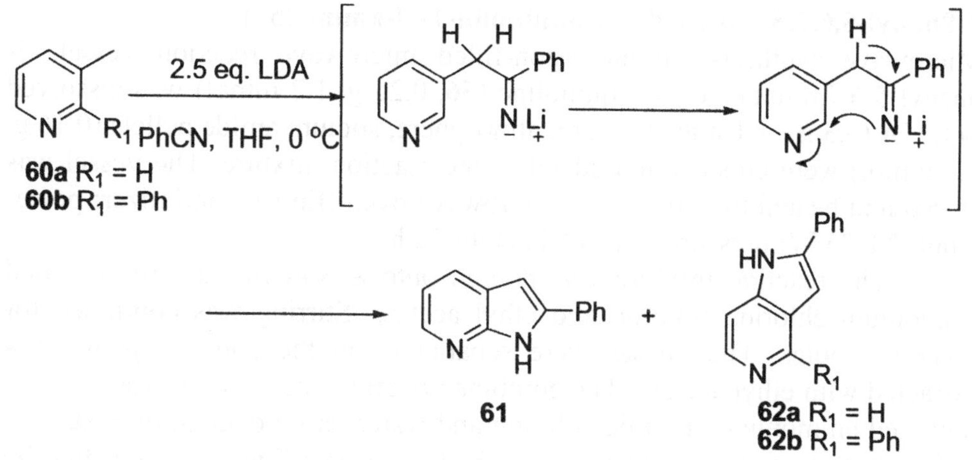

9.3.6 *Experimental*

2-Amino-4-chloro-3-nitropyridine (17a)

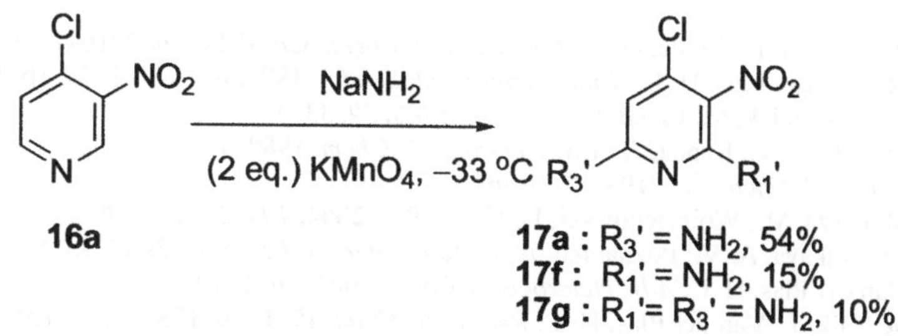

To 35–40 mL of liquid ammonia, (2.0–2.3 mmol) 4-chloro-3-nitropyridine and double the amount (4–4.6 mmol) of potassium permanganate were added, and the mixture was stirred for 5 h. After evaporation of ammonia, ~50 mL of water was added to the residue, and the mixture was continuously extracted with chloroform for 20 h.

The residue obtained after evaporation of the solvent from the combined extracts was dissolved in chloroform and separated by column chromatography. The first fraction gave, after washing with hexane, 2-amino-4-chloro-3-nitropyridine (**17f**) yield 15%, yellow crystals with mp 172–174 °C. The second fraction eluent was stripped off and the residue dissolved in boiling chloroform. The solution was concentrated and cooled to give 2-amino-4-chloro-5-nitropyridine (**17a**); yield 54%, light yellow crystals with mp 257–258 °C (sublime). The third fraction was eluted and the solution was concentrated to 20 mL and cooled to give 2,6-diamino-4-chloro-

3-nitropyridine (**17g**); yield 10%, yellow needles with mp 268–269 °C (decomposition).[14]

7-Phenyl-5,6,7,8-tetrahydro-isoquinolin-1-yl-amine (57)
Microwave synthesis: In an argon-filled microwave reaction vessel, 7-phenyl-5,6,7,8-tetrahydro-isoquinoline (**56**, 0.25 g, 1.2 mmol) was dissolved in DMA (6.3 mL). Under an argon atmosphere, sodium amide pellets (0.16 g, 4.1 mmol) were crushed and added to the reaction mixture. The vessel was closed and heated to 180 °C in a microwave oven (Emry's optimizer; power input: 20–25 W; pressure: 7.1–8.5 bar) for 12 h.

The reaction mixture was poured onto a cold mixture of saturated ammonium chloride solution and ethyl acetate. Stirring was continued for several minutes. The phases were separated and the aqueous phase was extracted with ethyl acetate. The combined organic phases were washed with saturated ammonium chloride solution and water, dried over sodium sulfate, and then evaporated to dryness. The obtained brown liquid was purified by column chromatography to give 160 mg (59% yield) of the title compound **57**.

9.3.7 References

1. Chichibabin, A. E.; Zeide, O. A. *J. Russ. Phys. Chem. Soc.* **1914,** *46*, 1216–1236.
2. (a) [R] Van der Plas, H. C. *Khim. Geterotsikl. Soedin.* **1987,** *8*, 1011–1027. (b) [R] Van der, P. H. C.; Wozniak, M. *Croat. Chem. Acta.* **1986,** *59*, 33–49.
3. [R] McGill, C. K.; Rappa, A. *Adv. Heterocyclic Chem.* **1988,** *44,* 1–79.
4. Marszak, I. *J. Chem. Soc.* **1946,** 760–761.
5. [R] Mąkosza, M.; Wojciechowski, K. *Chem. Rev.* **2004,** *104,* 2631–2666.
6. [R] Gulevskaya, A. V.; Pozharskii, A. F. *Adv. Heterocyclic Chem.* **2007**, *93*, 57–115.
7. [R] Van der Plas, H. C. *Adv. Heterocyclic Chem.* **2004**, *86*, 1–40.
8. (a) De Valk, J.; Van der Plas, H. C. *Rec. Trav. Chim.* **1971,** *90*, 1239–1245. (b) Kroon, A. P.; Van der Plas, H. C.; *Rec. Trav. Chim.* **1974,** *93*, 227–231. (c) Breuker, K.; Van der Plas, H. C. *Rec. Trav. Chim.* **1983,** *102*, 367–372. (d) Breuker, K.; Van der Plas, H. C. *J. Org. Chem.* **1979,** *44*, 4677–4680.
9. Zoltewicz, J. A. *Top. Curr. Chem.* **1975,** *59*, 33–64.
10. (a) Lont, P. J.; Van der Plas, H. C.; Verbeek, A. J. *Rec. Trav. Chim.* **1972,** *91*, 949–957. (b) De Valk, J.; Van der Plas, H. C. *Rec. Trav. Chim.* **1972,** *91*, 1414–1422. (c) Lont, P. J.; Van der Plas, H. C. *J. Rec. Trav. Chim.* **1973,** *92*, 311–316. (d) De Valk, J.; Van der Plas, H. C.; Jansen, F.; Koudijs, A. *Rec. Trav. Chim.* **1973,** *92*, 460–466.
11. Pathak, B. C.; Dutta, S. P.; Bardos, T. J. *J. Heterocyclic Chem.* **1969,** *6*, 447–452.
12. Knzie, M. G.; Felton, J. S. *Heterocycles* **1986,** *26*, 1815–1819.
13. [R] Pozharskii, A. F.; Simonov, A. M.; Doron'kin, V. N. *Russ. Chem. Rev.* **1978,** *47*, 1933–1969.
14. Pozharskii, A. F.; Kuz'menko, V. V.; Azimov, V. A.; Yakhontov, L. N. *Khim. Geterotsikl. Soedin.* **1973,** 1232–1239.
15. Wozniak, M.; Baranski, A.; Szpakiewicz, B. *Liebigs Ann. Chem.* **1991,** *3*, 875–878.
16. Zoltewicz, J. A.; Helmick, L. S. *J. Am. Chem. Soc.* **1972,** *94*, 682–683.
17. [R] Van der Plas, H. C. *Tetrahedron. Lett.* **1984,** *41*, 237–281.
18. (a) Van der Plas, H. C.; Charushin, V. N.; Veldhuizen, B. *J. Org. Chem.* **1983,** *48*, 1354–1357. (b) Pozharskii. A. F.; Gulevskaya, A. V. *Chem. Heterocyclic Compounds*. **2001,** *37*, 1461–1487.

19. Hara, H.; Van der Plas, H. C. *J. Heterocyclic Chem.* **1982,** *19*, 1285–1287.
20. (a) Rykowski, A.; Van der Plas, H. C. *J. Heterocyclic Chem.* **1984,** *21*, 433–434. (b) Rykowski, A.; Van der Plas, H. C. *J. Org. Chem.* **1980,** *45*, 881–885.
21. (a) Zoltewicz, J. A.; Hemlick, L. S.; Oestreich, T. M.; King, R. W.; Kandetzki, P. E. *J. Org. Chem.* **1973,** *38*, 1947–1949. (b) Kametani, T.; Kigasawa, K.; Iwabuchi, Y.; Hayasaka, T. *J. Heterocyclic Chem.* **1965,** *3*, 330.
22. Tondys, H.; Van der Plas, H. C.; Wozniak, M. *J. Heterocyclic Chem.* **1985,** *22*, 353–535.
23. Wozniak, M.; Baranski, A.; Nowak, K. *J. Org. Chem.* **1987,** *52*, 5643–5646.
24. Kobayashi, Y.; Kumadaki, I.; Taguchi, S.; Hanzawa, Y. *Tetrahedron. Lett.* **1970,** 3901–3902.
25. Wozniak, M.; Baranski, A.; Novak, K.; Poradowska, H. *Liebigs Ann. Chem.* **1990**, 653–657.
26. Chichibabin, A. E.; Oparina, M. P. *J. Russ. Phys. Chem. Soc.* **1920,** *50*, 543–548.
27. Bergstrom, F. W. *Liebigs Ann. Chem.* **1934,** 34–42.
28. (a) Bergstrom, F. W. *J. Amer. Chem. Soc.* **1934,** *56*, 1748–1751. (b) Bergstrom, F. W. *J. Org. Chem.* **1938,** *3*, 424–434.
29. Hara, H.; Van der Plas, H. C. *J. Heterocyclic Chem.* **1982,** *19*, 1527–1529.
30. (a) Palucki, M.; Hughes, D. L.; Yasuda, N.; Yang, C.; Reider, P. J. *Tetrahedron. Lett.* **2001,** *42*, 6811–6814. (b) Ciganek, E. *J. Org. Chem.* **1992,** *57*, 4521–4527.
31. Chrisman, W.; Knize, M. G.; Tang, M. J. *J. Heterocyclic Chem.* **2008,** 45, 1641–1649.
32. Palmer, A. M.; Grobbel, B.; Brehm, C.; Zimmermann, P. J.; Buhr, W.; Feth, M. P.; Holst, C.; Simon, W. A. *Bioorg. Med. Chem.* **2007,** *15*, 7647–7660.
33. Davis, M.; Wakefield, B. J.; Wardellt, J. A. *Tetrahedron* **1992,** *48*, 939–952.

9.4 Dimroth Rearrangement

Nicole L. Snyder and Taylor P. Adams

9.4.1 Description

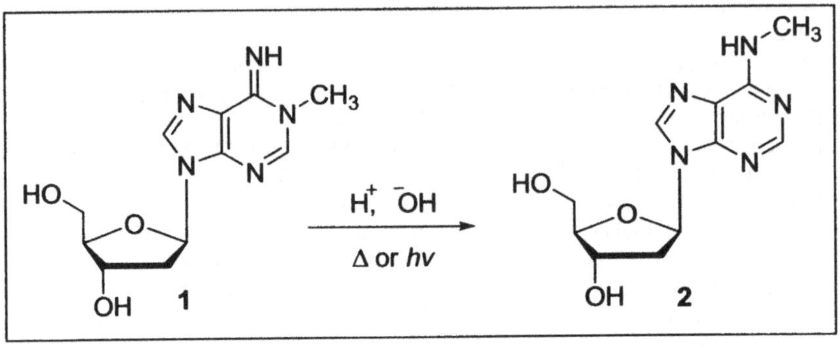

The Dimroth rearrangement is an isomerization process whereby atoms on a heterocyclic ring are translocated. The reaction can be catalyzed by a variety of acids and bases, with or without the presence of heat or light. Several factors may influence the rearrangement, including the number and types of heteroatoms in the ring, the number and types of substituents on the ring, and the kinetic and thermodynamic stabilities of the reactants and translocation products.

9.4.2 Historical Perspective

The Dimroth rearrangement was originally reported by Otto Dimroth in 1902[1] with the second report occurring seven years later in 1909.[2] The original reaction was discovered when 1,2,3-triazole derivatives rearranged under thermal conditions in the presence of base.

The most common rearrangements involve the translocation of exo- and endocyclic heteroatoms on a single heterocyclic ring. However, the Dimroth rearrangement has been shown to occur on a limited basis with heteroatoms in fused heterocyclic ring systems such as imidazo-, pyrazolo-, 1,2,3-triazolo- and 1,2,4-triazolo- and pyrimido-heterocycles.

The translocation of heteroatoms within rings of fused systems most often occurs with fused 5,6 aromatic systems and is accomplished through three major pathways as illustrated below.[3] In the first case, **A**, a heteroatom (denoted with a *) at the juncture between the two fused rings is translocated to a new location on one of the rings. In the second case, **B**, a heteroatom in one ring is translocated to form a new, differentially fused ring. In the third case, **C**, the presence of an amino, hydroxyl, or thiol group at the *ortho*-position of the heterocyclic ring translocates into the ring.

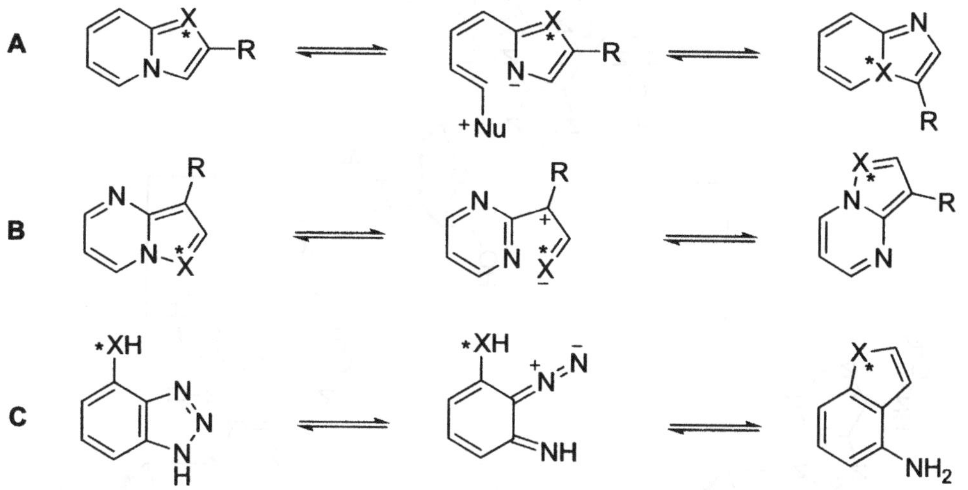

The translocation of exocyclic and endocyclic heteroatoms on a single heterocycle is shown below in **D**. In this case, the rearrangement is generally promoted by an exocyclic heteroatom on the ring. This is the most common example of a Dimroth rearrangement.

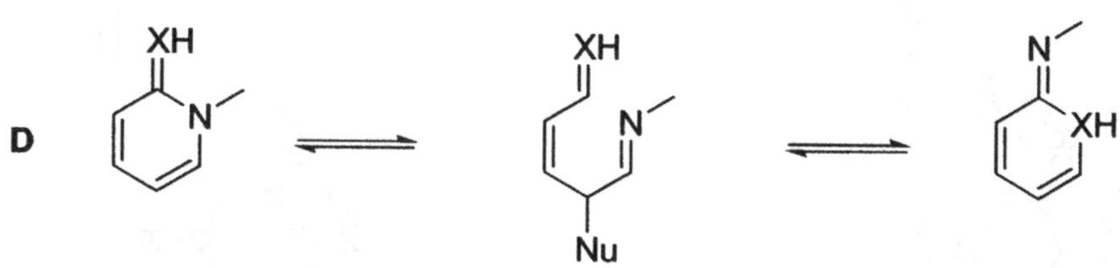

9.4.3 *Mechanism*

The Dimroth rearrangement has been shown to occur under both acidic and basic conditions, although classical Dimroth rearrangements usually occur with neutral molecules under alkaline conditions at a specific pH.[4] The general mechanism is illustrated using *N*-methyl adenosine **1**. Nucleophilic addition across the C–N bond of the purine ring system gives rise to intermediate **E**, which then undergoes ring opening and rearrangement to produce **F**. Rotation of the C–C imine bond to give **G**, followed by nucleophilic addition to recyclize the ring, gives **H**. Elimination of water followed by tautomerization generates the Dimroth product **2**.

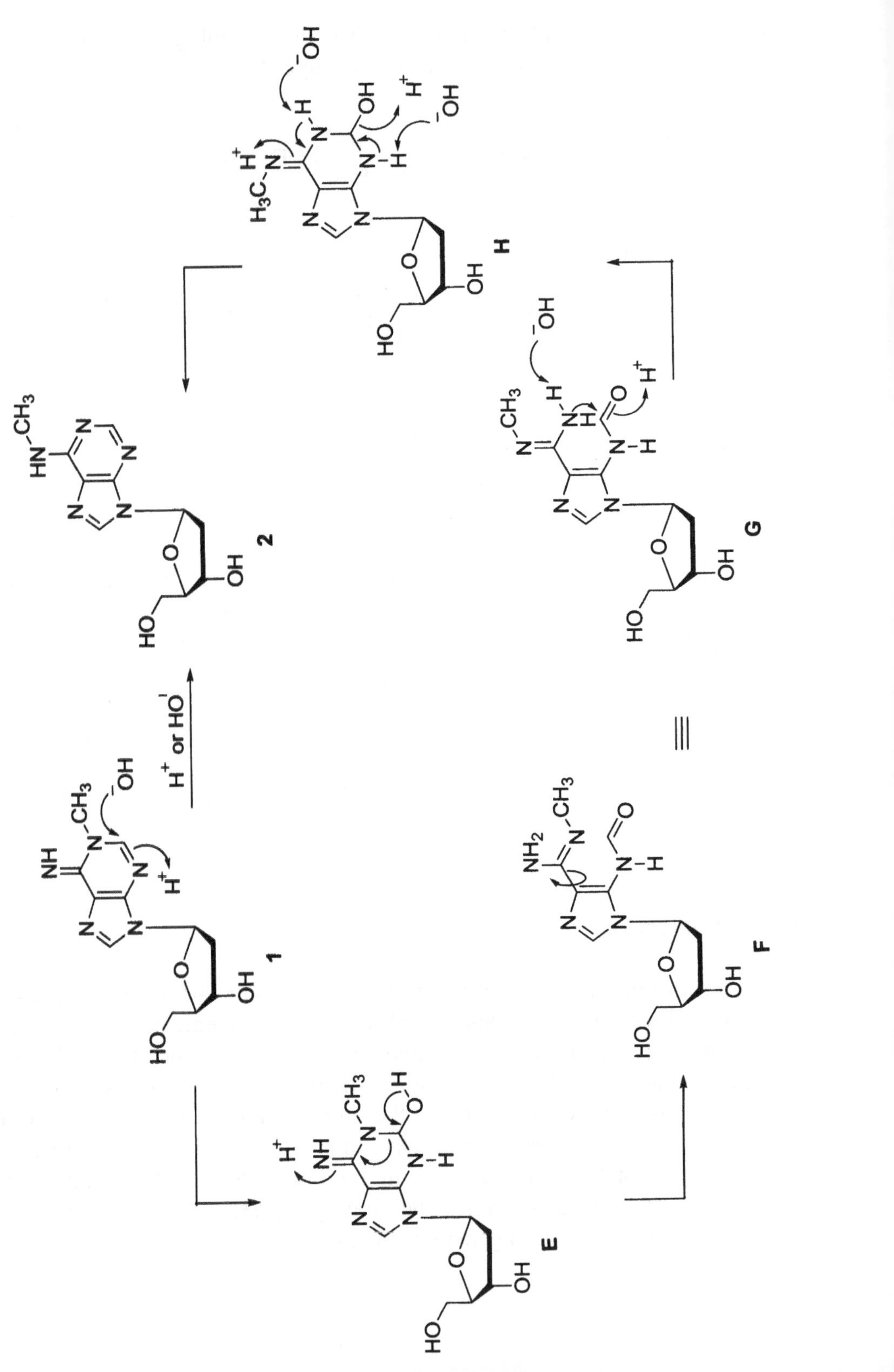
H+ or HO-
1
2
E
F
G
H

In general, it has been shown that increasing the number of nitrogen's in the ring increases the rate of the reaction.[5] The presence of an electron-withdrawing group can also have a favorable impact on the rate of the reaction.[6] In both cases, these substitutions increase the electrophilic nature of the ring, activating it toward nucleophilic attack and subsequent ring opening. Alkyl substituents have been shown to slow the rate of ring opening due to steric interactions and unfavourable electronic effects.[5] However, careful choice and placement of an alkyl group can also facilitate formation of the Dimroth product by establishing a favorable forward equilibrium.

Notably, the Dimroth rearrangement has been shown to occur in nature with the purine and pyrimidine bases of nucleosides and nucleotides upon exposure to certain chemical entities. For example, 3,4-epoxybutene,[7] styrene oxide[8] and other aromatic hydrocarbon based epoxides,[9] butadiene[10] and butadiene monoxide,[11] chloroethylene oxirane,[12] chlorambucil,[13] and acrolein,[14] among others, have been shown to facilitate Dimroth rearrangement, and in some cases subsequent cross-linking of DNA. While interesting from a mechanistic and biological perspective, these reactions will not be reviewed here.

9.4.4 Variations and Improvements

The Dimroth rearrangement is recognized as a general phenomenon in heterocyclic chemistry and occurs readily for activated heterocyclic ring systems under a variety of conditions. Initially, many Dimroth rearrangement products were discovered serendipitously as side products in the formation of highly nitrogenous or substituted heterocycles. In recent years, the Dimroth rearrangement has been exploited to generate a host of heterocyclic compounds for materials chemistry and biology.

In many cases, the Dimroth rearrangement is reversible, and so most of the variations and improvements for this reaction have focused on controlling either the substituents on the reactants or the reaction conditions in order to maximize the formation of the desired product. Specific examples, both historical and practical, are highlighted in the sections below. Examples are organized first by the type of rearrangement and then by the number of heteroatoms in the ring undergoing rearrangement.

9.4.5 *Synthetic Utility*

Heteroatom Translocation in Fused Heterocyclic Ring Systems

Rings Containing Two Heteroatoms

Imidazopyridines and imidazopyrimidines are two important classes of heterocycles. Derivatives of these compounds have been shown to have useful biological properties, including antifungal, antibacterial, and anticancer activity.

One of the first reports by Maury and co-workers of a Dimroth rearrangement occurring with imidazopyridines involved the rearrangement of functionalized imidazo[1,2-*a*]pyridines **5** and **7** in a refluxing solution of sodium hydroxide to give the corresponding isomers **6** and **8**.[6] It is interesting that 2-phenylimidazo[1,2-*a*]pyridine **3** did not undergo rearrangement to **4**, highlighting the importance of the electron-withdrawing nitro groups in activating the ring for rearrangement.[15] Maury and co-workers also showed that imidizopyrimidines (**9** and **10**), which contain an additional nitrogen atom in the pyridine ring, were similarly activated toward base catalyzed Dimroth rearrangement due to the enhancement of the electrophilic character at position 5 of the bicyclic system.[5]

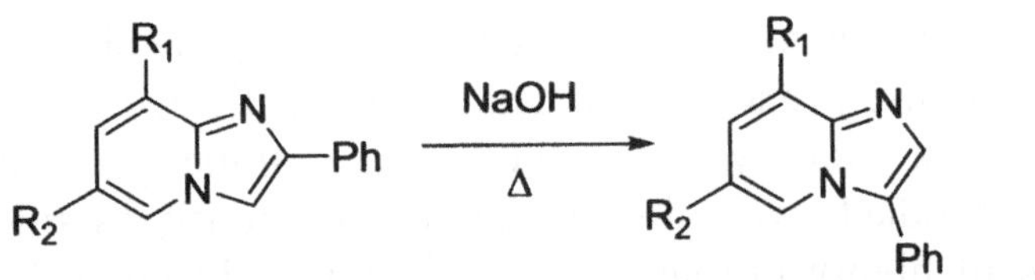

3, $R_1 = R_2 = H$
5, $R_1 = NO_2$, $R_2 = H$
7, $R_1 = H$, $R_2 = NO_2$

4, $R_1 = R_2 = H$ (NR)
6, $R_1 = NO_2$, $R_2 = H$
8, $R_1 = H$, $R_2 = NO_2$

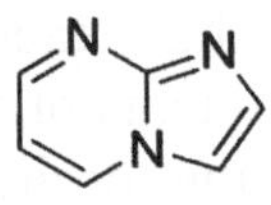

9

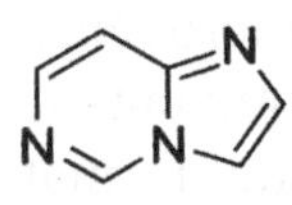

10

Carballares and co-workers employed a Dimroth rearrangement in their synthesis of a series of novel 3-substituted-2-aminoimidazo[1,2-*a*]pyrimidines with sterically bulky amino side chains.[16] Compounds **11** and **13**, when treated with a 5% solution of sodium methoxide in methanol at 80 °C, gave the corresponding Dimroth products **12** and **14** in 95% and 99% yield, respectively. In the same report, the authors prepared a number of analogous compounds substituted with aromatic and aliphatic functional groups (R) in high yields (91–99%). Their efforts to perform the Dimroth rearrangement with aromatic substituents on the exocyclic amine were equally successful.

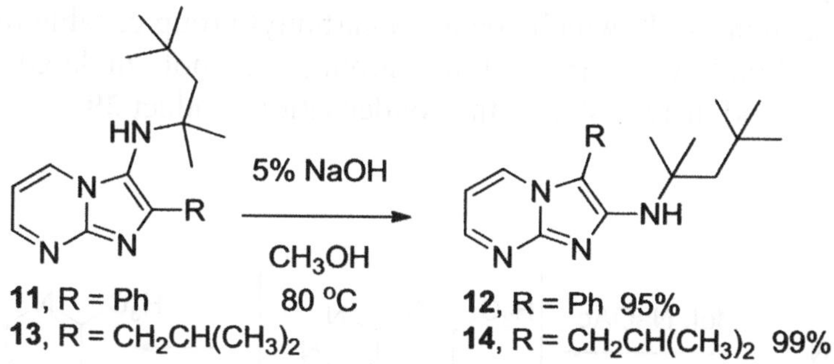

C-Nucleosides have also been prepared via Dimroth rearrangement. These compounds are desirable targets for the treatment of bacterial and viral infections due to their metabolic stability toward phosphorylase enzymes. Condensation of *C*-nucleoside **15** and 2-hydrazinopyrimidine **16** in refluxing ethanol led to the production of a nonisolatable purine analog intermediate, which underwent spontaneous Dimroth rearrangement to yield **17**.[17] This compound was isolated after treatment with methanolic ammonia to give the desired product in 45% yield.

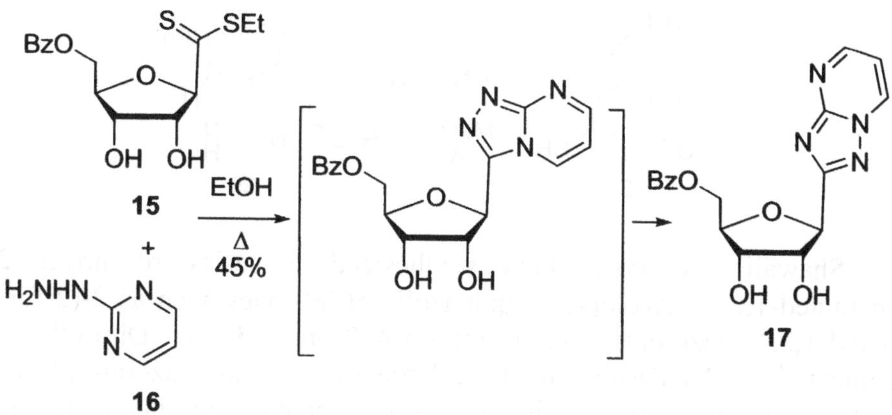

Rings Containing Three Heteroatoms

In their ongoing efforts to generate fused heterocyclic ring systems of potential physiological importance, Vas'kevich and co-workers employed the Dimroth rearrangement in their synthesis of a series of triazolo-pyrimidines.[18] Treatment of semicarbazides **18**, **20**, **22**, **24**, and **26** with methyl iodide in boiling methanol in the presence of sodium acetate resulted in formation of the corresponding 2-substituted-amino-5,7-dimethyl[1,2,4]-

triazolo[1,5-*a*]pyrimidines in good yields without formation of any sulfanyl-substituted thiazolopyrimidines. Notably, yields were improved for the more electronically deficient compounds **18**, **20** and **22**, which are activated toward base catalyzed substitution. The authors also found that analogous heterocycles, such as **28**, which contain a carbonyl group capable of forming a hydrogen bond with the amino group, do not undergo Dimroth rearrangement and instead stop at the condensation product **29**.

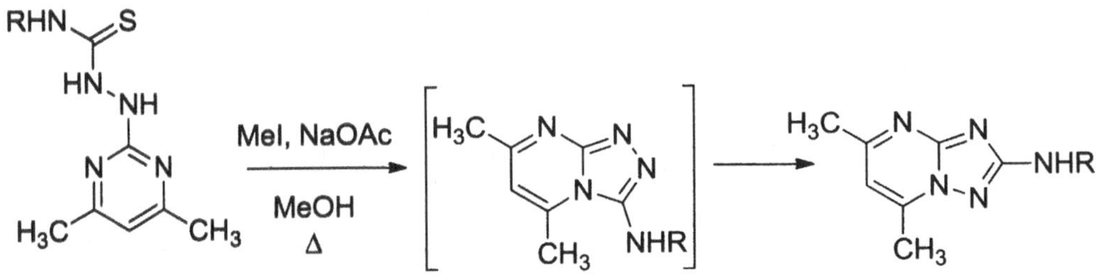

18, R = Ph
20, R = *p*-OCH_3-C_6H_4
22, R = *p*-CO_2Et-C_6H_4
24, R = CH_2CHCH_2
26, R = CH_3

19, R = Ph 73%
21, R = *p*-OCH_3-C_6H_4 77%
23, R = *p*-CO_2Et-C_6H_4 61%
25, R = CH_2CHCH_2 28%
27, R = CH_3 29%

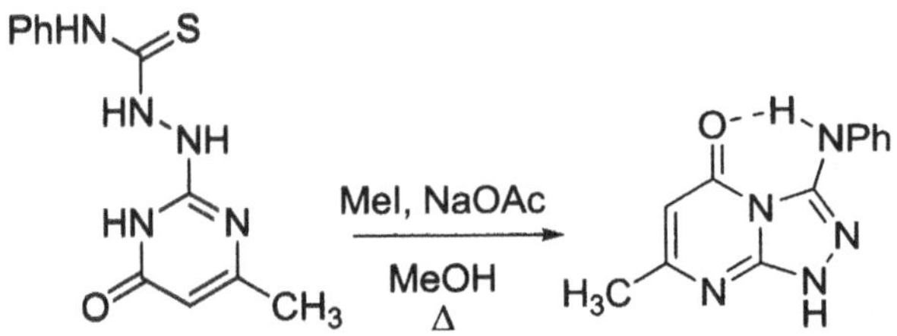

28 **29**

Shawali and co-workers synthesized a series of novel 2,4-disubstituted-1,2,4-triazolo[1,5-*a*]quinazolin-5(4*H*)-ones such as 2-(2-furyl)-4-phenyl-1,2,4-triazolo[1,5-*a*]quinazolin-5(4*H*)-one **31** via Dimroth rearrangement.[19] 2,4-Disubstituted-1,2,4-triazolo[1,5-*a*]quinazolin-5(4*H*)-one derivatives are of significant interest as therapeutics for the treatment of *Toxoplasmosis*. Quinazolinone **30**, when refluxed with ethanolic potassium hydroxide, readily formed the Dimroth product **31** in 78% yield. Several additional quinazolinones bearing different functional groups including phenyl, 4-bromophenyl, 4-nitrophenyl, 4-methoxyphenyl, naphthyl and thienyl derivatives (not shown) were also highlighted in their report. These compounds were prepared in upward of 86% yield with substitution having little to no impact on product yields. The authors employed a similar strategy in their later work on the synthesis of several substituted pyrazolotriazolopyrimidines.[20]

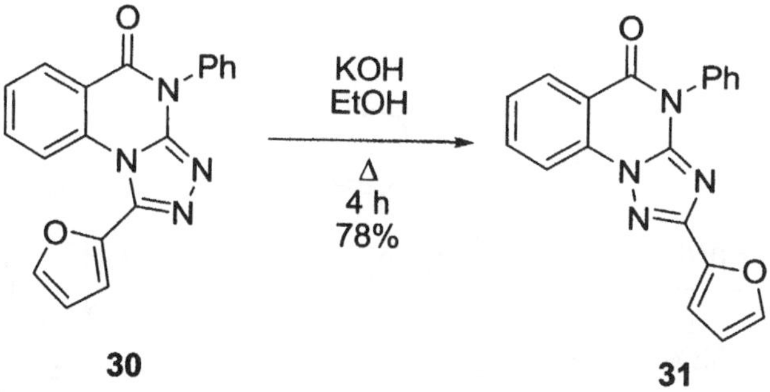

Loakes and co-workers observed the spontaneous Dimroth rearrangement of nucleosides **32**, **34** and **36** in the presence of a catalytic amount of sodium methoxide in methanol during hydrolysis of the acetate protecting groups.[21] The authors noted that a similar rearrangement did not occur in the presence of aqueous sodium hydroxide. The authors speculated that Dimroth rearrangement occurs in the former case, but not in the latter due to the formation of a methyl ester intermediate that can recyclize. In the latter case, an intermediate carboxylate ion, which does not recyclize, is formed.

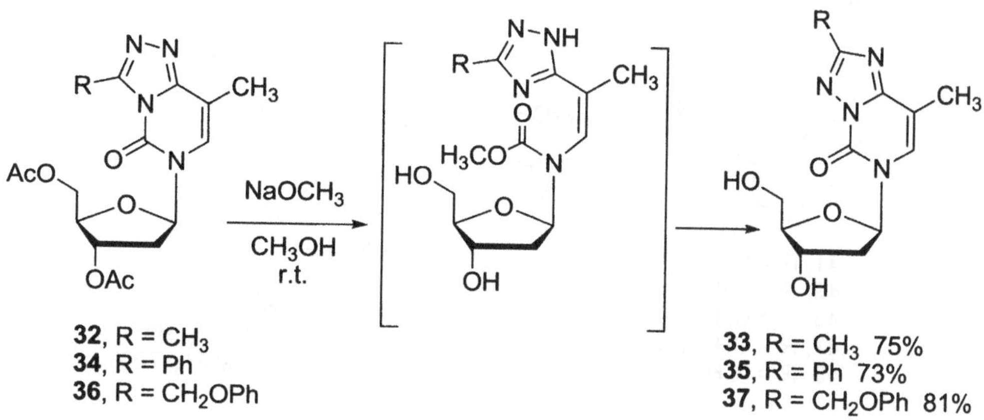

Rusinov and co-workers employed a Dimroth rearrangement in the final step of their novel synthesis of [1,2,4]triazolo[1,5-*d*][1,2,4]triazine derivatives.[22] These compounds, which are structural analogs of naturally occurring purine bases, have been shown to exhibit diverse biological properties, including antitumor and antiviral activity. Triazine **38**, when refluxed in nitrobenzene in the presence of **39**, underwent Dimroth rearrangement via a nonisolatable intermediate to afford 32% of triazinone **40**. Several additional 6-phenyl-1,2,4-triazin-3(2*H*)-ones were prepared in the same report, including methyl, phenyl, 3-nitrophenyl, and 3-pyridinyl analogs with yields of between 25 and 34% (not shown).

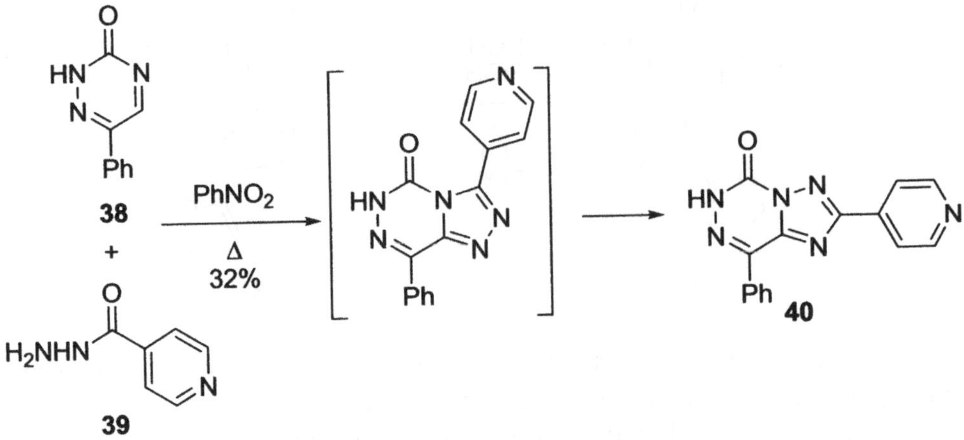

Vorob'ev and co-workers employed the Dimroth rearrangement in their preparation of a number of thieno[3,2-*e*][1,2,4]triazolo[1,5-*c*]pyrimidines.[23] Treatment of thienotriazolopyrimidines **41**, **43**, **45**, **47** and **49** with refluxing ethanolic sodium hydroxide gave the corresponding Dimroth products in yields ranging from 68–92%. The authors noted a slight decrease in yield when the thienotriazolopyrimidines were substituted with alkyl groups.

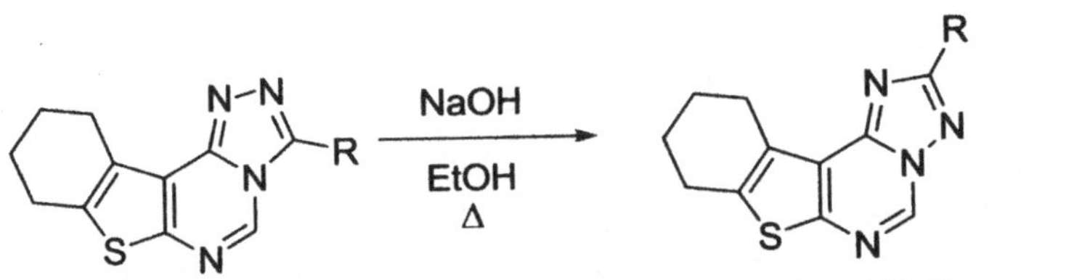

41, R = H
43, R = CH_3
45, R = CH_2CH_3
47, R = $CH(CH_3)_2$
49, R = $(CH_2)_3CH_3$

42, R = H 80%
44, R = CH_3 72%
46, R = CH_2CH_3 74%
48, R = $CH(CH_3)_2$ 68%
50, R = $(CH_2)_3CH_3$ 70%

Highly nitrogenous heterocycles such as 1,2,4-triazolo-1,3,5-triazines have been shown to rearrange upon treatment with excess base. 5,7-Bis(dimethylamino)-3-(methylthio)-1,2,4-triazolo[4,3-*a*][1,3,5]triazine **51** was isomerized to **53** via isolatable guanidine intermediate **52** in the presence of excess anhydrous dimethylamine, pyrrolidine or aniline in absolute ethanol at 85 °C.[24] This reaction was also shown to go forward under acidic conditions, although a mixture of **52** and **53** was recovered in the process.

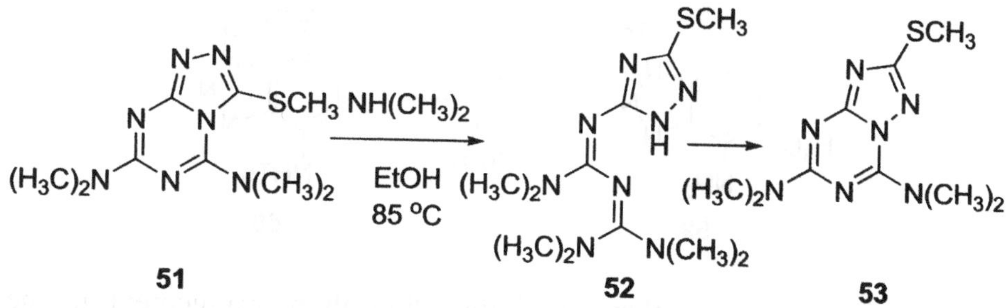

Tartakovsky and co-workers synthesized two novel tris[1,2,4]-triazolo[1,3,5]triazines via thermal Dimroth rearrangement.[25] These molecules are of interest due to their potential as components in thermally stable materials. [1,2,4]Triazolo[1,5-*a*:1′,5′-*c*:4″,3″-*e*][1,3,5]triazine **56** was obtained by heating either triphenyltris[1,2,4]triazolo[1,3,5]triazine **54** at 230 °C for 3 h or **55** at 350 °C for 4 h with yields of 80 and > 95%, respectively.

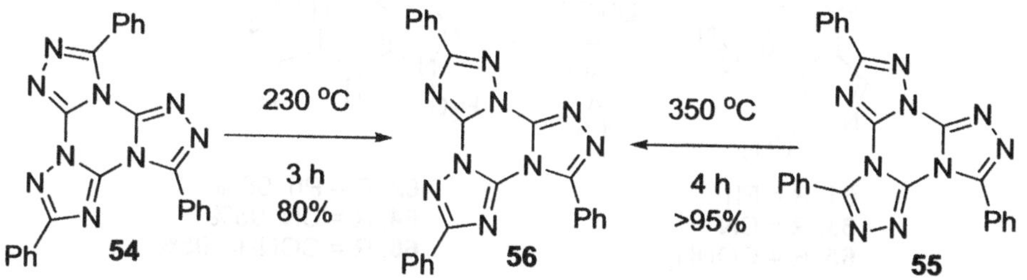

In their efforts to prepare purine ring analogs for biological studies, Temple and co-workers reported on the 85% conversion of diamino-1*H*-1,2,3-triazolo[4,5-*c*]pyridine **57** under basic condition to give the fused triazole diamino-3*H*-1,2,3-triazolo[4,5-*b*]pyridine **58**.[26] This reaction was shown to proceed through a diazo intermediate.

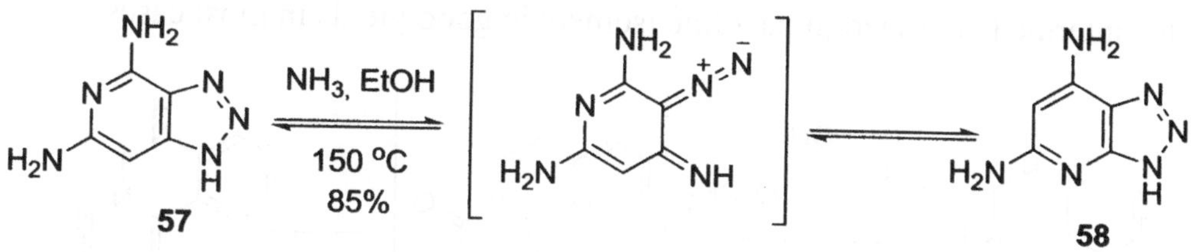

Fused 1,2,3-triazoles have also been shown to undergo thermal Dimroth rearrangement under neutral conditions. Benzotraizole **59** rearranged to the corresponding Dimroth product **60** in near quantitative yield after boiling for 3 h in toluene.[27] It is interesting to note that when **59** was refluxed in benzene under similar conditions, only a trace amount of **60** formed.

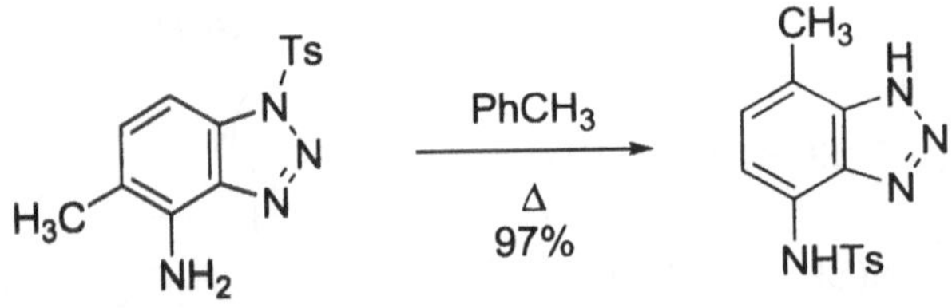

59 60

Laurina and co-workers used the Dimroth rearrangement in their synthesis of novel pyrazolo[3,4-*d*][1,2,3]triazolo[1,5-1]pyrimidine systems.[28] These compounds have the potential to serve as DNA intercalators in a similar fashion to acridines, anthracyclines, and actinomycins. Treatment of compounds **61**, **63** and **65** with refluxing sodium ethoxide in ethanol gave the corresponding Dimroth products **62**, **64** and **66** in high yield.

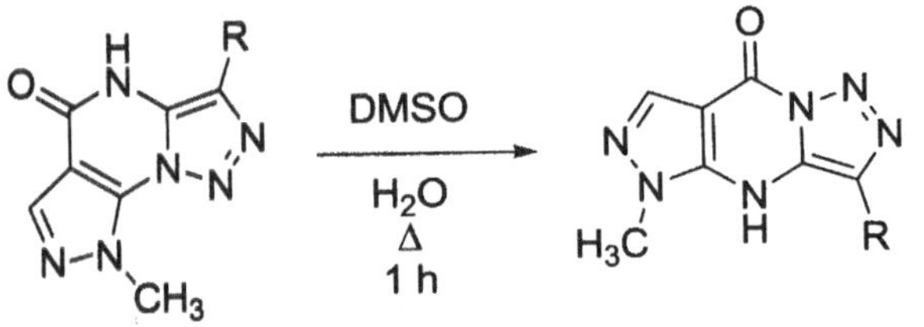

61, R = Ph
63, R = CN
65, R = $CONH_2$

62, R = Ph 98%
64, R = CN 95%
66, R = $CONH_2$ 95%

Karepenko and co-workers used the Dimroth rearrangement as a key step in their synthesis of intermediates toward novel spiro-fused (C5)-pyrazolino-(C6)-triazinones in an effort to explore the biological activity of these compounds.[29] Condensation of 4-hydrazinoquinazoline **67** with a number of (hetero)aryl substituted 2,4-diketoesters in glacial acetic acid led to the formation of the corresponding 3-acylmethyltriazoquinazolines **68** via the intermediate triazoquinazoline isomers in good yields in most cases.

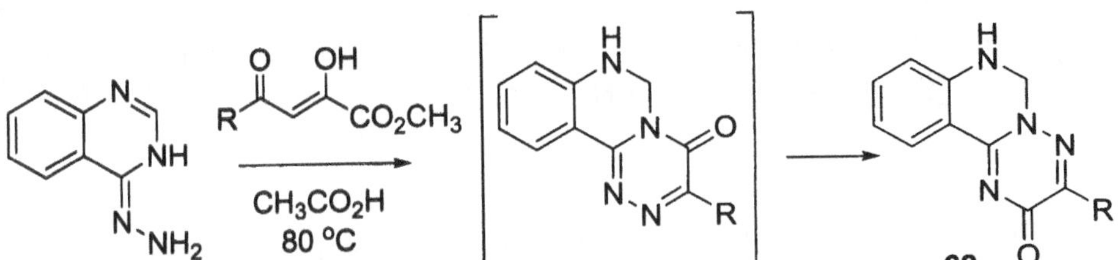

67 **68**

Thiones such as 1,2,3-triazolopyridinethione **69** underwent Dimroth rearrangement to form the corresponding amino-1,2,3-thiadiazolopyridine **70** in the presence of refluxing isopropanol.[30] The authors noted that triazole ring opening and subsequent rearrangement was controlled at least in part by the electron withdrawing inductive effects of the substituent in the pyridine ring (HNCS or NCSH > $NCNH_2$).

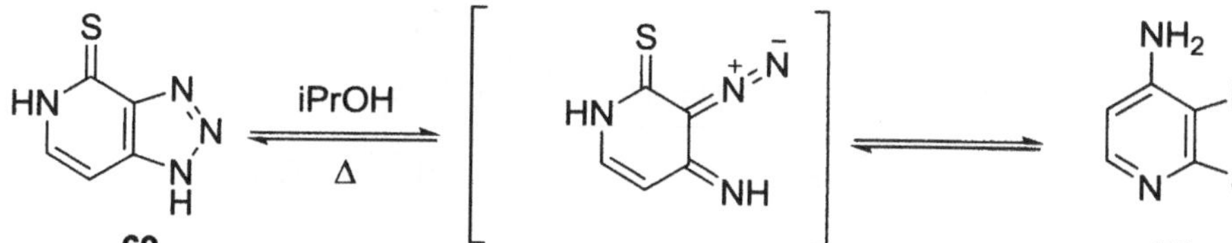

69 70

Rings Containing Four Heteroatoms

Georges and co-workers used the Dimroth rearrangement in their synthesis of substituted verdazyl compounds as templates for the diversity-oriented synthesis of heterocyclic compounds.[31] Verdazyl derivative **71** did not undergo rearrangement to form **72** when refluxed in ethyl acetate (not shown). However, treatment of **71** with either two equivalents of sodium hydride at room temperature or lithium diisopropyl amine in THF at 0 °C gave the corresponding Dimroth product **72** in 82% yield.

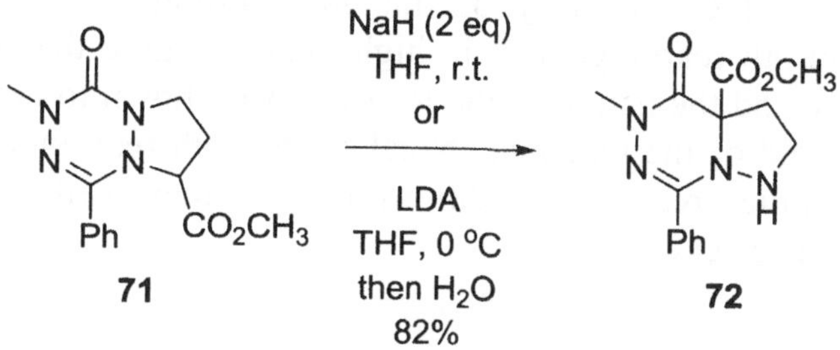

71 72

Heteroatom Translocation of Exocyclic and Endocyclic Heteroatoms in Heterocyclic Rings

Rings Containing One Heteroatom

The importance of electron-withdrawing groups on the Dimroth rearrangement was illustrated by Tschitschibabin and co-workers in their early work on the reaction of 1,2-dihydro-2-imino-1-methylpyridine derivatives **73** and **75** with aqueous sodium hydroxide. Analog **73** was slowly hydroylzed to produce the corresponding amide **74** without any rearrangement.[32] However, installation of a single nitro group on the pyridine ring (**75**) led to the generation of the rearranged product **76** under similar conditions.[33]

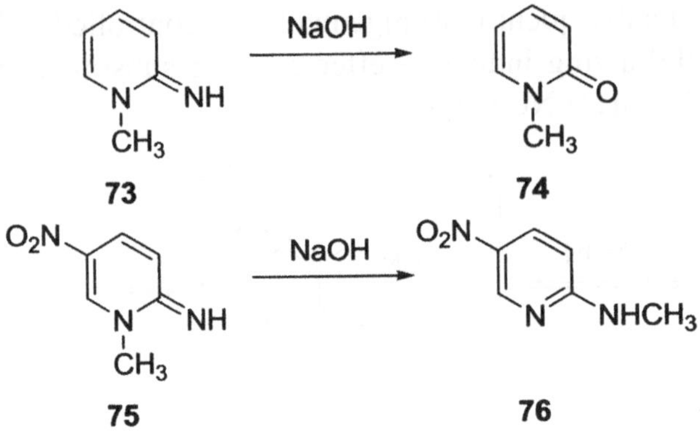

Nishiwaki and co-workers showed that nitroenamine **77** gave the corresponding aminopyridine **79** in quantitative yield via the intermediate iminopyridine when reacted with malonitrile **78** in chloroform at room temperature in the absence of base.[34] Excellent yields were also obtained using ethylcyanoacetate **80**, cyanoacetamide **82**, 2-pyridylnitrile **84**, phenylacetonitrile **86**, and benzoylacetonitrile **88** instead of **78**. However, the presence of a base was required for conversion of **82**, **84** and **86** to the corresponding aminopyridines. The authors also employed nitroenamines substituted with various R groups (*tert*-butyl, allyl, 2-ethoxycarbonyl) in an effort to explore the versatility of this methodology. Most derivatives underwent Dimroth rearrangement in high yields to give the corresponding aminopyridines, and the nature of the R group was shown to have little to no effect on the yield of the reaction. Notably Dimroth rearrangement did not occur under similar reaction conditions using either diethyl malonate or substituted amide esters.

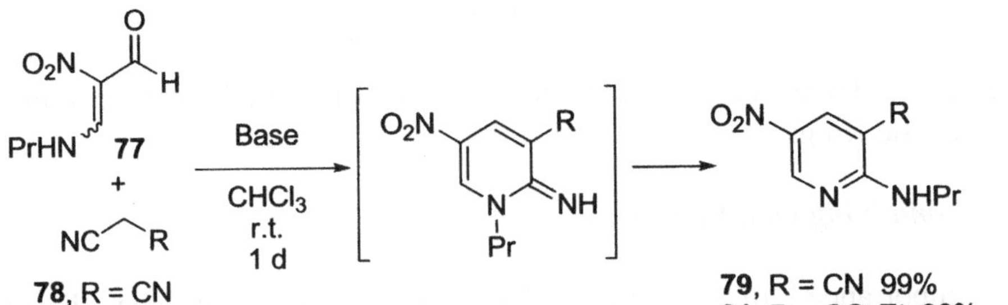

78, R = CN
80, R = CO_2Et
82, R = CO_2NH
84, R = 2-pyridyl
86, R = Ph
88, R = COPh

79, R = CN 99%
81, R = CO_2Et 93%
83, R = CO_2NH 99%
85, R = 2-pyridyl 90%
87, R = Ph 87%
89, R = COPh 97%

Hajós and co-workers observed an unexpected Dimroth rearrangement in their synthesis of linearly fused [1,2,4]triazolo[1,5-*b*]bisquinoline ring systems.[35] Diaminium salts substituted at the 4 position with methyl, ethyl, or benzyl functionalities (**90** where X = CH_3, CH_2CH_3 or Bn) gave the expected triazoles **99** in modest yields. However, treatment of unsubstituted diaminium salt **90** (X = H) with aldehydes **91, 93, 95** and **97** did not lead to the formation of the expected triazoles, but instead gave hydrazones **92, 94, 96** and **98** in 35–51% yield via a Dimroth rearrangement. The authors used ^{15}N-labeling studies to show that nucleophilic attack at position 1, which is required for the Dimroth rearrangement, is less likely to occur with substituted derivatives because the substitutents decrease the electropositive nature of that position.

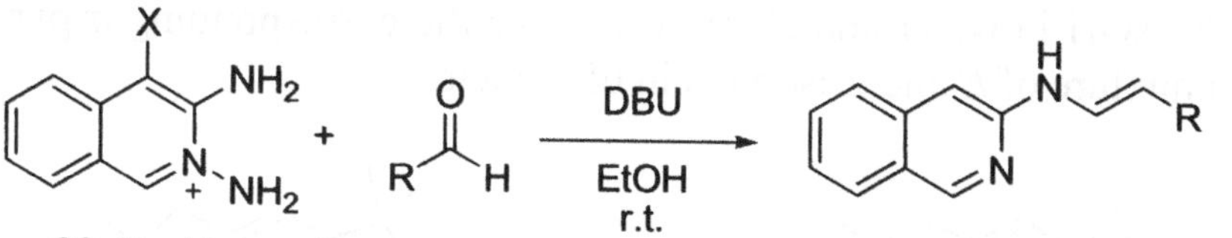

90, X = H

91, R = p-CH_3-C_6H_4
93, R = Ph
95, R = $CH_2CH_2CH_3$
97, R = CH_3

92, R = p-CH_3-C_6H_4 35%
94, R = Ph 51%
96, R = $CH_2CH_2CH_3$ 36%
98, R = CH_3 49%

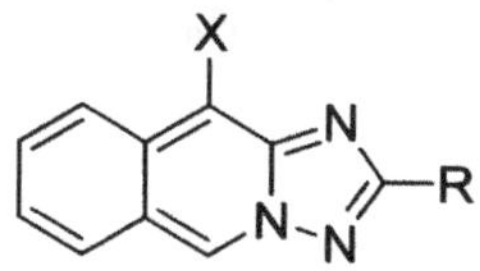

99, X = CH_3, $CH_2CH_2CH_3$, CH_2Ph

Langer and co-workers used an acid-catalyzed Dimroth rearrangement to prepare N-aryl-5-alkylildene-2,5-dihydropyrrol-2-ones.[36] These compounds are potent inhibitors of serine proteases, and are important intermediates in synthesis of γ-lactams and tetramic acid antibiotics such as pukeleimid A. Treatment of furan **100** in THF with and without the presence of lithium chloride gave the corresponding α,β-unsaturated ketone **101** in 56% and 12% yield, respectively, illustrating the importance of the acid catalyst for this system.

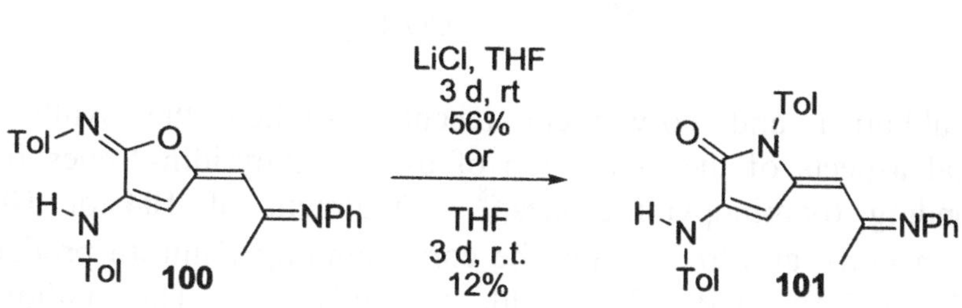

100 → **101**

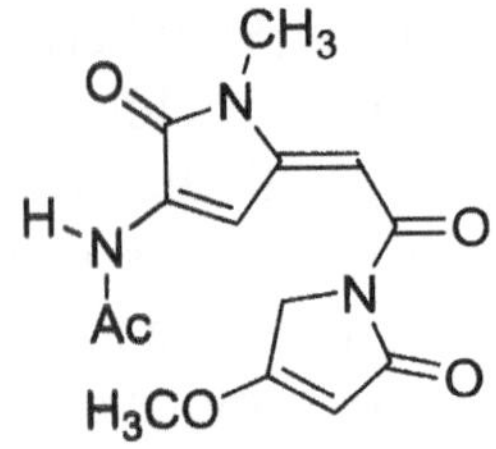

pukeleimid A

Jacobi and Liu[37] used the Dimroth rearrangement to generate structural motifs for semicorrins, tripyrrolines, and higher analogs such as cobyric acid, a functional analog of vitamin B_{12}. Treatment of iminolactone **102** with TsOH in water and chloroform gave the corresponding tripyrroline **103** as a mixture of *E* and *Z* isomers in 65% yield.

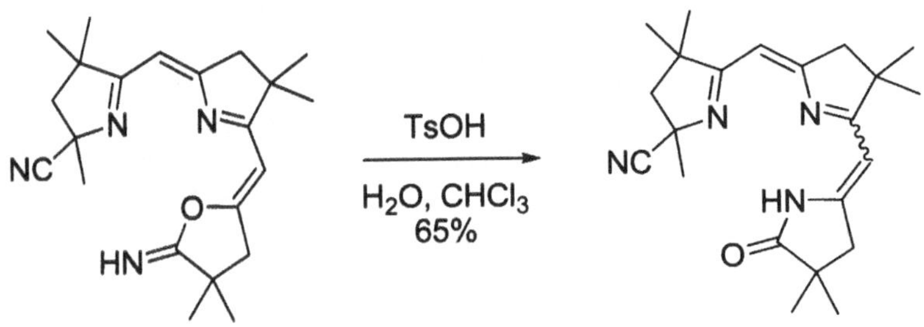

102 **103**

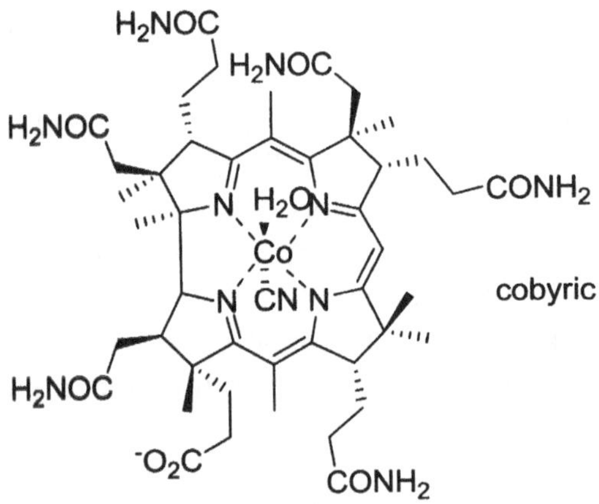

cobyric acid

Subbotina and co-workers recently studied the synthetic and theoretical aspects of the generation of 6-hydroxypyridin-2-ones from the corresponding 6-aminopyran-2-ones.[38] Treatment of **104** or **106** with pyridine in DMF at 120 °C gave the corresponding Dimroth products **105** and **107** in 70% and 57% yield, respectively. The authors used computational chemistry to rationalize the generation of a nonisolatable ketene intermediate in the reaction.

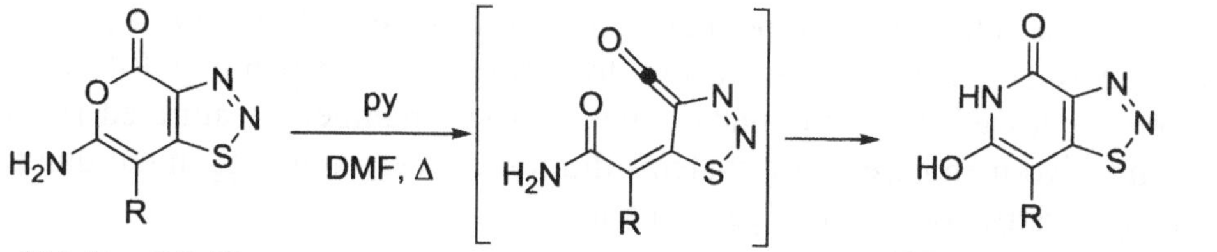

104, R = CO_2Et
106, R = CO_2CH_3

105, R = CO_2Et 70%
107, R = CO_2CH_3 57%

Rings Containing Two Heteroatoms

The effect of substituents in the Dimroth rearrangement of rings containing two heteroatoms was illustrated by Brown and co-workers in their earlier work with 1,2-dihydro-2-amino-1-methylpyrimidine **108**.[39] Treatment of **108** with anhydrous diethylamine gave the corresponding rearrangement product **109**. Replacement of the methyl group of **108** with higher alkane homologs (not shown) was shown to increase the rate of the rearrangement,[40] presumably due to increasing steric interactions, which hinder the reverse reaction. Multiple alkyl substitutions were also shown to have an impact on the rate of the reaction. For example, 1,2-dihydro-2-imino-1,6-dimethylpyrimidine **110** rearranged faster than 1,2-dihydro-2-imino-1,4-dimethylpyrimidine **111**. However, the yield for the Dimroth product of **111** was greater than that of **110**.

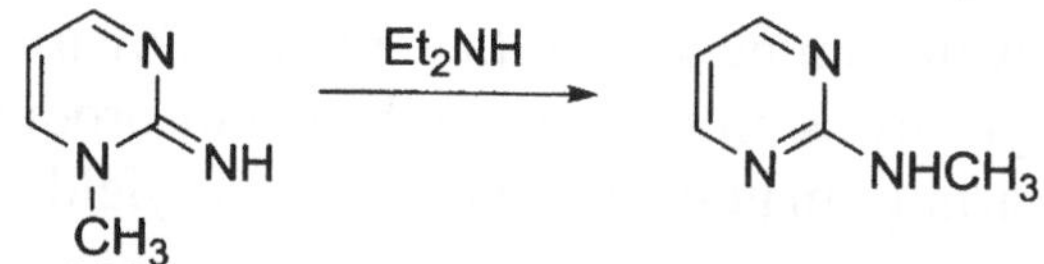

108 **109**

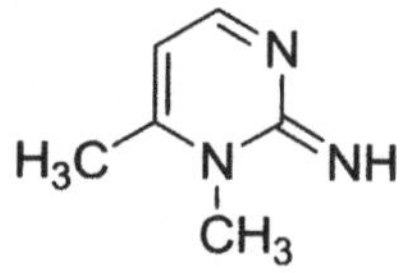

110

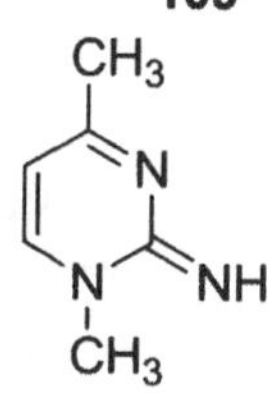

111

Brown and Ienaga also showed that 1,6-dihydro-6-imino-1,2-polymethylenepyrimidines such as **112** could be used to generate the corresponding β-bridged 6-aminopyrimidine derivatives **113** via Dimroth rearrangement, and that the rate of rearrangement was dependent on both substituent effects, as well as the number of carbons in the more saturated heterocycle.[41] For example, compounds with more electron-withdrawing subsitituents adjacent to the imino moiety rearranged faster than those with less electron-withdrawing subsitituents in the same position (**114** > **115** > **116**

> **112**). In addition, compounds containing seven atoms in the more saturated ring rearranged the fastest to form the corresponding β-bridged isomers, followed by compounds containing six-membered saturated rings. Four and five-membered ring systems did not undergo rearrangement due to the length restrictions for bridge formation.

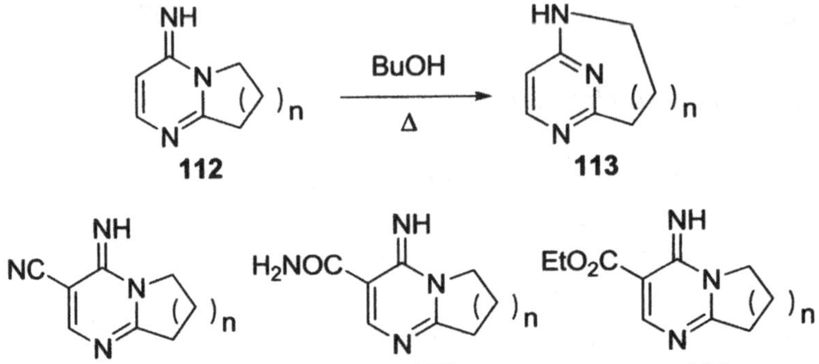

112 **113**

114 **115** **116**

Brown and Ienaga extended their work on the Dimroth rearrangement of 6-alkylaminopyrimidines and 2,*N*(6)-polymethylene-6-aminopyridines to 4-iminoquinazoline systems.[42] In particular, they showed that 6,6,7-tricyclic imine **117** and 6,6,9-tricyclic imine **119**, upon treatment with a refluxing solution of potassium hydroxide, rearranged to form the corresponding β-bridged isomers **118** and **120** in high yields. It is interesting to note that the analogous 6,6,8-tricyclic amine system (n = 6; not shown) gave a mixture of products with the hydrolysis product being the major product of the reaction. Larger tricyclic starting materials gave reduced yields of the Dimroth product.

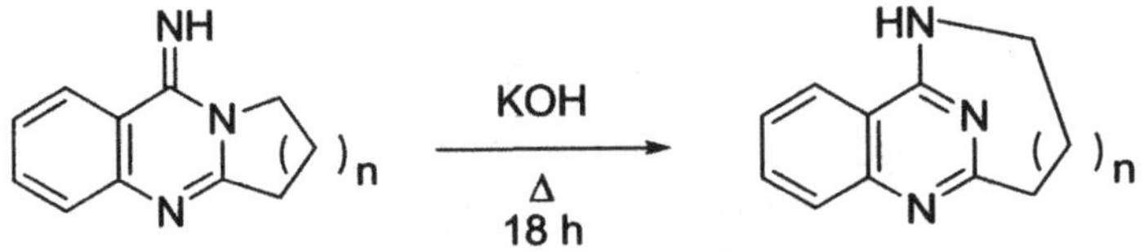

117, n = 5
119, n = 7

118, n = 5 91%
120, n = 7 85%

Narsaiah and co-workers employed the Dimroth rearrangement in their preparation of novel 4-substituted-amino-5-trifluormethyl-2,7-disubstituted pyrido[2,3-*d*]pyrimidines and evaluated their potential as antibiotics against Gram-positive and Gram-negative bacteria.[43] Treatment of iminoether **121** with methyl amine in ethanol at room temperature for 3 h gave rise to *N*-methyl-2-methyl-7-phenyl-5-(trifluoromethyl)pyrido[2,3-*d*]pyrimidin-4-amine **122** in 84% yield. This compound showed significant activity against *B. subtilis* and *S. aureus*. Several other derivatives were also prepared in similar yields using various amines, but unfortunately these compounds showed little efficacy as antibacterials.

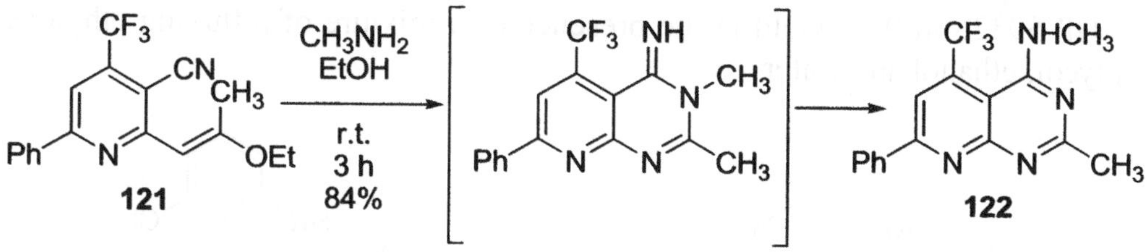

Reddy and co-workers[44] applied the Dimroth rearrangement to their syntheses of Iressa (gefitinib) and Tarceva (erlotinib), two 4-anilino-qinazoline compounds. These compounds have been shown to inhibit the epidermal growth factor receptor (EGFR), which is responsible for mediating cell division, motility, adhesion, and apoptosis. Reaction of imine **123** with 3-chloro-4-fluoroaniline in acetic acid at 130 °C gave Iressa in 70% yield. Similarly, reaction of imine **124** with 3-ethynyl aniline in refluxing acetic acid gave Tarceva in 92% yield. Besson and co-workers[45] recently used a similar, microwave-assisted Dimroth approach in their synthesis of the 4-aminoquinazoline Azixa, a homolog of Iressa and Tarceva (not shown).

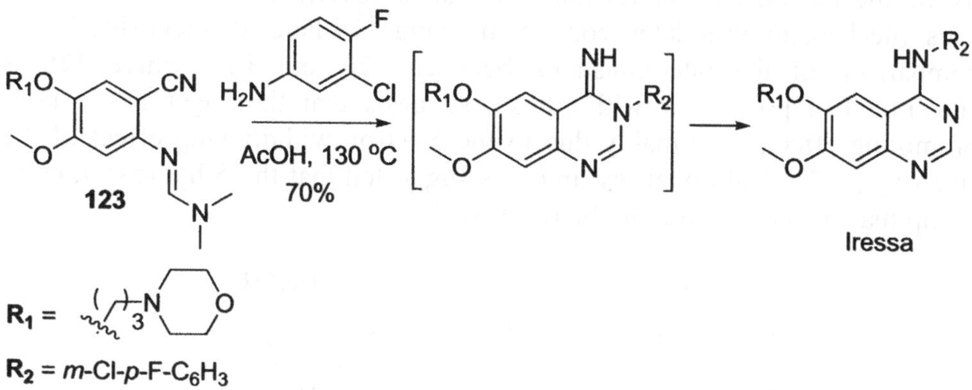

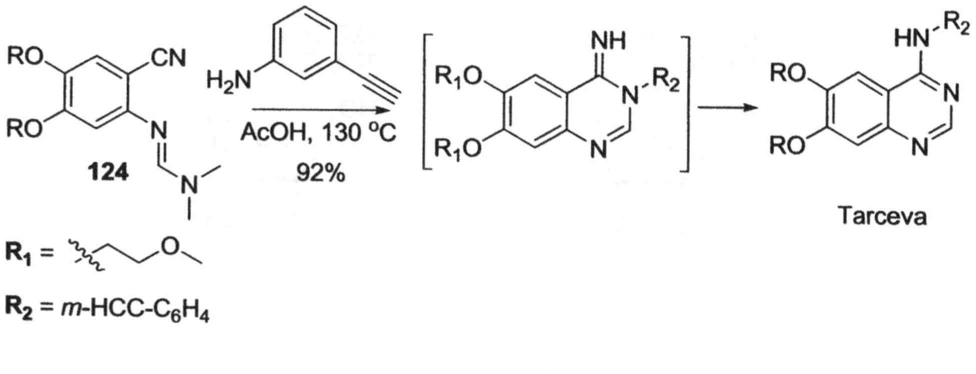

Fischer and Misun used the Dimroth rearrangement in their preparative synthesis of CGP-59326,[46] a pyrrolo[2,3-*d*]pyrimidine that has also been investigated as an EGFR inhibitor. Compound **125**, prepared in four steps from commercially available alanine, was readily converted to CGP-59326 in 92% yield in the presence of a mixture of refluxing ethylene glycol, ethanol and water.

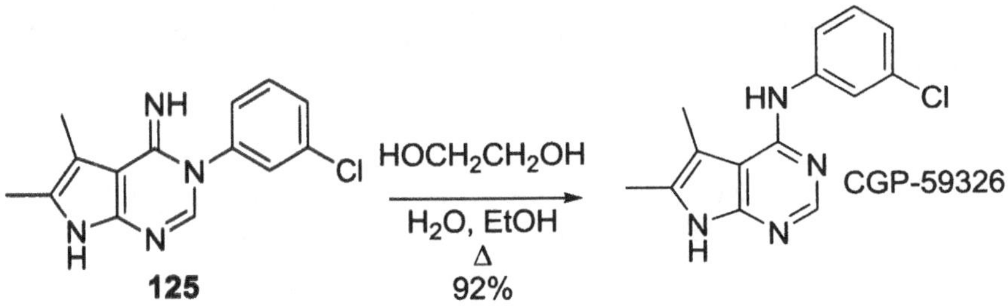

A number of adenine analogs have been prepared using the Dimroth rearrangement. Macon and Wolfenden[47] first reported on the conversion of 1-methyladenosine **126** to *N*-methyladenosine **127** when the former was treated with aqueous sodium hydroxide. The rate of conversion was shown to be proportional to the amount of hydroxide ion between pH values of 8 and 10, suggesting that the rate-determining step is the initial attack of the hydroxide ion on the neutral and protonated species at position 2 of **126**. This mechanism was later confirmed using ^{15}N-labeled adenosine.[48] A comparison of the rate constants between **126** and **128**, where **128** is substituted at position 9 with ribose, suggests that the sugar has a rate-promoting effect, presumably due to the electron-withdrawing effect of the ring oxygen.[49] Additional research has suggested that the 5-hydroxy methyl group may also play a role in the reaction.[50]

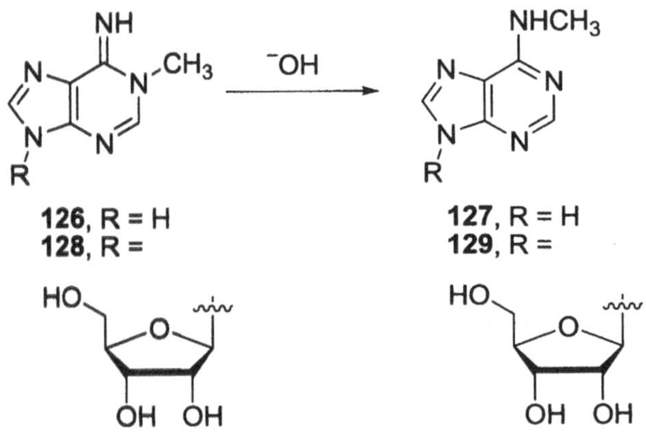

Kinetic studies have shown that the substituent at position 1 and pH also play an important role in the Dimroth rearrangement of adenine and adenosine derivatives.[50] For adenine derivatives substituted at position 1, the individual rates for methyl, ethyl and propyl derivatives were comparable, while the benzyl derivative rearranged faster (not shown). Above pH 10, methyl-substituted adenine derivatives rearranged fastest, while the benzyl-substituted derivatives rearranged the slowest. Subsequently, the authors found that adenosine **130** rearranged faster at pH values below 10, whereas adenoside **128** rearranged faster at pH values above 10, and in both cases the reaction rates were enhanced by the presence of ribose. The authors attributed these observations to the electronic effects of the substituents, which play a greater role at pH values less than 10 where the nucleoside remains protonated. At pH values above pH 10, the adenine/adenosine derivatives are neutral and sterics plays a more important role.

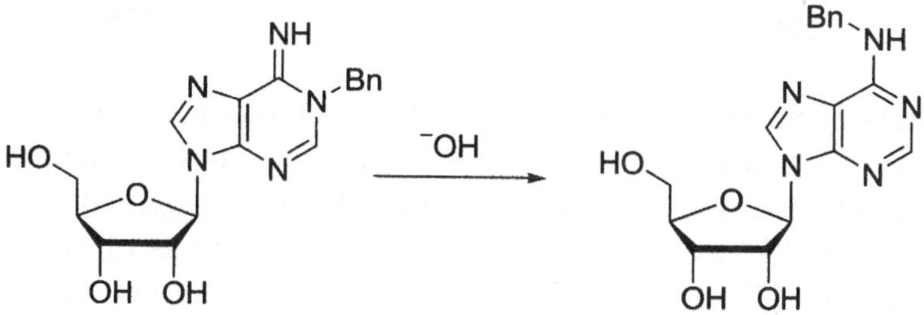

130 **131**

Later work by Fujii and co-workers on the Dimroth rearrangement of 1-(ω-hydroxyalkyl) adenine salts **132** revealed a rate increase over the corresponding alkyl substituted derivatives.[51] The authors also observed that the counterion played a significant role in the reaction rate; hydrobromide salts were shown to rearrange faster than the corresponding perchlorates to give the Dimroth products **133**. Interestingly, nucleoside analogs of these compounds failed to undergo Dimroth rearrangement to form the corresponding 6-*N* derivatives. Instead, a competing hydrolytic deamination reaction (not shown) resulted in the generation of the corresponding hypoxanthine derivatives.

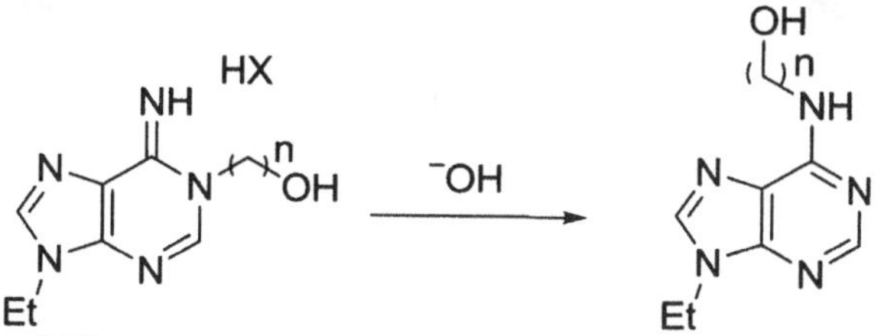

132, n = 2 or 3 **133**, n = 2 or 3

Ienaga and Pfleiderer demonstrated that polymethylene bridged nucleoside analogs **134** (where n = 5, 7 and 9) could undergo Dimroth rearrangement in boiling butanol to give the corresponding β-bridged isomers **135**.[52] For substrates were n < 5 the corresponding hydrolysis products (not shown) were isolated instead of the Dimroth products further illustrating the dependency of this reaction on the size of the saturated ring.

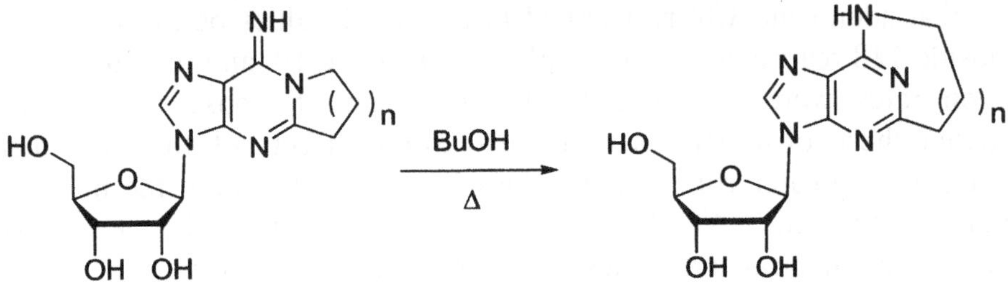

134 **135**

Carbonucleosides, which have similar biological properties to *C*-glycosides, have also been prepared via the Dimroth rearrangement. Borthwick and co-workers used a Dimroth rearrangement to prepare carbonucleoside **137** as a precursor to 4′-hydroxy analogs of carbocycle 2′-deoxy guanosine (2′-CDG),[53] which has shown good activity against herpes simplex virus types 1 and 2. Treatment of **136** with DBU in refluxing ethanol gave the corresponding purine derivative in 49% yield. The diastereomer (C5-epimer; not shown) was also prepared using a similar protocol in 35% yield.

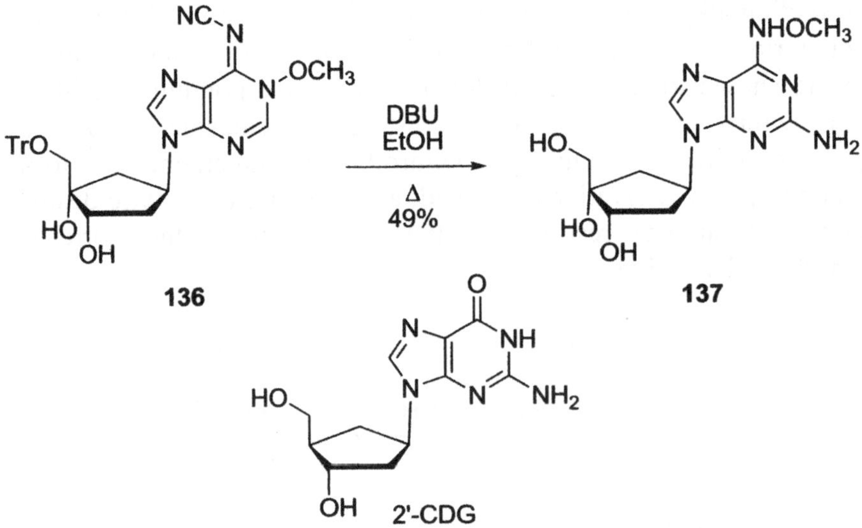

136 **137**

2'-CDG

Shuto and co-workers observed the Dimroth rearrangement of carbonucleoside **138** as an unwanted side product during the generation of a carbocyclic derivative of cyclic ADP-ribose (cADPR).[54] Treatment of carbonucleoside **138** with a hot solution of sodium hydroxide gave the corresponding Dimroth product **139** quantitatively.

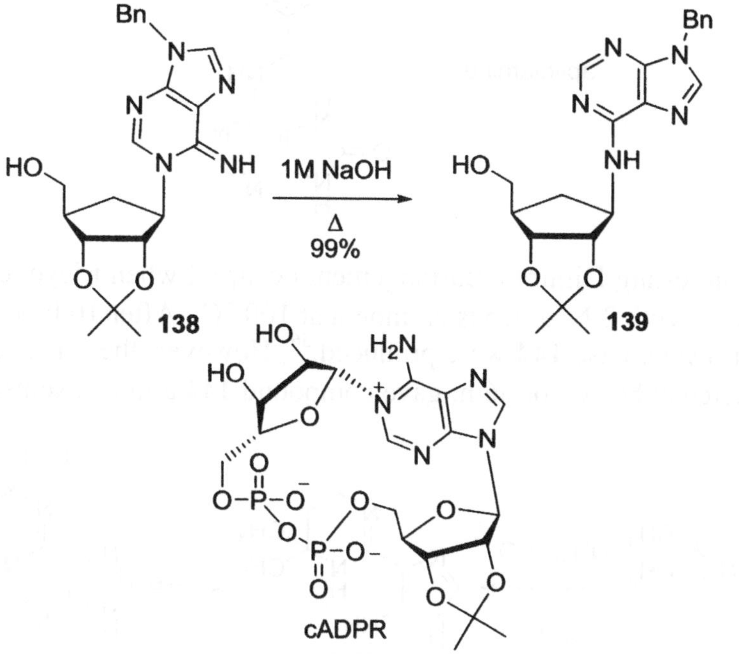

138 **139**

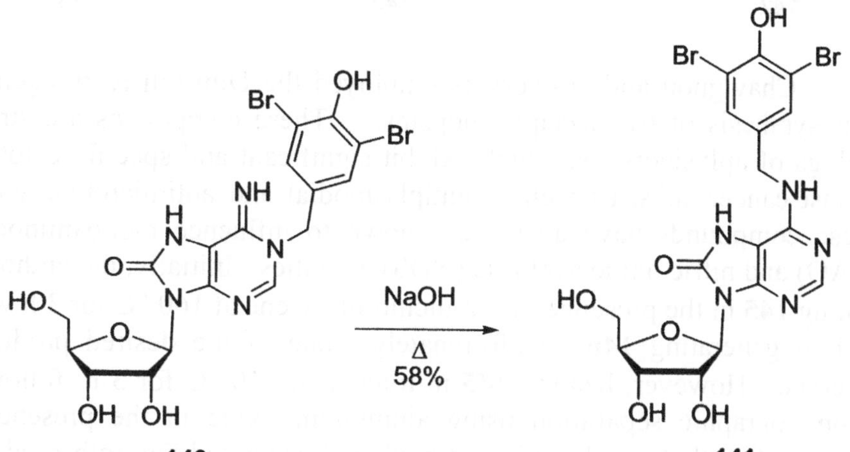

140 **141**

Itaya and co-workers used a Dimroth rearrangement to synthesize aplidiamine, a unique, naturally occurring 8-oxoadenine derivative, and its nucleoside analog.[55] Interest in aplidamine stems from its architectural homology to members of the phosmidosine family of antifungal antibiotic nucleotides. Treatment of **140** with boiling aqueous sodium hydroxide gave the corresponding nucleoside **141** in 58% overall yield.

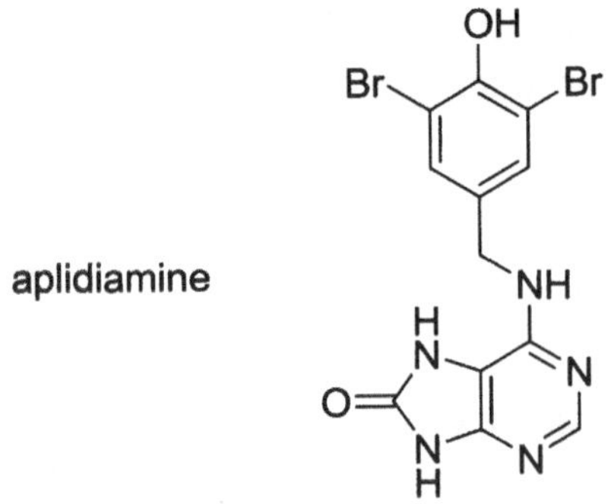

An interesting Dimroth rearrangement occurred when tricyclic purine **142** was treated with 2 N aqueous ammonia at 100 °C. After 10 h, a mixture of **143** and tricyclic base **144** were produced.[56] However, the same reaction, when conducted at 86 °C for 36 h, gave compound **144** almost exclusively in 60% yield.

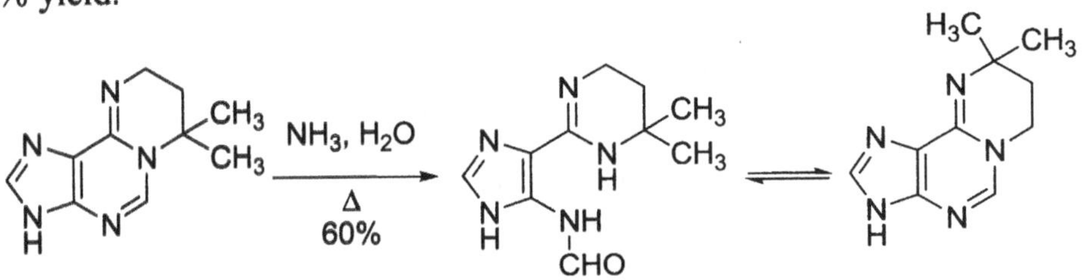

142 **143** **144**

Chavignon and co-workers employed the Dimroth rearrangement in their synthesis of two azaaplysinopsins.[57] These compounds are structural analogs of aplysinopsins, which exhibit significant and specific cytotoxicity against cancer cells, as well as antiplasmodial and antimicrobial activities. These compounds have also been shown to influence monoaminooxidase (MAO) and nitric oxide synthase (NOS) activities. Initially, the authors tried heating **145** in the presence of 1,2-dichlorobenzene at 160 °C for 3 h with the goal of generating **146**. Unfortunately, none of the desired product was observed. However, heating **145** in toluene at 110 °C for 3 h, followed by chromatographic separation using aluminium oxide in the presence of a mixture of methylene chloride and methanol generated Dimroth product **146** as a mixture of isomers.

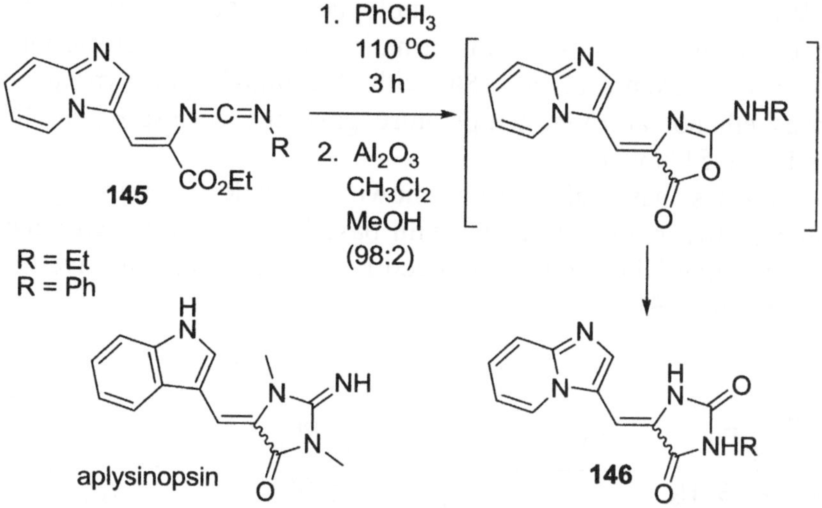

145

aplysinopsin

146

Uracil derivatives substituted at position 5 with an electron withdrawing group, or at position 1 with a phenyl group, have been shown to undergo Dimroth rearrangement.[58] For example, treatment of uracil derivative **147** with ethanolic potassium hydroxide gave the corresponding isomer **148** and 2-anilino-3-cyano-*N*-methylcrotonamide (not shown) as a mixture of products. Derivative **149** did not undergo rearrangement under similar conditions, illustrating the importance of substitution for this transformation.

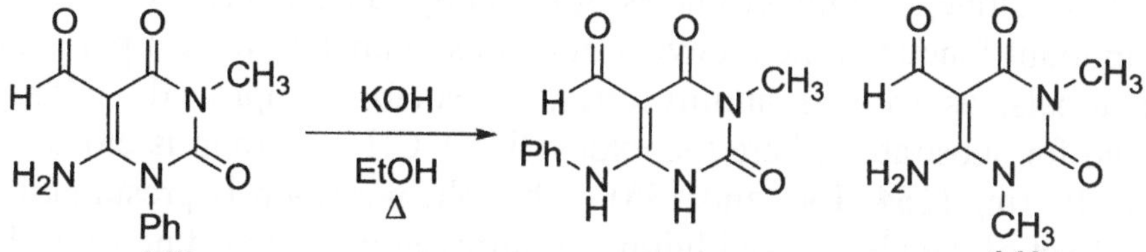

147 148 149

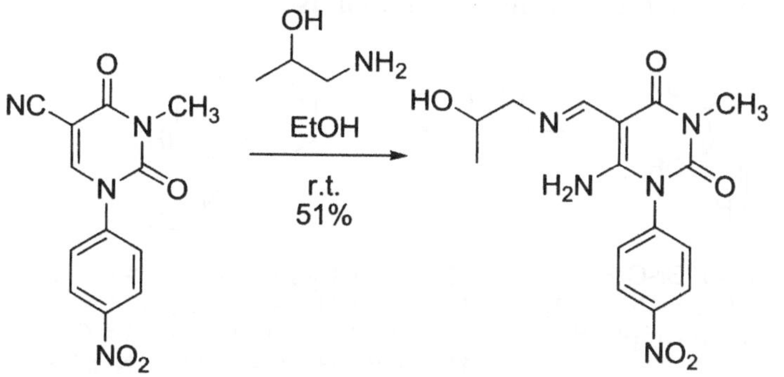

150 151

Walczak and co-workers employed an acid-catalyzed Dimroth rearrangement in their novel transformation of 5-cyanouracil derivatives.[59] Treatment of 5-cyano uracil **150** with 1-amino-2-hydroxypropanone in anhydrous ethanol at room temperature gave the corresponding Dimroth product **151** in 51% yield.

Oxazines have also been shown to undergo rearrangement as illustrated by the treatment of **152** with base.[60] In this case, elimination of the corresponding Dimroth product occurred spontaneously to give the more stable compound **153**.

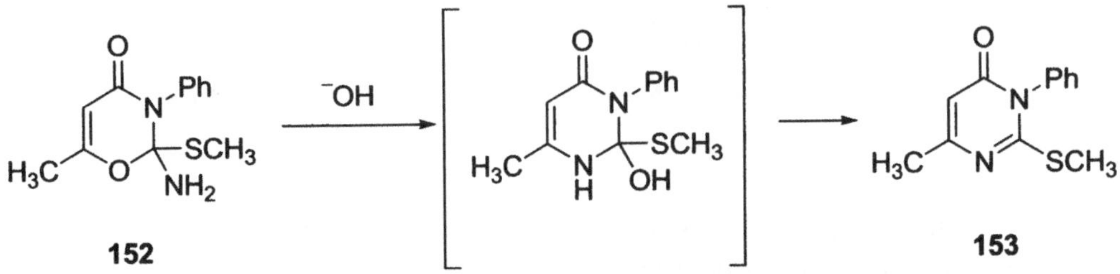

152 **153**

Kurz and co-workers used the Dimroth rearrangement to prepare a series of benzoxazinone analogs, which are of interest due to their high antimycotic activity.[61] Oxazines **154, 156, 158, 160, 162, 164** and **166** underwent spontaneous Dimroth rearrangement to produce the corresponding benzooxazinones (X = C) in modest yields. In this case it was shown that the *O*-substituent played little role in the yields of the final products. Kurz also applied this strategy to the synthesis of novel pyridooxazines (X = N), which are important functional heterocyclic moieties in analgesics, antipyretics and antibacterials, as well as antimycotics.[62] Notably, improved yields were achieved for substrates where the exocyclic O–CH_2–R group is closer to the heterocylic ring (**154, 156**, and **158**). This phemomenon is presumably due to the effect of the electron-withdrawing nitrogen in the pyridine ring. In the case of **162, 164**, and **166**, steric interactions most likely play a role in the reduced yields observed for these compounds.

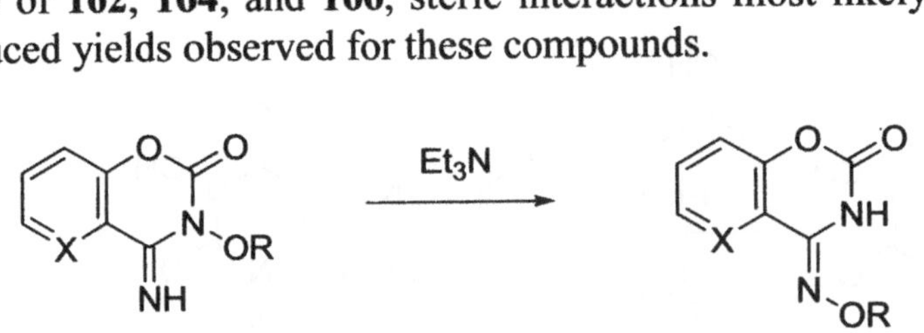

154, R = CH_2-*p*-Br-C_6H_4	**155**, R = CH_2-*p*-BrC_5H_4; X = C 55%, N 72%
156, R = CH_2-*p*-F-$CH_3C_6H_4$	**157**, R = CH_2-*p*-F$CH_3C_6H_4$; X = C ND, N 66%
158, R = CH_2-*p*-CH_3-C_6H_4	**159**, R = CH_2-*p*-$CH_3C_6H_4$; X = C 57%, N 68%
160, R = CH_3	**161**, R = CH_3; X = C 56%, N 51%
162, R = CH_2-C_6H_5	**163**, R = CH_2-C_6H_5; X = C 69%, N 62%
164, R = $(CH_2)_2$-C_6H_5	**165**, R = $(CH_2)_2$-C_6H_5; X = C 60%, N 55%
166, R = $(CH_3)_3$-C_6H_5	**167**, R = $(CH_2)_3$-C_6H_5; X = C 58%, N 58%

Zaleska and co-workers showed that thiazolidine rings can undergo Dimroth rearrangement under basic conditions.[63] Treatment of piperazin-3-one **168** with sodium hydride in DMF, followed by the addition of diiodomethane, gave the corresponding tricyclic imidizolidine ring **169** via the intermediate thiazolidine in 77% yield.

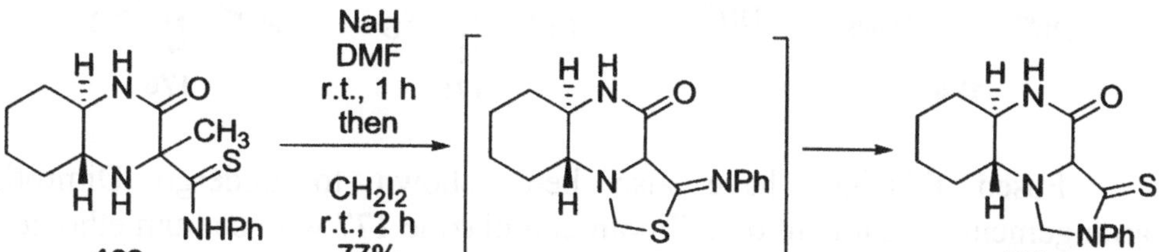

168 **169**

Treatment of piperazin-3-one **168** with sodium hydride in DMF, followed by the addition of diiodomethane gave the corresponding tricyclic imidizolidine ring **169** via the intermediate thiazolidine in 77% yield to give **171**.[64]

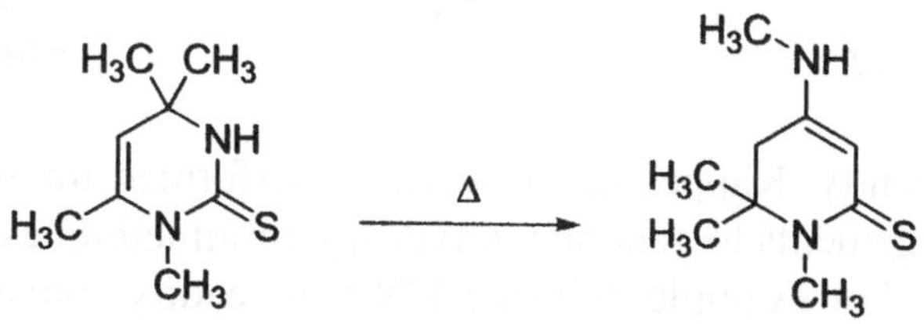

170 **171**

Similarly, Zigenuner and co-workers showed that thiazine **172** could undergo Dimroth rearrangement under thermal conditions to produce the corresponding pyrimidine **173** in 25% yield.[65]

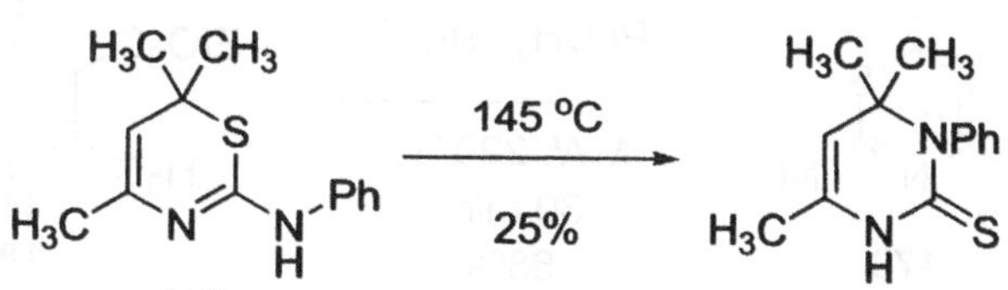

172 **173**

Thiazines also undergo Dimroth rearrangement under basic conditions to form the corresponding thiouracils. For example, *cis*-2-(4-substituted phenyl)amino-5-methyl-6-phenyl-5,6-dihydro-1,3-thiazin-4-one **174** rear-ranged in the presence of lithium hydride in DMF to give a mixture of the corresponding *cis*- and *trans*-thiouracils **175** and **176**.[66] This reaction was also succssfully demonstrated in the presence of triethylamine and absolute ethanol.

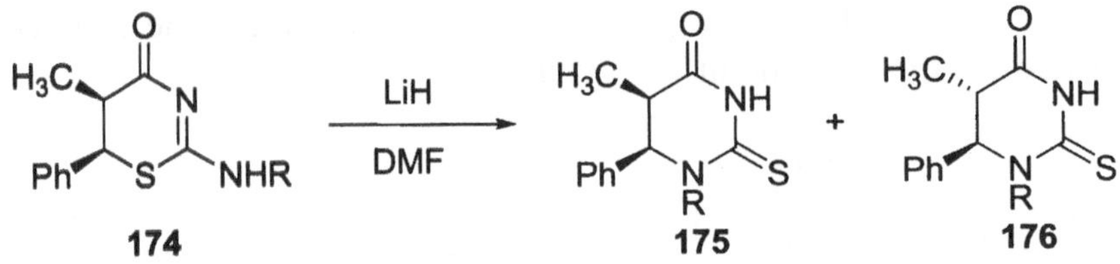

Fused thiazines have also been shown to undergo Dimroth rearrangement. Treatment of anilinothienothiazine **177** with sodium ethoxide in ethanol, followed by the addition of aqueous hydrochloric acid, led to the formation of the corresponding thioxopyrimidone **178** in 54% yield.[67]

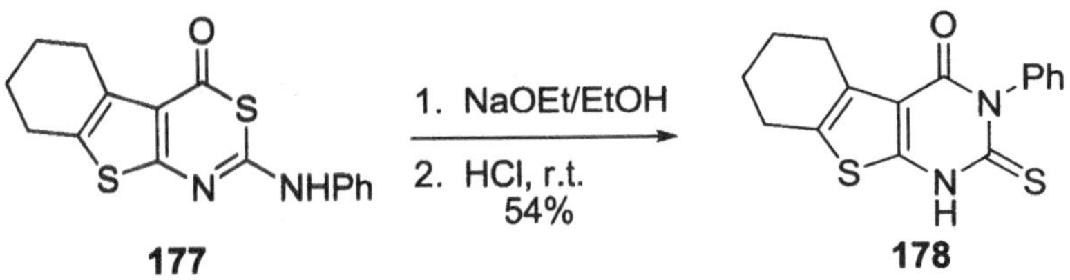

Most recently, Kappe and co-workers performed microwave-assisted Dimroth rearrangements to generate dihydropyrimidinethiones from thiazines in good yield.[68] For example, thiazine **179** was readily converted in 30 min to the corresponding dihydropyrimidinethione **180** in 68% yield in toluene heated to 220 °C with a microwave using a silicon-heating element (Si-HE). This work built on earlier work by Kappe and co-workers using a continuous flow microwave process to generate a series of dihydropyrimidinethiones.[69]

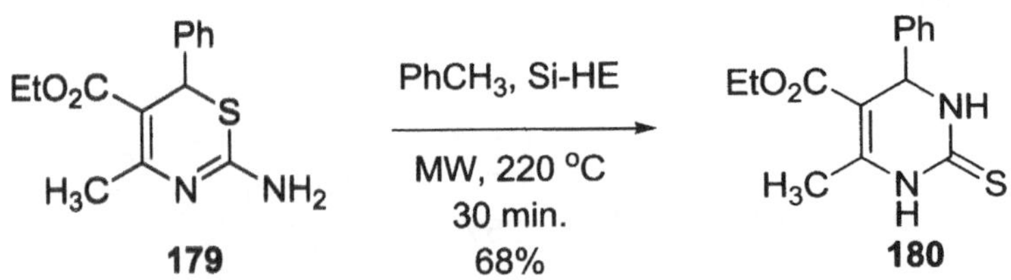

Rings Containing Three Heteroatoms

One of the first thermal rearrangements of 1,2,3-triazoles was reported in 1902 by Dimroth.[1,2] The original reaction involved heating 1,2,3-triazoles such as 5-amino-4-ethoxycarbonyl-1-phenyl-1,2,3-triazole **181** in absolute ethanol and benzene at 150 °C for 3 h, with the reaction favoring formation of the corresponding isomer **182**. This work was later expanded by Lieber and co-workers[70] who showed that the position of equilibrium shifts towards the more acidic isomer (**182**) as the electronegativity of the substituent R_1 is increased ($R_2 = C_6H_5$). Lieber and co-workers also revealed that a linear correlation existed between the logarithm of the equilibrium constant and Hammett's ρ-value for the individual R_1 groups employed in their study.

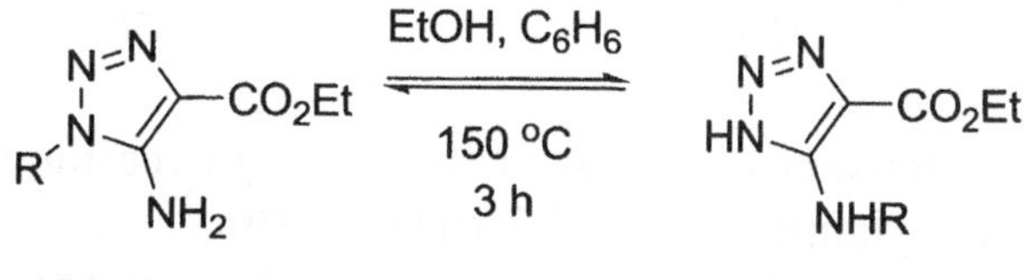

181, R_1= Ph, R_2 = CO_2Et **182**, R_1= Ph, R_2 = CO_2Et

Later, Ogata and colleagues investigated the photochemical-equilibrium between 1,4-diphenyl-5-amino-1,2,3-triazole **183** and its Dimroth product 4-phenyl-5-anilino-1,2,3-triazole **184**.[71] 1,4-Diphenyl-5-amino-1,2,3-triazole **183**, when irradiated with a 100-W high-pressure mercury lamp for 40 h, gave the corresponding 4-phenyl-5-anilino-1,2,3-triazole **186** through valence isomerisation via diarines **184** and **185** in 35% yield.

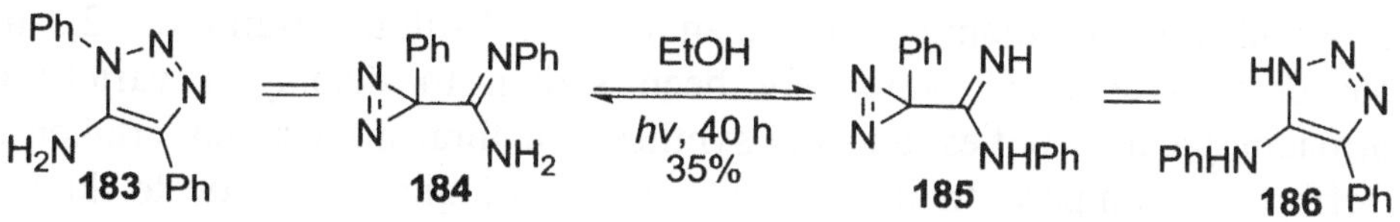

Albert and co-workers observed an unexpected Dimroth rearrangement in their efforts to improve the efficiency of 4-amino-3-benzyl-1,2,3-triazole synthesis.[72] The benzyl groups of compounds **187, 189, 191** and **193** partially migrated in the presence of hot bases such as potassium hydroxide to give complex mixtures of the starting materials (**187, 189, 191** and **193**) and products (**188, 190, 192** and **194**) in 30–45% yield. As the equilibrium of these rearrangements lay in the direction of the primary amines, retrogression of each reaction was made possible by heating in a neutral solvent such as ethanol. The authors also noted that by tuning the reaction conditions one could selectively produce the more basic triazoles (**187, 189, 191** and **193**) in good yield. Similar success was observed by Tennant and Sutherland[73] for the migration of 4-substituted-5-amino-1-phenyl-1,2,3-triazoles when heated with acetic acid and acetic anhydride, and by Vaughan and colleagues[74] in their synthesis of a number of 5-(arylamino)-1,2,3-triazoles.

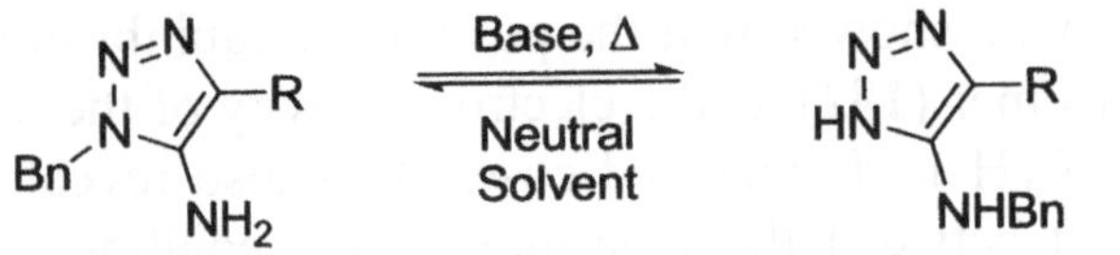

187, R = H
189, R =$CONH_2$
191, R = CO_2H
192, R = CN

188, R = H
190, R =$CONH_2$
192, R = CO_2H
194, R = CN

Hesse and Ognyanov employed the Dimroth rearrangement in their synthesis of 5-(phenylamino)-1*H*-1,2,3-triazole **196**.[75] This molecule was obtained in nearly quantitative yield by refluxing the corresponding 5-amino-1-phenyl-1*H*-1,2,3-triazole **195** in pyridine for 6 h.

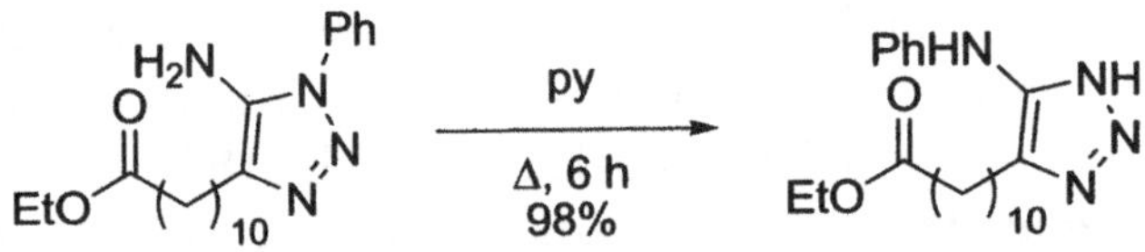

195 **196**

Batog and co-workers used the Dimroth rearrangement to prepare a series of novel monosubstituted *N*-(4-substituted-1*H*-1,2,3-triazol-5-yl)-1,2,5-oxadiazole-3,4-diamines (furazans).[76] Substituted-5-amino-1,2,3-triazole derivatives have previously been shown to display a variety of pharmacological activities, and the synthesis of furazans was undertaken in an effort to compare their reactivity and biological activity to known substituted-5-amino-1,2,3-triazole derivatives. 5-Amino-1,2,3-triazole derivatives **197, 199, 201** and **203** rearranged to form the mono-substituted diaminofurazans **198, 200, 202** and **204** upon continuous heating at 85–95 ºC in DMF for 1 h, with yields of between 52 and 87%.

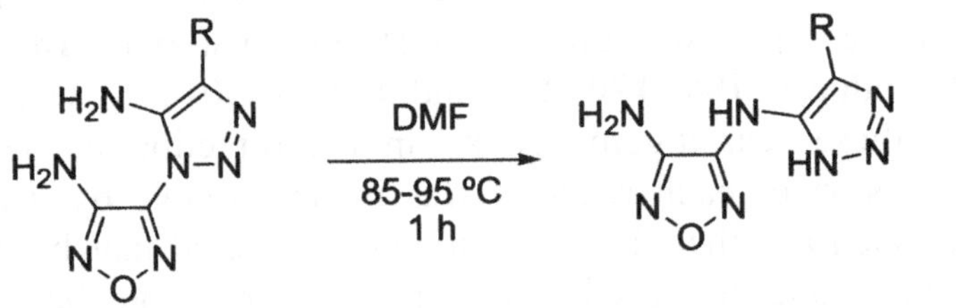

197, R = CN
199, R = $CONH_2$
201, R = *p*-NO_2-C_6H_4
203, R = *p*-Cl-C_6H_4

198, R = CN 87%
200, R = $CONH_2$ 52%
202, R = *p*-NO_2-C_6H_4 57%
204, R = *p*-Cl-C_6H_4 73%

Morzerin and co-workers made use of the Dimroth rearrangement in their synthesis of tetrakis(1,2,3-triazole-5-aminosulfonyl) calyx[4]arenes.[77]

These compounds are noteworthy because of their unique three-dimensional structures, extensive derivatization abilities, and tunable shapes, which makes them ideal molecular scaffolds for materials chemistry. Azidosulfonylcalix[4]arenes **205** and **207** were reacted with *N*-phenyl- and *N*-cyclohexyl-2-cyanoacetamides at 40 °C in the presence of sodium ethoxide for 15 h to give the tetrakis((1*H*-1,2,3-triazol-5-amine)-sulfonyl)calyx[4]arenes **206** and **208** in 38 and 68% yields, respectively. The reaction took place via the intermediate 1,2,3-triazole (not shown), which underwent spontaneous Dimroth rearrangement under the reported reaction conditions. The Dimroth rearrangement was also employed by Geide and colleagues in their synthesis of tetrakis [(1*H*-1,2,3-triazol-5-yl)-aminosulfonyl]calixarenes in 88% yield from the corresponding triazoles using excess triethylamine in boiling ethanol.[78]

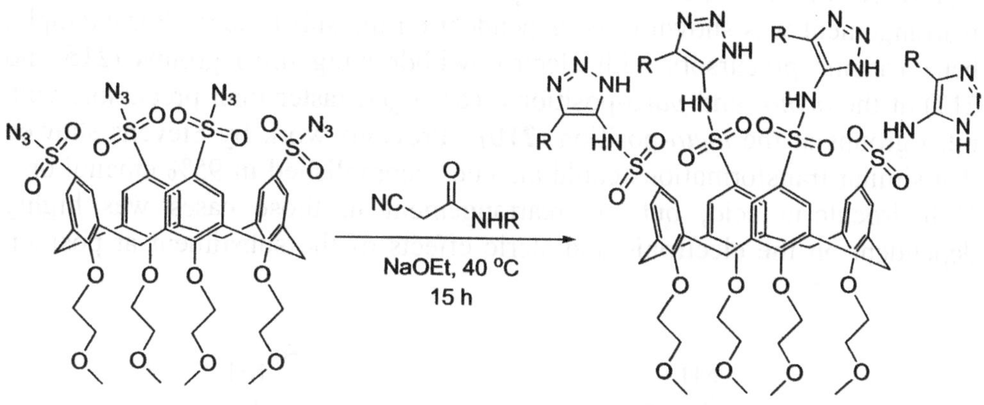

205, R = Ph
207, R = Cy

206, R = Ph 38%
208, R = Cy 68%

Bakulev and co-workers[79] reexamined the Dimroth rearrangement of 1,2,3-triazoles to 1,2,3-thiadiazoles initially developed by Regitz and Scherer[80] in their synthesis of 5-(*o*-aminophenyl)-1,2,3-thiadiazoles. Treatment of 1-(*o*-aminophenyl)-1,2,3-triazole-5-thiols **209**, **211** and **213** with nitrous acid and triethylorthoformate proved to be a facile route for the synthesis of the corresponding thiadiazoles **210**, **212** and **214**. The authors showed that the reverse reaction could be accomplished in near quantitative yield by refluxing for 1 h in triethylamine.[81] Bakulev`s group have employed this rearrangement in their syntheses of bis[1,2,3]triazolo[1,5-*b*;5′,1′-*f*][1,3,6]thiadiazepine derivatives[82] and tricyclic 1,3,6-thiadiazepines.[83]

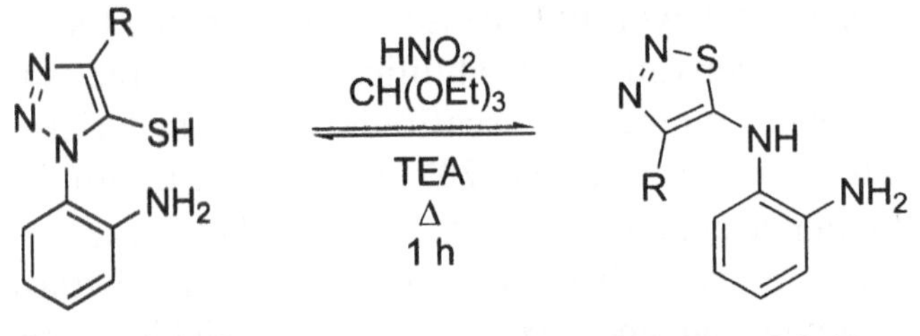

209, R = CO_2Et
211, R = $CONHCH_3$
213, R = $CONH_2$

210, R = CO_2Et
212, R = $CONHCH_3$
214, R = $CONH_2$

3-Substituted-3,4-dihydro-4-imino-1,2,3-benzotrazines such as **215**, **217**, and **219** have been shown to undergo Dimroth rearrangement in acetic acid in upward of 90% yield to give the corresponding Dimroth products regardless of the substituent at position 3.[84] However, the rate of rearrangement was shown to be dependent on the substituent. For example, benzotriazine precursors with electron-withdrawing nitro groups (**215** and **217**) at the *ortho*- and *para*-positions rearranged faster than precursors with nitro groups at the *meta*-position (**210**). Previous work by Stevens showed that similar transformations could also be accomplished in 95% ethanol or 2 N hydrochloric acid, but the rearrangement in those cases was highly dependent on the electronic and steric effects of the substituent at position 3.[85]

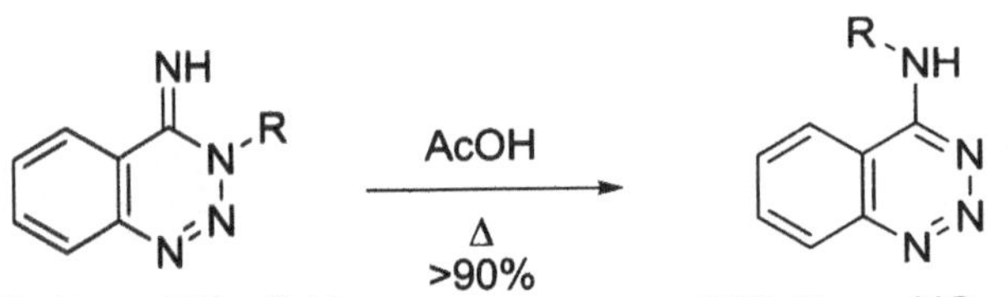

215, R = o-NO_2-C_6H_4
217, R = p-NO_2-C_6H_4
219, R = m-NO_2-C_6H_4

216, R = o-NO_2-C_6H_4
218, R = p-NO_2-C_6H_4
220, R = m-NO_2-C_6h_4

Reimer and co-workers showed that triazines such as 1-(nitrophenyl)-4,6-diamino-1,3,5-triazine **221** could undergo Dimroth rearrangement under basic conditions in hot ethanol to give the corresponding Dimroth product **222**.[86] Triazines such as **222** are important analogs of phenytoin and lamotrigine, two antiepileptic drugs that function by blocking neuronal sodium channels. In this example, the electron-withdrawing nitro group was shown to be critical for the stabilization of the transient dipolar acyclic intermediate formed during the rearrangement. Later work by Ma and co-workers showed that a series of 4,6-diamino-1,3,5-triazines could also undergo Dimroth rearrangement in refluxing ethanolic sodium hydroxide to produce the corresponding Dimroth products in 60–92% yield (not shown).[87]

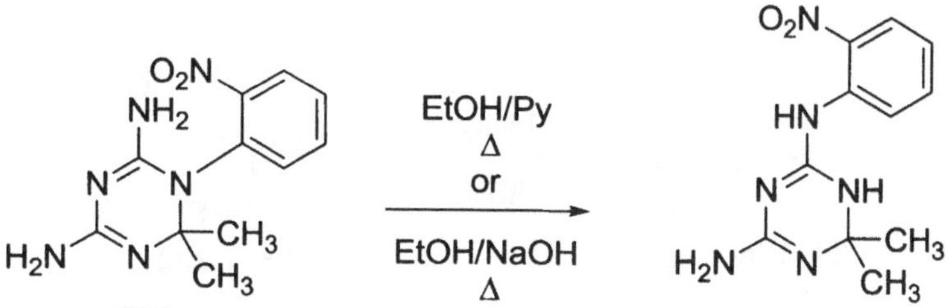

221 222

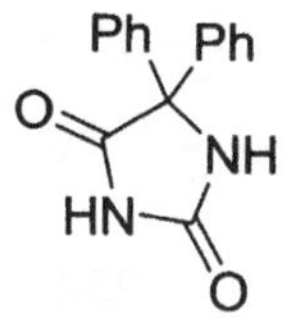

phenytoin

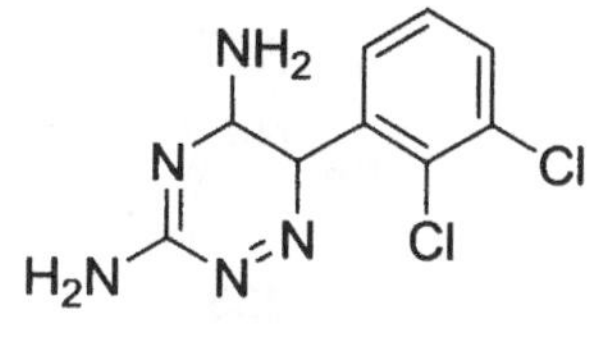

lamotrigine

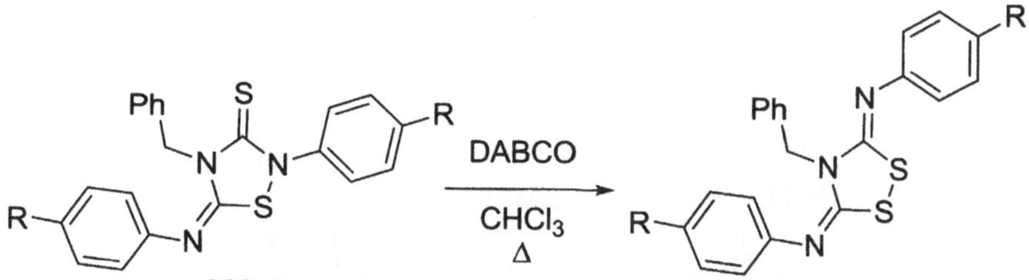

223, R = NO_2
225, R = Cl
227, R = H
229, R = CH_3

224, R = NO_2 99%
226, R = Cl 66%
228, R = H NR
230, R = CH_3 NR

L'abbe and co-workers reported on the Dimroth rearrangement of thiazolidines under basic conditions.[88] Treatment of **223** and **225** with diazabicyclo[2.2.2]octane in refluxing chloroform gave the corresponding Dimroth products **224** and **226** in 99 and 66% yields, respectively. The authors noted that the presence of an electron-withdrawing group was required for the formation of the desired product, as thiazolidenes **227** and **229** did not undergo Dimroth rearrangement.

In a later report, L'abbe and co-workers showed that dithianes could also undergo Dimroth rearrangement in the presence of Lewis acids such as benzoylisothiocyanate and benzoylchloride.[89] Treatment of **231** with benzoylisothiocyanate gave **232** in 80% yield. Similarly, treatment of **233** with benzoylchloride gave **234** in 55% yield. The authors hypothesized that these rearrangements occurred via a betaine intermediate.

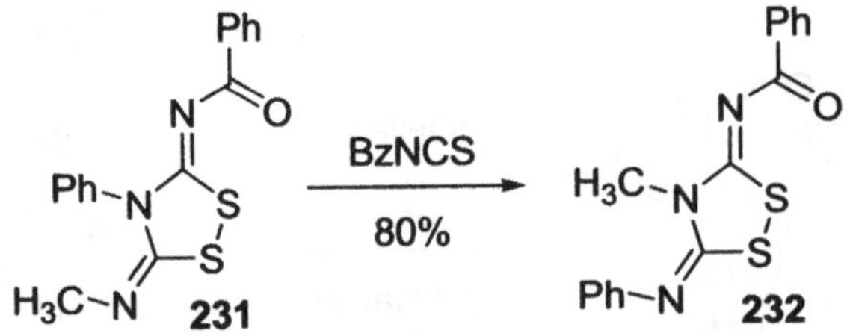

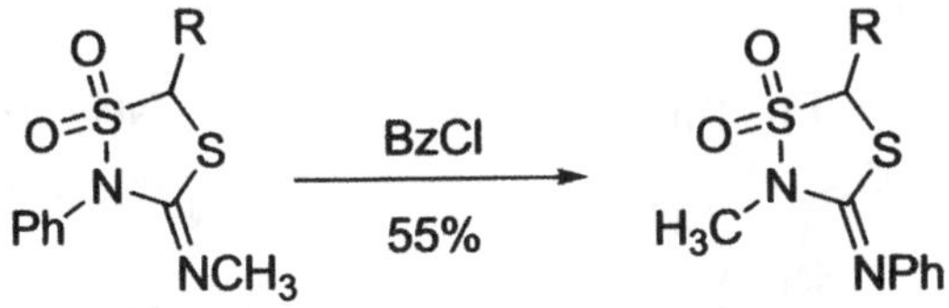

233 234

Rings Containing Four Heteroatoms

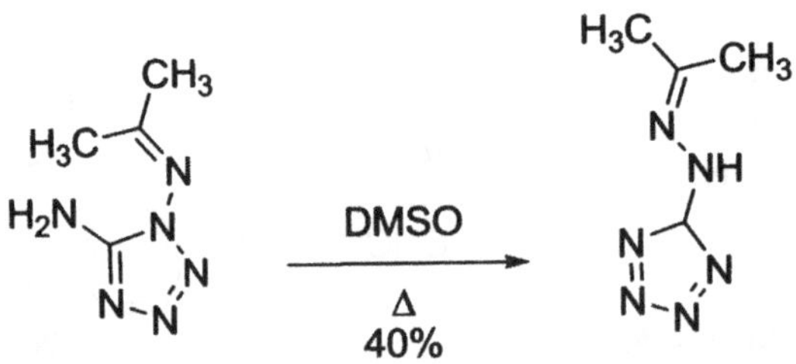

235 236

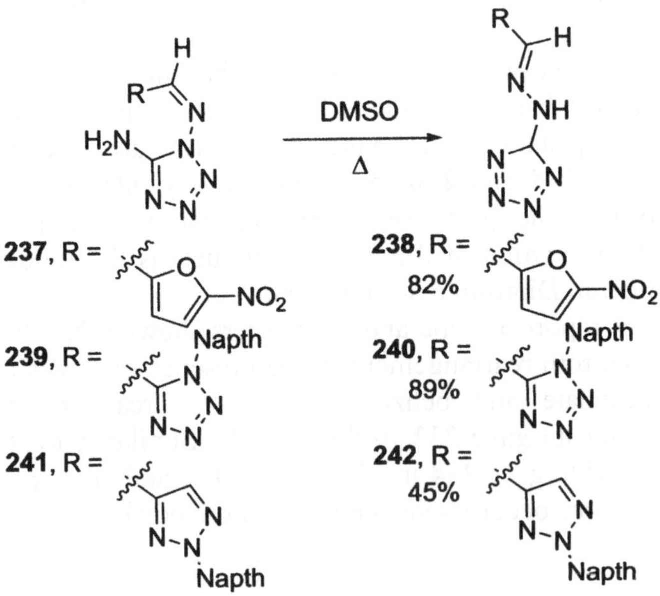

Tetrazoles have been shown to undergo Dimroth rearrangement under neutral conditions. For example, imines **235**, **237**, **239** and **241** underwent Dimroth rearrangement in DMSO to form the corresponding hydrazones.[90] Notably, higher yields were generally achieved for imines activated by electron-withdrawing groups.

9.4.6 Experimental

The first example by Shuto and co-workers[54] gives the standard reaction conditions for base catalyzed Dimroth rearrangement. In this particular example the yields were quantitative.

9-Benzyl-N6-[[(1R,2S,3R,4R)-2,3-(isopropylidenedioxy)-4-hydroxymethyl]-cyclopentyl]adenine

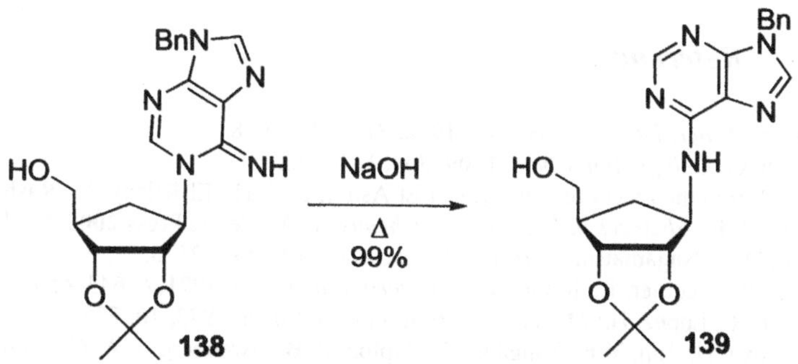

A suspension of **138** in 1 M NaOH (6 mL per mmol) was heated under reflux for 1 h, and then MeOH (3 mL per mmol) was added. The resulting solution was extracted with $CHCl_3$ (three times), and the organic layer was washed with brine, dried with Na_2SO_4, and evaporated. The residue was purified by silica column chromatography using 5% MeOH in $CHCl_3$ to give **139** as a white solid in 98% yield.

The following example by Lauria and co-workers[28] employs DMSO and water as a more neutral protocol for the Dimroth rearrangement. Lauria and co-workers performed several additional rearrangements in high yield using this protocol.

5-Methyl-3-phenyl-4,5-dihydro-8H-pyrazolo[3,4-d][1,2,3]triazolo[1,5-a]pyrimidin-8-one

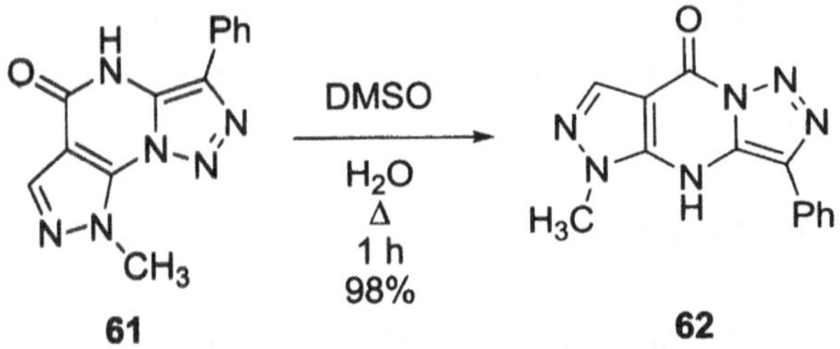

A solution of **61** was heated under reflux in dimethylsulfoxide (99.8%, 10 mL per 0.5 g) for 1 h. The cooled reaction mixture was then poured onto crushed ice and the solid was filtered and air dried to give **63** as a light yellow solid in 98% yield.

9.4.7 References

1 Dimroth, O. *Ber. Dtsch. Chem. Ges.* **1902,** *21*, 1029–1038.
2 Dimroth, O. *Leibigs Ann. Chem.* **1909,** *364*, 183–226.
3 Adapted from the recent review by: [R] El-Ashry, E. S. H.; El-Kilany, Y.; Rashed, N.; Assafir, H. In *Advances in Heterocyclic Chemistry*, Academic Press **2000**, 80–156.
4 Brown, D. J.; Nagamatsu, T. *Aust. J. Chem.* **1977,** *30*, 2515–2525.
5 Guerret, P.; Jacquier, R.; Maury, G. *J. Heterocyclic Chem.* **1971,** *8*, 643–650.
6 Jacquier, R.; Lopez, H.; Maury, G. *J. Heterocyclic Chem.* **1973,** *10*, 755–762.
7 Tretyakova, N.; Lin, Y-P.; Sangaiah, R.; Upton, P. B.; Swenberg, J. A. *Carcinogenesis* **1997,** *18*, 137–147.
8 Barlow, T.; Takeshita, J.; Dipple, A. *Chem. Res. Toxicol.* **1998,** *11*, 838–845.
9 Kim, H.-Y. H.; Finneman, J. I.; Harris, C. M.; Harris, T. M. *Chem. Res. Toxicol.* **2000,** *13*, 625–637.
10 (a) Kanuri, M.; Nechev, L. V.; Tamura, P. J.; Harris, C. M.; Harris, T. M.; Llyod, R. S. *Chem. Res. Toxicol.* **2002,** *15*, 1572–1580. (b) Goggin, M.; Anderson, C.; Park, S.; Swenberg, J.; Walker, V.; Tretyakova, N. *Chem. Res. Toxicol.* **2008,** *21*, 1163–1170.
11 Selzer, R. R.; Elfarra, A. A. *Carcinogenesis* **1999,** *20*, 285–292.
12 Munter, T.; Cottrell, L.; Hill, S.; Kronberg, L.; Watson, W. P.; Golding, B. T. *Chem. Res. Toxicol.* **2002,** *15*, 1549–1560.
13 Florea-Wang, D.; Haapala, E.; Mattinen, J.; Hakala, K.; Vilpo, J.; Hovinen, J. *Chem. Res. Toxicol.* **2003,** *16*, 403–408.
14 Pawlowicz, A. J.; Munter, T.; Zhao, Y.; Kronberg, L. *Chem. Res. Toxicol.* **2006,** *19*, 571–576.
15 Taylor, E. C.; Loeffler, P. K. *J. Am. Chem. Soc.* **1960,** *82*, 3147–3151.
16 Carballares, S.; Cifuentes, M. M.; Stephenson, G. A. *Tetrahedron Lett.* **2007,** *48*, 2041–2045.
17 El Khadem, H. S.; Kawai, J. *Carbohydr. Res.* **1989,** *189*, 149–160.
18 Vas'kevich, R. I.; Savitskii, P. V.; Zborovskii, Y. L.; Staninets, V. I.; Rusanov, E. B.; Chernega, A. N. *Russ. J. Org. Chem.* **2006,** *42*, 1403–1408.
19 Shawali, A.; Hassaneen, H.; Shurrab, N. *J. Heterocyclic Chem.* **2008,** *45*, 1825–1829.
20 Shawali, A. S.; Hassaneen, H. M.; Shurrab, N. K. *Tetrahedron* **2008,** *64*, 10339–10343.
21 Loakes, D.; Brown, D. M.; Salisbury, S. A. *J. Chem. Soc., Perkin Trans. 1* **1999,** 1333–1337.
22 Rusinov, V.; Zyryanov, G.; Egorov, I.; Ulomskii, E.; Aleksandrov, G.; Chupakhin, O. *Russ. J. Org. Chem.* **2004,** *40*, 85–89.
23 Vorob'ev, E. V.; Ketskii, M. E.; Krasnikov, V. V.; Mezheritskii, V. V.; Steglenko, D. V. *Russ. Chem. Bull., Intl. Ed.* **2006,** *55*, 2247–2255.
24 Singleton, J. L.; Coburn, M. D. *J. Heterocyclic Chem.* **1973,** *10*, 275.

25 Tartakovsky, V.; Frumkin, A.; Churakov. A.; Strelenko, Y. *Russ. Chem. Bull., Int. Ed.* **2005,** 54, 719–725.
26 Temple, C.; Smith, B. H.; Montgomery, J. A. *J. Org. Chem.* **1973,** *38*, 1095–1098.
27 Katritzky, A. R.; Ji, F. B.; Fan, W. Q.; Gallos, J. K.; Greenhill, J. V.; King, R. W.; Steel, P. J. *J. Org. Chem.* **1992,** *57*, 190–195.
28 Lauria, A.; Abbate, I.; Patella, C.; Gambino, N.; Silvestri, A.; Barone, G.; Almerico, A. M. *Tetrahedron Lett.* **2008,** *49*, 5125–5128.
29 Karpenko, A. V.; Kovalenko, S. I.; Shishkin, O. V. *Tetrahedron* **2009,** *65*, 5964–5972.
30 Temple, C.; Smith, B. H.; Montgomery, J. A. *J. Org. Chem.* **1972,** *37*, 3601–3604.
31 Chen, E. K .Y.; Bancerz, M.; Hamer, G. K.; Georges, M. K. *Eur. J. Org. Chem.* **2010,** 5681–5687.
32 Tschitschibabin, A. E.; Konowalowa, R. A.; Konowalowa, A. A. *Ber. Dtsch. Chem. Ges.* **1921,** *54*, 814–822.
33 Tschitschibabin, A. E.; Kirssanow, A. *Ber.* **1928,** *61*, 1223.
34 Nakaike, Y.; Hayashi, D.; Nishiwaki, N.; Tobe, Y.; Ariga, M. *Org. Biomol. Chem.* **2009,** *7*, 325–334.
35 Filák, L.; Reidl, Z.; Egyed, O.; Czugler, M.; Hoang, C. N.; Schantl, J. G.; Hajós, G. *Tetrahedron* **2008,** 1101–1113.
36 Anders, J. T.; Görls, H.; Langer, P. *Eur. J. Org. Chem.* **2004,** 1897–1910.
37 Jacobi, P. A.; Liu, H. *J. Am. Chem. Soc.* **1999,** *121*, 1958–1959.
38 Subbotina, J. O.; Fabian, W. M. F.; Tarasov, E. V.; Volkova, N. N.; Bakulev, V. A. *Eur. J. Org. Chem.* **2005,** 2914–2923.
39 Brown, D. J.; Ford, P. W.; Paddon-Row, M. N. *J. Chem. Soc. C* **1968,** 1452–1454.
40 Brown, D. J.; Ienega, K. *J. Chem. Soc., Perkins Trans. 1,* **1974,** 372–378.
41 Brown, D. J.; Ienaya, K. *Aust. J. Chem.* **1975,** *28*, 119–127.
42 Brown, D. J.; Ienaga, K. *J. Chem. Soc., Perkins 1* **1975,** 2182–2185.
43 Kanth, S. R.; Reddy, G. V.; Kishore, K. H.; Rao, P. S.; Narsaiah, B; Murthy, U. S. N. *Eur. J. Med. Chem.* **2006,** *41*, 1011–1016.
44 Chandregowda, V.; Rao, G. V.; Reddy, G. C. *Org. Process Res. Dev.* **2007,** *11*, 813–816.
45 Foucourt, A.; Dubouilh-Benard, C.; Chosson, E.; Corbiére, C.; Buquet, C.; Iannelli, M.; Leblond, B.; Marsais, F.; Besson, T. *Tetrahedron* **2010,** *66*, 4495–4502.
46 Fischer, R. W.; Misun, M. *Org. Process Res. Dev.* **2001,** *5*, 581–586.
47 Macon, J. B.; Wolfenden, R. *Biochemistry* **1968,** *7*, 3453–3458.
48 Engel, J. D. *Biochem. Biophys. Res. Commun.* **1975,** *64*, 581–586.
49 Itaya, T.; Tanka, F.; Fujii, T. *Tetrahedron* **1972,** *28*, 535–547.
50 Fujii, T.; Itaya, T.; Saito, T. *Chem. Pharm. Bull.* **1975,** *23*, 54–61.
51 Fujii, T.; Saito, T.; Terahara, N. *Chem. Pharm. Bull.* **1986,** *34*, 1094–1107.
52 Ienaga, K.; Pfleiderer, W. *Liebigs Ann. Chem.* **1979,** 1872–1880.
53 Borthwick, A. D.; Biggadike, K.; Paternoster, I. L. *Bioorg. Med. Chem. Lett.* **1993,** 2577–2580.
54 Shuto, S.; Fukuoka, M.; Manikowsky, A.; Ueno, Y.; Nakano, T.; Kuroda, R.; Kuroda, H.; Matsuda, A. *J. Am. Chem. Soc.* **2001,** *123*, 8750–8759.
55 (a) Itaya, T.; Hozumi, Y.; Kanai, T.; Ohta, T. *Tetrahedron Lett.* **1998,** *39*, 4695–696. (b) Itaya, T.; Hozumi, Y.; Kanai, T.; Ohta, T. *Chem. Pharm. Bull.* **1999,** *47*, 1297–1300.
56 Martin, D. M. G.; Reese, C. B. *J. Chem. Soc. (C)* **1968,** *14,* 1731–1738.
57 Chezal, J. M.; Delmas, G.; Mavel, S.; Elakmaoui, H.; Metin, J.; Diez, A.; Blanch, Y.; Gueiffier, A.; Rubiralta, M.; Teulade, J. C.; Chavignon, O. *J. Org. Chem.* **1997,** *62*, 4085–4087.
58 Hirota, K.; Ni, P. Z.; Suzuki, A.; Takasu, H.; Kitade, Y.; Maki, Y. *Chem. Pharm. Bull.* **1992,** *10*, 2839–2841.
59 Gondela, A.; Gabański, R.; Walczak, K. *Tetrahedron Lett.* **2004,** *45*, 8007–8009.
60 Lacey, R. N. *J. Chem. Soc.* **1954**, 839–849.
61 Kurz, T.; Widyan, K.; Wackendorff, C.; Schluter, K. *Synthesis* **2004,** *12*, 1987–1990.
62 Kurz, T. *Tetrahedron* **2005,** *46,* 3091–3096.
63 Zaleska, B.; Socha, R.; Karelus, M.; Grochowski, J.; Serda, P. *Synthesis* **2004,** *13*, 2169–2172.

64 (a) Sukumaran, P.; Rajasekharan, K. N. *Indian J. Chem., Sect. B* **1990,** 1070–1073. (b) Sukumaran, P.; Rajasekharan, K. N. *Indian J. Chem., Sect. B* **1990,** 1074–1076.

65 Zigeuner, G.; Strallhofer, T.; Wede, F.; Lintschinger, W.-B. *Monatsh. Chem.* **1975,** *106,* 1469–1477.

66 Dzurilla, M.; Kutschy, P.; Imrich, J.; Koscik, D.; Kraus, R. *Collect. Czech. Chem. Commun.* **1991,** *56,* 1287–1294.

67 Leistner, S.; Guetschow, M.; Wagner, G. *Synthesis* **1987,** 466–470.

68 Kremsner, J. M.; Kappe, C. O. *J. Org. Chem.* **2006,** *71,* 4651–4658.

69 Glasnov, T. N.; Vugts, D. J.; Koningstein, M. M.; Desai, B.; Fabian, W. M. F.; Orru, R. V. A. Kappe, C. O. *QSAR Comb. Sci.* **2006,** *25,* 509–518.

70 Lieber, E.; Chao, T-S.; Rao, C. N. R. *J. Org. Chem.* **1957,** *22,* 654–662.

71 Ogata, Y.; Takagi, K.; Hayashi, E. *B. Chem. Soc. Jpn.* **1977,** *50,* 2505–2506.

72 Adrien, Albert. *J. Chem. Soc.* **1970,** 230–235.

73 Tennant, G.; Sutherland, D. R. *J. Chem. Soc. C.* **1971,** 2156–2162.

74 Baines, K.; Rourke, T.; Vaughan, K.; Hooper, D. *J. Org. Chem.* **1981,** *46,* 856–859.

75 Ognyanov, V.; Hesse, M. *Helv. Chem. Acta.* **1991,** *74,* 899–903.

76 Rozhkov, V.; Batog, L.; Shevtsova, E.; Struchkova, M. *Mendeleev Commun.* **2004,** *14,* 76–77.

77 Morzerin, Y.; Pospelova, T.; Gluhareva, T.; Matern, A. *ARKIVOC* **2004,** *11,* 31–35.

78 Geide, I.; Glukhareva, T.; Matern, A.; Morzherin, Y. *Chem. Heterocyclic Compd.* **2006,** *42,* 121–122.

79 Volkova, N. N., Tarasov, E. V., Meervelt, L, Toppet, S., Dehaen, W., and Bakulev, V. A., *J. Chem. Soc., Perkin Trans. 1,* **2002,** 1574–1580.

80 Regitz, M.; Scherer, H. *Chem. Ber.* **1969,** *102,* 417–422.

81 Tarasov, E.; Volkova, N.; Morzherin, Y.; Bakulev, V. *Russ. J. Org. Chem.* **2004,** *40,* 870–873.

82 Volkova, N.; Tarasov, E.; Dehaen, W.; Bakulev, V. *Chem. Commun.* **1999,** 2273–2274.

83 Volka, N. N.; Tarasov, E. V.; Van Meervelt, L.; Toppet, S.; Dehaen, W.; Bakulev, V. A. *J. Chem. Soc., Perkin Trans. 1* **2002,** 1574–1580.

84 Siddiqui, M.; Stevens, M. F. G. *J. Chem. Soc., Perkin Trans 1* **1974,** 609–610.

85 Stevens, H. N. E.; Stevens, M. F. G. *J. Chem. Soc. C* **1970,** *6,* 765–771.

86 Stevens, M. F. G.; Chui, W. K.; Castro, M. A. *J. Heterocyclic Chem.* **1993,** *30,* 849–853.

87 Ma, X.; Poon, T.-Y.; Wong, P. T. H.; Chui, W.-K. *Bioorg. Med. Chem. Lett.* **2009,** *19,* 5644–5647.

88 L'abbe, G.; Verhelst, G.; Toppet, S. *J. Org. Chem.* **1977,** *42,* 1159–1163.

89 L'abbe, G.; Timmerman, A.; Martens, C.; Toppet, S. *J. Org. Chem.* **1978,** *43,* 4951–4955.

90 Moderhack, D.; Goos, K. H. Preu, L. *Chem. Ber.* **1990,** *123,* 1575–1577.

9.5 Hantzsch Synthesis

Nicole L. Snyder and Christopher J. Boisvert

9.5.1 Description

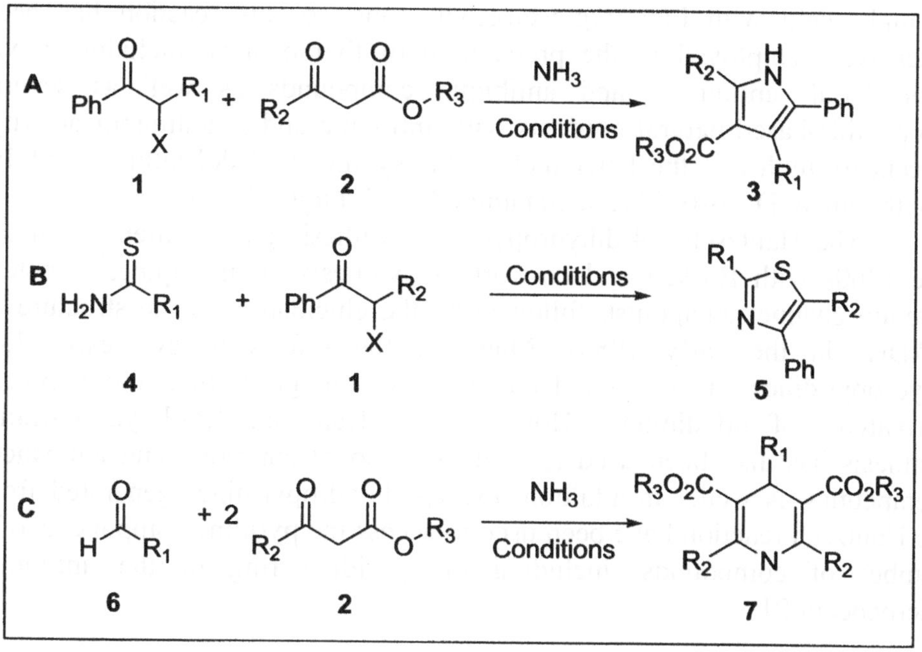

The Hantzsch synthesis has been used to generate pyrroles, thiazoles and dihydropyridine derivatives. Pyrroles (**3**) are generated from the reaction of β-ketoesters with ammonia, ammonia derivatives or primary amines, and α-haloketones (path **A**). Thiazoles (**5**) are generated from the reaction between α-haloketones and thiourea or thioamide derivatives (path **B**). Dihydropyridines (**7**) are generated from the reaction of aldehydes with β-ketoesters and ammonia or ammonia derivatives, or enamines derived from the reaction of ketones or β-ketoesters with amines (path **C**). Dihydropyridines can be readily converted to the corresponding pyridine derivatives and so this reaction is often termed the Hantzsch pyridine synthesis.

9.5.2 Historical Perspective

The Hantzsch dihydropyridine synthesis was first reported by Arthur Rudolf Hantzsch in 1881.[1] The Hantzsch pyrrole[2] and thiazole[3] reactions were reported in the years following Hantzsch's initial report on the synthesis of dihydropyridines.

The Hantzsch pyrrole synthesis remains useful for the preparation of *N*-substituted pyrroles from primary amines for materials chemistry and medicine. However, for pyrroles where *N* is unsubstituted, the Hantzsch pyrrole synthesis has largely been replaced by the Knorr pyrrole synthesis because the latter generally leads to higher product yields.

On the other hand, the Hantzsch thiazole synthesis generally proceeds in high yield with high regioselectivity, and so this reaction has been extensively employed in the preparation of therapeutics, including novel antimalarial, antitumor, and antibiotic compounds as well as several thiopeptide-based natural products with antibiotic and/or antitumor activity. Recent examples of the latter include the synthesis of dolstatins 3 and 10, amythiamicin D, IB-01211, bistratamide H and thuggacin A1.

The Hantzsch 1,4-dihydropyridine synthesis gained attention in the late 1960s with Bossert and co-workers synthesis of nifedipine,[4] a potent calcium channel antagonist, followed by the elucidation of the structure of NADH[5] in the early 1980s. Many of the efforts to synthesize 1,4-dhydropyridines since then have focused on generating more potent derivatives of nifedipine. However, the Hantzsch 1,4-dihydropyridine synthesis has also been used to synthesize novel antibiotic and anticancer therapeutics as well. In addition, the 1,4-dihydropyridines generated from the Hantzsch reaction have been used to access the pyridine components of a number of compounds, including the pyridine ring of the antibiotic micrococcin P1.

9.5.3 Mechanism

Pyrrole Synthesis

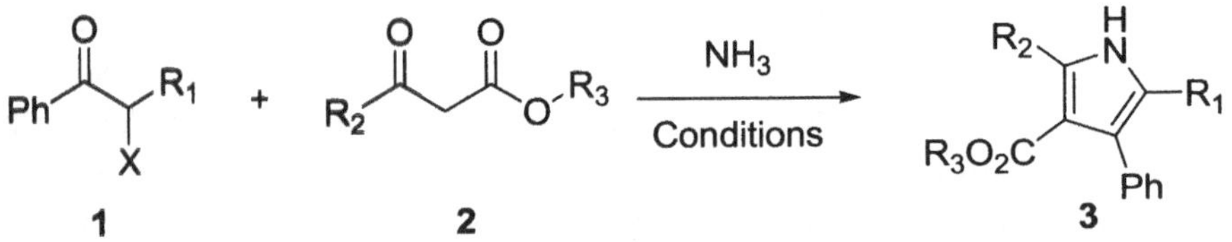

The mechanism for the Hantzsch pyrrole synthesis begins with enamine formation. Condensation of ammonia (or an ammonia surrogate) and β-ketoester **2** gives intermediate **A**. Intermediate **A** then undergoes dehydration and tautomerization (**B**) to produce enamine **C**. Michael addition of enamine **C** and α-haloketone **1** gives **D**, which forms **E** via β-elimination. Intramolecular nucleophilic substitution then generates **F**, which undergoes rapid isomerization to form the desired pyrrole **3**.

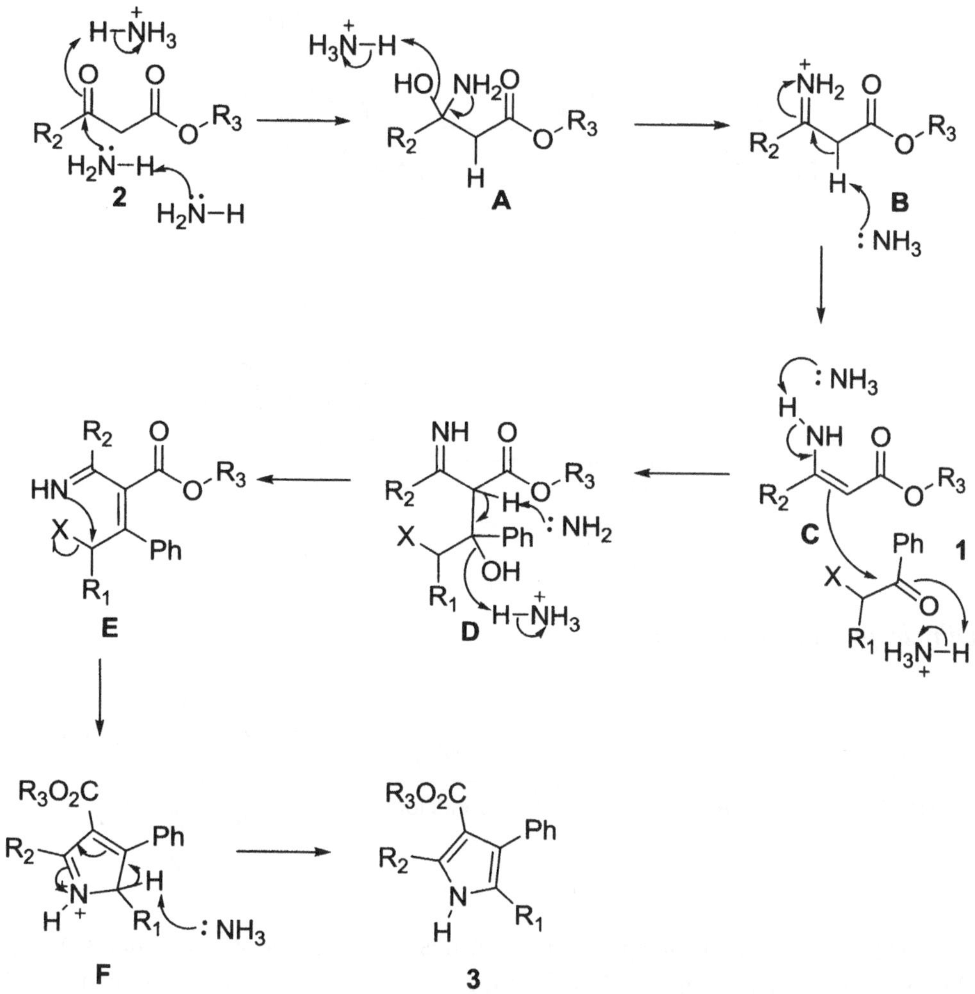

Thiazole Synthesis

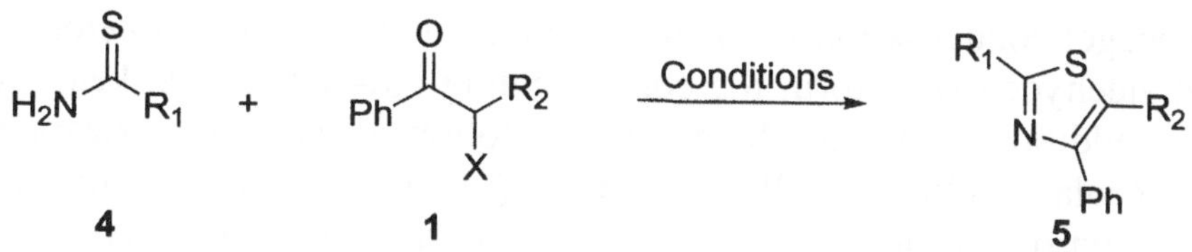

The mechanism for the Hantzsch thiazole synthesis begins with deprotonation of thioamide **4** (or thiourea) followed by nucleophilic substitution of the α-haloketone **1** to form **G**. **G** then undergoes intramolecular nucleophilic substitution at the ketone to form **H**, which undergoes rapid base-catalyzed elimination to give the desired thiazole **5**.

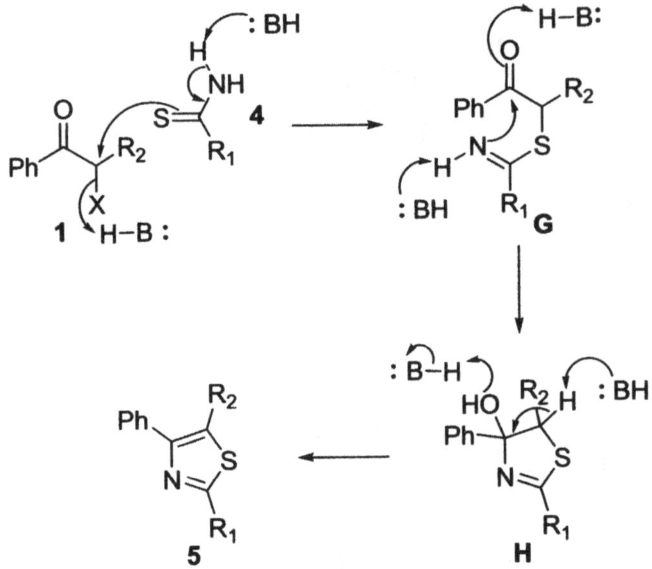

Dihydropyridine Synthesis

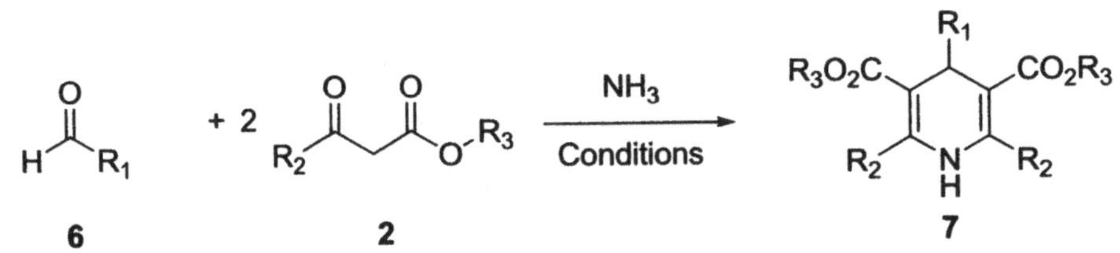

The Hantzsch 1,4-dihydopyridine synthesis was studied extensively by Berson and Brown using NMR spectroscopy.[6] One component of the Hantzsch 1,4-dihydropyridine is formed when one equivalent of a β-ketoester (**2**) or an activated methylene compound condenses with ammonia or an ammonia surrogate to give **K**. A second component is formed via Knoevenagel condensation between a second equivalent of a β-ketoester (**2**) and an aldehyde (**6**) to form the corresponding α,α-unsaturated ketoester **N**. Michael addition of **K** and **N** to give **O**, followed by tautomerization and intramolecular cyclization (**P**) gives **Q**. **Q** then undergoes elimination to form dihydropyridine **7**, generally as a mixture of stereoisomers. Dihydropyridine 7 can, in turn, undergo oxidation[7] to give the corresponding pyridine-3,5-dicarboxylate **R**, which can be decarboxylated to give the corresponding pyridine **S**.

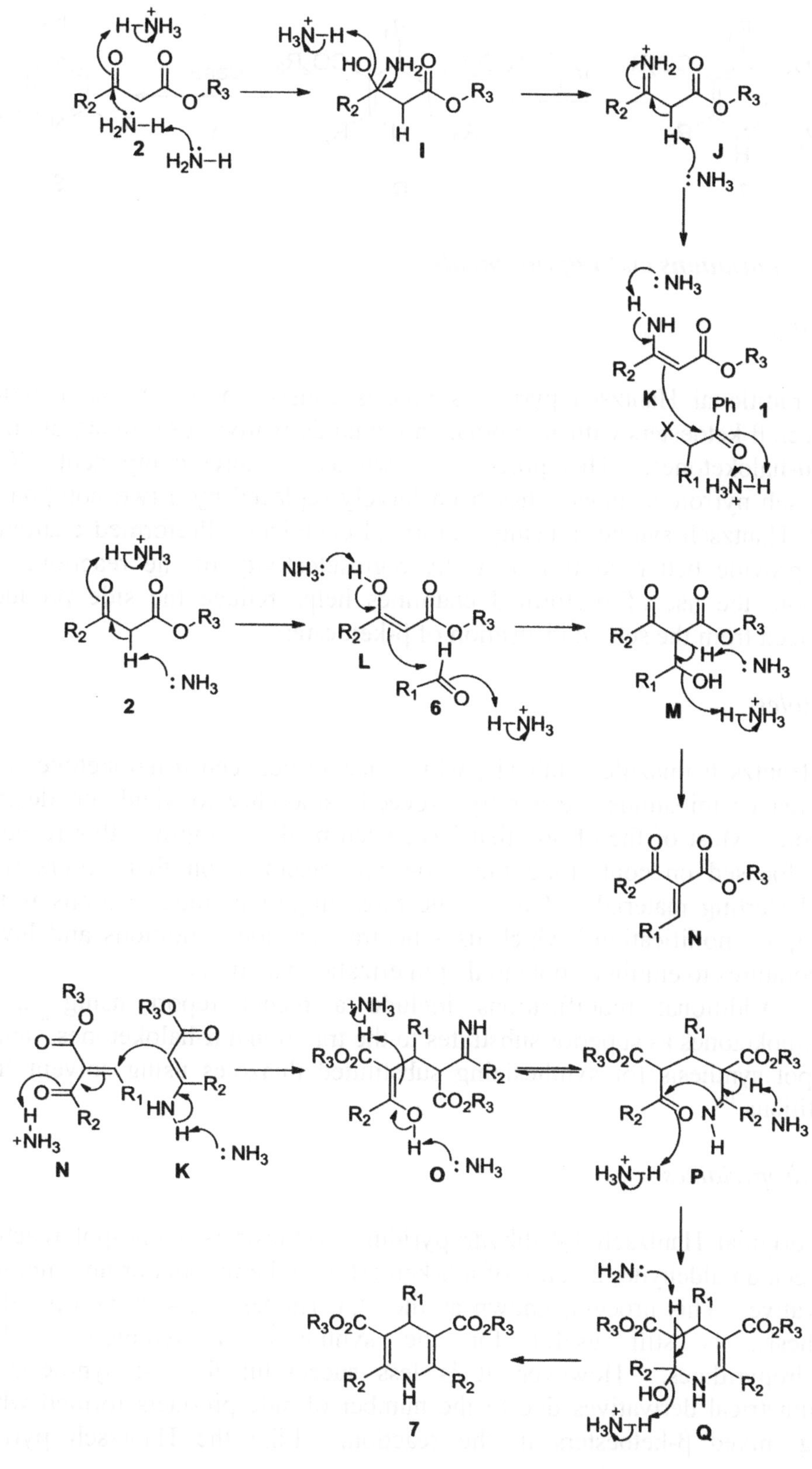

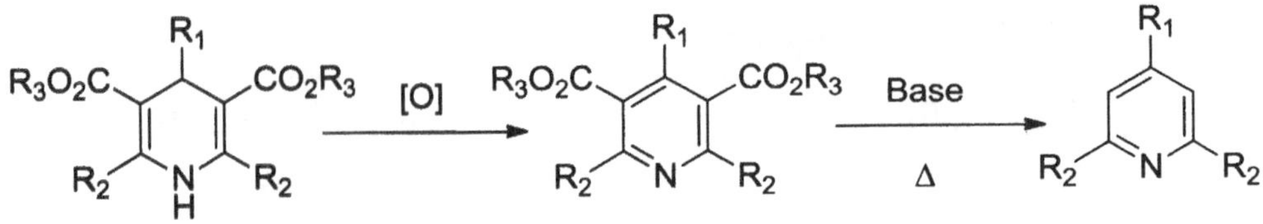

7 R S

9.5.4 *Variations and Improvements*

Pyrroles

The traditional Hantzsch pyrrole synthesis consists of a one-pot reaction between β-ketoesters with ammonia, ammonia derivatives or primary amines, and α-haloketones. This process, known as the three-component (3CP) Hantzsch pyrrole synthesis, has been largely replaced by a two component (2CP) Hantzsch synthesis using preformed enamines. Preformed enamines help provide better control over the regioselectivity of the reaction. In addition, the use of preformed enamines helps reduce the side products produced from the self condensation of β-ketoesters.

Thiazoles

The Hantzsch thiazole synthesis, which occurs between α-haloketones and thiourea or thioamides, generally proceeds smoothly to yield the desired thiazole. Most of the efforts that have been made to improve this reaction have focused on controlling the undesired racemization that occurs with chiral starting materials. One of the most important modifications is the Holzapfel modification,[8] which uses neutral reaction conditions and lower temperatures to eliminate potential epimerization reactions.

Additional modifications include a recent report using α, α-dibromoketones as superior substrates to the traditional α-haloketones,[9] and a one-pot synthesis for synthesizing substituted thiazoles using solvent free conditions.[10]

Dihydropyridines

The original Hantzsch 1,4-dihydropyridine synthesis is a one-pot reaction between an aldehyde, 2 equiv of a β-ketoester, and ammonia or an ammonia derivative. This process, known as the 3CP Hantzsch 1,4-dihydropyridine synthesis, is still useful for the synthesis of symmetrical 1,4-dihydropyridines. However, it is less successful for the synthesis of asymmetrical derivatives due to the number of side products formed when using mixed β-ketoesters in the reaction. Like the Hantzsch pyrrole synthesis, the 3CP 1,4-dihydropyridine synthesis has largely been replaced by a 2CP reaction system using preformed enamines, such as aminocrotonates and enaminoketones to provide better regiocontrol over the reaction products. For reactions where mixtures of stereoisomers are possible, the products are almost exclusively generated as racemates, although there have been a few reports where chiral auxiliaries have been used to control stereochemistry with varying success.[11]

Additional modifications to the Hantzsch 1,4-dihydropyridine synthesis generally involve the use of activated methylene compounds such as 1,3-diketones, ω-cyanoacetophenone, ω-phenylacetophenone, α, β-unsaturated ketones, and indane-1,3-diones.[7] A number of efforts to improve reaction yields using catalysts have also been reported, including the use of hydrotalcite materials,[12] triphenylphosphine[13] copper II triflate,[14] and covalently anchored sulfonic acid on silica gel.[15] Dihydropyridine synthesis has also recently been studied with high success using microwave[16] and solvent-free[17] reaction conditions.

9.5.5 Synthetic Utility

Pyrrole Synthesis

The Hantzsch pyrrole synthesis is a reaction used to synthesize pyrroles from a β-enaminoketone or ester generated by the reaction between ammonia or an ammonia surrogate with a β-diketone or β-ketoester and an α-haloketone. The initial three component reaction reported by Hantzsch involved the reaction between ethyl acetoacetate **8** and chloroacetone **9** in the presence of aqueous ammonia to give the corresponding pyrrole **10** in approximately 50% yield.[2]

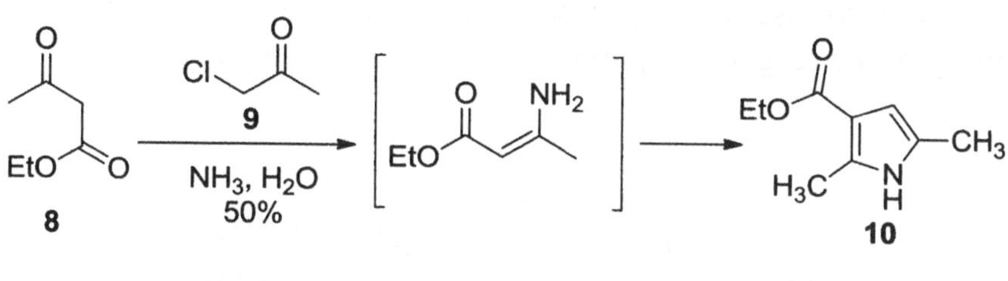

Nearly 70 years later, Roomi and MacDonald completed an extensive study on the scope and limitations of the Hantzsch pyrrole synthesis for the preparation of substituted pyrroles.[18] Over 25 pyrrole derivatives were prepared in modest to good yields, and the authors concluded that the Hantzsch pyrrole synthesis is most advantageous for the synthesis of pyrroles containing an alkyl substituent at position 3, with no substituent in position 4. Since their report, several groups have attempted to improve the Hantzsch pyrrole synthesis with the goal of generating novel pyrroles for use in materials chemistry and medicine.

Matiychuk and co-workers were able to synthesize several trisubstituted pyrroles containing aryl substituents at position 4 via Hantzsch pyrrole synthesis.[19] 3-Aryl-2-chloropropanals **11, 13, 15**, and **17** reacted with ethyl acetoacetate **8** in the presence of ammonia to form the desired pyrroles **12, 14, 16**, and **18** in modest yields. Modest yields were also obtained for benzyl acetoacetate derivatives in the same report using a similar protocol.

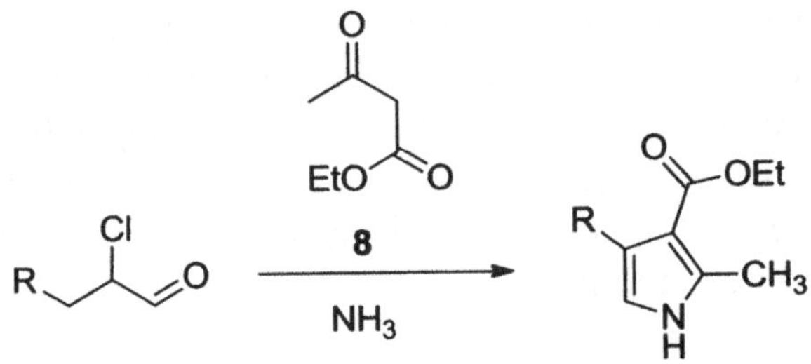

11, R = *o*-CH_3-C_6H_4
13, R = *p*-CH_3-C_6H_4
15, R = *o*-Cl-C_6H_4
17, R = *p*-Cl-C_6H_4

12, R = *o*-CH_3-C_6H_4 38%
14, R = *p*-CH_3-C_6H_4 48%
16, R = *o*-Cl-C_6H_4 42%
18, R = *p*-Cl-C_6H_4 44%

Skaddan and co-workers employed the Hantzsch synthesis in their preparation of isotopically labeled 1-*H*-pyrrole-2,3,5-tricarboxylic acid, a metabolic by-product of the breakdown of melatonin.[20] Labeled 1-*H*-pyrrole-2,3,5-tricarboxylic acid is currently under investigation as a tool to study the effectiveness of new hyper pigmentation drugs. ^{13}C-Labeled ethyl acetoacetate **19** and 2-chloroethanal **20**, when treated with ^{15}N-labeled aqueous ammonium hydroxide, produced pyrrole **21** in 35% yield. The authors chose to use this approach due to the relatively low cost and availability of the reagents required to prepare this compound.

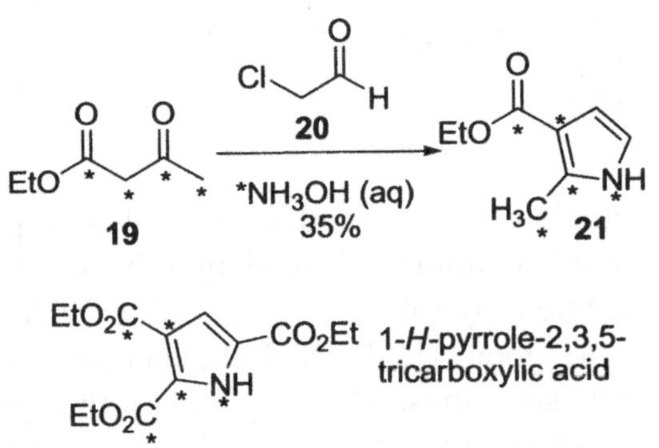

1-*H*-pyrrole-2,3,5-tricarboxylic acid

Kameswaran used a modified Hantzsch synthesis to prepare 2-aryl-5-trifluoromethylpyrrole derivatives[21] which are of interest as novel insecticides. Unfortunately, when the authors attempted the Hantzsch reaction using 3-bromo-1,1,1-trifluoropropanone **23** with β-oxonitriles, the corresponding furan was produced instead of the desired pyrrole (not shown). The authors concluded that the furan pathway was preferred due to the electrophilicity of the trifluoromethyl ketone and the enolizability of the oxo group. However, by modifying the reaction to include enamines such as **22**, **25** and **27**, they were able to produce the corresponding 2-aryl-5-trifluoromethylpyrroles in good yields.

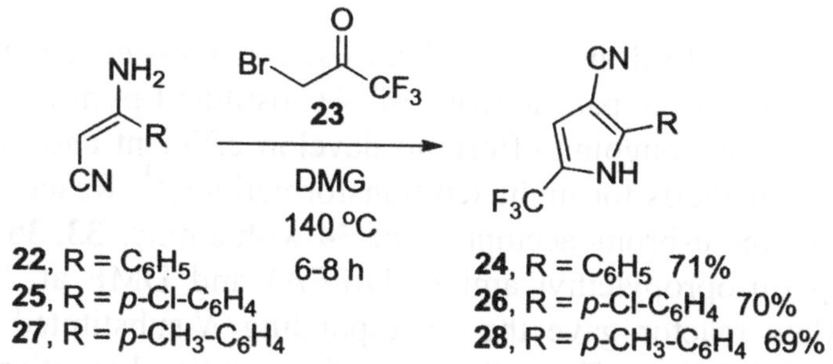

Trautwein and co-workers described an efficient method for the solid-phase synthesis of *N*-substituted pyrroles, including tetrasubstituted pyrroles.[22] Highly substituted pyrroles are functional components of compounds such as atorvastatin (Lipitor), an HMG-CoA reductase inhibitor used for lowering cholesterol. Reaction of β-ketoamides **29**, prepared from polymer bound acetoacetamide using a series of primary amines in trimethylorthoformate (not shown), with α-bromoketone derivatives **30** in the presence of 2,6-di-tertbutylpyridine (DTBP) and DMF yielded pyrroles **31** with diverse functionality in high purity. The authors found that polystyrene Rink amide resin was the best solid support because it was able to withstand the acetoacetylation conditions required to produce the polymer bound acetoacetamide.

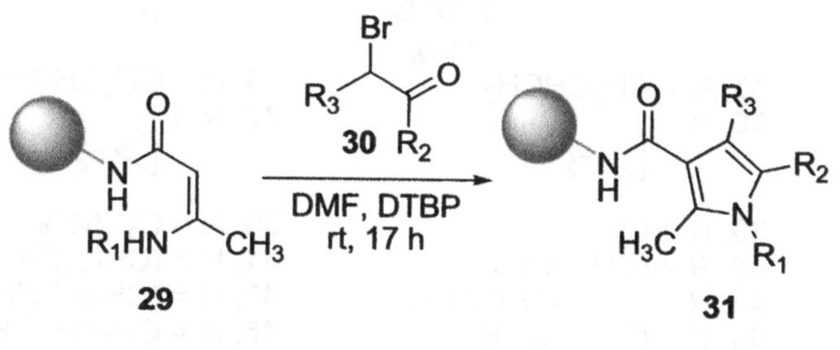

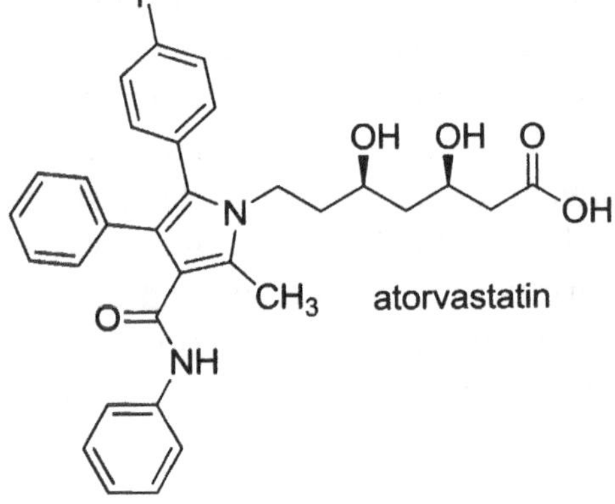

Recently Cosford and co-workers used a one-step continuous flow Hantzsch synthesis to prepare a series of *N*-substituted pyrrole-3-carboxylic acids as part of their ongoing efforts to develop efficient and high-yielding flow chemistry methods for multistep transformations.[23] Reaction of *t*-butyl acetoacetate **32** and α-bromoacetophenone **34** with amines **33**, **36**, **38**, **40**, **42**, and **44** using diisopropylethyl amine (DIPEA) and DMF at 200 °C in a continuous flow reactor gave the corresponding *N*-substituted pyrroles in modest yields (40-62%). The authors found the optimal reaction conditions were 2.2 equiv of *t*-butyl acetoacetate, 1 equiv of amine and 0.5 equiv of DIPEA in a 0.5M solution of α-bromoacetophenone predissolved in DMF. In addition, the authors noted that the HBr produced as a byproduct of the reaction simultaneously hydrolyzed the ester to give the free acid. Several additional β-ketoesters and α-bromoketones were explored in their report with similar success.

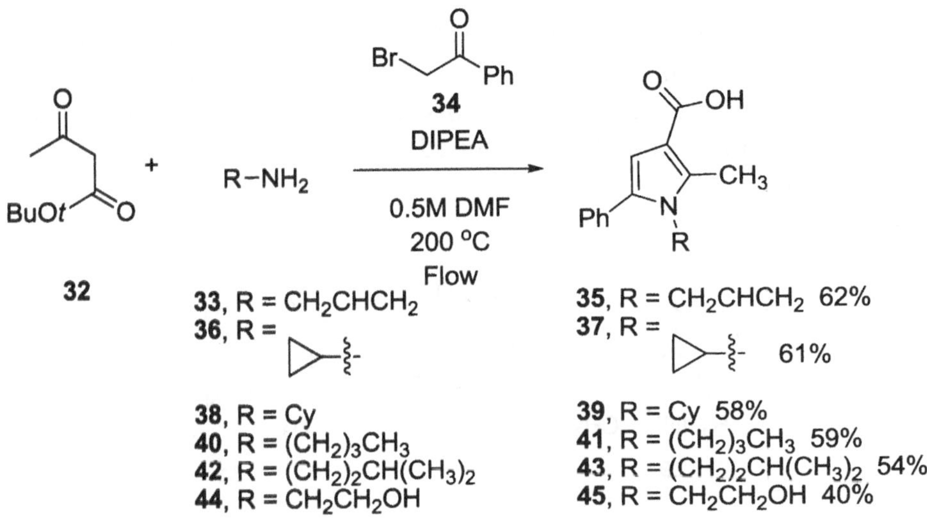

Calvo and co-workers reported on a method for producing a variety of 2,3-dihydropyrrolizines using a modified Hantzsch synthesis involving Weinreb 3-(pyrrolidin-2-ylidene)propionamides.[24] Pyrrolizines are important functional moieties in nonsteroidal antiinflammatory compounds such as ketorolac. The first step of their synthesis involved C-alkylation of an enamine derivative (**46**, **51**, or **55**) using *N*-methoxy-*N*-methyl-α-bromoacetamide **47** to produce amides **48**, **52**, or **56**. In the presence of organometallic compounds, the R_1 functionality is deactivated to form the corresponding carbonyl compounds **49**, **53** and **57** as isomeric mixtures of *E* and *Z* olefins. Refluxing ketones **49**, **53** and **57** in chloroform in the presence of silica gel led to the generation of pyrrolizines **50**, **54**, and **58** in good to excellent yields (40 to 99%).

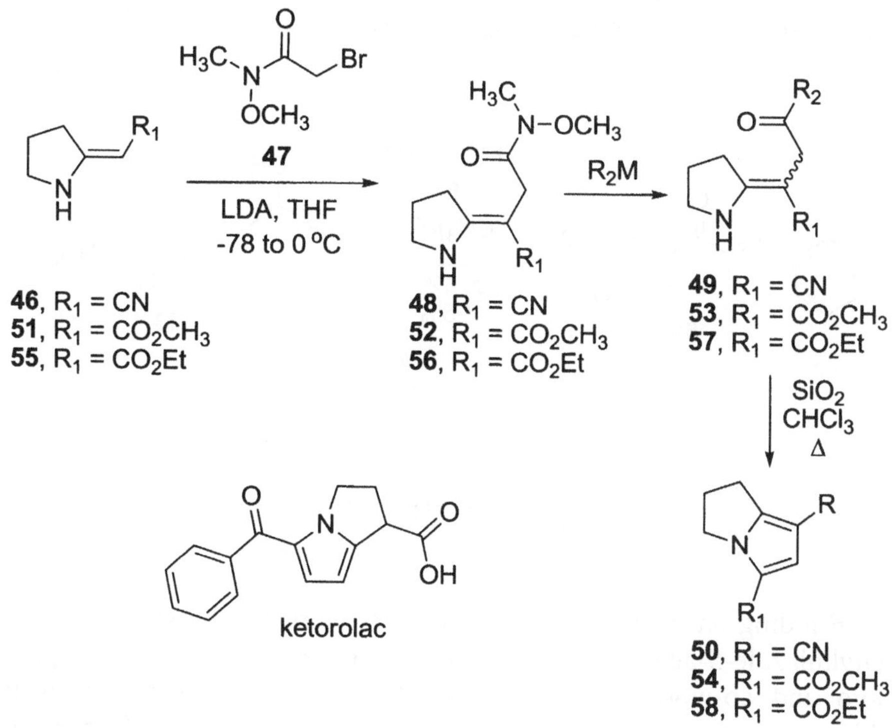

Thiazole Synthesis

The first reaction, reported by Hantzsch in the late 1880s, involved the one-pot condensation of chloroacetone **9** and thioacetamide **59** in refluxing benzene to produce thiazole **60** in near quantitative yeild.[3] Since Hantzsch's original report, the Hantzsch thiazole synthesis has been used to prepare a number of biologically relevant compounds and other functionally important thiazoles. Some of the more recent examples are highlighted below.

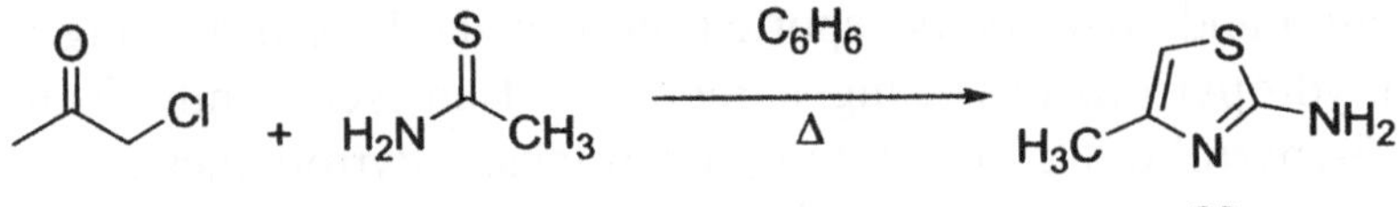

9 59 60

Avery and co-workers applied the Hantzsch thiazole synthesis to the synthesis of trisubstituted thiazole **64**, which showed good activity against falcipain-2 and faclipain-3, two cystein proteases of the malaria parasite *Plasmodium falciparum*.[25] Reaction of bromoester **61** with thiourea **62** in refluxing acetone for 12 h gave the corresponding thiazole **63** in near quantitative yield. Eight additional steps were required to reach compound **64**. Several other trisubstituted thiazole derivatives were also prepared by Avery and co-workers in the same report. However, most of the other derivatives showed little to no biological activity.

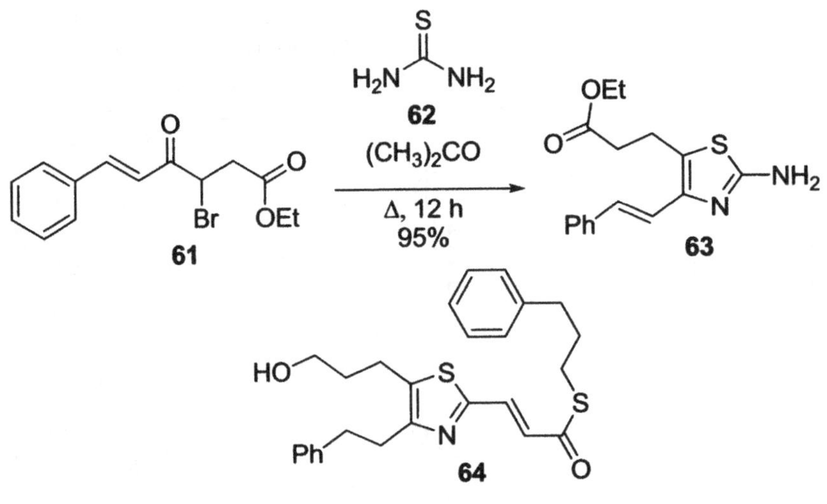

Building on Newbury and co-workers work on the synthesis of 2,4-diphenylthiazole-5-acetic acid derivatives as antiinflammatory agents,[26] Ikemoto and co-workers used the Hantzsch thiazole synthesis in their scalable synthesis of a pyridylethanolamine inhibitor of human β_3 adrenergic receptor inhibitor (β_3 ARI).[27] Treatment of α-bromoketone **65** with thioamide **66** gave the corresponding thiazole **67** in 89% yield. Four additional steps were required to generate the β_3 adrenergic receptor inhibitor in 66% overall yield.

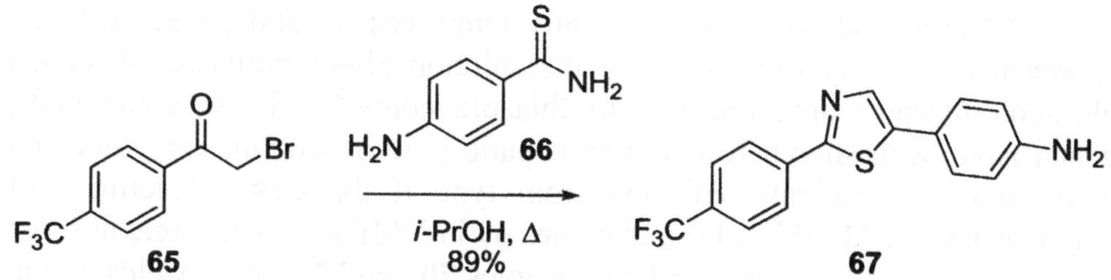

65
66
67
i-PrOH, Δ
89%

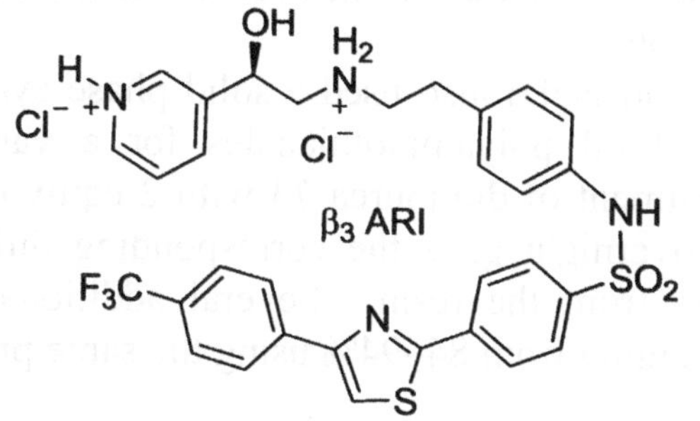

β3 ARI

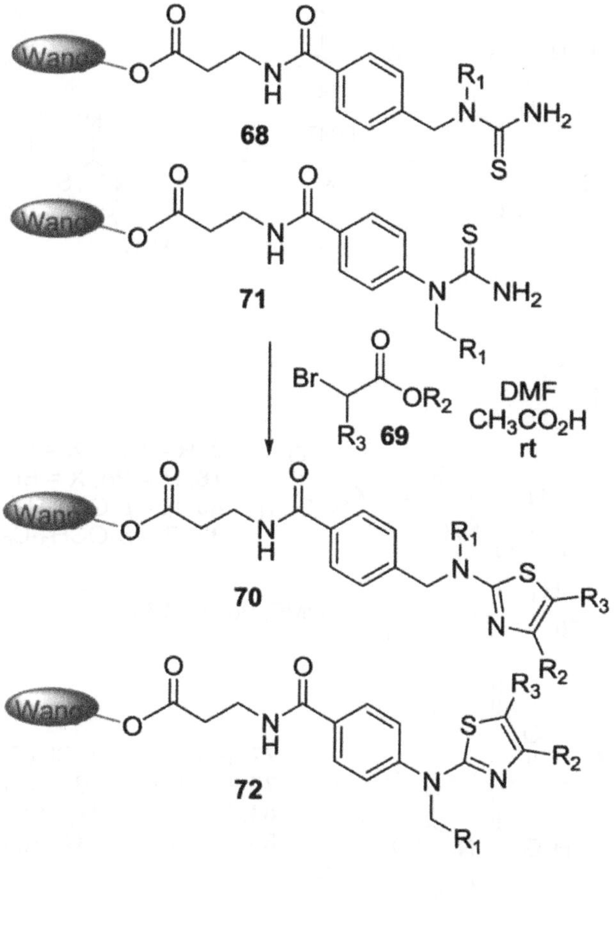

68
71
69
DMF
CH3CO2H
rt
70
72

Madsen and co-workers recently employed a solid phase synthetic approach as a competitive route to the solution phase synthesis of human glucagon receptor antagonists with thiazole cores.[28] These compounds, which have been shown to decrease hepatic glucose output, are promising therapeutics for patients suffering from type II diabetes. Treatment of thioureas **68** and **71** with α-bromoketones **69** in DMF and acetic acid at room temperature gave the corresponding thiazoles **70** and **72** in high yields and in good purity. The major advantage to this route was the ease of purifying the products of the reactions.

Arutyunyan and Nefzi also used a solid phase synthetic approach in their synthesis of chiral polyaminothiazoles for a variety of biological applications.[29] Treatment of dithiourea **73** with 2 equiv α-bromoacetone **74** in refluxing DMF overnight gave the corresponding dithiazole **75** in 88% yield after cleavage from the resin. Several additional derivatives were prepared in yields ranging from 84–94% using the same protocol.

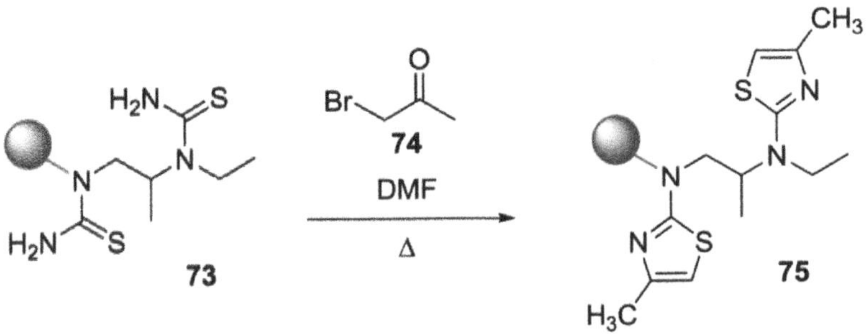

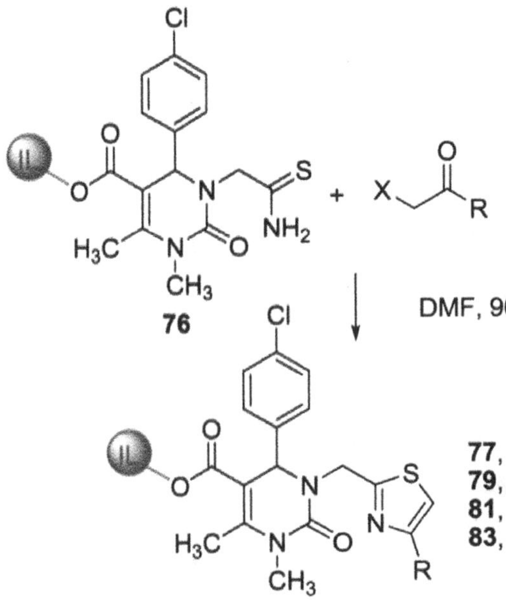

9, R = CH_3, X = Cl
78, R = Ph, X = Br
80, R = *p*-Cl-C_6H_4, X = Br
82, R = *p*-OCH_3-C_6H_4, X = Br

DMF, 90 °C, 16 h

77, R = CH_3, X = Cl 65%
79, R = Ph, X = Br 85%
81, R = *p*-Cl-C_6H_4, X = Br 88%
83, R = *p*-OCH_3-C_6H_4, X = Br 84%

Bazureau and co-workers used ionic liquid phase technology to prepare several new 3,4-dihydropyrimidine-2(1*H*)-ones.[30] Treatment of thioamide **76** (where IL = [HOC$_3$mim][PF$_6$]) with α-haloketones **9**, **78**, **80**, and **82** gave the corresponding thiazoles **77**, **79**, **81** and **83** in modest to high yields and high purity.

Fused thiazoles, especially compounds bearing an arylazo moiety at position 5, have been shown to have cytostatic and cytopatic activities. El-Meligie and El-Awady were able to prepare four fused arylazothiazole compounds through a common thiazole intermediate in good yields using the Hantzsch synthesis as one of the initial steps in their synthesis.[31] Treatment of **84** with iodine and thiourea **62** using a steam bath gave the corresponding thiazole **85** in 70% yield. Three additional steps were required to reach the target compound.

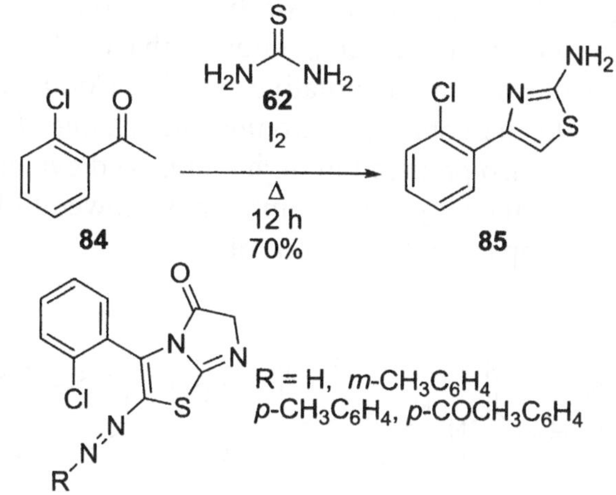

Velezheva and co-workers studied the influence of thioamide structures on the Hantzsch synthesis of thiazoles and thiazoloindoles.[32] These compounds serve as interesting templates for the synthesis of complex fused thiazole containing heterocycles. Reaction of 1-acetyl-2-bromo-3-indolinone **86** with thiourea derivatives **59**, **88**, **90**, and **92** gave the corresponding thiazoloindoles **87**, **89**, **91**, and **93**. The methylene unit was shown to be important for forming the thiazole, as thioamides lacking a methylene unit gave the corresponding hydroxythiazolines (not shown).

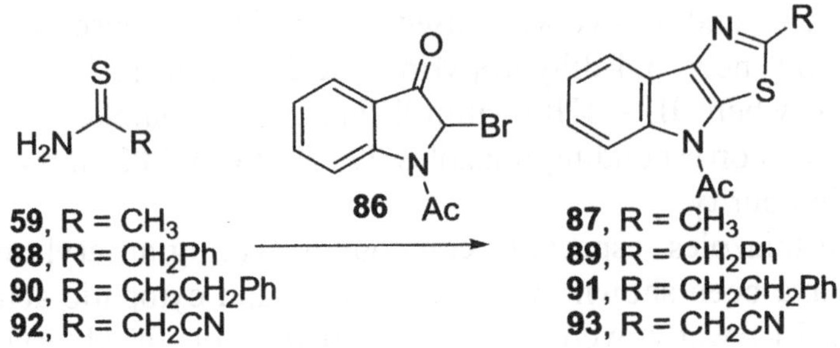

Coumarins are an important class of therapeutics. Several coumarin-thiazole conjugates displaying activity against *Helicobacter pylori* were recently prepared by Chimenti and co-workers[33] based on methodology developed by Mashraqui and co-workers for the synthesis of coumarin-based probes for imaging.[34] Reaction of 3-α-bromoacetyl coumarin **94** with thiosemicarbazone **95** in room temperature ethanol for 4 h gave the corresponding thiazole **96** in quantitative yield. Compound **96** showed comparable activity to the control, metronidazole, against *H. pylori*. Other derivatives of **96** were also prepared in high yields, and evaluated for activity against *H. pylori*. Unfortunately, most compounds showed little to no activity against *H. pylori* in comparison to the control.

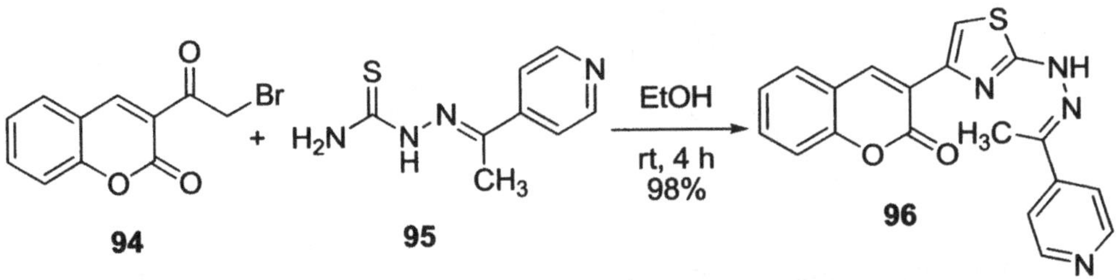

Lin and co-workers used the Hantzsch synthesis to prepare coumarin-based fluorescent "turn-on" thiourea probes.[35] Probes were composed of a coumarin dye, a carbonyl group, and a bromide group as illustrated by compound **97**. The carbonyl group was designed to decrease the fluorescence of the coumarin dye through intersystem crossing. In addition, the bromide, which is an important functional group for the Hantzsch reaction, serves a secondary role as a fluorescence quencher. These two qualities combined rendered the dye relatively inactive until treated with a thiourea derivative. Reaction of **97** with thiourea derivatives **62**, **99**, and **101** in refluxing DMF gave the corresponding thiazoles **98**, **100**, and **102**. No yields for the reaction products were reported; however, the authors noted a significant increase in fluorescence from compounds **98**, **100**, and **102**, showing proof of concept using the Hantzsch synthesis for the production of "turn-on" probes.

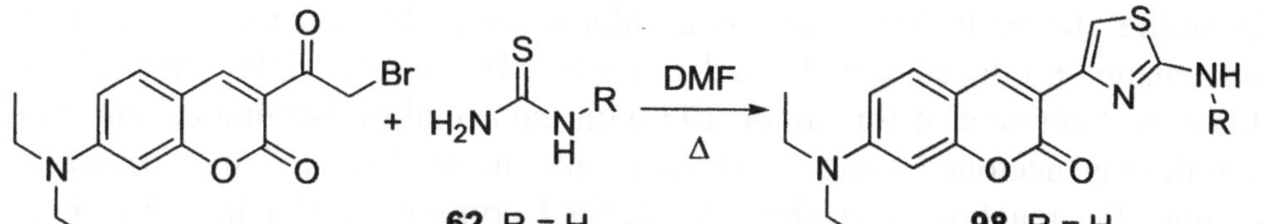

97

62, R = H
99, R = Ph
101, R = Napth

98, R = H
100, R = Ph
102, R = Napth

Kovacs and co-workers used the Hantzsch thiazole synthesis to prepare two thiazole C-nucleosides as pyranose analogs of tiazofurin, an antitumor compound, which functions by inhibiting inosine monophosphate dehydrogenase.[36] Their synthesis began with compounds **103**, which were prepared in four steps from the corresponding acetylated sugars. Treatment of **103** with ethyl bromopyruvate **104** in ethanol or acetonitrile, at ambient or elevated temperatures, led to the formation of the thiazole **105** in consistently low yield. The predominant product(s) of these reactions were shown to be furans **106** and **107** in ethanol and **107** in acetonitrile, with **108** prevailing in the presence of ethanol under longer reaction times. Epimerization of C1 was also noted. The authors were unable to increase yields of the desired thiazole **105**, despite using a variety of reaction conditions.

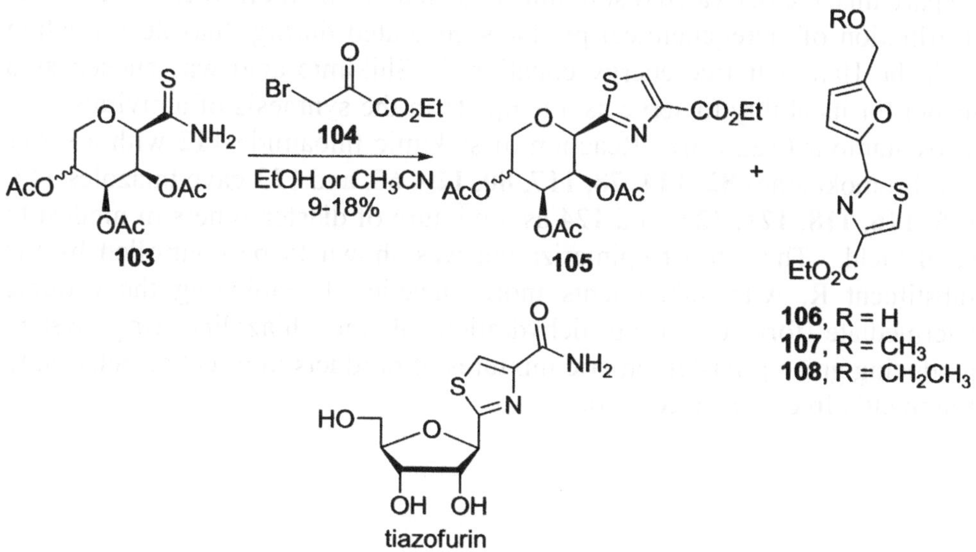

103

105

106, R = H
107, R = CH_3
108, R = CH_2CH_3

Hammer and co-workers used the Hantzsch thiazole synthesis to prepare thiazole and thiazole-*N*-oxide nucleobases with the goal of producing nucleobase analogs with pronounced directional dipoles and the capacity to base pair.[37] Their overall goal was to use these compounds as models to explore the biochemical and biophysical properties of DNA. Treatment of thioamide **109** with 50% aqueous 2-chloroethanal **20** in a mixture of THF and ethanol produced the desired thiazole **110** in low yield (only 28%). However, treatment of thioamide **109** with the acetal of 2-bromoethanal **111** in refluxing acetone, using a catalytic amount of 4M HCl in dioxane to promote the reaction, furnished the desired compound **110** in 70% yield, presenting a marked improvement.

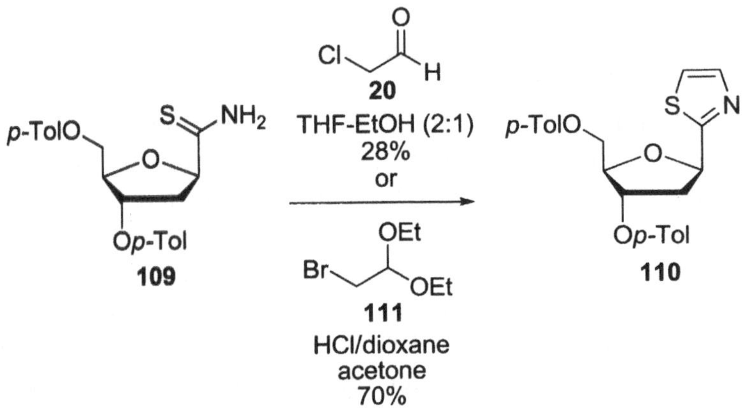

Goodwin and co-workers used the Hantzsch thiazole synthesis to prepare thiazole derivatized shikimic acid analogs in an effort to correlate the distribution of stereochemical products generated during thiazole formation with the Hammett free-energy equation.[38] Shikimic acid was chosen as a model for its ability to serve as a template in the synthesis of antivirals such as oseltamivir (Tamiflu). Reaction of shikimic thioamide **112** with a series of α-bromoketones **82, 114, 78, 117, 80, 120, 65**, and **123** gave thiazoles **113, 115, 116, 118, 121, 122**, and **124** as a mixture of diastereomers in modest to good yield. The rate of epimerization was shown to be controlled by the substituent R, with substituents more capable of stabilizing the cationic intermediate formed during dehydration of the thiazoline ring system producing greater diastereomeric mixtures of products in direct correlation to Hammett's free-energy equation.

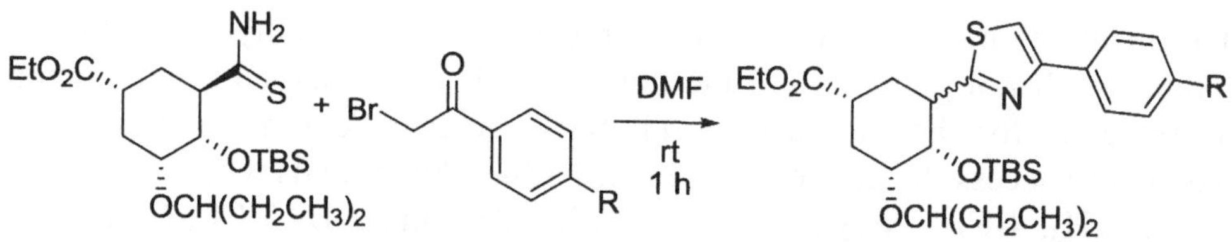

112

82, R = OCH_3
114, R = CH_3
78, R = H
117, R = F
80, R = Cl
120, R = OCF
65, R = CF_3
123, R = CN

113, R = OCH_3 91% (β:α 20:1)
115, R = CH_3 95% (β:α 20:1)
116, R = H 84% (β:α 7.8:1)
118, R = F 82% (β:α 6.3:1)
119, R = Cl 64% (β:α 4.8:1)
121, R = OCF_3 64% (β:α 2:1)
122, R = CF_3 66% (β:α 1:1.1)
124, R = CN 71% (β:α 1:1.5)

Urbanský and Drašar employed the Hantzsch thiazole synthesis in their preparation of semisynthetic steroids incorporating thiazoles. Semisynthetic steroids have shown importance as antiarrhythmic, cardiotonic and cytostatic agents.[39] Treatment of thioamide **125** with α-bromoderivatives **78**, **74**, **128**, **130**, and **133** gave the corresponding thiazolyl steroid derivatives in good to modest yields. For the production of **126** and **127**, benzene was shown to be an excellent solvent. For the less reactive methoxy, ethoxy, and isopropyloxy α-bromoderivatives (products **129**, **131**, and **133**, respectively), the corresponding alcohols (methanol, ethanol, and isopropanol) were found to be the best solvents. In addition, the HBr liberated in the reactions giving rise to **129**, **131**, and **133** was shown to catalyze the hydrolysis of the acetate group for these compounds to give mixtures of products.

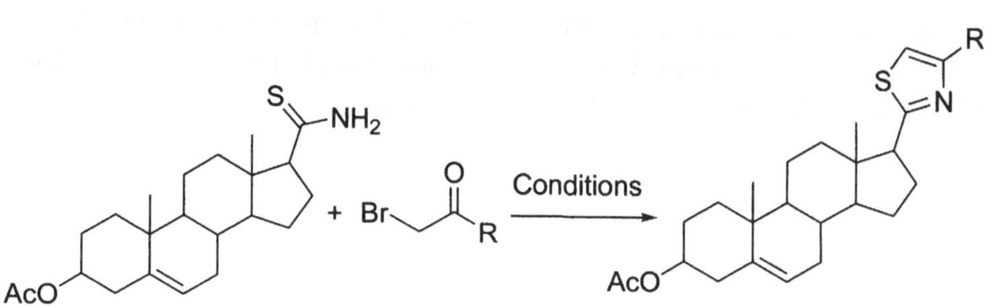

125

78, R = Ph
74, R = CH_3
128, R = OCH_3
130, R = OEt
132, R = O*i*Pr

126, R = Ph 76%
127, R = CH_3 64%
129, R = OCH_3 27%, 35%-OH
131, R = OEt 47%, 22%-OH
133, R = O*i*Pr 32%, 44%-OH

Sun and co-workers used the Hantzsch thiazole synthesis to prepare a series of novel diaryl urea derivatives **136** based on sorafenib as potential candidates for the treatment of nonsmall cell lung cancer and breast cancer.[40]

Treatment of thioamide **134** with bromoacetaldehyde diethylacetal **111** in refluxing ethanol gave 72% of the thiazole **135**, which served as a common intermediate for the synthesis of 11 diaryl ureas. Four more steps were required to access the desired diaryl ureas **136**. Compounds where R = 4-Cl-3-CF_3-C_6H_3, 4-Cl-3-NO_2-C_6H_3, 4-F-3-CF_3-C_6H_3, and 3-Br-5-CF_3-C_6H_3 were shown to be more potent than sorafenib.

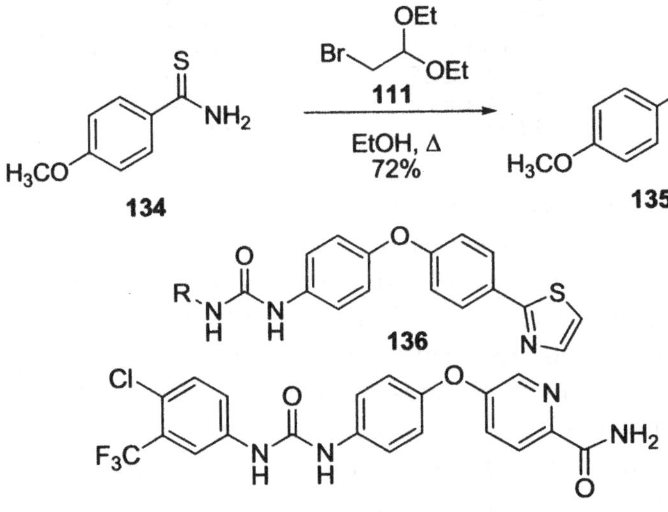

sorafenib

Sušnik and co-workers used the Hantzsch synthesis in their preparation of several novel 4-(2-amino-5-thiazolyl)-pyrimidine-2-amines[41] (**142**) as derivatives of imatinib (Gleevec), a protein kinase inhibitor used to treat leukemia. Reaction of α-bromoketone **137** with thioacetamide **138** or *N,N*-dimethyl thioacetamide **140** gave the corresponding thiazole derivatives **139** and **141** in 98% and 61% yields, respectively.

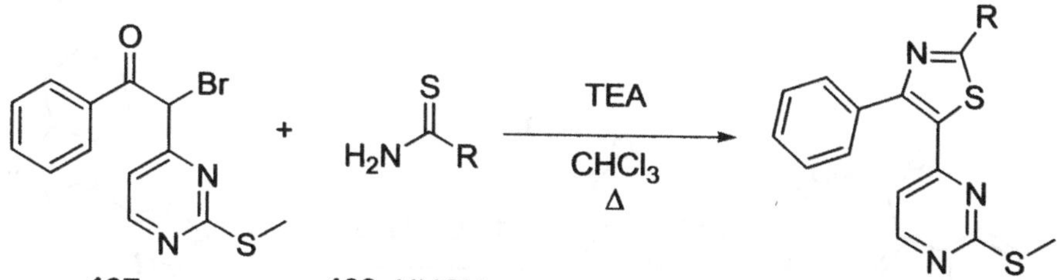

137 | 138, $NHCH_3$; 140, $N(CH_3)_2$ | 139, $NHCH_3$ 98%; 141, $N(CH_3)_2$ 61%

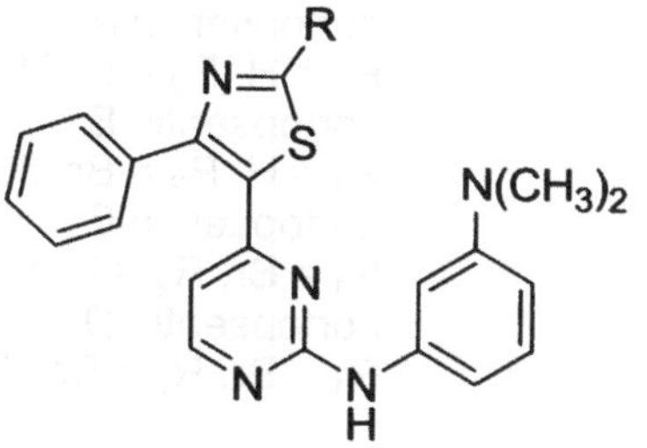

142

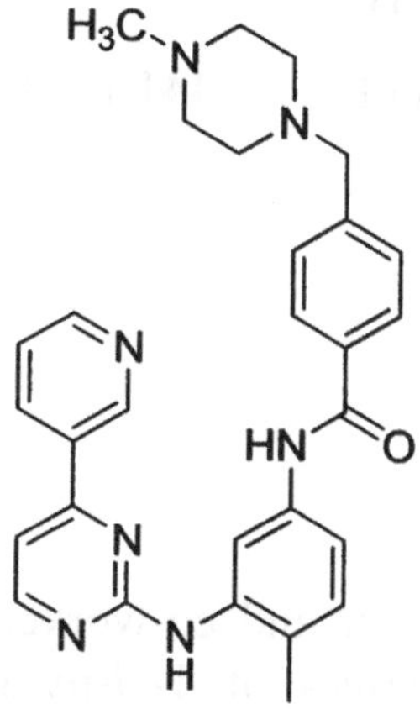

imatinib

Jiang and co-workers applied the Hantzsch thiazole synthesis to the preparation of several bis(indole)alkaloid nortopsentins alkaloids.[42] This family of compounds, which contain a characteristic 2,4-bis(3-indolyl)imidazole skeleton, exhibit bacteriostatic, antiviral and cytotoxic activities. Nortopsentins A–D, the focus of Jiang and co-workers work, showed significant *in vitro* cytotoxicity against P388 cells, and antifungal activity against *Candida albicans*. Refluxing **143** and **144** in absolute ethanol, followed by hydrolysis of the tosylate using sodium methoxide in methanol, gave the corresponding nortopsentins A–D in high yields.

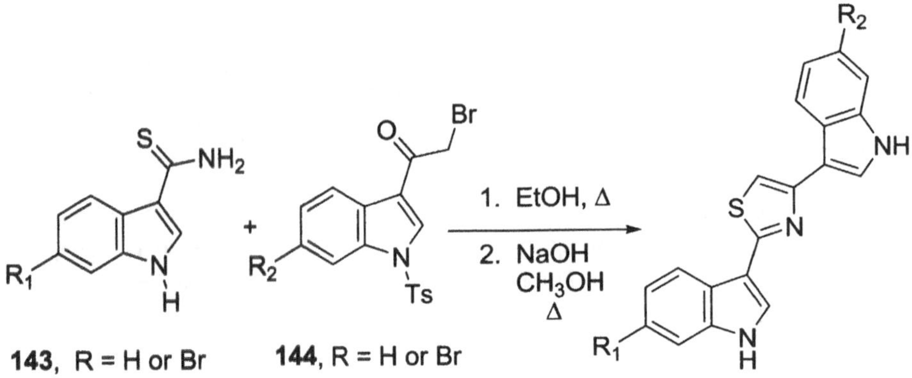

143, R = H or Br **144**, R = H or Br

nortopsentin A,
R_1 = H, R_2 = H 85.2%
nortopsentin B
R_1 = H, R_2 = Br 80.7%
nortopsentin C
R_1 = Br, R_2 = H 82.5%
nortopsentin D
R_1 = Br, R_2 = Br 78.6%

Kleinpeter and co-workers used the Hantzsch thiazole synthesis to prepare derivatives of 1-*H*-pyrazoloquinozalines[43] to study the structure activity relationships of these compounds. Compounds of this class are known to have tuberculostatic, bacteriostatic and fungistatic activity. Treatment of 2-acetylquinoxaline thiosemicarbazone **145** with ethyl bromopyruvate **104** in ethanol gave thiazole **146** exclusively in 59% yield. On the other hand, treatment of **145** with ethyl bromopyruvate **105** in 2-methoxylethanol gave thiazole **147** exclusively in 46% yield. The authors suggested that because **146** crystallizes from ethanol as it is formed, it does not undergo further redox reactions with the hydrogen halogenides formed in the course of the reaction. On the other hand, **146** remains in solution in 2-methoxyethanol and thus can further react with the hydrogen halogenides to form **147**. The authors prepared several additional derivatives of **147**, including dimeric derivatives, such as **148**, in good yield.

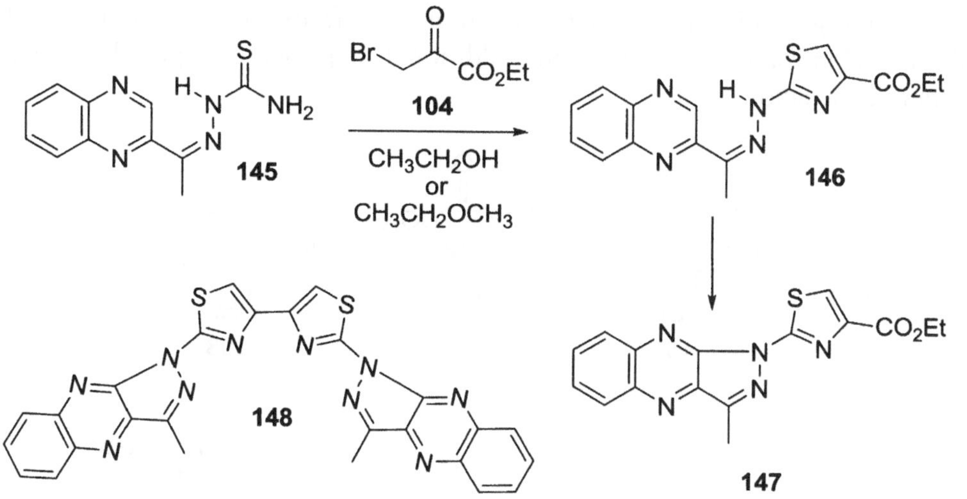

Poupat and co-workers employed the Hantzsch thiazole synthesis as a key step in their synthesis of an aminothiazole derivative of the antitumor alkaloid girolline.[44] Reaction of α-bromoketone **149**, prepared in four steps from commercially available D-(+)-arabitol, with *N*-Boc protected thiourea **150** in the presence of sodium bicarbonate in DMF at 50 °C gave the corresponding aminothiazole **151** in 83%. Five additional steps were required to generate the target compound 5-deazathiogirrolline.

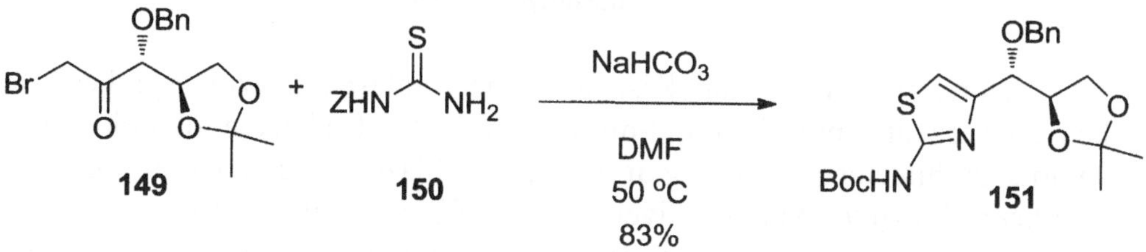

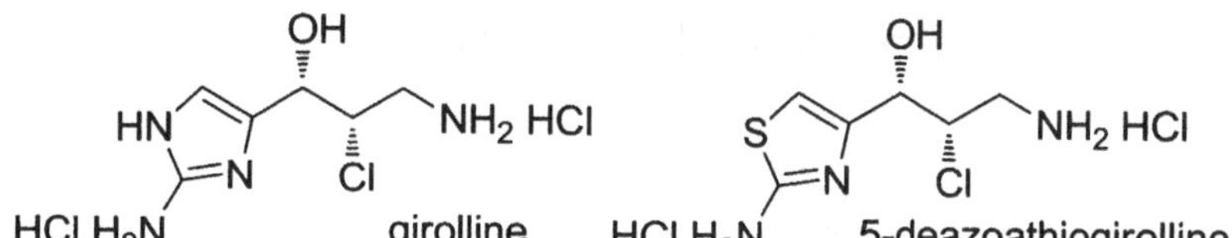

Bagley and co-workers used the Hantzsch thiazole synthesis to generate saramycetic acid,[45] a primary component of many thiopeptide antibiotics including micrococcin P1, the thiocillins, amythiamicins and bleomycin, as well as cyclothiazomycin, a selective inhibitor of human plasma renin. Treatment of thioamide **152** with ethyl bromopyruvate **104** in refluxing ethanol for one hour gave the corresponding thiazole **153** in quantitative yield. The carboxyethyl group of **153** was then converted in three steps to thioamide **154**, which was then subjected to a second Hantzsch reaction to give **155** in 54% yield over four steps from **153** (quantitative yield for the Hantzsch reaction). Four additional steps were required to reach saramycetic acid.

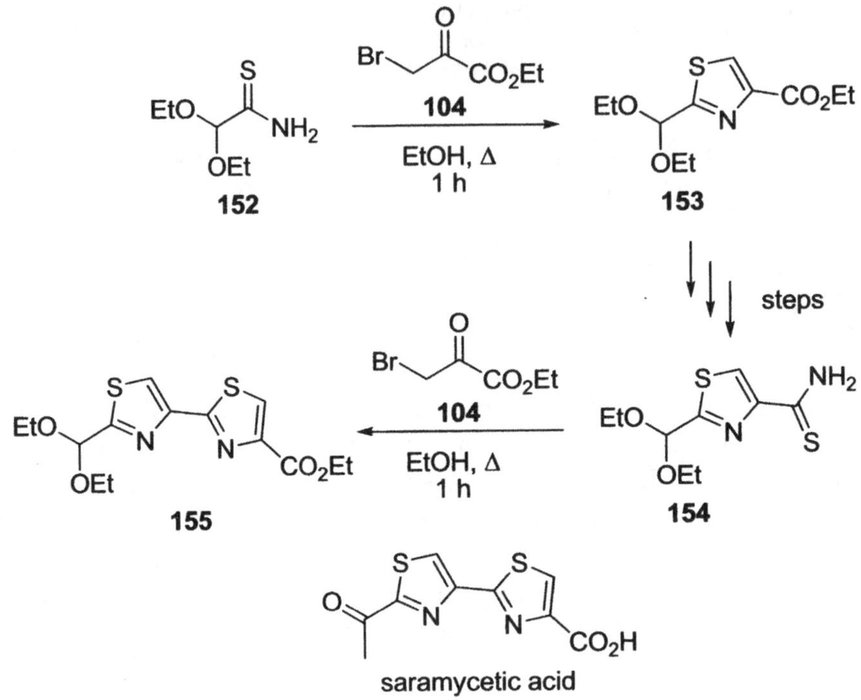

Kuriyama, Akaji, and Kiso used the Hantzsch thiazole synthesis in their convergent synthesis of (–)-mirabazole B.[46] Mirabazole alkaloids have a unique architecture consisting of thiazoline, thiazole and oxazole rings, and have been shown to exhibit a diverse array of biological activities. Reaction of **156**, prepared in 12 steps from D-alanine with chloroacetone **9** in refluxing ethanol, gave (–)-mirabazole B in 57% yield.

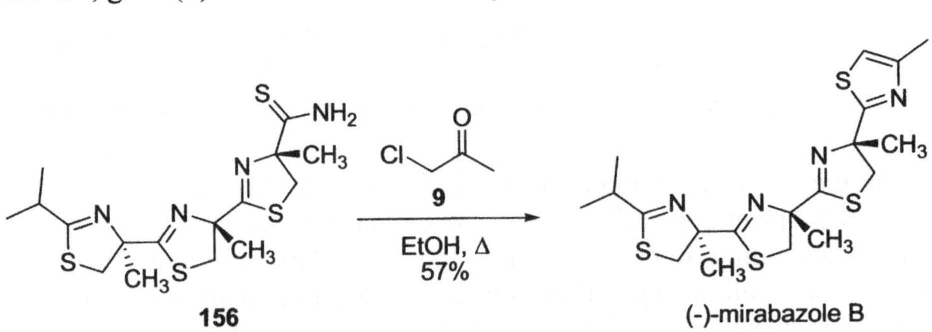

The Hantzsch synthesis has been employed by several groups in the syntheses of the thiazole rings of dolastatins 3 and 9. The dolastatin family of peptides is important due to their potent cytotoxic activity. One of the first efforts toward the synthesis of dolastatin 3 was by Pettit and Holzapfel who used the Hantzsch synthesis to synthesize an amino acid-based thiazole as a key component of this compound.[47] Thioamide **157**, prepared in seven steps from readily available glutamic acid, gave the corresponding thiazole **158** as as a racemic mixture in high yield when treated with ethyl bromopyruvate **104** in refluxing ethanol. Several additional amino acid-based thiazoles were also described in the same report. Pettit and co-workers later employed a Hantzsch strategy in their synthesis of amino acid based thiazole **159**,[48] thiazoles **158** and **159** were ultimately used to construct analogs of dolastain 3 for structural studies. Henichart and co-workers also used the Hantzsch thiazole synthesis in their preparation of several amino acid-based thiazole derivatives for the generation of dolastatin 3.[49]

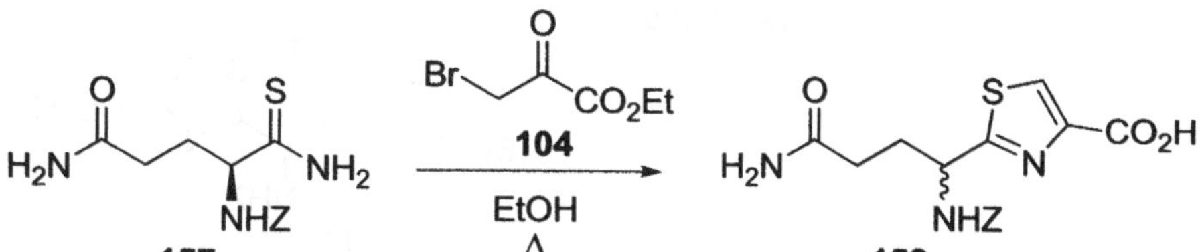

157 158

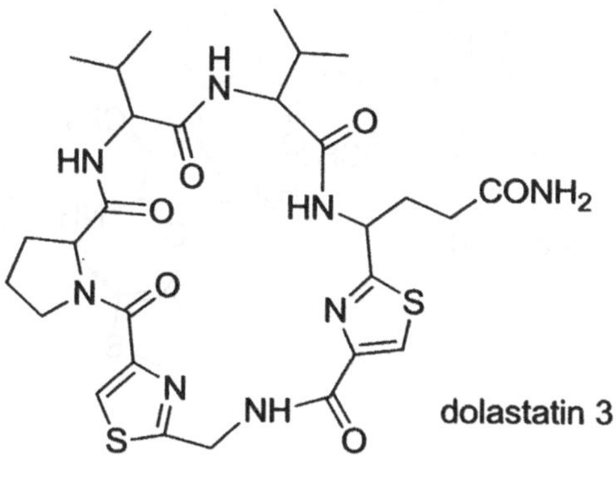

dolastatin 3

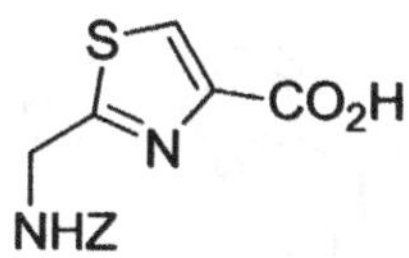

159

One of the challenges of generating amino acid-based thiazoles is the loss of stereochemistry during the reaction due to racemization. Holzapfel and co-workers were able to generate amino acid based thiazoles without the loss of stereochemistry by carefully controlling the reaction conditions.[8] Thioamide **160**, generated in five steps from L-glutamic acid, was treated first with ethyl bromopyruvate **104** under buffered conditions to generate the hydroxythiazoline intermediate that, when treated with trifluoroacetic anhydride (TFAA) in the presence of pyridine, furnished the corresponding thiazole **161** in greater than 90% yield over two steps with retention of configuration. The authors noted that maintenance of neutral conditions was important for suppressing imine-enamine interconversion and racemization from the HBr produced as a side product of the reaction. Holzapfel later applied a similar strategy to the synthesis of *N-t*-Boc-*S*-dolaphenine **162**, an amino acid based thiazole important for the synthesis of dolastatin 10.[50] A later report by Agullar and Meyers found that bulkier bases such as 2,6-dimethylpyridine were required to avoid racemisation with valine and alanine derivatives.[51]

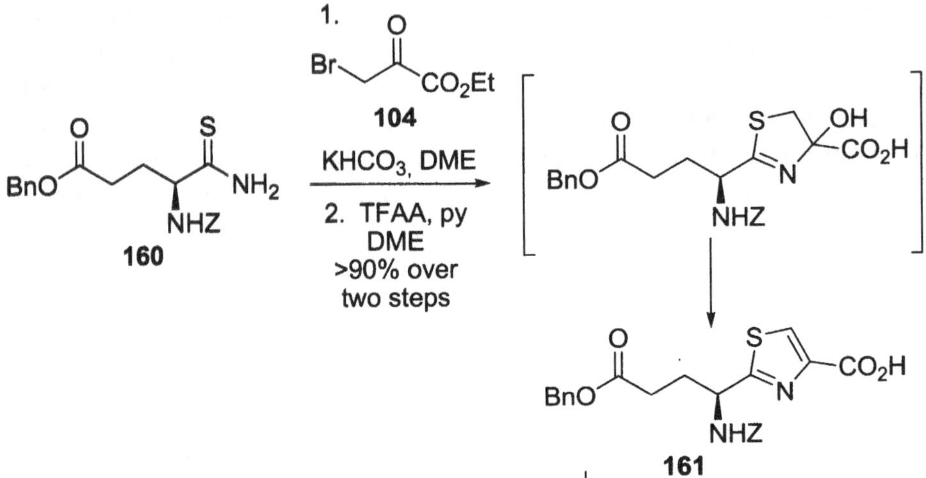

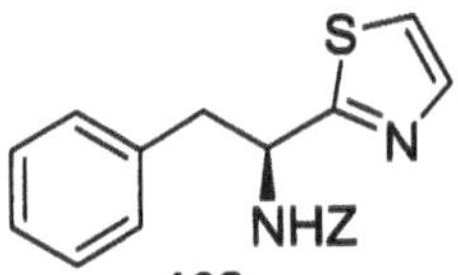

162

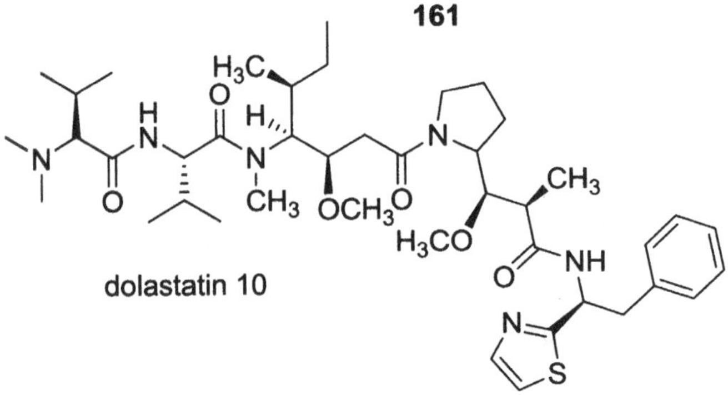

dolastatin 10

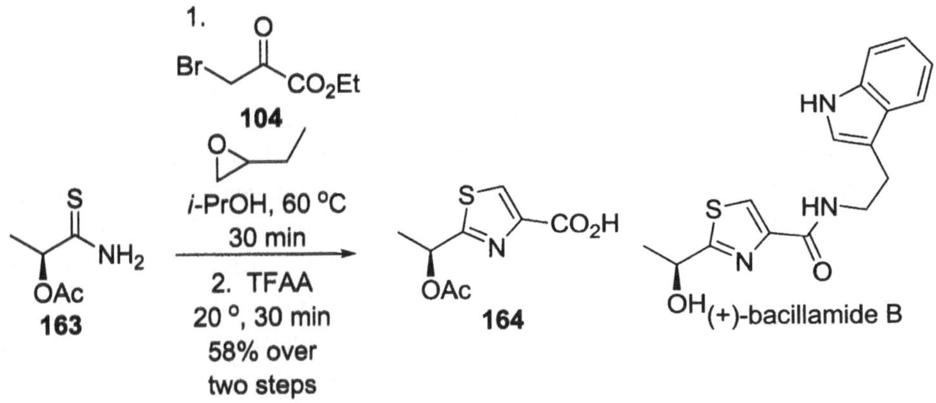

Bray and Olasoji employed the Hantzsch reaction in their total synthesis of (+) bacillamide B,[52] a compound with potential antibiotic activity. The authors used Schmidt's modification[53] of Holzapfel's method in order to avoid racemisation. Treatment of thioamide **163** with ethyl bromopyruvate **104** and ethyloxirane in isopropanol at 60 °C for 30 min, followed by careful addition of TFAA gave the chiral thiazole **164** in 58% yield. Bray and Olasoji's synthesis of bacillamide B follows on the heels of work by Xu and co-workers on the total synthesis of bacillamide C.[55]

The Hantzsch synthesis has been used to prepare a number of natural and synthetic cyclic peptides incorporating thiazoles. Pattenden and co-workers used the Hantzsch thiazole synthesis in their preparation of an *N*-Boc protected L-proline thiazole based amino acid, which was subsequently used to generate cyclic hexaeptides and octapeptides.[55] Thioamide **165**, prepared from commercially available Boc-L-proline, was subjected to Holzapfel's modified conditions[8] to generate the desired thiazole **166** in 61% yield. The biological evaluation of these compounds is currently under investigation.

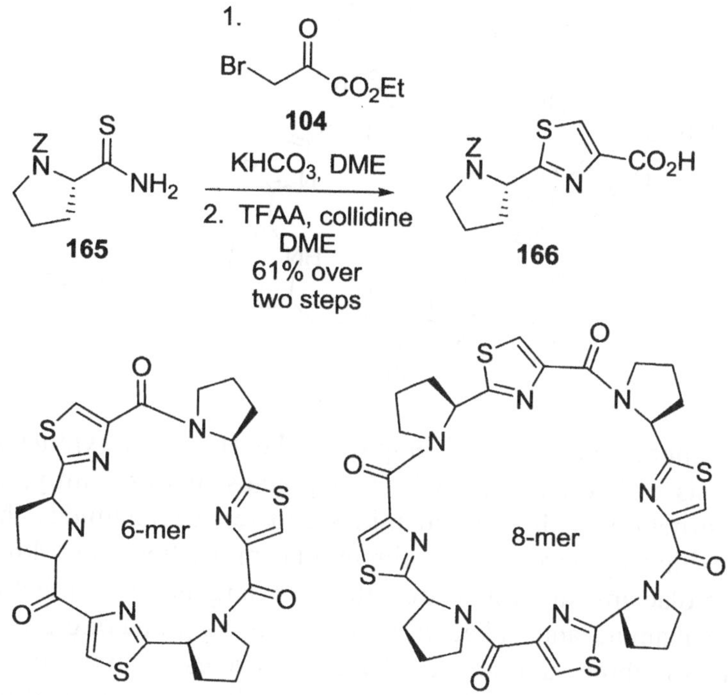

Moody and co-workers employed the Hantzsch thiazole synthesis on several occasions in their total synthesis of the thiopeptide antibiotic amythiamicin D.[56] Amythiamicin D functions by inhibiting the GTP-dependent elongation factors that are important for bacteria protein synthesis. Thiazole **167**, generated from *N*-Boc-valine in 65% on a 20-g scale using

Holzapfel's modified conditions (not shown),[8] was subject to a second Hantzsch synthesis using Holzapfel's conditions to produce thiazole **168** in 94% yield over two steps. The Hantzsch thiazole synthesis was also used in the same report in a later step in their synthesis of amythiamicin D.

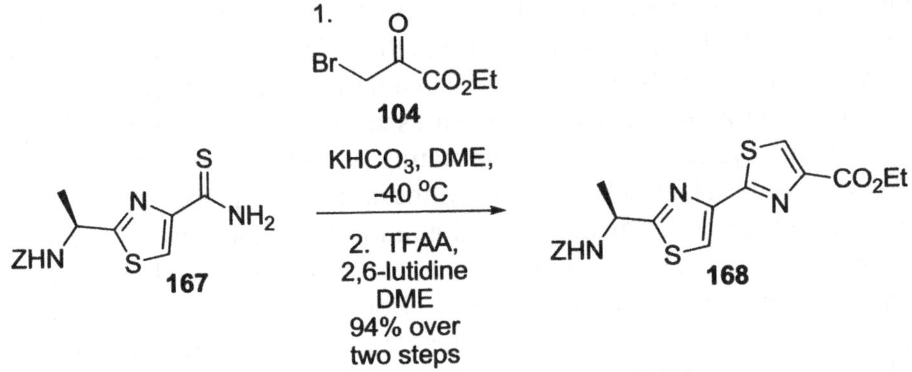

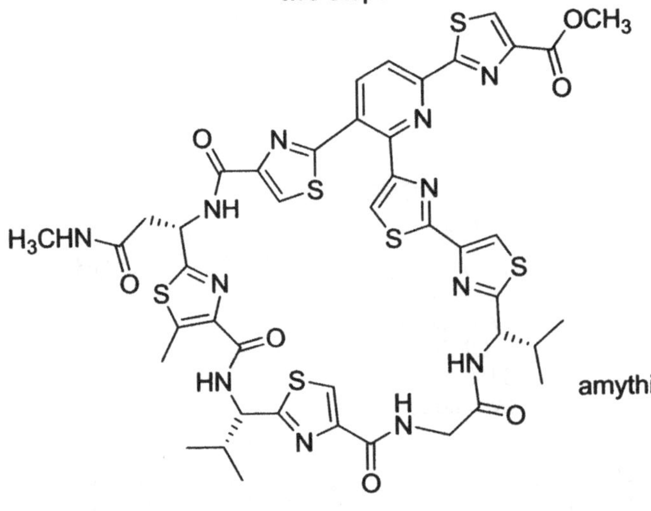

amythiamicin D

Nakamura and Takeuchi employed the Hantzsch synthesis in their total synthesis of bistratamide H.[57] Compounds of this family, which are similar in structure to the didmalamides and tenuecylclamides, have been shown to exhibit cytotoxic and antimicrobial properties. In addition, their unique macrolactam structures are attractive targets for synthetic organic chemists. Nakamura and Takeuchi used a fluorous protecting group in their preparation of three amino acid-based thiazoles commonly found as functional moieties in the bistratamide family of cyclic thiopeptides. Treatment of N^F-Teoc protected thioamides **169**, **171** and **173**, prepared in four steps from the commercially available amino acids, with ethyl bromopyruvate **104** using Holzapfel's modified conditions[8] gave the corresponding thiazoles **170**, **172**, and **173** in modest to high yields with excellent enantioselectivities. In addition, the fluorous-protecting group allowed for easy isolation of the desired products using fluorous liquid–liquid extraction. Compound **170** was then used in the total synthesis of bistratamide H. Meyers,[51] Katrizky,[58] and Smith[59] also employed the Hantzsch thiazole synthesis in the generation of amino acid-based thiazoles in their earlier work on the bistratamide family.

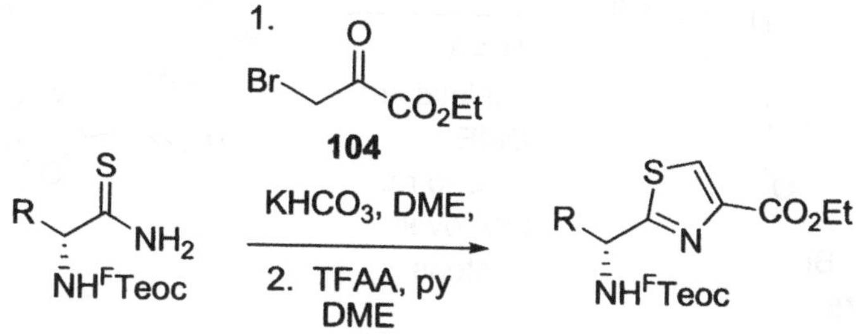

169, R = CH_3
171, R = *i*-Pr
173, R = CH_2Ph

170, R = CH_3 69% (92% *ee*)
172, R = *i*-Pr 80% (99% *ee*)
174, R = CH_2Ph 84% (95% *ee*)

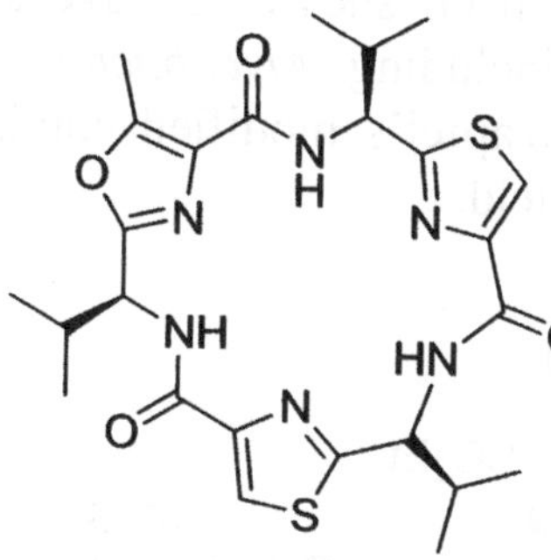

bistratamide H

Alvarez and co-workers employed the Hantzsch thiazole reaction as the final step of their total synthesis of IB-01211,[60] a cyclic peptide with cytotoxic properties effective against several tumor cell lines. In addition to the biological activity, the chemical architecture of this compound, which contains four oxazoles, one thiazole, and a tripeptide containing a didehydroamino acid residue, presents a significant challenge from a synthetic standpoint. Compound **175**, which contains both a thioamide and a masked α-bromoketone, was subjected to Holzapfel's modified Hantzsch synthesis[8] after unmasking of the acetal using refluxing formic acid to produce IB-01211 in 11% yield. Notably, in addition to macrocyclization via thiazole formation, the reaction conditions led to simultaneous elimination of the *t*-butoxy group to provide the exocyclic double bond. More recently, Nefzi and co-workers developed a two-step Hantzsch-based macrocylization approach for the synthesis of thiazole-containing cyclopeptides similar to IB-01211.[61] Their protocol was shown to produce cyclic peptides with yields ranging from 29–45% yield.

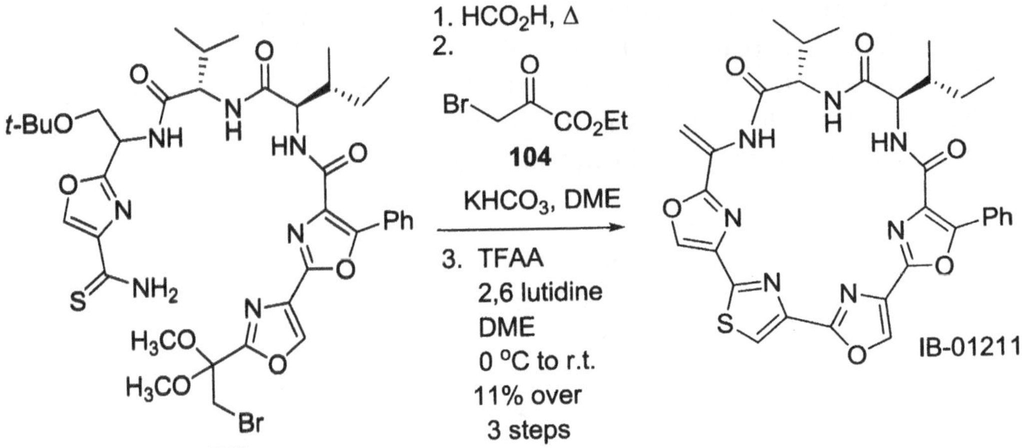

175

Ye and co-workers used the Hantzsch synthesis in their recent synthesis of the C1–C12 fragment common to the thuggacin family of natural products.[62] These compounds have been shown to have strong antibiotic activity against many organisms including *Mycobacterium tuberculosis*. Treatment of alkyne **176** using Holzapfel's modified conditions[8] gave the corresponding thiazole **177** in 68% yield.

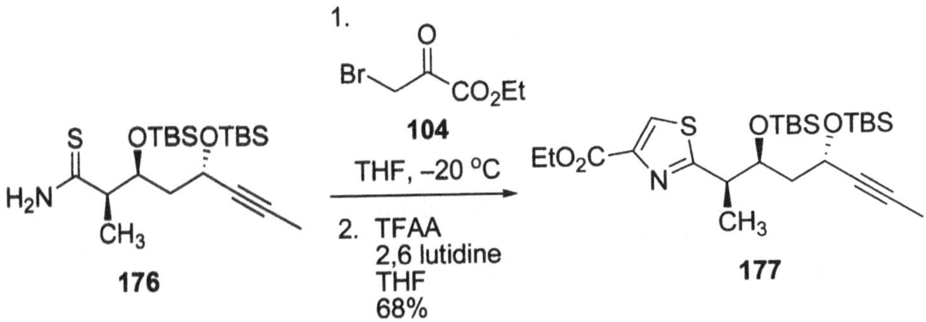

176 177

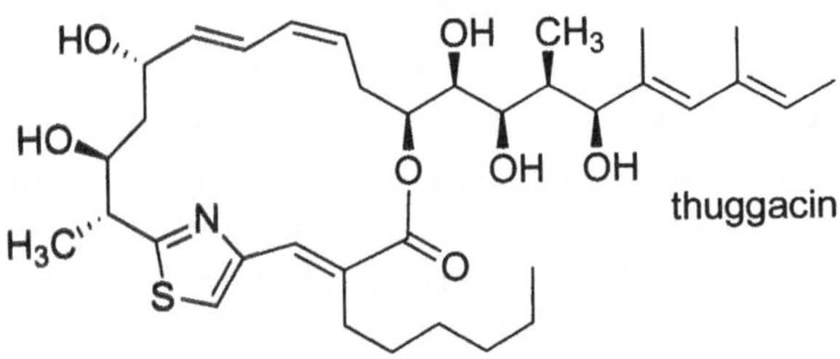

thuggacin A1

Dihydropyridine/Pyridine Synthesis

Hantzsch's original 1,4-dihydropyridine synthesis involved the one-pot reaction between 2 equiv of ethyl acetoacetate **8** and acetaldehyde **178** in refluxing aqueous ammonia to give the corresponding 1,4-dihydropyridine **179**.

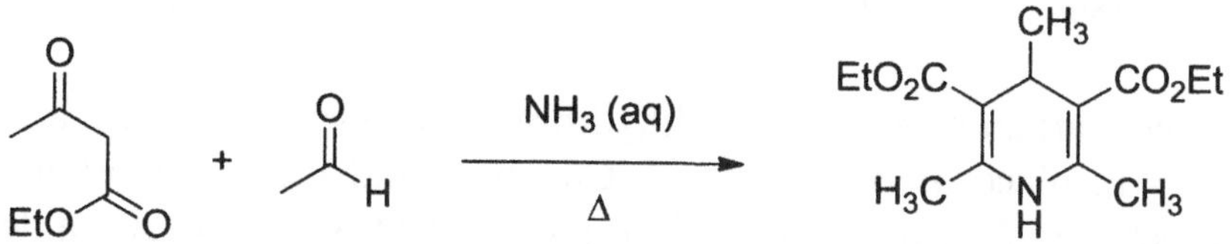

8 **178** **179**

Although the Hantzsch 1,4-dihydropyridine synthesis was originally reported in the late 1800s, interest in this reaction developed only in the past 50 years with Bossert and co-workers[4] synthesis of nifedipine, a potent calcium channel antagonist. The elucidation of the structure of NADH, which contains a dihydropyridine moiety, in the 1980s led to a second surge of research in this area.[5]

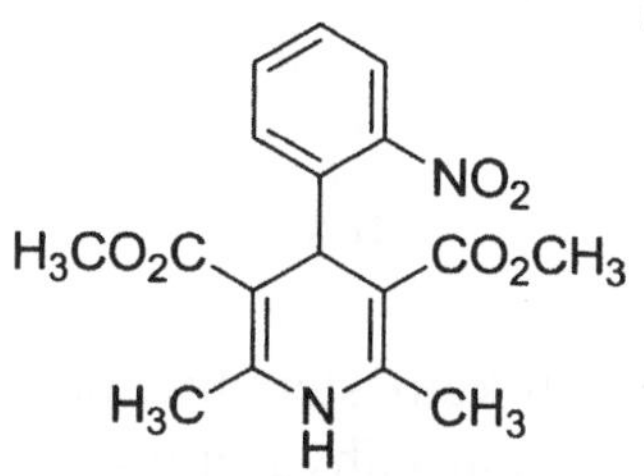

nifedipine

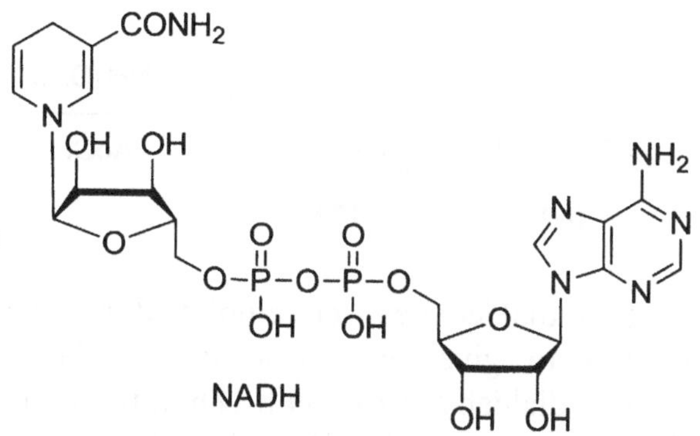

NADH

Bossert and co-workers original three-component Hantzsch synthesis of nifedipine used 1 equiv of 2-ntirobenzaldehyde **181**, 2 equiv of methyl acetoacetate **180**, and ammonia to give nifedipine in 72% yield.[4] Since their report, a number of variations on the Hantzsch 1,4-dihydropyridine synthesis have been reported. Many of the initial reports on the synthesis of 1,4-dihydropyridine derivatives were aimed at studying the structure activity relationships of these compounds with the goal of developing more potent and specific analogs of nifedipine.

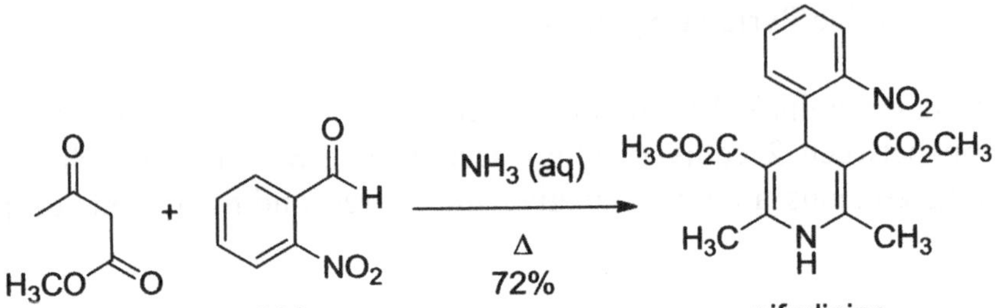

180 181

One variation of the 1,4-dihydropyridine synthesis uses acetylene derivatives in place of β-ketoesters. Chennant and Eisner reported on the reaction between 2 equiv of methyl propionate **182** with aromatic aldehydes such as **183** and ammonium acetate in acetic acid to produce a series of dihydropyridine derivatives **184** in good yield.[63] Unfortunately, differentially substituted acetylene derivatives and alkyl or nitrosubstituted aromatic aldehydes gave little to no yield of the desired 1,4-diydropyridine derivatives under the reported conditions. However, this method remains the best way to prepare of 2,6-unsubsitituted 1,4-dihydropyridines.

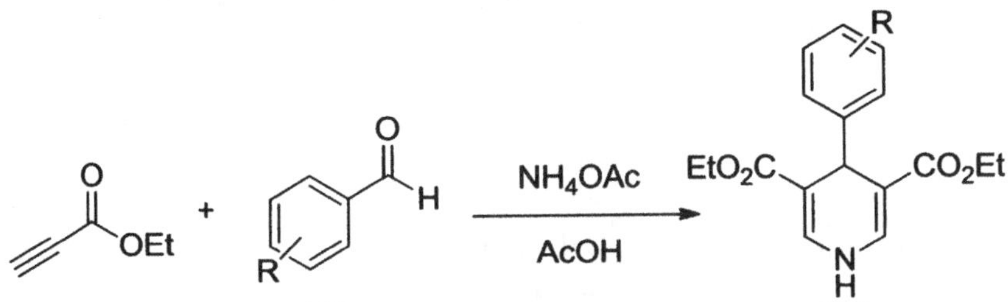

182 183 184

Spirodialdehydes were employed in place of β-ketoesters by Foos and co-workers to give rise to a unique class of spirodihydropyridines.[64] Reaction of dialdehyde **185** with **186** in refluxing benzene for 1.5 h, followed by the addition of pTsOH and additional refluxing for 1 h gave the corresponding spirodihydropyridine derivative **187** in near quantitative yield.

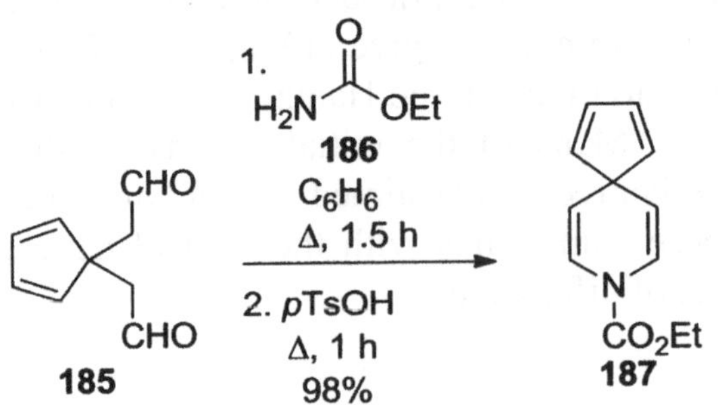

185 187

Meyer and co-workers were one of the first groups to use preformed enamines to synthesize 1,4-dihydropyridines in their synthesis of unsymmetrical fused nifedipine analogs.[65] Michael addition of alkylidene acetoacetic esters substituted with various aryl groups **188** and enaminocarbonyls **189** (where $n = 1$ or 2) in the presence of refluxing ethanol led to the corresponding fused systems **190** in good yield.

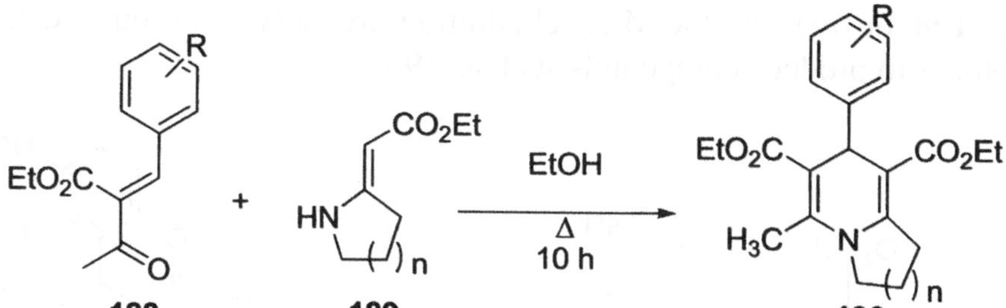

188 **189** **190**

Meyer and co-workers also used the active methylene compound 1,3-cyclohexanone **191** in their synthesis of highly substituted tetracyclic nifedipine derivatives. Reaction of **191** with 3-nitrobenzaldehyde **192** and enamines **189** gave the corresponding tricyclic systems **193** in modest yields ($n = 1$, 50% yield; $n = 3$, 46% yield). Compound **195** was prepared in a similar manner from enamine **194** in 59% yield.

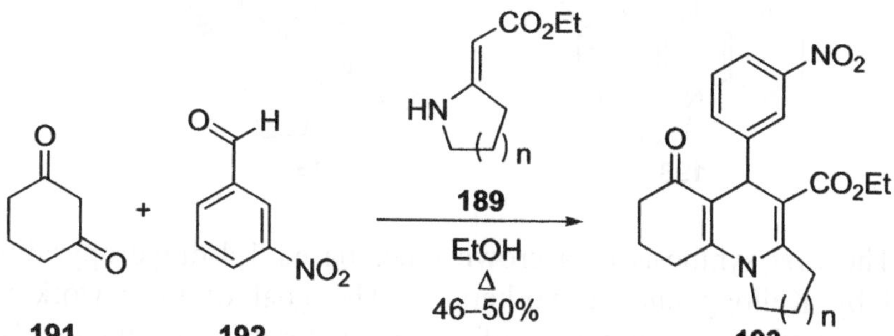

191 **192** **193**

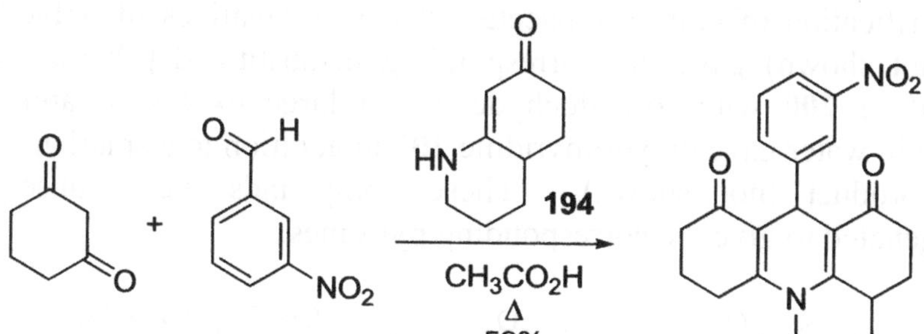

191 **192** **195**

In a later report, Meyers used ethyl acetoacetate **8**, aromatic aldehydes such as 3-nitrobenzaldehyde **192**, and 2,2′-methylenediimidazoline hydrochloride **196** to produce the corresponding 1,4-dihydropyridine **197** as a more complex derivative of nifedipine in 65% yield.[66] A series of aromatic aldehydes were employed to generate additional 1,4-dihydropyridines of this nature in 14–74% yield (not shown). Meyers and co-workers also prepared several novel analogs by varying the ketone source to produce compounds such as **198**, or by using the Michael addition products of various aldehydes and ketones to produce compounds such as **199**.

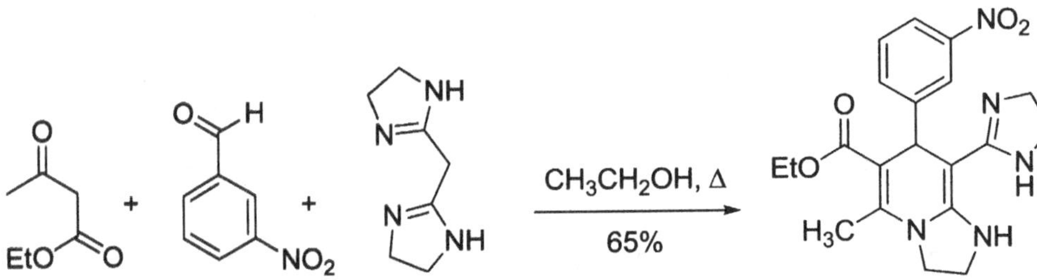

8 **192** **196** **197**

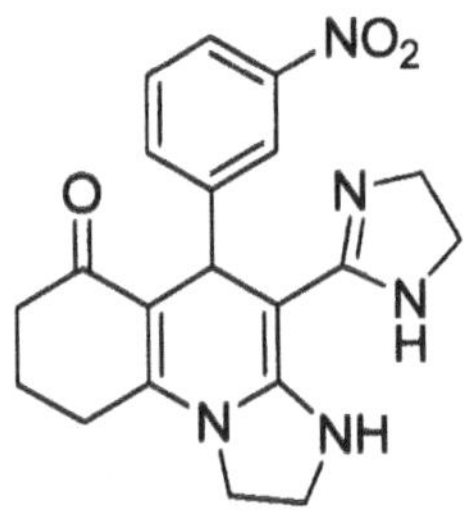

198

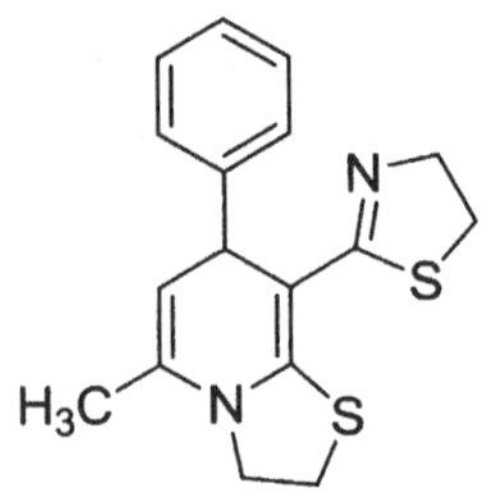

199

The first synthesis of a crown ether based 1,4-dihydropyridine was reported by Kellogg and co-workers.[67] The goal of their work was to generate synthetic versions of the natural coenzyme NADH. Transesterification of ethyl acetoacetate 8 with derivatives of polyethylene glycol (not shown) gave the corresponding disubstituted β-ketoester **200**. Treatment of **200** with formaldehyde and a large excess of ammonium carbonate in water gave dihydropyridine **201** in addition to a small amount of dimeric product (not shown). These compounds were subsequently dehydrogenated to give the corresponding pyridines.

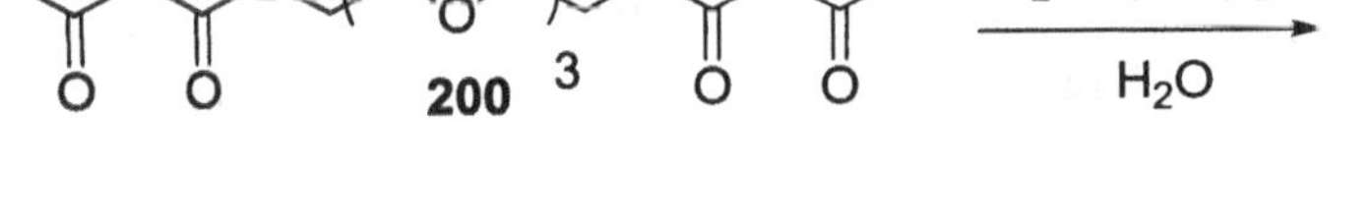

200

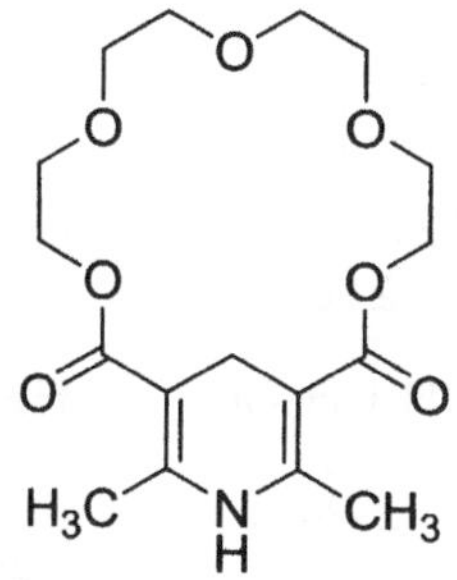

201

Natale and Quincy used a Hantzsch approach to prepare 4-(4′-isoxazyl)-1,4-dihydropyridines to study their potential as calcium channel antagonists.[68] Treatment of aldehydes **202**, **204** and **206**, prepared by oxidation of the readily available alcohol, with two equivalents of ethyl acetoacetate **8** in ammonium hydroxide and refluxing ethanol overnight gave the corresponding symmetrical dihydropyridine derivatives **203**, **205**, and **207** in modest yields. Notably, the more activated, electron withdrawing aldehyde performed better under the reported reaction conditions.

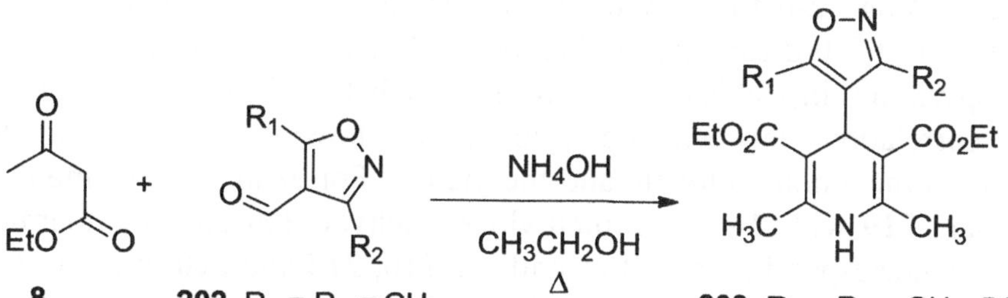

8

202, $R_1 = R_2 = CH_3$
204, R_1 = Et, $R_2 = CH_3$
206, $R_1 = CH_3$, R_2 = Ph

203, $R_1 = R_2 = CH_3$ 51%
205, R_1 = Et, $R_2 = CH_3$ 50%
207, $R_1 = CH_3$, R_2 = Ph 78%

Taylor and co-workers used the Hantzsch synthesis to prepare 2-(2′-aryl-2-oxoethylidene)-1,2,3,4-tetrahdyropyrimides and related compounds as derivatives of the novel calcium channel blocker Bay K 8644.[69] Treatment of β-ketoesters **208** and **212** with 2-trifluoromethylbenzaldehyde **209** and β-aminocrotonate **210** in refluxing ethanol for 12–24 h gave the corresponding dihydropyridine derivatives **211** and **213** in low yield. Several additional β-ketoesters, aromatic aldehydes and aminocrotonates were explored with yields ranging from 3–24% overall. It is interesting that only the exocyclic isomers of the dihydropyridine derivatives were isolated in the course of these reactions except in the cases where cyano substituted aminocrotonates were employed (not shown), and where mixtures of products were obtained. The authors went on to show that the yields of many of these reactions could be improved up to four-fold using a three-pot synthesis employing isoxazole intermediates.

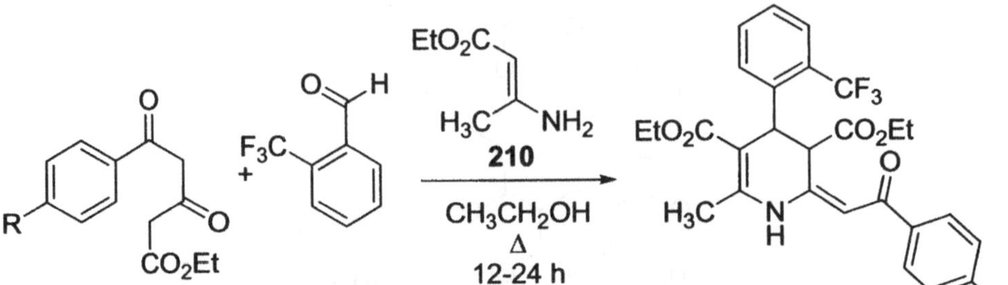

208, R = H
212, R = N(CH3)2
209
211, R = H 8%
213, R = N(CH3)2 8%

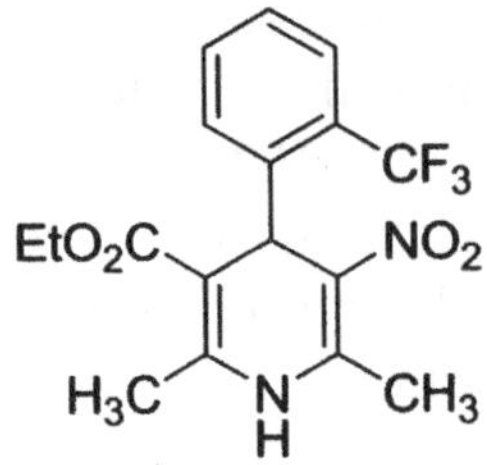

Bay K 8644

Iqbal, Trigger, and Knaus prepared a series of 1,4-dihydropyridine derivatives using 2-trifluoromethylbenzadehyde **209**, β-aminocrotonate derivatives, and 1,3-β-dicarbonyl and carbonyl derivatives to produce a series of dihydropyridines in yields ranging from 11 to 56%.[70] The authors noticed a direct correlation between the electron-withdrawing capacity of the substituent on the aminocrotonate and the yields. For example, treatment of **214**, **217** and **219** with 2-trifluoromethylbenzadehyde **209** and β-ketoamide **215** gave the corresponding dihydropyridines **216**, **218** and **220** in modest to low yields with the lowest yields obtained when the cyano and nitro substituted aminocrotonates were employed. This trend held for a number of derivatives prepared by Iqbal and co-workers.

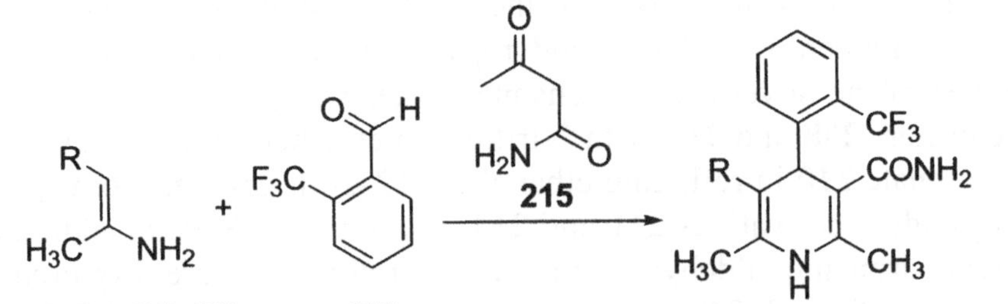

214, R = CO_2CH_3
217, R = CN
219, R = NO_2
209
216, R = CO_2CH_3 56%
218, R = CN 37%
220, R = NO_2 47%

Hermecz and co-workers used the Hantzsch 1,4-dihydropyridine synthesis in their preparation of naphthyridines.[71] Reaction of ethyl acetoacetate **8** and hexamethylenetetramine **221** as a novel source of nitrogen in acetic acid at 100 °C gave the desired pyridinedicarboxylate **222** (via the dihydropyridine) in 36.5% yield in addition to 2.3% of a product obtained in situ from the 4 + 2 cycloaddition between a heterodiene intermediate formed in the course of the reaction and the 1,4-dihydropyridine product (not shown).

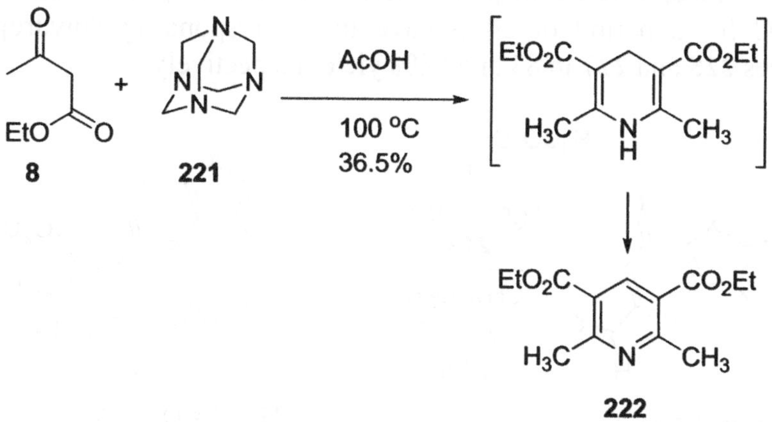

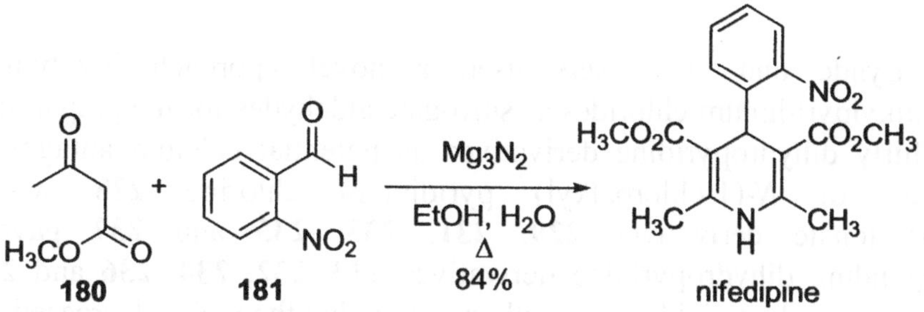

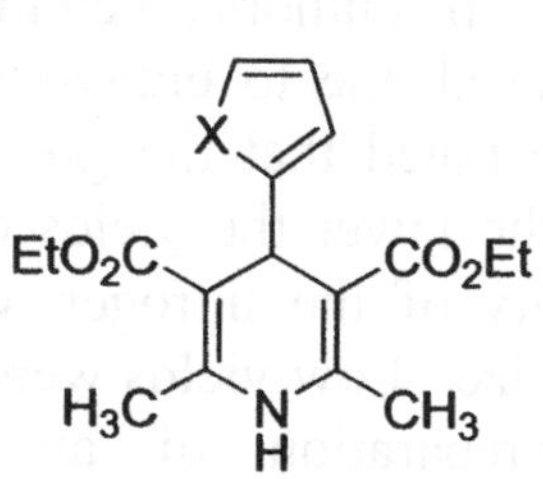

223, X = S or O

Ley and co-workers recently used magnesium nitride as a source of ammonia for the Hantzsch synthesis of 1,4-dihydropyridines in high yield.[72] Optimized conditions involved the reaction of 4.5 equiv ethyl acetoacetate **180** and 2 equiv of 2-nitrobenzaldehyde **181** with 1 equiv of magnesium nitride (0.9 M NH_3 concentration) in a refluxing solution of ethanol and water (6.5 equiv) for 16 h. The corresponding 1,4-dihydropyridine, in this case nifedipine, was produced in 84%. A number of other 1,4-

dihydropyridines such as thiophene and furan derivatives **223** were also produced in high yield using this methodology.

Görlitzer and Vogt employed the Hantzsch 1,4-dihydropyridine synthesis in their preparation of novel thiopyrone derivatives for use as potential leukotriene antagonists for the treatment of asthma.[73] Reaction of aldehydes **224** and **226** with β-aminocrotonate **214** in glacial acetic acid at 50–60 °C for a period of 24 h gave the corresponding dihydropyridine derivatives **225** and **227** in 83 and 86% yield, respectively.

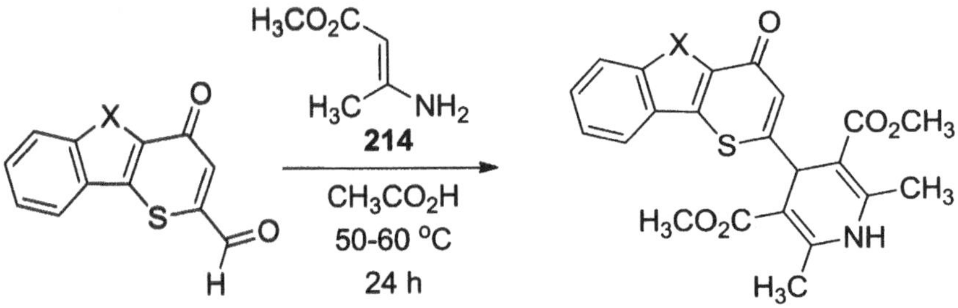

224, X = O
226, X = S

225, X = O 83%
227, X = S 86%

Eynde and co-workers used a novel approach involving *N*-chloroalkylpyridinium chlorides as surrogate aldehydes for the preparation of over thirty dihydropyridine derivatives as potential calcium antagonists.[74] Reaction of *N*-(1-chloroalkyl) pyridinium chloride **228** with β-aminocrotonate derivatives **229**, **231**, **233**, **235** and **237** gave the corresponding dihydropyridine derivatives **230**, **232**, **234**, **236** and **238** in modest to excellent yields. The authors noted that the yields decreased as the size of the R group increased due to unfavorable entropic affects for ring closure. The authors also noted that the greater the electron-withdrawing capacity of the R group, the lower the yields of the reaction. In the latter case, the decreased basicity of the nitrogen was shown to have a direct impact on its ability to cyclize. Low yields were also reported by Rimoli and co-workers[75] in their preparation of an *N*-benzyl substituted 1,4-dihydropyridine derivative from ethyl acetoacetate, benzaldehyde, and benzyl amine (not shown) supporting the work by Enyde and co-workers.

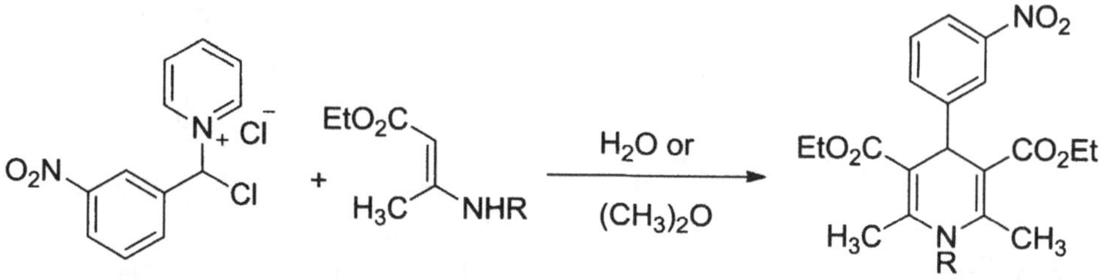

228

229, R = CH_3
231, R = CH_2Ph
233, R = *p*-OCH_3-C_6H_4
235, R = Ph
237, R = *p*-Cl-C_6H_4

230, R = CH_3 95%
232, R = CH_2Ph 50%
234, R = *p*-OCH_3-C_6H_4 45%
236, R = Ph 35%
238, R = *p*-Cl-C_6H_4 30%

Suh and co-workers prepared a pyrimidyl-1,4-dihydropyridine derivative as a novel derivative of nifedipine.[76] Reaction of methyl acetoacetate **180**, 5-formyluracil **239** and β-aminocrotonate **214** in refluxing isopropanol for 24 h furnished the corresponding 1,4-dihydropyridine derivative **240** in 52.5% yield. The ethoxyester derivative was also prepared in 49% yield, as were mixed esters in similar yields (not shown). Knaus and co-workers also prepared a symmetric pyrimidyl-1,4-dihydropyridine derivative using a traditional Hantzsch approach in higher yields (68% versus 52.5% for the methyl ester derivative), as well as asymmetric pyrimidyl-1,4-dihydropyridine derivatives bearing nitro groups on the 1,4-dihydropyridine ring, with the later prepared in only 10–18% yield (not shown).[77]

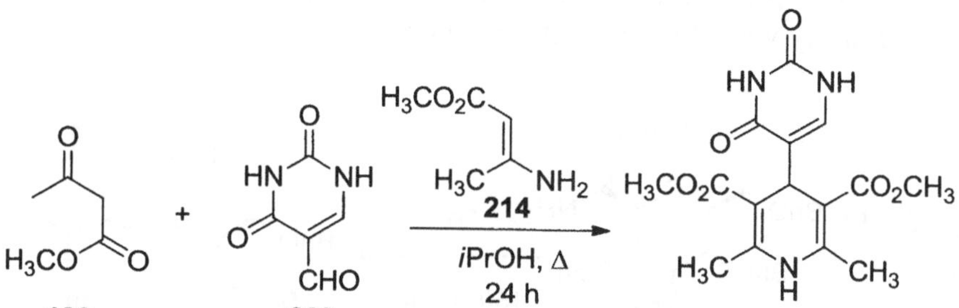

180 **239** **240**

Dondoni and co-workers used a three-component Hantzsch synthesis to prepare several symmetrical carbohydrate-based 1,4-dihydropyridine derivatives.[78] Their goal was to evaluate the corresponding *C*-glycosides as nucleoside mimetics of monastrol, an inhibitor of microtubule-associated protein Eg5. Reaction of ethyl acetoacetate **8**, galactosyl aldehyde **241** and β-aminocrotonate **214** in refluxing ethanol gave **242** in 88%. Ribose and mannose based 1,4-dihydopyridine glycoconjugates were also prepared in 71% and 70% yields, respectively (not shown).

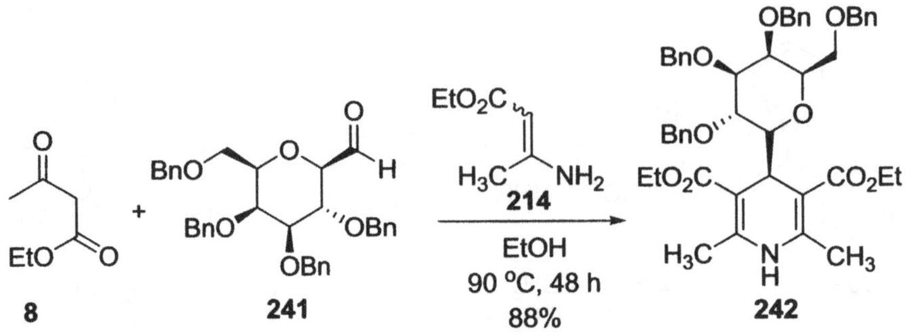

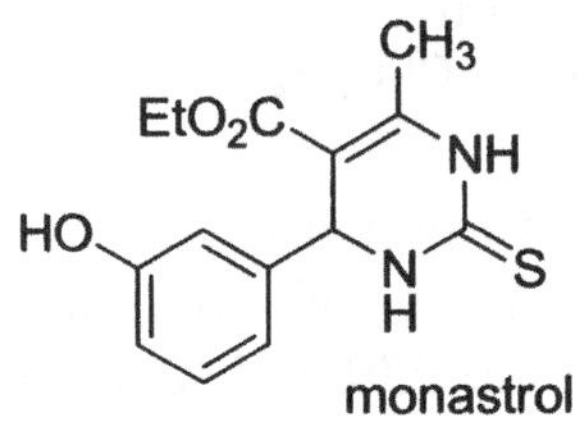

Dondoni and co-workers also prepared unsymmetrical carbohydrate-based 1,4-dihydropyridines using a two component Hantzsch synthesis. Treatment of unsaturated ketone **243** with β-aminocrotonate derivative **244** in DMF at 100 °C for 48 h gave the corresponding 1,4-dihydropyridine derivative **245** in 61% with only 30% diastereoselectivity. Ribose and mannose analogs were also prepared in 71% and 66% yields, respectively (not shown). A similar diastereoselectivity of 20% was observed when the ribose based β-aminocrotonate was employed. However, no enantioselectivity was obtained with mannose.

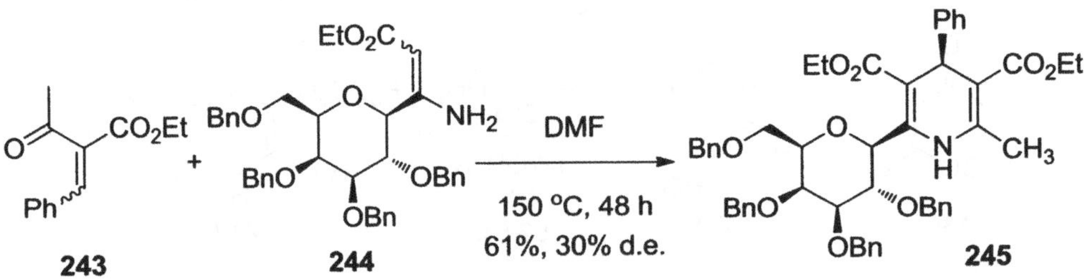

Enders, Müller and Demir showed that metallated chiral alkyl acetoacetate hydrazones could be used to generate select 1,4-dihydropyridine derivatives in modest yields with high enantioselectivities.[11a] For example, treatment of SADP-hydrazone **246**, prepared from the condensation of alkyl acetoacetates and *S*-(–)-1-amino-2-(dimethylmethoxymethyl)pyrrolidine (SADP), with *n*-butyllithium in the presence of acceptors such as **247** gave the corresponding 1,4-dihydropyridine **248** in 64% yield (96% enantioselectivity) after two-stage treatment with ammonium chloride.

Several additional compounds were prepared in similar yields with similar enantioselectivities in the same report.

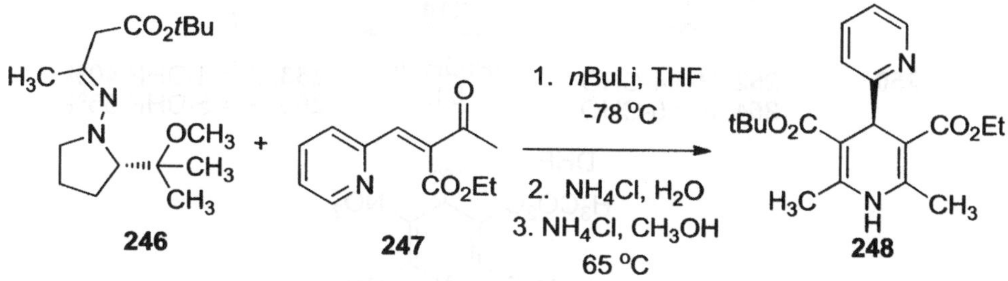

246 247 248

Shan and Knaus also used a chiral auxiliary in their modified Hantzsch approach to the synthesis of two new 1,4-dihydropyridine derivatives.[116] Unfortunately, reaction of 3-aminocrotonate ester **249** with 2-trifluoromethylbenzadehyde **209** and 3-nitropropionate **250** gave **251** as a mixture of diastereomers in low yield (22%) with nearly equal mixtures of the *R* and desired *S* enantiomers (*R*,*S*,*R*(10%) and *S*,*S*,*R*(12%)).

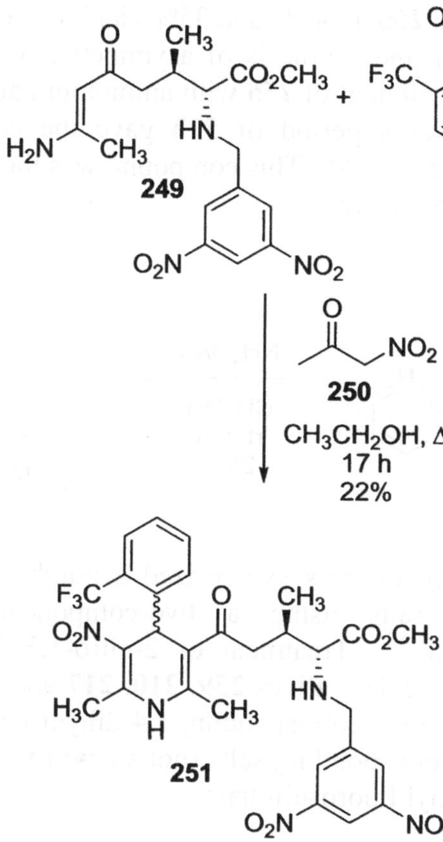

249 209 250 251

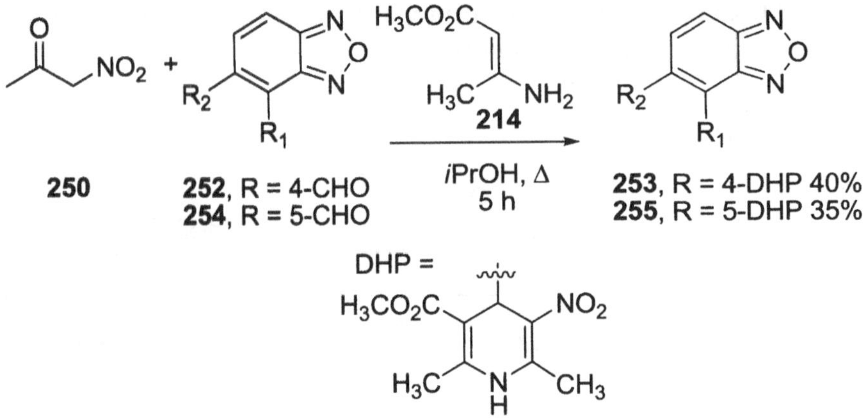

Building on their previous synthesis of benzofurazanyl-1,4-dihydropyridine based calcium antagonists using the Hantzsch synthesis,[79] Gasco and co-workers used a modified approach to prepare several nitrobenzofurazanyl-1,4-dihydropyridine derivatives.[80] Treatment of either 4-benzofurazanaldehyde **252** or 5-benzofurazanaldehyde **254** with 3-nitropropionate **250** and β-aminocrotonate **214** gave the corresponding 1,4-dihydropyridines **253** and **255** in 40% and 35% yield, respectively.

Beley reported on the synthesis of asymmetric terpyridines using a Hantzsch approach.[81] Treatment of **256** with ammonium acetate in methanol at room temperature over a period of 5 h gave the corresponding 1,4-dihydropyridine **257** in 25% yield. This compound was then oxidized to give the desired terpyridine derivative.

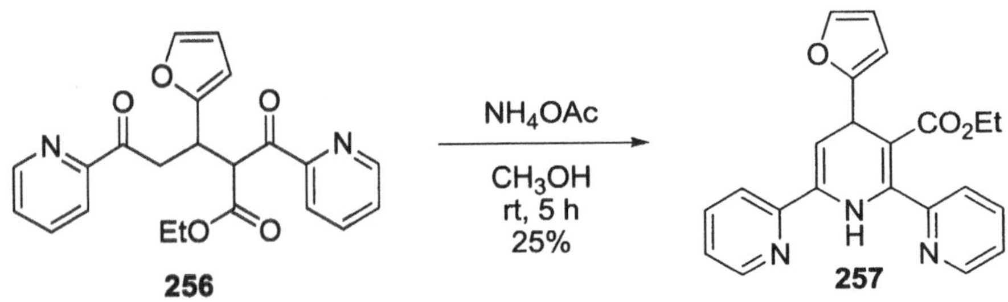

Sagitullina and co-workers synthesized a number of nitropyridines and their quaternary salts using a two-component Hantzsch 1,4-dihydropyridine synthesis.[82] Treatment of 2-nitro-1,3-diphenylpropenone **258** with β-aminocrotonate derivatives **259**, **210**, **217** and **263** in acetic acid at room temperature gave the corresponding 1,4-dihydropyridine derivatives in modest yields. The corresponding salts (not shown) were produced using dimethylsulfate and methyl fluorosulfonate.

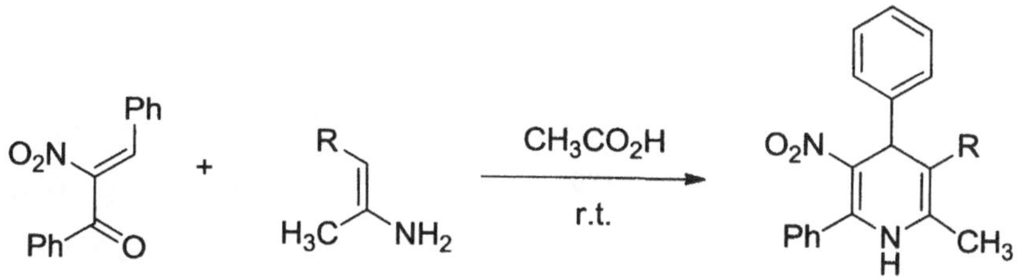

258 **259**, R = $COCH_3$
210, R = $CO_2CH_2CH_4$
217, R = CN
263, R = COPh

260, R = $COCH_3$ 84%
261, R = $CO_2CH_2CH_4$ 74%
262, R = CN 77%
264, R = COPh 64%

Bhandari and co-workers used a solid-phase approach to synthesize pyrrolo[3,4-*b*]pyridines via 1,4-dihydropyridines.[83] These bioactive chemotypes have been shown to exhibit activities at peripheral and central benzodiazepine receptors, as well as calcium-dependent potassium channels. AgaroGel or Tentagel linked compounds such as **265**, when treated with β-aminocrotonates **266** in DMF and trimethylorthoformate at 80 °C for 12 h gave the corresponding dihydropyridines **267**, which were readily converted to the corresponding pyrrolo[3,4-*b*]pyridines **268** in high yields in just three additional steps including cleavage from the resin.

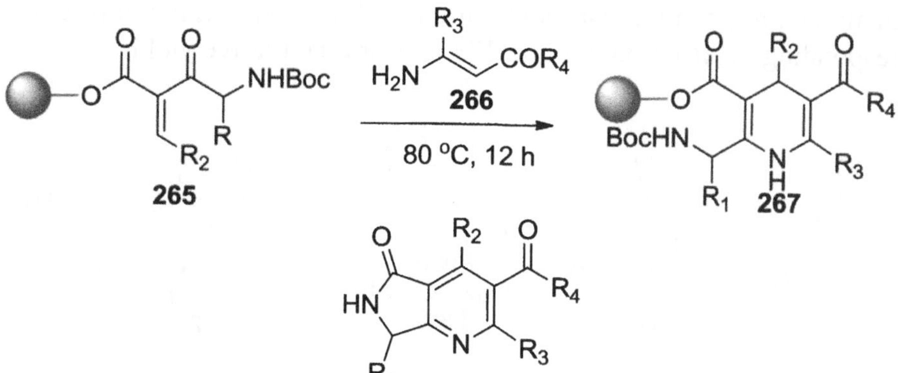

265 **267** **268**

Bazureau and co-workers employed a microwave-assisted, ionic liquid phase protocol to generate 1,4-dihydropyridines.[84] Reaction of ionic-liquid supported (IL) aldehyde **269**, methyl acetoacetate **180**, and either dimedone **270** in sodium acetate or β-aminocrotonate **214** with a 150W microwave at 50% power and 120 °C gave the corresponding 1,4-dihydropyridines **271** and **272**, respectively, in near quantitative yields. Notably, the authors were able to oxidize the IL-bound 1,4-dihydropyridines to the corresponding pyridine derivatives.

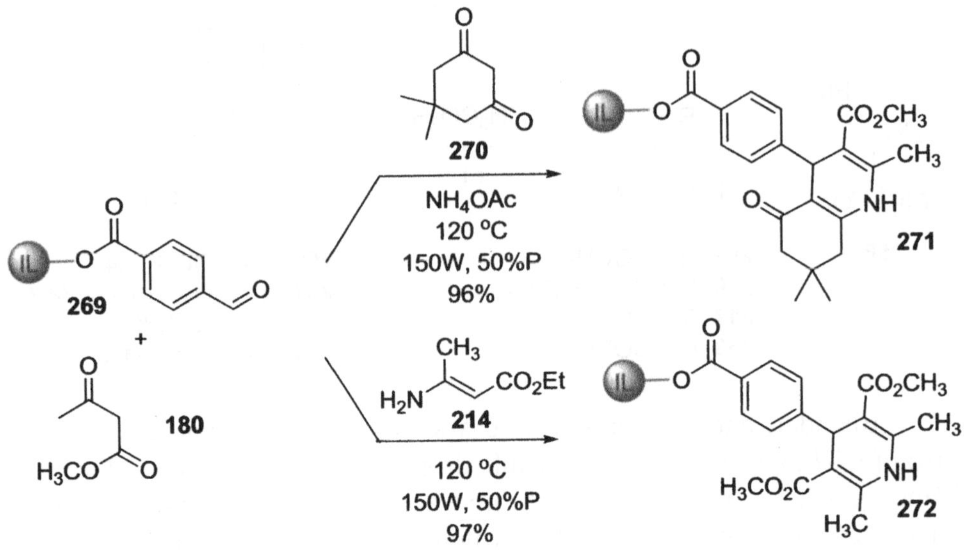

Su, Li and Li used 5-pyrrolidin-2yl-tetrazole as a promoter in their one-pot Hantzsch synthesis of polyhydroquinoline **273**.[85] These compounds are found as structural components in a number of natural products, and therefore are of significant interest to synthetic organic chemists. Reaction of ethyl acetoacetate **8** and dimedone **270** in the presence of 5-pyrrolidin-2yl-tetrazole and ammonium carbonate in ethanol at room temperature gave the corresponding polyhydroquinoline **273** in near quantitative yield.

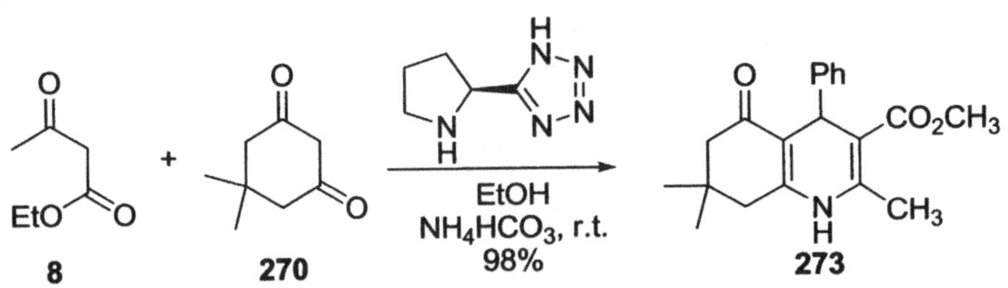

Štetinová and co-workers used the Hantzsch synthesis to prepare 1,2,4-trisubstituted pyrroles as part of their ongoing efforts to produce novel pyrrole derivatives with antihypertensive activity.[86] Reaction of aldehyde **274** with methyl acetoacetate **180** and glycine methyl ester hydrochloride **275** in sodium methoxide in methanol gave the corresponding 1,4-dihydropyridine **276** in 43% yield.

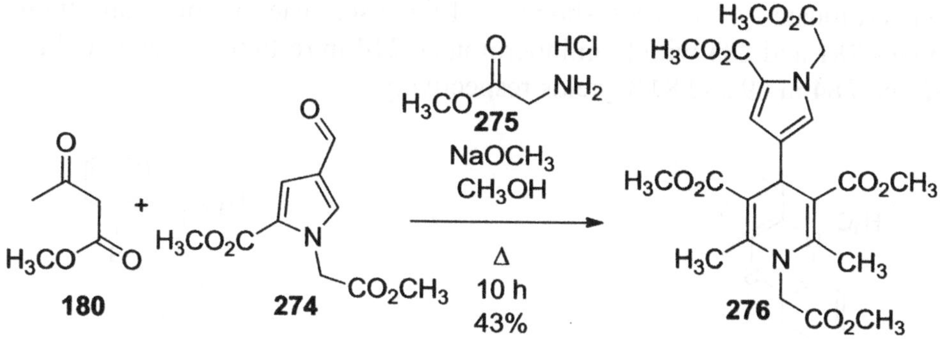

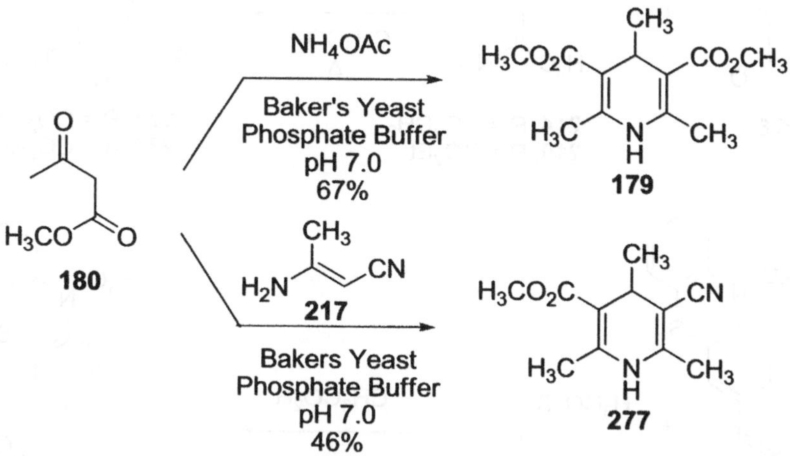

Lee used baker's yeast in his enzymatic approach to the synthesis of 1,4-dihydropyridines.[87] The traditional Hantzsch synthesis was explored whereby 2 equiv of methyl acetoacetate **180** and ammonium acetate were treated with baker's yeast in a phosphate buffer to produce the corresponding dihydropyridine **179** in 67% yield. A modified Hantzsch was also employed in the synthesis of asymmetrical derivative **277**. Treatment of 1 equiv of methyl acetoacetate **180** with 1 equiv of crotonitrile **217** gave the corresponding dihydropyridine derivative **277** in 46% yields. The traditional and modified approaches were also conducted using ethyl acetoacetate with similar yields for the reported reaction products.

Vigante and co-workers used the Hantzsch synthesis to prepare a series of novel symmetrical and unsymmetrical 1,4-dihydropyridine analogs to evaluate their antioxidant ability and therapeutic potential for the treatment of cancer.[88] Reaction of β-unsaturated ketone **278** with β-aminocrotonate derivatives **214** and **210** in refluxing acetic acid gave the corresponding 1,4-dihydropyridine derivatives **279** and **280** in 70% and 65% yield, respectively. This method proved more effective than traditional Hantzsch syntheses involving phenothiazine and either 2 equiv methyl acetoacetate (37%) or β-aminocrotonate (45%) (not shown). Likewise, reaction of β-unsaturated ketones **281** and **283** with β-aminocrotonate **214** in refluxing acetic acid gave **282** and **284** in 69 and 84% yields respectively.

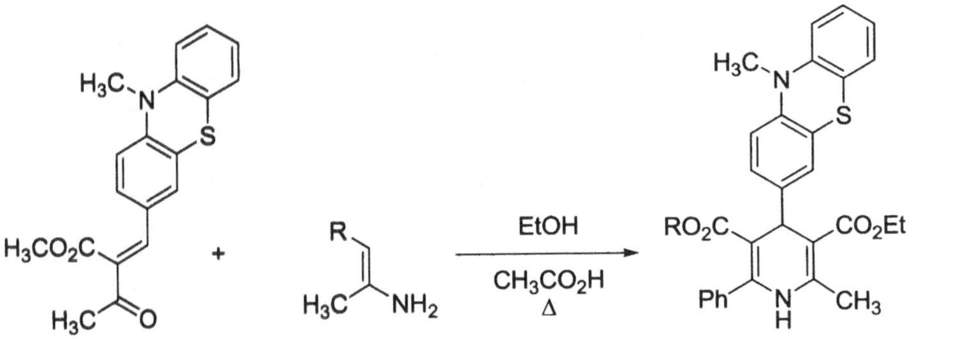

278

214, R = CO_2CH_3
210, R = CO_2Et

279, R = CO_2CH_3 70%
280, R = CO_2Et 65%

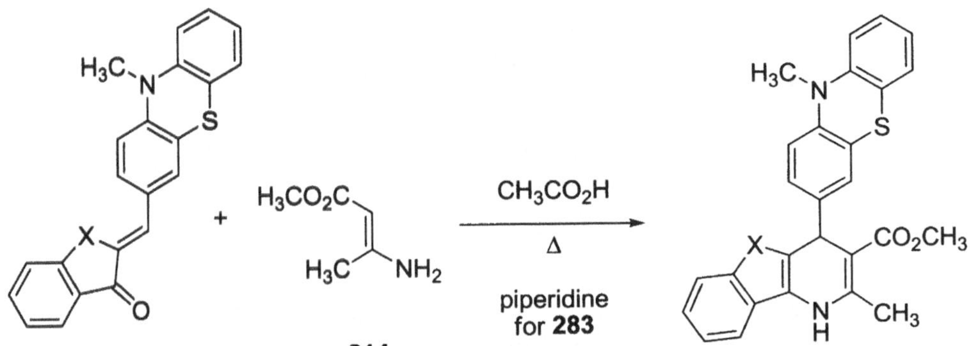

281, X = CO
283, X = SO_2

214

282, X = CO 69%
284, X = SO_2 84%

Kumar and co-workers synthesized a series of novel Hantzsch dihydropyridines to evaluate their potential as treatments for tuberculosis and cancer.[89] Reaction of 2 equiv of *p*-chloroacetoacetanline **285** with benzaldehyde **286** and aqueous ammonia in refluxing ethanol for 12 h gave the corresponding 1,4-dihydropyridine **287** in 30% yield. Several additional derivatives bearing various functional groups on the phenyl ring were prepared in yields ranging from 29–36% yield. Many compounds showed preliminary activity against Vero cells.

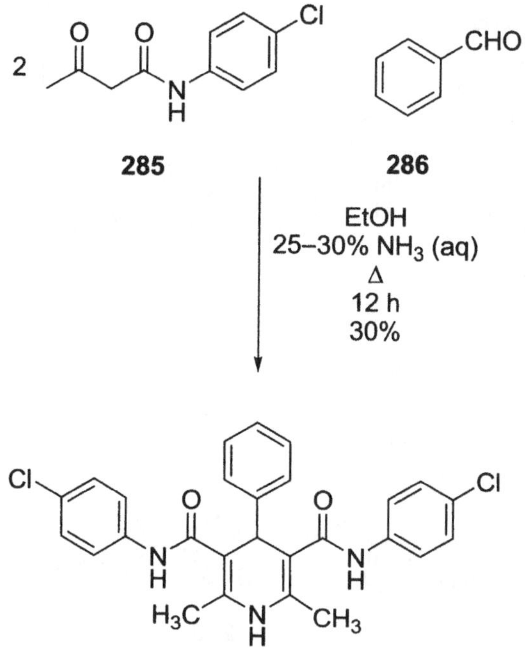

285 286

287

Lefranc and Ciufolini employed the Hantzsch 1,4-dihydropyridine synthesis in their total synthesis of micrococcin P1.[90] Micrococcin P1, a potent antibiotic against the malaria parasite *Plasmodium falicparum*, functions by disrupting protein biosynthesis by binding to ribosomes. Reaction of diketone **288** with ammonium acetate in ethanol gave the corresponding 1,4-dihydropyridine derivative, which was subsequently treated with DDQ in toluene to give pyridine **289** in near quantitative yield.

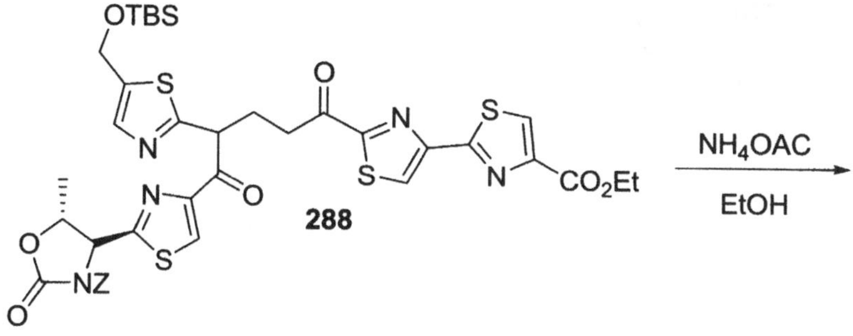

288

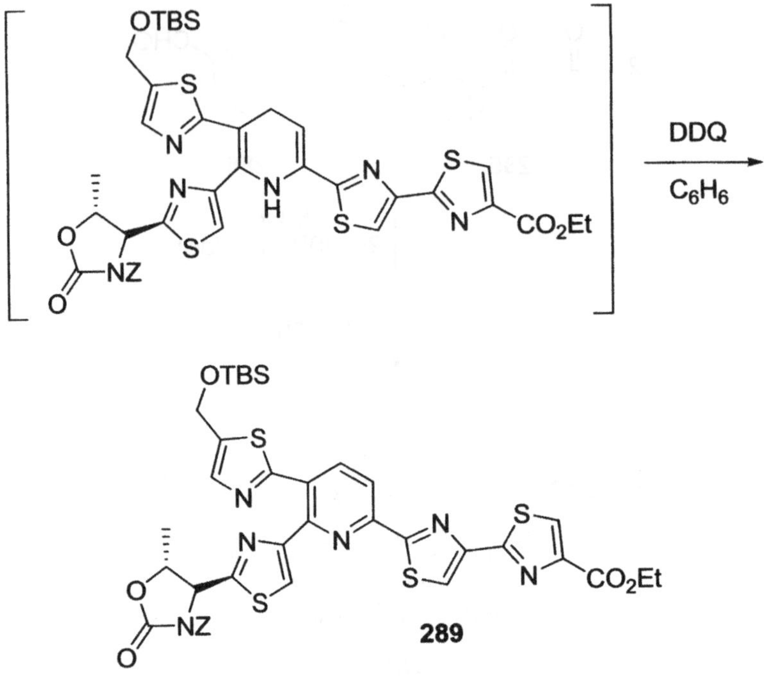

289

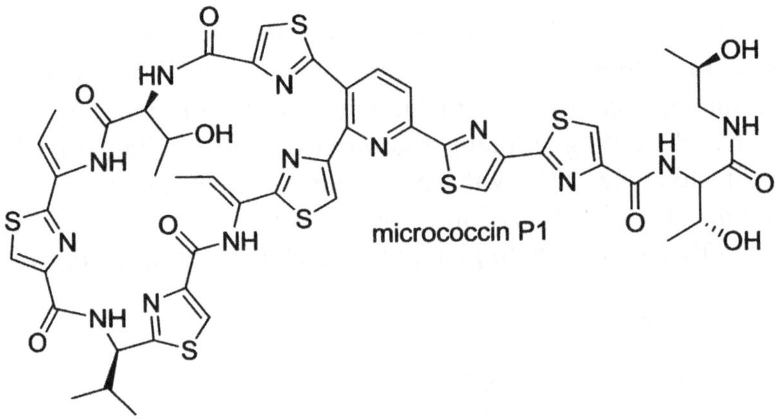

micrococcin P1

9.5.6 Experimental

The experimental examples provided below are given for the Hantzsch pyrrole, thiazole, and 1,4-dihydropyridine synthesis. The traditional method for preparing each type of heterocycle is presented first, followed by a modified protocol.

Pyrrole (General): ***Ethyl 2,5-dimethyl-1H-pyrrole-3-carboxylate***[2]

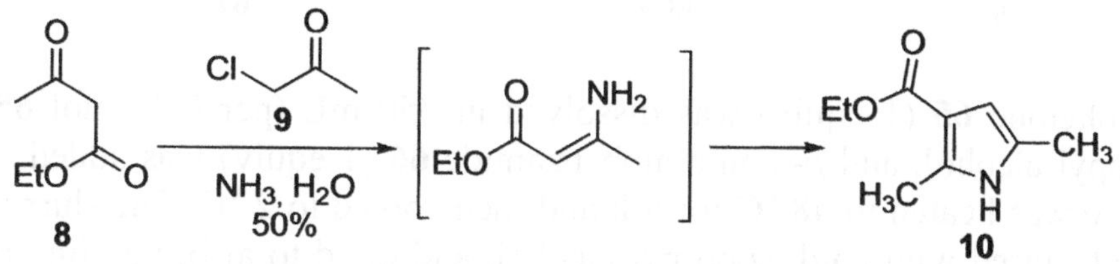

Aqueous ammonia (50 mL of 28 % NH_3 (aq) per 0.1 mol, 50 mL of H_2O per 0.1 mol) was added to the ethyl acetoacetate **8** (1 equiv) and chloroacetone **9** (1 equiv) and the mixture was stirred at 60 °C for 2 h and then left at room temperture overnight. The product of the reaction was extracted into ether and the extract was washed with 10% NaOH, water, 5% HCl, and again with water. The ether layer was dried (Na_2SO_4) and evaporated to give **9**. Recrystallization of **10** from ether-pentane gave the corresponding pyrrole in 50% yield.

Pyrrole (Modified): ***2-Phenyl-5-(trifluoromethyl)-1H-pyrrole-3-carbonitrile***[21]

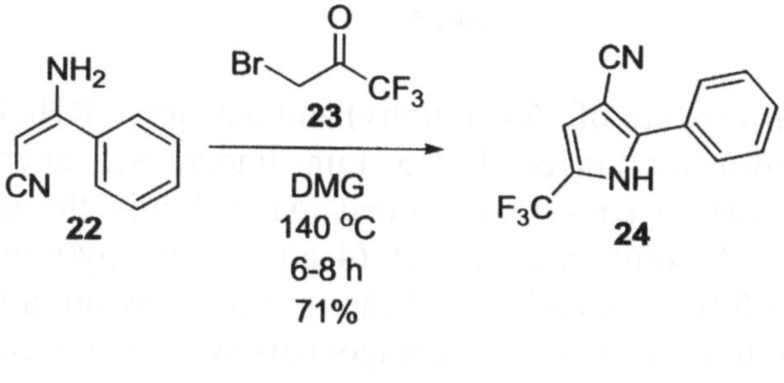

A solution of 3-bromo-1,1,1-trifluoropropanone **23** (1.2 equiv) and enamine **22** (1 equiv) in DGM (10 mL per mmol **23**) was heated at 140 °C for 6–8 h. The mixture was poured into a saturated $NaHCO_3$ solution (10 mL), extracted with EtOAc (3 × 5 mL), washed with brine, and dried with Na_2SO_4. The solvent was evaporated and the residue was purified by flash column chromatography on silica gel using 1:9 EtOAc/petroleum ether to give **24** in 71% yield.

Thiazole (General): ***4-[4-(4-Trifluoromethyl-phenyl)-thiazol-2-yl]-phenylamine hydrobromide salt***[27]

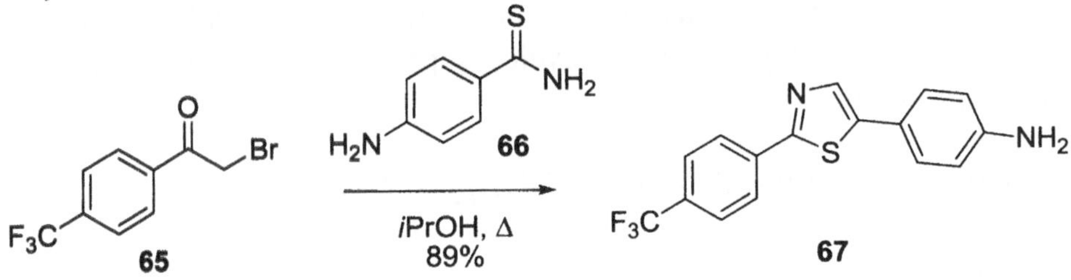

Bromoketone **65** (1 equiv) was dissolved in 750 mL (per 0.25 mol **65**) of isopropyl alcohol, and *p*-aminothiobenzamide **66** (1 equiv) was added. The mixture was heated to 48 °C for 4 h and then cooled to 2 °C. The slurry was filtered, rinsed with cold isopropyl alcohol, and dried to afford aniline **67** as the hydrobromide salt in 89% yield.

Thiazole (Modified): ***Ethyl-L-2-[3-(benzyloxycarbonyl)-1-(t-butoxy-carbonylamino) propyl]thiazole-4-carboxylate***[8]

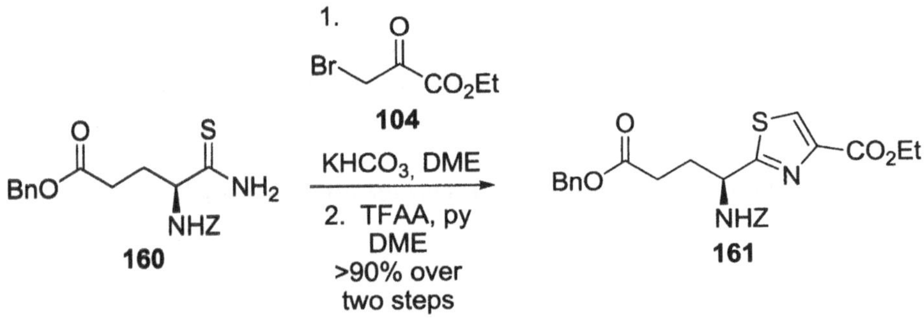

A heterogeneous mixture of **160** (1 equiv) and powdered $KHCO_3$ (9 equiv) in DME was vigorously stirred for 5 min under N_2, after which ethyl bromopyruvate **104** (3 equiv) was added. After 1 min the suspension was cooled to 0 °C. A solution of TFAA (4 equiv) and pyridine (8 equiv) in DME (1 mL) at 0 °C was added, and the reaction mixture allowed to reach ambient temperature. The volatile components were removed *in vacuo* and the residue resuspended in CH_3Cl and extracted with water. The organic phase was dried with Na_2SO_4, filtered and the filtrate condensed *in vacuo*. The resulting residue was purified using silica chromatography 3:4 EtOAc/hexanes to give optically pure **161** in 92% yield.

1,4-Dihydropyridine (General): ***Nifedipine***[4]

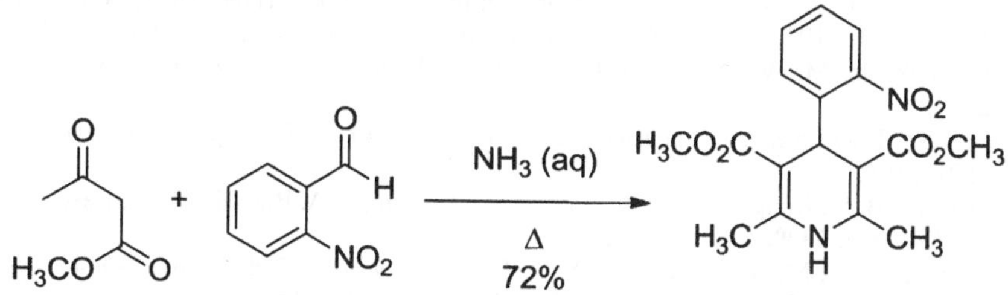

180 **181** nifedipine

2-Nitrobenzaldehyde **181** (1 equiv), methyl acetoacetate **180** (1.6 equiv), methanol (75 mL per equiv 2-nitrobenzaldehyde), and ammonia (32 mL per 290 mmol 2-nitrobenzaldehyde) were refluxed for several hours until the reaction was complete. The reaction was cooled to room temperature, and the resulting crystals were collected by vacuum filtration to produce nifedipine in 72% yield.

1,4-Dihydropyridine (Modified): ***Dimethyl-2,6-dimethyl-4-(4-oxo-4H-benzo[b]thiopyrano[2,3-d]furan-2-yl)-1,4-dihydropyridine-3,5-dicarboxylate***[73]

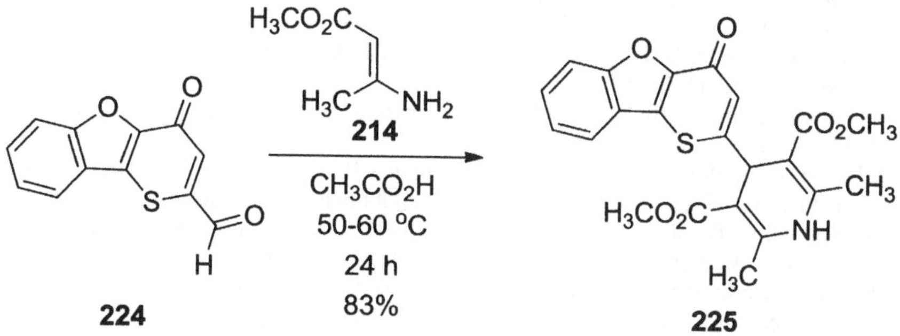

224 **225**

4-Thiopyron-2-carbaldehyde **224** (1 equiv) and aminocrotonate methyl ester **214** (2.5 equiv) were dissolved in 20 mL of acetic acid (per mmol **224**) and heated for 5 h at 50–60 °C. The reaction mixture was concentrated *in vacuo* and recrystallized from toluene to give **225** in 83% yield.

9.5.7 *References*

1 (a) Hantzsch, A. *Chem. Ber.* **1881,** *14*, 1637–1638. (b) Hatzsch, A. *Ann. Chem.* **1892,** 1–81.
2 Hantzsch, A. *Ber. Dtsch. Chem. Ges.* **1890,** *23*, 1474.
3 (a) Hantzsch, A.; Weber, H. J. *Ber. Dtsch. Chem. Ges.* **1887,** *20*, 3118–3132. (b) Hantzsch, A. *Ber. Dtsch. Chem. Ges.* **1888,** *21*, 942–946. (c) Hantzsch, A. *Liebigs Ann. Chem.* **1889,** *250*, 257–273.
4 US Pat. 3485847.
5 Schramm, M.; Thomas, G.; Franckowiak, G. *Nature* **1983,** *303*, 535.

6 Berson, J. A.; Brown, E. *J. Am. Chem. Soc.* **1955,** *77*, 444–447.
7 For a recent review on the scope and limitations of the Hantzsch 1,4-dihydropyridine synthesis see: [R] Saini, A.; Kumar, S.; Sandhu, J. S. *J. Sci. Indian Res.* **2008,** 95–111.
8 Bredenkamp, M. W.; Holzapfel, C. W.; van Zyl, W. J. *Synth. Commun.* **1990,** *20*, 2235–2249.
9 Prakash, R.; Kumar, A.; Aggarwal, R.; Prakash, O.; Singh, S. P. *Synth. Commun.* **2007,** *37*, 2501–2505.
10 Dawane, B. S.; Kongda, S. G.; Kamble, V. T.; Chavan, S. A. Bhosale, R. B.; Baseer, S. M. E. *J. Chem. Soc.* **2009,** *6*, S358–S362
11 (a) Enders, D.; Muller, S.; Demir, A. S. *Tetrahedron Lett.* **1998,** *29*, 6437–6440. (b) Shan, R.; Knaus, E. E. *Bioorg. Med. Chem. Lett.* **1999,** *9*, 2613–2614.
12 Antonyraj, C. A.; Kannan, S. *Appl. Catal. A: General* **2008,** *338*, 121–129.
13 Debache, A.; Ghalem, W.; Boulcina, R.; Belfaitah, A.; Rhouati, S.; Carboni, B. *Tetrahedron Lett.* **2009,** *50*, 5248–5250.
14 Paraskar, A. S.; Sudalai, A. *Indian J. Chem.* **2008,** *46B*, 331–335.
15 Gupta, R.; Gupta, R.; Paul, S.; Loupy, A. *Synthesis* **2007,** *18,* 2835–2838.
16 Öhbert, L.; Westman, J. *Synlett* **2002,** *8*, 1296–1298.
17 (a) Nasr-Esfhani, M.; Karami, B.; Behzadi, M. *J. Heterocyclic Chem.* **2009,** *46*, 931–935. (b) Pei, W.; Qin, W.; Li, W.; Sun, L. *Chin. J. Chem.* **2010,** *28*, 483–486.
18 Roomi, M. W.; MacDonald, S. F. *Can. J. Chem.* **1970,** *48*, 1689–1697.
19 Matiychuk, V. S.; Martyak, R. L.; Obushak, N. D. ; Ostapiuk, Y. V.; Pidlypnyi, N. I. *Chem. Heterocycl. Comp.* **2004,** *40,* 1218–1219.
20 Skaddan, M. B. *J. Labelled Compd. Radiopharm.* **2010,** *53*, 73–77.
21 Kameswaran, V.; Jiang, B. *Synthesis* **1996,** 530–532.
22 Trautwein, A. W.; Süβmuth, R. D.; Jung, G. *Bioorg. Med. Chem. Lett.* **1998,** *8*, 2381–2384.
23 Herath, A.; Cosford, N. D. P. *Org. Lett.* **2010,** *12,* 5182–5185.
24 Clavo, L.; González-Ortega, A.; Sañudo, M. C. *Synthesis* **2002,** *16*, 2450–2456.
25 Goud, P. M.; Sheri, A.; Desai, P. V.; Watkins, E. B.; Tekwani, B.; Sabnis, Y.; Gut, J.; Rosenthal, P. J.; Avery, M. A. *Med. Chem. Res.* **2005,** 74–105.
26 Brown, K.; Carter, D. P.; Cavalla, J. F.; Green, D.; Newberry, R. A.; Wilson, A. B. *J. Med. Chem.* **1974,** *17*, 1177–1181.
27 Ikemoto, N.; Liu, J.; Brands, K. M. J.; McNamara, J. M.; Reider, P. J. *Tetrahedron* **2003,** *59*, 1317–1325.
28 Madsen, P.; Kodra, J. T.; Behrens, C.; Nishimura, E.; Jeppesen, C. B.; Pridal, L.; Andersen, B.; Knudsen, L. B.; Valcarce-Aspergen, C.; Guldbrandt, M.; Christensen, I. T.; Jørgensen, A. S.; Ynddal, L.; Brand, C. L.; Bagger, M. A.; Lau, J. *J. Med. Chem.* **2009,** *52*, 2989–3000.
29 Arutyunyan, A.; Nefzi, A. *J. Comb. Chem.* **2010,** *72*, 315–317.
30 Legeay, J. C.; Eynde, J. J. V.; Bazureau, J. P. *Tetrahedron* **2008,** *64*, 5328–5335.
31 El-Meligie, S.; El-Awady, R. A. *J. Heterocyclic Chem.* **2002,** *39*, 1133–1138.
32 Lepeshkin, A. Y.; Turchin, K. F.; Sedov, A. L.; Velezheva, V. S. *Russ. Chem. Bull., Int. Ed.* **2007,** *56*, 1441–1446.
33 Chimenti, F.; Bizzarii, B.; Bolasco, A.; Secci, D.; Chimenti, P.; Granese, A.; Carrodori, S.; D'Ascenzio, M.; Scaltrito, M. M.; Sisto, F. *J. Heterocyclic Chem.* **2010,** *47*, 1269–1273.
34 Mashraqui, S. H.; Mstry, H.; Sundaram, S. *J. Heterocyclic Chem.* **2006,** *43*, 917–923.
35 Lin, W.; Cao, X.; Yuan, L.; Sing, Y. *Chem. Eur. J.* **2010,** *16*, 6454–6457.
36 Kovacs, L.; Herczegh, P.; Batta, G.; Farkas, I. *Tetrahedron* **1991,** *47*, 5539–5548.
37 Miller, T. J.; Farquar, H. D.; Sheybani, A.; Tallini, C. E.; Saurage, A. S.; Fronczek, F. R.; Hammer, R. P. *Org. Lett.* **2002,** *4*, 877–880.
38 Qiao, W.; So, S.-S.; Goodnow, R. A. *Org. Lett.* **2001,** *3*, 3655–3658.
39 Urbanský, M.; Drašar, P. *Synth. Commun.* **1993,** *23*, 829–845.
40 Sun, M.; Wu, X.; Chen J.; Cai, J.; Cao, M.; Ji, M. *Eur. J. Med. Chem.* **2010,** *45*, 2299–2306.
41 Sušnik, M. P.; Schnürch, M.; Mihovilovic, M. D.; Mereiter, K.; Stanetty, P. *Monatsch. Chem.* **2009,** *140*, 423–430.
42 Gu, X.-H.; Wan, X.-Z.; Jiang, B. *Bioorg. Med. Chem. Lett.* **1999,** *9*, 569–572.
43 Sarodnick, G.; Heydenreich, M.; Linker, T.; Kleinpeter, E. *Tetrahedron* **2003,** *59*, 6311–6321.
44 Nay, B.; Schiavi, B.; Ahond, A.; Poupat, C.; Potier, P. *Synthesis* **2005,** *1*, 97–101.

45 Glover, C.; Meritt, E. A.; Bagley, M. C. *Tetrahedron Lett.* **2007**, *48*, 7027–7030.
46 Kuriyama, N.; Akaji, K.; Kiso, Y. *Tetrahedron* **1997**, *53*, 8323–8334.
47 Holzapfel, C. W.; Pettit, G. R. *J. Org. Chem.* **1985**, *50*, 2323–2327.
48 Pettit, G. R.; Nelson, P. S.; Holzapfel, C. W. *J. Org. Chem.* **1985**, *50*, 2654–2659.
49 Houssin, R.; Lohez, M.; Bernier, J.-L.; Henichart, J.-P. *J. Org. Chem.* **1985**, *50*, 2787–2788.
50 Bredenkamp, M. W.; Holzapfel, C. W.; Snyman, R. M.; van Zyl, W. J. *Synth. Commun.* **1992**, *22*, 3029–3039.
51 Agullar, E.; Meyers, A. I. *Tetrahedron Lett.* **1994**, *35*, 2473–2476.
52 Bray, C. D.; Olasoji, J. *Synlett* **2010**, *4*, 599–601.
53 Schmidt, U.; Gleich, P.; Griesser, H.; Utz, R. *Synthesis* **1986**, 992–998.
54 Li, D.; Yang, H. S.; Cui, Q.; Mao, S. J.; Xu, X. H. *Chin. Chem. Lett.* **2009**, *20*, 1195–1197.
55 Jayaprakash, S.; Pattenden, G.; Viljoen, M. S.; Wilson, C. *Tetrahedron* **2003**, *59*, 6637–6646.
56 Hughes, R. A.; Thompson, S. P.; Alcaraz, L.; Moody, C. J. *J. Am. Chem. Soc.* **2005**, *127*, 15644–15651.
57 Nakamura, Y.; Takeuchi, S. *QSAR Comb. Sci.* **2006**, *25*, 703–708.
58 Katritzky, A. R.; Chen, J.; Yang, Z. *J. Org. Chem.* **1995**, *60*, 5638–5642.
59 Xia, Z.; Smith, C. D. *J. Org. Chem.* **2001**, *66*, 3459–3466.
60 (a) Hernandez, D.; Vilar, G.; Riego, E.; Canedo, L. M.; Cuevas, C.; Albericio, F.; Alvarez, M. *Org. Lett.* **2007**, *9*, 809–811. (b) Hernandez, D.; Riego, E.; Francesch, A.; Cuevas, C.; Albericio, F.; Alvarez, M. *Tetrahedron* **2007**, *63*, 9862–9870.
61 Nefzi, A.; Arutyunyan, S.; Fenwick, J. E. *J. Org. Chem.* **2010**, *75*, 7939–7941.
62 Tang, S.; Xu, Z.; Ye, T. *Tetrahedron: Asymm.* **2009**, *20*, 2027–2032.
63 Chennat, T.; Eisner, V. *J. Chem. Soc., Perkins Trans. 1* **1975**, 926.
64 Foos, J.; Steel, F.; Rizvi, S.; Fraenkel, G. *J. Org. Chem.* **1979**, *44*, 2522–2529.
65 Meyer, H.; Bossert, G.; Horstmann, H. *Liebigs Ann. Chem.* **1977**, 1888–1895.
66 Meyer, H. *Liebigs Ann. Chem.* **1981**, 1523–1533.
67 (a) van Bergen, T. J.; Kellogg, R. M. *J. Chem. Soc., Chem. Commun.* **1976**, 964–967. (b) Kellogg, R. M.; van Bergen, T. J.; van Doren, H.; Hedstrand, D.; Kooi, J.; Kruizinga, W. H.; Troostwijk, C. B. *J. Org. Chem.* **1980**, *45*, 2854–2861.
68 Natale, N. R.; Quincy, D. A. *Synth. Commun.* **1983**, *13*, 817–822.
69 Taylor, M. D.; Badger, E. W.; Steffen, R. P.; Haleen, S. J.; Pugsley, T. A.; Shih, Y. H.; Weishaar, R. E. *J. Med. Chem.* **1988**, *31*, 1659–1664.
70 Iqbal, N.; Triggle, C. R.; Knaus, E. E. *Drug. Dev. Res.* **1997**, *42*, 120–130.
71 Balough, M.; Hermecz, I.; Náray-Szabó, G.; Simon, K.; Mészáros, Z. *J. Chem. Soc., Perkin Trans. 1* **1986**, 753–757.
72 Bridgewood, K. L.; Veitch, G. E.; Ley, S. V. *Org. Lett.* **2008**, *10*, 3627–3629.
73 Görlitzer, K.; Vogt, R. *Arch. Pharm.* **1990**, *323*, 853–856.
74 Eynde, J.-J.; Mayence, A.; Maquestiau, A.; Anders, A. *Synth. Commun.* **1992**, *22*, 3291–3304.
75 Rimoli, M. G.; Avallone, L.; Zanarone, S.; Abignente, E.; Mangoni, A. *J. Heterocyclic Chem.* **2002**, *39*, 1117–1121.
76 Suh, J.; Hong, Y.; Bae, M. *Arch. Pharm. Res.* **1990**, *13*, 310–313.
77 Fassihi, A.; Velazquez, C.; Knaus, E. E. *J. Heterocyclic Chem.* **2004**, *41*, 263–266.
78 Dondoni, A.; Massi, A.; Minghini, E.; Bertolasi, V. *Helv. Chim. Acta* **2002**, *85*, 3331–3348.
79 Gasco, A. M.; Ermondi, G.; Fruttero, R.; Gasco, A. *Eur. J. Med. Chem.* **1996**, *31*, 3–9.
80 Visentin, S.; Amiel, P.; Fruttero, R.; Boschi, D.; Rousseil, C.; Giusta, L.; Carbone, E.; Gasco, A. *J. Med. Chem.* **1999**, *42*, 1422–1427.
81 Beley, M. *J. Heterocyclic Chem.* **2000**, *37*, 1077.
82 Sagitullina, G. P.; Glizdinskaya, L. V.; Sitnikov, G. V.; Sagitullin, R. S. *Chem. Heterocycl. Cmpds.* **2002**, *38*, 1336–1340.
83 Bhandari, A.; Li, B.; Gallop, M. A. *Synthesis* **1999**, *11*, 1951–1960.
84 Legeay, J.-C.; Eynde, J. J. V.; Bazureau, J. P. *Tetrahedron* **2005**, *61*, 12386–12397.
85 Su, W.; Li, J.; Li, J. *Aust. J. Chem.* **2008**, *61*, 860–863.
86 Štetinová, J.; Milata, V.; Prónayová, N.; Petrov, O.; Bartovič, A. *ARIVOC* **2005**, 127–139.
87 Lee, J. H. *Tetrahedron Lett.* **2005**, *46*, 7329–7330.
88 Vigante, B.; Tirzitis, G.; Tirzite, D.; Chekavichus, B.; Uldrkis, J. *Chem. Heterocycl. Comp.* **2007**, *43*, 225–232.

89 Kumar, B. R. P.; Masih, P.; Karthikeyan, E.; Bansal, A.; Vijayan, P. *Med. Chem. Res.* **2010,** *19*, 344–363.
90 Lefranc, D.; Ciufolini, M. A. *Angew. Chem., Int. Ed.* **2009,** *48*, 4198–4201.

9.6 Ortoleva–King Reaction

Ke Chen

9.6.1 Description

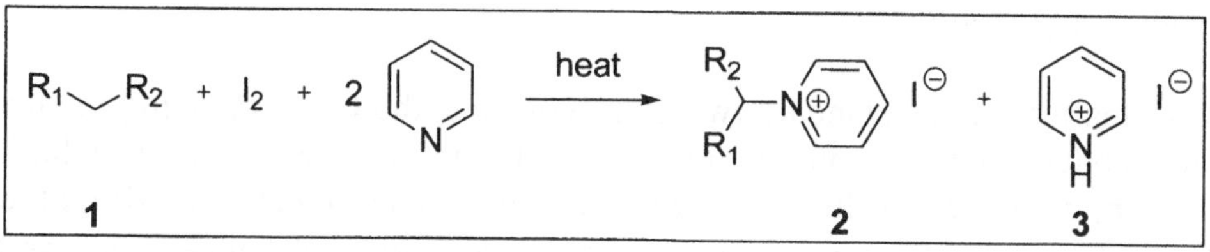

Ortoleva–King reaction[1,2] refers to the direct reaction of an active methyl or methylene compound **1** with iodine and pyridine to generate the corresponding pyridinium iodide **2** along with an equivalent of pyridine hydroiodide salt **3**.

9.6.2 Historical Perspective

Giovanni Ortoleva was the first to recognize the possibility of preparing pyridinium salts by heating cinnamic acid, iodine and pyridine in 1899,[3] and the titled reaction was later reported by L. Carroll King in 1944.[4] This elegant method offers direct access to synthetically useful pyridinium iodide salts. Even today, its potential remains underdeveloped.

9.6.3 Mechanism

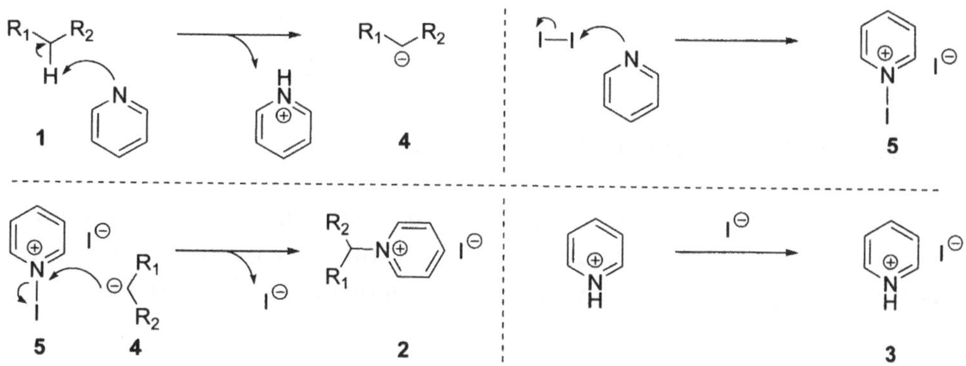

The mechanism of the Ortoleva–King reaction is proposed as following: Deprotonation of methylene compound **1** by one equivalent of pyridine affords its anion **4** along with pyridinium ion. A second equivalent of pyridine was activated by iodine to generate iodopyridinium intermediate **5**.

Displacement of the iodine from **5** by anion **4** gives desired pyridinium iodide **2**, while recombination of pyridinium ion and iodine ion leads to by-product pyridine hydroiodide salt **3**.

9.6.4 *Synthetic Utility*

General Application

In comparison to the standard two-step sequence, the Ortoleva–King method offers convenience as a one-pot procedure and, in most cases, better yield. In addition, it is very valuable when the halide intermediate is difficult to obtain. For example, converting α-tetralone **6** to its acyl halides such as **8** would be complicated by over-oxidation at the electron-rich phenyl ring. However, the preparation of its pyridinium salt **7** succeeds using the Ortoleva–King protocol due to its mild condition.[4]

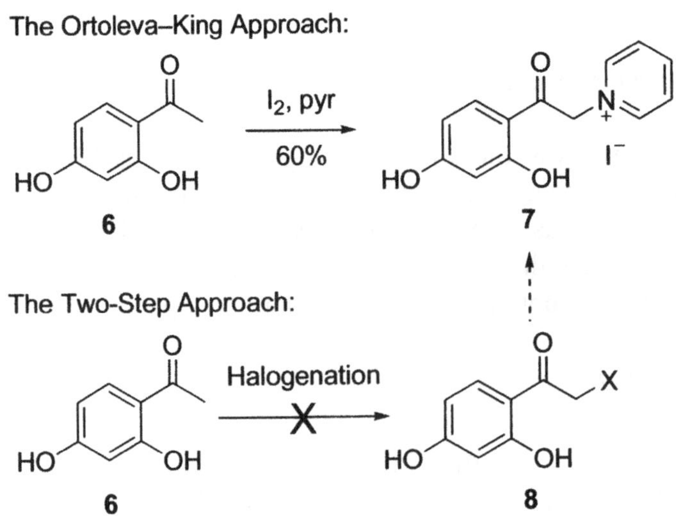

Application in Combination with Other Reactions

After a simple filtration or sometimes without isolating the pyridinium salt product, the Ortoleva–King reaction may be followed by further chemical transformation, giving rise to a variety of synthetically useful compounds.

Ortoleva–King Hydrolysis

The phenylacylpyridinium salt, obtained from the Ortoleva–King reaction, could undergo facile hydrolysis to give rise to the corresponding carboxylic acid. Pratt et al. reported an efficient synthesis of 6-carboxydehydroabietic acid **11** from methyl acetyldehydroabietate **9** via a one-pot procedure.[5] The intermediate Ortoleva–King product **10** was not isolated. After removal of pyridine, the crude was treated with potassium hydroxide in wet ethylene glycol, whereas the acid hydrolysis gave rise to **11** in excellent yield.

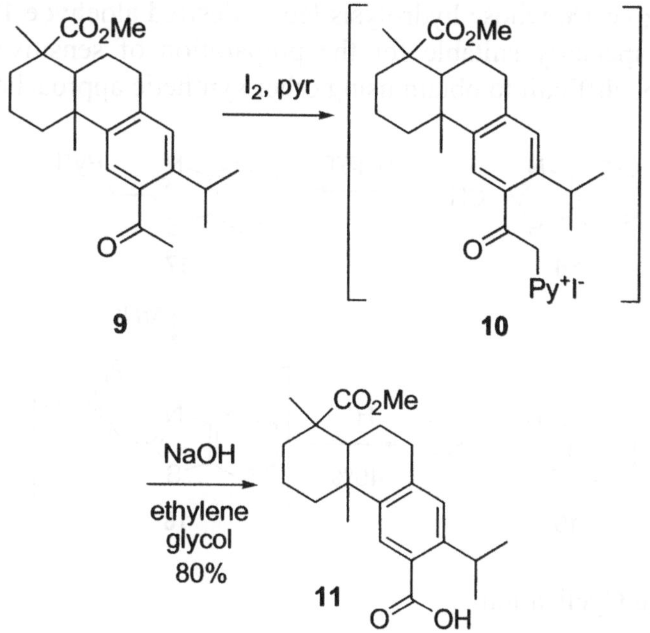

Research by Berson *et al.* showed that 2-pyridone **15** could be accessed via the basic hydrolysis of pyridinium diiodide **13**, which was obtained from the Ortoleva–King reaction of picolinium iodide **12** and used without isolation.[6] The reaction proceeded even at room temperature, therefore providing an extremely mild way to access 2-pyridones.

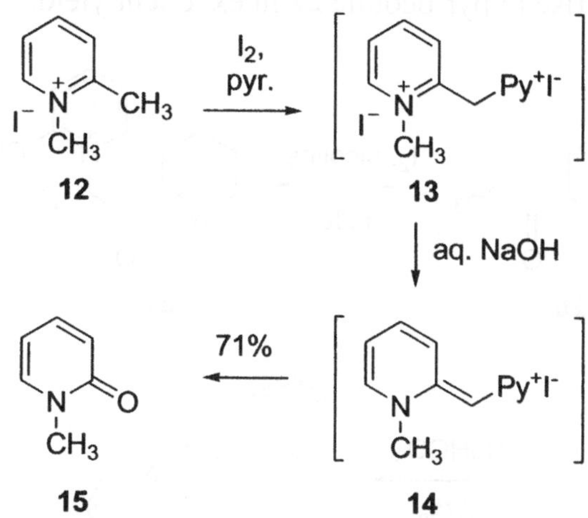

The Ortoleva–King reaction also found application in the following sequence where a methyl or methylene compound such as 2-methylthiazole **16** was converted to the corresponding aldehyde **19**.[7] The preparation of pyridinium salt **17** via Ortoleva–King reaction of **16** was followed by nitrone formation to give **18**, whose hydrolysis led to desired aldehyde **19**. This mild sequence is especially suitable for the preparation of sensitive aldehydes, which might be difficult to obtain using other synthetic approaches.

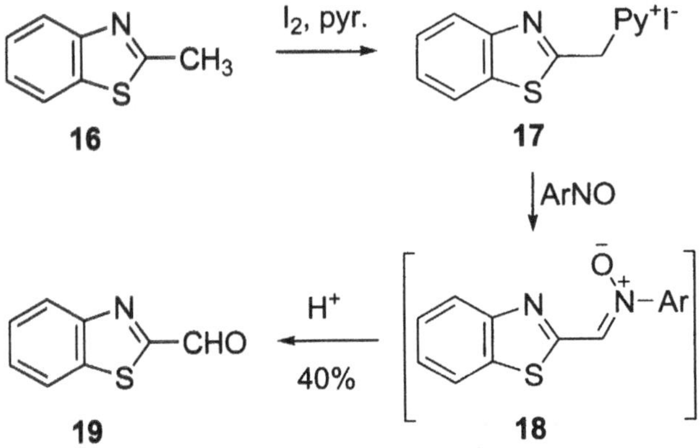

Ortoleva–King Cyclization

With well-positioned substituents, the product from the Ortoleva–King reaction may be followed by readily ring closures to yield a variety of heterocycles. Case in point was made by an elegant synthesis of 2-pyridyl-pyrrocoline **22** reported by Kröhnke *et al.*[8] First, picoline salt **21** was prepared from the Ortoleva–King reaction of acylpyridine **20** and picoline. Upon treatment of saturated sodium bicarbonate **21** underwent rapid cyclization to give rise to pyrrocoline **22** in excellent yield.

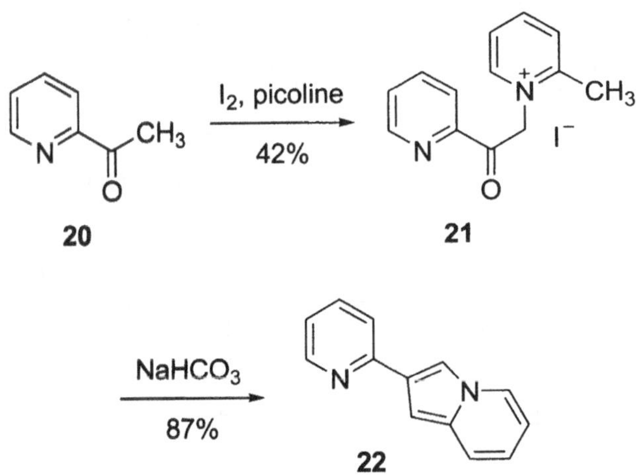

According to King *et al.*, reaction of quinaldine **23** and quinoline took a different course from the analogous reaction with pyridine or isoquinoline.[9] A structurally unique diquinoimidazolium iodide species **26** was obtained in nearly quantitative yield via the proposed *in situ* cyclization of intermediate **24** (**24**→**25**→**26**).

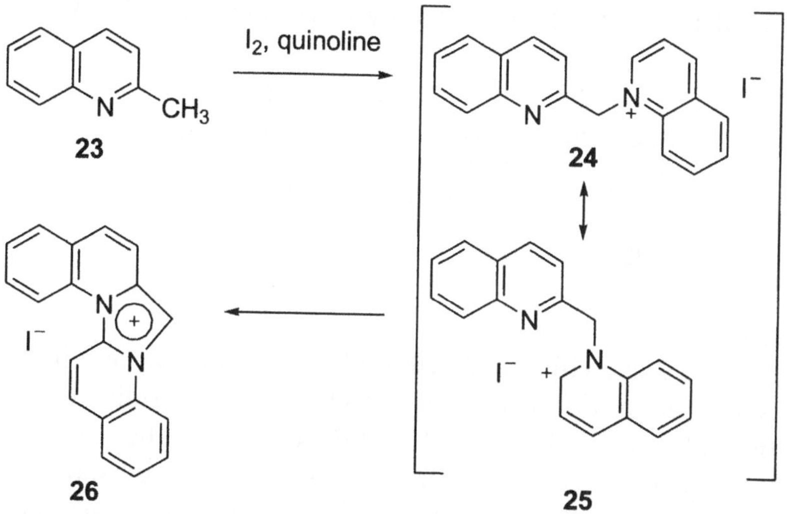

Ortoleva–King–Kröhnke sequence

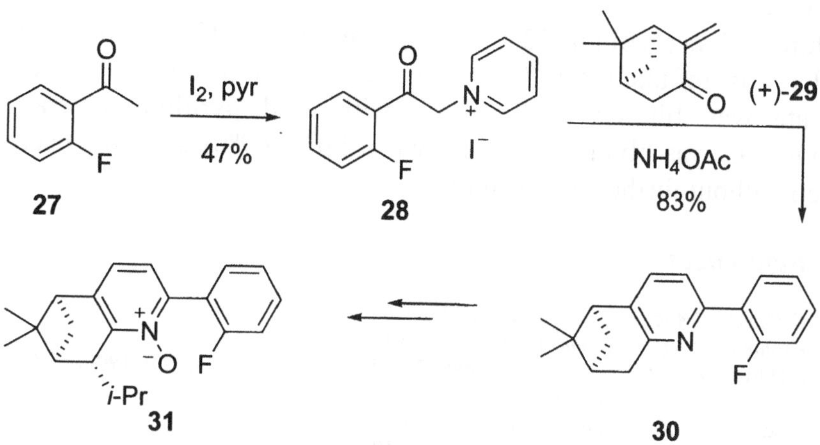

The Ortoleva–King–Kröhnke sequence was frequently used to prepare multisubstituted pyridines.[10–13] In fact, the acylpyridinium salt, obtained from the Ortoleva-King reaction, was often referred to the "Kröhnke salt". For example, a pyridine *N*-oxide based chiral ligand **31** was synthesized by Malkov *et al.*[11] Prepared from 2-fluoroacetophenone **27**, Kröhnke salt **28** underwent condensation with pinocarvone **29** and ammonium acetate, affording pyridine derivative **30**. Further elaboration of **30** led to the formation of desired *N*-oxide **31**, which served as a catalyst in asymmetric allylation of aldehydes.

9.6.5 *Experimental*

Preparation of 1-(2-furan-2-yl-2-oxo-ethyl)-pyridinium iodide **33**

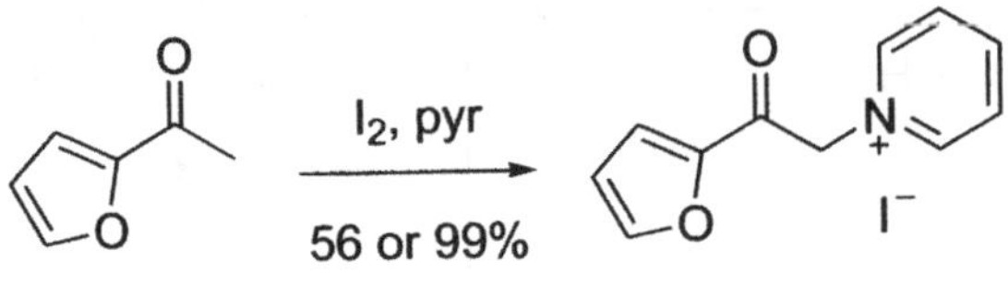

32 **33**

Precedure A[10]
To a solution of 2-acetylfuran **32** (10.0 mL, 100 mmol) in pyridine (25 mL) was added a solution of iodine (25.4 g, 100 mmol) in pyridine (75 mL), and the resulting solution was heated at 100–110 °C for 3 h. The reaction solution was left to stand overnight, filtered, washed with ethanol, and then recrystallized twice in ethanol. The crystal was dried under vacuum to yield compound **33** (17.7 g, 56.3 mmol). Yield 56%.

Precedure B[14]
A mixture of 2-acetylfuran **32** (20 mmol) and iodine (20 mmol) in pyridine (25 mL) was refluxed for 3 h. After cooling to room temperature, the precipitate was filtered and washed with cold pyridine several times. Compound **33** was obtained as a brown solid (99.9%) and was used in the next step without further purification.

9.6.6 *References*

1. Kröhnke, F. *Angew. Chem.* **1963,** *2*, 225–276.
2. Kröhnke, F. *Angew. Chem.* **1963,** *2*, 380–393.
3. (a) Ortoleva, G. *Gazz. Chim. Ital.* **1899,** *29 I*, 503. (b) *Gazz. Chim. Ital.* **1900,** *30 I*, 509.
4. King, L. C. *J. Am. Chem. Soc.* **1944,** *66*, 894–895.
5. Pratt, Y. T. *J. Am. Chem. Soc.* **1951,** *73*, 3803.
6. Berson, J. A.; Cohen, T. *J. Am. Chem. Soc.* **1956,** *78*, 416.
7. Ried, W.; Bender, H. *Chem. Ber.* **1956,** *89*, 1893.
8. Kröhnke, F.; Gross, K. F. *Chem. Ber.* **1959,** *92*, 22.
9. King, L. C.; Abramo, S. V. *J. Org. Chem.* **1958,** *23*, 1926.
10. Chen, C.-P.; Chen, Y.-J.; Lin. R.-X. *Tetrahedron: Asymmetry* **2004,** *15*, 3561–3571.
11. Markov, A. V.; Bell, M.; Vassieu, M.; Bugatti, V.; Kocovsky, P. *J. Mol. Catal. A: Chem.* **2003,** *196*, 179–186.
12. Lewis, J.; Constable, E. *Polyhedron* **1982,** *1*, 303–306.
13. [R] Kröhnke, F. *Synthesis* **1976,** 1–24.
14. Lee. E.-S.; *et al. Eur. J. Med. Chem.* **2008,** *43*, 675–682.

Appendixes

Appendix 1
Table of Contents for Volume 1: *Name Reactions in Heterocyclic Chemistry*
Published in 2005

Appendix 2
Table of Contents for Volume 2:
Name Reactions for Functional Group Transformations

Appendix 3
Table of Contents for Volume 3:
Name Reactions for Homologations—Part I
Published in 2009

Appendix 4
Table of Contents for Volume 3:
Name Reactions for Homologations—Part II
Published in 2009

Appendix 5
Table of Contents for Volume 5: *Name Reactions for Carbocyclic Ring Formations*

Published in 2010

Subject Index

Printed and bound by CPI Group (UK) Ltd, Croydon, CR0 4YY
16/09/2026
14963461-0006